Lecture Notes in Computer Science

# Lecture Notes in Artificial Intelligence  **13716**

Founding Editor

Jörg Siekmann

Series Editors

Randy Goebel, *University of Alberta, Edmonton, Canada*
Wolfgang Wahlster, *DFKI, Berlin, Germany*
Zhi-Hua Zhou, *Nanjing University, Nanjing, China*

The series Lecture Notes in Artificial Intelligence (LNAI) was established in 1988 as a topical subseries of LNCS devoted to artificial intelligence.

The series publishes state-of-the-art research results at a high level. As with the LNCS mother series, the mission of the series is to serve the international R & D community by providing an invaluable service, mainly focused on the publication of conference and workshop proceedings and postproceedings.

Massih-Reza Amini · Stéphane Canu ·
Asja Fischer · Tias Guns · Petra Kralj Novak ·
Grigorios Tsoumakas
Editors

# Machine Learning and Knowledge Discovery in Databases

European Conference, ECML PKDD 2022
Grenoble, France, September 19–23, 2022
Proceedings, Part IV

 Springer

*Editors*
Massih-Reza Amini
Grenoble Alpes University
Saint Martin d'Hères, France

Stéphane Canu
INSA Rouen Normandy
Saint Etienne du Rouvray, France

Asja Fischer
Ruhr-Universität Bochum
Bochum, Germany

Tias Guns
KU Leuven
Leuven, Belgium

Petra Kralj Novak
Central European University
Vienna, Austria

Grigorios Tsoumakas
Aristotle University of Thessaloniki
Thessaloniki, Greece

ISSN 0302-9743     ISSN 1611-3349 (electronic)
Lecture Notes in Artificial Intelligence
ISBN 978-3-031-26411-5     ISBN 978-3-031-26412-2 (eBook)
https://doi.org/10.1007/978-3-031-26412-2

LNCS Sublibrary: SL7 – Artificial Intelligence

This Springer imprint is published by the registered company Springer Nature Switzerland AG
The registered company address is: Gewerbestrasse 11, 6330 Cham, Switzerland

# Preface

The European Conference on Machine Learning and Principles and Practice of Knowledge Discovery in Databases (ECML–PKDD 2022) in Grenoble, France, was once again a place for in-person gathering and the exchange of ideas after two years of completely virtual conferences due to the SARS-CoV-2 pandemic. This year the conference was hosted for the first time in hybrid format, and we are honored and delighted to offer you these proceedings as a result.

The annual ECML–PKDD conference serves as a global venue for the most recent research in all fields of machine learning and knowledge discovery in databases, including cutting-edge applications. It builds on a highly successful run of ECML–PKDD conferences which has made it the premier European machine learning and data mining conference.

This year, the conference drew over 1080 participants (762 in-person and 318 online) from 37 countries, including 23 European nations. This wealth of interest considerably exceeded our expectations, and we were both excited and under pressure to plan a special event. Overall, the conference attracted a lot of interest from industry thanks to sponsorship, participation, and the conference's industrial day.

The main conference program consisted of presentations of 242 accepted papers and four keynote talks (in order of appearance):

- Francis Bach (Inria), Information Theory with Kernel Methods
- Danai Koutra (University of Michigan), Mining & Learning [Compact] Representations for Structured Data
- Fosca Gianotti (Scuola Normale Superiore di Pisa), Explainable Machine Learning for Trustworthy AI
- Yann Le Cun (Facebook AI Research), From Machine Learning to Autonomous Intelligence

In addition, there were respectively twenty three in-person and three online workshops; five in-person and three online tutorials; two combined in-person and one combined online workshop-tutorials, together with a PhD Forum, a discovery challenge and demonstrations.

Papers presented during the three main conference days were organized in 4 tracks, within 54 sessions:

- Research Track: articles on research or methodology from all branches of machine learning, data mining, and knowledge discovery;
- Applied Data Science Track: articles on cutting-edge uses of machine learning, data mining, and knowledge discovery to resolve practical use cases and close the gap between current theory and practice;
- Journal Track: articles that were published in special issues of the journals *Machine Learning* and *Data Mining and Knowledge Discovery*;

- Demo Track: short articles that propose a novel system that advances the state of the art and include a demonstration video.

We received a record number of 1238 abstract submissions, and for the Research and Applied Data Science Tracks, 932 papers made it through the review process (the remaining papers were withdrawn, with the bulk being desk rejected). We accepted 189 (27.3%) Research papers and 53 (22.2%) Applied Data science articles. 47 papers from the Journal Track and 17 demo papers were also included in the program. We were able to put together an extraordinarily rich and engaging program because of the high quality submissions.

Research articles that were judged to be of exceptional quality and deserving of special distinction were chosen by the awards committee:

- Machine Learning Best Paper Award: "*Bounding the Family-Wise Error Rate in Local Causal Discovery Using Rademacher Averages*", by Dario Simionato (University of Padova) and Fabio Vandin (University of Padova)
- Data-Mining Best Paper Award: "*Transforming PageRank into an Infinite-Depth Graph Neural Network*", by Andreas Roth (TU Dortmund), and Thomas Liebig (TU Dortmund)
- Test of Time Award for highest impact paper from ECML–PKDD 2012: "*Fairness-Aware Classifier with Prejudice Remover Regularizer*", by Toshihiro Kamishima (National Institute of Advanced Industrial Science and Technology AIST), Shotaro Akashi (National Institute of Advanced Industrial Science and Technology AIST), Hideki Asoh (National Institute of Advanced Industrial Science and Technology AIST), and Jun Sakuma (University of Tsukuba)

We sincerely thank the contributions of all participants, authors, PC members, area chairs, session chairs, volunteers, and co-organizers who made ECML–PKDD 2022 a huge success. We would especially like to thank Julie from the Grenoble World Trade Center for all her help and Titouan from Insight-outside, who worked so hard to make the online event possible. We also like to express our gratitude to Thierry for the design of the conference logo representing the three mountain chains surrounding the Grenoble city, as well as the sponsors and the ECML–PKDD Steering Committee.

October 2022

Massih-Reza Amini
Stéphane Canu
Asja Fischer
Petra Kralj Novak
Tias Guns
Grigorios Tsoumakas
Georgios Balikas
Fragkiskos Malliaros

# Organization

## General Chairs

Massih-Reza Amini      University Grenoble Alpes, France
Stéphane Canu      INSA Rouen, France

## Program Chairs

Asja Fischer      Ruhr University Bochum, Germany
Tias Guns      KU Leuven, Belgium
Petra Kralj Novak      Central European University, Austria
Grigorios Tsoumakas      Aristotle University of Thessaloniki, Greece

## Journal Track Chairs

Peggy Cellier      INSA Rennes, IRISA, France
Krzysztof Dembczyński      Yahoo Research, USA
Emilie Devijver      CNRS, France
Albrecht Zimmermann      University of Caen Normandie, France

## Workshop and Tutorial Chairs

Bruno Crémilleux      University of Caen Normandie, France
Charlotte Laclau      Telecom Paris, France

## Local Chairs

Latifa Boudiba      University Grenoble Alpes, France
Franck Iutzeler      University Grenoble Alpes, France

## Proceedings Chairs

Wouter Duivesteijn            Technische Universiteit Eindhoven,
                                     the Netherlands
Sibylle Hess                   Technische Universiteit Eindhoven,
                                     the Netherlands

## Industry Track Chairs

Rohit Babbar                  Aalto University, Finland
Françoise Fogelmann        Hub France IA, France

## Discovery Challenge Chairs

Ioannis Katakis              University of Nicosia, Cyprus
Ioannis Partalas             Expedia, Switzerland

## Demonstration Chairs

Georgios Balikas           Salesforce, France
Fragkiskos Malliaros       CentraleSupélec, France

## PhD Forum Chairs

Esther Galbrun             University of Eastern Finland, Finland
Justine Reynaud            University of Caen Normandie, France

## Awards Chairs

Francesca Lisi              Università degli Studi di Bari, Italy
Michalis Vlachos           University of Lausanne, Switzerland

## Sponsorship Chairs

Patrice Aknin               IRT SystemX, France
Gilles Gasso                INSA Rouen, France

# Web Chairs

| | |
|---|---|
| Martine Harshé | Laboratoire d'Informatique de Grenoble, France |
| Marta Soare | University Grenoble Alpes, France |

# Publicity Chair

| | |
|---|---|
| Emilie Morvant | Université Jean Monnet, France |

# ECML PKDD Steering Committee

| | |
|---|---|
| Annalisa Appice | University of Bari Aldo Moro, Italy |
| Ira Assent | Aarhus University, Denmark |
| Albert Bifet | Télécom ParisTech, France |
| Francesco Bonchi | ISI Foundation, Italy |
| Tania Cerquitelli | Politecnico di Torino, Italy |
| Sašo Džeroski | Jožef Stefan Institute, Slovenia |
| Elisa Fromont | Université de Rennes, France |
| Andreas Hotho | Julius-Maximilians-Universität Würzburg, Germany |
| Alípio Jorge | University of Porto, Portugal |
| Kristian Kersting | TU Darmstadt, Germany |
| Jefrey Lijffijt | Ghent University, Belgium |
| Luís Moreira-Matias | University of Porto, Portugal |
| Katharina Morik | TU Dortmund, Germany |
| Siegfried Nijssen | Université catholique de Louvain, Belgium |
| Andrea Passerini | University of Trento, Italy |
| Fernando Perez-Cruz | ETH Zurich, Switzerland |
| Alessandra Sala | Shutterstock Ireland Limited, Ireland |
| Arno Siebes | Utrecht University, the Netherlands |
| Isabel Valera | Universität des Saarlandes, Germany |

# Program Committees

# Guest Editorial Board, Journal Track

| | |
|---|---|
| Richard Allmendinger | University of Manchester, UK |
| Marie Anastacio | Universiteit Leiden, the Netherlands |
| Ira Assent | Aarhus University, Denmark |
| Martin Atzmueller | Universität Osnabrück, Germany |
| Rohit Babbar | Aalto University, Finland |

| | |
|---|---|
| Jaume Bacardit | Newcastle University, UK |
| Anthony Bagnall | University of East Anglia, UK |
| Mitra Baratchi | Universiteit Leiden, the Netherlands |
| Francesco Bariatti | IRISA, France |
| German Barquero | Universität de Barcelona, Spain |
| Alessio Benavoli | Trinity College Dublin, Ireland |
| Viktor Bengs | Ludwig-Maximilians-Universität München, Germany |
| Massimo Bilancia | Università degli Studi di Bari Aldo Moro, Italy |
| Ilaria Bordino | Unicredit R&D, Italy |
| Jakob Bossek | University of Münster, Germany |
| Ulf Brefeld | Leuphana University of Lüneburg, Germany |
| Ricardo Campello | University of Newcastle, UK |
| Michelangelo Ceci | University of Bari, Italy |
| Loic Cerf | Universidade Federal de Minas Gerais, Brazil |
| Vitor Cerqueira | Universidade do Porto, Portugal |
| Laetitia Chapel | IRISA, France |
| Jinghui Chen | Pennsylvania State University, USA |
| Silvia Chiusano | Politecnico di Torino, Italy |
| Roberto Corizzo | Università degli Studi di Bari Aldo Moro, Italy |
| Bruno Cremilleux | Université de Caen Normandie, France |
| Marco de Gemmis | University of Bari Aldo Moro, Italy |
| Sebastien Destercke | Centre National de la Recherche Scientifique, France |
| Shridhar Devamane | Global Academy of Technology, India |
| Benjamin Doerr | Ecole Polytechnique, France |
| Wouter Duivesteijn | Technische Universiteit Eindhoven, the Netherlands |
| Thomas Dyhre Nielsen | Aalborg University, Denmark |
| Tapio Elomaa | Tampere University, Finland |
| Remi Emonet | Université Jean Monnet Saint-Etienne, France |
| Nicola Fanizzi | Università degli Studi di Bari Aldo Moro, Italy |
| Pedro Ferreira | University of Lisbon, Portugal |
| Cesar Ferri | Universität Politecnica de Valencia, Spain |
| Julia Flores | University of Castilla-La Mancha, Spain |
| Ionut Florescu | Stevens Institute of Technology, USA |
| Germain Forestier | Université de Haute-Alsace, France |
| Joel Frank | Ruhr-Universität Bochum, Germany |
| Marco Frasca | Università degli Studi di Milano, Italy |
| Jose A. Gomez | Universidad de Castilla-La Mancha, Spain |
| Stephan Günnemann | Institute for Advanced Study, Germany |
| Luis Galarraga | Inria, France |

| | |
|---|---|
| Erik Schultheis | Aalto-yliopisto, Finland |
| Thomas Seidl | Ludwig-Maximilians-Universität München, Germany |
| Moritz Seiler | University of Münster, Germany |
| Kijung Shin | KAIST, South Korea |
| Shinichi Shirakawa | Yokohama National University, Japan |
| Marek Smieja | Jagiellonian University, Poland |
| James Edward Smith | University of the West of England, UK |
| Carlos Soares | Universidade do Porto, Portugal |
| Arnaud Soulet | Université de Tours, France |
| Gerasimos Spanakis | Maastricht University, the Netherlands |
| Giancarlo Sperli | University of Campania Luigi Vanvitelli, Italy |
| Myra Spiliopoulou | Otto von Guericke Universität Magdeburg, Germany |
| Jerzy Stefanowski | Poznan University of Technology, Poland |
| Giovanni Stilo | Università degli Studi dell'Aquila, Italy |
| Catalin Stoean | University of Craiova, Romania |
| Mahito Sugiyama | National Institute of Informatics, Japan |
| Nikolaj Tatti | Helsingin Yliopisto, Finland |
| Alexandre Termier | Université de Rennes 1, France |
| Luis Torgo | Dalhousie University, Canada |
| Leonardo Trujillo | Tecnologico Nacional de Mexico, Mexico |
| Wei-Wei Tu | 4Paradigm Inc., China |
| Steffen Udluft | Siemens AG Corporate Technology, Germany |
| Arnaud Vandaele | Université de Mons, Belgium |
| Celine Vens | KU Leuven, Belgium |
| Herna Viktor | University of Ottawa, Canada |
| Marco Virgolin | Centrum Wiskunde en Informatica, the Netherlands |
| Jordi Vitria | Universität de Barcelona, Spain |
| Jilles Vreeken | CISPA Helmholtz Center for Information Security, Germany |
| Willem Waegeman | Universiteit Gent, Belgium |
| Markus Wagner | University of Adelaide, Australia |
| Elizabeth Wanner | Centro Federal de Educacao Tecnologica de Minas, Brazil |
| Marcel Wever | Universität Paderborn, Germany |
| Ngai Wong | University of Hong Kong, Hong Kong, China |
| Man Leung Wong | Lingnan University, Hong Kong, China |
| Marek Wydmuch | Poznan University of Technology, Poland |
| Guoxian Yu | Shandong University, China |
| Xiang Zhang | University of Hong Kong, Hong Kong, China |

| Ye Zhu | Deakin University, USA |
| Arthur Zimek | Syddansk Universitet, Denmark |
| Albrecht Zimmermann | Université de Caen Normandie, France |

## Area Chairs

| Fabrizio Angiulli | DIMES, University of Calabria, Italy |
| Annalisa Appice | University of Bari, Italy |
| Ira Assent | Aarhus University, Denmark |
| Martin Atzmueller | Osnabrück University, Germany |
| Michael Berthold | Universität Konstanz, Germany |
| Albert Bifet | Université Paris-Saclay, France |
| Hendrik Blockeel | KU Leuven, Belgium |
| Christian Böhm | LMU Munich, Germany |
| Francesco Bonchi | ISI Foundation, Turin, Italy |
| Ulf Brefeld | Leuphana, Germany |
| Francesco Calabrese | Richemont, USA |
| Toon Calders | Universiteit Antwerpen, Belgium |
| Michelangelo Ceci | University of Bari, Italy |
| Peggy Cellier | IRISA, France |
| Duen Horng Chau | Georgia Institute of Technology, USA |
| Nicolas Courty | IRISA, Université Bretagne-Sud, France |
| Bruno Cremilleux | Université de Caen Normandie, France |
| Jesse Davis | KU Leuven, Belgium |
| Gianmarco De Francisci Morales | CentAI, Italy |
| Tom Diethe | Amazon, UK |
| Carlotta Domeniconi | George Mason University, USA |
| Yuxiao Dong | Tsinghua University, China |
| Kurt Driessens | Maastricht University, the Netherlands |
| Tapio Elomaa | Tampere University, Finland |
| Sergio Escalera | CVC and University of Barcelona, Spain |
| Faisal Farooq | Qatar Computing Research Institute, Qatar |
| Asja Fischer | Ruhr University Bochum, Germany |
| Peter Flach | University of Bristol, UK |
| Eibe Frank | University of Waikato, New Zealand |
| Paolo Frasconi | Università degli Studi di Firenze, Italy |
| Elisa Fromont | Université Rennes 1, IRISA/Inria, France |
| Johannes Fürnkranz | JKU Linz, Austria |
| Patrick Gallinari | Sorbonne Université, Criteo AI Lab, France |
| Joao Gama | INESC TEC - LIAAD, Portugal |
| Jose Gamez | Universidad de Castilla-La Mancha, Spain |
| Roman Garnett | Washington University in St. Louis, USA |
| Thomas Gärtner | TU Wien, Austria |

| | |
|---|---|
| Aristides Gionis | KTH Royal Institute of Technology, Sweden |
| Francesco Gullo | UniCredit, Italy |
| Stephan Günnemann | Technical University of Munich, Germany |
| Xiangnan He | University of Science and Technology of China, China |
| Daniel Hernandez-Lobato | Universidad Autonoma de Madrid, Spain |
| José Hernández-Orallo | Universität Politècnica de València, Spain |
| Jaakko Hollmén | Aalto University, Finland |
| Andreas Hotho | Universität Würzburg, Germany |
| Eyke Hüllermeier | University of Munich, Germany |
| Neil Hurley | University College Dublin, Ireland |
| Georgiana Ifrim | University College Dublin, Ireland |
| Alipio Jorge | INESC TEC/University of Porto, Portugal |
| Ross King | Chalmers University of Technology, Sweden |
| Arno Knobbe | Leiden University, the Netherlands |
| Yun Sing Koh | University of Auckland, New Zealand |
| Parisa Kordjamshidi | Michigan State University, USA |
| Lars Kotthoff | University of Wyoming, USA |
| Nicolas Kourtellis | Telefonica Research, Spain |
| Danai Koutra | University of Michigan, USA |
| Danica Kragic | KTH Royal Institute of Technology, Sweden |
| Stefan Kramer | Johannes Gutenberg University Mainz, Germany |
| Niklas Lavesson | Blekinge Institute of Technology, Sweden |
| Sébastien Lefèvre | Université de Bretagne Sud/IRISA, France |
| Jefrey Lijffijt | Ghent University, Belgium |
| Marius Lindauer | Leibniz University Hannover, Germany |
| Patrick Loiseau | Inria, France |
| Jose Lozano | UPV/EHU, Spain |
| Jörg Lücke | Universität Oldenburg, Germany |
| Donato Malerba | Università degli Studi di Bari Aldo Moro, Italy |
| Fragkiskos Malliaros | CentraleSupelec, France |
| Giuseppe Manco | ICAR-CNR, Italy |
| Wannes Meert | KU Leuven, Belgium |
| Pauli Miettinen | University of Eastern Finland, Finland |
| Dunja Mladenic | Jožef Stefan Institute, Slovenia |
| Anna Monreale | Università di Pisa, Italy |
| Luis Moreira-Matias | Finiata, Germany |
| Emilie Morvant | University Jean Monnet, St-Etienne, France |
| Sriraam Natarajan | UT Dallas, USA |
| Nuria Oliver | Vodafone Research, USA |
| Panagiotis Papapetrou | Stockholm University, Sweden |
| Laurence Park | WSU, Australia |

| Andrea Passerini | University of Trento, Italy |
| Mykola Pechenizkiy | TU Eindhoven, the Netherlands |
| Dino Pedreschi | University of Pisa, Italy |
| Robert Peharz | Graz University of Technology, Austria |
| Julien Perez | Naver Labs Europe, France |
| Franz Pernkopf | Graz University of Technology, Austria |
| Bernhard Pfahringer | University of Waikato, New Zealand |
| Fabio Pinelli | IMT Lucca, Italy |
| Visvanathan Ramesh | Goethe University Frankfurt, Germany |
| Jesse Read | Ecole Polytechnique, France |
| Zhaochun Ren | Shandong University, China |
| Marian-Andrei Rizoiu | University of Technology Sydney, Australia |
| Celine Robardet | INSA Lyon, France |
| Sriparna Saha | IIT Patna, India |
| Ute Schmid | University of Bamberg, Germany |
| Lars Schmidt-Thieme | University of Hildesheim, Germany |
| Michele Sebag | LISN CNRS, France |
| Thomas Seidl | LMU Munich, Germany |
| Arno Siebes | Universiteit Utrecht, the Netherlands |
| Fabrizio Silvestri | Sapienza, University of Rome, Italy |
| Myra Spiliopoulou | Otto-von-Guericke-University Magdeburg, Germany |
| Yizhou Sun | UCLA, USA |
| Jie Tang | Tsinghua University, China |
| Nikolaj Tatti | Helsinki University, Finland |
| Evimaria Terzi | Boston University, USA |
| Marc Tommasi | Lille University, France |
| Antti Ukkonen | University of Helsinki, Finland |
| Herke van Hoof | University of Amsterdam, the Netherlands |
| Matthijs van Leeuwen | Leiden University, the Netherlands |
| Celine Vens | KU Leuven, Belgium |
| Christel Vrain | University of Orleans, France |
| Jilles Vreeken | CISPA Helmholtz Center for Information Security, Germany |
| Willem Waegeman | Universiteit Gent, Belgium |
| Stefan Wrobel | Fraunhofer IAIS, Germany |
| Xing Xie | Microsoft Research Asia, China |
| Min-Ling Zhang | Southeast University, China |
| Albrecht Zimmermann | Université de Caen Normandie, France |
| Indre Zliobaite | University of Helsinki, Finland |

# Program Committee Members

| | |
|---|---|
| Amos Abbott | Virginia Tech, USA |
| Pedro Abreu | CISUC, Portugal |
| Maribel Acosta | Ruhr University Bochum, Germany |
| Timilehin Aderinola | Insight Centre, University College Dublin, Ireland |
| Linara Adilova | Ruhr University Bochum, Fraunhofer IAIS, Germany |
| Florian Adriaens | KTH, Sweden |
| Azim Ahmadzadeh | Georgia State University, USA |
| Nourhan Ahmed | University of Hildesheim, Germany |
| Deepak Ajwani | University College Dublin, Ireland |
| Amir Hossein Akhavan Rahnama | KTH Royal Institute of Technology, Sweden |
| Aymen Al Marjani | ENS Lyon, France |
| Mehwish Alam | Leibniz Institute for Information Infrastructure, Germany |
| Francesco Alesiani | NEC Laboratories Europe, Germany |
| Omar Alfarisi | ADNOC, Canada |
| Pegah Alizadeh | Ericsson Research, France |
| Reem Alotaibi | King Abdulaziz University, Saudi Arabia |
| Jumanah Alshehri | Temple University, USA |
| Bakhtiar Amen | University of Huddersfield, UK |
| Evelin Amorim | Inesc tec, Portugal |
| Shin Ando | Tokyo University of Science, Japan |
| Thiago Andrade | INESC TEC - LIAAD, Portugal |
| Jean-Marc Andreoli | Naverlabs Europe, France |
| Giuseppina Andresini | University of Bari Aldo Moro, Italy |
| Alessandro Antonucci | IDSIA, Switzerland |
| Xiang Ao | Institute of Computing Technology, CAS, China |
| Siddharth Aravindan | National University of Singapore, Singapore |
| Héber H. Arcolezi | Inria and École Polytechnique, France |
| Adrián Arnaiz-Rodríguez | ELLIS Unit Alicante, Spain |
| Yusuf Arslan | University of Luxembourg, Luxembourg |
| André Artelt | Bielefeld University, Germany |
| Sunil Aryal | Deakin University, Australia |
| Charles Assaad | Easyvista, France |
| Matthias Aßenmacher | Ludwig-Maximilians-Universität München, Germany |
| Zeyar Aung | Masdar Institute, UAE |
| Serge Autexier | DFKI Bremen, Germany |
| Rohit Babbar | Aalto University, Finland |
| Housam Babiker | University of Alberta, Canada |

| | |
|---|---|
| Antonio Bahamonde | University of Oviedo, Spain |
| Maroua Bahri | Inria Paris, France |
| Georgios Balikas | Salesforce, France |
| Maria Bampa | Stockholm University, Sweden |
| Hubert Baniecki | Warsaw University of Technology, Poland |
| Elena Baralis | Politecnico di Torino, Italy |
| Mitra Baratchi | LIACS - University of Leiden, the Netherlands |
| Kalliopi Basioti | Rutgers University, USA |
| Martin Becker | Stanford University, USA |
| Diana Benavides Prado | University of Auckland, New Zealand |
| Anes Bendimerad | LIRIS, France |
| Idir Benouaret | Université Grenoble Alpes, France |
| Isacco Beretta | Università di Pisa, Italy |
| Victor Berger | CEA, France |
| Christoph Bergmeir | Monash University, Australia |
| Cuissart Bertrand | University of Caen, France |
| Antonio Bevilacqua | University College Dublin, Ireland |
| Yaxin Bi | Ulster University, UK |
| Ranran Bian | University of Auckland, New Zealand |
| Adrien Bibal | University of Louvain, Belgium |
| Subhodip Biswas | Virginia Tech, USA |
| Patrick Blöbaum | Amazon AWS, USA |
| Carlos Bobed | University of Zaragoza, Spain |
| Paul Bogdan | USC, USA |
| Chiara Boldrini | CNR, Italy |
| Clément Bonet | Université Bretagne Sud, France |
| Andrea Bontempelli | University of Trento, Italy |
| Ludovico Boratto | University of Cagliari, Italy |
| Stefano Bortoli | Huawei Research Center, Germany |
| Diana-Laura Borza | Babes Bolyai University, Romania |
| Ahcene Boubekki | UiT, Norway |
| Sabri Boughorbel | QCRI, Qatar |
| Paula Branco | University of Ottawa, Canada |
| Jure Brence | Jožef Stefan Institute, Slovenia |
| Martin Breskvar | Jožef Stefan Institute, Slovenia |
| Marco Bressan | University of Milan, Italy |
| Dariusz Brzezinski | Poznan University of Technology, Poland |
| Florian Buettner | German Cancer Research Center, Germany |
| Julian Busch | Siemens Technology, Germany |
| Sebastian Buschjäger | TU Dortmund Artificial Intelligence Unit, Germany |
| Ali Butt | Virginia Tech, USA |

| | |
|---|---|
| Narayanan C. Krishnan | IIT Palakkad, India |
| Xiangrui Cai | Nankai University, China |
| Xiongcai Cai | UNSW Sydney, Australia |
| Zekun Cai | University of Tokyo, Japan |
| Andrea Campagner | Università degli Studi di Milano-Bicocca, Italy |
| Seyit Camtepe | CSIRO Data61, Australia |
| Jiangxia Cao | Chinese Academy of Sciences, China |
| Pengfei Cao | Chinese Academy of Sciences, China |
| Yongcan Cao | University of Texas at San Antonio, USA |
| Cécile Capponi | Aix-Marseille University, France |
| Axel Carlier | Institut National Polytechnique de Toulouse, France |
| Paula Carroll | University College Dublin, Ireland |
| John Cartlidge | University of Bristol, UK |
| Simon Caton | University College Dublin, Ireland |
| Bogdan Cautis | University of Paris-Saclay, France |
| Mustafa Cavus | Warsaw University of Technology, Poland |
| Remy Cazabet | Université Lyon 1, France |
| Josu Ceberio | University of the Basque Country, Spain |
| David Cechák | CEITEC Masaryk University, Czechia |
| Abdulkadir Celikkanat | Technical University of Denmark, Denmark |
| Dumitru-Clementin Cercel | University Politehnica of Bucharest, Romania |
| Christophe Cerisara | CNRS, France |
| Vítor Cerqueira | Dalhousie University, Canada |
| Mattia Cerrato | JGU Mainz, Germany |
| Ricardo Cerri | Federal University of São Carlos, Brazil |
| Hubert Chan | University of Hong Kong, Hong Kong, China |
| Vaggos Chatziafratis | Stanford University, USA |
| Siu Lun Chau | University of Oxford, UK |
| Chaochao Chen | Zhejiang University, China |
| Chuan Chen | Sun Yat-sen University, China |
| Hechang Chen | Jilin University, China |
| Jia Chen | Beihang University, China |
| Jiaoyan Chen | University of Oxford, UK |
| Jiawei Chen | Zhejiang University, China |
| Jin Chen | University of Electronic Science and Technology, China |
| Kuan-Hsun Chen | University of Twente, the Netherlands |
| Lingwei Chen | Wright State University, USA |
| Tianyi Chen | Boston University, USA |
| Wang Chen | Google, USA |
| Xinyuan Chen | Universiti Kuala Lumpur, Malaysia |

| | |
|---|---|
| Laurens Devos | KU Leuven, Belgium |
| Bhaskar Dhariyal | University College Dublin, Ireland |
| Nicola Di Mauro | University of Bari, Italy |
| Aissatou Diallo | University College London, UK |
| Christos Dimitrakakis | University of Neuchatel, Switzerland |
| Jiahao Ding | University of Houston, USA |
| Kaize Ding | Arizona State University, USA |
| Yao-Xiang Ding | Nanjing University, China |
| Guilherme Dinis Junior | Stockholm University, Sweden |
| Nikolaos Dionelis | University of Edinburgh, UK |
| Christos Diou | Harokopio University of Athens, Greece |
| Sonia Djebali | Léonard de Vinci Pôle Universitaire, France |
| Nicolas Dobigeon | University of Toulouse, France |
| Carola Doerr | Sorbonne University, France |
| Ruihai Dong | University College Dublin, Ireland |
| Shuyu Dong | Inria, Université Paris-Saclay, France |
| Yixiang Dong | Xi'an Jiaotong University, China |
| Xin Du | University of Edinburgh, UK |
| Yuntao Du | Nanjing University, China |
| Stefan Duffner | University of Lyon, France |
| Rahul Duggal | Georgia Tech, USA |
| Wouter Duivesteijn | TU Eindhoven, the Netherlands |
| Sebastijan Dumancic | TU Delft, the Netherlands |
| Inês Dutra | University of Porto, Portugal |
| Thomas Dyhre Nielsen | AAU, Denmark |
| Saso Dzeroski | Jožef Stefan Institute, Ljubljana, Slovenia |
| Tome Eftimov | Jožef Stefan Institute, Ljubljana, Slovenia |
| Hamid Eghbal-zadeh | LIT AI Lab, Johannes Kepler University, Austria |
| Theresa Eimer | Leibniz University Hannover, Germany |
| Radwa El Shawi | Tartu University, Estonia |
| Dominik Endres | Philipps-Universität Marburg, Germany |
| Roberto Esposito | Università di Torino, Italy |
| Georgios Evangelidis | University of Macedonia, Greece |
| Samuel Fadel | Leuphana University, Germany |
| Stephan Fahrenkrog-Petersen | Humboldt-Universität zu Berlin, Germany |
| Xiaomao Fan | Shenzhen Technology University, China |
| Zipei Fan | University of Tokyo, Japan |
| Hadi Fanaee | Halmstad University, Sweden |
| Meng Fang | TU/e, the Netherlands |
| Elaine Faria | UFU, Brazil |
| Ad Feelders | Universiteit Utrecht, the Netherlands |
| Sophie Fellenz | TU Kaiserslautern, Germany |

| Stefano Ferilli | University of Bari, Italy |
| Daniel Fernández-Sánchez | Universidad Autónoma de Madrid, Spain |
| Pedro Ferreira | Faculty of Sciences University of Porto, Portugal |
| Cèsar Ferri | Universität Politècnica València, Spain |
| Flavio Figueiredo | UFMG, Brazil |
| Soukaina Filali Boubrahimi | Utah State University, USA |
| Raphael Fischer | TU Dortmund, Germany |
| Germain Forestier | University of Haute Alsace, France |
| Edouard Fouché | Karlsruhe Institute of Technology, Germany |
| Philippe Fournier-Viger | Shenzhen University, China |
| Kary Framling | Umeå University, Sweden |
| Jérôme François | Inria Nancy Grand-Est, France |
| Fabio Fumarola | Prometeia, Italy |
| Pratik Gajane | Eindhoven University of Technology, the Netherlands |
| Esther Galbrun | University of Eastern Finland, Finland |
| Laura Galindez Olascoaga | KU Leuven, Belgium |
| Sunanda Gamage | University of Western Ontario, Canada |
| Chen Gao | Tsinghua University, China |
| Wei Gao | Nanjing University, China |
| Xiaofeng Gao | Shanghai Jiaotong University, China |
| Yuan Gao | University of Science and Technology of China, China |
| Jochen Garcke | University of Bonn, Germany |
| Clement Gautrais | Brightclue, France |
| Benoit Gauzere | INSA Rouen, France |
| Dominique Gay | Université de La Réunion, France |
| Xiou Ge | University of Southern California, USA |
| Bernhard Geiger | Know-Center GmbH, Germany |
| Jiahui Geng | University of Stavanger, Norway |
| Yangliao Geng | Tsinghua University, China |
| Konstantin Genin | University of Tübingen, Germany |
| Firas Gerges | New Jersey Institute of Technology, USA |
| Pierre Geurts | University of Liège, Belgium |
| Gizem Gezici | Sabanci University, Turkey |
| Amirata Ghorbani | Stanford, USA |
| Biraja Ghoshal | TCS, UK |
| Anna Giabelli | Università degli studi di Milano Bicocca, Italy |
| George Giannakopoulos | IIT Demokritos, Greece |
| Tobias Glasmachers | Ruhr-University Bochum, Germany |
| Heitor Murilo Gomes | University of Waikato, New Zealand |
| Anastasios Gounaris | Aristotle University of Thessaloniki, Greece |

| | |
|---|---|
| Antoine Gourru | University of Lyon, France |
| Michael Granitzer | University of Passau, Germany |
| Magda Gregorova | Hochschule Würzburg-Schweinfurt, Germany |
| Moritz Grosse-Wentrup | University of Vienna, Austria |
| Divya Grover | Chalmers University, Sweden |
| Bochen Guan | OPPO US Research Center, USA |
| Xinyu Guan | Xian Jiaotong University, China |
| Guillaume Guerard | ESILV, France |
| Daniel Guerreiro e Silva | University of Brasilia, Brazil |
| Riccardo Guidotti | University of Pisa, Italy |
| Ekta Gujral | University of California, Riverside, USA |
| Aditya Gulati | ELLIS Unit Alicante, Spain |
| Guibing Guo | Northeastern University, China |
| Jianxiong Guo | Beijing Normal University, China |
| Yuhui Guo | Renmin University of China, China |
| Karthik Gurumoorthy | Amazon, India |
| Thomas Guyet | Inria, Centre de Lyon, France |
| Guillaume Habault | KDDI Research, Inc., Japan |
| Amaury Habrard | University of St-Etienne, France |
| Shahrzad Haddadan | Brown University, USA |
| Shah Muhammad Hamdi | New Mexico State University, USA |
| Massinissa Hamidi | PRES Sorbonne Paris Cité, France |
| Peng Han | KAUST, Saudi Arabia |
| Tom Hanika | University of Kassel, Germany |
| Sébastien Harispe | IMT Mines Alès, France |
| Marwan Hassani | TU Eindhoven, the Netherlands |
| Kohei Hayashi | Preferred Networks, Inc., Japan |
| Conor Hayes | National University of Ireland Galway, Ireland |
| Lingna He | Zhejiang University of Technology, China |
| Ramya Hebbalaguppe | Indian Institute of Technology, Delhi, India |
| Jukka Heikkonen | University of Turku, Finland |
| Fredrik Heintz | Linköping University, Sweden |
| Patrick Hemmer | Karlsruhe Institute of Technology, Germany |
| Romain Hérault | INSA de Rouen, France |
| Jeronimo Hernandez-Gonzalez | University of Barcelona, Spain |
| Sibylle Hess | TU Eindhoven, the Netherlands |
| Fabian Hinder | Bielefeld University, Germany |
| Lars Holdijk | University of Amsterdam, the Netherlands |
| Martin Holena | Institute of Computer Science, Czechia |
| Mike Holenderski | Eindhoven University of Technology, the Netherlands |
| Shenda Hong | Peking University, China |

| Yupeng Hou | Renmin University of China, China |
| Binbin Hu | Ant Financial Services Group, China |
| Jian Hu | Queen Mary University of London, UK |
| Liang Hu | Tongji University, China |
| Wen Hu | Ant Group, China |
| Wenbin Hu | Wuhan University, China |
| Wenbo Hu | Tsinghua University, China |
| Yaowei Hu | University of Arkansas, USA |
| Chao Huang | University of Hong Kong, China |
| Gang Huang | Zhejiang Lab, China |
| Guanjie Huang | Penn State University, USA |
| Hong Huang | HUST, China |
| Jin Huang | University of Amsterdam, the Netherlands |
| Junjie Huang | Chinese Academy of Sciences, China |
| Qiang Huang | Jilin University, China |
| Shangrong Huang | Hunan University, China |
| Weitian Huang | South China University of Technology, China |
| Yan Huang | Huazhong University of Science and Technology, China |
| Yiran Huang | Karlsruhe Institute of Technology, Germany |
| Angelo Impedovo | University of Bari, Italy |
| Roberto Interdonato | CIRAD, France |
| Iñaki Inza | University of the Basque Country, Spain |
| Stratis Ioannidis | Northeastern University, USA |
| Rakib Islam | Facebook, USA |
| Tobias Jacobs | NEC Laboratories Europe GmbH, Germany |
| Priyank Jaini | Google, Canada |
| Johannes Jakubik | Karlsruhe Institute of Technology, Germany |
| Nathalie Japkowicz | American University, USA |
| Szymon Jaroszewicz | Polish Academy of Sciences, Poland |
| Shayan Jawed | University of Hildesheim, Germany |
| Rathinaraja Jeyaraj | Kyungpook National University, South Korea |
| Shaoxiong Ji | Aalto University, Finland |
| Taoran Ji | Virginia Tech, USA |
| Bin-Bin Jia | Southeast University, China |
| Yuheng Jia | Southeast University, China |
| Ziyu Jia | Beijing Jiaotong University, China |
| Nan Jiang | Purdue University, USA |
| Renhe Jiang | University of Tokyo, Japan |
| Siyang Jiang | National Taiwan University, Taiwan |
| Song Jiang | University of California, Los Angeles, USA |
| Wenyu Jiang | Nanjing University, China |

| Zhen Jiang | Jiangsu University, China |
| Yuncheng Jiang | South China Normal University, China |
| François-Xavier Jollois | Université de Paris Cité, France |
| Adan Jose-Garcia | Université de Lille, France |
| Ferdian Jovan | University of Bristol, UK |
| Steffen Jung | MPII, Germany |
| Thorsten Jungeblut | Bielefeld University of Applied Sciences, Germany |
| Hachem Kadri | Aix-Marseille University, France |
| Vana Kalogeraki | Athens University of Economics and Business, Greece |
| Vinayaka Kamath | Microsoft Research India, India |
| Toshihiro Kamishima | National Institute of Advanced Industrial Science, Japan |
| Bo Kang | Ghent University, Belgium |
| Alexandros Karakasidis | University of Macedonia, Greece |
| Mansooreh Karami | Arizona State University, USA |
| Panagiotis Karras | Aarhus University, Denmark |
| Ioannis Katakis | University of Nicosia, Cyprus |
| Koki Kawabata | Osaka University, Tokyo |
| Klemen Kenda | Jožef Stefan Institute, Slovenia |
| Patrik Joslin Kenfack | Innopolis University, Russia |
| Mahsa Keramati | Simon Fraser University, Canada |
| Hamidreza Keshavarz | Tarbiat Modares University, Iran |
| Adil Khan | Innopolis University, Russia |
| Jihed Khiari | Johannes Kepler University, Austria |
| Mi-Young Kim | University of Alberta, Canada |
| Arto Klami | University of Helsinki, Finland |
| Jiri Klema | Czech Technical University, Czechia |
| Tomas Kliegr | University of Economics Prague, Czechia |
| Christian Knoll | Graz, University of Technology, Austria |
| Dmitry Kobak | University of Tübingen, Germany |
| Vladimer Kobayashi | University of the Philippines Mindanao, Philippines |
| Dragi Kocev | Jožef Stefan Institute, Slovenia |
| Adrian Kochsiek | University of Mannheim, Germany |
| Masahiro Kohjima | NTT Corporation, Japan |
| Georgia Koloniari | University of Macedonia, Greece |
| Nikos Konofaos | Aristotle University of Thessaloniki, Greece |
| Irena Koprinska | University of Sydney, Australia |
| Lars Kotthoff | University of Wyoming, USA |
| Daniel Kottke | University of Kassel, Germany |

| | |
|---|---|
| Anna Krause | University of Würzburg, Germany |
| Alexander Kravberg | KTH Royal Institute of Technology, Sweden |
| Anastasia Krithara | NCSR Demokritos, Greece |
| Meelis Kull | University of Tartu, Estonia |
| Pawan Kumar | IIIT, Hyderabad, India |
| Suresh Kirthi Kumaraswamy | InterDigital, France |
| Gautam Kunapuli | Verisk Inc, USA |
| Marcin Kurdziel | AGH University of Science and Technology, Poland |
| Vladimir Kuzmanovski | Aalto University, Finland |
| Ariel Kwiatkowski | École Polytechnique, France |
| Firas Laakom | Tampere University, Finland |
| Harri Lähdesmäki | Aalto University, Finland |
| Stefanos Laskaridis | Samsung AI, UK |
| Alberto Lavelli | FBK-ict, Italy |
| Aonghus Lawlor | University College Dublin, Ireland |
| Thai Le | University of Mississippi, USA |
| Hoàng-Ân Lê | IRISA, University of South Brittany, France |
| Hoel Le Capitaine | University of Nantes, France |
| Thach Le Nguyen | Insight Centre, Ireland |
| Tai Le Quy | L3S Research Center - Leibniz University Hannover, Germany |
| Mustapha Lebbah | Sorbonne Paris Nord University, France |
| Dongman Lee | KAIST, South Korea |
| John Lee | Université catholique de Louvain, Belgium |
| Minwoo Lee | University of North Carolina at Charlotte, USA |
| Zed Lee | Stockholm University, Sweden |
| Yunwen Lei | University of Birmingham, UK |
| Douglas Leith | Trinity College Dublin, Ireland |
| Florian Lemmerich | RWTH Aachen, Germany |
| Carson Leung | University of Manitoba, Canada |
| Chaozhuo Li | Microsoft Research Asia, China |
| Jian Li | Institute of Information Engineering, China |
| Lei Li | Peking University, China |
| Li Li | Southwest University, China |
| Rui Li | Inspur Group, China |
| Shiyang Li | UCSB, USA |
| Shuokai Li | Chinese Academy of Sciences, China |
| Tianyu Li | Alibaba Group, China |
| Wenye Li | The Chinese University of Hong Kong, Shenzhen, China |
| Wenzhong Li | Nanjing University, China |

| | |
|---|---|
| Xiaoting Li | Pennsylvania State University, USA |
| Yang Li | University of North Carolina at Chapel Hill, USA |
| Zejian Li | Zhejiang University, China |
| Zhidong Li | UTS, Australia |
| Zhixin Li | Guangxi Normal University, China |
| Defu Lian | University of Science and Technology of China, China |
| Bin Liang | UTS, Australia |
| Yuchen Liang | RPI, USA |
| Yiwen Liao | University of Stuttgart, Germany |
| Pieter Libin | VUB, Belgium |
| Thomas Liebig | TU Dortmund, Germany |
| Seng Pei Liew | LINE Corporation, Japan |
| Beiyu Lin | University of Nevada - Las Vegas, USA |
| Chen Lin | Xiamen University, China |
| Tony Lindgren | Stockholm University, Sweden |
| Chen Ling | Emory University, USA |
| Jiajing Ling | Singapore Management University, Singapore |
| Marco Lippi | University of Modena and Reggio Emilia, Italy |
| Bin Liu | Chongqing University, China |
| Bowen Liu | Stanford University, USA |
| Chang Liu | Institute of Information Engineering, CAS, China |
| Chien-Liang Liu | National Chiao Tung University, Taiwan |
| Feng Liu | East China Normal University, China |
| Jiacheng Liu | Chinese University of Hong Kong, China |
| Li Liu | Chongqing University, China |
| Shengcai Liu | Southern University of Science and Technology, China |
| Shenghua Liu | Institute of Computing Technology, CAS, China |
| Tingwen Liu | Institute of Information Engineering, CAS, China |
| Xiangyu Liu | Tencent, China |
| Yong Liu | Renmin University of China, China |
| Yuansan Liu | University of Melbourne, Australia |
| Zhiwei Liu | Salesforce, USA |
| Tuwe Löfström | Jönköping University, Sweden |
| Corrado Loglisci | Università degli Studi di Bari Aldo Moro, Italy |
| Ting Long | Shanghai Jiao Tong University, China |
| Beatriz López | University of Girona, Spain |
| Yin Lou | Ant Group, USA |
| Samir Loudni | TASC (LS2N-CNRS), IMT Atlantique, France |
| Yang Lu | Xiamen University, China |
| Yuxun Lu | National Institute of Informatics, Japan |

| | |
|---|---|
| Massimiliano Luca | Bruno Kessler Foundation, Italy |
| Stefan Lüdtke | University of Mannheim, Germany |
| Jovita Lukasik | University of Mannheim, Germany |
| Denis Lukovnikov | University of Bonn, Germany |
| Pedro Henrique Luz de Araujo | University of Brasília, Brazil |
| Fenglong Ma | Pennsylvania State University, USA |
| Jing Ma | University of Virginia, USA |
| Meng Ma | Peking University, China |
| Muyang Ma | Shandong University, China |
| Ruizhe Ma | University of Massachusetts Lowell, USA |
| Xingkong Ma | National University of Defense Technology, China |
| Xueqi Ma | Tsinghua University, China |
| Zichen Ma | The Chinese University of Hong Kong, Shenzhen, China |
| Luis Macedo | University of Coimbra, Portugal |
| Harshitha Machiraju | EPFL, Switzerland |
| Manchit Madan | Delivery Hero, Germany |
| Seiji Maekawa | Osaka University, Japan |
| Sindri Magnusson | Stockholm University, Sweden |
| Pathum Chamikara Mahawaga | CSIRO Data61, Australia |
| Saket Maheshwary | Amazon, India |
| Ajay Mahimkar | AT&T, USA |
| Pierre Maillot | Inria, France |
| Lorenzo Malandri | Unimib, Italy |
| Rammohan Mallipeddi | Kyungpook National University, South Korea |
| Sahil Manchanda | IIT Delhi, India |
| Domenico Mandaglio | DIMES-UNICAL, Italy |
| Panagiotis Mandros | Harvard University, USA |
| Robin Manhaeve | KU Leuven, Belgium |
| Silviu Maniu | Université Paris-Saclay, France |
| Cinmayii Manliguez | National Sun Yat-Sen University, Taiwan |
| Naresh Manwani | International Institute of Information Technology, India |
| Jiali Mao | East China Normal University, China |
| Alexandru Mara | Ghent University, Belgium |
| Radu Marculescu | University of Texas at Austin, USA |
| Roger Mark | Massachusetts Institute of Technology, USA |
| Fernando Martínez-Plume | Joint Research Centre - European Commission, Belgium |
| Koji Maruhashi | Fujitsu Research, Fujitsu Limited, Japan |
| Simone Marullo | University of Siena, Italy |

| | |
|---|---|
| Elio Masciari | University of Naples, Italy |
| Florent Masseglia | Inria, France |
| Michael Mathioudakis | University of Helsinki, Finland |
| Takashi Matsubara | Osaka University, Japan |
| Tetsu Matsukawa | Kyushu University, Japan |
| Santiago Mazuelas | BCAM-Basque Center for Applied Mathematics, Spain |
| Ryan McConville | University of Bristol, UK |
| Hardik Meisheri | TCS Research, India |
| Panagiotis Meletis | Eindhoven University of Technology, the Netherlands |
| Gabor Melli | Medable, USA |
| Joao Mendes-Moreira | INESC TEC, Portugal |
| Chuan Meng | University of Amsterdam, the Netherlands |
| Cristina Menghini | Brown University, USA |
| Engelbert Mephu Nguifo | Université Clermont Auvergne, CNRS, LIMOS, France |
| Fabio Mercorio | University of Milan-Bicocca, Italy |
| Guillaume Metzler | Laboratoire ERIC, France |
| Hao Miao | Aalborg University, Denmark |
| Alessio Micheli | Università di Pisa, Italy |
| Paolo Mignone | University of Bari Aldo Moro, Italy |
| Matej Mihelcic | University of Zagreb, Croatia |
| Ioanna Miliou | Stockholm University, Sweden |
| Bamdev Mishra | Microsoft, India |
| Rishabh Misra | Twitter, Inc, USA |
| Dixant Mittal | National University of Singapore, Singapore |
| Zhaobin Mo | Columbia University, USA |
| Daichi Mochihashi | Institute of Statistical Mathematics, Japan |
| Armin Moharrer | Northeastern University, USA |
| Ioannis Mollas | Aristotle University of Thessaloniki, Greece |
| Carlos Monserrat-Aranda | Universität Politècnica de València, Spain |
| Konda Reddy Mopuri | Indian Institute of Technology Guwahati, India |
| Raha Moraffah | Arizona State University, USA |
| Pawel Morawiecki | Polish Academy of Sciences, Poland |
| Ahmadreza Mosallanezhad | Arizona State University, USA |
| Davide Mottin | Aarhus University, Denmark |
| Koyel Mukherjee | Adobe Research, India |
| Maximilian Münch | University of Applied Sciences Würzburg, Germany |
| Fabricio Murai | Universidade Federal de Minas Gerais, Brazil |
| Taichi Murayama | NAIST, Japan |

| | |
|---|---|
| Stéphane Mussard | CHROME, France |
| Mohamed Nadif | Centre Borelli - Université Paris Cité, France |
| Cian Naik | University of Oxford, UK |
| Felipe Kenji Nakano | KU Leuven, Belgium |
| Mirco Nanni | ISTI-CNR Pisa, Italy |
| Apurva Narayan | University of Waterloo, Canada |
| Usman Naseem | University of Sydney, Australia |
| Gergely Nemeth | ELLIS Unit Alicante, Spain |
| Stefan Neumann | KTH Royal Institute of Technology, Sweden |
| Anna Nguyen | Karlsruhe Institute of Technology, Germany |
| Quan Nguyen | Washington University in St. Louis, USA |
| Thi Phuong Quyen Nguyen | University of Da Nang, Vietnam |
| Thu Nguyen | SimulaMet, Norway |
| Thu Trang Nguyen | University College Dublin, Ireland |
| Prajakta Nimbhorkar | Chennai Mathematical Institute, Chennai, India |
| Xuefei Ning | Tsinghua University, China |
| Ikuko Nishikawa | Ritsumeikan University, Japan |
| Hao Niu | KDDI Research, Inc., Japan |
| Paraskevi Nousi | Aristotle University of Thessaloniki, Greece |
| Erik Novak | Jožef Stefan Institute, Slovenia |
| Slawomir Nowaczyk | Halmstad University, Sweden |
| Aleksandra Nowak | Jagiellonian University, Poland |
| Eirini Ntoutsi | Freie Universität Berlin, Germany |
| Andreas Nürnberger | Magdeburg University, Germany |
| James O'Neill | University of Liverpool, UK |
| Lutz Oettershagen | University of Bonn, Germany |
| Tsuyoshi Okita | Kyushu Institute of Technology, Japan |
| Makoto Onizuka | Osaka University, Japan |
| Subba Reddy Oota | IIIT Hyderabad, India |
| María Óskarsdóttir | University of Reykjavík, Iceland |
| Aomar Osmani | PRES Sorbonne Paris Cité, France |
| Aljaz Osojnik | JSI, Slovenia |
| Shuichi Otake | National Institute of Informatics, Japan |
| Greger Ottosson | IBM, France |
| Zijing Ou | Sun Yat-sen University, China |
| Abdelkader Ouali | University of Caen Normandy, France |
| Latifa Oukhellou | IFSTTAR, France |
| Kai Ouyang | Tsinghua University, France |
| Andrei Paleyes | University of Cambridge, UK |
| Pankaj Pandey | Indian Institute of Technology Gandhinagar, India |
| Guansong Pang | Singapore Management University, Singapore |
| Pance Panov | Jožef Stefan Institute, Slovenia |

| | |
|---|---|
| Apostolos Papadopoulos | Aristotle University of Thessaloniki, Greece |
| Evangelos Papalexakis | UC Riverside, USA |
| Anna Pappa | Université Paris 8, France |
| Chanyoung Park | UIUC, USA |
| Haekyu Park | Georgia Institute of Technology, USA |
| Sanghyun Park | Yonsei University, South Korea |
| Luca Pasa | University of Padova, Italy |
| Kevin Pasini | IRT SystemX, France |
| Vincenzo Pasquadibisceglie | University of Bari Aldo Moro, Italy |
| Nikolaos Passalis | Aristotle University of Thessaloniki, Greece |
| Javier Pastorino | University of Colorado, Denver, USA |
| Kitsuchart Pasupa | King Mongkut's Institute of Technology, Thailand |
| Andrea Paudice | University of Milan, Italy |
| Anand Paul | Kyungpook National University, South Korea |
| Yulong Pei | TU Eindhoven, the Netherlands |
| Charlotte Pelletier | Université de Bretagne du Sud, France |
| Jaakko Peltonen | Tampere University, Finland |
| Ruggero Pensa | University of Torino, Italy |
| Fabiola Pereira | Federal University of Uberlandia, Brazil |
| Lucas Pereira | ITI, LARSyS, Técnico Lisboa, Portugal |
| Aritz Pérez | Basque Center for Applied Mathematics, Spain |
| Lorenzo Perini | KU Leuven, Belgium |
| Alan Perotti | CENTAI Institute, Italy |
| Michaël Perrot | Inria Lille, France |
| Matej Petkovic | Institute Jožef Stefan, Slovenia |
| Lukas Pfahler | TU Dortmund University, Germany |
| Nico Piatkowski | Fraunhofer IAIS, Germany |
| Francesco Piccialli | University of Naples Federico II, Italy |
| Gianvito Pio | University of Bari, Italy |
| Giuseppe Pirrò | Sapienza University of Rome, Italy |
| Marc Plantevit | EPITA, France |
| Konstantinos Pliakos | KU Leuven, Belgium |
| Matthias Pohl | Otto von Guericke University, Germany |
| Nicolas Posocco | EURA NOVA, Belgium |
| Cedric Pradalier | GeorgiaTech Lorraine, France |
| Paul Prasse | University of Potsdam, Germany |
| Mahardhika Pratama | University of South Australia, Australia |
| Francesca Pratesi | ISTI - CNR, Italy |
| Steven Prestwich | University College Cork, Ireland |
| Giulia Preti | CentAI, Italy |
| Philippe Preux | Inria, France |
| Shalini Priya | Oak Ridge National Laboratory, USA |

| | |
|---|---|
| Ricardo Prudencio | Universidade Federal de Pernambuco, Brazil |
| Luca Putelli | Università degli Studi di Brescia, Italy |
| Peter van der Putten | Leiden University, the Netherlands |
| Chuan Qin | Baidu, China |
| Jixiang Qing | Ghent University, Belgium |
| Jolin Qu | Western Sydney University, Australia |
| Nicolas Quesada | Polytechnique Montreal, Canada |
| Teeradaj Racharak | Japan Advanced Institute of Science and Technology, Japan |
| Krystian Radlak | Warsaw University of Technology, Poland |
| Sandro Radovanovic | University of Belgrade, Serbia |
| Md Masudur Rahman | Purdue University, USA |
| Ankita Raj | Indian Institute of Technology Delhi, India |
| Herilalaina Rakotoarison | Inria, France |
| Alexander Rakowski | Hasso Plattner Institute, Germany |
| Jan Ramon | Inria, France |
| Sascha Ranftl | Graz University of Technology, Austria |
| Aleksandra Rashkovska Koceva | Jožef Stefan Institute, Slovenia |
| S. Ravi | Biocomplexity Institute, USA |
| Jesse Read | Ecole Polytechnique, France |
| David Reich | Universität Potsdam, Germany |
| Marina Reyboz | CEA, LIST, France |
| Pedro Ribeiro | University of Porto, Portugal |
| Rita P. Ribeiro | University of Porto, Portugal |
| Piera Riccio | ELLIS Unit Alicante Foundation, Spain |
| Christophe Rigotti | INSA Lyon, France |
| Matteo Riondato | Amherst College, USA |
| Mateus Riva | Telecom ParisTech, France |
| Kit Rodolfa | CMU, USA |
| Christophe Rodrigues | DVRC Pôle Universitaire Léonard de Vinci, France |
| Simon Rodríguez-Santana | ICMAT, Spain |
| Gaetano Rossiello | IBM Research, USA |
| Mohammad Rostami | University of Southern California, USA |
| Franz Rothlauf | Mainz Universität, Germany |
| Celine Rouveirol | Université Paris-Nord, France |
| Arjun Roy | Freie Universität Berlin, Germany |
| Joze Rozanec | Josef Stefan International Postgraduate School, Slovenia |
| Salvatore Ruggieri | University of Pisa, Italy |
| Marko Ruman | UTIA, AV CR, Czechia |
| Ellen Rushe | University College Dublin, Ireland |

| | |
|---|---|
| Dawid Rymarczyk | Jagiellonian University, Poland |
| Amal Saadallah | TU Dortmund, Germany |
| Khaled Mohammed Saifuddin | Georgia State University, USA |
| Hajer Salem | AUDENSIEL, France |
| Francesco Salvetti | Politecnico di Torino, Italy |
| Roberto Santana | University of the Basque Country (UPV/EHU), Spain |
| KC Santosh | University of South Dakota, USA |
| Somdeb Sarkhel | Adobe, USA |
| Yuya Sasaki | Osaka University, Japan |
| Yücel Saygın | Sabancı Universitesi, Turkey |
| Patrick Schäfer | Humboldt-Universität zu Berlin, Germany |
| Alexander Schiendorfer | Technische Hochschule Ingolstadt, Germany |
| Peter Schlicht | Volkswagen Group Research, Germany |
| Daniel Schmidt | Monash University, Australia |
| Johannes Schneider | University of Liechtenstein, Liechtenstein |
| Steven Schockaert | Cardiff University, UK |
| Jens Schreiber | University of Kassel, Germany |
| Matthias Schubert | Ludwig-Maximilians-Universität München, Germany |
| Alexander Schulz | CITEC, Bielefeld University, Germany |
| Jan-Philipp Schulze | Fraunhofer AISEC, Germany |
| Andreas Schwung | Fachhochschule Südwestfalen, Germany |
| Vasile-Marian Scuturici | LIRIS, France |
| Raquel Sebastião | IEETA/DETI-UA, Portugal |
| Stanislav Selitskiy | University of Bedfordshire, UK |
| Edoardo Serra | Boise State University, USA |
| Lorenzo Severini | UniCredit, R&D Dept., Italy |
| Tapan Shah | GE, USA |
| Ammar Shaker | NEC Laboratories Europe, Germany |
| Shiv Shankar | University of Massachusetts, USA |
| Junming Shao | University of Electronic Science and Technology, China |
| Kartik Sharma | Georgia Institute of Technology, USA |
| Manali Sharma | Samsung, USA |
| Ariona Shashaj | Network Contacts, Italy |
| Betty Shea | University of British Columbia, Canada |
| Chengchao Shen | Central South University, China |
| Hailan Shen | Central South University, China |
| Jiawei Sheng | Chinese Academy of Sciences, China |
| Yongpan Sheng | Southwest University, China |
| Chongyang Shi | Beijing Institute of Technology, China |

| | |
|---|---|
| Zhengxiang Shi | University College London, UK |
| Naman Shukla | Deepair LLC, USA |
| Pablo Silva | Dell Technologies, Brazil |
| Simeon Simoff | Western Sydney University, Australia |
| Maneesh Singh | Motive Technologies, USA |
| Nikhil Singh | MIT Media Lab, USA |
| Sarath Sivaprasad | IIIT Hyderabad, India |
| Elena Sizikova | NYU, USA |
| Andrzej Skowron | University of Warsaw, Poland |
| Blaz Skrlj | Institute Jožef Stefan, Slovenia |
| Oliver Snow | Simon Fraser University, Canada |
| Jonas Soenen | KU Leuven, Belgium |
| Nataliya Sokolovska | Sorbonne University, France |
| K. M. A. Solaiman | Purdue University, USA |
| Shuangyong Song | Jing Dong, China |
| Zixing Song | The Chinese University of Hong Kong, China |
| Tiberiu Sosea | University of Illinois at Chicago, USA |
| Arnaud Soulet | University of Tours, France |
| Lucas Souza | UFRJ, Brazil |
| Jens Sparsø | Technical University of Denmark, Denmark |
| Vivek Srivastava | TCS Research, USA |
| Marija Stanojevic | Temple University, USA |
| Jerzy Stefanowski | Poznan University of Technology, Poland |
| Simon Stieber | University of Augsburg, Germany |
| Jinyan Su | University of Electronic Science and Technology, China |
| Yongduo Sui | University of Science and Technology of China, China |
| Huiyan Sun | Jilin University, China |
| Yuwei Sun | University of Tokyo/RIKEN AIP, Japan |
| Gokul Swamy | Amazon, USA |
| Maryam Tabar | Pennsylvania State University, USA |
| Anika Tabassum | Virginia Tech, USA |
| Shazia Tabassum | INESCTEC, Portugal |
| Koji Tabata | Hokkaido University, Japan |
| Andrea Tagarelli | DIMES, University of Calabria, Italy |
| Etienne Tajeuna | Université de Laval, Canada |
| Acar Tamersoy | NortonLifeLock Research Group, USA |
| Chang Wei Tan | Monash University, Australia |
| Cheng Tan | Westlake University, China |
| Feilong Tang | Shanghai Jiao Tong University, China |
| Feng Tao | Volvo Cars, USA |

| | |
|---|---|
| Youming Tao | Shandong University, China |
| Martin Tappler | Graz University of Technology, Austria |
| Garth Tarr | University of Sydney, Australia |
| Mohammad Tayebi | Simon Fraser University, Canada |
| Anastasios Tefas | Aristotle University of Thessaloniki, Greece |
| Maguelonne Teisseire | INRAE - UMR Tetis, France |
| Stefano Teso | University of Trento, Italy |
| Olivier Teste | IRIT, University of Toulouse, France |
| Maximilian Thiessen | TU Wien, Austria |
| Eleftherios Tiakas | Aristotle University of Thessaloniki, Greece |
| Hongda Tian | University of Technology Sydney, Australia |
| Alessandro Tibo | Aalborg University, Denmark |
| Aditya Srinivas Timmaraju | Facebook, USA |
| Christos Tjortjis | International Hellenic University, Greece |
| Ljupco Todorovski | University of Ljubljana, Slovenia |
| Laszlo Toka | BME, Hungary |
| Ancy Tom | University of Minnesota, Twin Cities, USA |
| Panagiotis Traganitis | Michigan State University, USA |
| Cuong Tran | Syracuse University, USA |
| Minh-Tuan Tran | KAIST, South Korea |
| Giovanni Trappolini | Sapienza University of Rome, Italy |
| Volker Tresp | LMU, Germany |
| Yu-Chee Tseng | National Yang Ming Chiao Tung University, Taiwan |
| Maria Tzelepi | Aristotle University of Thessaloniki, Greece |
| Willy Ugarte | University of Applied Sciences (UPC), Peru |
| Antti Ukkonen | University of Helsinki, Finland |
| Abhishek Kumar Umrawal | Purdue University, USA |
| Athena Vakal | Aristotle University, Greece |
| Matias Valdenegro Toro | University of Groningen, the Netherlands |
| Maaike Van Roy | KU Leuven, Belgium |
| Dinh Van Tran | University of Freiburg, Germany |
| Fabio Vandin | University of Padova, Italy |
| Valerie Vaquet | CITEC, Bielefeld University, Germany |
| Iraklis Varlamis | Harokopio University of Athens, Greece |
| Santiago Velasco-Forero | MINES ParisTech, France |
| Bruno Veloso | Porto, Portugal |
| Dmytro Velychko | Carl von Ossietzky Universität Oldenburg, Germany |
| Sreekanth Vempati | Myntra, India |
| Sebastián Ventura Soto | University of Cordoba, Portugal |
| Rosana Veroneze | LBiC, Brazil |

| Jan Verwaeren | Ghent University, Belgium |
|---|---|
| Vassilios Verykios | Hellenic Open University, Greece |
| Herna Viktor | University of Ottawa, Canada |
| João Vinagre | LIAAD - INESC TEC, Portugal |
| Fabio Vitale | Centai Institute, Italy |
| Vasiliki Voukelatou | ISTI - CNR, Italy |
| Dong Quan Vu | Safran Tech, France |
| Maxime Wabartha | McGill University, Canada |
| Tomasz Walkowiak | Wroclaw University of Science and Technology, Poland |
| Vijay Walunj | University of Missouri-Kansas City, USA |
| Michael Wand | University of Mainz, Germany |
| Beilun Wang | Southeast University, China |
| Chang-Dong Wang | Sun Yat-sen University, China |
| Daheng Wang | Amazon, USA |
| Deng-Bao Wang | Southeast University, China |
| Di Wang | Nanyang Technological University, Singapore |
| Di Wang | KAUST, Saudi Arabia |
| Fu Wang | University of Exeter, UK |
| Hao Wang | Nanyang Technological University, Singapore |
| Hao Wang | Louisiana State University, USA |
| Hao Wang | University of Science and Technology of China, China |
| Hongwei Wang | University of Illinois Urbana-Champaign, USA |
| Hui Wang | SKLSDE, China |
| Hui (Wendy) Wang | Stevens Institute of Technology, USA |
| Jia Wang | Xi'an Jiaotong-Liverpool University, China |
| Jing Wang | Beijing Jiaotong University, China |
| Junxiang Wang | Emory University, USA |
| Qing Wang | IBM Research, USA |
| Rongguang Wang | University of Pennsylvania, USA |
| Ruoyu Wang | Shanghai Jiao Tong University, China |
| Ruxin Wang | Shenzhen Institutes of Advanced Technology, China |
| Senzhang Wang | Central South University, China |
| Shoujin Wang | Macquarie University, Australia |
| Xi Wang | Chinese Academy of Sciences, China |
| Yanchen Wang | Georgetown University, USA |
| Ye Wang | Chongqing University, China |
| Ye Wang | National University of Singapore, Singapore |
| Yifei Wang | Peking University, China |
| Yongqing Wang | Chinese Academy of Sciences, China |

| | |
|---|---|
| Yuandong Wang | Tsinghua University, China |
| Yue Wang | Microsoft Research, USA |
| Yun Cheng Wang | University of Southern California, USA |
| Zhaonan Wang | University of Tokyo, Japan |
| Zhaoxia Wang | SMU, Singapore |
| Zhiwei Wang | University of Chinese Academy of Sciences, China |
| Zihan Wang | Shandong University, China |
| Zijie J. Wang | Georgia Tech, USA |
| Dilusha Weeraddana | CSIRO, Australia |
| Pascal Welke | University of Bonn, Germany |
| Tobias Weller | University of Mannheim, Germany |
| Jörg Wicker | University of Auckland, New Zealand |
| Lena Wiese | Goethe University Frankfurt, Germany |
| Michael Wilbur | Vanderbilt University, USA |
| Moritz Wolter | Bonn University, Germany |
| Bin Wu | Beijing University of Posts and Telecommunications, China |
| Bo Wu | Renmin University of China, China |
| Jiancan Wu | University of Science and Technology of China, China |
| Jiantao Wu | University of Jinan, China |
| Ou Wu | Tianjin University, China |
| Yang Wu | Chinese Academy of Sciences, China |
| Yiqing Wu | University of Chinese Academic of Science, China |
| Yuejia Wu | Inner Mongolia University, China |
| Bin Xiao | University of Ottawa, Canada |
| Zhiwen Xiao | Southwest Jiaotong University, China |
| Ruobing Xie | WeChat, Tencent, China |
| Zikang Xiong | Purdue University, USA |
| Depeng Xu | University of North Carolina at Charlotte, USA |
| Jian Xu | Citadel, USA |
| Jiarong Xu | Fudan University, China |
| Kunpeng Xu | University of Sherbrooke, Canada |
| Ning Xu | Southeast University, China |
| Xianghong Xu | Tsinghua University, China |
| Sangeeta Yadav | Indian Institute of Science, India |
| Mehrdad Yaghoobi | University of Edinburgh, UK |
| Makoto Yamada | RIKEN AIP/Kyoto University, Japan |
| Akihiro Yamaguchi | Toshiba Corporation, Japan |
| Anil Yaman | Vrije Universiteit Amsterdam, the Netherlands |

| | |
|---|---|
| Hao Yan | Washington University in St Louis, USA |
| Qiao Yan | Shenzhen University, China |
| Chuang Yang | University of Tokyo, Japan |
| Deqing Yang | Fudan University, China |
| Haitian Yang | Chinese Academy of Sciences, China |
| Renchi Yang | National University of Singapore, Singapore |
| Shaofu Yang | Southeast University, China |
| Yang Yang | Nanjing University of Science and Technology, China |
| Yang Yang | Northwestern University, USA |
| Yiyang Yang | Guangdong University of Technology, China |
| Yu Yang | The Hong Kong Polytechnic University, China |
| Peng Yao | University of Science and Technology of China, China |
| Vithya Yogarajan | University of Auckland, New Zealand |
| Tetsuya Yoshida | Nara Women's University, Japan |
| Hong Yu | Chongqing Laboratory of Comput. Intelligence, China |
| Wenjian Yu | Tsinghua University, China |
| Yanwei Yu | Ocean University of China, China |
| Ziqiang Yu | Yantai University, China |
| Sha Yuan | Beijing Academy of Artificial Intelligence, China |
| Shuhan Yuan | Utah State University, USA |
| Mingxuan Yue | Google, USA |
| Aras Yurtman | KU Leuven, Belgium |
| Nayyar Zaidi | Deakin University, Australia |
| Zelin Zang | Zhejiang University & Westlake University, China |
| Masoumeh Zareapoor | Shanghai Jiao Tong University, China |
| Hanqing Zeng | USC, USA |
| Tieyong Zeng | The Chinese University of Hong Kong, China |
| Bin Zhang | South China University of Technology, China |
| Bob Zhang | University of Macau, Macao, China |
| Hang Zhang | National University of Defense Technology, China |
| Huaizheng Zhang | Nanyang Technological University, Singapore |
| Jiangwei Zhang | Tencent, China |
| Jinwei Zhang | Cornell University, USA |
| Jun Zhang | Tsinghua University, China |
| Lei Zhang | Virginia Tech, USA |
| Luxin Zhang | Worldline/Inria, France |
| Mimi Zhang | Trinity College Dublin, Ireland |
| Qi Zhang | University of Technology Sydney, Australia |

| | |
|---|---|
| Qiyiwen Zhang | University of Pennsylvania, USA |
| Teng Zhang | Huazhong University of Science and Technology, China |
| Tianle Zhang | University of Exeter, UK |
| Xuan Zhang | Renmin University of China, China |
| Yang Zhang | University of Science and Technology of China, China |
| Yaqian Zhang | University of Waikato, New Zealand |
| Yu Zhang | University of Illinois at Urbana-Champaign, USA |
| Zhengbo Zhang | Beihang University, China |
| Zhiyuan Zhang | Peking University, China |
| Heng Zhao | Shenzhen Technology University, China |
| Mia Zhao | Airbnb, USA |
| Tong Zhao | Snap Inc., USA |
| Qinkai Zheng | Tsinghua University, China |
| Xiangping Zheng | Renmin University of China, China |
| Bingxin Zhou | University of Sydney, Australia |
| Bo Zhou | Baidu, Inc., China |
| Min Zhou | Huawei Technologies, China |
| Zhipeng Zhou | University of Science and Technology of China, China |
| Hui Zhu | Chinese Academy of Sciences, China |
| Kenny Zhu | SJTU, China |
| Lingwei Zhu | Nara Institute of Science and Technology, Japan |
| Mengying Zhu | Zhejiang University, China |
| Renbo Zhu | Peking University, China |
| Yanmin Zhu | Shanghai Jiao Tong University, China |
| Yifan Zhu | Tsinghua University, China |
| Bartosz Zieliński | Jagiellonian University, Poland |
| Sebastian Ziesche | Bosch Center for Artificial Intelligence, Germany |
| Indre Zliobaite | University of Helsinki, Finland |
| Gianlucca Zuin | UFM, Brazil |

## Program Committee Members, Demo Track

| | |
|---|---|
| Hesam Amoualian | WholeSoft Market, France |
| Georgios Balikas | Salesforce, France |
| Giannis Bekoulis | Vrije Universiteit Brussel, Belgium |
| Ludovico Boratto | University of Cagliari, Italy |
| Michelangelo Ceci | University of Bari, Italy |
| Abdulkadir Celikkanat | Technical University of Denmark, Denmark |

Zhirong Yang     Norwegian University of Science and Technology, Norway

Xiangyu Zhao     City University of Hong Kong, Hong Kong, China

**Sponsors**

# Contents – Part IV

## Multi-agent Reinforcement Learning

## Bandits and Online Learning

## Active and Semi-supervised Learning

## Private and Federated Learning

# Reinforcement Learning

# Coupling User Preference with External Rewards to Enable Driver-centered and Resource-aware EV Charging Recommendation

Chengyin Li, Zheng Dong, Nathan Fisher, and Dongxiao Zhu[✉]

Department of Computer Science, Wayne State University, Detroit, MI 48201, USA
{cyli,dong,fishern,dzhu}@wayne.edu

**Abstract.** Electric Vehicle (EV) charging recommendation that both accommodates user preference and adapts to the ever-changing external environment arises as a cost-effective strategy to alleviate the range anxiety of private EV drivers. Previous studies focus on centralized strategies to achieve optimized resource allocation, particularly useful for privacy-indifferent taxi fleets and fixed-route public transits. However, private EV driver seeks a more personalized and resource-aware charging recommendation that is tailor-made to accommodate the user preference (when and where to charge) yet sufficiently adaptive to the spatiotemporal mismatch between charging supply and demand. Here we propose a novel Regularized Actor-Critic (RAC) charging recommendation approach that would allow each EV driver to strike an optimal balance between the user preference (historical charging pattern) and the external reward (driving distance and wait time). Experimental results on two real-world datasets demonstrate the unique features and superior performance of our approach to the competing methods.

**Keywords:** Actor critic · Charging recommendation · Electric Vehicle (EV) · User preference · External reward

## 1 Introduction

Electric Vehicles (EVs) are becoming popular due to their decreased carbon footprint and intelligent driving experience over conventional internal combustion vehicles [1] in personal transportation tools. Meanwhile, the miles per charge of an EV is limited by its battery capacity, together with sparse allocations of charging stations (CSs) and excessive wait/charge time, which are major driving factors for the so-called range anxiety, especially for private EV drivers. Recently, developing intelligent driver-centered charging recommendation algorithms are emerging as a cost-effective strategy to ensure sufficient utilization of the existing charging infrastructure and satisfactory user experience [2,3].

Existing charging recommendation studies mainly focus on public EVs (e.g., electric taxis and buses) [3,4]. With relatively fixed schedule routines, and no

M.-R. Amini et al. (Eds.): ECML PKDD 2022, LNAI 13716, pp. 3–19, 2023.
https://doi.org/10.1007/978-3-031-26412-2_1

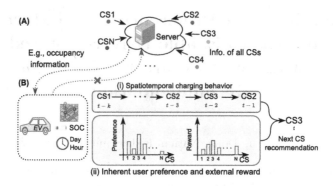

**Fig. 1.** Driver-centered and resource-aware charging recommendation. (A) Centralized charging recommendation enables optimized resource allocation, where bi-directional information sharing between the sever and EVs is assumed. (B) Driver-centered charging recommendation considers user preference and external reward, where only monodirectional information (e.g., the occupancy information of all CSs) sharing from the sever to each EV is required (green dotted line). Therefore, private information of an EV, like GPS location, is not uploaded to the server (pink dotted line). (Color figure online)

privacy or user preference consideration, the public EV charging recommendation for public transits can be made completely to optimize CS resource utilization. In general, these algorithms often leverage a global server, which monitors all the CSs in a city (Fig. 1A). Charging recommendation can be fulfilled upon requests for public EVs by sending their GPS locations and state of charge (SOC). This kind of recommendation gives each EV an optimal driving and wait time before charging. Instead of using one single global server, many servers can be distributed across a city [4,5] to reduce the recommendation latency for public EVs.

Although server-centralized methods have an excellent resource-aware property for the availability of charging for CSs, for private EVs, they rarely accommodate individual user preferences of charging and even have the risk of private data leakage (e.g., GPS location). Thus, the centralized strategy would also impair the trustworthiness [6–8] of the charging recommendation. A driver-centered instead of a server-centralized charging recommendation strategy would be preferred for a private EV to follow its user preference without leaking private information. In this situation (Fig. 1B), there would be a sequence of on-EV charging events records (when and which CS) that reflect the personal preference of charging patterns for a private EV driver. To enable the resource-aware property for a driver-centered charging recommendation, creating a public platform for sharing availability of CSs is needed.

Motivated by the success of recent research on the next POI (Point Of Interest) recommendation centered on each user, these studies can also be adapted to solve the charging recommendation problem for private EVs when viewing each CS as a POI. Different from collaborative filtering, based on the general recommendation that learns similarities between users and items [9], the following POI recommendation algorithms attempt to predict the most likely next POI that a user will visit based on the historical trajectory [10–14]. Although these methods indeed model user preferences, they are neither resource-aware nor adapted to the ever-changing external environment.

As such, a desirable charging recommender for a private EV requires: (1) learning the user preference from its historical charging patterns for achieving driver-centered recommendation, and (2) having a good external reward (optimal driving and wait time before charging) to achieve resource-aware recommendation (Fig. 1 B). By treating the private EV charging recommendation as the next POI recommendation problem, maximizing external rewards (with a shorter time of driving and wait before charging) by exploring possible CSs for each recommendation, reinforcement learning can be utilized. To leverage user preference and external reward, we propose a novel charging recommendation framework, Regularized Actor-Critic (RAC), for private EVs. The critic is based on a resource-saving over all CSs to give a evaluation value over the prediction of actor representing external reward, and the actor is reinforced by the reward and simultaneously regularized by the driver's user preference. Both actor and critic are based on deep neural networks (DNNs).

We summarize the main contributions of this work as follows: (1) we design and develop a novel framework RAC to give driver-centered and resource-aware charging recommendations on-EV recommendation; (2) RAC is tailor-made for each driver, allowing each to accommodate inherent user preference and also adapt to ever-changing external reward; and (3) we propose a warm-up training technique to solve the cold-start recommendation problem for new EV drivers.

## 2   Related Work

Next POI recommendation has attracted much attention recently in location-based analysis. There are two lines of POI recommendation methods: (1) following user preference from sequential visiting POIs regularities, and (2) exploiting external incentive via maximizing the utility (reward) of recommendations.

For the first line of research, the earlier works primarily attempt to solve the sequential next-item recommendation problem using temporal features. For example, [11] introduces Factorizing Personalized Markov Chain (FPMC) that captures sequential dependency between the recent and next items as well as the general taste of a user using a combination of matrix factorization and Markov chains for next-basket recommendation. [12] proposes a time-related Long-Short Term Memory (LSTM) network to capture both long- and short-term sequential influence for next item recommendation. [15] attempts to model user' preference drift over time to achieve a better user experience in next item recommendation. These next-item recommendation approaches only use temporal features whereas next POI recommendation would need to use both temporal and geospatial features.

More recent studies of next POI recommendation not only model temporal relations but also consider geospatial context, such as ST-RNN [16] and ATST-LSTM [17]. [13] proposes a hierarchical extension of LSTM to code spatial and temporal contexts into the LSTM for general location recommendation.

[14] introduces a spatiotemporal gated network model where they leverage time gate and distance gate to control the effect of the last visited POI on next POI recommendation. [10] extends the gates with a power-law attention mechanism with more attention on the nearby POIs and explores the subsequence patterns for next POI recommendation. [18] develops a long and short-term preference learning model considering sequential and context information for next POI recommendation. User preference-based methods can achieve significant performance for the following users' previous experience; however, they are restricted from making novel recommendations beyond users' previous experience.

Although few studies exploit external incentive, these methods can help explore new possibilities for next POI recommendation. Charging Recommendation with multi-agent reinforcement learning is applied for public EVs [19,20], in which private information from each EV is inevitably required. [21] proposes an inverse reinforcement learning method for next visit action recommendation by maximizing the reward that the user gains when discovering new, relevant, and non-popular POIs. This study utilizes the optimal POI selection policy (the POI visit trajectory of a similar group users) as the guidance. As such, it is only applicable for the centralized charging recommendation for privacy-indifferent public transit fleets where charging events are aggregated to the central server to learn the user group. However, this approach is not applicable to the driver-centered EV charging recommendation problem that we are tackling since the individual charging pattern is learned without data sharing across drivers. Besides the inverse reinforcement learning approach, [22] introduces deep reinforcement learning for news recommendation, and [23] proposes supervised reinforcement learning for treatment recommendation. These methods are also based on learning similar user groups thus not directly applicable to the driver-centered EV charging recommendation task, the latter is further subject to resource and geospatial constraints.

Despite the existing approaches utilized spatiotemporal, social network, and/or contextual information for effective next POI recommendations, they do not possess the desirable features for CS recommendation, which are (1) driver-centered: the trade-off between the driver's charging preference and the external reward is tuned for each driver, particularly for new drivers, and (2) resource-aware: there is usually capacity constraint on a CS but not on a social check-in POI.

## 3   Problem Formulation

Each EV driver is considered as an agent, and the trustworthy server that collects occupancy information of all the CSs represent the external ever-changing environment. We considered our charging recommendation as a finite-horizon MDP problem where a stochastic policy consists of a state space $\mathcal{S}$, an action space $\mathcal{A}$, and a reward function $r \colon \mathcal{S} \times \mathcal{A} \to \mathbb{R}$. At each time point $t$, an EV driver with the current state $s_t \in \mathcal{S}$, chooses an action $a_t$, i.e., the one-hot encoding of a CS, based on a stochastic policy $\pi_\theta(a|s)$ where $\theta$ is the set of parameters, and receives a reward $r_t$ from the spatiotemporal environment. Our objective is to learn such a stochastic policy $\pi_\theta(a|s)$ to select an action $a_t \sim \pi_\theta(a|s)$ by maximizing the sum of discounted rewards (return $R$) from the time point $t$, which is defined as $R_t = \sum_{i=t}^{T} \gamma^{(i-t)} r(s_i, a_i)$, and simultaneously minimizing the difference from the EV driver's decision $\hat{a}_t$. $\gamma \in [0, 1]$, e.g., 0.99, is a discount factor to balance the importance of immediate and future rewards. $T$ is the furthermost time point we use.

The charging recommendation task is a process to learn a good policy for next CS recommendation for an EV driver. By modeling user behaviors with situation awareness, two types of methods can be designed to learn the policy: value based Reinforcement Learning (RL) to maintain a greedy policy, and policy gradient based RL to learn a parameterized stochastic policy $\pi_\theta(a|s)$ or a deterministic policy $\mu_\theta(s)$, where $\theta$ represents the set of parameters of the policy. For the discrete property of CSs, we focus on learning a personalized stochastic policy $\pi_\theta(a|s)$ for each EV driver using DNNs.

## 4  Our Approach

### 4.1  Background

Q-learning [24] is an off-policy learning strategy for solving RL problems that finds a greedy policy $\mu(s) = \arg\max_a Q^\mu(s, a)$, where $Q^\mu(s, a)$ is $Q$ value or action-value, and it is usually used for a small discrete action space. For any finite Markov decision process, Q-learning finds an optimal policy in the sense of maximizing the expected value of the total reward over any successive steps, starting from the current state. The value of $Q^\mu(s, a)$ can be calculated with dynamic programming. With the introduction of DNNs, a deep Q network (DQN) is used to learn such $Q$ function $Q_w(s, a)$ with parameter $w$, and DNN is incapable of handling a high dimension action space. During training, a replay buffer is introduced for sampling, and DQN asynchronously updates a target network $Q_w^{tar}(s, a)$ to minimize the expectation of square loss.

Policy gradient [25] is another approach to solve RL problems and can be employed to handle continuous or high-dimensional discrete actions, and it targets modeling and optimizing the policy directly. The policy is usually modeled with a parameterized function respect to $\theta$, $\pi(s, a)$. The value of the reward (objective) function depends on this policy and then various algorithms can be applied to optimize $\theta$ for the best reward. To learn the parameter $\theta$ of $\pi_\theta(a|s)$, we maximize the expectation of state-value function $V^{\pi_\theta}(s) = \sum_a \pi_\theta(a|s) Q^{\pi_\theta}(s, a)$, where $Q^{\pi_\theta}(s, a)$ is the state-value function. Then we need to maximize $J(\pi_\theta) = E_{s \sim \rho^{\pi_\theta}}[V^{\pi_\theta}(s_1)]$, where $\rho^{\pi_\theta}$ represents the discounted state distribution. Policy gradient learns the parameter $\theta$ by the gradient $\nabla_\theta J(\pi_\theta)$, which is calculated with the policy gradient theorem: $\nabla_\theta J(\pi_\theta) = \mathbb{E}_{s \sim \rho^{\pi_\theta}, a}[\nabla_\theta \log \pi_\theta(a|s) Q^{\pi_\theta}(s, a)]$. These calculations are guaranteed by the policy gradient theorem.

Actor-critic [26] method combines the advantages of Q-learning and policy gradient to accelerate and stabilize the learning process in solving RL problems.

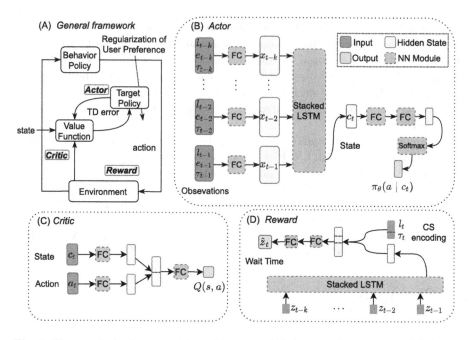

**Fig. 2.** Our regularized actor-critic architecture. (A) A general framework. (B) Actor network decides which action (CS) to take (charge). (C) Critic network tells the actor how good the action is and how it should be adjusted. (D) Reward network estimates wait time estimation at each CS.

It has two components: a) an actor to learn the parameter $\theta$ of $\pi_\theta$ in the direction of the gradient $\nabla J(\pi_\theta)$ to maximize $J(\pi_\theta)$, and b) a critic to estimate the parameter $w$ in an action-value function $Q_w(s, a)$.

In this paper, we use an off-policy actor-critic [23,26], where the actor updates the policy weights. The critic learns an off-policy estimate of the value function for the current actor policy, different from the (fixed) behavior policy. The actor then uses this estimate to update the policy. Actor-critic methods consist of two models, which may optionally share parameters. Critic updates the value function parameters $w$ for state action-value $Q_w(s, a)$. Actor updates the policy parameters $\theta$ for $\pi_\theta(a|s)$, in the direction suggested by the critic. $\pi_\theta(a|s)$ is obtained by averaging the state distribution of behavior policy $\beta(a|s)$. $\beta(a|s)$ for collecting samples is a known policy (predefined just like a hyperparameter). The objective function sums up the reward over the state distribution defined by this behavior policy: $J(\pi_\theta) = E_{s \sim \rho^{\pi_\beta}}[Q^\pi(s, a)\pi_\theta(a|s)]$, where $\pi_\beta$ is the stationary distribution of the behavior policy $\beta(a|s)$; and $Q^\pi$ is the action-value function estimated with regard to the target policy $\pi$.

## 4.2 The Regularized Actor-Critic (RAC) Method

To find an optimal policy for the MDP problem also with following user preference, we use the regularized RL method, specifically with a regularized actor-critic model [23], which combines the advantages of Q-learning and policy gradient. Since the computation cost becomes intractable with many states and actions when using policy iteration and value iteration, we introduce a DNN-based actor-critic model to reduce the computation cost and stabilize the learning. While the traditional actor-critic model aims to maximize the reward without considering a driver's preference, we also use regularization to learn the user's historical charging behavior as a representation of user preference. Our proposed general regularized actor-critic framework is shown in Fig. 2A.

The actor network learns a policy $\pi_\theta(a|s)$ with a set of parameters $\theta$ to render charging recommendation for each EV driver, where the input is $s_t$ and the output is the probabilities of all actions in $\mathcal{A}$ of transitioning to a CS $a_t$. By optimizing the two learning tasks simultaneously, we maximize the following objective function:

$$J(\theta) = (1 - \epsilon)J_{RL}(\theta) + \epsilon(-J_R(\theta)),$$

where $\epsilon$ is tuning parameter to weigh between inherent user preference and external reward (return) when making recommendation. The RL objective $J_{RL}$ aims to maximize the expected return via learning the policy $\pi_\theta(a|s)$ by maximizing the state value of an action that is averaged over the state distribution of the CS selection for each EV driver, i.e.,

$$J_{RL}(\theta) = \mathbb{E}_{s \sim \rho^{\pi_\theta}, a \ \pi_\theta(a|s)}[Q_w(s, a)].$$

The regularization objective $J_R$ aims to minimize the discrepancy between the recommended CS and preferred CS for each user via minimizing the difference between CS recommended by $\pi_\theta(a|s)$ and CS given by each EV driver's previous selection, in terms of the cross entropy loss, i.e.,

$$J_R(\theta) = \mathbb{E}_{s \sim \rho^{\hat{\mu}(s)}}[-\frac{1}{K} \sum_{k=1}^{K} \hat{a}_{t,k} \log \pi_\theta^k(a|s) - (1 - \hat{a}_{t,k}) \log(1 - \pi_\theta^k(a|s))].$$

Using DNNs, $\theta$ can be learned with stochastic gradient decedent (SGD) algorithms.

The critic network is jointly learned with the actor network, where the inputs are the current and previous states (i.e., CSs) of each EV driver, actions, and rewards. The critic network uses a DNN to learn the action-value function $Q_w(s, a)$, which is used to update the parameters of the actor in the direction of reward improvement. The critic network is only needed for guiding the actor during training whereas only actor network is required at test stage. We update the parameter $w$ via minimizing

$$J(w) = \mathbb{E}_{r_t, s_t \sim \rho^{\hat{\mu}(s)}}[(Q_w(s_t, a_t) - y^t)^2],$$

in which $y^t = r(s_t, a_t) + \gamma Q_w^{tar}(s_{t+1}, a_{t+1})$, $a_{t+1} \sim \pi_\theta(s_{t+1})$ is the charging action recommended by the actor network, and $\delta = (Q_w(s_t, a_t) - y^t)$ is Temporal Difference (TD) error, which is used for learning the Q-function.

## 4.3   The RAC Framework for EV Charging Recommendation

In the previous formulation, we assume the state of an EV driver is fully observable. However, we are often unable to observe the full states of an EV driver. Here we reformulate the environment of RAC as Partially Observable Markov Decision Process (POMDP). In POMDP, $O$ is used to denote the observation set, and we obtain each observation $o \in O$ directly from $p(o_t|s_t)$. For simplicity, we use a stacked LSTM together with the previous Fully Connected (FC) layers for each input step (Fig. 2B), to summarize previous observations to substitute the partially observable state $s_t$ with $c_t = f_{\phi_1}(o_{t-k}, ..., o_{t-2}, o_{t-1})$. Each $o = (l, e, \tau)$ represents a observation in different time points, and $\phi_1$ is the set of parameters of $f$. $l$ denotes the CS location context information, $e$ presents charging event related features (e.g., SOC), and $\tau$ represents the time point (e.g., day of a week and hour of a day). $l$ is a combination context with the geodesic distance from previous CS (calculated by the latitudes and longitudes), one-hot encoding of this CS and the POI distribution around this CS. $o$ is a concatenation of $l$, $e$ and $\tau$ vectors. The samples for training the actor model is generated from the behavior actor $\beta(s|a)$ (i.e., from the real world charging trajectories) via a buffer in an off-policy setting.

Our RAC consists of three main DNN modules for estimating the actor, the critic, and the reward, as shown in Fig. 2. Actor DNN (Fig. 2B) captures each driver's charging preference. We take a subsequence of the most recent CSs as input to extract the hidden state $c_t$ through a stacked-LSTM. With the following fully connected layers, we recommend the CS to go next for an EV driver. During training, the actor is supervised with the TD from the critic network to maximize the expected reward and the actual CS selection from this driver with cross-entropy loss to minimize the difference (Fig. 2A). Since the actor is on each EV and takes private charging information as input, it is a driver-centered charging recommendation model.

To enable resource-awareness, we use a one-way information transmission scheme, shown in Fig. 1. We train a resource-aware actor for each EV driver via estimating $Q$ value from the critic DNN with addition of the immediate reward $r$ estimated with a reward DNN. Figure 2C shows the prediction of $Q_w$ value of state $s$ and action $a$, and the state here would be substituted with $c$ in POMDP setting. Figure 2D describes how to estimate the wait time $\hat{z}$ in all CSs. We can calculate the immediate reward for each pair of $(s, a)$ by combining with the estimated drive time. To tackle the cold start problem for new EV drivers, we introduce a warm-up training technique to update the model and will illustrate the details in the experiment section.

## 4.4   Timely Estimation of Reward

In stead of using traditional static reward, we dynamically estimate the reward from external environment. Since the drive time to and wait time at the CS play a key role for private EV driver's satisfaction, we estimate rewards based on these two factors. Specifically, we directly use the geodesic distance from map to represent the drive time and use a DNN (Fig. 2D) to timely estimate the wait time for each charging. Therefore, a timely estimation of reward for choosing each CS can be given by a simple equation: $\hat{r}_t = -100(\frac{\hat{z}_t}{\tilde{z}} + \zeta\frac{\hat{d}_t}{\tilde{d}})$ where $\hat{z}_t = g_{\phi_2}(l_t, \tau_t, z_{t-k}, \dots, z_{t-1})$ is the predicted wait time through reward network, in which $\phi_2$ is the parameters of the reward DNN (LSTM).$z_{t-k}$ is the wait time in $k$ steps before the current time step, and it is directly summarized from the dataset we used. $\hat{d}_t$ is the estimated driving distance to the corresponding CS. Further, $\tilde{d}$ and $\tilde{z}$ represent statistically averaged driving distance and wait time in each CS, and they are constant values for a specific CS. $\zeta$ is a coefficient, which usually has an inverse relationship with an EV driver's familiarity with the routes (visiting frequency of each CS). For simplicity, we set $\zeta$ as 0.8 for the most visited CS, and 1 for other situations. To make the predicted wait time and predicted driving distance to be additive, we do normalization for the predicted values by the averaged wait time $\hat{z}$ and $\hat{d}$ respectively for each CS. Since the wait time and drive time are estimated by each CS, our RAC framework is resource-aware to make CS recommendation for each EV driver.

Putting all the components as mentioned above together, the training algorithm of RAC is shown in Algorithm 1.

## 4.5   Geospatial Feature Learning

The POI distribution within the neighborhood of each CS is what we used to learn the geospatial features from each CS. With this information, we can infer the semantic relationships among the CSs to assist in recommending CSs for each driver. Google Map defines 76 types (e.g., schools, restaurants, and hospitals) of POIs. Specifically, for each CS, we use its latitude and longitude information together with a geodesic radius of 600 m to pull the surrounding POIs. We count the number of POIs for each type to obtain a 76-dimension vector (e.g., $POI \in \mathbb{R}^{76}$) as the POI distribution. We concatenate this vector with other information, i.e., geodesic distances to CSs and one-hot encoding of the CS. With the charging event features and the timestamp-related features, we learn a unified embedding through an MLP for each input step of the stacked-LSTM.

## 5   Experiments

### 5.1   Experimental Setup

All the experiments are implemented on two real world charging events datasets from Dundee city[1] and Glasgow city[2]. The POI distribution for each CS is

---

[1] https://data.dundeecity.gov.uk/dataset/.
[2] http://ubdc.gla.ac.uk/dataset/.

**Algorithm 1.** The RAC training algorithm

**Input:** Actions $A$, observations $O$, reward function $r$, # of CSs $M$, historical wait time at each hour $(z_1^j, ..., z_T^j)$, and coordinates (latitude$^j$, longitude$^j$) in $j$-th CS

**Hyper-parameters:** Learning rate $\alpha = 0.001$, $\epsilon = 0.5$, the finite-horizon step $T = 10$, number of episodes $I$, and $\gamma = 0.99$

**Output:** $\theta, \phi_1, \phi_2, w$

1: Store sequences $(o_1, a_1, r_1, ..., o_T, a_T, r_T)$ by behavior policy $\beta(a|s)$ in buffer $D$, each $o = (l, e, \tau)$, and # of epochs $N$;
2: Random initialize actor $\pi_\theta$, critic $Q_w$, target critic $Q_w^{tar}$, TD error $\delta = 0$, and reward network $f_\phi$;
3: **for** $n = 1$ to $N$ **do**
4:     Sample $(o_1^i, a_1^i, r_1^i, ..., o_T^i, a_T^i, r_T^i) \subset D$, $i = 1, ..., I$
5:     $c_t^i \leftarrow f_{\phi_1}(o_{t-k}^i, o_{t-k}^i, ..., o_{t-1}^i)$
6:     $a_t^i, a_t^{i+1} \leftarrow$ sampled by $\pi_\theta$
7:     $\hat{z}_t^i \leftarrow g_{\phi_2}(l_t, \tau_t, z_{t-k}, ..., z_{t-1})$
8:     $\hat{r}_t^i \leftarrow -100(\frac{\hat{z}_t^i}{\bar{z}^i} + \zeta\frac{\hat{d}_t^i}{\bar{d}^i})$
9:     $y_t^i \leftarrow \hat{r}_t^i + \gamma Q_w^{tar}(c_t^{i+1}, a_t^{i+1})$
10:     $\hat{a}_t^i \leftarrow$ given by the EV driver's selection
11:     $\delta_t^i \leftarrow Q_w(c_T^i, \hat{a}_t^i) - y_t^i$
12:     $w \leftarrow w - \alpha\frac{1}{IT}\sum_i\sum_t \delta_t^i \nabla_w Q_w(c_t^i, a_t^i)$
13:     $\phi_1 \leftarrow \phi_1 - \alpha\frac{1}{IT}\sum_i\sum_t \delta_t^i \nabla_{\phi_1} f_{\phi_1}$
14:     $\phi_2 \leftarrow \phi_2 - \alpha\frac{1}{IT}\sum_i\sum_t \delta_t^i \nabla_{\phi_2} g_{\phi_2}$
15:     $\nabla_w Q_w(c_t^i, a_t^i) \leftarrow$ given by $Q_w(c_t^i, a_t^i)$
16:     $\eta_t^i = \frac{1}{M}\sum_{k=1}^M \frac{\hat{a}_{t,k}^i - a_{t,k}^i}{(1-a_{t,k}^i)a_{t,k}^i}$
17:     $\theta \leftarrow \theta + \alpha\frac{1}{IT}\sum_i\sum_t[(1-\epsilon)\nabla_w Q_w(c_t^i, a_t^i) + \epsilon\eta_t^i]$
18: **end for**

obtained from Google Place API[3]. The code of our method is publicly available on this link: https://github.com/cyli2019/RAC-for-EV-Charging-Rec.

**Datasets and Limitations.** For Dundee city, we select the charging events from the time range of 6/6/2018-9/6/2018, in which there are 800 unique EV drivers, 44 CSs and 19, 115 charging events. For Glasgow city, in the time range of 9/1/2013-2/14/2014, we have 47 unique EV drivers, 8 CSs and 507 charging events. For each charging event, the following variables are available: CS ID, charging event ID, EV charging date, time, and duration, user ID, and consumed energy (in kWh) for each transaction. For each user ID, we observe a sequence of charging events in chronological order to obtain the observations $O$. For each

---

[3] https://developers.google.com/maps/documentation/places/web-service.

CS ID, we learn the geospatial feature to determine their semantic similarity according to POI types.

To model an EV driver preference, we train a model using the CS at each time point as the outcome and the previous charging event sequence as the input. To enable situation awareness, for a specific CS, there is a chronically ordered sequence of wait time, and we use the wait time corresponding to each time point as the outcome and that of previous time points as inputs in our reward network to forecast hourly wait time for all CS's. Combined with the estimated drive time that are inverse proportional to familiarity adjusted geodesic distance, we determine the timely reward for each EV driver's charging event.

To our knowledge, these two datasets are the only publicly available driver-level charging event data for our driver-centered charging recommendation task, though with relatively small size and unavailability of certain information. Due to the privacy constraints, the global positioning system (GPS) information of each driver and the corresponding timestamp are not publicly available as well as traffic information in these two cities during the time frame. As such, we have no choice but having to assume the EV driver transits from CS to CS and using driving distance between CSs combined with estimated wait time at each CS to calculate the external reward. Another assumption we made is using the time interval of each charging event to approximate the SOC of the EV since all EVs in the data sets are of the same model. The method developed in this paper is general that does not rely on the aforementioned assumption; when GPS, timestamp and SOC information become available, our method is ready to work without change.

**Evaluation Metrics.** Similar to POI recommendation, we treat the earlier 80% sequences of each driver as a training set, the middle 10% as a validation set, and the latter 10% as a test set. Two standard metrics are adopted to evaluate methods' performance, namely, Precision (P@K) and Recall (R@K) on the test set. To quantify the external reward for making a charging recommendation, we also use a Mean Average Reward (MAR) as an evaluation metric. Each reward is calculated based on familiarity-adjusted geodesic distance and projected wait time at the recommended CS, and MAR is the average value over all users across all time points in the test set. To solve the cold-start problem for EV drivers who have few charging events, we use 5% of data in the earlier sequences from all users (with more than 10 charging events) to train a model as warm-up, the rest 95% following the same data splitting strategy described above followed by training with each driver's private data. We assume that for the earliest 5% of data can be shared without privacy issues when the user related information is eliminated.

**Baselines.** We compare RAC with the following baseline methods, including two classic methods (i.e., MC, and FPMC [11]), three DNN-based state-of-the-art methods (i.e., Time-LSTM [12] , ST-RNN [16], and ATST-LSTM [17]). We select these methods as the baselines for method comparison, instead of other

**Table 1.** Performance comparison with different learning methods. Results of the best-performing RAC model are boldfaced; the runner-up is labeled with '*'; 'Improvement' refers to the percentage of improvement that RAC achieves relative to the runner-up results.

| Dataset | Metrics | MC | FPMC | Time-LSTM | ST-RNN | ATST-LSTM | RAC-zero | RAC | Improvement |
|---------|---------|------|------|-----------|--------|-----------|----------|-------|-------------|
| Dundee | $P@1$ | 0.204 | 0.242 | 0.313 | 0.326 | 0.368* | 0.385 | **0.424** | 15.2% |
| | $P@3$ | 0.256 | 0.321 | 0.367 | 0.402 | 0.435* | 0.463 | **0.509** | 17.0% |
| | $P@5$ | 0.321 | 0.363 | 0.436 | 0.437 | 0.484* | 0.528 | **0.577** | 19.2% |
| | $R@1$ | 0.146 | 0.195 | 0.203 | 0.216 | 0.247* | 0.285 | **0.292** | 18.2% |
| | $R@3$ | 0.153 | 0.226 | 0.236 | 0.278 | 0.298* | 0.344 | **0.368** | 23.5% |
| | $R@5$ | 0.192 | 0.237 | 0.245 | 0.325 | 0.375* | 0.427 | **0.479** | 27.7% |
| | $MAR$ | −327.8 | −265.9 | −210.4 | −195.4 | −164.5* | −133.2 | **−114.6** | 30.3% |
| Glasgow | $P@1$ | 0.163 | 0.207 | 0.264 | 0.252 | 0.294* | 0.313 | **0.364** | 23.8% |
| | $P@3$ | 0.226 | 0.262 | 0.325 | 0.356 | 0.375* | 0.40.9 | **0.458** | 22.1% |
| | $P@5$ | 0.285 | 0.301 | 0.398 | 0.405 | 0.428* | 0.482 | **0.497** | 16.1% |
| | $R@1$ | 0.108 | 0.093 | 0.122 | 0.128 | 0.133* | 0.13.1 | **0.164** | 23.3% |
| | $R@3$ | 0.126 | 0.135 | 0.174 | 0.182 | 0.216* | 0.224 | **0.253** | 17.1% |
| | $R@5$ | 0.173 | 0.167 | 0.263 | 0.323 | 0.334* | 0.395 | **0.406** | 21.5% |
| | $MAR$ | −456.3 | −305.4 | −232.2 | −210.9 | −196.4* | −164.2 | **−154.3** | 21.4% |

general POI recommendation methods (e.g., multi-step or sequential POI recommendation problem), because they directly address the next POI recommendation problem. One variant of our RAC (i.e., RAC-zero) is trained from scratch without warm-up training. The description of the baselines are: (1) MC: first-order Markov Chain utilizes sequential data to predict a driver's next action based on the last actions via learning a transition matrix. (2) FPMC: Matrix factorization method learns the general taste of a driver by factorizing the matrix over observed driver-item preferences. Factorization Personalized Markov Chains model is a combination of MC and MF approaches for the next-basket recommendation. (3) Time-LSTM: Time-LSTM is a state-of-the-art variant of LSTM model used in recommender systems. Time-LSTM improves the modeling of sequential patterns by explicitly capturing the multiple time structures in the check-in sequence. We used the best-performing version reported in their paper. (4) ST-RNN : It is a RNN-based method that incorporates spatiotemporal contexts for next location prediction. (5) ATST-LSTM: It utilizes POIs and spatiotemporal contexts in a multi-modal manner for next POI prediction. In addition, to evaluate the effect of warm-up training on solving the cold-start problem, we compare our RAC with its a variant, RAC-zero, which is trained from scratch.

## 5.2  Performance Comparison

The parameter tuning information during the training are described above, and after that we make comparison for our approach with the baselines methods. Table 1 presents the performance (R@K, P@K, and MAR) of all methods across the two datasets. We test $K$ with 1, 3, and 5, and based on the parameter tuning results, we use the setting of two-layer stacked-LSTM for both actor and

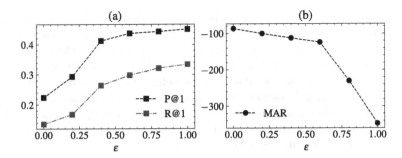

**Fig. 3.** The tension between maximizing (a) inherent user preference and (b) external reward on the averaged user.

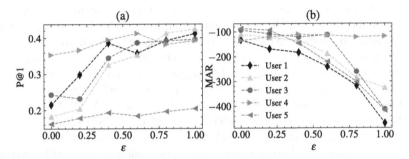

**Fig. 4.** A case study of five individual EV drivers (a) inherent user preference, and (b) external reward.

reward networks, embedding/hidden sizes of (100, 100), $\epsilon$ of 0.5, and learning rate of 0.001. The feeding steps for LSTMs in actor and reward networks are set to 5 and 10 respectively. In terms of charging recommendation task, the RNN based methods (Time-LSTM, ST-RNN, ATST-LSTM, and RAC) generally outperforms non-RNN based competitors (MC, and FPMC) owing to the leverage of spatiotemporal features. For the former, ATST-LSTM is better than ST-RNN possibly due to the effective use of attention mechanism. ST-RNN has slightly better performance over Time-LSTM due to the incorporation of spatial features. Overall, our proposed RAC consistently achieves the best performance not only on precision/recall but also over MAR, in which the improvement column are the comparisons between RAC and the runner-up model (ATST-LSTM). This is translated into the fact that overall RAC is capable of accommodating inherent user preference and ensuring the external rewards to a maximum extent in rendering charging recommendations.

To demonstrate the influence of warm-up training in RAC, we compare it with the training-from-scratch-approach RAC-zero. From Table 1, RAC demonstrates a better overall performance on the Dundee dataset for the relative abundance of samples for warm-up training; in the meanwhile, due to the limited number of warm-up training samples in the Glasgow dataset, this improvement is relatively slight. Conventionally, the Glasgow dataset with fewer charging stations might have better recommendation accuracy than the Dundee dataset. However, we should know that most (over 80%) EVs are revisiting no more than eight charging stations for both datasets. Therefore, for driver-centered charging pattern, the number of possible CSs is similar for these two datasets, resulting in even worse performance for the Glasgow dataset than the Dundee dataset. Overall, our proposed RAC consistently achieves the best performance not only on precision/recall but also over MAR.

## 5.3   Driver-centered CS Recommendation

Figure 3 illustrates the effect of personalization tuning parameter $\epsilon$ on precision/recall and reward of the recommendation. Since RAC is a driver-centered recommendation method, each driver can experiment with the parameter $\epsilon$ to weigh more on inherent user preference or on external award when seeking driver-centered charging recommendations. In Fig. 3A, the P@1 and R@1 of RAC climb up as $\epsilon$ increases, and becomes stable at around 0.5, indicating a larger value would not further improve the performance. In Fig. 3(b), MAR first decreases slightly before 0.5 and then drops quickly afterwards. Collectively, it appears an average driver can get the best of both worlds when $\epsilon$ is around 0.5.

Figure 4 shows that drivers 1–3 follow a very similar pattern to the average driver in Fig. 3 where $\epsilon$ is around 0.5, representing a good trade-off to balance between the inherent user preference and the external reward. Driver 4 represents a special case where the driver preference aligns well with the external reward; in this case the charging recommendation is invariant to the choice of $\epsilon$. Hence the recommendation can be made either based on user preference or external reward since they are consistent to each other. Driver 5 represents a new driver with low precision and recall due to the lack of historical charging data. As such, the recommendation can simply be made based mostly on the external reward via setting $\epsilon$ to a low value, e.g., 0.2. In sum, tuning $\epsilon$ indeed enables an individual driver to be more attentive to his/her preference or to the external reward when seeking EV charging recommendation.

Figure 5 demonstrates the award (e.g., wait time and driving distance) for three representative EV drivers, User 3, User 4, and User 5, under two different values of $\epsilon$, i.e., 0.2 and 0.8. Recall the latter denotes the weight on an EV driver to follow historical charging pattern. Therefore, an increase of the $\epsilon$ value from 0.2 to 0.8 indicates that the charging recommendation is rendered based more on the driver's previous charging pattern than the reward from external environment. In Fig. 5, we describe three types of drivers demonstrated by different trade-offs: (1) For User 3, the wait time and driving distance are both increasing, resulting in a smaller reward, whereas a better prediction accuracy. (2) For User 4, the wait time and driving distance remains shorter yet stable across the two values of $\epsilon$, demonstrating both a larger reward and higher prediction accuracy.

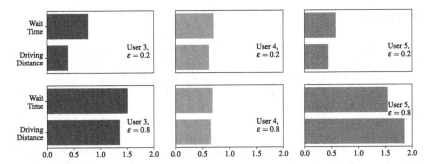

**Fig. 5.** Examples of explanations for the recommendations. We use the mean value of the normalized wait time and driving distance to make the comparison fair.

(3) For User 5 who is a newer driver, the reward increases similarly to User 3. However, the prediction accuracy stays low regardless of the choice of $\epsilon$ due to the limited information on historical charging pattern of the new driver. In summary, for drivers such as User 3 whose charging patterns are vastly deviated from what would be recommended by the external award, tuning $\epsilon$ would allow the drivers to be more attentive to either historical charging patterns or the external award. For drivers such as User 4 whose historical charging pattern is consistent with the more rewarding charging option as determined by shorter wait time and driving distance, the choice of $\epsilon$ does not matter, representing an optimal charging recommendation scenario. For new drivers such as User 5, a charging recommendation that is largely based on the external reward may be more appropriate.

## 6  Conclusion

In this paper, we propose a resource-aware and driver-centered charging recommendation method for private EVs. We devise a flexible regularized actor-critic framework, i.e., using RL to maximize external reward as the regularization to model inherent user preference for each driver. Our approach is sufficiently flexible for a wide range of EV drivers including new drivers with limited charging pattern data. Experimental results on real-world datasets demonstrate the superior performance of our approach over the state-of-the-arts in the driver-centered EV charging recommendation task.

**Acknowledgements.** This work is supported by the National Science Foundation under grant no. IIS-1724227.

# References

1. O'Donovan, C., Frith, J.: Electric buses in cities: driving towards cleaner air and lower co 2. Tech. Rep, Bloomberg New Energy Finance, NY, USA (2018)
2. Wang, X., Yuen, C., Hassan, N.U., An, N., Wu, W.: Electric vehicle charging station placement for urban public bus systems. IEEE Trans. Intell. Transp. Syst. **18**(1), 128–139 (2016)
3. Wang, G., Xie, X., Zhang, F., Liu, Y., Zhang, D.: Bcharge: data-driven real-time charging scheduling for large-scale electric bus fleets. In: RTSS, pp. 45–55 (2018)
4. Guo, T., You, P., Yang, Z.: Recommendation of geographic distributed charging stations for electric vehicles: a game theoretical approach. In: 2017 IEEE Power & Energy Society General Meeting, pp. 1–5 (2017)
5. Cao, Y., Kaiwartya, O., Zhuang, Y., Ahmad, N., Sun, Y., Lloret, J.: A decentralized deadline-driven electric vehicle charging recommendation. IEEE Syst. J. (2018)
6. Qiang, Y., Li, C., Brocanelli, M., Zhu, D.: Counterfactual interpolation augmentation (CIA): a unified approach to enhance fairness and explainability of DNN. In: IJCAI (2022)
7. Li, X., Li, X., Pan, D., Zhu, D.: Improving adversarial robustness via probabilistically compact loss with logit constraints. In: AAAI, vol. 35, pp. 8482–8490 (2021)
8. Pan, D., Li, X., Zhu, D.: Explaining deep neural network models with adversarial gradient integration. In: IJCAI, pp. 2876–2883 (2021)
9. Pan, D., Li, X., Li, X., Zhu, D.: Explainable recommendation via interpretable feature mapping and evaluation of explainability. In: IJCAI, pp. 2690–2696 (2020)
10. Zhao, K., et al.: Discovering subsequence patterns for next poi recommendation. In: IJCAI, pp. 3216–3222 (2020)
11. Rendle, S., Freudenthaler, C., Schmidt-Thieme, L.: Factorizing personalized markov chains for next-basket recommendation. In: WWW 2010, pp. 811–820 (2010)
12. Zhu, Y., et al.: What to do next: modeling user behaviors by time-LSTM. In: IJCAI, vol. 17, pp. 3602–3608 (2017)
13. Kong, D., Wu, F.: HST-LSTM: a hierarchical spatial-temporal long-short term memory network for location prediction. In: IJCAI, vol. 18, pp. 2341–2347 (2018)
14. Zhao, P., et al.: Where to go next: a spatio-temporal gated network for next poi recommendation. In: AAAI, vol. 33, pp. 5877–5884 (2019)
15. Li, L., Zheng, L., Yang, F., Li, T.: Modeling and broadening temporal user interest in personalized news recommendation. Expert Syst. Appl. **41**(7), 3168–3177 (2014)
16. Liu, Q., Wu, S., Wang, L., Tan, T.: Predicting the next location: a recurrent model with spatial and temporal contexts. In: AAAI, vol. 30 (2016)
17. Huang, L., Ma, Y., Wang, S., Liu, Y.: An attention-based spatiotemporal LSTM network for next poi recommendation. IEEE Trans. Serv. Comput. (2019)
18. Wu, Y., Li, K., Zhao, G., Qian, X.: Long-and short-term preference learning for next poi recommendation. In: CIKM 2019, pp. 2301–2304 (2019)
19. Wang, E., et al.: Joint charging and relocation recommendation for e-taxi drivers via multi-agent mean field hierarchical reinforcement learning. IEEE Trans. Mobile Comput. (2020)
20. Zhang, W., et al.: Intelligent electric vehicle charging recommendation based on multi-agent reinforcement learning. In: WWW 2021, pp. 1856–1867 (2021)
21. Massimo, D., Ricci, F.: Harnessing a generalised user behaviour model for next-poi recommendation. In: RecSys 2018, pp. 402–406 (2018)

22. Zheng, G., et al.: DRN: a deep reinforcement learning framework for news recommendation. In: WWW 2018 (2018)
23. Wang, L., Zhang, W., He, X., Zha, H.: Supervised reinforcement learning with recurrent neural network for dynamic treatment recommendation. In: KDD 2018, pp. 2447–2456 (2018)
24. Watkins, C.J., Dayan, P.: Q-learning. Mach. Learn. **8**(3–4), 279–292 (1992)
25. Sutton, R.S., McAllester, D.A., Singh, S.P., Mansour, Y., et al.: Policy gradient methods for reinforcement learning with function approximation. In: NIPS, vol. 99, pp. 1057–1063 (1999)
26. Schmitt, S., Hessel, M., Simonyan, K.: Off-policy actor-critic with shared experience replay. In: ICML (2020)

# Multi-Objective Actor-Critics for Real-Time Bidding in Display Advertising

Haolin Zhou[1], Chaoqi Yang[2], Xiaofeng Gao[1(✉)], Qiong Chen[3], Gongshen Liu[4], and Guihai Chen[1]

[1] MoE Key Lab of Artificial Intelligence, Department of Computer Science and Engineering, Shanghai Jiao Tong University, Shanghai, China
koziello@sjtu.edu.cn, {gao-xf,gchen}@cs.sjtu.edu.cn
[2] University of Illinois at Urbana-Champaign, Illinois, USA
chaoqiy2@illinois.edu
[3] Tencent Ads, Beijing, China
evechen@tencent.com
[4] School of Electronic Information and Electrical Engineering, Shanghai Jiao Tong University, Shanghai, China
lgshen@sjtu.edu.cn

**Abstract.** Online Real-Time Bidding (RTB) is a complex auction game among which advertisers struggle to bid for ad impressions when a user request occurs. Considering display cost, Return on Investment (ROI), and other influential Key Performance Indicators (KPIs), large ad platforms try to balance the trade-off among various goals in dynamics. To address the challenge, we propose a **Multi-ObjecTive Actor-Critics** algorithm based on reinforcement learning (RL), named `MoTiAC`, for the problem of bidding optimization with various goals. In `MoTiAC`, objective-specific agents update the global network asynchronously with different goals and perspectives, leading to a robust bidding policy. Unlike previous RL models, the proposed `MoTiAC` can simultaneously fulfill multi-objective tasks in complicated bidding environments. In addition, we mathematically prove that our model will converge to Pareto optimality. Finally, experiments on a large-scale real-world commercial dataset from Tencent verify the effectiveness of `MoTiAC` versus a set of recent approaches.

**Keywords:** Real-time bidding · Reinforcement learning · Display advertising · Multiple objectives

This work was supported by the National Key R&D Program of China [2020YFB 1707903]; the National Natural Science Foundation of China [61872238, 61972254]; Shanghai Municipal Science and Technology Major Project [2021SHZDZX0102]; the CCF-Tencent Open Fund [RAGR20200105] and the Tencent Marketing Solution Rhino-Bird Focused Research Program [FR202001]. Chaoqi Yang completed this work when he was an undergraduate student at Shanghai Jiao Tong University. Xiaofeng Gao is the corresponding author.
H. Zhou and C. Yang—Equal contribution.

**Supplementary Information** The online version contains supplementary material available at https://doi.org/10.1007/978-3-031-26412-2_2.

# 1  Introduction

The rapid development of the Internet and smart devices has created a decent environment for the advertisement industry. As a result, real-time bidding (RTB) has gained continuous attention in the past few decades [23]. A typical RTB setup consists of publishers, supply-side platforms (SSP), data management platforms (DMP), ad exchange (ADX), and demand-side platforms (DSP). When an online browsing activity triggers an ad request in one bidding round, the SSP sends this request to the DSP through the ADX, where eligible ads compete for the impression. The bidding agent, DSP, represents advertisers to come up with an optimal bid and transmits the bid back to the ADX (e.g., usually within less than 100ms [23]), where the winner is selected to be displayed and charged by a generalized second price (GSP).

In the RTB system, *bidding optimization* in DSP is regarded as the most critical problem [24]. Unlike Sponsored Search (SS) [25], where advertisers make keyword-level bidding decisions, DSP in the RTB setting needs to calculate the optimal impression-level bidding under the basis of user/customer data (e.g., income, occupation, purchase behavior, gender, etc.), target ad (e.g., content, click history, budget plan, etc.) and auction context (e.g., bidding history, time, etc.) in every single auction [24].

Thus, our work focuses on DSP, where *bidding optimization* happens. In real-time bidding, two fundamental challenges need to be addressed. Firstly, the RTB environment is highly dynamic. In [20,24,26], researchers make a strong assumption that the bidding process is stationary over time. However, the sequence of user queries (e.g., incurring impressions, clicks, or conversions) is time-dependent and mostly unpredictable [25], where the outcome influences the next auction round. Traditional algorithms usually learn an independent predictor and conduct fixed optimization that amounts to a greedy strategy, often not leading to the optimal return [3]. Agents with reinforcement learning (RL) address the aforementioned challenge to some extent [7,12,25]. RL-based methods can alleviate the instability by learning from immediate feedback and long-term reward. However, these methods are limited to either *Revenue* or *ROI*, which is only one part of the overall utility. In the problem of RTB, we assume that the utility is two-fold, as outlined: (i) the cumulative cost should be kept within the budget; (ii) the overall revenue should be maximized. Therefore, the second challenge is that the real-world RTB industry needs to consider multiple objectives, which are not adequately addressed in the existing literature.

To address the challenges mentioned above, we propose a *Multi-Objective Actor-Critic* model, named MoTiAC. We generalize the popular asynchronous advantage actor-critic (A3C) [13] reinforcement learning algorithm for multiple objectives in the RTB setting. Our model employs several local actor-critic networks with different objectives to interact with the same environment and then updates the global network asynchronously according to different reward signals. Instead of using a fixed linear combination of different objectives, MoTiAC can decide on adaptive weights over time according to how well the current situation conforms with the agent's prior. We evaluate our model on click data collected from the Tencent ad bidding system. The experimental results verify the effectiveness of our approach versus a set of baselines.

The contributions in this paper can be summarized as follows:

- We identify two critical challenges in RTB and are well motivated to use multi-objective RL as the solution.
- We propose a novel multi-objective actor-critic model MoTiAC for optimal bidding and prove the superiority of our model from the perspective of Pareto optimality.
- Extensive experiments on a real industrial dataset collected from the Tencent ad system show that MoTiAC achieves state-of-the-art performance.

## 2   Preliminaries

### 2.1   Definition of oCPA and Bidding Process

In the online advertising scenario, there are three main ways of pricing. Cost-per-mille (CPM) [7] is the first standard, where revenue is proportional to *impression*. Cost-per-click (CPC) [24] is a performance-based model, i.e., only when users *click* the ad can the platform get paid. In the cost-per-acquisition (CPA) model, the payment is attached to each *conversion* event. Regardless of the pricing ways, ad platforms always try to maximize revenue while simultaneously maintaining the overall cost within the budget predefined by advertisers.

In this work, we focus on one pricing model that is currently used in Tencent online ad bidding systems, called optimized cost-per-acquisition (oCPA), in which **advertisers are supposed to set a target CPA price, denoted by** $\text{CPA}_{\text{target}}$ **for each conversion while the charge is based on each click.** The critical point for the bidding system is to make an optimal strategy to allocate overall impressions among ads properly, such that (i) the real click-based cost is close to the estimated cost calculated from $\text{CPA}_{\text{target}}$, specifically,

$$\#\text{clicks} \times \text{CPC}_{\text{next}} \approx \#\text{conversions} \times \text{CPA}_{\text{target}}, \tag{1}$$

where $\text{CPC}_{\text{next}}$ is the cost charged by the second highest price and $\text{CPA}_{\text{target}}$ is pre-defined for each conversion; (ii) more overall conversions. In the system, the goal of our bidding agent is to generate an optimal **$\text{CPC}_{\text{bid}}$** price, adjusting the winner of each impression. We denote $\mathcal{I} = \{1, 2, ..., n\}$ as bidding iterations, $\mathcal{A} = \{ad_1, ad_2, ...\}$ as a set of all advertisements. For each iteration $i \in \mathcal{I}$, $ad_j \in \mathcal{A}$, our bidding agent will decide on a $\text{CPC}_{\text{bid}}^{(i,j)}$ to play auction. Then the ad with the highest $\text{CPC}_{\text{bid}}^{(i,j)}$ wins the impression and then receives possible $\#\text{clicks}^{(i,j)}$ (charged by $\text{CPC}_{\text{next}}^{(i,j)}$ per click) and $\#\text{conversions}^{(i,j)}$ based on user engagements.

### 2.2   Optimization Goals in Real-Time Bidding

On the one hand, when $\text{CPC}_{\text{bid}}$ is set higher, ads are more likely to win this impression to get clicks or later conversions, and vice versa. However, on the other hand, higher $\text{CPC}_{\text{bid}}$ means lower opportunities for other ad impressions. Therefore, to determine the appropriate bidding price, we define the two optimization objectives as follows:

**Objective 1: Minimize Overall CPA.** The first objective in RTB bidding problem is to allocate impression-level bids in every auction round, so that each ad will get reasonble opportunities for display and later get clicks or conversions, which makes $\text{CPA}_{\text{real}}$ close to $\text{CPA}_{\text{target}}$ pre-defined by the advertisers:

$$\text{CPA}_{\text{real}}^{(j)} = \frac{\sum_{i \in \mathcal{I}} \#\text{clicks}^{(i,j)} \times \text{CPC}_{\text{next}}^{(i,j)}}{\sum_{i \in \mathcal{I}} \#\text{conversions}^{(i,j)}}, \quad \forall ad_j \in \mathcal{A}. \tag{2}$$

To achieve the goal of minimizing overall CPA, i.e., be in line with the original budget, a lower ratio between $\text{CPA}_{\text{real}}^{(j)}$ and $\text{CPA}_{\text{target}}^{(j)}$ is desired. Precisely, when the ratio is smaller than 1, the agent will receive a positive feedback. On the contrary, when the ratio is greater than 1, it means that the actual expenditure exceeds the budget and the agent will be punished by a negative reward.

**Objective 2: Maximize Conversions.** The second objective is to enlarge conversions as much as possible under the condition of a reasonable $\text{CPA}_{\text{real}}$, so that platform can stay competitive and run a sustainable business:

$$\#\text{conversions}^{(j)} = \sum_{i \in \mathcal{I}} \#\text{conversions}^{(i,j)}, \quad \forall ad_j \in \mathcal{A}, \tag{3}$$

where $\#\text{conversions}^{(j)}$ is a cumulative value until the current bidding auction. Obviously, relatively high #conversions will receive a positive reward. When the policy network gives fewer conversions, the agent will be punished with a negative reward.

Note that in the real setting, optimization objectives used by advertising platforms can be adjusted based on actual business needs. In the implementation and evaluation of MoTiAC, we use **ROI** (Return on Investment) and **Revenue**, corresponding to the two objectives for optimization, i.e., minimizing overall CPA and maximizing the number of conversions. Their definition will be detailed in Sec. 4.1.

## 3   Methodology

As shown in Sec. 2.1, the RTB problem is a multi-objective optimization problem. We need to control advertisers' budgets and make profitable decisions for the ad platform. Traditional RTB control policy or RL agent can hardly handle these challenges. In this work, we design MoTiAC to decouple the training procedure of multiple objectives into disentangled worker groups of actor-critics. We will elaborate on the technical details of MoTiAC in the following subsections.

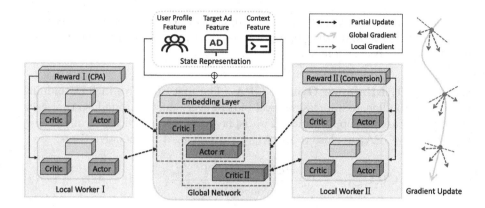

**Fig. 1.** Framework of the proposed MoTiAC in RTB.

## 3.1   Asynchronous Advantage Actor-Critic Model in RTB

An actor-critic reinforcement learning setting [8] in our RTB scenario consists of:

- **state** $s$: each state is composed of anonymous feature embeddings extracted from the user profile and bidding environment, indicating the current bidding state.
- **action** $a$: action is defined as the bidding price for each ad based on the input state. Instead of using discrete action space [20], our model outputs a distribution so that action can be sampled based on probability.
- **reward** $r$: obviously, the reward is a feedback signal from the environment to evaluate how good the previous action is, which guides the RL agent towards a better policy. In our model, we design multiple rewards based on different optimization goals. Each actor-critic worker group deals with one type of reward from the environment and later achieves multiple objectives together.
- **policy** $\pi_\theta(\cdot)$: policy is represented as $\pi_\theta(a_t|s_t)$, which denotes the probability to take action $a_t$ under state $s_t$. In an actor-critic thread, actor works as a policy network, and critic stands for value function $V(s; \theta_v)$ of each state. The parameters are updated according to the experience reward obtained during the training process.

For each policy $\pi_\theta$, we define the utility function as

$$U(\pi_\theta) = E_{\tau \sim p_\theta(\tau)}[R(\tau)], \qquad (4)$$

where $p_\theta(\tau)$ denotes the distribution of trajectories under policy $\pi_\theta$, and $R(\tau)$ is a return function over trajectory $\tau$, calculated by summing all the reward signals in the trajectory. The utility function is used to evaluate the quality of an action taken in a specific state. We also introduce value function from critic to reduce the varaition that may occur when updating parameters in real time.

After collecting a number of tuples $(s_t, a_t, r_t)$ from each trajectory $\tau$, the policy network $\pi_\theta(\cdot)$ is updated by

$$\theta \leftarrow \theta + \eta_{actor} \sum_{t=1}^{T} (R(s_t) - V(s_t)) \nabla_\theta \log \pi_\theta(a_t \mid s_t), \tag{5}$$

where $\eta_{actor}$ represents the learning rate of policy network, $T$ is a preset maximum step size in a trajectory, $R(s_t) = \sum_{n=t}^{T} \gamma^{n-t} r_n$ denotes the cumulative discounted reward, and $\gamma$ is a decaying factor. The critic network, $V(s; \theta_v)$, could also be updated by:

$$\theta_v \rightarrow \theta_v + \eta_{critic} \frac{\partial (R_t - V_{\theta_v}(s_t))^2}{\partial \theta_v}, \tag{6}$$

where $\eta_{critic}$ represents the learning rate of value function.

## 3.2  Adaptive Reward Partition

In this subsection, we consider the general $K$-objective case, where $K$ is the total number of objectives. As stated in Sec. 2.2, multiple objectives should be considered in modeling the RTB problem. One intuitive way [14] of handling multiple objectives is to integrate them into a single reward function linearly, and we call it *Reward Combination*: (i) A linear combination of rewards is firstly computed, where $w_k$ quantifies the relative importance of the corresponding objective $R_k(\cdot)$:

$$R(s) = \sum_{k=1}^{K} w_k \times R_k(s). \tag{7}$$

(ii) A single-objective agent is then defined with the expected return equal to value function $V(s)$. However, a weighted combination is only valid when objectives do not compete [17]. In the RTB setting, the relationship between objectives can be complicated, and they usually conflict on different sides. The intuitive combination might flatten the gradient for each objective, and thus the agent is likely to limit itself within a narrow boundary of search space. Besides, a predefined combination may not be flexible in the dynamic bidding environment. Overall, such a *Reward Combination* method is unstable and inappropriate for the RTB problem, as we will show in the experiments.

**Reward Partition.** We now propose the *Reward Partition* scheme in MoTiAC. Specifically, we design reward for each objective and employ one group of actor-critic networks with the corresponding reward. There is one global network with an actor and multiple critics in our model. At the start of one iteration, each local network copies parameters from a global network. Afterward, local networks from each group will begin to explore based on their objective and apply weighted gradients to the actor and one of the critics (partial update) in the global network asynchronously, as shown in Fig. 1. Formally, we denote the total utility and value function of the $k^{th}$ group ($k = 1, \cdots, K$) as $U^k(\pi_\theta)$ and $V_k(s; \theta_v)$, respectively. Different from the original Eqn. (5), the parameter updating policy network in one actor-critic group of MoTiAC is formulated as

$$\theta \leftarrow \theta + \eta_{actor} w_k \sum_{t=1}^{T} (R_k(s_t) - V_k(s_t)) \nabla_\theta \log \pi_\theta(a_t \mid s_t), \qquad (8)$$

where $w_k$ is an objective-aware customized weight for optimization in range (0,1) and is tailored for each $adj \in A$. We can simply set $w_k$ as

$$w_k = \frac{R_k(s_t) - V_k(s_t)}{\sum_{l=1}^{K}(R_l(s_t) - V_l(s_t))}, \qquad (9)$$

while dynamically adjusting the value of $w_k$ by giving higher learning weights to the local network that contributes more to the total reward. Motivated by Bayesian RL [5], we can generalize the customized weight and parameterize $w_k$ by introducing a latent multinomial variable $\phi$ with $w_k = p(\phi = k|\tau)$ under trajectory $\tau$, named as *agent's prior*. We set the initial prior as

$$p(\phi = k|\tau_0) = \frac{1}{K}, \quad \forall\, k = 1, 2, \ldots, K, \qquad (10)$$

where $\tau_0$ indicates that the trajectory just begins. When $\tau_t$ is up to state $s_t$, i.e., $\tau_t = \{s_1, a_1, r_1, s_2, a_2, r_2, \ldots s_t\}$, we update the posterior by

$$p(\phi = k|\tau_t) = \frac{p(\tau_t|\phi = k)p(\phi = k)}{\sum_k p(\tau_t|\phi = k)p(\phi = k)}, \qquad (11)$$

where $p(\tau_t|\phi = k)$ tells how well the current trajectory agrees with the utility of objective $k$. Based on priority factor $w_k$, together with the strategy of running different exploration policies in different groups of workers, the overall changes being made to the global actor parameters $\theta$ are likely to be less correlated and more objective-specific in time, which means our model can make wide exploration and achieve a balance between multiple objectives with a global overview.

In addition, we present some analysis for the two reward aggregation methods in terms of parameters update and value function approximation. If we attach the weights of *Reward Combination* to the gradients in *Reward Partition*, the parameters updating strategy should be identical on average. For *Reward Combination*, global shared actor parameters $\theta$ is updated by

$$\theta \leftarrow \theta + \eta_{actor} \sum_t \left( \left( \sum_{k=1}^{K} w_k \times R_k(s_t) - V_k(s_t) \right) \times \nabla_\theta \log \pi_\theta(a_t \mid s_t) \right),$$

while in *Reward Partition*, the expected global gradient is given as

$$\theta \leftarrow \theta + \eta_{actor} \sum_t \left( \left( \sum_{k=1}^{K} (w_k \times R_k(s_t) - w_k \times V_k(s_t)) \right) \times \nabla_\theta \log \pi_\theta(a_t \mid s_t) \right).$$

The difference between the two reward aggregating methods lies in the advantage part. Thus the effect of parameter updates heavily depends on how well and precisely the critic can learn from its reward. By learning in a decomposed manner, the proposed *Reward Partition* advances the *Reward Combination* by using easy-to-learn functions to approximate single rewards, thus yielding a better policy.

### 3.3   Optimzation and Training Procedure

In the framework of `MoTiAC`, the policy network explores continuous action space and outputs action distribution for inference. Therefore, loss for a single actor-critic worker (objective-$k$) is gathered from actor $\theta$, critic $\theta_v$, and action distribution entropy $H$ to improve exploration by discouraging premature convergence to sub-optimal [13],

$$L_{\theta,\theta_v} = \eta_{actor}E[R(\tau)] + \eta_{critic}\sum_{s_t \in \tau}\|V_{\theta_v}(s_t) - R(s_t)\|^2 + \beta\sum_{s_t \in \tau}H(\pi(s_t)), \quad (12)$$

where $\beta$ represents the strength of entropy regularization.

   After one iteration (e.g., 10-minute bidding simulation), we compute gradients for each actor-critic network and push the weighted gradients to the global network. With multiple actor-learners applying online updates in parallel, the global network could explore to achieve a robust balance between multiple objectives. The training procedure of `MoTiAC` is shown in Algorithm 1.

### 3.4   Convergence Analysis of `MoTiAC`

In this section, we use a toy demonstration to provide insights into the convergence property for the proposed `MoTiAC`. As illustrated in Fig. 2, the solid black line is the gradient contour of *objective 1*, and the black dash line is for *objective 2*. The yellow area within their intersection is the area of the optimal strategy, where both advertisers and publishers satisfy with their benefits. Due to the highly dynamic environment of RTB [3], the optimal bidding strategy will change dramatically.

   Traditionally in a multi-objective setting, when people use linear combinations or other more complex transformations [11], like policy votes [18] of reward functions. They implicitly assume that the optimal solution is fixed, as shown in the upper part of Fig. 2. Consequently, their models can only learn the *initial optimal* and fail to characterize the dynamics. However, according to the dynamic environment in RTB, our `MoTiAC` adjusts the gradient w.r.t each possible situation towards a new optimal based on each objective separately and will easily be competent for real-world instability. Each gradient w.r.t the objectives forces the agent closer to the optimal for compensation rather than conflicts. Finally, the agent would reach the area of *new optimal* and tunes its position in the micro-level, called convergence.

---

**Algorithm 1:** Training for each actor-critic thread in `MoTiAC`

---

1 // Assume global shared parameters $\theta$ and $\theta_v$;
2 // Assume objective-specific parameters $\theta_k'$ and $\theta_{v,k}'$, $k \in \{1, 2, \ldots, K\}$;
3 Initialize step counter $t \leftarrow 1$; epoch $T$; discounted factor $\gamma$;
4 **while** $t < T_{max}$ **do**
5      Reset gradients: $d\theta \leftarrow 0$ and $d\theta_v \leftarrow 0$ ;
6      Synchronize specific parameters $\theta_k' = \theta$ and $\theta_{v,k}' = \theta_v$;
7      Get state $s_t$ extracted from user profile features and bidding environment;
8      // Assume ad set $\mathcal{A} = \{ad_1, ad_2, \ldots\}$;
9      **for** $ad_j \in \mathcal{A}$ **do**
10          **repeat**
11              Determine bidding price $a_t$ according to policy $\pi(a_t \mid s_t; \theta_k')$;
12              Receive reward $r_t$ w.r.t *objective* $k$;
13              Reach new state $s_{t+1}$;
14              $t \leftarrow t + 1$;
15          **until** *terminal state*;
16          **for** $n \in \{t - 1, \ldots, 1\}$ **do**
17              $r_n \leftarrow r_n + \gamma \times r_{n+1}$;
18              // Accumulative gradient w.r.t $\theta_k'$;
19              $d\theta_k' \leftarrow d\theta_k' + \eta_{actor} \sum (r_n - V(s_n; \theta_{v,k}')) \nabla_{\theta_k'} \log \pi(a_n|s_n) + \beta \sum \nabla_{\theta_k'} H(\pi(a_n|s_n))$;
20              // Accumulative gradient w.r.t $\theta_{v,k}'$;
21              $d\theta_{v,k}' \leftarrow d\theta_{v,k}' + \eta_{critic} \sum \partial \|r_n - V(s_n; \theta_{v,k}')\|^2 / \partial \theta_{v,k}'$;
22          **end**
23          // Asynchronously update $\theta$ and $\theta_v$ with $d\theta_k'$ and $d\theta_{v,k}'$;
24          // Compute $w_k = p(\phi = k|\tau)$ w.r.t objective $k$;
25          $\theta \leftarrow \theta + w_k \times d\theta_k'$ and $\theta_v \leftarrow \theta_v + w_k \times d\theta_{v,k}'$;
26      **end**
27 **end**

---

We further prove that the global policy will converge to the Pareto optimality between these objectives. The utility expectation of the objective $k$ is denoted as $E[U^k(\pi_\theta)]$. We begin the analysis with Theorem 1 [10],

**Theorem 1.** *(Pareto Optimality). If $\pi^*$ is a Pareto optimal policy, then for any other policy $\pi$, one can at least find one $k$, so that $0 < k \leq K$ and,*

$$E[U^k(\pi^*)] \geq E[U^k(\pi)]. \tag{13}$$

The multi-objective setting assumes that the possible policy set $\Pi$ spans a convex space ($K$-simplices). The optimal policy of any affine interpolation of objective utility will be also optimal [4]. We restate in Theorem 2 by only considering the non-negative region.

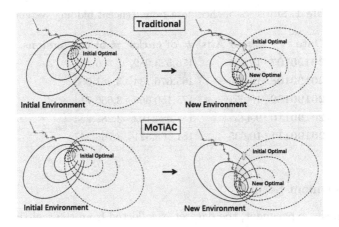

**Fig. 2.** Convergence illustration of `MoTiAC`.

**Theorem 2.** $\pi^*$ *is Pareto optimal iff there exits* $\{l_k > 0 : \sum_k l_k = 1\}$ *such that,*

$$\pi^* \in \arg\max_\pi \left[ \sum_k l_k E[U^k(\pi)] \right]. \tag{14}$$

*Proof.* We derive the gradient by aggregating Eqn. (8) as,

$$\nabla = \sum_{\tau_t} \sum_k p(\phi = k|\tau_t) \nabla_\theta U^k(\tau_t; \pi_\theta) \propto \sum_k p(\phi = k) \sum_{\tau_t} p(\tau_t|\phi = k) \nabla_\theta U^k(\tau_t; \pi_\theta)$$

$$= \sum_k p(\phi = k) \nabla_\theta E_{\tau_t}[U^k(\tau_t; \pi_\theta)] = \nabla_\theta \left[ \sum_k p(\phi = k) E_{\tau_t}[U^k(\tau_t; \pi_\theta)] \right]. \tag{15}$$

By making $l_k = p(\phi = k)$ (Note that $\sum_k p(\phi = k) = 1$), we find that the overall gradient conform with the definition of Pareto optimality in Eqn. (14). Therefore, we conclude that `MoTiAC` converges to Pareto optimal, indicating that it can naturally balance different objectives.

## 4   Experiments

In the experiment, we use real-world industrial data to answer the following three research questions:

- **RQ1:** How does `MoTiAC` perform compared with other baseline methods?
- **RQ2:** What is the best way to aggregate multiple objectives?
- **RQ3:** How does `MoTiAC` balance the exploration of different objectives?

**Table 1.** Statistics of click data from Tencent bidding system.

| Date | # of Ads | # of clicks | # of conversions |
|---|---|---|---|
| **20190107** | 10,201 | 176,523,089 | 3,886,155 |
| **20190108** | 10,416 | 165,676,734 | 3,661,060 |
| **20190109** | 10,251 | 178,150,666 | 3,656,714 |
| **20190110** | 9,445 | 157,084,102 | 3,287,254 |
| **20190111** | 10,035 | 181,868,321 | 3,768,247 |

## 4.1 Experiment Setup

*Dataset.* In the experiment, the dataset is collected from the real-time commercial ads bidding system of Tencent. There are nearly 10,000 ads daily with a huge volume of click and conversion logs. According to real-world business, the bidding interval is set to be 10 min (144 bidding sessions for a day), which is much shorter than one hour [7]. Basic statistics can be found in Table 1.

*Compared Baselines.* We carefully select related methods for comparison and adopt the same settings for all the compared methods with 200 iterations. Details about implementation can be seen in Appendix A.2.

- **Proportional-Integral-Derivative (PID):** [2] is a widely used feedback control policy, which produces the control signal from a linear combination of proportional, integral, and derivative factors.
- **Advantage Actor-Critic (A2C):** [13] makes the training process more stable by introducing an advantage function. [7] generalizes the actor-critic structure in the RTB setting.
- **Deep Q-Network (DQN):** [20] uses DQN with a single objective under the assumption of consistent state transition in the RTB problem, while the similar structure can also be coupled with a dynamic programming approrach [3].
- **Aggregated A3C (Agg-A3C):** Agg-A3C [13] is proposed to disrupt the correlation of training data by introducing an asynchronous update mechanism.

We linearly combine multiple rewards (following *Reward Combination*) for all the baselines. Besides, we adopt two variants of our model: *Objective1-A3C (O1-A3C)* and *Objective2-A3C (O2-A3C)*, by only considering one of the objectives. We use four days of data for training and another day for testing and then use the cross-validation strategy on the training set for hyper-parameter selection. Similar settings can be found in literature [20,26].

*Evaluation Metrics.* We clarify the objectives of our problem based on the collected data. In Sec. 2.2, we claim that our two objectives are: (1) *minimize overall CPA*; (2) *maximize conversions*. We refer to the industrial convention and redefine our goals in the experiments. *Revenue* is a common indicator for platform

Table 2. Comparative results based on PID.

| Model | Relative Cost | Relative ROI | Relative Revenue | R-score |
|---|---|---|---|---|
| PID | 1.0000 | 1.0000 | 1.0000 | 1.0000 |
| A2C | 1.0366 (+3.66%) | 0.9665 (−3.35%) | 1.0019 (+0.19%) | 0.9742 |
| DQN | 0.9765 (−2.35%) | 1.0076 (+0.76%) | 0.9840 (−1.60%) | 0.9966 |
| Agg-A3C | 1.0952 (+9.52%) | 0.9802 (−1.98%) | 1.0625 (+6.25%) | 0.9929 |
| O1-A3C | 0.9580 (−4.20%) | 1.0170 (+1.70%) | 0.9744 (−2.56%) | 1.0070 |
| O2-A3C | 1.0891 (+8.91%) | 0.9774 (−2.26%) | 1.0645 (+6.45%) | 0.9893 |
| MoTiAC | 1.0150 (+1.50%) | 1.0267 (+2.67%) | 1.0421 (+4.21%) | 1.0203 |

earnings, which turns out to be proportional to conversions. *Cost* is the money paid by advertisers, which also appears to be a widely accepted factor in online advertising. Therefore, without loss of generality, we reclaim our two objectives to be:

$$\text{Revenue}^{(j)} = \text{conversions}^{(j)} \times \text{CPA}_{\text{target}}^{(j)}, \quad \text{Cost}^{(j)} = \#\text{clicks}^{(j)} \times \text{CPC}_{\text{next}}^{(j)}, \tag{16}$$

$$\max \mathbf{ROI} \leftarrow \max \sum_{Ad_j \in A} \frac{\text{Revenue}^{(j)}}{\text{Cost}^{(j)}}, \tag{17}$$

which corresponds to the first objective: *CPA goal*, and

$$\max \mathbf{Revenue} \leftarrow \max \sum_{Ad_j \in A} \text{Revenue}^{(j)}, \tag{18}$$

related to the second objective: *Conversion goal*.

For the two variants of MoTiAC, O1-A3C corresponds to maximizing ROI, while O2-A3C is related to maximizing Revenue. In addition to directly comparing these two metrics, we also use *R-score* proposed in [12] to evaluate the model performance. The higher the *R-score*, the more satisfactory the advertisers and platform will be. In the real-world online ad system, PID is currently used to control bidding. We employ it as a standard baseline, and most of the comparison results will be based on PID, i.e., $value \rightarrow \frac{value}{value_{PID}}$, except for Sec. 4.4.

## 4.2  RQ1: Comparison with Recent Baselines

We perform the comparison of MoTiAC with other approaches. The results are shown in Table 2. The values in the parentheses represent the percentage of improvement or reduction towards PID. An optimal method is expected to improve both metrics (ROI & Revenue) compared with the current PID baseline.

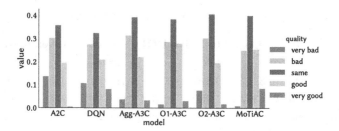

**Fig. 3.** Bidding quality distribution of compared methods over PID.

**Objective Comparison.** We find that MoTiAC best balances the trade-off between two objectives (ROI & Revenue) based on the above considerations. Also, it has the highest R-score. Specifically, A2C is the worst since it gains a similar revenue (conversion goal) but a much lower ROI (CPA goal) than PID. The result proves that the A2C structure cannot fully capture the dynamics in the RTB environment. Based on a hybrid reward, DQN has a similar performance as O1-A3C, with relatively fewer conversions than other methods. We suspect the discrete action space may limit the policy to a local and unstable optimal. By solely applying the weighted sum in a standard A3C (Agg-A3C), the poor result towards ROI is not surprising. As the RTB environment varies continuously, fixing the formula of reward aggregation cannot capture the dynamic changes. It should be pointed out that two ablation models, O1-A3C and O2-A3C, present two extreme situations. O1-A3C performs well in the first ROI objective but performs poorly for the Revenue goal and vice versa for O2-A3C. By shifting the priority of different objectives over time, our proposed MoTiAC uses the agent's prior as a reference to make the decision in the future, precisely capturing the dynamics of the RTB sequence. Therefore, it outperforms all the other baselines.

Comparing *Reward Partition* and *Reward Combination*, the advantages of MoTiAC over other baselines show that our proposed method of accumulating rewards overall reduces the difficulty of agent learning and makes it easier for the policy network to converge around the optimal value.

**Bidding Quality Analysis.** To further verify the superiority of MoTiAC compared to other methods, we analyze the relative bidding quality of these methods over PID. We group all the ads into five categories based on their bidding results. The detailed evaluation metrics can be found in Appendix A.1. As shown in Fig. 3, both A2C and O2-A3C present more bad results compared than good ones, indicating that these two models could not provide a gain for the existing bidding system at a finer granularity. O1-A3C has a relatively similar performance as PID, as they both aims at minimizing real *CPA*. We also find that DQN tends to make bidding towards either very good or very bad, once again demonstrating the instability of the method. Agg-A3C shares the same distribution pattern with O1-A3C and vanilla PID, which indicates that the combined reward does not work in our scenario. The proposed MoTiAC turns out to have a desirable improvement over PID with more ads on the right *good* side and fewer

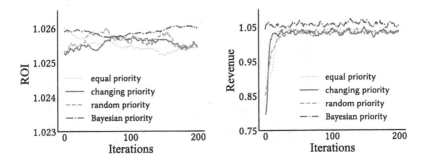

**Fig. 4.** Results under different priority functions.

ads on the left *bad* side. Note that the negative transfer of multi-objective tasks makes some bidding results inevitably worse. However, we can still consider that MoTiAC can achieve the best balance among all the compared methods.

### 4.3   RQ2: Variants of $w_k$

To give a comprehensive view of MoTiAC, we perform different ways to aggregate objectives. Four different variants of $w_k$ are considered in the experiment. Since we have two objectives, we use $w_1(t)$ for the first objective and $1 - w_1(t)$ for the second:

- equal priority: $w_1(t) = \frac{1}{2}$;
- changing priority: $w_1(t) = \exp(-\alpha \cdot t)$ with a scalar $\alpha$;
- random priority: $w_1(t) = \text{random}([0, 1])$;
- Bayesian priority: One can refer to Eqn. (11).

As shown in Fig. 4, we present the training curves for ROI and Revenue. The first three strategies are designed before training and will not adjust to the changing environment. It turns out that they perform similarly in both objectives and could gain a decent improvement over the PID case by around +2.5% in ROI and +3% in Revenue. However, in *equal priority*, the curve of ROI generally drops when the iteration goes up, which stems from the fact that fixed equal weights cannot fit the dynamic environment. For *changing priority*, it is interesting that ROI first increases then decreases for priority shifting, as different priority leads to different optimal. In *random priority*, curves dramatically change in a small range since the priority function outputs the weight randomly. The *Bayesian priority* case, on the contrary, sets priority based on the conformity of the agent's prior and current state. Reward partition with agent prior dominates the first three strategies by an increasingly higher ROI achievement by +2.7% and better Revenue by around +4.2%.

### 4.4   RQ3: Case Study

In this section, we try to investigate how MoTiAC balances the exploration of multiple objectives and achieves the optimal globally. We choose one typical ad

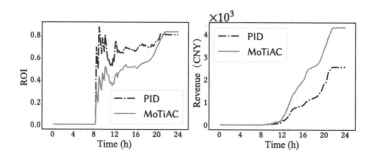

**Fig. 5.** ROI and Revenue curves of the target ad's reponse.

**Table 3.** Numerical results of the target ad using PID and MoTiAC.

| Models | Revenue (CNY) | Cost (CNY) | ROI |
|---|---|---|---|
| **PID** | $3.184 \times 10^3$ | $2.548 \times 10^3$ | 0.8003 |
| MoTiAC | $4.298 \times 10^3$ | $5.199 \times 10^3$ | 0.8267 |

with large conversions and show the bidding process within 24 h. As PID is the current model in the real ad system, we use PID to compare with MoTiAC and draw the results of ROI and Revenue curve in Fig. 5. We also collect the final numerical results in Table 3.

Figure 5 shows a pretty low ROI initially. For the target ad, both models first try to lift the ROI. Based on the figure presented on the left, the red dashed curve rises from 0 to about 0.7 sharply for PID at 8h. The potential process should be that PID has given up most of the bid chances and only concentrates on those with a high conversion rate (CVR) so that we have witnessed a low Revenue gain of the PID model in the right figure from 8h to around 21h. Though the ROI curve remains relatively low, our MoTiAC can select good impression-level chances while considering the other objective. At 24h, MoTiAC finally surpasses PID in ROI because of the high volume of pre-gained Revenue. With long-term consideration, MoTiAC beats PID on both the cumulative ROI and Revenue. We can conclude that PID is greedy out of the immediate feedback mechanism. It is always concerned with the current situation and never considers further benefits. When the current state is under control, PID will appear conservative and give a shortsighted strategy, resulting in a seemingly good ROI and poor Revenue (like the red curve in Fig. 5). However, MoTiAC has a better overall view. It foresees the long-run benefit and will keep exploration even temporarily deviating from the right direction or slowing down the rising pace (ROI curve for the target ad at 8h). Under a global overview, MoTiAC can finally reach better ROI and Revenue than PID.

# 5   Related Work

*Real-time Bidding.* Researchers have proposed static methods [15] for optimal biddings, such as constraint optimization [24], to perform an impression-level evaluation. However, traditional methods inevitably ignore that real-world situations in RTB are often dynamic [21] due to the unpredictability of user behavior [7] and different marketing plans [22] from advertisers. Furthermore, the auction process of optimal bidding is formulated as a Markov decision process (MDP) in recent study [7,12]. Considering the various goals of different players in RTB, a robust framework is required to balance these multiple objectives. Therefore, we are motivated to propose a novel multi-objective RL model to maximize the overall utility of RTB.

*Reinforcement Learning.* Significant achievements have been made by the emergence of RL algorithms, such as policy gradient [19] and actor-critic [8]. With the advancement of GPU and deep learning (DL), more successfully deep RL algorithms [9,13] have been proposed and applied to various domains. Meanwhile, there are previous attempts to address the multi-objective reinforcement learning (MORL) problem [6], where the objectives are combined mainly by static or adaptive linear weights [1,14] or captured by a set of policies and evolving preferences [16].

# 6   Conclusion and Future Directions

In this paper, we propose **Multi-ObjecTive Actor-Critics** for real-time bidding in display advertising. MoTiAC utilizes objective-aware actor-critics to solve the problem of multi-objective bidding optimization. Our model can follow adaptive strategies in a dynamic RTB environment and outputs the optimal bidding policy by learning priors from historical data. We conduct extensive experiments on the real-world industrial dataset. Empirical results show that MoTiAC achieves state-of-the-art on the Tencent advertising dataset. One future direction could be extending multi-objective solutions with priors in the multi-agent reinforcement learning area.

# References

1. Abels, A., Roijers, M.D., Lenaerts, T., Nowé, A., Stechkelmacher, D.: Dynamic weights in multi-objective deep reinforcement learning. In: International Conference on Machine Learning (ICML), pp. 11–20 (2019)
2. Bennett, S.: Development of the PID controller. IEEE Control Syst., 58–62 (1993)
3. Cai, H., et al.: Real-time bidding by reinforcement learning in display advertising. In: Proceedings of the 10th ACM International Conference on Web Search and Data Mining (WSDM), pp. 661–670 (2017)
4. Critch, A.: Toward negotiable reinforcement learning: shifting priorities in pareto optimal sequential decision-making. arXiv:1701.01302 (2017)

5. Ghavamzadeh, M., Mannor, S., Pineau, J., Tamar, A., et al.: Bayesian reinforcement learning: a survey. Foundations and Trends® in Machine Learning 8(5–6), 359–483 (2015)
6. Hayes, C.F., et al.: A practical guide to multi-objective reinforcement learning and planning. arXiv preprint arXiv:2103.09568 (2021)
7. Jin, J., Song, C., Li, H., Gai, K., Wang, J., Zhang, W.: Real-time bidding with multi-agent reinforcement learning in display advertising. In: Proceedings of the 27th ACM International Conference on Information and Knowledge Management (CIKM) (2018)
8. Konda, V.R., Tsitsiklis, J.N.: Actor-critic algorithms. In: Advances in Neural Information Processing Systems (NeurIPS), pp. 1008–1014 (2000)
9. Lillicrap, T.P., Hunt, J.J., Pritzel, A., Heess, N., Erez, T., Tassa, Y., Silver, D., Wierstra, D.: Continuous control with deep reinforcement learning. arXiv:1509.02971 (2015)
10. Lin, X., Zhen, H.L., Li, Z., Zhang, Q.F., Kwong, S.: Pareto multi-task learning. In: Advances in Neural Information Processing Systems (NeurIPS), pp. 12060–12070 (2019)
11. Lizotte, D.J., Bowling, M.H., Murphy, S.A.: Efficient reinforcement learning with multiple reward functions for randomized controlled trial analysis. In: International Conference on Machine Learning (ICML), pp. 695–702 (2010)
12. Lu, J., Yang, C., Gao, X., Wang, L., Li, C., Chen, G.: Reinforcement learning with sequential information clustering in real-time bidding. In: Proceedings of the 28th ACM International Conference on Information and Knowledge Management (CIKM), pp. 1633–1641 (2019)
13. Mnih, V., et al.: Asynchronous methods for deep reinforcement learning. In: International Conference on Machine Learning (ICML), pp. 1928–1937 (2016)
14. Pasunuru, R., Bansal, M.: Multi-reward reinforced summarization with saliency and entailment. arXiv:1804.06451 (2018)
15. Perlich, C., Dalessandro, B., Hook, R., Stitelman, O., Raeder, T., Provost, F.: Bid optimizing and inventory scoring in targeted online advertising. In: Proceedings of the 18th ACM SIGKDD International Conference on Knowledge Discovery and Data Mining (KDD), pp. 804–812 (2012)
16. Pirotta, M., Parisi, S., Restelli, M.: Multi-objective reinforcement learning with continuous pareto frontier approximation. In: 29th AAAI Conference on Artificial Intelligence (AAAI) (2015)
17. Sener, O., Koltun, V.: Multi-task learning as multi-objective optimization. In: Advances in Neural Information Processing Systems (NeurIPS), pp. 525–536 (2018)
18. Shelton, C.R.: Balancing multiple sources of reward in reinforcement learning. In: Advances in Neural Information Processing Systems (NeurIPS), pp. 1082–1088 (2001)
19. Sutton, R.S., McAllester, D.A., Singh, S.P., Mansour, Y.: Policy gradient methods for reinforcement learning with function approximation. In: Advances in Neural Information Processing Systems (NeurIPS), pp. 1057–1063 (2000)
20. Wang, Y., et al.: Ladder: a human-level bidding agent for large-scale real-time online auctions. arXiv:1708.05565 (2017)
21. Wu, D., et al.: Budget constrained bidding by model-free reinforcement learning in display advertising. In: Proceedings of the 27th ACM International Conference on Information and Knowledge Management (CIKM) (2018)
22. Xu, J., Lee, K.C., Li, W., Qi, H., Lu, Q.: Smart pacing for effective online ad campaign optimization. In: Proceedings of the 21th ACM SIGKDD International Conference on Knowledge Discovery and Data Mining (KDD), pp. 2217–2226 (2015)

23. Yuan, S., Wang, J., Zhao, X.: Real-time bidding for online advertising: measurement and analysis. In: Proceedings of the 7th International Workshop on Data Mining for Online Advertising (ADKDD), p. 3 (2013)
24. Zhang, W., Yuan, S., Wang, J.: Optimal real-time bidding for display advertising. In: Proceedings of the 20th ACM SIGKDD International Conference on Knowledge Discovery and Data Mining (KDD), pp. 1077–1086 (2014)
25. Zhao, J., Qiu, G., Guan, Z., Zhao, W., He, X.: Deep reinforcement learning for sponsored search real-time bidding. arXiv preprint arXiv:1803.00259 (2018)
26. Zhu, H., et al.: Optimized cost per click in Taobao display advertising. In: Proceedings of the 23rd ACM SIGKDD International Conference on Knowledge Discovery and Data Mining (KDD), pp. 2191–2200 (2017)

# Batch Reinforcement Learning from Crowds

Guoxi Zhang[1(✉)] and Hisashi Kashima[1,2]

[1] Graduate School of Informatics, Kyoto University, Kyoto, Japan
guoxi@ml.ist.i.kyoto-u.ac.jp, kashima@i.kyoto-u.ac.jp
[2] RIKEN Guardian Robot Project, Kyoto, Japan

**Abstract.** A shortcoming of batch reinforcement learning is its requirement for rewards in data, thus not applicable to tasks without reward functions. Existing settings for the lack of reward, such as behavioral cloning, rely on optimal demonstrations collected from humans. Unfortunately, extensive expertise is required for ensuring optimality, which hinder the acquisition of large-scale data for complex tasks. This paper addresses the lack of reward by learning a reward function from preferences between trajectories. Generating preferences only requires a basic understanding of a task, and it is faster than performing demonstrations. Thus, preferences can be collected at scale from non-expert humans using crowdsourcing. This paper tackles a critical challenge that emerged when collecting data from non-expert humans: the noise in preferences. A novel probabilistic model is proposed for modelling the reliability of labels, which utilizes labels collaboratively. Moreover, the proposed model smooths the estimation with a learned reward function. Evaluation on Atari datasets demonstrates the effectiveness of the proposed model, followed by an ablation study to analyze the relative importance of the proposed ideas.

**Keywords:** Preference-based reinforcement learning · Crowdsourcing

## 1 Introduction

Batch Reinforcement Learning (RL) (Lange et al. 2012) is a setting for RL that addresses its limitation of data acquisition. Online RL needs to generate new data during learning, either via simulation or physical interaction. However, simulators with high fidelity are not always available, and real-world interactions raise safety and ethical concerns. Batch RL instead reuses existing vastly available data, so it has received increasing attention in recent years (Pavse et al. 2020, Agarwal et al. 2020, Gelada and Bellemare 2019, Kumar et al. 2019, Fujimoto et al. 2019).

In batch RL, the data consists of observations, actions, and rewards. For example, in recommender systems, the observations are user profiles, and the actions are items to be recommended. The rewards are scalars, evaluating actions for their consequences on achieving a given task. The mapping from observations and actions to rewards is called a reward function. Often, the sequence of observations, actions, and rewards generated during interaction is called a trajectory.

M.-R. Amini et al. (Eds.): ECML PKDD 2022, LNAI 13716, pp. 38–51, 2023.
https://doi.org/10.1007/978-3-031-26412-2_3

While batch RL is a promising data-driven setting, its dependence on rewards can be problematic. Many tasks of interest lack a reward function. Consider an application to StarCraft, a famous real-time strategy game, for example. Typically, evaluative feedback is given for the final result of a series of action sequences (i.e., the entire trajectory), not for individual actions about their contributions to the result. In the RL literature, the lack of reward has been addressed by inverse RL (Abbeel and Ng 2004) or Behavioral Cloning (BC) (Schaal 1996), which eliminate the need for rewards by leveraging expert demonstrations. However, their assumption on demonstrations can be hard to satisfy. Optimal demonstrations require extensive expertise. In practice, the competency of human demonstrators differs (Mandlekar et al. 2019), causing RL and BC algorithms to fail (Mandlekar et al. 2021).

This paper addresses the lack of reward signals by learning a reward function from preferences. A *preference* is the outcome of a comparison between two trajectories for the extent the task is solved. Compared to demonstrations, providing preferences requires less human expertise. For example, demonstrating the moves of professional sports players is difficult, but with general knowledge a sports fan can still appreciate the moves in games. Hence, preferences can be collected at scale from a large group of non-expert humans, possibly via crowdsourcing (Vaughan 2017), which is the use of the vast amount of non-expert human knowledge and labor that exists on the Internet for intellectual tasks. In the RL literature, learning from preferences is discussed as Preference-based RL (PbRL) (Wirth et al. 2017). Recent advances show that agents can solve complex tasks using preferences in an online RL setting (Christiano et al. 2017, Ibarz et al. 2018). This paper extends PbRL to a batch RL setting and focuses on the following challenge: How to learn a reward function from noisy preferences?

Denoising becomes a major requirement when preferences are collected using crowdsourcing. Crowd workers can make mistakes due to the lack of ability or motivation, thus data generated with crowdsourcing is often very noisy. Similar observation is made when collecting demonstrations from the crowd. Mandlekar et al. (2018; 2019) discovers that collected demonstrations hardly facilitate policy learning due to noise in demonstrations. However, this challenge has been overlooked by the PbRL community. The reason is that, in an online setting, preferences are often collected from recruited collaborators, so their quality is assured. Meanwhile, the present study assumes preferences are collected from non experts, and little is known or can be assured about the annotators.

This paper proposes a probabilistic model, named Deep Crowd-BT (DCBT), for learning a reward function from noisy preferences. DCBT assumes that each preference label is potentially unreliable, and it estimates the label reliability from data. As shown in Fig. 1, the idea behind DCBT is to model the correlation of the reliability of a label with its annotator, other labels for the same query, and the estimated reward function. The conditional dependency on the annotator models the fact that unreliable annotators tend to give unreliable labels. Meanwhile, as it is a common practice to solicit multiple labels for the same query, the proposed model collaboratively utilizes labels from different annotators for the same query. Yet in practice each query can only be labelled by a small group of annotators given a fixed labelling budget, so DCBT also utilizes the estimated reward function effectively smooths the label reliability.

A set of experiments on large scale offline datasets (Agarwal et al. 2020) for Atari 2600 games verifies the effectiveness of DCBT. The results show that DCBT facilitates fast convergence of policy learning and outperforms reward learning algorithms that ignore the noise. Furthermore, an ablation study is also performed to analyze the relative importance of collaboration and smoothing. The contributions of this paper are summarized as follows:

- This paper addresses the lack of reward in batch RL setting by learning a reward function from noisy preferences.
- A probabilistic model is proposed to handle the noise in preferences, which collaboratively models the reliability of labels and smooths it with the estimated reward function.
- Experiments on Atari games, accompanied by an ablation study, verify the efficacy of the proposed model.

## 2   Related Work

Batch RL is a sub-field of RL that learns to solve tasks using pre-collected data instead of online interaction. Efforts have been dedicated to issues that emerge when learning offline, such as the covariate shift (Gelada and Bellemare 2019) and the overestimation of Q function (Kumar et al. 2019). This paper addresses the lack of reward in batch RL setting, which complements policy learning algorithms.

In the literature of RL, the lack of reward is canonically addressed by either inverse RL (Abbeel and Ng 2004) or BC (Schaal 1996). Inverse RL does not apply as it requires online interactions. BC suffers from the inefficiency of data acquisition and the imperfectness of collected demonstrations. While data acquisition can be scaled up using crowdsourced platforms such as the RoboTurk (Mandlekar et al. 2018; 2019), the imperfectness of demonstrations remains an issue for BC (Mandlekar et al. 2021). Recent results show that BC can work on mixtures of optimal and imperfect demonstrations via collecting confidence scores for trajectories (Wu et al. 2019) or learning an ensemble model for behavioral policies (Sasaki and Yamashina 2021). These methods still require large amount of optimal trajectories.

Instead, this paper addresses the lack of reward by collecting preferences from humans. As preferences require less expertise than optimal demonstrations, they can be collected using methods such as crowdsourcing (Vaughan 2017). In the literature of RL, PbRL is shown to be successful for complex discrete and continuous control tasks (Christiano et al. 2017, Ibarz et al., 2018). Interested readers may refer to the detailed survey from Wirth et al. (2017). However, the existing work on PbRL is restricted to clean data in the online RL setting, while crowdsourced data are often noisy (Zheng et al. 2017, Rodrigues and Pereira

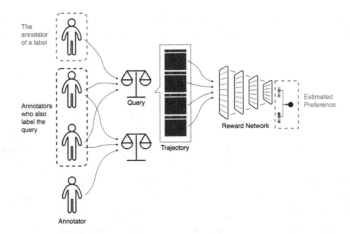

**Fig. 1.** This diagram illustrates information utilized by the proposed DCBT model. An arrow pointing to a query means that the annotator gives a label for the query. When determining the reliability of a label, the DCBT model utilizes: (a) the ID of the annotator of the label (the blue box), (b) the IDs of annotators who also label the query and the labels they give (the red box), and (c) an estimate for preference computed using the learned reward function. Part (b) is our idea of utilizing other labels for the same query to assist in modeling label reliability. Meanwhile, an estimate for preference can be computed with (c), which is also helpful in determining label reliability. (Color figure online)

2018). While being detrimental to reward learning (Ibarz et al. 2018), noise in preferences remains a overlooked issue. This paper extends PbRL to a batch setting and overcomes the issue of noise in preferences. It extends the propabilistic model proposed by Chen et al. (2013) to effectively model label reliability while learning a reward functions from preferences.

# 3 Problem Setting

## 3.1 Markov Decision Process

A sequential decision-making problem is modeled as interactions between an agent and an environment, described as a discounted infinite-horizon MDP: $\langle \mathcal{S}, \mathcal{A}, R, P, \gamma \rangle$, where $\mathcal{S}$ refers to the state space (we interchangeably use *states* and *observations* in this paper), $\mathcal{A}$ is a finite and discrete set of actions, and $R(s, a)$ is the reward function. $P(s'|s, a)$ is the transition probability that characterizes how states transits depending on the action, which is not revealed to the agent. $\gamma \in \mathbb{R}$ is the discount factor.

Interactions roll out in discrete time steps. At step $t$, the agent observes $s_t \in \mathcal{S}$ and selects action $a_t$ according to a policy $\pi : \mathcal{S} \to \mathcal{A}$. Based on $(s_t, a_t)$, the environment decides next state $s_{t+1}$ according to $P(s_{t+1}|s_t, a_t)$, and the reward

$r_t \in \mathbb{R}$ is determined according to $R(s, a)$. The objective of policy learning is to find a policy $\pi$ that maximizes $\sum_{t=1}^{\infty} \gamma^{t-1} R(s_t, a_t)$, the discounted sum of the rewards.

## 3.2   Reward Learning Problem

In this paper, trajectories are assumed to be missing from trajectories. A trajectory can be written as $\eta = (s_1, a_1, \ldots, s_{T_c}, a_{T_c})$, where $T_c$ is the length of this trajectory. A learning agent is provided with a set of trajectories and preferences over these trajectories generated by a group of annotators. The $i^{\text{th}}$ sample can be written as a four tuple: $(\eta_{i,1}, \eta_{i,2}, y_i, w_i)$. The pair $(\eta_{i,1}, \eta_{i,2})$ is the preference query, and $y_i$ is the preference label. $y_i = $ "$\succ$" if in $\eta_{i,1}$ the task is solved better than in $\eta_{i,2}$, $y_i = $ "$\approx$" if the two clips are equally good, and $y_i = $ "$\prec$" if in $\eta_{i,1}$ the task is solved worse than in $\eta_{i,2}$. $w_i$ is the ID of the annotator who gave $y_i$. Let $N$ be the number of preferences, and let $M$ be the number of annotators.

The preferences are assumed to be noisy, as the annotators may lack of expertise of commitment. Yet no information other than annotator IDs is revealed to the learning agent. From the preferences, the learning agent aims at learning a reward function $\hat{R}$. The learning problem is summarized as follows:

- Input: A set of noisy preferences $D = \{(\eta_{i,1}, \eta_{i,2}, y_i, w_i)\}_{i=1}^{N}$, where $\eta_{i,1}$ and $\eta_{i,2}$ are two trajectories, $w_i$ is the identity of the annotator, and $y_i \in \{\text{"}\succ\text{"}, \text{"}\approx\text{"}, \text{"}\prec\text{"}\}$ is the preference label.
- Output: An estimated reward function $R : \mathcal{S} \times \mathcal{A} \rightarrow \mathbb{R}$.

After learning $R$, the trajectories are augmented with estimated rewards given by $R$. They are now in the standard format for off-policy policy learning, and in principle any algorithm of the kind is applicable.

## 4   Proposed Method

This section presents the proposed Deep Crowd-BT (DCBT) model. A diagram for DCBT is shown in Fig. 1, and a pseudocode for reward learning with DCBT is provided in Algorithm 1.

### 4.1   Modeling Preferences

Given a sample $(\eta_{i,1}, \eta_{i,2}, y_i, w_i)$, DCBT first computes the rewards of $\eta_{i,1}$ and $\eta_{i,2}$. For each state-action pair $(s, a)$ in $\eta_{i,1}$ and $\eta_{i,2}$, the reward network outputs a scalar $R(s, a)$. This is done by the reward network shown in the upper part of Fig. 1. Let $\theta_R$ be the parameters of $R$. For image input, $R$ is parameterized with convolutional neural networks followed by feedforward networks.

Then the probability of the event "$\eta_{i,1} \succ \eta_{i,2}$" is modeled with the Bradly-Terry model (BT) (Bradley and Terry 1952):

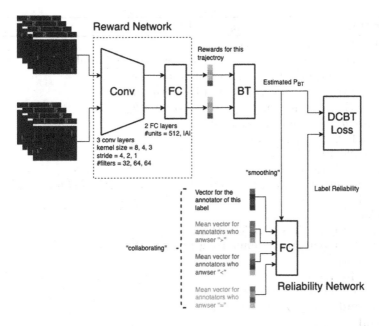

**Fig. 2.** This figure illustrates the architecture for learning reward function with the proposed DCBT model. After learning, the reward function (shown in dashed box) can be used to infer rewards using states and actions of trajectories.

$$P_{BT}\left(\eta_{i,1} \succ \eta_{i,2}\right) = \frac{\exp\left(G(\eta_{i,1})\right)}{\exp\left(G(\eta_{i,1})\right)) + \exp\left(G(\eta_{i,2})\right)},$$

$$G(\eta_{i,1}) = \frac{1}{T_c} \sum_{(s,a) \in \eta_{i,1}} R(s,a),$$
(1)

$$G(\eta_{i,2}) = \frac{1}{T_c} \sum_{(s,a) \in \eta_{i,2}} R(s,a).$$

The larger $G(\eta_{i,1})$ is, the larger $P_{BT}(\eta_{i,1} \succ \eta_{i,2})$ becomes. $P_{BT}(\eta_{i,1} \succ \eta_{i,2}) > 0.5$ if $G(\eta_{i,1}) > G(\eta_{i,2})$, that is, the trajectory with a larger sum of rewards is preferred. If the preferences are consistent with the task of interest, then maximizing the loglikelihood of $P_{BT}\left(\eta_{i,1} \succ \eta_{i,2}\right)$ will generate reward function that facilitate learning of competitive policies (Christiano et al., 2017,Ibarz et al., 2018).

### 4.2   Handling Noise in Preferences

When the preferences are noisy, preference labels are not entirely consistent with the underlying task. Ibarz et al. (2018) point out that policies degenerate for noisy preference data. In literature, the Crowd-BT model (Chen et al. 2013) is a probabilistic model for modeling noisy pairwise comparisons. It assumes

---

**Algorithm 1:** Reward Learning with DCBT

---

**Input:** $D$, a noisy preference dataset

$\quad\quad\quad\quad T_{\text{INIT}}$, the number of graident steps for the initialization phase

$\quad\quad\quad\quad T_{\text{TOTAL}}$, the total number of gradient steps

$\quad\quad\quad\quad T_{\text{ALT}}$, the period of alternative optimization

$\quad\quad\quad\quad \beta T_{\text{ALT}}$ is the number of steps to train $\theta_R$

Initialize $\theta_R$, $\theta_W$ and $\theta_\alpha$ randomly

**for** $t = 1, 2, \ldots, T_{\text{TOTAL}}$ **do**

$\quad$ Sample a batch of preference data

$\quad$ **if** $t \leq T_{\text{INIT}}$ **then**

$\quad\quad|\quad$ update $\theta_R, \theta_W, \theta_\alpha$ by minimizing $L_{\text{DCBT}-\text{INIT}} + \lambda_1 L_{\text{IDF}} + \lambda_2 L_{\ell_1, \ell_2}$

$\quad$ **end**

$\quad$ **else**

$\quad\quad$ **if** $t \bmod T_{\text{ALT}} < \beta T_{\text{ALT}}$ **then**

$\quad\quad\quad|\quad$ update $\theta_R$ by minimizing Equation 8.

$\quad\quad$ **end**

$\quad\quad$ **else**

$\quad\quad\quad|\quad$ update $\theta_W, \theta_\alpha$ by minimizing Equation 8.

$\quad\quad$ **end**

$\quad$ **end**

**end**

---

each annotator makes errors with an annotator-specific probability. Denote by $\alpha_w = P(\eta_{i,1} \succ_w \eta_{i,2} \mid \eta_{i,1} \succ \eta_{i,2})$ the probability that annotator $w$ gives "$\succ$" for $(\eta_{i,1}, \eta_{i,2})$, when the groundtruth is $\eta_{i,1} \succ \eta_{i,2}$. With the Crowd-BT model,

$$P_{\text{Crowd}-\text{BT}}(y_i = \text{``}\succ\text{''}) = \alpha_{w_i} P_{\text{BT}}(\eta_{i,1} \succ \eta_{i,2}) + (1 - \alpha_{w_i})(1 - P_{\text{BT}}(\eta_{i,1} \succ \eta_{i,2})). \tag{2}$$

In other words, the Crowd-BT model assumes that the reliability of labels from the same annotator is the same and fixed, in regardless of queries being compared. This assumption, however, is inadequate in our case and suffers from at least two reasons. In our case, each query is labeled by multiple annotators. The other labels for the same query and the credibility of their annotators are informative when modeling the reliability of a label, but they are not utilized in the Crowd-BT model. Meanwhile, in practice rich coverage of possible trajectories are required to ensure the generalization of $R$. Thus, under limited labeling budget, each preference query is labeled by a tiny group of annotators, which incurs high variance in labels.

The proposed DCBT extends Crowd-BT by explicitly overcoming the above-mentioned issues. Instead of per-annotator reliability parameter, it learns a per-sample reliability defined as:

$$\alpha_i = P(y_i = \text{``}\succ\text{''} \mid w_i, C_i, P_{\text{BT}}(\eta_{i,1} \succ \eta_{i,2})). \tag{3}$$

The probability of $y_i = $ "$\succ$" can be expressed as

$$P_{DCBT}(y_i = "\succ") = \alpha_i P_{BT}(\eta_{i,1} \succ \eta_{i,2}) + (1 - \alpha_i)(1 - P_{BT}(\eta_{i,1} \succ \eta_{i,2})). \tag{4}$$

For each sample, the reliability network shown in the bottom part of Fig. 1 outputs label reliability $\alpha_i$. This network is parameterized with the fully-connected layer with sigmoidal activation. Let the parameter of this network be $\theta_\alpha$. This network addresses the drawbacks of the Crowd-BT model by taking as input $w_i$, $C_i$ and $P_{BT}(\eta_{i,1} \succ \eta_{i,2})$.

$C_i$ is a set that contains labels and their annotators for the same preference query $(\eta_{i,1}, \eta_{i,2})$. Specifically,

$$C_i = \{(y_j, w_j) \mid \eta_{j,1} = \eta_{i,1}, \eta_{j,2} = \eta_{i,2}, j \neq i, (\eta_{j,1}, \eta_{j,2}, y_j, w_j) \in D\}. \tag{5}$$

Using $C_i$, the reliability network utilizes crowdsourced preferences *collaboratively*. The intuition is that, for the same query, labels from other annotators and the credibility of these annotators provide useful information for modeling $\alpha_i$. A label might be reliable if it is consistent with other labels, especially with those from credible annotators. Moreover, the label values matter. For example, suppose ("$\prec$", $w_j$) $\in C_i$, and $w_j$ is a credible annotator. Then while both "$\succ$" and "$\approx$" are inconsistent with "$\prec$", the latter should be a more reliable one. The reliability network relies on a worker embedding matrix $\theta_W \subset \mathbb{R}^d$. It takes as input the embedding vector of $w_i$. For $C_i$, it groups the label-annotator pairs into three groups by labels values. Then it computes the mean vector of worker embedding vectors in each group and concatenates these mean vectors together. For example, the red vector in the bottom of Fig. 1 corresponds to the mean vector of annotators who answer "$\prec$". Zero vector is used when a group does not contain any annotators.

Meanwhile, the reliability network utilizes $P_{BT}(\eta_1 \succ \eta_2)$ to address the variance in labels. The present study claims that label reliability also depends on the difficulty of queries. A label $y_i = $ "$\approx$" is less reliable if $\eta_1$ is significantly better than $\eta_2$, when compared to the case in which $\eta_1$ is only slightly better than $\eta_2$. Thus the reliability network utilizes $P_{BT}(\eta_1 \succ \eta_2)$ in determining the reliability of $y_i$. As the $R$ summarizes information from all annotators, utilizing $P_{BT}(\eta_1 \succ \eta_2)$ effectively *smooths* the labels collected for each queries.

### 4.3   Learning

The parameter $\theta_R$, $\theta_W$, and $\theta_\alpha$ can be learned by minimizing the following objective function:

$$L_{DCBT}(\theta_R, \theta_W, \theta_\alpha) = -\frac{1}{N} \sum_{i=1}^{N} [\tilde{y}_i \log(P_{DCBT}) + (1 - \tilde{y}_i) \log(1 - P_{DCBT})], \tag{6}$$

where $P_{DCBT}$ is a short-hand notation for $P_{DCBT}(y_i = $ "$\succ$"). $\tilde{y}_i$ equals to 1 if $y_i = $ "$\succ$", 0.5 if $y_i = $ "$\approx$" and 0 if $y_i = $ "$\prec$". Note that $P_{DCBT}$ is a function of

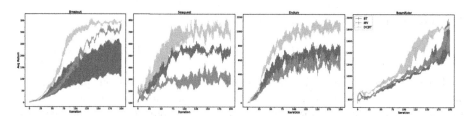

**Fig. 3.** Performance comparison among the proposed DCBT model, BT model, and MV. The curves are for the average returns obtained during training a QR-DQN agent. The QR-DQN agents trained with the proposed method performs better than those trained with BT and MV for game *Seaquest* and *Enduro*. Moreover, for all of the games, the proposed DCBT model enables faster policy learning convergence, which is also desirable in practice.

$\theta_R, \theta_W$ and $\theta_\alpha$ for a fixed dataset $D$. This dependency is omitted in notations for simplicity.

As mentioned by Chen et al. (2013), there is an identifiability issue for learning from pairwise comparisons. For a query $(\eta_{i,1}, \eta_{i,2})$, $\mathrm{P}_{BT}(\eta_{i,1} \succ \eta_{i,2})$ does not change when adding an arbitrary constant $C \in \mathbb{R}$ to $G(\eta_{i,1})$ and $G(\eta_{i,2})$. Preliminary experiments show that reward network tends to output large values, a similar situation with the over-fitting problem in supervised learning. Following Chen et al. (2013), our learning algorithm utilizes a regularization term $L_{reg}$ to restrict reward values around zero. Moreover, $\ell_1$ and $\ell_2$ regularization are also helpful to reduce over-fitting.

$$L_{reg}(\theta_R) =$$
$$-\frac{1}{2N} \sum_{i=1}^{N} \sum_{k=1}^{2} \left[ \log \left( \frac{\exp\left(G(\eta_{i,k})\right)}{\exp\left(G(\eta_{i,k})\right) + 1} \right) + \log \left( \frac{1}{\exp\left(G(\eta_{i,k})\right) + 1} \right) \right]. \tag{7}$$

The overall objective function can be compactly written as

$$L(\theta_R, \theta_W, \theta_\alpha) = L_{DCBT} + \lambda_1 L_{reg} + \lambda_2 L_{\ell_1, \ell_2}, \tag{8}$$

Besides, an initialization phase is required before optimizing Eq. 8. This phases initializes $R$ by maximizing the loglikelihood of the BT model. The intuition is that by regarding all labels as correct, the reward network can attain intermediate ability in modeling preferences. It also initializes $\theta_\alpha$ and $\theta_W$ by minimizing the cross entropy between $\alpha_i$ and a Bernoulli distribution with parameter $\bar{\alpha}$. This follows the initialization procedure for the Crowd-BT model. Formally, the objective function for initialization phase, $L_{DCBT-INIT}$, is defined as:

$$L_{DCBT-INIT}(\theta_R, \theta_W, \theta_\alpha) =$$
$$-\frac{1}{N} \sum_{i}^{N} [\tilde{y}_i \log \mathrm{P}_{BT} + (1 - \tilde{y}_i) \log(1 - \mathrm{P}_{BT}) + \bar{\alpha} \log(\alpha_i) + (1 - \bar{\alpha}) \log(1 - \alpha_i)], \tag{9}$$

where $P_{BT}$ is a short-hand notation for $P_{BT}(y_i = \text{``}\succ\text{''})$. Note that it is a function of $\theta_R$, and $\alpha_i$ is a function of $\theta_R, \theta_W$ and $\theta_\alpha$. $\bar{\alpha}$ is a hyper-parameter, which is set to 0.99. The gradients from the latter two terms of $L_{DCBT-INIT}$ to $\theta_R$ are blocked, as they use pseudo labels instead of true labels.

After initialization, the objective function in Eq. 8 is minimized with an alternative scheme, similar to the method described by Chen et al. (2013). At each round, $\theta_R$ is first optimized for several gradient steps while keeping $\theta_W$ and $\theta_\alpha$ fixed. Then, $\theta_W$ and $\theta_\alpha$ are optimized for several steps while $\theta_R$ is fixed. The entire algorithm is given in Algorithm 1.

## 5    Evaluation

### 5.1    Setup

The present study utilizes the benchmark datasets published by Agarwal et al. (2020) for experimental evaluations; they contain trajectories for Atari 2600 games collected during training a DQN agent. Due to limited computation resources, four of the games used in existing work for online PbRL (Ibarz et al. 2018) are selected. In practice, trajectories are too long to be processed, so they are truncated to clips of length 30 ($T_c = 30$). From each game 50,000 clips of trajectories are randomly sampled. Queries are sampled randomly from these clips.

To have precise control for error rates, preferences are generated by 2,500 simulated annotators. Each annotator generates a correct preference label with a fixed probability sampled from $\mathrm{Beta}(7, 3)$. When making mistakes, an annotator selects one of the two incorrect labels uniformly at random. Each annotator labels at most 20 queries, which means the number of preferences does not exceed 50, 000. In our experiments, each query is annotated by at most ten different annotators.

Algorithms for reward learning are evaluated using the performance of the same policy policy-learning algorithm. Reward functions are learned using different reward learning algorithms on the same set of noisy preferences. Then, the reward functions are utilized to compute rewards for learning policies using the same policy-learning algorithm, which means that the performance of the policies reflects the performance of the reward-learning algorithms. For the policy learning algorithm, the quantile-regression DQN algorithm (QR-DQN) (Dabney et al. 2018) is adopted due to its superior performance. A recent empirical analysis shows that this algorithm yields the state-of-the-art performance in batch RL settings (Agarwal et al. 2020). Obtained policies are evaluated in terms of the average return obtained per episode. Experiments are repeated three times on three different sets of trajectories of a game. The mean values of returns and their standard error are reported.

### 5.2    Implementation Details

Figure 2 shows the network structure used for reward learning using the DCBT model. The other reward learning algorithms utilize the same reward network.

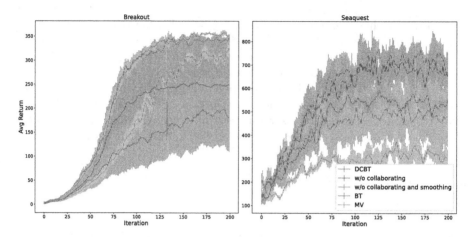

**Fig. 4.** The results of the ablation study for DCBT. For *Breakout*, removing annotator collaborating decreases performance, but further removing smoothing improves performance. For *Seaquest*, removing collaborating results in little difference, but further removing smoothing decreases performance. These results show that combining the two ideas can effectively handle noise in preferences and overcome drawbacks of the two ideas.

The convolutional network part of this architecture is the same as that of the QR-DQN agent released by (Agarwal et al. 2020).

The QR-DQN agent is trained for 200 training iteration. In each training iteration, to speed up training, the agent is continuously trained for 62,500 **gradient steps**. In its original implementation (Agarwal et al. 2020), agents are trained for 250,000 **environment steps**, during which parameters are updated every four environment steps. Therefore, in our evaluations for each iteration, agents are trained for the same number of gradient steps as (Agarwal et al. 2020) performed. Except for this difference, the other hyper-parameters are not altered.

## 5.3    Alternative Methods

For all of the four games, the proposed DCBT model is compared with the following two baseline methods.

*BT Model.* This method regards all preferences as correct ones, and utilizes the BT model to learn $\hat{R}$. It is the method used in the recent work for online PbRL setting (Christiano et al. 2017, Ibarz et al., 2018).

*Majority Voting (MV).* This method counts the occurrence of different labels for the same query. The label with the maximum count is chosen as the estimated label. Ties are broken randomly. Using the estimated labels, $\hat{R}$ is learned with the BT model. The estimated labels generated by this method still contain noise, but they are less noisy than the original labels used by the BT method.

An ablation study is carried out to analyze the effect of annotator collaborating and smoothing in modeling label reliability. For game *Breakout* and *Seaquest*, the DCBT model is compared with the following two methods.

*w/o Collaborating.* This is a variant of the proposed method that ignores other annotators who also label the same query.

*w/o Collaborating and Smoothing.* Not only *w/o collaborating*, the estimated rewards are also ignored. Note that this method only considers the identity of the annotator of a query, so it is equivalent to the Crowd-BT model.

### 5.4   Results

Figure 3 shows the results for all the four games. For the two games, *Seaquest* and *Enduro*, the proposed DCBT achieves the best final performance, which means that DCBT successfully generates reward functions that align with the tasks of interest. Meanwhile, for the other two games, *Breakout* and *BeamRider*, using MV results in slower convergence but close final performance. Thus, for these two games, the MV method can generate reward functions aligned with the tasks of interest, but such reward functions hinder fast convergence. Only for *BeamRider*, the BT method has a similar final performance as the DCBT model. In addition to improved final performance, faster convergence is also a desirable property in practice, especially in scenarios with limited resources. From this perspective, the proposed DCBT model also outperforms its alternatives.

Figure 4 shows the ablation study results for DCBT on *Breakout* and *Seaquest*. For *Breakout*, removing the annotator collaborating (shown in orange) decreases its performance, although it is still better than the BT method (shown in red). Hence for this game, label collaborating plays an important roll for DCBT. Interestingly, further removing the smoothing (shown in green) boosts the performance, which is even slightly better than the DCBT model. Since this is a rather simple model compared to DCBT, the increase in performance might be due to less overfitting.

For *Seaquest*, removing annotator collaborating hardly affects the performance. Furthermore, removing smoothing significantly decreases the performance. So for this game, the idea of smoothing might be important for the performance of the proposed DCBT model.

Our experimental results show that, the efficacy of annotator collaborating and smoothing might be task specific, but their drawbacks can be overcome by combining them together. Further investigation of the reasons is left as future work.

## 6   Conclusion

This paper address the lack of reward function in batch RL setting. Existing settings for this problem rely on optimal demonstrations provided by humans, which is unrealistic for complex tasks. So even though data acquisition is scaled up with crowdsourcing, effective policy learning is still challenging. This paper tackles this problem by learning reward functions from noisy preferences. Generating preferences requires less expertise than generating demonstrations. Thus they can be solicited from vastly available non-expert humans. A critical challenge lies in the noise of preferences, which is overlooked in the literature of PbRL. This challenge is addressed with a novel probabilistic model called DCBT. DCBT collaboratively models the correlation between label reliability and annotators. It also utilizes the estimated reward function to compute preference estimates, which effectively smooths labels reliability. Evaluations on Atari 2600 games show the efficacy of the proposed model in learning reward functions from noisy preferences, followed by an ablation study for annotator collaborating and smoothing. Overall, this paper explores a novel methodology for harvesting human knowledge to learn policies in batch RL setting.

Our ablation study indicates the occurrence of over-fitting, which means the reward model might overly fit states and actions in preferences. How to inject induction bias to overcome such an effect is an interesting future work. Moreover, while there are 50 million states and actions for each game, only 50,000 are covered in queries. Under a fixed labelling budget, how to more effectively generate queries is an important issue.

**Acknowledgment.** This work was partially supported by JST CREST Grant Number JPMJCR21D1.

# References

Abbeel, P., Ng, A.Y.: Apprenticeship learning via inverse reinforcement learning. In: Proceedings of the 21st International Conference on Machine Learning (ICML) (2004)

Agarwal, R., Schuurmans, D., Norouzi, M.: An optimistic perspective on offline reinforcement learning. In: III, H.D., Singh, A. (eds.) Proceedings of the 37th International Conference on Machine Learning. Proceedings of Machine Learning Research, vol. 119, pp. 104–114. PMLR (2020)

Bradley, R.A., Terry, M.E.: Rank analysis of incomplete block designs: I. the method of paired comparisons. Biometrika **39**(3/4), 324–345 (1952)

Chen, X., Bennett, P.N., Collins-Thompson, K., Horvitz, E.: Pairwise ranking aggregation in a crowdsourced setting. In: Proceedings of the Sixth ACM International Conference on Web Search and Data Mining (WSDM), pp. 193–202 (2013)

Christiano, P.F, Leike, J., Brown, T., Martic, M., Legg, S., Amodei, D.: Deep reinforcement learning from human preferences. In: Guyon, I., et al. (eds.) Advances in Neural Information Processing Systems 30, pp. 4302–4310. Curran Associates Inc. (2017)

Dabney, W., Rowland, M., Bellemare, M.G., Munos, R.: Distributional reinforcement learning with quantile regression. In: Proceedings of the 32nd AAAI Conference on Artificial Intelligence, pp. 2892–2901 (2018)

Fujimoto, S., Conti, E., Ghavamzadeh, M., Pineau, J.: Benchmarking batch deep reinforcement learning algorithms (2019)

Gelada, C., Bellemare, M.G.: Off-policy deep reinforcement learning by bootstrapping the covariate shift. In: Proceedings of the 33rd AAAI Conference on Artificial Intelligence (AAAI), pp. 3647–3655 (2019)

Ibarz, B., Leike, J., Pohlen, T., Irving, G., Legg, S., Amodei, D.: Reward learning from human preferences and demonstrations in atari. In: Advances in Neural Information Processing Systems 31, pp. 8022–8034. Curran Associates Inc. (2018)

Kumar, A., Fu, J., Tucker, G., Levine, S.: Stabilizing off-policy q-learning via bootstrapping error reduction. In: Advances in Neural Information Processing Systems 32 (2019)

Lange, S., Gabel, T., Riedmiller, M.: Batch Reinforcement Learning, pp. 45–73. Springer, Heidelberg (2012)

Mandlekar, A., et al.: Roboturk: a crowdsourcing platform for robotic skill learning through imitation. In: Proceedings of the Second Conference on Robot Learning (CoRL), pp. 879–893 (2018)

Mandlekar, A., et al.: Scaling robot supervision to hundreds of hours with roboturk: robotic manipulation dataset through human reasoning and dexterity. In: Proceedings of the 2019 IEEE/RSJ International Conference on Intelligent Robots and Systems, pp. 1048–1055 (2019)

Mandlekar, A., et al.: What matters in learning from offline human demonstrations for robot manipulation (2021)

Pavse, B., Durugkar, I., Hanna, J., Stone, P.: Reducing sampling error in batch temporal difference learning. In: Proceedings of the 37th International Conference on Machine Learning (ICML), pp. 7543–7552 (2020)

Rodrigues, F., Pereira, F.: Deep learning from crowds. In: Proceedings of the 32nd AAAI Conference on Artificial Intelligence (AAAI) (2018)

Sasaki, F., Yamashina, R.: Behavioral cloning from noisy demonstrations. In: Proceeding of the International Conference on Learning Representations (ICLR) (2021)

Schaal, S.: Learning from demonstration. In: Advances in Neural Information Processing Systems 9 (1996)

Vaughan, J.W.: Making better use of the crowd: how crowdsourcing can advance machine learning research. J. Mach. Learn. Res. 18(1), 7026–7071 (2017). ISSN 1532–4435

Wirth, C., Akrour, R., Neumann, G., Fürnkranz, J.: A survey of preference-based reinforcement learning methods. J. Mach. Learn. Res. 18(136), 1–46 (2017)

Wu, Y.-H. , Charoenphakdee, N., Bao, H., Tangkaratt, V., Sugiyama, M.: Imitation learning from imperfect demonstration. In: Proceedings of the 36th International Conference on Machine Learning (ICML), pp. 6818–6827 (2019)

Zheng, Y., Li, G., Li, Y., Shan, C., Cheng, R.: Truth inference in crowdsourcing: is the problem solved? Proc. VLDB Endowment 10(5), 541–552 (2017)

# Oracle-SAGE: Planning Ahead in Graph-Based Deep Reinforcement Learning

Andrew Chester$^{(\boxtimes)}$ ⓘ, Michael Dann ⓘ, Fabio Zambetta ⓘ,
and John Thangarajah ⓘ

School of Computing Technologies, RMIT University, Melbourne, Australia
{andrew.chester,michael.dann,fabio.zambetta,john.thangarajah}@rmit.edu.au

**Abstract.** Deep reinforcement learning (RL) commonly suffers from high sample complexity and poor generalisation, especially with high-dimensional (image-based) input. Where available (such as some robotic control domains), low dimensional vector inputs outperform their image based counterparts, but it is challenging to represent complex dynamic environments in this manner. Relational reinforcement learning instead represents the world as a set of objects and the relations between them; offering a flexible yet expressive view which provides structural inductive biases to aid learning. Recently relational RL methods have been extended with modern function approximation using graph neural networks (GNNs). However, inherent limitations in the processing model for GNNs result in decreased returns when important information is dispersed widely throughout the graph. We outline a hybrid learning and planning model which uses reinforcement learning to propose and select subgoals for a planning model to achieve. This includes a novel action selection mechanism and loss function to allow training around the non-differentiable planner. We demonstrate our algorithms effectiveness on a range of domains, including MiniHack and a challenging extension of the classic taxi domain.

**Keywords:** Reinforcement learning · GNNs · Symbolic planning

## 1 Introduction

Despite the impressive advances of deep reinforcement learning (RL) over the last decade, most methods struggle to generalise effectively to different environments [15]. A potential explanation for this is that deep RL agents find it challenging to create meaningful abstractions of their input. Humans conceptualise the world in terms of distinct objects and the relations between them, which grants us the ability to respond effectively to novel situations by breaking them down

**Supplementary Information** The online version contains supplementary material available at https://doi.org/10.1007/978-3-031-26412-2_4.

into familiar components [19,29]. Within the field of reinforcement learning, this approach is best exemplified by relational RL [7]. It has been argued that we now have the tools to combine the power of deep learning with a relational perspective through the use of graph neural networks (GNNs) [2].

While some work exists in this area [14,17,20], much of it is performed on domains (e.g. block tower stability predictions) which have a particular specialised graph representation. Navigational domains are both important in real-world applications (self-driving cars, robot locomotion), and naturally suited to being represented as a graph. For example, a road network can be modelled as a graph with nodes for each intersection and edges for each road. This representation can naturally handle bridges, tunnels, and one-way streets in a way that would be challenging for a standard image based representation to capture accurately.

We hypothesise that naively applying GNNs in RL will be challenging due to their architecture limiting their ability to synthesise information dispersed across long distances in the input graph. A GNN operates by applying *message passing* steps on its input, which limits the effective receptive field for each node to its local neighbourhood [39]. We provide evidence for this hypothesis through a series of experiments on a targeted synthetic domain. Increasing the number of message passing steps indefinitely is not a feasible solution due to increased memory requirements and instability in training [31].

We propose a solution in the RL context by drawing inspiration from recent work in hybrid symbolic planning and RL methods [16,26]. Augmenting the learner with a planning system allows it to integrate data from beyond its receptive field. We illustrate this idea though the following taxi domain which serves as a running example throughout this paper. Imagine being a taxi driver (the GNN-based learner) in a busy city, deciding where to drive and which passengers to take while trying to maximise your earnings for the day. You have available to you a GPS mapping system (the planner), which allows you to plan routes and get time estimates for any destination of your choosing. Interacting with the GPS may change your decisions. For example, you may decide to head to the airport (a subgoal) where you are likely to find paying customers. The GPS however notifies you of a crash on the way and so it will take much longer than usual to get there (feedback). With that additional information, you may change your mind and decide to drive around the nearby streets looking for a passenger instead (the alternate subgoal).

This paper outlines Oracle-SAGE, a hybrid learning and planning approach in which the reinforcement learner proposes multiple subgoals for the planner to evaluate. Once the planner has generated plans to accomplish these subgoals, it returns the projected symbolic states that will result from executing each plan. The discriminator then makes a final decision by ranking these future states in order of desirability, allowing it to revise its subgoals in light of feedback from the planner. The planner thus functions as an oracle, attempting to predict the results of achieving the learner's subgoals. An advantage of this approach is that graph based input combines easily with symbolic planning models; PDDL domains are naturally similar to graphs as they are both object oriented [34].

This proposed architecture poses a number of novel challenges. First, the discriminator needs a way of ranking the projected states; we define the *path-value* as a variant of the value function for future states. Second, the learner has two components: the meta-controller and the discriminator, separated by a non-differentiable planner. In order to train both of these from learned experience, we need to define a *path-value loss function* which allows gradients to propagate around the non-differentiable component. Finally, communication between learner and planner requires both components to share a common symbolic state representation, which we model with graphs. In order to address these, our core contributions in this paper are:

- A novel subgoal (and action) selection mechanism which integrates feedback from a non-differentiable planner.
- A path-value loss function to enable training the action selection network through the non-differentiable planning component.
- A comprehensive evaluation of Oracle-SAGE's effectiveness.

We demonstrate Oracle-SAGE's general purpose effectiveness on a set of complex domains including an extended taxi domain [6] and a subset of the challenging roguelike game Nethack [32]. We benchmark against state-of-the-art learning and hybrid approaches, and show that Oracle-SAGE outperforms all competing methods. We also compare against an ablation which demonstrates the importance of all components of our proposed method.

## 2   Preliminaries

### 2.1   Reinforcement Learning

A Markov Decision Process (MDP) $\mathcal{M}$ is a tuple $(\mathcal{S}, \mathcal{A}, T, R, \gamma)$, where $s \in \mathcal{S}$ are the environment states, $a \in \mathcal{A}$ are the actions, $T : (\mathcal{S} \times \mathcal{A} \times \mathcal{S}) \to [0,1]$ is the transition function specifying the environment transition probabilities, and $R : (\mathcal{S} \times \mathcal{A} \times \mathbb{R}) \to [0,1]$ gives the probabilities of rewards. The agent maximises the total *return* $U$, exponentially discounted by a *discount rate* $\gamma \in [0,1]$: at a given step $t$, $U_t = \sum_{k=0}^{T} \gamma^k r_{t+k+1}$, where $T$ is the remaining episode length. The probability of taking an action in a state is given by the *policy* $\pi(a|s)$ : $(\mathcal{S} \times \mathcal{A}) \to [0,1]$.

### 2.2   Symbolic Planning

We use the planning domain definition language (PDDL) to model symbolic planning problems [27]. A planning task consists of a *domain* $D$, and *instance* $N$. The domain is $(\mathcal{T}, \mathcal{P}, \mathcal{F}, \mathcal{O})$, where $\mathcal{T}$ are object types, $\mathcal{P}$ are Boolean predicates $P(o_1, \ldots, o_k)$, and $\mathcal{F}$ are numeric functions $f(o_1, \ldots, o_k)$, where $o_1, \ldots, o_k$ are object variables. $\mathcal{O}$ are planning operators, with *preconditions* and *postconditions* (changes to predicates and functions caused by the operator). The instance comprises $(B, I, G)$, where $B$ defines the objects present, $I$ is a conjunction of predicates and functions which describes the initial state, and $G$ is the goal which similarly describes the desired end state. In our domains, the grounded planning operators are equivalent to the set of actions in the underlying MDP, i.e. $\mathcal{A} = \mathcal{O}$, so for simplicity we simply refer to these grounded operators as actions.

We sketch here our transformation of PDDL problems into graphs; it may be applied to any PDDL domain where all predicates and numeric functions have arity no greater than 2. Each object is represented as a node, with object types represented as one-hot encoded node attributes. Unary predicates are also represented as binary node attributes, while unary functions are represented as real-valued attributes. Binary predicates and functions are represented as edge attributes between their two objects. Actions modify the graph by changing node and edge attributes, as well as edges themselves according to the postconditions of the action; for any given action $a$ and current state $s$, PDDL semantics defines a transition function $\Delta$ to compute the next state: $s' = \Delta(s, a)$. See Fig. 1a for a visual example of the taxi domain as a graph.

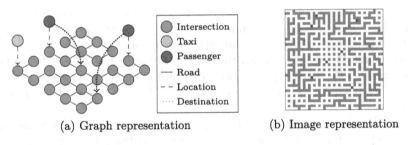

(a) Graph representation      (b) Image representation

**Fig. 1.** Taxi domain representations. In the image, passenger destinations are in green, the roads are in white, the taxi is red, and passengers are blue. (Color figure online)

### 2.3 Graphs and Graph Neural Networks

In this article we assume the state is encoded as a directed multi-graph: $s = (\mathcal{V}, \mathcal{E}, u)$, where $\mathcal{V}$ is the set of nodes, $\mathcal{E}$ is the set of edges, and $u$ is the global state; represented as a special node with no edges. All nodes $v_i \in \mathcal{V}$ and $u$ have types and attributes associated with them, represented as a real-valued vector as described above. Similarly edges $e_j \in \mathcal{E}$ have types and attributes, and there may be multiple edges of different types between the same two nodes: $e = (v_s, v_d, e_a)$, where $v_s, v_d$ are the source and destination nodes respectively. The length of the attribute vectors for nodes and edges depends on the domain. Computation over these graphs is done with a Graph Network (GN) [2], a framework which generalises the most common GNN architectures. As our GN functions are implemented by neural networks, we use the term GNN throughout. We denote the output of the GNN for each node, edge, and global state by $v_i', e_j', u'$ respectively, and refer to the entire set of nodes as $\boldsymbol{v'}$.

# 3  GNN Information Horizon Problem

As mentioned in the introduction, we hypothesise that GNNs perform poorly when important information is distributed widely throughout the graph. In this section, we formalise this problem after providing intuition on why it manifests. We introduce a domain that is targeted to display this failure mode, and empirically demonstrate that RL methods using standard GNNs do not effectively learn, but Oracle-SAGE does.

A GNN is composed of blocks which perform a *message passing* step on their input, producing updated embeddings for edges and nodes. Formally; each block consists of edge, node and global update operations:

$$e'_j = \Phi^e(v_{s_j}, v_{d_j}, e_j, u) \tag{1}$$

$$v'_i = \Phi^v(v, \rho^{e \to v}(e'_j), u) \tag{2}$$

$$u' = \Phi^u(\rho^{v \to u}(v'_i), \rho^{e \to u}(e'_j), u) \tag{3}$$

where $\Phi^{\{e,v,u\}}$ are arbitrary functions; in this work we use a single FC layer which takes the concatenated arguments as input. Similarly, $\rho^{x \to y}$ are aggregation functions which operate over an arbitrary number of inputs; we use elementwise max. It should be noted that $\rho^{e \to v}$ only aggregates over edges which are adjacent to the node $v_i$; $\rho^{e \to u}$ and $\rho^{v \to u}$ are global and aggregate over all edges/nodes respectively. Multiple blocks of this form are stacked to form the GNN.

From the above update equations, it can be seen that in any single message passing step, each node can only process information from its immediate neighbours. This limits the effective *receptive field* of any node to other nodes within $h$ steps, where $h$ is the number of blocks. In a navigational context, this limits the ability to determine connectivity between points greater than a distance of $h$ away, or to aggregate information along a path greater than length $h$. This is the *information horizon problem*; for a given GNN architecture there is a limited horizon beyond which node level information cannot propagate. The planner in our model is not subject to this information horizon, and should therefore be able to accurately predict future states using the entire state information.

## 3.1  Synthetic Domain

We construct a synthetic bandit-like domain [4] designed to test this hypothesis. An agent chooses which of $c$ corridors should be traversed (Fig. 2). Along each corridor are $l$ spaces, each of which contains a number of green (positive) or red (negative) tokens. The agent collects all tokens in rooms it traverses, and the final reward is equal to the number of green tokens minus the number of red tokens in the agents possession. Unlike in a traditional bandit problem, the number of tokens in every room is randomly generated at the start of each episode; to allow learning, the environment is fully observable. The agent selects a corridor by applying a softmax layer to the output of the final node in each corridor.

Since there are no further choices to make after choosing a corridor, the agent moves directly to the end in a single action. Conceptually, the optimal policy is simple: sum the tokens for each path and select the one which has the highest number of green minus red.

We hypothesise, as per the information horizon problem, that a GNN should be able to learn an optimal policy as long as the number of message passing steps ($h$) in the GNN is at least $l$. For $h < l$, the informa-tion from the entire corridor will be unable to propagate through the network to be inte-grated into a single node for action selection. Instead, the best the GNN can do is to choose

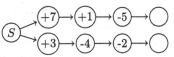

**Fig. 2.** Synthetic domain with $c = 2$ and $l = 3$. This is the input state graph given to the agents, not a rep-resentation of the MDP itself.

the most promising path based on the first $h$ steps. To demonstrate this, we use SR-DRL as an example of an RL agent with GNN-based processing which is subject to the information horizon problem [17]. By contrast Oracle-SAGE (full details in Sect. 4) is provided with a planning model that, when given a goal to reach the end of the corridor, can project the agent to the end of the corridor with the number of red and green tokens it would collect on the way. Crucially however, the planning model does not know that the final reward is equal to green minus red tokens and so cannot by itself be used to choose the best action.

## 3.2  Synthetic Domain Results

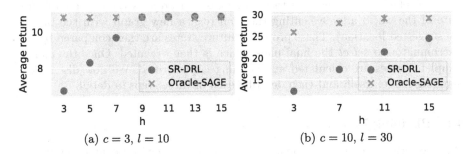

(a) $c = 3, l = 10$

(b) $c = 10, l = 30$

**Fig. 3.** Converged scores for small (a) and large (b) synthetic domain

Figure 3 shows the performance of both methods as the number of message-passing steps ($h$) is varied. The results in the small setting provide compelling evidence for our hypothesis. For $h \geq l$, SR-DRL performs well as it can inte-grate information over the entire path. However, for $h < l$ the return degrades gradually as a path that may look promising in the first $h$ steps may then have large numbers of red tokens later. On the other hand, Oracle-SAGE performs near optimally regardless of the GNN horizon as it sends the top 3 choices to the planner to evaluate, and so when $c = 3$ the discriminator can choose directly between all options. While this shows that Oracle-SAGE addresses the information horizon problem, it may not reflect results for most environments where the number of possible goals exceeds 3.

In the large setting we see that Oracle-SAGE is no longer optimal with a short horizon, as the planner can only evaluate a fraction of the total subgoal possibilities. Nevertheless, it still outperforms SR-DRL for a given horizon length since it evaluates the top 3 promising paths. For $h = 3$; the meta-controllers first choice (which would be SR-DRL's action) is only chosen 41% of the time, in the remainder the increased information from the planner results in a revision to the chosen subgoal. This performance gap suggests that Oracle-SAGE may perform better than SR-DRL in more complex environments, even when the number of possible subgoals is large. With this promise in mind, we now describe it in greater detail.

## 4   Oracle-SAGE

Oracle-SAGE (Fig. 4) combines reinforcement learning with symbolic planning using a shared symbolic graph representation. At a high level, the meta-controller *proposes* $k$ subgoals to the planner. These represent an initial guess of the most promising subgoals to pursue, prior to planning. In our taxi example, a single subgoal could be: "move to location $x$" or "deliver passenger $y$". The planner then creates a plan to reach each of these subgoals and *projects* the expected future state of the world after executing the plan (e.g. a new graph with the taxi in the suggested location). These projected future states are then compared by the discriminator to *select* the final plan, which is then executed. Once the plan is complete, losses are calculated and training is performed. We now discuss the planning model itself, and then describe each of these steps in detail.

### 4.1   Planning Model

Our algorithm requires sufficient knowledge of the environment dynamics to predict future states in *partial* detail. More concretely, we assume that the model is suitable for short-term planning, but not necessarily for long-term planning. Such models are referred to as *myopic* planning models in [5]. For example, in the taxi domain, the model provided might only encompass the local actions of the taxi, e.g. "if you drive along this road, you get to this intersection" and "if you pick up a passenger in your current location, that passenger will be in the taxi". Critically though, it does not know anything about where passengers will appear or what their destinations will be. This makes it feasible to apply our method in domains where the full environment dynamics are complex and unknown, so long as some action consequences are easily specified, such as those introduced in Sect. 5.

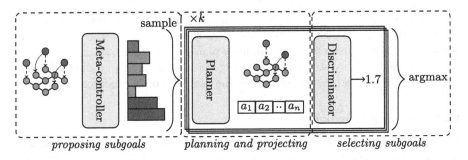

*proposing subgoals*          *planning and projecting*          *selecting subgoals*

**Fig. 4.** Oracle-SAGE action selection mechanism. The meta-controller generates a distribution over subgoals, then samples $k$ to be proposed to the planner. For each of the $k$ subgoals, the planner generates a plan and a projected final state. The discriminator then predicts the values of these states and selects the plan with the highest value.

## 4.2    Proposing Subgoals

The idea of proposing multiple subgoals is motivated by the information horizon problem. While some subgoals (say, delivering a passenger) may look promising to the GNN-based meta-controller, it may be unable to determine the total cost of achieving that subgoal (the time taken to reach their destination). By proposing multiple subgoals to the planner, the discriminator can avoid subgoals which have unexpectedly large costs according to their projected outcome.

Concretely, the meta-controller is an actor-critic RL system that operates at a higher level than the base MDP. The action space for the meta-controller is the set of possible planning subgoals as defined by the planning model. Further details of the semantics of each environment representation can be found in the experiments.

The meta-controller is comprised of a GNN $\mathcal{G}$, the actor $\pi$ and the critic $V$. The GNN converts the input state graph into the set of node embeddings and the global embedding: $v', u' = \mathcal{G}(s; \theta_G)$, where $\theta_G$ are the parameters of the GNN. The critic is implemented by a fully-connected layer parameterised by $\theta_v$ and takes only the global state as input: $V(u'; \theta_v)$. Finally the actor is implemented by a fully-connected layer parameterised by $\theta_a$, followed by a softmax layer, which gives the policy as a probability distribution over subgoals: $\pi(g|v', u'; \theta_a)$. Unlike a standard RL agent, the meta-controller samples $k$ subgoals from $\pi$ without replacement, which are then passed to the planner.

## 4.3    Planning and Projecting

The planner uses its partial knowledge of the environment dynamics to "look ahead" and predict what would happen if the proposed subgoal were to be achieved. This may include consequences that were not taken into account by the meta-controller due to the information horizon problem. This future state may look less (or more) promising to the discriminator, and it can then select one of the proposed subgoals accordingly.

Specifically, the planner receives $k$ subgoals from the meta-controller and processes them in parallel. For each subgoal $g$, the planner constructs a planning problem: $(s, g)$ and from this determines a plan $p = [a_0, a_1, \ldots, a_{n-1}]$. The plan is then applied step by step to the starting state $s_0$ with $s_{i+1} = \Delta(s_i, a_i)$. This gives $\tilde{s} = s_n$ as the projected state after the plan has been executed.

## 4.4   Selecting Subgoal

The role of the discriminator is to select which of the projected future states is best, and hence select the corresponding plan to execute. To do so the discriminator first applies a GNN $\mathcal{G}$ to $\tilde{s}$ to obtain the embedded global state vector: $\tilde{u}' = \mathcal{G}(\tilde{s}; \theta_G)$. This ensures there is a fixed-size representation regardless of the size of the state graph. This GNN shares parameters with the meta-controller; this is optional but improved performance in our experiments.

It then ranks the projected states in order of desirability, taking previously accumulated rewards into account. To explain the intuition here, imagine you have just delivered a passenger to a remote destination and received a large reward. The state immediately following this seems unpromising - you are stranded in the middle of nowhere with no passengers in sight - but the plan itself is a good choice due to the accumulated reward. To address this issue, we define a path value function, $V_p(u', \tilde{u}'; \theta_p)$ which takes as input the (embedded) current state $u'$ and projected future state $\tilde{u}'$. It is trained to predict the expected future reward for being in state $s$ and then (after some number of steps) being in state $\tilde{s}$. The path value function is implemented as a fully-connected layer parameterised by $\theta_p$, which is trained using the path value loss defined below. This process is repeated for each of the $k$ projected states and the plan with the highest path value is chosen for execution.

## 4.5   Executing Plan

In our environments, the symbolic actions in the plan correspond to the actions in the base MDP, so these are simply executed in sequence until the plan terminates. If the planning model is abstract (i.e. operates at a higher level than the MDP actions), then a low-level RL controller could be trained to achieve each symbolic planning step as in [5,26]. The meta-controller operates on a temporally extended scale; one subgoal might correspond to dozens of atomic actions in the underlying MDP. Consequently the experience tuples we store for an $n$ step plan are of the form $(s_t, \tilde{s}_t, U_{t:t+n}, s_{t+n}, g_t)$, where $U_{t:t+n} = \sum_{i=0}^{n} \gamma^i r_{t+i}$ is the accumulated discounted reward for the plan.

## 4.6   Training

We use A2C [28] to train our agent, but any policy gradient method could be applied. The overall loss is the sum of four components: the policy loss, value loss, and entropy loss as usual for A2C, as well as the novel path value loss.

$$\mathcal{L} = \mathcal{L}_P + \kappa_1 \cdot \mathcal{L}_V + \kappa_2 \cdot \mathcal{L}_E + \kappa_3 \cdot \mathcal{L}_{PV} \qquad (4)$$

where:

- Policy loss: $\mathcal{L}_P(\theta_G, \theta_v, \theta_a) = -\ln(\pi(g_t | \boldsymbol{v}'_t, u'_t; \theta_a)) \cdot A(u'_t, a; \theta_v)$
- Value loss: $\mathcal{L}_V(\theta_G, \theta_v) = (V(u'_t; \theta_v) - (U_{t:t+n} + \gamma^n \cdot V(u'_{t+n})))^2$
- Entropy loss: $\mathcal{L}_E(\theta_G, \theta_a) = H(\pi(g_t | \boldsymbol{v}'_t; \theta_a))$
- Path value loss: $\mathcal{L}_{PV}(\theta_G, \theta_p) = (V_p(u'_t, \tilde{u}'_t; \theta_p) - (U_{t:t+n} + \gamma^n \cdot V(u'_{t+n})))^2$

and $\kappa_i$ are hyperparameters, $H$ is the entropy, and $A$ is the advantage function.

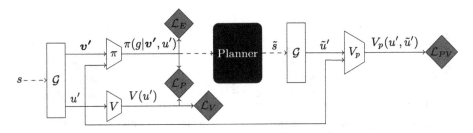

**Fig. 5.** Visualisation of loss calculations and gradient flows. Rectangles represent GNNs; trapeziums represent FC layers. Blue diamonds are loss components, and the black rectangle is a non-differentiable symbolic planner. Solid lines show computation with gradient flow in the opposite direction, dashed lines do not admit gradients. (Color figure online)

## 5  Experiments

We empirically test two hypotheses: 1. Does Oracle-SAGE mitigate the information horizon problem in complex navigation tasks? 2. Are the state projection and discriminator critical to Oracle-SAGE's success, or is planning sufficient?

To evaluate the first hypothesis we compare against SR-DRL, a graph based RL model, which suffers from the information horizon problem [17]. To evaluate the second, we compare against an ablation of Oracle-SAGE which proposes a single subgoal to the planner and therefore has no discriminator. This is similar to SAGE, albeit it leverages a graph-based instead of image-based input [5]. For completeness, we also show results for standard RL approaches using image-based representations of our domains; for these, the information horizon problem does not apply, but they do not have the benefit of the semantically richer graph-based representation or planning model [28,32]. Results are averaged over 5 random seeds, and the number of proposed subgoals ($k$) is set to 3.[1]

### 5.1  Taxi Domain

We extend the classic Taxi domain [6] as follows. A single taxi operates in a randomly generated $20 \times 20$ grid world comprised of zones with different levels of

---

[1] Code available at https://github.com/AndrewPaulChester/oracle-sage.

connectivity (Fig. 1b). During each episode, passengers appear in random cells with random destinations, with up to 20 present at once. The taxi receives a reward of 1 for every passenger that is delivered to their destination. As described previously, the planning model can predict the movements of the taxi, but does not know where future passengers will appear, so constructing a single optimal plan for the entire episode is impossible. The subgoal space of the meta-controller is to deliver any passenger or to move to any square. The image based benchmark is a standard CNN-based A2C agent [28].

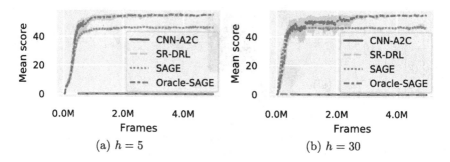

(a) $h = 5$                    (b) $h = 30$

**Fig. 6.** Taxi domain results. Each line is averaged over 5 seeds, with 95% CI shaded.

Figure 6 shows the results in the taxi domain, we first discuss results with $h = 5$. Neither CNN-A2C or graph-based SR-DRL learn to reliably deliver passengers in this environment. The large grid makes it challenging to randomly deliver passengers, resulting in a very sparse reward environment. Even worse, the randomisation of the maze-like road network at every episode prevents memorisation of a lucky action sequence.

By contrast, SAGE performs quite well, quickly learning to deliver around 46 passengers per episode. This success can be attributed to its ability to construct plans to reach far off subgoals; the planner can handle the low-level navigation reliably. It still falls short of Oracle-SAGE though, which delivers about 20% more passengers on average. By investigating the model choices in more detail, we can see why this occurs. Oracle-SAGE assigns nearly equal probabilities to its top 3 choices, and after receiving the projected future state is approximately equally likely to choose any one of them. However the average length of a plan in Oracle-SAGE is 34 compared to 42 for SAGE. This indicates that Oracle-SAGE's discriminator is learning to choose the shorter plans; i.e. deliver passengers that are closer to the taxi, thereby getting the same reward in a shorter time. The meta-controller is unable to learn to distinguish between these subgoals since the passenger destinations are beyond the information horizon.

The results with a horizon of 30 are largely similar to those with the smaller horizon. We start to see instability in the policies with these deep GNNs, but no sign of any benefit, even though the horizon length is comparable to the average plan length. This may indicate that even when the information is not strictly outside of the GNN horizon, the relevant features are too hard to learn in a reasonable time when compared to the more semantically compact projected state representation. We were unable to train with a horizon longer than 30 due to instability and memory constraints.

## 5.2    MiniHack Domain

We also show results using MiniHack [32], an environment suite built on top of the dungeon crawling game Nethack [22]. Our custom MiniHack domain consists of a number of rooms connected by randomised corridors, with a staircase in all rooms except the one the agent starts in. Each room contains a couple of stationary traps, and a random subset of rooms in each episode contain deadly monsters. The agent receives a reward at each step of -0.1, with +10 for successfully descending a staircase and -10 for being killed by a trap or monster. The maximum length of each episode is 100 steps. The planning model provided to the agent is similar to that in the taxi domain; it is restricted to movement actions, it has no knowledge of monsters, traps or staircases. The subgoal space of the meta-controller is to move to any visible square, or to fire rocks with the sling present in the players starting inventory. We use RND as described in the original MiniHack paper [32] as the image based benchmark for this domain.

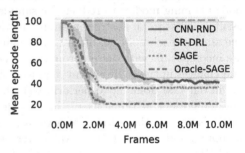

The results in Fig. 7 show that Oracle-SAGE outperforms SAGE in this domain. While SAGE can learn to set subgoals of reaching staircases in rooms without monsters, it fails to distinguish between near and far staircases as the distances are beyond the information horizon. As such, it essentially chooses a safe staircase at random, resulting in a higher average episode length and hence lower average score. By contrast Oracle-SAGE can propose a number of different

**Fig. 7.** MiniHack results; deaths count as a length of 100

staircases as potential options, and then evaluate the time taken to get to each one according to the projected future state. This allows it to choose the closest safe staircase reliably, improving performance.

SR-DRL fails to learn in this environment. The rooms with stairs often have monsters in them, which makes them very dangerous for agents early in training that are still acting largely randomly. The monsters are likely to kill the agent before it stumbles upon the stairs, and so it learns to avoid being killed by monsters at the cost of never entering any of the rooms with stairs. Due to its intrinsic exploration bonus, RND continues to explore and does eventually learn to reach the closest set of stairs; the other rooms are too challenging to reach even with intrinsic exploration. As a result, it reliably converges to a suboptimal policy: aim for the closest stairs regardless of whether there is a monster present.

## 6    Related Work

There are two broad paradigms for addressing sequential decision problems: planning and RL [10]. A wide variety of prior work integrates these two strategies, such as model-based RL [12,13,33], learning based planners [9], and hierarchical hybrid planning/learning architectures [16,21,26]. These approaches vary in the amount of information provided to the agent. Model-based RL frequently assumes no information except direct environment interaction, instead learning the models of the environment from scratch. This requires the least effort for human designers, but accurate models often require large amounts of experience to learn. At the other end of the spectrum, planning based approaches often assume access to a perfect environment model [9]. While this can reduce the sample complexity dramatically, it is impractical for many domains of interest as the true environment dynamics are unknown. This work is aimed at a middle ground, where we assume access to a *myopic* [5] model of the world which is suitable for short term but not long term planning. We contrast this with an *abstract* planning model, which is suitable for long term planning but does not contain the necessary details to act directly in the environment.

**RL-Planning Hybrids.** Much recent work augments RL systems with symbolic planning models to reduce the sample complexity [16,21,23,26]. These all assume access to an abstract planning model and generate a single fixed plan at the start of each episode, often from a human provided goal. These techniques are incompatible with our environments where only myopic models are available, and changes in the environment require replanning during an episode. For example in our taxi domain, to perform well the agent must deliver passengers that are not present at the start of the episode, which is outside of these methods' capabilities. Another branch of work assumes access to an abstract state space mapping, and applies tabular value iteration at the high level to guide a low-level RL policy [30,37,38]. These methods assume that the abstract state space is small, and cannot scale to our domains which require function approximation at the high level. Most similar to our current approach, SAGE assumes only a myopic planning model [5], but lacks feedback from the planner. Finally we note that none of the approaches in this section operate on graph-structured input domains.

**Graph-Based RL.** We can categorise work that combines GNNs with RL based on the source of the graphs used. Some work assumes this graph takes a special form and is provided separately to the state observations of the agent, such as a representation of the agent's body [36], or a network of nearby agents in a multi-agent setting [18]. In others, the graph is derived directly from the observation itself using a domain-specific algorithm, such as text based adventure games [1]. Finally are those methods in which the graph forms the state observation itself. Some of these have specialised graph representations or action selection mechanisms which restrict their applicability to a single application domain, such as block stacking [14, 24] or municipal maintenance planning [20]. Concurrently to our work, Beeching et al. perform navigation over a graph in a realistic 3D environment [3]. They explicitly target scenarios where a symbolic planning model is not applicable, but restrict themselves to pure navigation tasks to reach a provided endpoint, rather than the general reward maximisation objective in our work. Symnet [8] uses GNN based RL to solve relational planning tasks. While this is domain independent, it requires a complete description of the environment dynamics in RDDL (a probabilistic PDDL variant) and so is not applicable in our domains where some environment dynamics are unknown. The most closely related approach to ours is SR-DRL which uses a similar problem set-up, but does not have access to a planning model and so is subject to the information horizon problem [17].

**GNN Receptive Field.** Our description of the GNN information horizon problem draws on prior work regarding the receptive fields of GNNs outside of the RL context [35]. Some authors have tried to address this by using spectral convolution methods, which aggregate information from a wider neighborhood in a single GNN block [25]. These approaches are computationally intensive and do not generalise well across graphs with different structures [39]. Another approach is to deepen the GNN to expand the receptive field [11, 31]. While this is promising, it requires commensurately more resources to train, and as demonstrated in our taxi experiments does not improve performance in our RL setting.

## 7   Conclusion

GNNs show promise in extending relational RL algorithms with modern function approximation, allowing for object-centric reasoning. In this paper we formalised the GNN information horizon problem in deep RL, and showed empirically on a synthetic domain that it leads to degraded performance for our benchmarks on large graphs. This motivated Oracle-SAGE; a graph-based hybrid learning and planning algorithm which incorporates planning predictions into its decision process to mitigate such a problem. We demonstrated its effectiveness against a range of benchmarks and ablations on an extended taxi and a MiniHack domain.

A limitation of this work is that the top-$k$ action selection and planning projection requires additional computational resources when compared to competing approaches. We have also assumed that the provided PDDL actions map directly onto the base MDP. Including a low-level goal-directed RL controller that works as a layer under the planning model would allow this approach to be applied to a wider set of domains [16, 21, 26]. Finally, our future work may also investigate interim experience augmentation [5] to increase sample efficiency in domains with long plans.

# References

1. Ammanabrolu, P., Riedl, M.: Playing text-adventure games with graph-based deep reinforcement learning. In: NAACL (2019)
2. Battaglia, P., et al.: Relational inductive biases, deep learning, and graph networks. arXiv preprint arXiv:1806.01261 (2018)
3. Beeching, E., et al.: Graph augmented deep reinforcement learning in the GameR-Land3D environment. arXiv preprint arXiv:2112.11731 (2021)
4. Berry, D.A., Fristedt, B.: Bandit problems: sequential allocation of experiments (Monographs on Statistics and Applied Probability) (1985)
5. Chester, A., Dann, M., Zambetta, F., Thangarajah, J.: SAGE: generating symbolic goals for myopic models in deep reinforcement learning. arXiv preprint arXiv2203.05079 (2022)
6. Dietterich, T.G.: Hierarchical reinforcement learning with the MAXQ value function decomposition. JAIR **13**, 227–303 (2000)
7. Džeroski, S., De Raedt, L., Driessens, K.: Relational reinforcement learning. Mach. Learn. **43**(1), 7–52 (2001)
8. Garg, S., Bajpai, A., Mausam, M.: Size independent neural transfer for RDDL planning. In: ICAPS (2019)
9. Garg, S., Bajpai, A., Mausam, M.: Symbolic network: generalized neural policies for relational MDPs. In: ICML (2020)
10. Geffner, H.: Model-free, model-based, and general intelligence. arXiv preprint arXiv:1806.02308 (2018)
11. Godwin, J., et al.: Simple GNN regularisation for 3D molecular property prediction and beyond. In: ICLR (2022)
12. Ha, D., Schmidhuber, J.: Recurrent world models facilitate policy evolution. In: NeurIPS (2018)
13. Hafner, D., Lillicrap, T., Norouzi, M., Ba, J.: Mastering atari with discrete world models. In: ICLR (2021)
14. Hamrick, J.B., et al.: Relational inductive bias for physical construction in humans and machines. In: Proceedings of the Annual Meeting of the Cognitive Science Society (2018)
15. Henderson, P., Islam, R., Bachman, P., Pineau, J., Precup, D., Meger, D.: Deep reinforcement learning that matters. In: AAAI (2018)
16. Illanes, L., Yan, X., Icarte, R.T., McIlraith, S.A.: Symbolic plans as high-level instructions for reinforcement learning. In: ICAPS (2020)
17. Janisch, J., Pevný, T., Lisý, V.: Symbolic relational deep reinforcement learning based on graph neural networks. arXiv preprint arXiv:2009.12462 (2020)
18. Jiang, J., Dun, C., Huang, T., Lu, Z.: Graph convolutional reinforcement learning. In: ICLR (2019)
19. Kemp, C., Tenenbaum, J.B.: The discovery of structural form. PNAS **105**(31), 10687–10692 (2008)
20. Kerkkamp, D., Bukhsh, Z., Zhang, Y., Jansen, N.: Grouping of maintenance actions with deep reinforcement learning and graph convolutional networks. In: ICAART (2022)
21. Kokel, H., Manoharan, A., Natarajan, S., Ravindran, B., Tadepalli, P.: RePReL: integrating relational planning and reinforcement learning for effective abstraction. In: ICAPS (2021)
22. Küttler, H., et al.: The NetHack learning environment. In: NeurIPS (2020)

23. Leonetti, M., Iocchi, L., Stone, P.: A synthesis of automated planning and reinforcement learning for efficient, robust decision-making. AIJ **241**, 103–130 (2016)
24. Li, R., Jabri, A., Darrell, T., Agrawal, P.: Towards practical multi-object manipulation using relational reinforcement learning. In: ICRA (2020)
25. Liu, Z., Chen, C., Li, L., Zhou, J., Li, X., Song, L., Qi, Y.: GeniePath: graph neural networks with adaptive receptive paths. In: AAAI (2019)
26. Lyu, D., Yang, F., Liu, B., Gustafson, S.: SDRL: interpretable and data-efficient deep reinforcement learning leveraging symbolic planning. In: AAAI (2019)
27. McDermott, D., et al.: PDDL - the planning domain definition language. Technical Report, Yale Center for Computational Vision and Control (1998)
28. Mnih, V., et al.: Human-level control through deep reinforcement learning. Nature **518**(7540), 529 (2015)
29. Navon, D.: Forest before trees: the precedence of global features in visual perception. Cogn. Psychol. **9**(3), 353–383 (1977)
30. Roderick, M., Grimm, C., Tellex, S.: Deep abstract Q-networks. In: AAMAS (2018)
31. Rong, Y., Huang, W., Xu, T., Huang, J.: DropEdge: towards deep graph convolutional networks on node classification. In: ICLR (2020)
32. Samvelyan, M., et al.: MiniHack the planet: a sandbox for open-ended reinforcement learning research. In: NeurIPS Track on Datasets and Benchmarks (2021)
33. Schrittwieser, J., et al.: Mastering Atari, Go, chess and shogi by planning with a learned model. Nature **588**(7839), 604–609 (2020)
34. Sievers, S., Röger, G., Wehrle, M., Katz, M.: Theoretical foundations for structural symmetries of lifted PDDL tasks. In: ICAPS (2019)
35. Topping, J., Di Giovanni, F., Chamberlain, B.P., Dong, X., Bronstein, M.M.: Understanding over-squashing and bottlenecks on graphs via curvature. arXiv preprint arXiv:2111.14522 (2021)
36. Wang, T., Liao, R., Ba, J., Fidler, S.: NerveNet: learning structured policy with graph neural networks. In: ICLR (2018)
37. Winder, J., et al.: Planning with abstract learned models while learning transferable subtasks. In: AAAI (2020)
38. Wöhlke, J., Schmitt, F., van Hoof, H.: Hierarchies of planning and reinforcement learning for robot navigation. In: ICRA (2021)
39. Zhou, J., et al.: Graph neural networks: a review of methods and applications. AI Open **1**, 57–81 (2020)

# Reducing the Planning Horizon Through Reinforcement Learning

Logan Dunbar[1(✉)] [iD], Benjamin Rosman[2] [iD], Anthony G. Cohn[1,3,4,5,6] [iD],
and Matteo Leonetti[7] [iD]

[1] School of Computing, University of Leeds, Leeds, UK
{sclmd,a.g.cohn}@leeds.ac.uk
[2] University of the Witwatersrand, Johannesburg, South Africa
benjamin.rosman1@wits.ac.za
[3] Tongji University, Shanghai, China
[4] Alan Turing Institute, London, UK
[5] Qingdao University of Science and Technology, Qingdao, China
[6] Shandong University, Jinan, China
[7] Department of Informatics, King's College London, London, UK
matteo.leonetti@kcl.ac.uk

**Abstract.** Planning is a computationally expensive process, which can limit the reactivity of autonomous agents. Planning problems are usually solved in isolation, independently of similar, previously solved problems. The depth of search that a planner requires to find a solution, known as the planning horizon, is a critical factor when integrating planners into reactive agents. We consider the case of an agent repeatedly carrying out a task from different initial states. We propose a combination of classical planning and model-free reinforcement learning to reduce the planning horizon over time. Control is smoothly transferred from the planner to the model-free policy as the agent compiles the planner's policy into a value function. Local exploration of the model-free policy allows the agent to adapt to the environment and eventually overcome model inaccuracies. We evaluate the efficacy of our framework on symbolic PDDL domains and a stochastic grid world environment and show that we are able to significantly reduce the planning horizon while improving upon model inaccuracies.

**Keywords:** Planning · Planning horizon · Reinforcement learning

## 1 Introduction

Planning is a notoriously complex problem, with propositional planning shown to be PSPACE-complete [5]. The planning horizon is the maximum depth that a planner must search before finding a solution, and the number of paths through the search graph grows exponentially with a deepening planning horizon. This exponential growth makes graphs with an even moderate branching factor slow to traverse. Furthermore, frequent replanning when solving such problems limits

© The Author(s), under exclusive license to Springer Nature Switzerland AG 2023
M.-R. Amini et al. (Eds.): ECML PKDD 2022, LNAI 13716, pp. 68–83, 2023.
https://doi.org/10.1007/978-3-031-26412-2_5

the reactivity of agents, making it challenging to incorporate planning into real-time applications.

Machine learning is increasingly used to take advantage of previously solved planning instances to guide the search and plan faster [12]. Much of the work has focused on using plans from simpler problems to learn generalised heuristics for larger problems [20,30,31]; however, even with improved heuristics the planning horizon is left untouched, leaving the problem prohibitively expensive in general. Methods that learn a policy from a set of plans [1,3,23] have been studied to compile the deliberative behaviour of the planner into a reactive policy. Such methods can be highly effective, but rely on large training sets of plans, and are not suitable for online learning.

Our work is inspired by psychological experiments, which have shown that humans have two distinct decision making systems, commonly known as the habit/stimulus-response system and the goal-directed system [6,14,25]. Early work thought these two systems were in competition for resources, but Gershman et al. [8] showed that these systems exhibit a more cooperative architecture, where a model-based system was used to train a model-free system that acted on the environment. With this inspiration, we model the goal-directed system as a classical planner with access to a model, and the habit-based system as a model-free reinforcement learning (RL) algorithm.

In this paper, we propose the Plan Compilation Framework (PCF), a framework that uses model-free RL to compile the plans of a classical planner into a reactive policy. We target an agent that repeatedly executes a task, learning to plan less and less as the agent accumulates experience. Our method can be used online as the agent faces the task, does not require a training set to be initialised, and is agnostic to the type of planner used. Eventually, the behaviour becomes completely reactive and model-free, with the added benefit that it can leverage reinforcement learning to further optimize the policy beyond what can be planned on the model.

## 2   Background

In this section we introduce the notation and background on planning and RL that we will use throughout the paper.

### 2.1   Planning

We consider discrete planning problems, consisting of a tuple $\langle \mathcal{S}, \mathcal{A}, \tilde{\mathcal{T}}, \tilde{\mathcal{C}}, s_0, \mathcal{S}_{\mathcal{G}} \rangle$ where $\mathcal{S}$ is a finite set of states, $\mathcal{A}$ is a finite set of actions, $\tilde{\mathcal{T}} : \mathcal{S} \times \mathcal{A} \times \mathcal{S} \rightarrow \{0, 1\}$ is a deterministic transition function that models the environment, $\tilde{\mathcal{C}} : \mathcal{S} \times \mathcal{A} \times \mathcal{S} \rightarrow \mathbb{R}$ is the model's cost function, $s_0$ is an initial state and $\mathcal{S}_{\mathcal{G}} \subseteq \mathcal{S}$ is a set of goal states. Planning produces a plan $P(s_0) \doteq [a_0, a_1, \ldots a_n]$, a sequence of actions that transforms $s_0$ into a goal state $s_{n+1} \in \mathcal{S}_{\mathcal{G}}$ when applied sequentially.

Planners can be seen as executing a simple abstract algorithm: from the initial state, choose a next state $s_i$ according to some strategy, check if $s_i \in \mathcal{S}_{\mathcal{G}}$ and exit with the plan if true, otherwise expand the state into its successors using the model, and loop until a goal is found. Our work only requires changing the goal test logic of existing planners, which affords extensive integration opportunities.

## 2.2   Reinforcement Learning

A reinforcement learning problem is modelled as a Markov Decision Process (MDP), a tuple $\langle \mathcal{S}, \mathcal{A}, \mathcal{T}, \mathcal{R}, \gamma \rangle$ where $\mathcal{S}$ is a finite set of states, $\mathcal{A}$ is a finite set of actions, $\mathcal{T} : \mathcal{S} \times \mathcal{A} \times \mathcal{S} \rightarrow [0, 1]$ is the environment's transition function, $\mathcal{R} : \mathcal{S} \times \mathcal{A} \times \mathcal{S} \rightarrow [r_{min}, r_{max}]$ is a reward function and $\gamma$ is a discount factor.

A policy $\pi (a \mid s)$ is a probability distribution over actions conditioned on the current state $s \in \mathcal{S}$. The return $G_t = \sum_{k=0}^{\infty} \gamma^k r_{t+k+1}$ is the cumulative discounted reward obtained from time step $t$. The expected return when starting in state $s$, taking action $a$ and following policy $\pi$ is known as the state-action value function:

$$Q_\pi (s, a) \doteq \mathbb{E}_\pi \left[ \sum_{k=0}^{\infty} \gamma^k r_{t+k+1} \,\middle|\, s_t = s, a_t = a \right]. \tag{1}$$

The value function is bounded by $[q_{min}, q_{max}]$ where $q_{min} = r_{min}/(1 - \gamma)$ and $q_{max} = r_{max}/(1 - \gamma)$. The goal of RL is to find an optimal policy $\pi^*$ that maximises $Q_\pi (s, a), \forall s, a$. Two central methods for learning a value function are Monte-Carlo methods and temporal difference learning [27].

**Monte-Carlo Methods.** Constant-$\alpha$ Monte Carlo [27] uses the actual return $G_t$ as the target when updating the value function:

$$Q (s_t, a_t) \leftarrow Q (s_t, a_t) + \alpha \left[ G_t - Q (s_t, a_t) \right]. \tag{2}$$

This produces an unbiased but high variance estimate of the value function, which we leverage to allow our agent to initially favour actions specified by the planner, as discussed in Sec 4.1.

**Temporal Difference Learning.** Q-learning [28] is a common single-step algorithm, whose update target is $r_{t+1} + \gamma \max_a Q (s_{t+1}, a)$, which replaces the return in Eq. 2. Methods can also look further than one step ahead, with the $n$-step return defined as

$$G_{t:t+n} \doteq r_{t+1} + \gamma r_{t+2} + \cdots + \gamma^{n-1} r_{t+n} + \gamma^n V_{t+n-1}(s_{t+n}), \tag{3}$$

where $V_{t+n-1}(s_{t+n})$ is the value of state $s_{t+n}$ and can be approximated by $\max_a Q (s_{t+n}, a)$. This allows the agent to blend actual returns generated by environmental interactions with bootstrapping estimates further into the future.

## 2.3   Distances Between Distributions

In this work we use the concept of a distance between distributions to categorise when the policy for a particular state has stabilised. A common way to quantify the difference between two probability distributions $P(x)$ and $Q(x)$ is the relative entropy, also known as the Kullback-Leibler Divergence (KLD) [17]:

$$\mathcal{D}_{KL} (P \| Q) = \sum_{x \in \mathcal{X}} P(x) \log \left( \frac{P(x)}{Q(x)} \right). \tag{4}$$

A related measure, which, unlike the KLD, is a distance metric, is known as the Jenson-Shannon Divergence (JSD): [19]

$$\mathcal{D}_{JS}\left(P\,\|\,Q\right) = \frac{\mathcal{D}_{KL}\left(P\,\|\,M\right)}{2} + \frac{\mathcal{D}_{KL}\left(Q\,\|\,M\right)}{2} \tag{5}$$

where $M = \frac{1}{2}\left(P+Q\right)$. This measure allows us to quantify the change in our policy when we update the value function for a state, as detailed in Sec 4.2.

## 3   Related Work

Gershman et al. [8] showed that a simple implementation of the DYNA architecture [26] was able to replicate their psychological findings, lending credence to the idea of a cooperative model-based and model-free system. In DYNA, a model-free algorithm chooses actions, and a model-based algorithm trains the model-free values. While a model can be specified as prior knowledge, generally the agent learns a model of the environment through interactions. Even if a model is specified, DYNA would be unable to initially prefer the actions from this model, as our work does. This negates some of the benefits of pre-specified models, namely fast and reliable goal achievement. DYNA also requires that the model be able to generate accurate rewards, as they are incorporated into the value function from which the model-free algorithm chooses its actions. We require only that the planner generate a plan to achieve the goal, which allows us to use classical planning.

Other techniques for reducing the planning time include Lifelong Planning A* (LPA*) and D* Lite [15], which are incremental heuristic search methods. LPA* repeatedly calculates the shortest distances from the start state to the goal state as the edge costs of the graph change. D* Lite considers the opposing problem and searches from the goal to the current state. This formulation allows the start state to change without needing to recompute the entire search graph. In D* Lite the planning horizon is effectively shortened by the agent moving towards the goal during plan enactment, but for any particular state the horizon does not change. Both LPA* and D* Lite require access to a predecessor model which is often harder to specify than a successor model. We could leverage these planners in our work, but we find the dependence on a predecessor model too limiting.

Our concept of a *learnt* state (Sect. 4.2) is similar to the *known* state from Explicit Explore or Exploit (E³) [13]. In E³ a state becomes known when it has been visited and the actions tried sufficiently often to produce an estimate of the transitions and pay-offs with high probability. They use the concept of the *known* state to partition the MDP into known and unknown regions, learning to exploit the current known states or to explore the unknown states. They need to explicitly learn the model of the transition and reward probabilities, which can be slow and require visiting many states that are irrelevant to the current goal. We similarly partition the MDP into *learnt* and *unlearnt* states, but we leverage the prior knowledge of the planner to initialise the *learnt* states, thereby reducing the environmental interactions while making sure to achieve the goal, as well as providing a more relevant exploration frontier. We also reduce our dependence on the planner over time, moving to efficient model-free learning, while $E^3$ continues to maintain and update its model estimates, which could become costly in the long-term.

The work by Grounds and Kudenko [9] and Grzes and Kudenko [10] use planning to shape [22,29] the rewards received by learning algorithms. Grounds and Kudenko [9] learn a low-level Q-learning behaviour for each STRIPS operator by using the STRIPS planner to generate plans from which they derive a shaping value. Grzes and Kudenko [10] use a similar technique, but instead use the generated high-level plan to shape the reward of a single Q-learner. They use the step number of the plan to provide a potential field for shaping. This guides the Q-learner to choose actions that would lead the agent along the planner's path. Although shaping provides suggestions to the agent, it is unable to enforce which actions are chosen. This means the agent can very quickly start behaving like pure Q-learning if the shaping and environmental reward are mismatched. Our goal in this work is to utilise the information contained in the planner as effectively as possibly, preventing Q-learning-like exploration which can leave the agent goal-impoverished while trying to explore the entire state space.

The work most similar to ours is Learning Real-Time-A* (LRTA*) [16], which is generalised by Real-Time Dynamic Programming (RTDP) [2]. LRTA* aims to bring planning to real-time applications by performing local search within a limited horizon and stores the results in an evaluation function which is updated over successive trials. RTDP was subsequently developed as a form of asynchronous dynamic programming [4] that uses Bellman backups to obtain the values of states. RTDP operates under the same horizon and time constraints as LRTA*, making it also suitable for real-time applications. Both LRTA* and RTDP require a perfect model to backup values accurately through the state space. Where our work differs is that LRTA* and RTDP will eventually converge to the optimal values for the possible policies afforded by their models. Through repeated trials they will converge to the model-optimal values, but the search horizon is never reduced, nor is actual environmental information incorporated into the search. Our framework instead leverages the information contained in the planner's model during the initial phase of operation, and gradually cedes control to the model-free learning algorithm. This means the planner will at some point never be called, and the model-free learner will be completely responsible for choosing actions, incorporating real environmental feedback and adapting as necessary.

## 4    Plan Compilation Framework

We now propose the Plan Compilation Framework with the Plan Compiler (PC) agent, the goal of which is to leverage reinforcement learning to reduce the planning time of a classical planner when repeatedly solving a problem from different initial states, eventually becoming purely reactive. The classical planner produces plans that take the agent to the goal, while model-free RL learns to compile the planner's behaviour into a value function online. Once the model-free policy of a state has stabilised through successive value function updates, we consider it *learnt* and the planner is no longer required for that state, ceding control to the model-free RL algorithm. We then augment the goal set of the planner with the *learnt* states, generating plans to the goal, or to a *learnt* state, where the model-free RL algorithm can take over and react accordingly. This process reduces the planning horizon through repeated interactions, as well as leverages the ability of model-free RL to explore and improve upon the performance of the planner's model.

The PC agent's algorithm is shown in Algorithm 1. It is a classical RL agent computing actions based on the current state and learning from the observed next state and reward. The key elements are described in the following sections.

## 4.1   Compiling the Planner's Policy

The planner induces a policy $\pi_{pln}$ over the state space by virtue of being able to generate a plan $P(s)$ and returning the first action from that plan. We use model-free RL to compile this policy into a value function $Q$ by learning to ascribe higher value to the actions chosen by $\pi_{pln}$ in the initial phases of operation. Q-value initialisation, as shown by Matignon et al. [21], plays a crucial role in the initial behaviour of the agent. We want to replicate the behaviour of the planner, not explore every action. We therefore pessimistically initialise our Q-values, $q_{init} \leftarrow q_{min} - \delta$, driving the agent to prefer states and actions it has previously visited. This requires knowing the lower bound of return we expect to receive while following the planner, which is always possible when $r_{min}$ is known.

The biased nature of bootstrapping TD updates learn to favour actions *not* in $\pi_{pln}$ in the early phases of operation. The total change to the state-action value during update is

$$\delta Q(s, a) \leftarrow \alpha [r' + \gamma \max_{a'} Q(s', a') - Q(s, a)], \qquad (6)$$

which, when all Q-values are uniformly initialised to $q_{init}$, is dominated by $r'$ due to $\gamma \max_{a'} Q(s', a') - Q(s, a) = \gamma q_{init} - q_{init} \approx 0$ on the first update. Any negative reward therefore lowers an action's value, pushing the policy away from $\pi_{pln}$. We want, initially, the actions suggested by the planner, regardless of reward sign, to be preferred over all other actions. To do this, we leverage the unbiased Monte Carlo update (Sect. 2.2), which uses the return as the target rather than a biased bootstrapped value. This, coupled with the pessimistic initialisation, preserves the desired action preferences when learning to replicate $\pi_{pln}$, as seen in Algorithm 1, Ln. 25, 28.

## 4.2   Relinquishing Control to the Learner

Once the agent has learnt to replicate the planner's policy in a state, it can query the model-free learner for the action rather than the computationally expensive

---

**Algorithm 1:** Plan Compiler Agent

---

**Input:**

$\epsilon,\ \epsilon_{exp},\ \alpha,\ \alpha_l,\ \tau_D,\ \tau_l,\ \zeta$

$\forall s, a,\quad Q(s, a) \leftarrow q_{min} - \delta$

$\forall s, a,\quad Q_{exp}(s, a) \leftarrow q_{max} + \delta$

$e \leftarrow 0$

```
1  for each episode do
2  |   Initialise s
3  |   repeat
4  |   |   a ← act (s)                                          #Ln. 10
5  |   |   Take action a, observe r', s'
6  |   |   learn (s, a, r', s')                                 #Ln. 21
7  |   |   s ← s'
8  |   until terminal (s)
9  end
10 def act (s):
11 |   if e ≤ 0 then                                            #Not Exploring
12 |   |   if !learnt (s) then
13 |   |   |   return action from planner
14 |   |   else if rand() < ε_exp then                          #Start Exploring
15 |   |   |   v ← argmax_a Q(s, a)
16 |   |   |   e ← ξ|v|
17 |   |   else
18 |   |   |   return ε-greedy action from Q
19 |   return ε-greedy action from Q_exp                        #Exploring
20 end
21 def learn (s, a, r', s'):
22 |   if learnt (s) and learnt (s') then
23 |   |   bootstrap (Q, (s, a, r', s'))
24 |   else
25 |   |   traj.append(s, a, r', s')
26 |                                                            } Update Q
27 |   if terminal (s') then
28 |   |   monte_carlo (Q, traj)
29 |   else if learnt (s') then
30 |   |   n_step (Q, traj)
31 |   bootstrap (Q_exp, (s, a, r', s'))                        } Update Q_exp
32 |   if e > 0 then
33 |   |   e ← e - |r'|                                         #Eq. 11
34 end
```

planner. We define a state as being *learnt* when its policy stabilises through successive Q-value updates. We compute the JSD distance (Eq. 5) between the state's policy before $\pi_{pre}$ and after $\pi_{post}$ performing the Q-value update. We introduce a function $l : \mathcal{S} \rightarrow [0,1]$ that quantifies how stable the policy is for a particular state, akin to anomaly detection [7]. We track the stability estimate with an exponential recency-weighted average

$$l(s) \leftarrow l(s) + \alpha_l \left[ u_l - l(s) \right] \tag{7}$$

where $\alpha_l$ is the stability learning rate, and $u_l$ is the update target. The target is binary, and computes whether the update caused a policy change above a threshold $\tau_{\mathcal{D}}$[1]:

$$u_l = \begin{cases} 1, & \text{if } \mathcal{D}_{JS} \left( \pi_{pre} \,\|\, \pi_{post} \right) < \tau_{\mathcal{D}} \\ 0, & \text{otherwise.} \end{cases} \tag{8}$$

When $l(s)$ is above a threshold $\tau_l \in (0,1)$, meaning the policy for state $s$ has stabilised, we consider $s$ learnt:

$$learnt(s) = \begin{cases} True, & \text{if } l(s) > \tau_l \\ False, & \text{otherwise.} \end{cases} \tag{9}$$

Once a state is considered learnt, and the agent next encounters it, the learner chooses an action from $Q$, without the need to invoke the planner. We prevent the state from reverting to unlearnt by setting $u_l(s) \leftarrow 1$ for all subsequent stability updates. The learner can now exploit the planner's policy or can choose to explore alternate actions to potentially improve upon the planner's policy.

Once states are considered learnt, the negative effect of the bootstrapping bias is reduced, and the agent can update the value function with TD learning, which allows online updates during the episode. The update procedure is detailed in Algorithm 1, Ln. 22–30. The control flow of action selection is shown in Fig. 1 and detailed in Algorithm 1, Ln. 10–20.

### 4.3   Exploration

We facilitate local exploration around the paths provided by the planner with the aim to improve upon its performance without incurring drastic exploration costs. The pessimistically initialised Q-values continuously drive the agent back onto the paths preferred by the planner, preventing exploration.

To enable controlled exploration while deviating from the planner's behaviour we take two steps: (1) we introduce a second value function $Q_{exp}$, which is optimistically initialised; (2) we establish an exploration budget based on state values and interrupt exploration when the budget is spent. $Q_{exp}$ is updated online with bootstrapping as seen in Algorithm 1, Ln. 31.

When in a learnt state, with some small probability $\epsilon_{exp}$ the agent begins exploring, choosing actions from the optimistic value function $Q_{exp}$. The state's

---

[1] This can be prespecified or adapted online as per the work of De Klerk et al. [7].

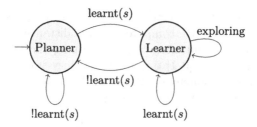

**Fig. 1.** Control flow of Plan Compiler action selection: the planner has initial control while states are !learnt($s$), once a state becomes learnt($s$), control passes to the learner. The learner maintains control while successive states are learnt($s$), transferring control back to the planner if a state is !learnt($s$). The learner also maintains control while exploring, either keeping control or giving control back to the planner once the quota is depleted, depending on the state's learnt status.

current pessimistic value $V(s) \leftarrow \max_a Q(s, a)$ is an estimate of how much return the agent expects to receive from that point onwards, which we use to bound the amount of exploration the agent is allowed. Using this value, we calculate an exploration quota $e_t$, which is proportional to $V(s)$

$$e_t \leftarrow \xi |V(s_t)|, \tag{10}$$

where $\xi \geq 0$ determines how far, in terms of value, we are willing to explore away from the planner's paths. The agent then exclusively chooses actions from $Q_{exp}$ which takes the agent into unexplored regions of the state space. The exploration behaviour is shown in Algorithm 1, Ln. 14–16, 19.

While exploring, we reduce the exploration quota by the reward received

$$e_{t+1} \leftarrow e_t - |r_{t+1}|, \tag{11}$$

shown in Algorithm,1, Ln. 33. When the quota drops to or below zero, exploration ends, at which point the agent will be in an unlearnt state, choosing an action from the planner, or in a learnt state, choosing an action from $Q$. The quota has the effect of preventing the agent from straying too far from the learnt regions, thereby maintaining reactivity, preventing excessive exploration, and making sure the agent achieves the goal regularly and reliably. The two parameters $\epsilon_{exp}$ and $\xi$ are tunable to specify how often and how far the agent is allowed to travel from the paths generated by the planner.

### 4.4   Reducing the Planning Horizon

Once a state has become learnt, we can reduce the planning horizon by augmenting the goal set of the planner:

$$\mathcal{S}'_{\mathcal{G}} \leftarrow \mathcal{S}_{\mathcal{G}} \cup \{s \mid learnt(s)\}. \tag{12}$$

The planner is then able to plan to states for which the learner has a stable policy, from which it can react accordingly. Through repeated interactions with the environment, more states become learnt, shortening the planning horizon until the planner is rarely invoked.

Initially, we may pay a path-length cost due to planning to *learnt* states as opposed to goal states, or due to pernicious initial state distributions. Over time, as the agent explores locally around the planner's paths, the model-free algorithm learns the true state-action values and undoes the reward penalty. Additionally, decoupling the model-free learner from the planner, in the way PCF does, allows the use of deterministic planners in stochastic domains as an alternative to computationally expensive sampling-based stochastic planners. Learning overcomes the inaccurate model used for planning, as shown in the DARLING system [18].

## 5  Experiments

We designed two experiments to evaluate the properties of our method. In the first experiment we show the trade off between the reduction in the number of states expanded by the planner and its cost in terms of sub-optimality of the reward. We use three PDDL domains from the International Planning Competition (IPC)[2] and FastDownward [11] as the planner. In the second experiment we show how, through reinforcement learning, our system can improve over planning with an inaccurate model. For this experiment we cannot use IPC domains, since these do not incorporate any modeling error. Therefore, we use a grid world whose parameters we can control so as to diverge from the model.[3]

### 5.1  PDDL Domains

Depot is a logistics-like domain where crates are trucked between depots and stacked in specific orders using hoists, with resource constraints on the hoists and trucks. 15-Puzzle is a classic planning domain with 15 unique tiles in a $4 \times 4$ grid that can exchange places with a single blank tile, the objective being to arrange the tiles in ascending order. 15-Blocks is a domain with 15 blocks that can be stacked atop a table or one another, and the goal is to create a particular pattern of stacked blocks. For all three domains, the agent receives -1 reward for every action taken in the environment.

We use two settings of FastDownward, called FD-G and FD-GFF, using the Context Enhanced Additive heuristic and the FastForward heuristic respectively, both of which are non-admissible and produce sub-optimal plans. We chose the faster setting per environment. We show that we can integrate our Plan Compiler agent with these planners, which we label PC-FD-G and PC-FD-GFF. Our $Q$ and $Q_{exp}$ are tables using hashed PDDL states and actions to store the values. The results are averaged over 5 runs of 20k episodes with random initial states. The PC parameters used were: $\epsilon = 0.1$, $\alpha = 1$, $\alpha_l = 1$, $\tau_D = 0.01$, $\tau_l = 0.9$, $\xi = 0$, $\epsilon_{exp} = 0$.

---

[2] Implemented by PDDLGym [24].
[3] Code at https://github.com/logan-dunbar/plan_compilation_framework.

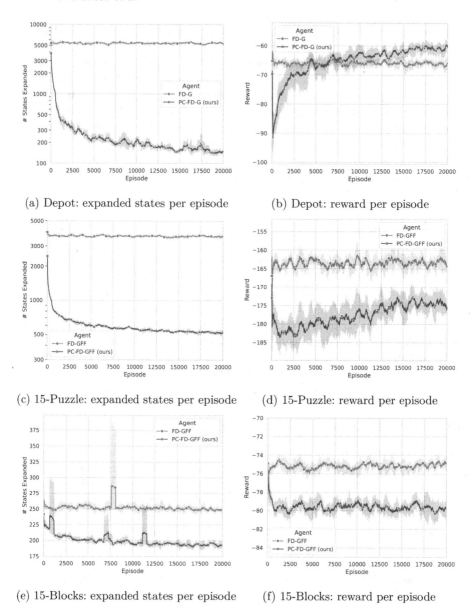

(a) Depot: expanded states per episode    (b) Depot: reward per episode

(c) 15-Puzzle: expanded states per episode    (d) 15-Puzzle: reward per episode

(e) 15-Blocks: expanded states per episode    (f) 15-Blocks: reward per episode

**Fig. 2.** PDDL domains results

**Results.** The results for the PDDL experiments are shown in Fig. 2. The number of states expanded per episode is shown for each domain in Figs. 2a,c,e. The graphs clearly show a large reduction in the number of states expanded during each episode, noting the logarithmic scale for Depot and 15-Puzzle. For Depot, the number of states expanded is reduced by an order of magnitude after approximately 1k episodes, with 25x fewer expansions after 5k episodes. The results are similar for 15-Puzzle, with 5x fewer expansions after 2.5k episodes. This justifies the potentially exponential reduction in computation achieved when reducing the planning horizon, as alluded to in the introduction. For 15-Blocks, the number of state expansions is reduced by 20% after 5k episodes. The FastForward heuristic performs extremely well in this domain, averaging only 250 states expanded for solutions with average path length of 75, which is already close to optimal.

Reward per episode is shown for each domain in Figs. 2b,d,f. In Depot, the reward drops sharply in the very early episodes. However, the model-free RL algorithm quickly begins exploring locally around the paths generated by the planner, bringing the reward obtained to parity after 7.5k episodes, and begins to outperform the planner from that point onward. This means that after 20k episodes the system is both outperforming the planner in terms of reward and has reduced the computational burden of the planner by over an order of magnitude. For 15-Puzzle, the characteristic drop-off in reward is present in the initial episodes, after which there is a clear trend of improvement. In this domain, the reward does not reach parity with the planner after 20k episodes, meaning that the advantage in terms of planning time has a long-lasting cost in terms of reward. However, the trend suggests that longer runtime will once again bring the performance in line with the planner, and the near order of magnitude reduction in number of states expanded might be considered a worthwhile trade-off for a roughly 6% loss of reward performance. If optimality can be foregone, and a slight reduction in reward tolerated, a large reduction in computational requirements can be gained. In 15-Blocks, the reward received initially drops sharply, but then stabilises 7% worse than just using the planner. The huge space and branching factor seems to prevent the model-free algorithm from finding improved plans, but this could also be due to FastForward providing excellent paths very close to optimal, coupled with our constant cost of $\epsilon$-greedy exploration.

## 5.2   Grid World

The grid world is a stochastic $50 \times 50$ maze-like world with walls, quicksand, random initialisation and a fixed goal location. The grid for each run is randomly generated to contain 20% walls, and 25% of the remaining free space is allocated to be quicksand. Even rows and columns are twice as likely to receive quicksand. This has the effect of creating maze-like paths which the agent can learn to traverse. We operate under the assumption that the goal is reachable, and therefore ensure that at least 50% of the non-wall locations can reach the goal, otherwise we regenerate the grid. In each episode a random initial state is chosen from the set of reachable states. Moving into a wall receives -5 reward, stepping into quicksand receives -100 reward, and every other action receives -1 reward. The agent can move in the 4 cardinal directions, and choosing to move in a direction succeeds 80% of the time, with 20% chance to move in either of the neighbouring directions. $Q$ and $Q_{exp}$ are tables of hashed states and actions storing the values.

The model used for planning is deterministic, and therefore incorrect. Furthermore, the model has no cost, such that the planner aims to compute the shortest path, which may not be optimal in terms of reward. We test our framework against vanilla Q-learning (Q in Fig. 3), A* (A*) and RTDP (RTDP), and we use both A* (PC-A*) and RTDP (PC-RTDP) as our Plan Compiler planners. The results are averaged over 5 runs of 10k episodes. The PC parameters used were: $\epsilon = 0.1$, $\alpha = 0.1$, $\alpha_l = 0.1$, $\tau_D = 0.01$, $\tau_l = 0.9$, $\xi = 0.5$, and $\epsilon_{exp}$ is linearly reduced from 0.03 to 0 in 8k episodes.

The second part of this experiment showcases the effect of the exploration quota $\xi$. Higher values of $\xi$ should result in more exploration in the early phases of operation with a commensurate loss in reward, while a lower $\xi$ should remain more faithful to the plans generated by the planner, achieving the goal more regularly in the early phase but possibly losing out on finding better paths in the long run. We use A* as our planner, vary the quota for six different settings $\xi = \{0.05, 0.1, 0.15, 0.3, 0.5, 1\}$, fix the remaining parameters to: $\epsilon = 0.1$, $\alpha = 0.1$, $\alpha_l = 0.1$, $\tau_D = 0.01$, $\tau_l = 0.9$, and $\epsilon_{exp}$ is linearly reduced from 0.03 to 0 in 8k episodes.

**Results.** The results for the gridworld domain are shown in Fig. 3. Figure 3a shows a large reduction in the number of states expanded, where both PC-A* and PC-RTDP have essentially reduced the planning horizon to 0 after 750 episodes. This is possible due to the small state space allowing the agent to visit and learn every state. RTDP also drops off rapidly, but as the planner has no way of reducing the planning horizon to 0, it maintains a constant planning cost. This could become prohibitive in larger environments with an open-loop planning cycle such as implemented here, whereas our method will continue to reduce its computational requirements over time. A* maintains a fairly large computational burden as it has no way of incorporating the previous solutions.

The reward per episode is shown in Fig. 3b. Q-learning is known to be sample inefficient and this can be seen in the asymptotic reward in the first few episodes. This makes Q-learning impractical in real-world settings, because it would be too costly for an agent or robot to explore as much as Q-learning requires. However, the large amount of exploration does mean that Q-learning can find excellent paths and we see this with the agent averaging $-700$ reward after 10k episodes. A* and RTDP have no knowledge of the true transition function, nor of the real reward function, nor any way to incorporate experience. This means they will only be able to achieve what their models afford them. RTDP initially has behaviour similar to that of Q-learning, as it computes its value function using dynamic programming updates. It manages this quickly in this small world and then reaches a steady state of reward of approximately $-950$. A* is much the same, with the random initialisation and grids providing some small fluctuations. PC-A* starts at the same performance level as A*, drops slightly until some states become learnt and it can begin exploring. This can be seen in the drop off between episodes 0–1k. The exploration begins to pay dividends at this point and the agent improves until reaching parity with Q-learning, at $-700$ reward.

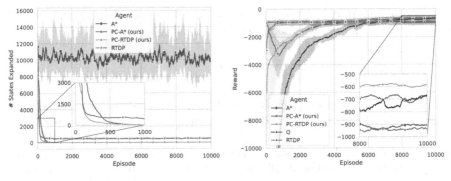

(a) Expanded states per episode          (b) Reward per episode

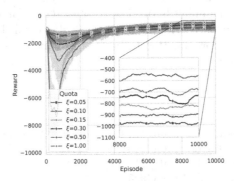

(c) Reward per episode for varying quota values

**Fig. 3.** GridWorld results

This is remarkable considering the amount of exploration Q-learning required versus that of PC-A*. This confirms that even suboptimal plans can be excellent exploratory guides. PC-RTDP starts with RTDP's characteristic Q-learning like exploratory behaviour, with large variability in the early phases of operation. But it too puts that exploration to good use and quickly outperforms all other agents with an average of $-600$ reward after 10k episodes. This shows that we can leverage the strengths of other approaches and even improve upon them.

The final experiment in Fig. 3c shows the effect the quota parameter $\xi$ has on exploration. A lower $\xi$ results in less exploration, and this can clearly be seen in the first 2k episodes. $\xi = 1$ is the most exploratory, receiving large negative reward as it traverses the state space, while $\xi = 0.05$ is the least exploratory, remaining truer to the planner's paths. The rest are properly ordered between the two extreme values of $\xi$. This reverses as the exploratory behaviour finds better paths than what the planner could suggest, resulting in $\xi = 1$ achieving the highest long term reward, $\xi = 0.05$ the lowest, and the rest ordered in between. This demonstrates the flexibility a designer has when using the Plan Compilation Framework. Depending on the situation, one can vary $\xi$ to either engage in exploratory behaviour or to be faithful to the guiding planner and prefer reducing the computational requirements of the planner.

## 6    Conclusions

In this paper we have designed and implemented the novel Plan Compilation Framework that combines classical planning with model-free reinforcement learning in such a way as to compile the planner's policy into a model-free RL value function, dramatically reducing the planning horizon over time and improving the long term computational efficiency of the system. We show by experiment in PDDL domains that we are able to reduce the number of states expanded by the FastDownward planner, and we show in a stochastic grid world that we are able to reduce the planning horizon of A* and RTDP planners whilst also improving upon an imperfect model. In future work we will look to introduce the power of function approximation to be able to handle infinite state spaces and robotic applications.

## References

1. Argall, B.D., Chernova, S., Veloso, M., Browning, B.: A survey of robot learning from demonstration. Robot. Auton. Syst. **57**(5) (2009)
2. Barto, A.G., Bradtke, S.J., Singh, S.P.: Learning to act using real-time dynamic programming. Artif. Intell. **72**(1) (1995)
3. Bejjani, W., Dogar, M.R., Leonetti, M.: Learning physics-based manipulation in clutter: combining image-based generalization and look-ahead planning. In: IEEE/RSJ International Conference on Intelligent Robots and Systems (2019)
4. Bertsekas, D.P.: Distributed asynchronous computation of fixed points. Math. Program. **27**(1) (1983)
5. Bylander, T.: Complexity results for planning. In: 12th International Joint Conference on Artificial Intelligence (1991)
6. Daw, N.D., Niv, Y., Dayan, P.: Uncertainty-based competition between prefrontal and dorsolateral striatal systems for behavioral control. Nat. Neurosci. **8** (2005)
7. De Klerk, M., Venter, P.W., Hoffman, P.A.: Parameter analysis of the Jensen-Shannon divergence for shot boundary detection in streaming media applications. SAIEE Africa Res. J. **109**(3) (2018)
8. Gershman, S.J., Markman, A.B., Otto, A.R.: Retrospective revaluation in sequential decision making: a tale of two systems. J. Exp. Psychol. General **143**(1) (2014)
9. Grounds, M., Kudenko, D.: Combining reinforcement learning with symbolic planning. In: Tuyls, K., Nowe, A., Guessoum, Z., Kudenko, D. (eds.) AAMAS/ALAMAS 2005-2007. LNCS (LNAI), vol. 4865, pp. 75–86. Springer, Heidelberg (2008). https://doi.org/10.1007/978-3-540-77949-0_6
10. Grzes, M., Kudenko, D.: Plan-based reward shaping for reinforcement learning. In: 4th International IEEE Conference Intelligent Systems, vol. 2. IEEE (2008)
11. Helmert, M.: The fast downward planning system. J. Artif. Intell. Res. **26** (2006)
12. Jiménez, S., De La Rosa, T., Fernández, S., Fernández, F., Borrajo, D.: A review of machine learning for automated planning. Knowl. Eng. Rev. **27**(4) (2012)

13. Kearns, M., Singh, S.: Near-optimal reinforcement learning in polynomial time. Mach. Learn. **49**(2–3) (2002)
14. Keramati, M., Dezfouli, A., Piray, P.: Speed/accuracy trade-off between the habitual and the goal-directed processes. PLoS Comput. Biol. **7**(5) (2011)
15. Koenig, S., Likhachev, M.: Fast replanning for navigation in unknown terrain. IEEE Trans. Robot. **21**(3) (2005)
16. Korf, R.E.: Real-time heuristic search. Artif. Intell. **42**(2) (1990)
17. Kullback, S., Leibler, R.A.: On information and sufficiency. Ann. Math. Stat. **22**(1) (1951)
18. Leonetti, M., Iocchi, L., Stone, P.: A synthesis of automated planning and reinforcement learning for efficient, robust decision-making. Artif. Intell. **241** (2016)
19. Lin, J.: Divergence measures based on the Shannon entropy. IEEE Trans. Inf. Theory **37**(1) (1991)
20. Marom, O., Rosman, B.: Utilising uncertainty for efficient learning of likely-admissible heuristics. In: Proceedings of the International Conference on Automated Planning and Scheduling, vol. 30 (2020)
21. Matignon, L., Laurent, G.J., Le Fort-Piat, N.: Reward function and initial values: better choices for accelerated goal-directed reinforcement learning. In: International Conference on Artificial Neural Networks (2006)
22. Ng, A.Y., Harada, D., Russell, S.: Policy invariance under reward transformations: theory and application to reward shaping. In: Proceedings of the Sixteenth International Conference on Machine Learning (1999)
23. Pérez-Higueras, N., Caballero, F., Merino, L.: Learning robot navigation behaviors by demonstration using a RRT* planner. In: Agah, A., Cabibihan, J.-J., Howard, A.M., Salichs, M.A., He, H. (eds.) ICSR 2016. LNCS (LNAI), vol. 9979, pp. 1–10. Springer, Cham (2016). https://doi.org/10.1007/978-3-319-47437-3_1
24. Silver, T., Chitnis, R.: PDDLGym: gym environments from PDDL problems. In: International Conference on Automated Planning and Scheduling (ICAPS) PRL Workshop (2020)
25. Solway, A., Botvinick, M.M.: Goal-directed decision making as probabilistic inference: a computational framework and potential neural correlates. Psychol. Rev. **119**(1) (2012)
26. Sutton, R.S.: Dyna, an integrated architecture for learning, planning, and reacting. ACM SIGART Bull. **2**(4) (1991)
27. Sutton, R.S., Barto, A.G.: Reinforcement Learning: An Introduction. MIT Press, Cambridge (2018)
28. Watkins, C.J.C.H.: Learning from delayed rewards. Ph.D. thesis, King's College, Cambridge (1989)
29. Wiewiora, E., Cottrell, G.W., Elkan, C.: Principled methods for advising reinforcement learning agents. In: Proceedings of the 20th International Conference on Machine Learning (2003)
30. Yoon, S.W., Fern, A., Givan, R.: Learning heuristic functions from relaxed plans. In: Proceedings of the Sixteenth International Conference on Automated Planning and Scheduling, vol. 2 (2006)
31. Yoon, S., Fern, A., Givan, R.: Learning control knowledge for forward search planning. J. Mach. Learn. Res. **9**(4) (2008)

# State Representation Learning for Goal-Conditioned Reinforcement Learning

Lorenzo Steccanella[✉] and Anders Jonsson

Department of Information and Communication Technologies,
Universitat Pompeu Fabra, Barcelona, Spain
{lorenzo.steccanella,anders.jonsson}@upf.edu

**Abstract.** This paper presents a novel state representation for reward-free Markov decision processes. The idea is to learn, in a self-supervised manner, an embedding space where distances between pairs of embedded states correspond to the minimum number of actions needed to transition between them. Compared to previous methods, our approach does not require any domain knowledge, learning from offline and unlabeled data. We show how this representation can be leveraged to learn goal-conditioned policies, providing a notion of similarity between states and goals and a useful heuristic distance to guide planning and reinforcement learning algorithms. Finally, we empirically validate our method in classic control domains and multi-goal environments, demonstrating that our method can successfully learn representations in large and/or continuous domains.

**Keywords:** Representation learning · Goal-conditioned reinforcement learning · Reward shaping · Reinforcement learning

## 1 Introduction

In reinforcement learning, an agent attempts to learn useful behaviors through interaction with an unknown environment. By observing the outcome of actions, the agent has to learn from experience which action to select in each state in order to maximize the expected cumulative reward.

In many applications of reinforcement learning, it is useful to define a metric that measures the similarity of two states in the environment. Such a metric can be used, e.g., to define equivalence classes of states in order to accelerate learning, or to perform transfer learning in case the domain changes according to some parameters but retains part of the structure of the original domain. A metric can also be used as a heuristic in goal-conditioned reinforcement learning, in which the learning agent has to achieve different goals in the same environment. A goal-conditioned policy for action selection has to reason not only about the current

**Supplementary Information** The online version contains supplementary material available at https://doi.org/10.1007/978-3-031-26412-2_6.

state, but also on a known goal state that the agent should reach as quickly as possible.

In this work, we propose a novel algorithm for computing a metric that estimates the minimum distance between pairs of states in reinforcement learning. The idea is to compute an embedding of each state into a Euclidean space (see Fig. 1), and define a distance between pairs of states equivalent to the norm of their difference in the embedded space. We formulate the problem of computing the embedding as a constrained optimization problem, and relax the constraints by transforming them into a penalty term of the objective. An embedding that minimizes the objective can then be estimated via gradient descent.

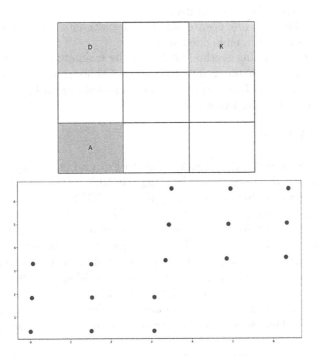

**Fig. 1.** Top: a simple gridworld where an agent has to pick up a key and open a door (key and door positions are fixed). Bottom: the learned state embedding $\phi$ on $\mathbb{R}^2$. The state is composed of the agent location and whether or not it holds the key.

The proposed metric can be used as a basis for goal-conditioned reinforcement learning, and has an advantage over other approaches such as generalized value functions. The domain of a generalized value function includes the goal state in addition to the current state, which intuitively increases the complexity of learning and hence the effort necessary to properly estimate a goal-conditioned policy. In contrast, the domain of the proposed embedding is just the state itself, and the distance metric is estimated by comparing pairs of embedded states.

In addition to the novel distance metric, we also propose a model-based approach to reinforcement learning in which we learn a transition model of actions directly in the embedded space. By estimating how the embedding will change after taking a certain action, we can predict whether a given action will take the agent closer to or further from a given target state. We show how to use the transition model to plan directly in embedded space. As an alternative, we also show how to use the proposed distance metric as a heuristic in the form of reward shaping when learning to reach a particular goal state.

The contributions of this work can be summarized as follows:

1. We propose a self-supervised training scheme to learn a distance function by embedding the state space into a low-dimensional Euclidean space $\mathcal{R}^e$ where a chosen p-norm distance between embedded states approximates the minimum distance between the actual states.
2. Once an embedding has been computed, we estimate a transition model of the actions directly in embedded space.
3. We propose a planning method that uses the estimated transition model to select actions, and a potential-based reward shaping mechanism that uses the learned distance function to provide immediate reward to the agent in a reinforcement learning framework.

## 2    Related Work

Our work relies on self-supervised learning to learn an embedding space useful for goal-conditioned reinforcement learning (GCRL).

Goal-Conditioned Supervised Learning, or GCSL [4], learns a goal-conditioned policy using supervised learning. The algorithm iteratively samples a goal from a given distribution, collects a suboptimal trajectory for reaching the goal, relabels the trajectory to add expert tuples to the dataset, and performs supervised learning on the dataset to update the policy via maximum likelihood.

Similar to our work, Dadashi et al. [3] learn embeddings and define a pseudo-metric between two states as the Euclidean distance between their embeddings. Unlike our work, an embedding is computed both for the state-action space and the state space. The embeddings are trained using loss functions inspired by bisimulation.

Tian et al. [12] also learn a predictive model and a distance function from a given dataset. However, unlike our work, the predictive model is learned for the original state space rather than the embedded space, and the distance function is in the form of a universal value function that takes the goal state as input in addition to the current state-action pair. Moreover, in their work they use "negative" goals assuming extra domain knowledge in the form of proprioceptive state information from the agent (e.g. robot joint angles). Schaul et al. [10] also learn universal value functions by factoring them into two components $\phi : S \to \mathbb{R}$ and $\varphi : G \to \mathbb{R}$, where $G$ is the set of goal states. For a more comprehensive survey of goal-conditioned reinforcement learning, we refer to Liu et al. [6].

## 3    Background

In this section we introduce necessary background knowledge and notation.

## 3.1   Markov Decision Processes

A Markov decision process (MDP) [9] is a tuple $\mathcal{M} = \langle S, A, P, r \rangle$, where $S$, $A$ denote the state space and action space, $P : S \times A \to \Delta(S)$ is a transition kernel and $r : S \times A \to \mathbb{R}$ is a reward function. At time $t$, the learning agent observes a state $s_t \in S$, takes an action $a_t \in A$, obtains a reward $r_t$ with expected value $\mathbb{E}[r_t] = r(s_t, a_t)$, and transitions to a new state $s_{t+1} \sim P(\cdot|s_t, a_t)$.

A stochastic policy $\pi : S \to \Delta(A)$ is a mapping from states to probability distributions over actions. The aim of reinforcement learning is to compute a policy $\pi$ that maximizes some notion of expected future reward.

In this work, we consider the discounted reward criterion, for which the expected future reward of a policy $\pi$ can be represented using a value function $V^\pi$, defined for each state $s \in S$ as

$$V^\pi(s) = \mathbb{E}\left[ \sum_{t=1}^{\infty} \gamma^{t-1} r(S_t, A_t) \,\middle|\, S_1 = s \right].$$

Here, random variables $S_t$ and $A_t$ model the state and action at time $t$, respectively, and the expectation is over the action $A_t \sim \pi(\cdot|S_t)$ and next state $S_{t+1} \sim P(\cdot|S_t, A_t)$. The discount factor $\gamma \in (0, 1]$ is used to control the relative importance of future rewards, and to ensure $V^\pi$ is bounded.

As an alternative to the value function $V^\pi$, one can instead model expected future reward using an action-value function $Q^\pi$, defined for each state-action pair $(s, a) \in S \times A$ as

$$Q^\pi(s, a) = \mathbb{E}\left[ \sum_{t=1}^{\infty} \gamma^{t-1} r(S_t, A_t) \,\middle|\, S_1 = s, A_1 = a \right].$$

The value function $V^\pi$ and action-value function $Q^\pi$ are related through the well-known Bellman equations:

$$V^\pi(s) = \sum_{a \in A} \pi(a|s) Q^\pi(s, a),$$
$$Q^\pi(s, a) = r(s, a) + \gamma \sum_{s' \in S} P(s'|s, a) V^\pi(s').$$

The aim of learning is to find an optimal policy $\pi^*$ that maximizes the value in each state, i.e. $\pi^*(s) = \arg\max_\pi V^\pi$. The optimal value function $V^*$ and action-value function $Q^*$ satisfy the Bellman optimality equations:

$$V^*(s) = \max_{a \in A} Q^*(s, a),$$
$$Q^*(s, a) = r(s, a) + \gamma \sum_{s' \in S} P(s'|s, a) V^*(s').$$

## 3.2   Goal-Conditioned Reinforcement Learning

Standard RL only requires the agent to complete one task defined by the reward function. In Goal-Conditioned Reinforcement Learning (GCRL) the observation is augmented with an additional goal that the agent is require to achieve when taking a decision in an episode [2, 10]. GCRL augments the MDP tuple $\mathcal{M}$ with a set of goal states and a desired goal distribution $\mathcal{M}_G = \langle S, G, p_g, A, P, r \rangle$, where G is a subset of the state space $G \subseteq S$, $p_g$ is the goal distribution and the reward function $r : S \times A \times G \rightarrow \mathbb{R}$ is defined on goals G. Therefore the objective of GCRL is to reach goal states via a goal-conditioned policy $\pi : S \times G \rightarrow \Delta(A)$ that maximizes the expectation of the cumulative return over the goal distribution.

**Self-Imitation Learning.** When we consider the goal space to be equal to the state space $G = S$ we can treat any trajectory $t = \{s_0, a_0, ..., a_{n-1}, s_n\}$ and any sub-trajectory $t_{i,j} \in t$, as a successful trial for reaching their final states. Goal Conditioned Supervised Learning (GCSL) [4] iteratively performs behavioral cloning on sub-trajectories collected in a dataset $\mathcal{D}$ by learning a policy $\pi$ conditioned on both the goal and the number of timesteps to reach the goal h.

$$J(\pi) = \mathbb{E}_{\mathcal{D}}[log\pi(a \mid s, g, h)].$$

### 3.3   Reward Shaping

An important challenge in reinforcement learning is solving domains with sparse rewards, i.e. when the immediate reward signal is almost always zero.

Reward Shaping attempts to solve this issue by augmenting a sparse reward signal r with a reward shaping function F, $\bar{r} = r + F$. Based on this idea, Ng et al. [8] proposed Potential-based reward shaping (PBRS) as an approach to guarantee policy invariance while reshaping the environment reward r. If the reward is constructed from a potential function, policy invariance guarantees to unalter the optimal policy. Formally PBRS defines F as:

$$F = \gamma \Phi(s') - \Phi(s)$$

where $\Phi : S \rightarrow \mathcal{R}$ is a real-valued potential function.

## 4   Contribution

In this section we present our main contribution, a method for learning a state representation of an MDP that can be leveraged to learn goal-conditioned policies. We first introduce notation that will be used throughout, then present our method for learning an embedding, and finally show how to integrate the embedding in algorithms for planning and learning.

We first define the Minimum Action Distance (MAD) $d_{MAD}(s, s')$ as the minimum number of actions necessary to transition from state s to state s'.

**Definition 1.** *(Minimum Action Distance) Let $T(s' \mid \pi, s)$ be the random variable denoting the first time step in which state s' is reached in the MDP when starting from state s and following policy $\pi$. Then $d_{MAD}(s, s')$ is defined as:*

$$d_{MAD}(s, s') := \min_{\pi} min \left[ T(s' \mid \pi, s) \right].$$

The Minimum Action Distance between states is a priori unknown, and is not directly observable in continuous and/or noisy state spaces where we cannot simply enumerate the states and keep statistics about the MAD metric. Instead, we will approximate an upper bound using the distances between states observed on trajectories. We introduce the notion of Trajectory Distance (TD) as follows:

**Definition 2.** *(Trajectory Distance) Given any trajectory* $t = s_0, ..., s_n \sim \mathcal{M}$ *collected in an MDP* $\mathcal{M}$ *and given any pair of states along the trajectory* $(s_i, s_j) \in t$ *such that* $0 \leq i \leq j \leq n$*, we define* $d_{TD}(s_i, s_j \mid t)$ *as*

$$d_{TD}(s_i, s_j \mid t) = (j - i),$$

*i.e. the number of decision steps required to reach* $s_j$ *from* $s_i$ *on trajectory* $t$*.*

### 4.1   State Representation Learning

Our goal is to learn a parametric state embedding $\phi_\theta : S \to \mathcal{R}^e$ such that the distance $d$ between any pair of embedded states approximates the Minimum Action Distance from state $s$ to state $s'$ or vice versa.

$$d(\phi_\theta(s), \phi_\theta(s')) \approx min(d_{MAD}(s, s'), d_{MAD}(s', s)). \tag{1}$$

We favour symmetric embeddings since it allows us to use norms as distance functions, e.g. the L1 norm $d(z, y) = ||z - y||_1$. Later we discuss possible ways to extend our work to asymmetric distance functions.

To learn the embedding $\phi_\theta$, we start by observing that given any state trajectory $t = \{s_0, ..., s_n\}$, choosing any pair of states $(s_i, s_j) \in t$ with $0 \leq i \leq j \leq n$, their distance along the trajectory represents an upper bound of the MAD.

$$d_{MAD}(s_i, s_j) \leq d_{TD}(s_i, s_j \mid t). \tag{2}$$

Inequality (2) holds for any trajectory sampled by any policy and allows to estimate the state embedding $\phi_\theta$ offline from a dataset of collected trajectories $\mathcal{T} = \{t_1, ..., t_n\}$. We formulate the problem of learning this embedding as a constrained optimization problem:

$$\min_\theta \sum_{t \in \mathcal{T}} \sum_{(s,s') \in t} (||\phi_\theta(s) - \phi_\theta(s')||_l - d_{TD}(s, s' \mid t))^2,$$
$$\text{s.t.} \quad ||\phi_\theta(s) - \phi_\theta(s')||_l \leq d_{TD}(s, s' \mid t) \quad \forall t \in \mathcal{T}, \forall (s, s') \in t. \tag{3}$$

Intuitively, the objective is to make the embedded distance between pairs of states as close as possible to the observed trajectory distance, while respecting the upper bound constraints. Without constrains, the objective is minimized when the embedding matches the expected Trajectory Distance $\mathbb{E}[d_{TD}]$ between all pairs of states observed on trajectories in the dataset $\mathcal{T}$. In contrast, constraining the solution to match the minimum TD with the upper-bound constrains $||\phi_\theta(s) - \phi_\theta(s')||_l \leq d_{TD}(s, s' \mid t)$ allows us to approximate the MAD.

Evidently, the precision of this approximation depends to the quality of the given trajectories.

To make the constrained optimization problem tractable, we relax the hard constrains in (3) and convert them into a penalty term in order to retrieve a simple unconstrained formulation that is solvable with gradient descent and fits within the optimization scheme of neural networks.

$$\min_\theta \ \sum_{t \in \mathcal{T}} \sum_{(s,s') \in t} \left[ \frac{1}{d_{TD}(s,s' \mid t)^2} (\|\phi_\theta(s) - \phi_\theta(s')\|_l - d_{TD}(s,s' \mid t))^2 \right] + C,$$

(4)

where $C$ is our penalty term defined as

$$C = \sum_{t \in \mathcal{T}} \sum_{(s,s') \in t} \left[ \frac{1}{d_{TD}(s,s' \mid t)^2} \max \left(0, \|\phi_\theta(s) - \phi_\theta(s')\|_l - d_{TD}(s,s' \mid t) \right)^2 \right].$$

The penalty term $C$ introduce a quadratic penalization of the objective for violating the upper-bound constraints $\|\phi_\theta(s) - \phi_\theta(s')\|_l <= d_{TD}(s,s' \mid t)$, while the term $\frac{1}{d_{TD}(s,s'|t)^2}$ normalizes each sample loss to be in the range $[0,1]$. The normalizing term also has the effect of prioritizing pairs of states that are close together on a trajectory, while giving less weight to pairs of states that are further apart. Intuitively, this makes sense since there is more uncertainty regarding the MAD of pairs of states that are further apart on a trajectory.

## 4.2   Learning Transition Models

In the previous section we showed how to learn a state representation that encodes a distance metric between states. This distance allows us to identify states $s_t$ that are close to a given goal state, i.e. $d(\phi_\theta(s_t), \phi_\theta(s_{goal})) < \epsilon$, or to measure how far we are from the goal state, i.e. $d(\phi_\theta(s_t), \phi_\theta(s_{goal}))$. However, on its own, the distance metric does not directly give us a policy for reaching the desired goal state.

In this section we propose a method to learn a transition model of actions, that combined with our state representation allows us to plan directly in the embedded space and derive policies to reach any given goal state. Given a dataset of trajectories $\mathcal{T}$ and a state embedding $\phi_\theta(s)$, we seek a parametric transition model $\rho_\zeta(\phi_\theta(s), a)$ such that for any triple $(s, a, s') \in \mathcal{T}$, $\rho_\zeta(\phi_\theta(s), a) \approx \phi_\theta(s')$.

We propose to learn this model simply by minimizing the squared error as

$$\min_\zeta \ \sum_t^{\mathcal{T}} \sum_{s,a,s'}^t \left[ (\rho_\zeta(\phi_\theta(s), a) - \phi_\theta(s'))^2 \right].$$

(5)

Note that in this minimization problem, the parameters $\theta$ of our state representation are fixed, since they are considered known and are thus not optimized at this stage.

## 4.3   Latent Space Planning

The functions $\rho_\zeta$ and $\phi_\theta$ together represent an approximate model of the underlying MDP.

We propose a Model Predictive Control algorithm that we call Plan-Dist, which computes a policy to reach a given desired goal state $s_{goal} \in S$ by unrolling trajectories for a fixed horizon $H$ in the embedded space. Plan-Dist uses the negative distance between the actual state $s_t$ and the goal state $s_{goal}$ as the desired reward function to be maximized, i.e. $r(s) = -d(\phi_\theta(s_t), \phi_\theta(s_{goal}))$. Our algorithm considers discrete action spaces and discretizes the action space otherwise. Plan-Dist samples a number $N$ of action trajectories $T_{N,H}$ from the set of all possible action sequences of length $H$, $T_{N,H} \subset A_H$. The trajectories are then unrolled recursively in the latent space starting from our actual state $s_t$ and using the transition model $\phi_\theta(s_{t+1}) \approx \rho_\zeta(\phi_\theta(s_t), a_t)$. At time step $t$, the first action of the trajectory that minimizes the distance to the goal is performed and this process is repeated at each time step until a terminal state is reached (cf. Algorithm 1).

---

**Algorithm 1.** PLAN-DIST

---

1: **Input**: environment $e$, state embedding $\phi_\theta$, transition model $\rho_\zeta$, horizon $H$,
    number $N$ of trajectories to evaluate
2: $s \leftarrow initial\,state$
3: $s_{goal} \leftarrow goal\,state$
4: $z_{goal} \leftarrow \phi_\theta(s_{goal})$
5: **while** within budget **do**
6:     $T_{N,H} \leftarrow$ sample $N$ action sequences of length $H$
7:     $t_{MaxReward} \leftarrow None$
8:     $r_{max} \leftarrow MinReward$
9:     **for** $t_a \in T_{N,H}$ **do**
10:         $z = \phi_\theta(s)$
11:         $r = r - d(z, z_{goal})$
12:         **for** $a_t \in t_a$ **do**
13:             $z_{t+1} = \rho_\zeta(z, a_t)$
14:             $r = r - d(z_{t+1}, z_{goal})$
15:         **end for**
16:         **if** $r > r_{max}$ **then**
17:             $r_{max} = r$
18:             $t_{MaxReward} \leftarrow t_a$
19:         **end if**
20:     **end for**
21:     $s' \leftarrow$ apply action $t_{MaxReward}[0]$ in state $s$
22:     $s = s'$
23: **end while**

---

## 4.4   Reward Shaping

Our last contribution is to show how to combine prior knowledge in the form of goal states and our learned distance function to guide existing reinforcement learning algorithms.

We assume that a goal state is given and we augment the environment reward $r(s, a)$ observed by the reinforcement learning agent with Potential-based Reward Shaping [8] of the form:

$$\bar{r}(s, a) = r(s, a) + F(s, \gamma, s'), \tag{6}$$

where $F$ is our potential-based reward:

$$F(s, \gamma, s') = -\gamma d(\phi_\theta(s'), \phi_\theta(s_{goal})) + d(\phi_\theta(s), \phi_\theta(s_{goal})).$$

Here, $d(\phi_\theta(\cdot), \phi_\theta(s_{goal}))$ represents our estimated Minimum Action Distance to the goal $s_{goal}$. Note that for a fixed goal state $s_{goal}$, $-d(\phi_\theta(\cdot), \phi_\theta(s_{goal}))$ is a real-valued function of states which is maximized when $d = 0$.

Intuitively our reward shaping schema is forcing the agent to reach the goal state as soon as possible while maximizing the environment reward $r(s, a)$. By using potential-based reward shaping $F(s, \gamma, s')$ we are ensuring that the optimal policy will be invariant [8].

## 5   Experimental Results

In this section we present results from experiments where we learn a state embedding and transition model offline from a given dataset of trajectories[1] We then use the learned models to perform experiments in two settings:

1. Offline goal-conditioned policy learning: Here we evaluate the performance of our Plan-Dist algorithm against GCSL [4].
2. Reward Shaping: In this setting we use the learned MAD distance to reshape the reward of a DDQN [13] agent (DDQN-PR) for discrete action environments and DDPG [5] for continuos action environment (DDPG-PR), and we compare it to their original versions.

---

[1] The code to reproduce the experimental results is available at: https://github.com/lorenzosteccanella/SRL.

*Classic Control Environments*

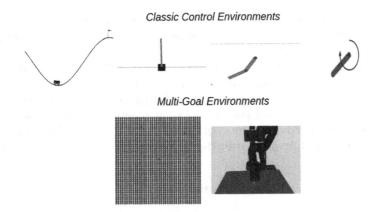

*Multi-Goal Environments*

**Fig. 2.** Evaluation Tasks. Top row: MountainCar-v0, CartPole-v0, AcroBot-v1 and Pendulum-v0. Bottom row: GridWorld and SawyerReachXYZEnv-v1.

### 5.1   Dataset Collection and Domain Description

We test our algorithms on the classic RL control suite (cf. Figure 2). Even though termination is often defined for a range of states, we fix a single goal state among the termination states. These domains have complex dynamics and random initial states, making it difficult to reach the goal state without dedicated exploration. The goal state selected for each domain is:

- MountainCar-v0: [0.50427865, 0.02712902]
- CartPole-v0: [0, 0, 0, 0]
- AcroBot-v1: [-0.9661, 0.2581, 0.8875, 0.4607, -1.8354, -5.0000]
- Pendulum-v0: [1, 0, 0]

Additionally, we test our model-based algorithm Plan-Dist in two multi-goal domains(see. Figure 2):

- A $40 \times 40$ GridWorld.
- The multiworld domain SawyerReachXYZEnv-v1, where a multi-jointed robotic arm has to reach a given goal position.

In each episode, a new goal $s_{goal}$ is sampled at random, so the set of possible goal states $G$ equals the entire state space $S$. These domains are challenging for reinforcement learning algorithms, and even previous work on goal-conditioned reinforcement learning usually considers a small fixed subset of goal states.

In each of these domains we collect a dataset that approximately covers the state space, since we want to be able to use any state as a goal state. Collecting these datasets is not trivial. As an example, consider the MountainCar domain where a car is on a one-dimensional track, positioned between two mountains. A simple random trajectory will not be enough to cover all the state space since it will get stuck in the valley without being able to move the cart on top of the mountains. Every domain in the classic control suite presents this exploration difficulty and for these environments we rely on collecting trajectories performed by the algorithms DDQN [13] and DDPG [5] while learning a policy for these domains. Note that we use DDPG only in the Pendulum domain, which is characterized by a continuous action space.

In Table 1 we report the size, the algorithm/policy used to collect the trajectories, the average reward and the maximum reward of each dataset. Note that the average reward is far from optimal and that both Plan-Dist (our offline algorithm) and GCSL improve over the dataset performance (cf. Figure 3).

**Table 1.** Dataset description.

| Environments | Trajectories Dataset | Algorithm to Collect Trajectories | Avg Reward Dataset | Max Reward Dataset |
|---|---|---|---|---|
| MountainCar-v0 | 100 | DDQN | $-164.26$ | $-112$ |
| CartPole-v0 | 200 | DDQN | $+89.42$ | $+172$ |
| AcroBot-v1 | 100 | DDQN | $-158.28$ | -92.0 |
| Pendulum-v0 | 100 | DDPG | $-1380.39$ | -564.90 |
| GridWorld | 100 | RandomPolicy | – | – |
| SawyerReach-XYZEnv-v1 | 100 | RandomPolicy | – | – |

## 5.2   Learning a State Embedding

The first step of our procedure consists in learning a state embedding $\phi_\theta$ from a given dataset of trajectories $\mathcal{T}$. From each trajectory $t_i = \{s_0, ..., s_n\} \in \mathcal{T}$ we collect all samples $(s_{i|t_i}, s_{j|t_i}, d_{TD}(s_{i|t_i}, s_{j|t_i} \mid t_i))$, $0 \leq i \leq j \leq n$, and populate a Prioritized Experience Replay (PER) memory [11]. We use PER to prioritize the samples based on how much they violate our penalty function in (4).

We used mini-batches $B$ of size 512 with the AdamW optimizer [7] and a learning rate of $5 * 10^{-4}$ for 100,000 steps to train a neural network $\phi_\theta$ by minimizing the following loss derived from (4):

$$\mathcal{L}(\mathcal{B}) = \sum_{(s,s',d_{TD}) \in B} \left[ \frac{1}{d_{TD}^2} (\|\phi_\theta(s) - \phi_\theta(s')\|_1 - d_{TD})^2 \right] + C,$$

where $C$ is the penalty term defined as:

$$C = \sum_{(s,s',d_{TD}) \in B} \left[ \frac{1}{d_{TD}^2} \max(0, \|\phi_\theta(s) - \phi_\theta(s')\|_1 - d_{TD})^2 \right]$$

We use an embedding dimension of size 64 with an L1 norm as the metric to approximate the MAD distance. Empirically, the L1 norm turns out to perform better than the L2 norm in high-dimensional embedding spaces. These findings are in accordance with theory [1].

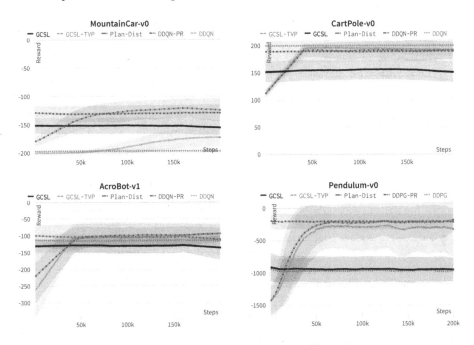

**Fig. 3.** Results in the classic RL control suite.

## 5.3  Learning Dynamics

We use the same dataset of trajectories $\mathcal{T}$ to learn a transition model. We collect all the samples $(s, a, s')$ in a dataset $\mathcal{D}$ and train a neural network $\rho_\zeta$ using mini-batches $\mathcal{B}$ of size 512 with the AdamW optimizer [7] and a learning rate of $5 * 10^{-4}$ for 10,000 steps by mimizing the following loss derived from (5):

$$\mathcal{L}(\mathcal{B}) = \sum_{s, s', d_{TD}}^{B} \left[ (\rho_\zeta(\phi_\theta(s), a) - \phi_\theta(s'))^2 \right]$$

## 5.4  Experiments

We compare our algorithm Plan-Dist against an offline variant of GCSL, where GCSL is trained from the same dataset of trajectories as our models $\phi_\theta$ and $\rho_\zeta$. The GCSL policy and the models $\phi_\theta$ and $\rho_\zeta$ are all learned offline and frozen at test time.

Ghosh et al. [4] propose two variants of the GCSL algorithm, a Time-Varying Policy where the policy is conditioned on the remaining horizon $\pi(a|s, g, h)$ (in our experiments we refer to this as GCSL-TVP) and a horizon-less policy $\pi(a|s, g)$ (we refer to this as GCSL).

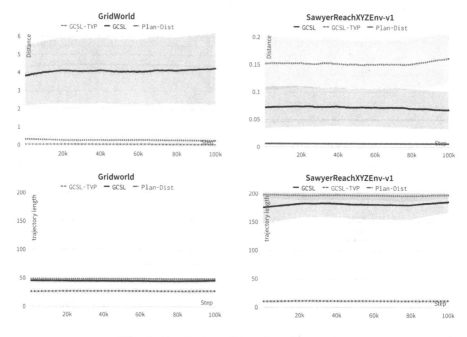

**Fig. 4.** Results in multi-goal environments.

We refer to our reward shaping algorithms as DDQN-PR/DDPG-PR and their original counterpart without reward shaping as DDQN/DDPG. DDQN is used in domains in which the action space is discrete, while DDPG is used for continuous action domains.

For all the experiments we report results averaged over 10 seeds where the shaded area represents the standard deviation and the results are smoothed using an average window of length 100. All the hyper-parameters used for each algorithm are reported in the appendix.

In the multi-goal environments in Fig. 4 we report two metrics: the distance to the goal with respect to the state reached at the end of the episode, and the length of the performed trajectory. In both domains, the episode terminates either when we reach the goal state or when we reach the maximum number of steps (50 steps for GridWorld, and 200 steps for SawyerReachXYZEnv-v1). We evaluate the algorithms for 100,000 environment steps.

We can observe that Plan-Dist is able to outperform GCSL, being able to reach the desired goal state with better precision and by using shorter paths. We do not compare to reinforcement learning algorithms in these domains since they struggle to generalize when the goal changes so frequently.

On the classic RL control suite in Fig. 3 we report the results showing the total reward achieved at the end of each episode. Here we compare both goal-conditioned algorithms and state-of-the-art reinforcement learning algorithms for 200,000 environment steps. Plan-Dist is still able to outperform GCSL in almost all domains, while performing slightly worse than GCSL-TVP in CartPole-v0. Compared to DDQN-PR/DDPG-PR, Plan-Dist is able to reach similar total reward, but in MountainCar-v0, DDQN-PR is eventually able to achieve higher reward.

The reward shaping mechanism of DDQN-PR/DDPG-PR is not helping in the domains CartPole-v0, Pendulum-v0 and Acrobot-v0. In these domains, it is hard to define a single state as the goal to reach in each episode. As an example, in CartPole-v0 we defined the state $[0, 0, 0, 0]$ as our goal state and we reshape the reward accordingly, but this is not in line with the environment reward that instead cares only about balancing the pole regardless of the position of the cart. While in these domains we do not observe an improvement in performance, it is worth noticing that our reward shaping scheme is not adversely affecting DDQN-PR/DDPG-PR, and they are able to achieve results that are similar to those of their original counterparts.

Conversely, in MountainCar-v0 where the environment reward resembles a goal reaching objective, since the goal is to reach the peak of the mountain as fast as possible, our reward shaping scheme is aligned with the environment objective and DDQN-PR outperforms DDQN in terms of learning speed and total reward on the fixed evaluation time of 200,000 steps.

## 6    Discussion and Future Work

We propose a novel method for learning a parametric state embedding $\phi_\theta$ where the distance between any pair of states $(s, s')$ in embedded space approximates the Minimum Action Distance, $d(\phi_\theta(s), \phi_\theta(s')) \approx d_{MAD}(s, s')$. One limitation of our approach is that we consider symmetric distance functions, while in general the MAD in an MDP could be asymmetric, $d_{MAD}(s, s') \neq d_{MAD}(s', s)$. Schaul et al. [10] raise a similar issue in the context of learning Universal Value Functions, and propose an asymmetric distance function on the following form:

$$d_A(s, s') = \|\sigma(\psi_1(s'))(\phi(s) - \psi_2(s'))\|_l,$$

where $\sigma$ is a the logistic function and $\psi_1$ and $\psi_2$ are two halves of the same embedding vector. In their work they show similar performance using the symmetric and asymmetric distance functions. Still, an interesting future direction would be to use this asymmetric distance function in the context of our self-supervised training scheme.

While our work focuses on estimating the MAD between states and empirically shows the utility of the resulting metric for goal-conditioned reinforcement learning, the distance measure could be uninformative in highly stochastic environment where the expected shortest path distance better measures the distance between states. One possible way to approximate this measure using our self-supervised training scheme would be to minimize a weighted version of our objective in (3):

$$\min_{\theta} \sum_{t \in \mathcal{T}} \sum_{(s,s') \in t} 1/d_{TD}^{\alpha}(\|\phi_\theta(s) - \phi_\theta(s')\|_l - d_{TD}(s, s' \mid t))^2.$$

Here, the term $1/d_{TD}$ is exponentiated by a factor $\alpha$ which decides whether to favour the regression over shorter or longer Trajectory Distances. Concretely, when $\alpha < 1$ we favour the regression over shorter Trajectory Distances, approximating a Shortest Path Distance.

In our work we learn a distance function offline from a given dataset of trajectories, and one possible line of future research would be to collect trajectories while simultaneously exploring the environment in order to learn the distance function.

In this work we focus on single goal reaching tasks, in order to have a fair comparison with goal-conditioned reinforcement learning agents in the literature. However, the use of our learned distance function is not limited to this setting and we can consider multi-goal tasks, such as reaching a goal while maximizing the distance to forbidden (obstacle) states, reaching the nearest of two goals, and in general any linear and non-linear combination of distances to states given as input.

Lastly, it would be interesting to use this work in the contest of Hierarchical Reinforcement Learning, in which a manager could suggest subgoals to our Plan-Dist algorithm.

**Acknowledgments.** Anders Jonsson is partially funded by the Spanish grant PID2019-108141GB-I00 and the European project TAILOR (H2020, GA 952215).

# References

1. Aggarwal, C.C., Hinneburg, A., Keim, D.A.: On the surprising behavior of distance metrics in high dimensional space. In: Van den Bussche, J., Vianu, V. (eds.) ICDT 2001. LNCS, vol. 1973, pp. 420–434. Springer, Heidelberg (2001). https://doi.org/10.1007/3-540-44503-X_27
2. Andrychowicz, M., et al.: Hindsight experience replay. Adv. Neural Inf. Process. Syst. 30 (2017)
3. Dadashi, R., Rezaeifar, S., Vieillard, N., Hussenot, L., Pietquin, O., Geist, M.: Offline reinforcement learning with pseudometric learning. In: Meila, M., Zhang, T. (eds.) Proceedings of the 38th International Conference on Machine Learning, ICML 2021, 18–24 July 2021, Virtual Event. Proceedings of Machine Learning Research, vol. 139, pp. 2307–2318. PMLR (2021)
4. Ghosh, D., et al.: Learning to reach goals via iterated supervised learning. In: International Conference on Learning Representations (2020)
5. Lillicrap, T.P., et al.: Continuous control with deep reinforcement learning. arXiv preprint arXiv:1509.02971 (2015)
6. Liu, M., Zhu, M., Zhang, W.: Goal-conditioned reinforcement learning: Problems and solutions (2022). https://doi.org/10.48550/ARXIV.2201.08299, https://arxiv.org/abs/2201.08299
7. Loshchilov, I., Hutter, F.: Decoupled weight decay regularization. arXiv preprint arXiv:1711.05101 (2017)

8. Ng, A.Y., Harada, D., Russell, S.: Policy invariance under reward transformations: theory and application to reward shaping. In: ICML, vol. 99, pp. 278–287 (1999)
9. Puterman, M.L.: Markov decision processes: discrete stochastic dynamic programming. John Wiley & Sons (2014)
10. Schaul, T., Horgan, D., Gregor, K., Silver, D.: Universal value function approximators. In: International Conference on Machine Learning, pp. 1312–1320. PMLR (2015)
11. Schaul, T., Quan, J., Antonoglou, I., Silver, D.: Prioritized experience replay. arXiv preprint arXiv:1511.05952 (2015)
12. Tian, S., et al.: Model-based visual planning with self-supervised functional distances. In: International Conference on Learning Representations (2021)
13. Van Hasselt, H., Guez, A., Silver, D.: Deep reinforcement learning with double q-learning. In: Proceedings of the AAAI Conference on Artificial Intelligence, vol. 30 (2016)

# Bootstrap State Representation Using Style Transfer for Better Generalization in Deep Reinforcement Learning

Md Masudur Rahman[✉] and Yexiang Xue

Department of Computer Science, Purdue University, West Lafayette, IN 47907, USA
{rahman64,yexiang}@purdue.edu

**Abstract.** Deep Reinforcement Learning (RL) agents often overfit the training environment, leading to poor generalization performance. In this paper, we propose Thinker, a bootstrapping method to remove adversarial effects of confounding features from the observation in an unsupervised way, and thus, it improves RL agents' generalization. Thinker first clusters experience trajectories into several clusters. These trajectories are then bootstrapped by applying a style transfer generator, which translates the trajectories from one cluster's style to another while maintaining the content of the observations. The bootstrapped trajectories are then used for policy learning. Thinker has wide applicability among many RL settings. Experimental results reveal that Thinker leads to better generalization capability in the Procgen benchmark environments compared to base algorithms and several data augmentation techniques.

**Keywords:** Deep reinforcement learning · Generalization in reinforcement learning

## 1 Introduction

Deep reinforcement learning has achieved tremendous success. However, deep neural networks often overfit to confounding features in the training data due to their high flexibility, leading to poor generalization [6,7,14,33]. These confounding features (e.g., background color) are usually not connected to the reward; thus, an optimal agent should avoid focusing on them during the policy learning. Even worse, confounding features lead to incorrect state representations, which prevents deep RL agents from performing well even in slightly different environments.

Many approaches have been proposed to address this challenges including data augmentation approaches such as random cropping, adding jitter in image-based observation [7,20–22,27], random noise injection [17], network randomization [4,23,25], and regularization [7,17,20,32] have shown to improve generalization. The common theme of these approaches is to increase diversity in the training data so as the learned policy would better generalize. However, this perturbation is primarily done in isolation of the task semantic, which might change an essential aspect of the observation, resulting in sub-optimal policy learning.

M.-R. Amini et al. (Eds.): ECML PKDD 2022, LNAI 13716, pp. 100–115, 2023.
https://doi.org/10.1007/978-3-031-26412-2_7

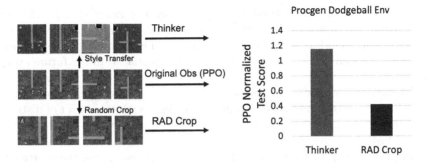

**Fig. 1.** Comparison between style transfer-based Thinker and random crop data augmentation-based RAD [21] agents on Procgen Dodgeball. [**Left**] We see that The random crop removes many essential aspects of the observation while the style transfer retains most game semantics and changes mainly the background and texture of objects. [**Right**] In generalization, the Thinker agent achieves better performance compared to PPO. In contrast, the RAD Crop agent significantly worsens the base PPO's performance.

Moreover, the random perturbation in various manipulations of observations such as cropping, blocking, or combining two random images from different environment levels might result in unrealistic observations that the agent will less likely observe during testing. Thus these techniques might work poorly in the setup where agents depend on realistic observation for policy learning.

For example, consider a RL maze environment where the agent takes the whole maze board image as input observation to learn a policy where the background color of maze varies in each episode. Thus, applying random cropping might hide essential part of the observation which eventually results in poor performance. Our proposed method tackle this issue by changing style of the observation (e.g., background color) while maintaining the maze board's semantic which eventually help RL agent to learn a better policy. It is also desirable to train the agent with realistic observations, which helps it understand the environments' semantics. Otherwise, the agent might learn unexpected and unsafe behaviors while entirely focusing on maximizing rewards even by exploiting flaws in environments such as imperfect reward design.

In this paper, we propose Thinker, a novel bootstrapping approach to remove the adversarial effects of confounding features from observations and boost the deep RL agent performance. Thinker automatically creates new training data via changing the visuals of the given observation. An RL agent then learns from various observation style instead of a single styled the original training data. Intuitively, this approach help the agent not to focus much on the style which assumed to be confounder and can change in future unseen environments. Compared to previous approaches, our proposed method focuses on transforming the visual style of observations realistically while keeping the semantics same. Thus, the transferred trajectories corresponds to those that possibly appear in testing environments, hence assisting the agent in adapting to the unseen scenarios. Design of our method is motivated by the *counterfactual thinking* nature of human - *"what if the background color of the image observation was Red instead of Blue?"*; thus the name is Thinker. This imagination-based thinking often beneficial for decision-making on similar scenarios in the future events [8, 28].

Our method uses a similar mechanism to disentangled confounding features. Figure 1 shows an overview of the Thinker module. It maintains a set of distributions (cluster of sample observations) of experience data, which can be learned using a clustering algorithm.

Our proposed approach consists of a style transfer-based observation translation method that considers content of the observation. Trajectory data from the agent's replay buffer is clustered into different categories, and then observation is translated from one cluster style to another cluster's style. Here the style is determined by the commonality of observation features in a cluster. Thus this style translation is targeted toward non-generalizable features. The agent should be robust toward changes of such features. Moreover, the translated trajectories correspond to those that possibly appear in testing environments, assisting the agent in adapting to unseen scenarios.

Thinker learns generators between each pair of clusters using adversarial loss [12] and cycle consistency loss [5, 35]. The generator can translate observations from one cluster to another; that means changing style to another cluster while maintaining the semantic of the observation in the underlying task. After training, all generators are available to the RL agent to use during its policy learning process.

During policy training, the agent can query the Thinker module with new observations and get back the translated observations. The agent can then use the translated observation for policy training. Here, the Thinker module bootstraps the observation data and tries to learn better state representation, which is invariant to the policy network's unseen environment. Intuitively, the observation translation process is similar to asking the counterfactual question; what if the new observation is coming from a different source (visually different distribution)?

Note that, Thinker works entirely in an unsupervised way and does not require any *additional* environment interactions. Thus the agent can learn policy without collecting more data in the environment, potentially improving sample efficiency and generalization in unseen environments.

We evaluated the effectiveness of Thinker module on Procgen [6] benchmark environments. We evaluated the usefulness of Thinker on the standard on-policy RL algorithm, Proximal Policy Optimization (PPO) [30]. We observe that Thinker often can successfully transfers style from one cluster to another, generating semantically equivalent observation data. Moreover, our agent performs better in generalization to unseen test environments than PPO. We further evaluate our method with two popularly used data augmentation approaches: random cropping and random cutout [21]. We demonstrate that these data augmentation method sometimes worsen the base PPO algorithm while our proposed approach improve the performance in both sample efficiency (Train Reward) and general-ization (Test Reward).

In summary, our contributions are listed as follows:

- We introduce Thinker, a bootstrapping method to remove adverse effects of confounding features from the observation in an unsupervised way.
- Thinker can be used with existing deep RL algorithms where experience trajectory is used for policy training. We provide an algorithm to leverage Thinker in different RL settings.
- We evaluate Thinker on Procgen environments where it often successfully translates the visual features of observations while keeping the game semantic intact. Overall, our Thinker agent performs better in sample efficiency and generalization than the base PPO [30] algorithm and two data augmentation-based approaches: random crop and random cutout [21].

The source code of our Thinker module is available at https://github.com/masud99r/thinker.

## 2   Background

**Markov Decision Process (MDP).** An MDP can be denoted as $\mathcal{M} = (\mathcal{S}, \mathcal{A}, \mathcal{P}, r)$ where $\mathcal{S}$ is a set states, $\mathcal{A}$ is a set of possible actions. At every timestep $t$, from an state $s_t \in \mathcal{S}$, the agent takes an action $a_t \in \mathcal{A}$ and the environment proceed to next state. The agent then receives a reward $r_t$ as the environment moves to a new state $s_{t+1} \in \mathcal{S}$ based on the transition probability $P(s_{t+1}|s_t, a_t)$.

**Reinforcement Learning.** In reinforcement learning, the agent interacts with the environment in discrete timesteps that can be defined as an MDP, denoted by $M = (\mathcal{S}, \mathcal{A}, \mathcal{P}, r)$, $\mathcal{P}$ is the transition probability between states after agent takes action, and $r$ is the immediate reward the agent gets. In practice, the state $(\mathcal{S})$ is unobserved, and the agent gets to see only a glimpse of the underlying system state in the form of observation $(\mathcal{O})$. The agent's target is to learn a policy $(\pi)$, which is a mapping from state to action, by maximizing collected rewards. In addition, to master skills in an environment, the agent needs to extract useful information from the observation, which helps take optimal actions. In deep reinforcement learning (RL), the neural network architecture is often used to represent the policy (value function, Q-function). In this paper, we use such a deep RL setup in image-based observation space.

**RL Agent Evaluation.** Traditionally, RL agent trained in an environment where it is evaluated how quickly it learns the policy. However, the evaluation is often done on the same environment setup. While this evaluation approach can measure policy learning efficiency, it critically misses whether the agent actually learned the necessary skill or just memorized some aspect of the environment to get the maximum reward in training. In this setup, the agent can often overfit to the scoreboard or timer in a game which can lead to the best reward; however, the agent can completely ignore other parts of the environment [31,34]. The agent can even memorize the training environment to achieve the best cumulative reward [34]. In contrast, in this paper, we use a zero-shot generalization [31] setup where the agent is trained and tested on different environment instances. Furthermore, the agent's performance is evaluated on unseen environment instances; thus, the agent must master skills during training to perform better in generalization.

**Generalization Issue in Deep RL.** The agent's goal is to use necessary information from the observation and learn behavior that maximizes reward. However, due to the lack of variability in observations, the agent might focus on spurious features. This problem becomes commonplace in RL training, especially if the observation space is large, such as the RGB image. In such cases, the agent might memorize the trajectory without actually learning the underlying task. This issue might be undetected if the agent trains and evaluates in the same environment. The agent trained in such a task (environment) might overfit to the trained environment and fail to generalize in the same task but with a slightly different environment. For example, background color might be irrelevant for a game, and the game might have different backgrounds at different episodes, but the game logic will remain the same. These unimportant features are the confounder that might mislead agents during training. The issue might be severe in deep reinforcement learning as the agent policy is often represented using high-capacity neural networks. If the agent focuses on these confounder features, it might overfit and fails to generalize.

**Style Transfer with Generative Adversarial Network.** The task of style transfer is to change particular features of a given image to another, where generative adversarial network (GAN) has achieved enormous success [5,18,19,35]. This setup often consists of images from two domains where models learn to style translate images from one domain to another. The shared features then define the style among images in a domain. A pairing between two domains images is necessary to make many translation methods work. However, such information is not available in the reinforcement learning setup. Nevertheless, an unpaired image-to-image translation method can be used, which does not require a one-to-one mapping of annotated images from two domains. In this paper, we leverage StarGAN [5] that efficiently learns mappings among various domains using a single generator and discriminator. In the RL setup, we apply a clustering approach which first separates the trajectory data into clusters. Then we train the StarGAN on these clusters that learn to style translate images among those clusters.

## 3   Bootstrap Observations with Thinker

Our proposed method, Thinker, focuses on removing the adverse confounding features, which helps the deep RL agent to learn invariant representations from the observations, which eventually help to learn generalizable policy. Figure 2 shows an overview of our method. Thinker maintains a set of distributions achieved by clustering observation data that come from the experience trajectory. We implemented our method on a high-dimensional RGB image observation space.

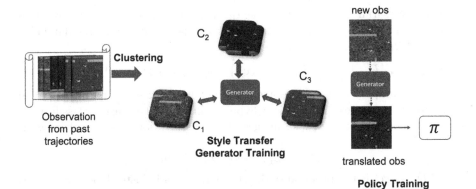

**Fig. 2.** Overview of style Thinker module. The experience trajectory observation data for the task environment are separated into different classes based on visual features using Gaussian Mixture Model (GMM) clustering algorithm. The task of the generator is then to translate the image from one class image to another classes images. In this case, the "style" is defined as the commonality among images in a single class. Given a new observation, it first infers into its (source) cluster using GMM, and then the generator translated it to (target) another cluster style. The target cluster is taken randomly from the rest of the cluster. The translated observations are used to train the policy.

**Clustering Trajectories.** The trajectory data is first clustered into several $(n)$ clusters. Though any clustering algorithms can be leveraged for this clustering process, in this paper, we describe a particular implementation of our method, where we use the Gaussian Mixture Model (GMM) for clustering, and ResNet [15] for feature extraction and dimension reduction. Furthermore, this clustering process focuses entirely on the visual aspect of the observation without necessarily concentrating on the corresponding reward structure. Therefore, images would be clustered based on these visual characteristics. In the next step, the observation dataset is clustered using the GMM algorithm. Images in these clusters' are then used to carry out style transfer training.

**Generator Training.** We train a single generator $G$ to translate image from one cluster to another. We build on the generator on previous works [5, 35] which is a unified framework for a multi-domain image to image translation. Given an input image $x$ from a source cluster the output translated image $x'$ conditioned on the target cluster number $c$, that is $x' \leftarrow G(x, c)$, where $c$ is a randomly chosen cluster number. A discriminator $D : x \rightarrow \{D_{src}(x), D_{cls}(x)\}$ is used to distinguish real image and fake image generated by $G$. Here $D_{src}$ distinguish between fake and real images of the source, and $D_{cls}$ determines the cluster number of the given input image $x$. Generator G tries to fool discriminator $D$ in an adversarial setting by generating a realistic image represented by the true image distribution.

The adversarial loss is calculated as the Wasserstein GAN [3] objective with gradient penalty [13] which stabilize the training compared to regular GAN objective [12]. This loss is defined as

$$\mathcal{L}_{adv} = \mathbb{E}_x[D_{src}] - \mathbb{E}_{x,c}[D_{src}(G(x,c))] - \lambda_{gp}\mathbb{E}_{\hat{x}}[(||\nabla_{\hat{x}}D_{src}(\hat{x})|| - 1)^2], \quad (1)$$

where $\hat{x}$ is sampled uniformly along a straight line between a pair of real and generated fake images and $\lambda_{gp}$ is a hyperparameter. The cluster classification loss is defined for real and fake images. The classification loss of real image is defined as

$$\mathcal{L}_{cls}^r = \mathbb{E}_{x,c'}[-\log D_{cls}(c'|x)], \quad (2)$$

where $D_{cls}(c'|x)$ is the probability distribution over all cluster labels. Similarly, the classification loss of fake generated image is defined as

$$\mathcal{L}_{cls}^f = \mathbb{E}_{x,c}[-\log D_{cls}(c|G(x,c))], \quad (3)$$

The full discriminator loss is

$$\mathcal{L}_D = -\mathcal{L}_{adv} + \lambda_{cls}\mathcal{L}_{cls}^r, \quad (4)$$

which consists of the adversarial loss $\mathcal{L}_{adv}$, and domain classification loss $\mathcal{L}_{cls}^r$ and $\lambda_{cls}$ is a hyperparameter. The discriminator detects a fake image generated by the generator G from the real image in the given class data.

To preserve image content during translation a reconstruction loss is applied

$$\mathcal{L}_{rec} = \mathbb{E}_{x,c,c'}[||x - G(G(x,c),c')||_1], \quad (5)$$

where we use the $L1$ norm.

The $\mathcal{L}_{rec}$ is the reconstruction loss which makes sure the generator preserves the content of the input images while changing the domain-related part of the inputs. This cycle consistency loss [5,35] $\mathcal{L}_{rec}$ makes sure the translated input can be translated back to the original input, thus only changing the domain related part and not the semantic. Thus, the generator loss is

$$\mathcal{L}_G = \mathcal{L}_{adv} + \lambda_{cls}\mathcal{L}_{cls}^f + \lambda_{rec}\mathcal{L}_{rec}, \quad (6)$$

where $\mathcal{L}_{adv}$ is adversarial loss, and $\mathcal{L}_{cls}^f$ is the loss of detecting fake image and the $\lambda_{rec}$ is a hyperparameter.

**Train Agent with Thinker.** During policy training, the agent can query the generator module with a new observation and get back translated observation (Algorithm 1). The agent can then use the translated observation for policy training. Intuitively, the observation translation process is similar to asking the counterfactual question; *"what if the new observation is coming from a visually different episode distribution)?"* The Thinker method can be applied to existing deep RL algorithms where experience data is used to train policy networks. In this paper, we evaluate Thinker with on-policy PPO [30]. Intuitively, Thinker maintains a counterfactual-based visual thinking component, which it can invoke at any learning timestep and translate the observation from one distribution to another. Algorithm 1 describes detailed steps of training deep RL agents with Thinker.

---

**Algorithm 1.** Thinker

---

Get PPO for policy learning RL agent

Collect observation trajectory $\mathcal{D}$ using initial policy

Cluster dataset $\mathcal{D}$ into $n$ clusters using GMM

Train Generator $G$ with the $n$ clusters by optimizing equation 4, and 6

**for** each iteration **do**

    **for** each environment step **do**

        $a_t \sim \pi_\theta(a_t|x_t)$

        $x_{t+1} \sim P(x_{t+1}|x_t, a_t)$

        $r_t \sim R(x_t, a_t)$

        $\mathcal{B} \leftarrow \mathcal{B} \cup \{(x_t, a_t, r_t, x_{t+1})\}$

        Translate all obs $x \in \mathcal{B}$ to get $\mathcal{B}'$ using Generator $G$

        Train policy $\pi_\theta$ on $\mathcal{B}'$ with PPO

    **end for**

**end for**

---

# 4 Experiments

## 4.1 Implementation

We implemented Thinker using Ray framework: Tune, and RLlib [24], which supports simple primitive and unified API to build scalable applications.

**Clustering.** We use Gaussian Mixture Model (GMM) implementation available in Scikit-learn [26]. We first pass observation through the pre-trained ResNet18 [15] model which is trained on ImageNet dataset [29]. The ResNet18 model[1] converts the RGB image into a 1000 dimensional vector. We use the layer just before the final softmax layer to get this vector. This dimensionality reduction step drastically reduces the training and inference time of the Gaussian mixture model. Given the number of cluster $n$, this model train on $n$ clusters. These $n$ clusters data are stored by the agent, which is later used for the generator training. The inference module takes input an observation, and it returns the cluster-ID to which it belongs, which is used to identify the target cluster for the style translation. The number of clusters is the hyperparameter, which can depend on the diversity of environment levels. However, for our method to be effective, at least two clusters are required. Therefore, unless otherwise mentioned, we reported comparison results using the number of clusters $n = 3$ (better performing in ablation).

**Learning Generator.** After clustering, all data is then feed to the generator module which learns a single generator that style translate between any pair of clusters. The agent can choose various cluster numbers (hyperparameter) during training time. Each time the clustering is trained, the generator must be updated with the new cluster samples. In our experiments, we train the generator once and at the beginning of the training. Note that the collected trajectory data

---

[1] https://pytorch.org/hub/pytorch_vision_resnet/.

should be representative enough to train a good generator. Thus initially, the agent has to sufficiently explore the environments to have a diverse observation in the buffer. In our experiment, we use an initial policy whose parameters are chosen randomly to allow exploration and enable diverse data collection for the cluster and generator training.

## 4.2   Experiment Setup

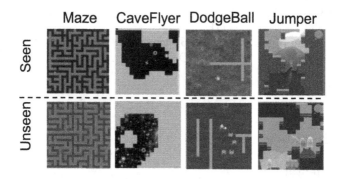

**Fig. 3.** Some snippets of different Procgen environments. The training (seen) levels vary drastically from the testing (unseen) environment. The agent must master the skill without overfitting irrelevant non-generalizable aspects of the environment to perform better in unseen levels.

For the StarGAN training, we use 500 iterations, where in each iteration, the data were sampled from the available clusters dataset. These sampled data were used to train the generators' networks. For the generator model, we use a ResNet-based CNN architecture with 6 residual blocks. The hyperparameters are set $\lambda_{cls} = 1$, $\lambda_{rec} = 10$, and $\lambda_{gp} = 10$.

**Environment.** We conducted experiments on four OpenAI Procgen [6] environments consisting of diverse procedurally-generated environments with different action sets: Maze, CaveFlyer, Dodgeball, and Jumper. These environments are chosen due to their relatively larger generalization gap [6]. We conduct experiments on these environments to measure how quickly (sample efficiency) a reinforcement learning agent learns generalizable policy. Some snippets of different Procgen environments are given in Fig. 3. All environments use a discrete 15 dimensional action space which generates $64 \times 64 \times 3$ RGB image observations.

**Settings.** As suggested in the Procgen benchmark paper [6], we trained the agents on 200 levels for *easy* difficulty levels and evaluated on the full distribution of levels. We report evaluation results on the full distribution (i.e., test), including unseen levels, focusing on generalization as well as training learning curve for sample efficiency. We used the standard Proximal Policy Optimization (PPO) [30] and data augmentation techniques for our baseline comparison. PPO learns policy in an on-policy approach by alternating between sampling data through interaction with the environment and optimizing a surrogate objective function, enabling multiple epochs of minibatch updates using stochastic gradient ascent.

On the other hand, RAD is a data augmentation technique [21] which shows effective empirical evidence in complex RL benchmarks including some Procgen environments. In particular, the *Cutout Color* augmentation technique which has shown better results in many Procgen environments compared in [21] thus we compare with this data augmentation technique. Additionally, we experimented on random crop augmentation. However, this augmentation fails to achieve any reasonable performance in the experimented environments. Thus, we do not report the results for random crop here in our experiments.

We used RLlib [24] to implement all the algorithms. For all the agents (Thinker, PPO, and RAD), to implement the policy network (model), we use a CNN architecture used in IMPALA [9], which also found to work better in the Procgen environments [6]. To account for the agents' performance variability, we run each algorithm with 5 random seeds. Policy learning hyperparameter settings (RLlib's default [24]) for Thinker, PPO, and RAD are set the same for a fair comparison. The hyperparameters are given in Table 1.

**Table 1.** Hyperparameters for experiments - RLlib

| Description | Hyperparameters | Description | Hyperparameters |
|---|---|---|---|
| Discount factor | 0.999 | The GAE (lambda) | 0.95 |
| Learning rate | $5.0e - 4$ | Epochs per train batch | 3 |
| SGD batch | 2048 | Training batch size | 16384 |
| KL divergence | 0.0 | Target KL divergence | 0.01 |
| Coeff. of value loss | 0.5 | Coeff. of the entropy | 0.01 |
| PPO clip parameter | 0.2 | Clip for the value | 0.2 |
| Global clip | 0.5 | PyTorch framework | *torch* |
| Settings for model | IMPALA CNN | Rollout fragment | 256 |

**Evaluation Metric.** It has been observed that a single measure in the form of mean or median can hide the uncertainty implied by different runs [2]. In this paper, we report the reward distribution of all 5 random seed runs in the form of a boxplot to mitigate the above issue.

**Computing Details.** We used the following machine configurations to run our experiments: 20 core-CPU with 256 GB of RAM, CPU Model Name: Intel(R) Xeon(R) Silver 4114 CPU @ 2.20 GHz, and an Nvidia A100 GPU. In our setup, for each run of a training of 25M timesteps, Thinker took approx. 14 h (including approx. 2 h of generator training), RAD-Random Crop took approx. 30 h, RAD-Cutout Color took approx. 9 h and PPO took approx. 8 h.

## 4.3   Results

We now discuss the results of our experiments. We first discuss the generalization results and then sample efficiency. Further, we evaluate how our agents perform in different hyperparameter values of the number of clusters. Finally, we demonstrate samples of the style transfer by our generator.

**Generalization on Unseen Environments.** We show how each agent achieves generalization after training for 25 million timesteps. This scenario is a zero-shot setting, which means we do not train the agent on the test environment levels (unseen to the trained agent). We report the reward in different random seed runs in a boxplot. The generalization results are computed by evaluating the trained agents on test levels (full distribution) for 128 random episode trials.

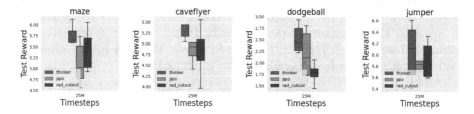

**Fig. 4.** Generalization (Test) results. Our agent Thinker performs better in all environments than the base PPO algorithm and RAD cutout data augmentation.

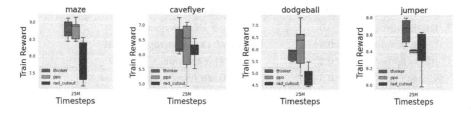

**Fig. 5.** Sample efficiency (Train) results. Thinker achieves better sample efficiency in the Jumper environment while performing comparably with the base PPO algorithm in other environments. Note that our agent Thinker still achieves competitive results during training despite being optimized for generalization.

Figure 4 shows the boxplot of the test performance at the end of the training. We observe that our agent Thinker performs better (in the median, 25th, and 75th reward) compared to the base PPO algorithm and RAD cutout data augmentation. On the other hand, the random cutout data augmentation approach sometimes worsens the performance compared to the base PPO. In all cases, Thinker performs better than the data augmentation-based approach. Random Crop performed worst and could not produce any meaningful reward in these environments. Thus, for brevity, we omit them from Fig. 4.

These results show the importance of our bootstrapped observations data during policy training, which could help us learn a policy that performs better across unseen levels of environments than baselines.

**Sample Efficiency During Training.** We further evaluate the sample efficiency of our method during training. We show in Fig. 5 the final train reward after training the agents for 25 million timesteps.

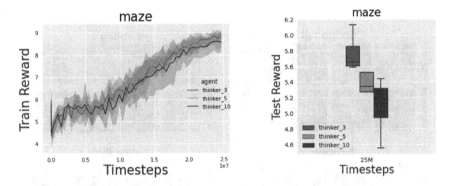

**Fig. 6.** Ablation results. Thinker's performance on different cluster numbers (3, 5, and 10) on the Maze Procgen environment. The results are averaged over 5 seeds. [**Left**] Thinker's learning curve during training. [**Right**] Thinker's generalization performance in boxplot on unseen levels after the training.

Thinker achieves better sample efficiency in the Jumper environment and performs comparably with the base PPO algorithm in other environments. However, the random cutout data augmentation mostly fails to improve (and sometimes worsens) the performance over the base PPO algorithm. Note that the ultimate goal of our agent Thinker is to perform better in test time. Despite that objective, it still achieves competitive results during training.

On the other hand, our agent Thinker performs better than the data augmentation-based approach in all the environments. We omit the random crop data augmentation result for brevity due to its poor performance. 6.

**Ablation Study.** The ablation results for different cluster numbers are shown in Fig. 6. We observe that the number of clusters has some effect on policy learning. The generalization (Test Reward) performance is dropping with the increase in clusters. When the number of clusters is large, that is 10; the generator might overfit each cluster's features and translate the essential semantic part of the observation, thus resulting in lower performance. However, the cluster number does not affect the train results (Train Reward). We see the best results at cluster number 3 in the Maze Procgen environment.

**Style Transfer Sample.** Figure 7 shows some sample style translations by our trained generator. Overall, the generator performs style transfer while mostly maintaining the game semantics. For example, in the Dodgeball environment, in the second column, we see that the background color of the observation is gray, while in the translated observation, it is mostly blue. Additionally, the game objects (e.g., small dots, horizontal and vertical bar) remain in place. These objects are the essential part where the agent needs to focus while solving the task.

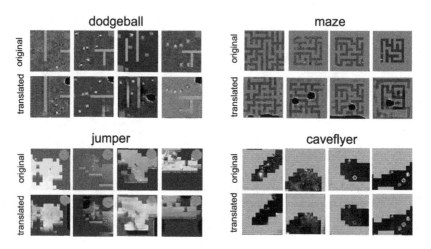

**Fig. 7.** Sample style translations by our trained generator on some Procgen environments. The top images are the original observations for each environment, and the corresponding bottom images are the translated images. We see that the contents of the translated images mostly remain similar to the original images while the style varies.

## 5   Related Work

Regularization has been used to improve RL generalization [7,10,17,20,32]. On the other hand, data augmentation has been shown promising results in generalization and in high-dimensional observation space [7,20,20,21,27]. Network randomization [4,23,25] and random noise injection [17], leveraging inherent sequential structure in RL [1] have been explored to improve RL robustness and generalization. In these cases, the idea is to learn invariant features and disentangled representation [16] robust to visual changes in the testing environment. In our work, we explicitly tackle this problem by generating semantically similar but visually different observation samples, which ideally cancel out unimportant features in the environment and thus learn invariant state representations. Our method focuses on performing realistic visual style transfer of observations while keeping the semantic same. Thus, the target observation corresponds to possible testing environments, aiming to prepare the agent for the unseen scenarios. A closely related paper of our style transfer approach is [11]. They require access to the annotated agent's trajectory data in both source and target domains for the GAN training. In our case, we do not need the information of levels and their sample beforehand; instead, we automatically cluster the trajectory data based on observation's visual features. Thus, the style transfer happens between learned clusters. Additionally, our approach uses visual-based clustering; thus, one cluster may have data from multiple levels, potentially preventing GAN from overfitting [11] to any particular environment levels.

## 6    Discussion

In conclusion, we proposed a novel bootstrapping method to remove the adverse effects of confounding features from the observation in an unsupervised way. Our method first clusters experience trajectories into several clusters; then, it learns StarGAN-based generators. These generators translate the trajectories from one cluster's style to another, which are used for policy training. Our method can be used with existing deep RL algorithms where experience trajectory is used for policy training. Evaluating on visually enriched environments, we demonstrated that our method improves the performance of the existing RL algorithm while achieving better generalization capacity and sample efficiency.

The impacts of Thinker on policy learning depends on the quality of the bootstrapped data generations. Thus our method is better suited for the cases where different levels of an environment vary visually (e.g., changing background color, object colors, texture). In the scenarios where different levels of an environment vary due to mostly its semantic logic differences (e.g., the structural difference in a maze), our method might face challenges. Lack of visual diversity in the clustering might lead the generator to overfit, impacting the its translation performance across these clusters. A possible alternative is to cluster observation data using other features that vary between clusters in addition to visual aspects. A large number of the cluster might place less diverse observation in individual cluster focusing on low-level objects' details, which might cause the generator to overfit. We suggest to reduce the number of clusters in such scenarios. During policy learning, the agent requires some time to train and infer the Thinker module. However, this additional time is negligible compared to deep RL agents' typical stretched running time. Additionally, as we are training a single generator for all the cluster pairs, we find the overhead of Thinker training time reasonable in the context of deep RL agent training.

**Acknowledgements.** This research was supported by NSF grants IIS-1850243, CCF-1918327.

# References

1. Agarwal, R., Machado, M.C., Castro, P.S., Bellemare, M.G.: Contrastive behavioral similarity embeddings for generalization in reinforcement learning. In: International Conference on Learning Representations (2021)
2. Agarwal, R., Schwarzer, M., Castro, P.S., Courville, A., Bellemare, M.G.: Deep reinforcement learning at the edge of the statistical precipice. In: Advances in Neural Information Processing Systems (2021)
3. Arjovsky, M., Chintala, S., Bottou, L.: Wasserstein generative adversarial networks. In: International Conference on Machine Learning, pp. 214–223. PMLR (2017)
4. Burda, Y., Edwards, H., Storkey, A., Klimov, O.: Exploration by random network distillation. arXiv preprint arXiv:1810.12894 (2018)
5. Choi, Y., Choi, M., Kim, M., Ha, J.W., Kim, S., Choo, J.: StarGAN: unified generative adversarial networks for multi-domain image-to-image translation. In: Proceedings of the IEEE Conference on Computer Vision and Pattern Recognition, pp. 8789–8797 (2018)
6. Cobbe, K., Hesse, C., Hilton, J., Schulman, J.: Leveraging procedural generation to benchmark reinforcement learning. arXiv preprint arXiv:1912.01588 (2019)
7. Cobbe, K., Klimov, O., Hesse, C., Kim, T., Schulman, J.: Quantifying generalization in reinforcement learning. In: International Conference on Machine Learning, pp. 1282–1289. PMLR (2019)
8. Epstude, K., Roese, N.J.: The functional theory of counterfactual thinking. Pers. Soc. Psychol. Rev. **12**(2), 168–192 (2008)
9. Espeholt, L., et al.: Impala: scalable distributed deep-RL with importance weighted actor-learner architectures. arXiv preprint arXiv:1802.01561 (2018)
10. Farebrother, J., Machado, M.C., Bowling, M.: Generalization and regularization in DQN. arXiv preprint arXiv:1810.00123 (2018)
11. Gamrian, S., Goldberg, Y.: Transfer learning for related reinforcement learning tasks via image-to-image translation. In: International Conference on Machine Learning, pp. 2063–2072. PMLR (2019)
12. Goodfellow, I., et al.: Generative adversarial nets. Adv. Neural. Inf. Process. Syst. **27**, 2672–2680 (2014)
13. Gulrajani, I., Ahmed, F., Arjovsky, M., Dumoulin, V., Courville, A.: Improved training of wasserstein gans. arXiv preprint arXiv:1704.00028 (2017)
14. Hardt, M., Recht, B., Singer, Y.: Train faster, generalize better: stability of stochastic gradient descent. In: International Conference on Machine Learning, pp. 1225–1234. PMLR (2016)
15. He, K., Zhang, X., Ren, S., Sun, J.: Deep residual learning for image recognition. In: Proceedings of the IEEE Conference on Computer Vision and Pattern Recognition, pp. 770–778 (2016)
16. Higgins, I., et al.: Darla: improving zero-shot transfer in reinforcement learning. arXiv preprint arXiv:1707.08475 (2017)
17. Igl, M., et al.: Generalization in reinforcement learning with selective noise injection and information bottleneck. In: Advances in Neural Information Processing Systems, pp. 13978–13990 (2019)
18. Isola, P., Zhu, J.Y., Zhou, T., Efros, A.A.: Image-to-image translation with conditional adversarial networks. In: Proceedings of the IEEE Conference on Computer Vision and Pattern Recognition, pp. 1125–1134 (2017)
19. Kim, T., Cha, M., Kim, H., Lee, J.K., Kim, J.: Learning to discover cross-domain relations with generative adversarial networks. In: International Conference on Machine Learning, pp. 1857–1865. PMLR (2017)

20. Kostrikov, I., Yarats, D., Fergus, R.: Image augmentation is all you need: regularizing deep reinforcement learning from pixels. arXiv preprint arXiv:2004.13649 (2020)
21. Laskin, M., Lee, K., Stooke, A., Pinto, L., Abbeel, P., Srinivas, A.: Reinforcement learning with augmented data. In: Advances in Neural Information Processing Systems (2020)
22. Laskin, M., Srinivas, A., Abbeel, P.: Curl: contrastive unsupervised representations for reinforcement learning. In: International Conference on Machine Learning, pp. 5639–5650. PMLR (2020)
23. Lee, K., Lee, K., Shin, J., Lee, H.: Network randomization: a simple technique for generalization in deep reinforcement learning. In: International Conference on Learning Representations (2020)
24. Liang, E., et al.: Rllib: abstractions for distributed reinforcement learning. In: International Conference on Machine Learning, pp. 3053–3062. PMLR (2018)
25. Osband, I., Aslanides, J., Cassirer, A.: Randomized prior functions for deep reinforcement learning. In: Advances in Neural Information Processing Systems, pp. 8617–8629 (2018)
26. Pedregosa, F., et al.: Scikit-learn: machine learning in Python. J. Mach. Learn. Res. **12**, 2825–2830 (2011)
27. Raileanu, R., Goldstein, M., Yarats, D., Kostrikov, I., Fergus, R.: Automatic data augmentation for generalization in deep reinforcement learning. arXiv preprint arXiv:2006.12862 (2020)
28. Roese, N.J.: The functional basis of counterfactual thinking. J. Pers. Soc. Psychol. **66**(5), 805 (1994)
29. Russakovsky, O., et al.: ImageNet large scale visual recognition challenge. Int. J. Comput. Vision **115**(3), 211–252 (2015). https://doi.org/10.1007/s11263-015-0816-y
30. Schulman, J., Wolski, F., Dhariwal, P., Radford, A., Klimov, O.: Proximal policy optimization algorithms. arXiv preprint arXiv:1707.06347 (2017)
31. Song, X., Jiang, Y., Tu, S., Du, Y., Neyshabur, B.: Observational overfitting in reinforcement learning. In: International Conference on Learning Representations (2020)
32. Wang, K., Kang, B., Shao, J., Feng, J.: Improving generalization in reinforcement learning with mixture regularization. arXiv preprint arXiv:2010.10814 (2020)
33. Zhang, A., Ballas, N., Pineau, J.: A dissection of overfitting and generalization in continuous reinforcement learning. arXiv preprint arXiv:1806.07937 (2018)
34. Zhang, C., Vinyals, O., Munos, R., Bengio, S.: A study on overfitting in deep reinforcement learning. arXiv preprint arXiv:1804.06893 (2018)
35. Zhu, J.Y., Park, T., Isola, P., Efros, A.A.: Unpaired image-to-image translation using cycle-consistent adversarial networks. In: Proceedings of the IEEE International Conference on Computer Vision, pp. 2223–2232 (2017)

# Imitation Learning with Sinkhorn Distances

Georgios Papagiannis[1(✉)] and Yunpeng Li[2]

[1] Imperial College London, London, UK
g.papagiannis21@imperial.ac.uk
[2] University of Surrey, Guildford, UK
yunpeng.li@surrey.ac.uk

**Abstract.** Imitation learning algorithms have been interpreted as variants of divergence minimization problems. The ability to compare occupancy measures between experts and learners is crucial in their effectiveness in learning from demonstrations. In this paper, we present tractable solutions by formulating imitation learning as minimization of the Sinkhorn distance between occupancy measures. The formulation combines the valuable properties of optimal transport metrics in comparing non-overlapping distributions with a cosine distance cost defined in an adversarially learned feature space. This leads to a highly discriminative critic network and optimal transport plan that subsequently guide imitation learning. We evaluate the proposed approach using both the reward metric and the Sinkhorn distance metric on a number of MuJoCo experiments. For the implementation and reproducing results please refer to the following repository https://github.com/gpapagiannis/sinkhorn-imitation.

## 1 Introduction

Recent developments in reinforcement learning (RL) have allowed agents to achieve state-of-the-art performance on complex tasks from learning to play games [18,33,38] to dexterous manipulation [24], provided with well defined reward functions. However, crafting such a reward function in practical scenarios to encapsulate the desired objective is often non-trivial. Imitation learning (IL) [20] aims to address this issue by formulating the problem of learning behavior through expert demonstration and has shown promises on various application domains including autonomous driving and surgical task automation [2,13,21,23,43].

The main approaches to imitation learning include that of behavioral cloning (BC) and inverse reinforcement learning (IRL). BC mimics the expert's behavior by converting the task into a supervised regression problem [23,30]. While simple to implement, it is known to suffer from low sample efficiency and poor generalization performance due to covariate shift and high sample correlations in the expert's trajectory [26,27]. Algorithms such as Dataset Aggregation (DAgger)

---

G. Papagiannis—Work done as a student at the University of Surrey.

© The Author(s), under exclusive license to Springer Nature Switzerland AG 2023
M.-R. Amini et al. (Eds.): ECML PKDD 2022, LNAI 13716, pp. 116–131, 2023.
https://doi.org/10.1007/978-3-031-26412-2_8

[26] and Disturbances for Augmenting Robot Trajectories (DART) [17] alleviate this issue. However, they require constantly querying an expert for the correct actions.

Inverse reinforcement learning instead aims to recover a reward function which is subsequently used to train the learner's policy [19,42]. IRL approaches have shown significantly better results [1,2,5,16,22,40] including being sample efficient in terms of expert demonstration. However, IRL itself is an ill-posed problem - multiple reward functions can characterize a specific expert behavior, therefore additional constraints need to be imposed to recover a unique solution [19,41,42]. In addition, the alternating optimization procedure between reward recovery and policy training leads to increased computational cost.

Adversarial imitation learning, on the other hand, bypasses the step of explicit reward inference as in IRL and directly learns a policy that matches that of an expert. Generative adversarial imitation learning (GAIL) [11] minimizes the Jensen-Shannon (JS) divergence between the learner's and expert's occupancy measures through a generative adversarial networks (GANs)-based training process. GAIL was developed as a variant of the reward regularized maximum entropy IRL framework [42], where different reward regularizers lead to different IL methods. GAIL has been extended by various other methods aiming to improve its sample efficiency in regard to environment interaction through off-policy RL [4,14,15,31,36]. Recent development [9] provides a unified probabilistic perspective to interpret different imitation learning methods as $f$-divergence minimization problems and showed that the state-marginal matching objective of IRL approaches is what contributes the most to their superior performance compared to BC. While these methods have shown empirical success, they inherit the same issues from $f$-divergence and adversarial training, such as training instability in GAN-based training [10] and mode-covering behavior in the JS and Kullback-Leibler (KL) divergences [9,12].

An alternative approach is to utilize optimal transport-based metrics to formulate the imitation learning problem. The optimal transport (OT) theory [37] provides a flexible and powerful tool to compare probability distributions through coupling of distributions based on the metric in the underlying spaces. The Wasserstein adversarial imitation learning (WAIL) [39] was proposed to minimize the dual form of the Wasserstein distance between the learner's and expert's occupancy measures, similar to the training of the Wasserstein GAN [3]. The geometric property of the Wasserstein distance leads to numerical stability in training and robustness to disjoint measures. However, the solution to the dual formulation is intractable; approximations are needed in the implementation of neural networks to impose the required Lipschitz condition [28]. [7] introduced Primal Wasserstein imitation learning (PWIL), that uses a reward proxy derived based on an upper bound to the Wasserstein distance between the state-action distributions of the learner and the expert. While PWIL leads to successful imitation, it is unclear how it inherits the theoretical properties of OT, since the transport map between occupancy measures is suboptimal, based on a greedy coupling strategy whose approximation error is difficult to quantify.

In this paper we present Sinkhorn imitation learning (SIL), a *tractable* solution to optimal transport-based imitation learning by leveraging the coupling of occupancy measures and the computational efficiency of the Sinkhorn distance [6], that inherits the theoretical properties of OT. Our main contributions include: (i) We propose and justify an imitation learning training pipeline that minimizes the Sinkhorn distance between occupancy measures of the expert and the learner; (ii) We derive a reward proxy using a set of trainable and highly discriminative optimal transport ground metrics; (iii) We demonstrate through experiments on the MuJoCo simulator [35] that SIL obtains comparable results with the state-of-the-art, outperforming the baselines on a number of experiment settings in regard to both the commonly used reward metric and the Sinkhorn distance.

The rest of this paper is organized as follows. In Sect. 2 we provide the necessary background for this work. Section 3 introduces the proposed Sinkhorn Imitation Learning (SIL) framework. Section 4 provides details of experiments to evaluate the performance of SIL on a number of MuJoCo environments. We conclude the paper and discuss future research directions in Sect. 5.

## 2   Background

### 2.1   Imitation Learning

**Notation.** We consider a Markov Decision Process (MDP) which is defined as a tuple $\{\mathcal{S}, \mathcal{A}, \mathcal{P}, r, \gamma\}$, where $\mathcal{S}$ is a set of states, $\mathcal{A}$ is a set of possible actions an agent can take on the environment, $\mathcal{P} : \mathcal{S} \times \mathcal{A} \times \mathcal{S} \to [0,1]$ is a transition probability matrix, $r : \mathcal{S} \times \mathcal{A} \to \mathbb{R}$ is a reward function and $\gamma \in (0,1)$ is a discount factor. The agent's behavior is defined by a stochastic policy $\pi : \mathcal{S} \to \text{Prob}(\mathcal{A})$ and $\Pi$ is the set of all such policies. We use $\pi_E, \pi \in \Pi$ to refer to the expert and learner policy respectively. The performance measure of policy $\pi$ is defined as $\mathcal{J} = \mathbb{E}_\pi[r(s,a)] = \mathbb{E}[\sum_{t=0}^\infty \gamma^t r(s_t, a_t)|\mathcal{P}, \pi]$ where $s_t \in \mathcal{S}$ is a state observed by the agent at time step $t$. With a slight abuse of notations, we also use $r((s,a)_\pi)$ to denote explicitly that $(s,a)_\pi \sim \pi$. $\tau_E$ and $\tau_\pi$ denote the set of state-action pairs sampled by an expert and a learner policy respectively during interaction with the environment, also referred to as trajectories. The distribution of state-action pairs generated by policy $\pi$ through environment interaction, also known as the occupancy measure $\rho_\pi : \mathcal{S} \times \mathcal{A} \to \mathbb{R}$, is defined as

$$\rho_\pi(s,a) = (1-\gamma)\pi(a|s)\sum_{t=0}^\infty \gamma^t P_\pi[s_t = s]$$ where $P_\pi[s_t = s]$ denotes the probability of a state being $s$ at time step $t$ following policy $\pi$.

*Generative Adversarial Imitation Learning.* Ho and Ermon [11] extended the framework of MaxEnt IRL by introducing a reward regularizer $\psi(r) : \mathcal{S} \times \mathcal{A} \to \mathbb{R}$:

$$\text{IRL}_\psi(\pi_E) := \arg\max_r -\psi(r) + \min_{\pi \in \Pi}\big(-\mathcal{H}^{causal}(\pi) - \mathbb{E}_\pi[r(s,a)]\big) + \mathbb{E}_{\pi_E}[r(s,a)] \,,$$

$$(1)$$

where $\mathcal{H}^{causal}(\pi) := \mathbb{E}_{\rho_\pi}[-\log \pi(a|s)]/(1-\gamma)$ [41]. The process of RL following IRL can be formulated as that of occupancy measure matching [11]:

$$\text{RL} \circ \text{IRL}_\psi(\pi_E) := \arg\min_{\pi \in \Pi} - \mathcal{H}^{causal}(\pi) + \psi^*(\rho_\pi - \rho_E) , \qquad (2)$$

where $\psi^*$ corresponds to the convex conjugate of the reward regularizer $\psi(r)$. The regularized MaxEnt IRL framework bypasses the expensive step of reward inference and learns how to imitate an expert by matching its occupancy measure. Different realizations of the reward regularizer lead to different IL frameworks. A specific choice of the regularizer leads to the Generative Adversarial Imitation Learning (GAIL) framework that minimizes the Jensen-Shannon divergence between the learner's and expert's occupancy measures [11].

*f-Divergence MaxEnt IRL.* Recently, Ghasemipour et al. [9] showed that training a learner policy $\pi$ to minimize the distance between two occupancy measures can be generalised to minimize any $f$-divergence between $\rho_E$ and $\rho_\pi$ denoted as $D_f(\rho_E \| \rho_\pi)$. Different choices of $f$ yield different divergence minimization IL algorithms [9] and can be computed as:

$$\max_{T_\omega} \mathbb{E}_{(s,a)\sim\rho_E}[T_\omega(s,a)] - \mathbb{E}_{(s,a)\sim\rho_\pi}[f^*(T_\omega(s,a)))] , \qquad (3)$$

where $T_\omega : \mathcal{S}\times\mathcal{A} \to \mathbb{R}$ and $f^*$ is the convex conjugate of the selected $f$-divergence. The learner's policy is optimized with respect to the reward proxy $f^*(T_\omega(s,a))$.

## 2.2 Optimal Transport

While divergence minimization methods have enjoyed empirical success, they are still difficult to evaluate in high dimensions [34], due to the sensitivity to different hyperparameters and difficulty in training depending on the distributions that are evaluated [28]. The optimal transport (OT) theory [37] provides effective methods to compare degenerate distributions by accounting for the underlying metric space. Consider $P_k(\Gamma)$ to be the set of Borel probability measures on a Polish metric space $(\Gamma, d)$ with finite $k$-th moment. Given two probability measures $p, q \in P_k(\Gamma)$, the $k$-Wasserstein metric is defined as [37]:

$$\mathcal{W}_k(p,q)_c = \left( \inf_{\zeta\in\Omega(p,q)} \int_\Gamma c(x,y)^k d\zeta(x,y) \right)^{\frac{1}{k}} , \qquad (4)$$

where $\Omega(p,q)$ denotes the set of joint probability distributions whose marginals are $p$ and $q$, respectively. $c(x,y)$ denotes the cost of transporting sample $x \sim p$ to $y \sim q$. The joint distribution $\zeta$ that minimizes the total transportation cost is referred to as the optimal transport plan.

*Sinkhorn Distances.* The solution to Eq. (4) is generally intractable for high dimensional distributions in practice. A regularized form of the optimal transport formulation was proposed by Cuturi [6] that can efficiently compute the Wasserstein metric. The Sinkhorn distance $\mathcal{W}_s^\beta(p,q)_c$ between $p$ and $q$ is defined as:

$$W_s^\beta(p,q)_c = \inf_{\zeta_\beta \in \Omega_\beta(p,q)} \mathbb{E}_{x,y \sim \zeta_\beta}[c(x,y)] , \qquad (5)$$

where $\Omega_\beta(p,q)$ denotes the set of all joint distributions in $\Omega(p,q)$ with entropy of at least $\mathcal{H}(p) + \mathcal{H}(q) - \beta$ and $\mathcal{H}(\cdot)$ computes the entropy of a distribution. The distance is evaluated on two distributions $p$ and $q$ where in the context of adversarial IL correspond to the state-action distributions of the learner and the expert policies.

## 3   SIL: Sinkhorn Imitation Learning

We consider the problem of training a learner policy $\pi$ to imitate an expert, by matching its state-action distribution $\rho_E$ in terms of minimizing their Sinkhorn distance. To facilitate the development of the learning pipeline, we begin by discussing how the Sinkhorn distance is used to evaluate similarity between occupancy measures.

Consider the case of a learner $\pi$ interacting with an environment and generating a trajectory of state-action pairs $\tau_\pi \sim \pi$ that characterizes its occupancy measure. A trajectory of expert demonstrations $\tau_E \sim \pi_E$ is also available as the expert trajectories. The optimal transport plan $\zeta_\beta$ between the samples of $\tau_\pi$ and $\tau_E$ can be obtained via the Sinkhorn algorithm [6]. Following Eq. (5) we can evaluate the Sinkhorn distance of $\tau_\pi$ and $\tau_E$ as follows:

$$W_s^\beta(\tau_\pi, \tau_E)_c = \sum_{(s,a)_\pi \in \tau_\pi} \sum_{(s,a)_{\pi_E} \in \tau_E} c\Big((s,a)_\pi, (s,a)_{\pi_E}\Big) \zeta_\beta\Big((s,a)_\pi, (s,a)_{\pi_E}\Big) . \quad (6)$$

**Reward Proxy.** We now introduce a reward proxy suitable for training a learner policy that minimizes $W_s^\beta(\tau_\pi, \tau_E)_c$ in order to match the expert's occupancy measure.

The reward function $v_c((s,a)_\pi)$ for each sample $(s,a)_\pi$ in the learner's trajectory is defined as:

$$v_c((s,a)_\pi) := - \sum_{(s,a)_{\pi_E} \in \tau_E} c\Big((s,a)_\pi, (s,a)_{\pi_E}\Big) \zeta_\beta\Big((s,a)_\pi, (s,a)_{\pi_E}\Big) . \quad (7)$$

The optimization objective of the learner policy $\mathcal{J} = \mathbb{E}_\pi[r((s,a)_\pi)]$ under $r((s,a)_\pi) := v_c((s,a)_\pi)$ corresponds to minimizing the Sinkhorn distance between the learner's and expert's trajectories defined in Eq. (6). Hence, by maximizing the optimization objective $\mathcal{J}$ with reward $v_c((s,a)_\pi)$, a learner is trained to minimize the Sinkhorn distance between the occupancy measures of the learner and the expert demonstrator.

**Adversarial Reward Proxy.** The reward specified in Eq. (7) can only be obtained after the learner has generated a complete trajectory. The optimal transport plan $\zeta_\beta((s,a)_\pi, (s,a)_{\pi_E})$ then weighs the transport cost of each sample

$(s, a)_\pi \in \tau_\pi$ according to the samples present in $\tau_\pi$ and $\tau_E$. The dependence of $v_c((s, a)_\pi)$ to all state-action pairs in $\tau_\pi$ and $\tau_E$ can potentially result in the same state-action pair being assigned significantly different rewards depending on the trajectory that it is sampled from. Such dependence can lead to difficulty in maximizing the optimization objective $\mathcal{J}$ (and equivalently in minimizing the Sinkhorn distance between the occupancy measures from the learner and the expert). Empirical evidence is provided in the ablation study in Sect. 4.

In order to provide a discriminative signal to the learner's policy and aid the optimization process, we consider adversarially training a critic to penalize non-expert state-action pairs by increasing their transport cost to the expert's distribution, drawing inspiration from the adversarially trained transport ground metric in the OT-GAN framework [29]. The critic $c_w((s, a)_\pi, (s, a)_{\pi_E})$ parameterized by $w$ is defined as follows:

$$c_w((s, a)_\pi, (s, a)_{\pi_E}) = 1 - \frac{f_w((s, a)_\pi) \cdot f_w((s, a)_{\pi_E})}{||f_w((s, a)_\pi)||_2 ||f_w((s, a)_{\pi_E})||_2} , \qquad (8)$$

where $\cdot$ denotes the inner product between two vectors. $f_w(\cdot) : \mathcal{S} \times \mathcal{A} \to \mathbb{R}^d$ maps the environment's observation space $\mathcal{S} \times \mathcal{A}$ to an adversarially learned feature space $\mathbb{R}^d$ where $d$ is the feature dimension. The adversarial reward proxy $v_{c_w}((s, a)_\pi)$ is obtained by substituting the transport cost $c(\cdot, \cdot)$ in Eq. (7) with $c_w(\cdot, \cdot)$ defined by Eq. (8). SIL learns $\pi$ by solving the following minimax optimization problem:

$$\arg \min_\pi \ \max_w \mathcal{W}_s^\beta (\rho_\pi, \rho_E)_{c_w} . \qquad (9)$$

*Remark 1.* For SIL, the adversarial training part of the transport cost is not part of the approximation procedure of the distance metric, as in GAIL [11] and WAIL [39]. The Sinkhorn distance is computed directly via the Sinkhorn iterative procedure [6] with the transport cost defined in Equation (8).

**Algorithm.** The pseudocode for the proposed Sinkhorn imitation learning (SIL) framework is presented in Algorithm 1. In each iteration we randomly match each of the learner's generated trajectories to one of the expert's and obtain their Sinkhorn distance. The reason behind this implementation choice is to maintain a constant computational complexity with respect to a potentially increasing number of demonstrations. We then alternate between one step of updating a critic network $c_w$ to maximize the Sinkhorn distance between the learner's and expert's trajectories and a policy update step to minimize the distance between occupancy measures with the learned reward proxy. As SIL depends on complete environment trajectories to compute the Sinkhorn distance, it is inherently an on-policy method. Hence, to train our imitator policy we use Trust Region Policy Optimization (TRPO) [32] for our experiments.

**Algorithm 1:** Sinkhorn imitation learning (SIL)
___

**Input**: Set of expert trajectories $\{\tau_E\} \sim \pi_E$, Sinkhorn regularization
parameter $\beta$, initial learner's policy parameters $\theta_0$, initial critic
network parameters $w_0$, number of training iterations $K$

1: **for** iteration $k = 0$ to $K$ **do**
2:     Sample a set of trajectories $\{\tau_{\pi_{\theta_k}}\}_k \sim \pi_{\theta_k}$.
3:     Create a set of trajectory pairs $\{(\tau_{\pi_{\theta_k}}, \tau_E)\}_k$ by randomly
       matching trajectories from the learner's set to the expert's.
4:     For each pair in $\{(\tau_{\pi_{\theta_k}}, \tau_E)\}_k$, calculate $\mathcal{W}_s^\beta(\tau_{\pi_{\theta_k}}, \tau_E)_{c_w}$ using the Sinkhorn
       algorithm (Eq. (5)) and transport cost as in Eq. (8), in order to update the
       reward proxy $v_{c_{w_k}}((s,a)_{\pi_{\theta_k}})$ for each state action pair.
5:     Update $w_k$ to maximize $\mathcal{W}_s^\beta(\tau_{\pi_{\theta_k}}, \tau_E)_{c_w}$ using gradient ascent with the gradient:

$$\nabla_{w_k} \frac{1}{m} \sum_{\{(\tau_{\pi_{\theta_k}}, \tau_E)\}_k} \mathcal{W}_s^\beta(\tau_{\pi_{\theta_k}}, \tau_E)_{c_w} , \qquad (10)$$

       where $m$ is the number of trajectory pairs.
6:     Update policy parameter $\theta_k$ using TRPO and reward $v_{c_{w_k}}((s,a)_{\pi_{\theta_k}})$ updated in
       Step 4.
7: **end for**

**Output**: Learned policy $\pi_{\theta_k}$.
___

### 3.1   Connection to Regularized MaxEnt IRL

We now show how SIL can be interpreted as a variant of the regularized MaxEnt
IRL framework [11] given a specific choice of $\psi(r)$.

**Definition 1.** *Consider a learner's policy and expert's demonstrations, as well
as their induced occupancy measures $\rho_\pi$ and $\rho_E$. We define the following reward
regularizer:*

$$\psi_{\mathcal{W}}(r) := -\mathcal{W}_s^\beta(\rho_\pi, \rho_E)_{c_w} + \mathbb{E}_{\rho_\pi}[r(s,a)] - \mathbb{E}_{\rho_E}[r(s,a)] . \qquad (11)$$

**Proposition 1.** *The reward regularizer $\psi_{\mathcal{W}}(r)$ defined in Eq. (11) leads to an
entropy regularized MaxEnt IRL algorithm. When $r((s,a)_\pi) = v_{c_w}((s,a)_\pi)$,*

$$RL \circ IRL_{\psi_{\mathcal{W}}}(\pi_E) = \arg\min_{\pi \in \Pi} -\mathcal{H}^{causal}(\pi) + \sup_w \mathcal{W}_s^\beta(\rho_\pi, \rho_E)_{c_w} . \qquad (12)$$

Equation (12) corresponds to the process of updating a critic network to
maximize the Sinkhorn distance between the learner's and expert's occupancy
measures, followed by the process of finding a policy $\pi$ to minimize it. The added
term $\mathcal{H}^{causal}(\pi)$ is treated as a regularization parameter.

*Proof.* Consider the set of possible rewards $\mathcal{R} := \{r : \mathcal{S} \times \mathcal{A} \to \mathbb{R}\}$ in finite state-action space as in [11] and [9]. The joint state-action distributions $\rho_\pi$ and $\rho_E$ are represented as vectors in $[0,1]^{\mathcal{S} \times \mathcal{A}}$.

Define $\psi_{\mathcal{W}}(r) := -\mathcal{W}_s^\beta(\rho_\pi, \rho_E) + \mathbb{E}_{\rho_\pi}[r(s,a)] - \mathbb{E}_{\rho_E}[r(s,a)]$, where $\mathcal{W}_s^\beta(\rho_\pi, \rho_E)$ is obtained with the transport cost $c_w$ defined in Equation (8). Given $r(s,a) = v_{c_w}(s,a)$ and recall that the convex conjugate of a function $g$ is $g^*(y) = \sup_{x \in \text{dom}(g)}(y^T x - g(x))$, we obtain

$$\psi_{\mathcal{W}}^*(\rho_\pi - \rho_E) = \sup_{r \in \mathcal{R}}[(\rho_\pi - \rho_E)^T r - \psi_{\mathcal{W}}(r)] = \sup_{r \in \mathcal{R}}\Big[\sum_{\mathcal{S} \times \mathcal{A}}(\rho_\pi(s,a) - \rho_E(s,a)) \cdot r(s,a)$$

$$+ \mathcal{W}_s^\beta(\rho_\pi, \rho_E) - \sum_{\mathcal{S} \times \mathcal{A}}(\rho_\pi(s,a) - \rho_E(s,a)) \cdot r(s,a)\Big] =$$

$$\sup_{r \in \mathcal{R}} \mathcal{W}_s^\beta(\rho_\pi, \rho_E) = \sup_{v_{c_w} \in \mathcal{R}} \mathcal{W}_s^\beta(\rho_\pi, \rho_E) = \sup_w \mathcal{W}_s^\beta(\rho_\pi, \rho_E) . \tag{13}$$

From Eq. (2),

$$\text{RL} \circ \text{IRL}_\psi(\pi_E) = \underset{\pi \in \Pi}{\arg\min} - \mathcal{H}^{causal}(\pi) + \psi_{\mathcal{W}}^*(\rho_\pi - \rho_E)$$

$$= \underset{\pi \in \Pi}{\arg\min} - \mathcal{H}^{causal}(\pi) + \sup_w \mathcal{W}_s^\beta(\rho_\pi, \rho_E) . \tag{14}$$

## 4 Experiments

To empirically evaluate the Sinkhorn imitation learning (SIL) algorithm, we benchmark SIL against BC in the four MuJoCo [35] environments studied in [9], namely Hopper-v2, Walker2d-v2, Ant-v2 and HalfCheetah-v2, as well as the Humanoid-v2 environment. Given that SIL is an on-policy method due to the requirement of complete trajectories, two on-policy adversarial IL algorithms, namely GAIL [11] and AIRL [8], are also included as baselines. All algorithms are evaluated against the true reward metric obtained through environment interaction, in addition to the Sinkhorn distance between the samples from the learned policy and the expert demonstrations.

Initially we train policies using TRPO [32] to obtain expert performance. The expert policies are used to generate sets of expert demonstrations. The performance of the obtained expert policies can be found in Table 1. To study the robustness of SIL in learning from various lengths of trajectory sets we train the algorithms on sets of $\{2, 4, 8, 16, 32\}$ and for Humanoid-v2 for $\{8, 16, 32\}$ sets. All trajectories are subsampled by a factor of 20 starting from a random offset, a common practice found in [8,9,11]. SIL, GAIL and AIRL are trained for 250 iterations allowing approximately $50,000$ environment interactions per iteration. For Humanoid-v2 we train the algorithms for 350 iterations. All reported results correspond to performance metrics obtained after testing the learner policies on 50 episodes.

**Table 1.** Performance of expert policies providing the demonstrations trained using TRPO.

| Environments | Expert Performance |
|---|---|
| Hopper-v2 | $3354.74 \pm 1.87$ |
| HalfCheetah-v2 | $4726.53 \pm 133.12$ |
| Walker2d-v2 | $3496.44 \pm 8.79$ |
| Ant-v2 | $5063.11 \pm 337.50$ |
| Humanoid-v2 | $6303.36 \pm 97.71$ |

### 4.1 Implementation Details

**Adversarial Critic.** The critic network consists of a 2-layer MLP architecture with 128 units each with ReLU activations. For each experiment we report the best performing result after training the critic with the following learning rates $\{0.0004, 0.0005, 0.0006, 0.0007, 0.0008, 0.0009\}$ and output dimensions $\{5, 10, 30\}$. Although different choices of the critic network output dimension may yield better results for the proposed SIL algorithm in different environments, no further attempt was made to fine-tune the output for the critic. We note that for most experiment settings a critic output dimension of 30 and learning rate of 0.0005 among the pool of candidate values yield the best results.

**Reward Proxy.** After obtaining the value of $v_{c_w}$ as defined in Equations (7) and (8), we add a value of $\frac{2}{L}$ where $L$ is the trajectory length and scale the reward by 2. By doing so we set the range of $v_{c_w}$ to be $0 \leq v_{c_w} \leq 4$ which proved to be effective for environments requiring a survival bonus. We keep track of a running standard deviation to normalize rewards.

**Policy Architecture and Training.** For both the expert and learner policies, we use the same architecture comprised of a 2-layer MLP architecture each with 128 units with ReLU activations. The same architecture is used amongst all imitation learning algorithms. For all adversarial IL algorithms, as well as obtaining expert performance, we train the policies using Trust Region Policy Optimization [32]. Finally, we normalize environment observations by keeping track of the running mean and standard deviation.

**GAIL and AIRL.** To aid the performance of the benchmarks algorithms GAIL and AIRL in the HalfCheetah-v2 environment, we initialize the policies with that from behavioural cloning.

**Computational Resource.** The experiments were run on a computer with an Intel (R) Xeon (R) Gold 5218 CPU 2.3 GHz and 16 GB of RAM, and a RTX 6000 graphic card with 22 GB memories.

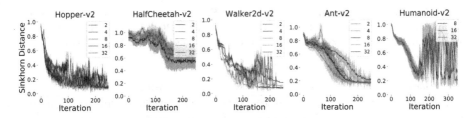

**Fig. 1.** Mean and standard deviation of the Sinkhorn distance evaluated during training of SIL using a fixed cosine transport cost by stochastically sampling an action from the learner's policy.

## 4.2   Results

**Sinkhorn Metric.** We begin by evaluating performance amongst IL methods using the Sinkhorn metric. Since our goal is to assess how well imitation learning algorithms match the expert's occupancy measure, the Sinkhorn distance offers a valid metric of similarity between learner's and expert's trajectories compared to the reward metric which is also often unavailable in practical scenarios. We report the Sinkhorn distance between occupancy measures computed with a fixed cosine distance-based transport cost during testing and evaluation:

$$c((s,a)_\pi, (s,a)_{\pi_E}) = 1 - \frac{[s,a]_\pi \cdot [s,a]_{\pi_E}}{||[s,a]_\pi||_2 ||[s,a]_{\pi_E}||_2} \ , \tag{15}$$

where $[s,a]_\pi$ denotes the concatenated vector of state-action of policy $\pi$ and $||\cdot||_2$ computes the L2 norm. Table 2 reports the Sinkhorn metric evaluated between the trajectories generated by the learned policies with the demonstrations provided by the expert. A smaller Sinkhorn distance corresponds to higher similarity between the learner's and expert's generated trajectories. SIL, AIRL and GAIL obtain comparable performance in most of the environments. The proposed SIL algorithm outperforms the baselines in almost all experiments on the environments of HalfCheetah-v2 and Ant-v2, while AIRL achieves superior performance on the environments of Hopper-v2 and Walker2d-v2. GAIL on the other hand obtains relatively poor performance with regard to the Sinkhorn distance when provided with only 2 expert trajectories on the environments of Hopper-v2, HalfCheetah-v2 and Ant-v2. As expected, behavioral cloning fails to obtain competitive performance in almost all experiment settings especially when provided with a small number of expert demonstrations.

In addition, SIL outperforms GAIL and AIRL on the Humanoid-v2 environment when provided with 8 and 16 trajectories, where SIL demonstrates significantly improved sample efficiency in terms of both expert demonstrations and environment interactions. GAIL outperforms the rest when trained with 32 trajectories on the Humanoid-v2 environment. Interestingly, BC obtains superior performance with regard to the Sinkhorn distance on the Humanoid-v2 environment when provided with 8 trajectories, but low performance regarding the reward metric as shown in Table 3.

**Table 2.** Mean and standard deviation of the Sinkhorn distance between the expert demonstrations and samples from imitator policies for BC, GAIL, AIRL and SIL. A fixed cosine transport cost is used only for evaluation (Smaller distance denotes better performance).

| Environments | Trajectories | BC | GAIL | AIRL | SIL |
|---|---|---|---|---|---|
| Hopper-v2 | 2 | $0.467 \pm 0.009$ | $0.098 \pm 0.003$ | $\mathbf{0.069 \pm 0.001}$ | $0.073 \pm 0.001$ |
| | 4 | $0.408 \pm 0.080$ | $0.120 \pm 0.010$ | $\mathbf{0.066 \pm 0.009}$ | $0.082 \pm 0.010$ |
| | 8 | $0.300 \pm 0.029$ | $0.074 \pm 0.004$ | $\mathbf{0.068 \pm 0.006}$ | $0.071 \pm 0.005$ |
| | 16 | $0.182 \pm 0.042$ | $0.106 \pm 0.008$ | $\mathbf{0.074 \pm 0.010}$ | $0.078 \pm 0.012$ |
| | 32 | $0.157 \pm 0.084$ | $\mathbf{0.071 \pm 0.008}$ | $0.072 \pm 0.009$ | $0.089 \pm 0.008$ |
| HalfCheetah-v2 | 2 | $1.043 \pm 0.058$ | $0.940 \pm 0.181$ | $0.577 \pm 0.157$ | $\mathbf{0.546 \pm 0.138}$ |
| | 4 | $0.791 \pm 0.096$ | $0.633 \pm 0.095$ | $0.630 \pm 0.091$ | $\mathbf{0.620 \pm 0.101}$ |
| | 8 | $0.841 \pm 0.071$ | $0.702 \pm 0.095$ | $0.708 \pm 0.054$ | $\mathbf{0.700 \pm 0.052}$ |
| | 16 | $0.764 \pm 0.166$ | $\mathbf{0.670 \pm 0.128}$ | $0.671 \pm 0.112$ | $0.688 \pm 0.131$ |
| | 32 | $0.717 \pm 0.129$ | $0.695 \pm 0.113$ | $0.699 \pm 0.091$ | $\mathbf{0.685 \pm 0.083}$ |
| Walker2d-v2 | 2 | $0.474 \pm 0.023$ | $0.067 \pm 0.008$ | $\mathbf{0.034 \pm 0.005}$ | $0.080 \pm 0.004$ |
| | 4 | $0.694 \pm 0.011$ | $0.067 \pm 0.006$ | $\mathbf{0.036 \pm 0002}$ | $0.079 \pm 0.005$ |
| | 8 | $0.335 \pm 0.004$ | $0.069 \pm 0.005$ | $\mathbf{0.036 \pm 0.003}$ | $0.063 \pm 0.003$ |
| | 16 | $0.199 \pm 0.013$ | $0.061 \pm 0.004$ | $\mathbf{0.037 \pm 0.005}$ | $0.102 \pm 0.007$ |
| | 32 | $0.196 \pm 0.098$ | $0.052 \pm 0.003$ | $\mathbf{0.042 \pm 0.004}$ | $0.147 \pm 0.003$ |
| Ant-v2 | 2 | $0.843 \pm 0.033$ | $0.344 \pm 0.068$ | $0.164 \pm 0.006$ | $\mathbf{0.158 \pm 0.008}$ |
| | 4 | $0.684 \pm 0.159$ | $0.165 \pm 0.119$ | $0.163 \pm 0.008$ | $\mathbf{0.157 \pm 0.014}$ |
| | 8 | $0.996 \pm 0.029$ | $0.159 \pm 0.016$ | $0.164 \pm 0.019$ | $\mathbf{0.155 \pm 0.012}$ |
| | 16 | $0.724 \pm 0.149$ | $0.225 \pm 0.106$ | $0.173 \pm 0.062$ | $\mathbf{0.165 \pm 0.022}$ |
| | 32 | $0.452 \pm 0094$ | $0.176 \pm 0.029$ | $\mathbf{0.172 \pm 0.020}$ | $0.173 \pm 0.018$ |
| Humanoid-v2 | 8 | $\mathbf{0.336 \pm 0.089}$ | $0.386 \pm 0.011$ | $1.015 \pm 0.015$ | $0.379 \pm 0.296$ |
| | 16 | $0.290 \pm 0.086$ | $0.428 \pm 0.027$ | $1.034 \pm 0.017$ | $\mathbf{0.182 \pm 0.011}$ |
| | 32 | $0.182 \pm 0.028$ | $\mathbf{0.162 \pm 0.144}$ | $1.026 \pm 0.015$ | $0.250 \pm 0.180$ |

**Reward Metric.** To better understand how performance changes in terms of the Sinkhorn distance metric translates to the true reward, Table 3 shows the reward obtained with the learned policies in the same experiments reported in Table 2. While all adversarial imitation learning algorithms exhibit similar reward values compared to the expert policies, we observe that SIL generally obtains lower reward compared to AIRL on Ant-v2. In addition, AIRL obtains lower reward compared to SIL and GAIL on Walker2d-v2. However, both SIL and AIRL yield superior performance in these environments when evaluated using the Sinkhorn distance as shown in Table 2. The result suggests that evaluating the performance of imitation learning algorithms with a true similarity metric, such as the Sinkhorn distance, can be more reliable since our objective is to match state-action distributions.

**Table 3.** Mean and standard deviation of the reward metric performance of imitator policies for BC, GAIL, AIRL and SIL.

| Environments | Trajectories | BC | GAIL | AIRL | SIL |
|---|---|---|---|---|---|
| Hopper-v2 | 2 | $391.38 \pm 42.98$ | $3341.27 \pm 38.96$ | $3353.33 \pm 2.05$ | $\mathbf{3376.70 \pm 2.45}$ |
| | 4 | $659.51 \pm 166.32$ | $3206.85 \pm 1.56$ | $\mathbf{3353.75 \pm 1.67}$ | $3325.66 \pm 4.24$ |
| | 8 | $1094.39 \pm 145.93$ | $3216.93 \pm 3.08$ | $\mathbf{3369.17 \pm 3.04}$ | $3335.31 \pm 2.66$ |
| | 16 | $2003.71 \pm 655.85$ | $\mathbf{3380.97 \pm 2.16}$ | $3338.07 \pm 2.14$ | $3376.55 \pm 2.65$ |
| | 32 | $2330.82 \pm 1013.71$ | $3333.93 \pm 1.47$ | $\mathbf{3361.56 \pm 1.93}$ | $3326.52 \pm 3.62$ |
| HalfCheetah-v2 | 2 | $-60.80 \pm 23.12$ | $764.91 \pm 546.47$ | $4467.83 \pm 61.13$ | $\mathbf{4664.65 \pm 91.73}$ |
| | 4 | $1018.68 \pm 236.13$ | $\mathbf{5183.67 \pm 118.74}$ | $4578.84 \pm 102.92$ | $4505.88 \pm 130.50$ |
| | 8 | $1590.73 \pm 279.05$ | $\mathbf{4902.46 \pm 721.43}$ | $4686.22 \pm 147.89$ | $4818.82 \pm 251.27$ |
| | 16 | $2434.30 \pm 733.29$ | $4519.49 \pm 157.99$ | $\mathbf{4783.79 \pm 197.27}$ | $4492.37 \pm 134.35$ |
| | 32 | $3598.98 \pm 558.70$ | $4661.17 \pm 147.21$ | $4633.48 \pm 116.89$ | $\mathbf{4795.68 \pm 191.90}$ |
| Walker2d-v2 | 2 | $591.92 \pm 32.77$ | $3509.37 \pm 8.08$ | $3497.80 \pm 9.64$ | $\mathbf{3566.32 \pm 16.11}$ |
| | 4 | $314.77 \pm 9.21$ | $\mathbf{3537.63 \pm 4.14}$ | $3496.61 \pm 10.94$ | $3523.73 \pm 21.91$ |
| | 8 | $808.37 \pm 5.28$ | $3394.15 \pm 4.74$ | $\mathbf{3488.68 \pm 10.67}$ | $3420.13 \pm 16.38$ |
| | 16 | $1281.80 \pm 81.11$ | $3444.96 \pm 23.99$ | $3459.84 \pm 8.25$ | $\mathbf{3557.51 \pm 11.67}$ |
| | 32 | $1804.74 \pm 1154.36$ | $3427.61 \pm 9.79$ | $\mathbf{3495.04 \pm 17.18}$ | $3203.32 \pm 23.65$ |
| Ant-v2 | 2 | $845.14 \pm 172.37$ | $3443.87 \pm 716.61$ | $\mathbf{5190.89 \pm 67.94}$ | $4981.70 \pm 50.89$ |
| | 4 | $897.54 \pm 2.14$ | $4912.92 \pm 606.99$ | $\mathbf{5182.42 \pm 65.70}$ | $5020.71 \pm 89.74$ |
| | 8 | $991.92 \pm 2.92$ | $5112.21 \pm 102.23$ | $5083.30 \pm 77.48$ | $\mathbf{5112.55 \pm 62.87}$ |
| | 16 | $1014.14 \pm 447.66$ | $4854.87 \pm 895.63$ | $\mathbf{5034.80 \pm 331.64}$ | $4935.33 \pm 87.15$ |
| | 32 | $2197.20 \pm 487.00$ | $5009.60 \pm 247.43$ | $\mathbf{5013.36 \pm 119.12}$ | $4581.27 \pm 123.75$ |
| Humanoid-v2 | 8 | $1462.47 \pm 1139.19$ | $1249.26 \pm 187.71$ | $3897.47 \pm 1047.03$ | $\mathbf{4456.09 \pm 2707.92}$ |
| | 16 | $2100.93 \pm 1116.79$ | $496.11 \pm 113.28$ | $4396.01 \pm 433.63$ | $\mathbf{6380.37 \pm 40.35}$ |
| | 32 | $4807.86 \pm 1903.08$ | $\mathbf{6252.73 \pm 570.72}$ | $1884.92 \pm 764.89$ | $5593.19 \pm 1967.86$ |

**Training Stability.** Table 2 showcases that SIL consistently minimizes the Sinkhorn distance while being robust to varying lengths of expert demonstrations. Figure 1 depicts the evolution of the Sinkhorn distance between occupancy measures of the learner and the expert in the training process of SIL. In spite of the training instability observed on Walker2d-v2 with 2 or 32 expert trajectories on the Humanoid-v2 environment, SIL still successfully learns to imitate the expert demonstrator. We speculate that training stability could be improved in these settings with further hyperparameter tuning as discussed in Sect. 5 which we leave for future work. Training stability of SIL is evident on the Hopper-v2, Ant-v2 and HalfCheetah-v2 environments.

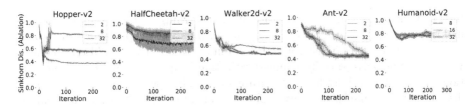

**Fig. 2.** *Ablation Study.* Mean and standard deviation of the Sinkhorn distance during training of SIL for three sets of varying number of trajectories. The critic network update has been replaced with a fixed cosine transport cost defined in Eq. (15).

**Table 4.** Mean and standard deviation of the reward and Sinkhorn metric performance after re-training SIL with a *fixed* cosine transport cost defined in Eq. (15).

| Environments | Metric | 2 | 8 | 32 |
|---|---|---|---|---|
| Hopper-v2 | Reward | $264.72 \pm 1.28$ | $520.88 \pm 29.83$ | $9.44 \pm 0.31$ |
| | Sinkhorn | $0.036 \pm 0.007$ | $0.552 \pm 0.008$ | $0.777 \pm 0.007$ |
| HalfCheetah-v2 | Reward | $-1643.98 \pm 198.31$ | $-844.52 \pm 267.42$ | $-1220.92 \pm 217.86$ |
| | Sinkhorn | $0.670 \pm 0.141$ | $0.841 \pm 0.035$ | $0.424 \pm 0.031$ |
| Walker2d-v2 | Reward | $60.64 \pm 7.92$ | $-2.39 \pm 14.05$ | $-11.38 \pm 1.22$ |
| | Sinkhorn | $0.538 \pm 0.006$ | $0.487 \pm 0.005$ | $0.466 \pm 0.009$ |
| Ant-v2 | Reward | $1482.03 \pm 480.99$ | $607.87 \pm 87.09$ | $114.22 \pm 123.24$ |
| | Sinkhorn | $0.398 \pm 0.090$ | $0.419 \pm 0.025$ | $0.424 \pm 0.031$ |
| | | **8** | **16** | **32** |
| Humanoid-v2 | Reward | $447.87 \pm 31.26$ | $505.47 \pm 67.62$ | $335.48 \pm 65.14$ |
| | Sinkhorn | $0.760 \pm 0.011$ | $0.789 \pm 0.013$ | $0.835 \pm 0.012$ |

**Ablation Study.** To study the effect of minimizing the Sinkhorn distance between occupancy measures using a fixed transport cost, we repeat our experiments on the environments Hopper-v2, HalfCheetah-v2, Walker2d-v2 and Ant-v2 with $\{2, 8, 32\}$ trajectory sets. For Humanoid-v2 we conduct the experiments on sets of $\{8, 16, 32\}$. In this ablation study, instead of training a critic network in an adversarially learned feature space, we assign a reward proxy defined by Equation (7) with a *fixed* cosine transport cost introduced in Equation (15).

Figure 2 depicts the evolution of the Sinkhorn distance between occupancy measures during training of SIL, after replacing the adversarial objective of the critic network with a fixed transport cost. While the training process is more stable, it fails to achieve good performance in terms both of the Sinkhorn distance metric (Figs. 1 and 2) and reward metric (see Table 4). The result suggests that the training objective of the critic network has been a crucial part of the proposed algorithm in providing sufficiently strong signals to the learner policy to match the expert's state-action distribution.

## 5   Conclusion

In this work we presented Sinkhorn imitation learning (SIL), a solution to optimal transport based imitation learning, by formulating the problem of matching an expert's state-action distribution as minimization of their Sinkhorn distance. We utilized an adversarially trained critic that maps the state-action observations to an adversarially learned feature space. The use of the critic provides a discriminative signal to the learner policy to facilitate the imitation of an expert demonstrator's behavior. Experiments on 5 MuJoCo environments demonstrate that SIL exhibits competitive performance compared to the baselines.

The Sinkhorn imitation learning framework can be extended in several directions to address current limitations which we aim to study in future work. Currently, SIL's formulation makes it compatible with only on-policy RL methods as computing the Sinkhorn distance necessitates complete trajectories. While SIL is efficient compared to other on-policy adversarial IL benchmarks, it still requires more environment interactions to learn compared to off-policy adversarial IL methods. Hence, it is an interesting future direction to extend SIL to be compatible with off-policy RL algorithms, in line with previous work [7,14,15,25] to yield a method that both inherits the theoretical benefits of OT while being sample efficient. Additionally, performance of SIL was reported with a fixed critic network structure in all studied experiments. Hence, it is unclear what is the effect of the network architecture in guiding imitation learning. It will be of practical significance to investigate the impact of different critic network architectures on training stability and computational efficiency, as well as its relationship to the dimension of state-action space. Another interesting research area is to extend the current framework to incorporate the temporal dependence of the trajectory in the construction of the optimal transport coupling and subsequently the reward proxy. We anticipate that this will be a promising direction for improving the sample efficiency and generalization performance of the optimal transport-based adversarial imitation learning framework.

# References

1. Abbeel, P., Dolgov, D., Ng, A.Y., Thrun, S.: Apprenticeship learning for motion planning with application to parking lot navigation. In: Proc. IEEE/RSJ International Conference on Intelligent Robots and Systems, pp. 1083–1090 (2008)
2. Abbeel, P., Coates, A., Ng, A.Y.: Autonomous helicopter aerobatics through apprenticeship learning. Int. J. Robot. Res. **29**(13), 1608–1639 (2010)
3. Arjovsky, M., Chintala, S., Bottou, L.: Wasserstein generative adversarial networks. In: Proceedings of the International Conference on Machine Learning (ICML), pp. 214–223. Sydney, Australia (2017)
4. Blondé, L., Kalousis, A.: Sample-efficient imitation learning via generative adversarial nets. In: Proceedings of the International Conference on Artificial Intelligence and Statistics (AISTATS). Okinawa, Japan (2019)
5. Coates, A., Abbeel, P., Ng, A.Y.: Learning for control from multiple demonstrations. In: Proceedings of the International Conference on Machine Learning (ICML), pp. 144–151. Helsinki, Finland (2008)
6. Cuturi, M.: Sinkhorn distances: lightspeed computation of optimal transport. In: Proceedings of the Advances in Neural Information Processing Systems (NeurIPS), pp. 2292–2300. Lake Tahoe, Nevada, USA (2013)
7. Dadashi, R., Hussenot, L., Geist, M., Pietquin, O.: Primal Wasserstein imitation learning. In: Proceedings of the International Conference on Learning Representations (ICLR) (2021)
8. Fu, J., Luo, K., Levine, S.: Learning robust rewards with adverserial inverse reinforcement learning. In: Proceedings of the International Conference on Learning Representations (ICLR). Vancouver, Canada (2018)

9. Ghasemipour, S.K.S., Zemel, R., Gu, S.: A divergence minimization perspective on imitation learning methods. In: Proceedings of the Conference on Robot Learning (CoRL). Osaka, Japan (2019)
10. Goodfellow, I., et al.: Generative adversarial nets. In: Proc. In: Advances in Neural Information Processing Systems (NeurIPS), pp. 2672–2680. Montréal, Canada (2014)
11. Ho, J., Ermon, S.: Generative adversarial imitation learning. In: Proceedings of the Advances in Neural Information Processing Systems (NeurIPS), pp. 4565–4573. Barcelona, Spain (2016)
12. Ke, L., Barnes, M., Sun, W., Lee, G., Choudhury, S., Srinivasa, S.S.: Imitation learning as f-divergence minimization. arXiv:1905.12888 (2019)
13. Kober, J., Peters, J.: Learning motor primitives for robotics. In: Proceedings of the IEEE International Conference on Robotics and Automation, pp. 2112–2118. Kobe, Japan (2009)
14. Kostrikov, I., Agrawal, K.K., Levine, S., Tompson, J.: Discriminator-actor-critic: addressing sample inefficiency and reward bias in adversarial imitation learning. In: Proceedings of the International Conference on Learning Representations (ICLR). New Orleans, USA (2019)
15. Kostrikov, I., Nachum, O., Tompson, J.: Imitation learning via off-policy distribution matching. In: Proceedings of the International Conference on Learning Representations (ICLR) (2020)
16. Kuderer, M., Kretzschmar, H., Burgard, W.: Teaching mobile robots to cooperatively navigate in populated environments. In: Proceedings of the IEEE/RSJ International Conference on Intelligent Robots and Systems, pp. 3138–3143 (2013)
17. Laskey, M., Lee, J., Fox, R., Dragan, A., Goldberg, K.: DART: noise injection for robust imitation learning. In: Proceedings of the Conference on Robot Learning (CoRL). Mountain View, USA (2017)
18. Mnih, V., Kavukcuoglu, K., Silver, D., et al.: Human-level control through deep reinforcement learning. Nature **518**, 529–533 (2015)
19. Ng, A.Y., Russell, S.: Algorithms for inverse reinforcement learning. In: Proceedings of the International Conference on Machine Learning (ICML), pp. 663–670. Stanford, CA, USA (2000)
20. Osa, T., Pajarinen, J., Neumann, G., Bagnell, J.A., Abbeel, P., Peters, J.: An algorithmic perspective on imitation learning. Now Foundations and Trends (2018)
21. Osa, T., Sugita, N., Mitsuishi, M.: Online trajectory planning in dynamic environments for surgical task automation. In: Proceedings of the Robotics: Science and Systems. Berkley, CA, USA (2014)
22. Park, D., Noseworthy, M., Paul, R., Roy, S., Roy, N.: Inferring task goals and constraints using Bayesian nonparametric inverse reinforcement learning. In: Proceedings of the Conference on Robot Learning (CoRL). Osaka, Japan (2019)
23. Pomerleau, D.A.: ALVINN: an autonomous land vehicle in a neural network. In: Proceedings of the Advances in Neural Information Processing Systems (NeurIPS), pp. 305–313 (1989)
24. Rajeswaran, A., et al.: Learning complex dexterous manipulation with deep reinforcement learning and demonstrations. In: Proceedings of the Robotics: Science and Systems (RSS). Pittsburgh, Pennsylvania (2018)
25. Reddy, S., Dragan, A.D., Levine, S.: SQIL: imitation learning via reinforcement learning with sparse rewards. In: Proceedings of the International Conference on Learning Representations (ICLR) (2020)

26. Ross, S., Gordon, G., Bagnell, J.: A reduction of imitation learning and structured prediction to no-regret online learning. In: Proceedings of the Conference on Artificial Intelligence and Statistics, pp. 627–635. Fort Lauderdale, FL, USA (2011)

27. Ross, S., Bagnell, A.: Efficient reductions for imitation learning. In: Proceedings of the International Conference on Artificial Intelligence and Statistics, pp. 661–668, 13–15 May 2010. Chia Laguna Resort, Sardinia, Italy

28. Salimans, T., et al.: Improved techniques for training GANs. In: Proceedings of the Advances in Neural Information Processing Systems (NeurIPS), pp. 2234–2242. Barcelona, Spain (2016)

29. Salimans, T., Zhang, H., Radford, A., Metaxas, D.N.: Improving GANs using optimal transport. In: Proceedings of the International Conference on Learning Representation (ICLR). Vancouver, Canada (2018)

30. Sammut, C., Hurst, S., Kedzier, D., Michie, D.: Learning to fly. In: Proceedings of the International Conference on Machine Learning (ICML), pp. 385–393. Aberdeen, Scotland, United Kingdom (1992)

31. Sasaki, F., Yohira, T., Kawaguchi, A.: Sample efficient imitation learning for continuous control. In: Proceedings of the International Conference on Learning Representations (ICLR) (2019)

32. Schulman, J., Levine, S., Abbeel, P., Jordan, M., Moritz, P.: Trust region policy optimization. In: Proceedings of the International Conference on Machine Learning (ICML), pp. 1889–1897. Lille, France (2015)

33. Silver, D., et al.: Mastering the game of go with deep neural networks and tree search. Nature **529**, 484–503 (2016)

34. Sriperumbudur, B.K., Gretton, A., Fukumizu, K., Lanckriet, G.R.G., Schölkopf, B.: A note on integral probability metrics and $\phi$-divergences. arXiv:0901.2698 (2009)

35. Todorov, E., Erez, T., Tassa, Y.: MuJoCo: a physics engine for model-based control. In: Proceedings of the IEEE/RSJ International Conference on Intelligent Robots and Systems, pp. 5026–5033 (2012)

36. Torabi, F., Geiger, S., Warnell, G., Stone, P.: Sample-efficient adversarial imitation learning from observation. arXiv:1906.07374 (2019)

37. Villani, C.: Optimal Transport: Old and New, vol. 338. Springer Science & Business Media, Berlin, Germany (2008). https://doi.org/10.1007/978-3-540-71050-9

38. Vinyals, O., et al.: Grandmaster level in StarCraft II using multi-agent reinforcement learning. Nature **575**, 350–354 (2019)

39. Xiao, H., Herman, M., Wagner, J., Ziesche, S., Etesami, J., Linh, T.H.: Wasserstein adversarial imitation learning. arXiv:1906.08113 (2019)

40. Zhu, Y., et al.: Reinforcement and imitation learning for diverse visuomotor skills. arXiv:1802.09564 (2018)

41. Ziebart, B., Bagnell, A.J.: Modeling interaction via the principle of maximum causal entropy. In: Proceedings of the International Conference on Machine Learning (ICML), pp. 1255–1262. Haifa, Israel (2010)

42. Ziebart, B., Mass, A., Bagnell, A.J., Dey, A.K.: Maximum entropy inverse reinforcement learning. In: Proceedings of the AAAI Conference on Artificial intelligence, pp. 1433–1438 (2008)

43. Zucker, M., et al.: Optimization and learning for rough terrain legged locomotion. Int. J. Robot. Res. **30**(2), 175–191 (2011)

# Safe Exploration Method
# for Reinforcement Learning Under
# Existence of Disturbance

Yoshihiro Okawa[1]([✉])⬤, Tomotake Sasaki[1]⬤, Hitoshi Yanami[1],
and Toru Namerikawa[2]⬤

[1] Artificial Intelligence Laboratory, Fujitsu Limited, Kawasaki, Japan
{okawa.y,tomotake.sasaki,yanami}@fujitsu.com
[2] Department of System Design Engineering, Keio University, Yokohama, Japan
namerikawa@keio.jp

**Abstract.** Recent rapid developments in reinforcement learning algorithms have been giving us novel possibilities in many fields. However, due to their exploring property, we have to take the risk into consideration when we apply those algorithms to safety-critical problems especially in real environments. In this study, we deal with a safe exploration problem in reinforcement learning under the existence of disturbance. We define the safety during learning as satisfaction of the constraint conditions explicitly defined in terms of the state and propose a safe exploration method that uses partial prior knowledge of a controlled object and disturbance. The proposed method assures the satisfaction of the explicit state constraints with a pre-specified probability even if the controlled object is exposed to a stochastic disturbance following a normal distribution. As theoretical results, we introduce sufficient conditions to construct conservative inputs not containing an exploring aspect used in the proposed method and prove that the safety in the above explained sense is guaranteed with the proposed method. Furthermore, we illustrate the validity and effectiveness of the proposed method through numerical simulations of an inverted pendulum and a four-bar parallel link robot manipulator.

**Keywords:** Reinforcement learning · Safe exploration · Chance constraint

## 1 Introduction

Guaranteeing safety and performance during learning is one of the critical issues to implement reinforcement learning (RL) in real environments [12,14]. To address this issue, RL algorithms and related methods dealing with safety have been studied in recent years and some of them are called "safe reinforcement learning" [10]. For example, Biyik et al. [4] proposed a safe exploration

---

**Supplementary Information** The online version contains supplementary material available at https://doi.org/10.1007/978-3-031-26412-2_9.

algorithm for a deterministic Markov decision process (MDP) to be used in RL. They guaranteed to prevent states from being unrecoverable by leveraging the Lipschitz continuity of its unknown transition dynamics. In addition, Ge et al. [11] proposed a modified Q-learning method for a constrained MDP solved with the Lagrange multiplier method so that their algorithm seeks for the optimal solution ensuring that the safety premise is satisfied. Several methods use prior knowledge of the controlled object (i.e., environment) for guaranteeing the safety [3,17]. However, few studies evaluated their safety quantitatively from a viewpoint of satisfying state constraints at each timestep that are defined explicitly in the problems. Evaluating safety from this viewpoint is often useful when we have constraints on a physical system and need to estimate the risk caused by violating those constraints beforehand.

Recently, Okawa et al. [19] proposed a safe exploration method that is applicable to existing RL algorithms. They quantitatively evaluated the above-mentioned safety in accordance with probabilities of satisfying the explicit state constraints. In particular, they theoretically showed that their proposed method assures the satisfaction of the state constraints with a pre-specified probability by using partial prior knowledge of the controlled object. However, they did not consider the existence of external disturbance, which is an important factor when we consider safety. Such disturbance sometimes makes the state violate the constraints even if the inputs (i.e., actions) used in exploration are designed to satisfy those constraints. Furthermore, they made a strong assumption regarding the controlled objects such that the state remains within the area satisfying the constraints if the input is set to be zero as a conservative input, i.e., an input that does not contain an exploring aspect.

In this study, we extend Okawa et al.'s work [19] and tackle the safe exploration problem in RL under the existence of disturbance[1]. Our main contributions are the following.

- We propose a novel safe exploration method for RL that uses partial prior knowledge of both the controlled object and disturbance.
- We introduce sufficient conditions to construct conservative inputs not containing an exploring aspect used in the proposed method. Moreover, we theoretically prove that our proposed method assures the satisfaction of explicit state constraints with a pre-specified probability under the existence of disturbance that follows a normal distribution.

We also demonstrate the validity and effectiveness of the proposed method with the simulated inverted pendulum provided in OpenAI Gym [6] and a four-bar parallel link robot manipulator [18] with additional disturbances.

The rest of this paper is organized as follows. In Sect. 2, we introduce the problem formulation of this study. In Sect. 3, we describe our safe exploration method. Subsequently, theoretical results about the proposed method are shown

---

[1] Further comparison with other related works is given in Appendix A (electronic supplementary material).

in Sect. 4. We illustrate the results of simulation evaluation in Sect. 5. We discuss the limitations of the proposed method in Sect. 6, and finally, we conclude this paper in Sect. 7.

## 2    Problem Formulation

We consider an input-affine discrete-time nonlinear dynamic system (environment) expressed by the following state transition equation:

$$\boldsymbol{x}_{k+1} = \boldsymbol{f}(\boldsymbol{x}_k) + \boldsymbol{G}(\boldsymbol{x}_k)\boldsymbol{u}_k + \boldsymbol{w}_k, \tag{1}$$

where $\boldsymbol{x}_k \in \mathbb{R}^n$, $\boldsymbol{u}_k \in \mathbb{R}^m$, and $\boldsymbol{w}_k \in \mathbb{R}^n$ stand for the state, input (action) and disturbance at timestep $k$, respectively, and $\boldsymbol{f} : \mathbb{R}^n \to \mathbb{R}^n$ and $\boldsymbol{G} : \mathbb{R}^n \to \mathbb{R}^{n \times m}$ are unknown nonlinear functions. We suppose that the state $\boldsymbol{x}_k$ is directly observable. An immediate cost $c_{k+1} \geq 0$ is given depending on the state, input and disturbance at each timestep $k$:

$$c_{k+1} = c(\boldsymbol{x}_k, \ \boldsymbol{u}_k, \ \boldsymbol{w}_k), \tag{2}$$

where the immediate cost function $c : \mathbb{R}^n \times \mathbb{R}^m \times \mathbb{R}^n \to [0, \infty)$ is unknown while $c_{k+1}$ is supposed to be directly observable. We consider the situation where the constraints that the state is desired to satisfy from the viewpoint of safety are explicitly given by the following linear inequalities:

$$\boldsymbol{H}\boldsymbol{x} \preceq \boldsymbol{d}, \tag{3}$$

where $\boldsymbol{d} = [d_1, \ldots, d_{n_c}]^\top \in \mathbb{R}^{n_c}$, $\boldsymbol{H} = [\boldsymbol{h}_1, \ldots, \boldsymbol{h}_{n_c}]^\top \in \mathbb{R}^{n_c \times n}$, $n_c$ is the number of constraints and $\preceq$ means that the standard inequality $\leq$ on $\mathbb{R}$ holds for all elements. In addition, we define $\mathcal{X}_s \subset \mathbb{R}^n$ as the set of safe states, that is,

$$\mathcal{X}_s := \{\boldsymbol{x} \in \mathbb{R}^n | \boldsymbol{H}\boldsymbol{x} \preceq \boldsymbol{d}\}. \tag{4}$$

Initial state $\boldsymbol{x}_0$ is assumed to satisfy $\boldsymbol{x}_0 \in \mathcal{X}_s$ for simplicity.

The primal goal of reinforcement learning is to acquire a policy (control law) that minimizes or maximizes an evaluation function with respect to the immediate cost or reward, using them as cues in its trial-and-error process [20]. In this study, we consider the standard discounted cumulative cost as the evaluation function to be minimized:

$$J = \sum_{k=0}^{T} \gamma^k c_{k+1}. \tag{5}$$

Here, $\gamma$ is a discount factor ($0 < \gamma \leq 1$) and $T$ is the terminal time.

Besides (5) for the cost evaluation, we define the safety in this study as satisfaction of the state constraints and evaluate its guarantee quantitatively. In detail, we consider the following chance constraint with respect to the satisfaction of the explicit state constraints (3) at each timestep $k$:

$$\Pr\{\boldsymbol{H}\boldsymbol{x}_k \preceq \boldsymbol{d}\} \geq \eta, \tag{6}$$

where $\Pr\{\boldsymbol{H}\boldsymbol{x}_k \preceq \boldsymbol{d}\}(= \Pr\{\boldsymbol{x}_k \in \mathcal{X}_s\})$ denotes the probability that $\boldsymbol{x}_k$ satisfies the constraints (3).

The objective of the proposed safe exploration method is to make the chance constraint (6) satisfied at every timestep $k = 1, 2, \ldots, T$ for a pre-specified $\eta$, where $0.5 < \eta < 1$ in this study.

Figure 1 shows the overall picture of the reinforcement learning problem in this study. The controller (agent) depicted as the largest red box generates an input (action) $\boldsymbol{u}_k$ according to a base policy with the proposed safe exploration method and apply it to the controlled object (environment) depicted as the green box, which is a discrete-time nonlinear dynamic system exposed to a disturbance $\boldsymbol{w}_k$. According to an RL algorithm, the base policy is updated based on the state $\boldsymbol{x}_{k+1}$ and immediate cost $c_{k+1}$ observed from the controlled object. In addition to updating the base policy to minimize the evaluation function, the chance constraint should be satisfied at every timestep $k = 1, 2, \ldots, T$. The proposed method is described in detail in Sects. 3 and 4.

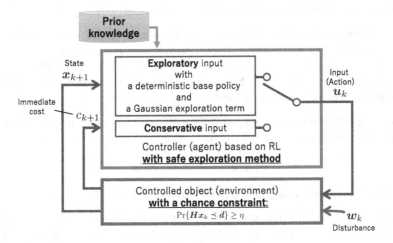

**Fig. 1.** Overview of controlled object (environment) under existence of disturbance and controller (agent) based on an RL algorithm with the proposed safe exploration method. The controller updates its base policy through an RL algorithm, while the proposed safe exploration method makes the chance constraint of controlled object satisfied by adjusting its exploration process online.

As the base policy, we consider a nonlinear deterministic feedback control law

$$\boldsymbol{\mu}(\,\cdot\,;\boldsymbol{\theta}) : \mathbb{R}^n \to \mathbb{R}^m$$
$$x \mapsto \boldsymbol{\mu}(\boldsymbol{x};\boldsymbol{\theta}), \tag{7}$$

where $\boldsymbol{\theta} \in \mathbb{R}^{N_\theta}$ is an adjustable parameter to be updated by an RL algorithm. When we allow exploration, we generate an input $\boldsymbol{u}_k$ by the following equation:

$$\boldsymbol{u}_k = \boldsymbol{\mu}(\boldsymbol{x}_k;\boldsymbol{\theta}_k) + \boldsymbol{\varepsilon}_k, \tag{8}$$

where $\varepsilon_k \in \mathbb{R}^m$ is a stochastic exploration term that follows an $m$-dimensional normal distribution (Gaussian probability density function) with mean $\mathbf{0} \in \mathbb{R}^m$ and variance-covariance matrix $\boldsymbol{\Sigma}_k \in \mathbb{R}^{m \times m}$, denoted as $\varepsilon_k \sim \mathcal{N}(\mathbf{0}, \boldsymbol{\Sigma}_k)$. In this case, as a consequence of the definition, $\boldsymbol{u}_k$ follows a normal distribution $\mathcal{N}(\boldsymbol{\mu}(\boldsymbol{x}_k; \boldsymbol{\theta}_k), \boldsymbol{\Sigma}_k)$.

We make the following four assumptions about the controlled object and the disturbance. The proposed method uses these prior knowledge to generate inputs, and the theoretical guarantee of satisfying the chance constraint is proven by using these assumptions.

**Assumption 1.** *Matrices $\boldsymbol{A} \in \mathbb{R}^{n \times n}$ and $\boldsymbol{B} \in \mathbb{R}^{n \times m}$ in the following linear approximation model of the nonlinear dynamics (1) are known:*

$$\boldsymbol{x}_{k+1} \simeq \boldsymbol{A}\boldsymbol{x}_k + \boldsymbol{B}\boldsymbol{u}_k + \boldsymbol{w}_k. \tag{9}$$

The next assumption is about the disturbance.

**Assumption 2.** *The disturbance $\boldsymbol{w}_k$ stochastically occurs according to an $n$-dimensional normal distribution $\mathcal{N}(\boldsymbol{\mu}_w, \boldsymbol{\Sigma}_w)$, where $\boldsymbol{\mu}_w \in \mathbb{R}^n$ and $\boldsymbol{\Sigma}_w \in \mathbb{R}^{n \times n}$ are the mean and the variance-covariance matrix, respectively. The mean $\boldsymbol{\mu}_w$ and variance-covariance matrix $\boldsymbol{\Sigma}_w$ are known, and the disturbance $\boldsymbol{w}_k$ and exploration term $\varepsilon_k$ are uncorrelated at each timestep $k$.*

We define the difference $\boldsymbol{e}(\boldsymbol{x}, \boldsymbol{u}) \in \mathbb{R}^n$ between the nonlinear system (1) and the linear approximation model (9) (i.e., approximation error) as below:

$$\boldsymbol{e}(\boldsymbol{x}, \boldsymbol{u}) := \boldsymbol{f}(\boldsymbol{x}) + \boldsymbol{G}(\boldsymbol{x})\boldsymbol{u} - (\boldsymbol{A}\boldsymbol{x} + \boldsymbol{B}\boldsymbol{u}). \tag{10}$$

We make the following assumption on this approximation error.

**Assumption 3.** *Regarding the approximation error $\boldsymbol{e}(\boldsymbol{x}, \boldsymbol{u})$ defined by (10), $\bar{\delta}_j < \infty$, $\bar{\Delta}_j < \infty$, $j = 1, \ldots, n_c$ that satisfy the following inequalities are known:*

$$\bar{\delta}_j \geq \sup_{\boldsymbol{x} \in \mathbb{R}^n, \ \boldsymbol{u} \in \mathbb{R}^m} |\boldsymbol{h}_j^\top \boldsymbol{e}(\boldsymbol{x}, \boldsymbol{u})|, \quad j = 1, 2, \ldots, n_c, \tag{11}$$

$$\bar{\Delta}_j \geq \sup_{\boldsymbol{x} \in \mathbb{R}^n, \ \boldsymbol{u} \in \mathbb{R}^m} |\boldsymbol{h}_j^\top \left(\boldsymbol{A}^{\tau-1} + \boldsymbol{A}^{\tau-2} + \cdots + \boldsymbol{I}\right) \boldsymbol{e}(\boldsymbol{x}, \boldsymbol{u})|, \quad j = 1, 2, \ldots, n_c, \tag{12}$$

*where $\tau$ is a positive integer.*

The following assumption about the linear approximation model and the constraints is also made.

**Assumption 4.** *The following condition holds for $\boldsymbol{B}$ and $\boldsymbol{H} = [\boldsymbol{h}_1, \ldots, \boldsymbol{h}_{n_c}]^\top$:*

$$\boldsymbol{h}_j^\top \boldsymbol{B} \neq \mathbf{0}, \quad \forall j = 1, 2, \ldots, n_c. \tag{13}$$

Regarding the above-mentioned assumptions, Assumptions 1 and 4 are similar to the ones used in [19], while we make a relaxed assumption on the approximation error in Assumption 3 and remove assumptions on the autonomous dynamics $\boldsymbol{f}$ and conservative inputs used in [19].

## 3  Safe Exploration Method with Conservative Inputs

Now, we propose the safe exploration method to guarantee the safety in the sense of satisfaction of the chance constraint (6). As shown in Fig. 1, the basic idea is to decide whether to explore or not by using the knowledge about the controlled object and disturbance. The detailed way is given as Algorithm 1 below. Here $\wedge$ is the logical conjunction, $\Phi$ is the normal cumulative distribution function,

---

**Algorithm 1.** Proposed safe exploration method

---

At every timestep $k \geq 0$, observe state $\boldsymbol{x}_k$ and generate input $\boldsymbol{u}_k$ as follows:

(i) if $\boldsymbol{x}_k \in \mathcal{X}_s \wedge \left( \left\| \boldsymbol{h}_j^\top \boldsymbol{\Sigma}_w^{\frac{1}{2}} \right\|_2 \leq \dfrac{1}{\Phi^{-1}(\eta_k')} (d_j - \boldsymbol{h}_j^\top \hat{\boldsymbol{x}}_{k+1} - \delta_j), \forall \delta_j \in \{\pm \bar{\delta}_j\}, \forall j = 1, \dots, n_c \right)$

$\boldsymbol{u}_k = \boldsymbol{\mu}(\boldsymbol{x}_k; \boldsymbol{\theta}_k) + \boldsymbol{\varepsilon}_k$, where $\boldsymbol{\varepsilon}_k \sim \mathcal{N}(\boldsymbol{0}, \boldsymbol{\Sigma}_k)$,

(ii) **elseif** $\boldsymbol{x}_k \in \mathcal{X}_s \wedge \left( \left\| \boldsymbol{h}_j^\top \boldsymbol{\Sigma}_w^{\frac{1}{2}} \right\|_2 > \dfrac{1}{\Phi^{-1}(\eta_k')} (d_j - \boldsymbol{h}_j^\top \hat{\boldsymbol{x}}_{k+1} - \delta_j), \text{ for some } \delta_j \in \{\pm \bar{\delta}_j\} \right)$

$\boldsymbol{u}_k = \boldsymbol{u}_k^{stay}$,

(iii) **else** (i.e., $\boldsymbol{x}_k \notin \mathcal{X}_s$)

$\boldsymbol{u}_k = \boldsymbol{u}_k^{back}$.

---

$$\hat{\boldsymbol{x}}_{k+1} := \boldsymbol{A}\boldsymbol{x}_k + \boldsymbol{B}\boldsymbol{\mu}(\boldsymbol{x}_k; \boldsymbol{\theta}_k) + \boldsymbol{\mu}_w, \qquad \eta_k' := 1 - \frac{1 - \left(\frac{\eta}{\xi^k}\right)^{\frac{1}{T}}}{n_c}, \qquad (14)$$

and $\xi$ is a positive real number that satisfies $\eta^{\frac{1}{T}} < \xi < 1$. The quantity $\hat{\boldsymbol{x}}_{k+1}$ is a one-step ahead predicted state based on the mean of the linear approximation model (9) with substitution of (8)[2]. In the case (i), the degree of exploration is adjusted by choosing the variance-covariance matrix $\boldsymbol{\Sigma}_k$ of the stochastic exploration term $\boldsymbol{\varepsilon}_k$ to satisfy the following inequality for all $j = 1, \dots, n_c$:

$$\left\| \boldsymbol{h}_j^\top \boldsymbol{B}' \begin{bmatrix} \boldsymbol{\Sigma}_k & \\ & \boldsymbol{\Sigma}_w \end{bmatrix}^{\frac{1}{2}} \right\|_2 \leq \frac{1}{\Phi^{-1}(\eta_k')} (d_j - \boldsymbol{h}_j^\top \hat{\boldsymbol{x}}_{k+1} - \delta_j), \forall \delta_j \in \{\pm \bar{\delta}_j\}, \qquad (15)$$

where $\boldsymbol{B}' = [\boldsymbol{B}, \boldsymbol{I}]$.

Note that the case $\boldsymbol{x}_k \in \mathcal{X}_s$ (i.e., the current state satisfies all constraints) is divided to (i) and (ii) depending on the one-step ahead predicted state $\hat{\boldsymbol{x}}_{k+1}$, and we use an exploratory input only when $\left\| \boldsymbol{h}_j^\top \boldsymbol{\Sigma}_w^{\frac{1}{2}} \right\|_2 \leq \frac{1}{\Phi^{-1}(\eta_k')}(d_j - \boldsymbol{h}_j^\top \hat{\boldsymbol{x}}_{k+1} - \delta_j), \forall \delta_j \in \{\pm \bar{\delta}_j\}$ holds for all $j$. Rough and intuitive meaning of this condition is that we allow exploration only when the next state probably stays in $\mathcal{X}_s$ even if we generate the input with $\boldsymbol{\varepsilon}_k$, given that $\boldsymbol{\Sigma}_k$ is a solution of (15).

---

[2] Note that the means of $\boldsymbol{\varepsilon}_k$ and $\boldsymbol{w}_k$ are assumed to be $\boldsymbol{0}$ and $\boldsymbol{\mu}_w$, respectively.

The inputs $\boldsymbol{u}_k^{stay}$ and $\boldsymbol{u}_k^{back}$ used in the cases (ii) and (iii) are defined as below. These inputs do not contain exploring aspects, and thus we call them conservative inputs.

**Definition 1.** *We call* $\boldsymbol{u}_k^{stay}$ *a conservative input of the first kind with which*
$\Pr\{\boldsymbol{H}\boldsymbol{x}_{k+1} \preceq \boldsymbol{d}\} \geq \left(\frac{\eta}{\xi^k}\right)^{\frac{1}{\tau}}$ *holds if* $\boldsymbol{x}_k = \boldsymbol{x} \in \mathcal{X}_s$ *occurs at timestep* $k \geq 0$.

**Definition 2.** *We call* $\boldsymbol{u}_k^{back}$, $\boldsymbol{u}_{k+1}^{back}$, $\dots$, $\boldsymbol{u}_{k+\tau-1}^{back}$ *a sequence of conservative inputs of the second kind with which for some* $j \leq \tau$, $\Pr\{\boldsymbol{x}_{k+j} \in \mathcal{X}_s\} \geq \xi$ *holds if* $\boldsymbol{x}_k = \boldsymbol{x} \notin \mathcal{X}_s$ *occurs at timestep* $k \geq 1$. *That is, using these inputs in this order, the state moves back to* $\mathcal{X}_s$ *within* $\tau$ *steps with a probability of at least* $\xi$.

We give sufficient conditions to construct these $\boldsymbol{u}_k^{stay}$ and $\boldsymbol{u}_k^{back}$ in Sect. 4.3. As shown in the examples in Sect. 5.1, the controllability index of the linear approximation model can be used as a clue to find the positive integer $\tau$.

Figure 2 illustrates how the proposed method switches the inputs differently in accordance with the three cases. In the case (i), the state constraints are satisfied and the input contains exploring aspect, (ii) the state constraints are satisfied but the input does not contain exploring aspect, and (iii) the state constraints are not satisfied and the input does not contain exploring aspect.

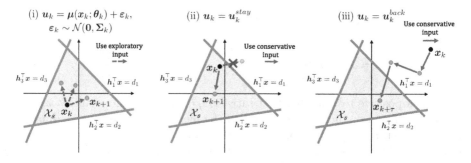

**Fig. 2.** Illustration of the proposed method for a case of $n = 2$ and $n_c = 3$. The proposed method switches two types of inputs in accordance with the current and one-step ahead predicted state information: exploratory inputs generated by a deterministic base policy and a Gaussian exploration term are used in the case (i), while the conservative ones that do not contain exploring aspect are used in the cases (ii) and (iii).

The proposed method, Algorithm 1, switches the exploratory inputs and the conservative ones in accordance with the current and one-step ahead predicted state information by using prior knowledge of both the controlled object and disturbance, while the previous work [19] only used that of the controlled object. In addition, this method adjusts the degree of exploration to an appropriate level by restricting $\boldsymbol{\Sigma}_k$ of the exploration term $\boldsymbol{\varepsilon}_k$ to a solution of (15), which also contains prior knowledge of both the controlled object and disturbance.

# 4    Theoretical Guarantee for Chance Constraint Satisfaction

In this section, we provide theoretical results regarding the safe exploration method we introduced in the previous section. In particular, we theoretically prove that the proposed method makes the state constraints satisfied with a pre-specified probability, i.e., makes the chance constraint (6) hold, at every timestep.

We consider the case (i) in Algorithm 1 in Subsect. 4.1 and the case (iii) in Subsect. 4.2, respectively. We provide Theorem 1 regarding the construction of conservative inputs used in the cases (ii) and (iii) in Subsect. 4.3. Then, in Subsect. 4.4, we provide Theorem 2, which shows that the proposed method makes the chance constraint (6) satisfied at every timestep $k$ under Assumptions 1–4. Proofs of the lemmas and theorems described in this section are given in Appendix B.

## 4.1    Theoretical Result on the Exploratory Inputs Generated with a Deterministic Base Policy and a Gaussian Exploration Term

First, we consider the case (i) in Algorithm 1 in which we generate an input containing exploring aspect according to (8) with a deterministic base policy and a Gaussian exploration term. The following lemma holds.

**Lemma 1.** *Let $q \in (0.5, 1)$. Suppose Assumptions 1, 2, and 3 hold. Generate input $u_k$ according to (8) when the state of the nonlinear system (1) at timestep $k$ is $x_k$. Then, the following inequality is a sufficient condition for $\Pr\{h_j^\top x_{k+1} \leq d_j\} \geq q$, $\forall j = 1, \ldots, n_c$:*

$$\left\| h_j^\top B' \begin{bmatrix} \Sigma_k & \\ & \Sigma_w \end{bmatrix}^{\frac{1}{2}} \right\|_2 \leq \frac{1}{\Phi^{-1}(q)} \left\{ d_j - h_j^\top (Ax_k + B\mu(x_k; \theta_k) + \mu_w) + \delta_j \right\},$$

$$\forall j = 1, 2, \ldots, n_c, \quad \forall \delta_j \in \{\bar{\delta}_j, -\bar{\delta}_j\}, \tag{16}$$

*where $B' = [B, I]$ and $\Phi$ is the normal cumulative distribution function.*

Proof is given in Appendix B.1. This lemma is proved with the equivalent transformation of a chance constraint into its deterministic counterpart [5, §4.4.2] and holds since the disturbance $w_k$ follows a normal distribution and is uncorrelated to the input $u_k$ according to Assumption 2 and (8). Furthermore, this lemma shows that, in the case (i), the state satisfies the constraints with an arbitrary probability $q \in (0.5, 1)$ by adjusting the variance-covariance matrix $\Sigma_k$ used to generate the Gaussian exploration term $\varepsilon_k$ so that the inequality (16) would be satisfied.

## 4.2  Theoretical Result on the Conservative Inputs of the Second Kind

Next, we consider the case (iii) in Algorithm 1 in which the state constraints are not satisfied. In this case, we use the conservative inputs of the second kind defined in Definition 2. Regarding this situation, the following lemma holds.

**Lemma 2.** *Suppose we use input sequence $\boldsymbol{u}_k^{back}$, $\boldsymbol{u}_{k+1}^{back}$, $\ldots$, $\boldsymbol{u}_{k+j-1}^{back}$ ($j < \tau$) given in Definition 2 when $\boldsymbol{x}_{k-1} \in \mathcal{X}_s$ and $\boldsymbol{x}_k = \boldsymbol{x} \notin \mathcal{X}_s$ occur. Also suppose $\boldsymbol{x}_k \in \mathcal{X}_s \Rightarrow \Pr\{\boldsymbol{x}_{k+1} \in \mathcal{X}_s\} \geq p$ holds with $p \in (0, 1)$. Then $\Pr\{\boldsymbol{x}_k \in \mathcal{X}_s\} \geq \xi^k p^\tau$ holds for all $k = 1, 2, \ldots, T$ if $\boldsymbol{x}_0 \in \mathcal{X}_s$.*

Proof is given in Appendix B.2. This lemma gives us a theoretical guarantee to make a state violating the constraints satisfy them with a desired probability after a certain number of timesteps if we use conservative inputs (or input sequence) defined in Definition 2.

## 4.3  Theoretical Result on How to Generate Conservative Inputs

As shown in Algorithm 1, our proposed method uses conservative inputs $\boldsymbol{u}_k^{stay}$ and $\boldsymbol{u}_k^{back}$ given in Definitions 1 and 2, respectively. Therefore, when we try to apply this method to real problems, we need to construct such conservative inputs. To address this issue, in this subsection, we introduce sufficient conditions to construct those conservative inputs, which are given by using prior knowledge of the controlled object and disturbance. Namely, regarding $\boldsymbol{u}_k^{stay}$ and $\boldsymbol{u}_k^{back}$ used in Algorithm 1, we have the following theorem.

**Theorem 1.** *Let $q \in (0.5, 1)$. Suppose Assumptions 1, 2 and 3 hold. Then, if input $\boldsymbol{u}_k$ satisfies the following inequality for all $j = 1, 2, \ldots, n_c$ and $\delta_j \in \{\bar{\delta}_j, -\bar{\delta}_j\}$, $\Pr\{\boldsymbol{x}_{k+1} \in \mathcal{X}_s\} \geq q$ holds:*

$$d_j - \boldsymbol{h}_j^\top (\boldsymbol{A}\boldsymbol{x}_k + \boldsymbol{B}\boldsymbol{u}_k + \boldsymbol{\mu}_w) - \delta_j \geq \Phi^{-1}(q') \left\| \boldsymbol{h}_j^\top \boldsymbol{\Sigma}_w^{\frac{1}{2}} \right\|_2, \qquad (17)$$

*where $q' = 1 - \frac{1-q}{n_c}$.*

*In addition, if input sequence $\boldsymbol{U}_k = [\boldsymbol{u}_k^\top, \boldsymbol{u}_{k+1}^\top, \ldots, \boldsymbol{u}_{k+\tau-1}^\top]^\top$ satisfies the following inequality for all $j = 1, 2, \ldots, n_c$ and $\Delta_j \in \{\bar{\Delta}_j, -\bar{\Delta}_j\}$, $\Pr\{\boldsymbol{x}_{k+\tau} \in \mathcal{X}_s\} \geq q$ holds:*

$$d_j - \boldsymbol{h}_j^\top \left( \boldsymbol{A}^\tau \boldsymbol{x}_k + \hat{\boldsymbol{B}}\boldsymbol{U}_k + \hat{\boldsymbol{C}}\hat{\boldsymbol{\mu}}_w \right) - \Delta_j \geq \Phi^{-1}(q') \left\| \boldsymbol{h}_j^\top \hat{\boldsymbol{C}} \begin{bmatrix} \boldsymbol{\Sigma}_w & & \\ & \ddots & \\ & & \boldsymbol{\Sigma}_w \end{bmatrix}^{\frac{1}{2}} \right\|_2, \qquad (18)$$

*where $\hat{\boldsymbol{\mu}}_w = [\boldsymbol{\mu}_w^\top, \ldots, \boldsymbol{\mu}_w^\top]^\top \in \mathbb{R}^{n\tau}$, $\hat{\boldsymbol{B}} = [\boldsymbol{A}^{\tau-1}\boldsymbol{B}, \boldsymbol{A}^{\tau-2}\boldsymbol{B}, \ldots, \boldsymbol{B}]$ and $\hat{\boldsymbol{C}} = [\boldsymbol{A}^{\tau-1}, \boldsymbol{A}^{\tau-2}, \ldots, \boldsymbol{I}]$.*

*Sketch of Proof.* First, from Bonferroni's inequality, the following relation holds for $q' = 1 - \frac{1-q}{n_c}$ and $\forall \delta_j$, $\forall j = 1, \ldots, n_c$:

$$\Pr\{\boldsymbol{H}\boldsymbol{x}_{k+1} \preceq \boldsymbol{d}\} \geq q \Leftarrow \Pr\left\{\boldsymbol{h}_j^\top (\boldsymbol{A}\boldsymbol{x}_k + \boldsymbol{B}\boldsymbol{u}_k + \boldsymbol{w}_k) + \delta_j \leq d_j\right\} \geq q'. \qquad (19)$$

Next, as input $\boldsymbol{u}_k$ and disturbance $\boldsymbol{w}_k$ follow normal distributions and are uncorrelated (Assumption 2 and (8)), the following relation holds [5, §4.4.2]:

$$\Pr\{\boldsymbol{h}_j^\top \left(\boldsymbol{A}\boldsymbol{x}_k + \boldsymbol{B}\boldsymbol{u}_k + \boldsymbol{w}_k\right) + \delta_j \leq d_j\} \geq q'$$
$$\Leftrightarrow d_j - \boldsymbol{h}_j^\top \left(\boldsymbol{A}\boldsymbol{x}_k + \boldsymbol{B}\boldsymbol{u}_k\right) - \delta_j - \boldsymbol{h}_j^\top \boldsymbol{\mu}_w \geq \Phi^{-1}(q') \left\|\boldsymbol{h}_j^\top \boldsymbol{\Sigma}_w^{\frac{1}{2}}\right\|_2. \tag{20}$$

Therefore, the first part of the theorem is proved. The second part of the theorem is proved in the same way. Full proof is given in Appendix B.3.    □

This theorem means that, if we find solutions of (17) with $q' = 1 - \frac{1-\left(\frac{\eta}{\xi^k}\right)^{\frac{1}{\tau}}}{n_c}$ and (18) with $q' = 1 - \frac{1-\xi}{n_c}$, they can be used as the conservative inputs $\boldsymbol{u}_k^{stay}$ and $\boldsymbol{u}_k^{back}$ in Definitions 1 and 2, respectively. Since (17) and (18) are linear w.r.t. $\boldsymbol{u}_k$ and $\boldsymbol{U}_k$, we can use linear programming solvers to find the solutions. Concrete examples of the conditions given in this theorem are shown in simulation evaluations in Sect. 5.

### 4.4 Main Theoretical Result: Theoretical Guarantee for Chance Constraint Satisfaction

Using the complementary theoretical results described so far, we show our main theorem that guarantees the satisfaction of the safety when we use our proposed safe exploration method, Algorithm 1, even with the existence of disturbance.

**Theorem 2.** *Let $\eta \in (0.5, 1)$. Suppose Assumptions 1 through 4 hold. Then, by generating input $\boldsymbol{u}_k$ according to Algorithm 1, chance constraint (6) are satisfied at every timestep $k = 1, 2, \ldots, T$.*

*Sketch of Proof.* First, consider the case of (i) in Algorithm 1. From Lemma 1, Assumptions 3 and 4, and Bonferroni's inequality,

$$\Pr\{\boldsymbol{H}\boldsymbol{x}_{k+1} \preceq \boldsymbol{d}\} \geq \left(\frac{\eta}{\xi^k}\right)^{\frac{1}{\tau}} \tag{21}$$

holds if the input $\boldsymbol{u}_k$ is generated by (8) with $\boldsymbol{\Sigma}_k$ satisfying (15), and thus, chance constraint (6) is satisfied for $k = 1, 2, \ldots, T$.

Next, in the case of (ii) in Algorithm 1, by generating an input as $\boldsymbol{u}_k = \boldsymbol{u}_k^{stay}$ that is defined in Definition 1, $\Pr\{\boldsymbol{H}\boldsymbol{x}_{k+1} \preceq \boldsymbol{d}\} \geq \left(\frac{\eta}{\xi^k}\right)^{\frac{1}{\tau}}$ holds when $\boldsymbol{x}_k \in \mathcal{X}_s$.

Finally, by generating input as $\boldsymbol{u}_k = \boldsymbol{u}_k^{back}$ in case (iii) of Algorithm 1, $\Pr\{\boldsymbol{H}\boldsymbol{x}_k \preceq \boldsymbol{d}\} \geq \eta$ holds for any $\boldsymbol{x}_k \in \mathbb{R}^n$, $k = 1, 2, \ldots, T$ from Lemma 2. Hence, noting $\left(\frac{\eta}{\xi^k}\right)^{\frac{1}{\tau}} > \eta$, $\Pr\{\boldsymbol{H}\boldsymbol{x}_k \preceq \boldsymbol{d}\} \geq \eta$ is satisfied for $k = 1, 2, \ldots, T$. Full proof is given in Appendix B.4.    □

The theoretical guarantee of safety proved in Theorem 2 is obtained with the equivalent transformation of a chance constraint into its deterministic counterpart under the assumption on disturbances (Assumption 4). That is, this theoretical result holds since the disturbance follows a normal distribution and is uncorrelated to the input. The proposed method, however, can be applicable to deal with other types of disturbance if the sufficient part holds with a certain transformation.

# 5    Simulation Evaluation

## 5.1    Simulation Conditions

We evaluated the validity of the proposed method with the inverted-pendulum provided as "Pendulum-v0" in OpenAI Gym [6] and the four-bar parallel link robot manipulator with two degrees of freedom dealt in [18]. Configuration figures of both problems are provided in Fig. C.1 in Appendix. We added external disturbances to these problems.

*Inverted-pendulum:* The discrete-time dynamics of this problem is given by

$$
\begin{bmatrix} \phi_{k+1} \\ \zeta_{k+1} \end{bmatrix} = \begin{bmatrix} \phi_k + T_s \zeta_k \\ \zeta_k - T_s \frac{3g}{2\ell} \sin(\phi_k + \pi) \end{bmatrix} + \begin{bmatrix} 0 \\ T_s \frac{3}{m\ell^2} \end{bmatrix} u_k + \boldsymbol{w}_k
$$
$$
=: \boldsymbol{f}(\boldsymbol{x}_k) + \boldsymbol{G} u_k + \boldsymbol{w}_k, \tag{22}
$$

where $\phi_k \in \mathbb{R}$ and $\zeta_k \in \mathbb{R}$ are the angle and rotating speed of the pendulum and $\boldsymbol{x}_k = [\phi_k, \zeta_k]^\top$. Further, $u_k \in \mathbb{R}$ is an input torque, $T_s$ is a sampling period, and $\boldsymbol{w}_k \in \mathbb{R}^2$ is the disturbance where $\boldsymbol{w}_k \sim \mathcal{N}(\boldsymbol{\mu}_w, \boldsymbol{\Sigma}_w)$, $\boldsymbol{\mu}_w = [\mu_{w,\phi}, \mu_{w,\zeta}]^\top \in \mathbb{R}^2$ and $\boldsymbol{\Sigma}_w = \mathrm{diag}(\sigma_{w,\phi}^2, \sigma_{w,\zeta}^2) \in \mathbb{R}^{2\times2}$. Concrete values of these and the other variables used in this evaluation are listed in Table C.1 in Appendix. We use the following linear approximation model of the above nonlinear system:

$$
\boldsymbol{x}_{k+1} \simeq \begin{bmatrix} 1 & T_s \\ 0 & 1 \end{bmatrix} \boldsymbol{x}_k + \begin{bmatrix} 0 \\ T_s \frac{3}{m\ell^2} \end{bmatrix} u_k + \boldsymbol{w}_k
$$
$$
=: \boldsymbol{A} \boldsymbol{x}_k + \boldsymbol{B} u_k + \boldsymbol{w}_k. \tag{23}
$$

The approximation errors $\boldsymbol{e}$ in (10) is given by

$$
\boldsymbol{e}(\boldsymbol{x}, u) = \boldsymbol{f}(\boldsymbol{x}) + \boldsymbol{G}u - (\boldsymbol{A}\boldsymbol{x} + \boldsymbol{B}u) = \begin{bmatrix} 0 \\ -T_s \frac{3g}{2\ell} \sin(\phi + \pi) \end{bmatrix}. \tag{24}
$$

We set constraints on $\zeta_k$ as $-6 \le \zeta_k \le 6$, $\forall k = 1, \dots, T$. This condition becomes

$$
\boldsymbol{h}_1^\top \boldsymbol{x}_k \le d_1, \ \boldsymbol{h}_2^\top \boldsymbol{x}_k \le d_2, \ \forall k = 1, \dots, T, \tag{25}
$$

where $\boldsymbol{h}_1^\top = [0, 1]$, $\boldsymbol{h}_2^\top = [0, -1]$, $d_1 = d_2 = 6$, and $n_c = 2$. Therefore, Assumption 4 holds since $\boldsymbol{h}_j^\top \boldsymbol{B} \ne 0$, $j \in \{1, 2\}$. Furthermore, the approximation model given by (23) is controllable because of its coefficient matrices $\boldsymbol{A}$ and $\boldsymbol{B}$, and its controllability index is 2. According to this result, we set $\tau = 2$ and we have

$$
\sup_{\boldsymbol{x} \in \mathbb{R}^2, \ u \in \mathbb{R}} |\boldsymbol{h}_j^\top \boldsymbol{e}(\boldsymbol{x}, u)| = T_s \frac{3g}{2\ell}, \ \ j \in \{1, 2\}, \tag{26}
$$

$$
\sup_{\boldsymbol{x} \in \mathbb{R}^2, \ u \in \mathbb{R}} |\boldsymbol{h}_j^\top (\boldsymbol{A} + \boldsymbol{I}) \boldsymbol{e}(\boldsymbol{x}, u)| = T_s \frac{3g}{\ell}, \ \ j \in \{1, 2\}, \tag{27}
$$

since $|\sin(\phi + \pi)| \le 1$, $\forall \phi \in \mathbb{R}$. Therefore we used in this evaluation $T_s \frac{3g}{2\ell}$ and $T_s \frac{3g}{\ell}$ as $\bar{\delta}_j$ and $\bar{\Delta}_j$, respectively, and they satisfy Assumption 3.

Regarding immediate cost, we let

$$c_{k+1} = \left( \{(\phi_k + \pi) \bmod 2\pi\} - \pi \right)^2 + 0.1\zeta_k^2 + 0.001u_k^2. \tag{28}$$

The first term corresponds to swinging up the pendulum and keeping it inverted. Furthermore, in our method, we used the following conservative inputs:

$$u_k^{stay} = -\frac{m\ell^2}{3T_s}(\zeta_k + \mu_{w,\phi}), \quad \begin{bmatrix} u_k^{back} \\ u_{k+1}^{back} \end{bmatrix} = \begin{bmatrix} -\frac{m\ell^2}{3T_s}(\zeta_k + 2\mu_{w,\phi}) \\ 0 \end{bmatrix}. \tag{29}$$

Both of these inputs satisfy the inequalities in Theorem 1 with the parameters in Table C.1, and can be used as conservative inputs defined in Definitions 1 and 2.

*Four-bar Parallel Link Robot Manipulator:* We let $\boldsymbol{x} = [q_1, q_2, \varpi_1, \varpi_2]^\top \in \mathbb{R}^4$ and $\boldsymbol{u} = [v_1, v_2]^\top \in \mathbb{R}^2$ where $q_1, q_2$ are angles of links of a robot, $\varpi_1, \varpi_2$ are their rotating speed and $v_1, v_2$ are armature voltages from an actuator. The discrete-time dynamics of a robot manipulator with an actuator including external disturbance $\boldsymbol{w}_k \in \mathbb{R}^4$ where $\boldsymbol{w}_k \sim \mathcal{N}(\boldsymbol{\mu}_w, \boldsymbol{\Sigma}_w)$, $\boldsymbol{\mu}_w = [\mu_{w,q_1}, \mu_{w,q_2}, \mu_{w,\varpi_1}, \mu_{w,\varpi_2}]^\top \in \mathbb{R}^4$ and $\boldsymbol{\Sigma}_w = \mathrm{diag}(\sigma_{w,q_1}^2, \sigma_{w,q_2}^2, \sigma_{w,\varpi_1}^2, \sigma_{w,\varpi_2}^2) \in \mathbb{R}^{4\times4}$ is given by

$$\boldsymbol{x}_{k+1} = \begin{bmatrix} q_{1_k} + T_s \varpi_{1_k} \\ q_{2_k} + T_s \varpi_{2_k} \\ \varpi_{1_k} - T_s \frac{\hat{d}_{11}}{\hat{m}_{11}} \varpi_{1_k} - T_s \frac{V_1}{\hat{m}_{11}} \cos q_{1_k} \\ \varpi_{2_k} - T_s \frac{\hat{d}_{22}}{\hat{m}_{22}} \varpi_{2_k} - T_s \frac{V_2}{\hat{m}_{22}} \cos q_{2_k} \end{bmatrix} + T_s \begin{bmatrix} 0 & 0 \\ 0 & 0 \\ \frac{\alpha}{\hat{m}_{11}} & 0 \\ 0 & \frac{\alpha}{\hat{m}_{22}} \end{bmatrix} \boldsymbol{u}_k + \boldsymbol{w}_k$$

$$=: \boldsymbol{f}(\boldsymbol{x}_k) + \boldsymbol{G}\boldsymbol{u}_k + \boldsymbol{w}_k, \tag{30}$$

where $\hat{m}_{ii} = \eta^2 J_{mi} + M_{ii}, \hat{d}_{ii} = \eta^2 \left( D_{mi} + \frac{K_t K_b}{R} \right), i \in \{1,2\}, \alpha = \frac{\eta K_a K_t}{R}$. The definitions of symbols in (30) and their specific values except the sampling period $T_s$ are given in [18]. Derivation of (30) is detailed in Appendix C.2. Similarly, we obtain the following linear approximation model of (30) by ignoring gravity term:

$$\boldsymbol{x}_{k+1} \simeq \begin{bmatrix} 1 & 0 & T_s & 0 \\ 0 & 1 & 0 & T_s \\ 0 & 0 & (1 - T_s \frac{\hat{d}_{11}}{\hat{m}_{11}}) & 0 \\ 0 & 0 & 0 & (1 - T_s \frac{\hat{d}_{22}}{\hat{m}_{22}}) \end{bmatrix} \begin{bmatrix} q_{1_k} \\ q_{2_k} \\ \varpi_{1_k} \\ \varpi_{2_k} \end{bmatrix} + T_s \begin{bmatrix} 0 & 0 \\ 0 & 0 \\ \frac{\alpha}{\hat{m}_{11}} & 0 \\ 0 & \frac{\alpha}{\hat{m}_{22}} \end{bmatrix} \boldsymbol{u}_k + \boldsymbol{w}_k$$

$$=: \boldsymbol{A}\boldsymbol{x}_k + \boldsymbol{B}\boldsymbol{u}_k + \boldsymbol{w}_k. \tag{31}$$

In the same way as the setting of the inverted pendulum problem described above, we set constraints on the upper and lower bounds regarding rotating speed $\varpi_1$ and $\varpi_2$ with $\boldsymbol{h}_1 = [0,0,1,0]^\top$, $\boldsymbol{h}_2 = [0,0,-1,0]^\top$, $\boldsymbol{h}_3 = [0,0,0,1]^\top$, $\boldsymbol{h}_4 = [0,0,0,-1]^\top$. Since $|\cos q_i| \le 1$, $i \in \{1,2\}$, we have the following relations:

$$\sup_{\boldsymbol{x}\in\mathbb{R}^4,\boldsymbol{u}\in\mathbb{R}^2}|\boldsymbol{h}_j^\top \boldsymbol{e}(\boldsymbol{x},\boldsymbol{u})| = \begin{cases} T_s\frac{V_1}{\hat{m}_{11}}, j \in \{1,2\} \\ T_s\frac{V_2}{\hat{m}_{22}}, j \in \{3,4\} \end{cases}, \tag{32}$$

$$\sup_{\boldsymbol{x}\in\mathbb{R}^4,\boldsymbol{u}\in\mathbb{R}^2}|\boldsymbol{h}_j^\top (\boldsymbol{A}+\boldsymbol{I})\boldsymbol{e}(\boldsymbol{x},\boldsymbol{u})| = \begin{cases} |2 - T_s\frac{\hat{d}_{11}}{\hat{m}_{11}}|T_s\frac{V_1}{\hat{m}_{11}}, j \in \{1,2\} \\ |2 - T_s\frac{\hat{d}_{22}}{\hat{m}_{22}}|T_s\frac{V_2}{\hat{m}_{22}}, j \in \{3,4\} \end{cases}. \tag{33}$$

We use them as $\bar{\delta}_j$ and $\bar{\varDelta}_j$, and therefore Assumption 3 holds. Assumption 4 also holds with $\boldsymbol{h}_1, \boldsymbol{h}_2, \boldsymbol{h}_3, \boldsymbol{h}_4$ and $\boldsymbol{B}$. In this setting, we used immediate cost

$$c_{k+1} = 2\Big(\{(q_{1_k}+\pi) \bmod 2\pi\} - \pi\Big)^2 + 2\Big(\{((q_{2_k}+\pi)-5\pi/6) \bmod 2\pi\} - \pi\Big)^2$$
$$+ 0.1(\varpi_{1_k}^2 + \varpi_{2_k}^2) + 0.001\boldsymbol{u}_k^\top \boldsymbol{u}_k. \tag{34}$$

The first two terms corresponds to changing the pose of manipulator to the one depicted on the right in Fig. C.2 in Appendix and keeping that pose. Furthermore, in our method, we used the following conservative inputs:

$$\boldsymbol{u}_k^{stay} = \begin{bmatrix} -\frac{1}{b_1}\{(1-a_1)\varpi_{1_k}+(1-a_1)\mu_{w,\varpi_1}\} \\ -\frac{1}{b_2}\{(1-a_2)\varpi_{2_k}+(1-a_2)\mu_{w,\varpi_2}\} \end{bmatrix}, \tag{35}$$

$$\boldsymbol{u}_k^{back} = \begin{bmatrix} -\frac{1}{(1-a_1)b_1}\{(1-a_1)^2\varpi_{1_k}+(2-a_1)\mu_{w,\varpi_1}\} \\ -\frac{1}{(1-a_2)b_2}\{(1-a_2)^2\varpi_{2_k}+(2-a_2)\mu_{w,\varpi_2}\} \end{bmatrix}, \quad \boldsymbol{u}_{k+1}^{back} = \begin{bmatrix} 0 \\ 0 \end{bmatrix}, \tag{36}$$

where $a_1$, $a_2$, $b_1$ and $b_2$ are derived from elements of $\boldsymbol{A}$ and $\boldsymbol{B}$ and they are $a_1 = T_s\hat{d}_{11}/\hat{m}_{11}$, $a_2 = T_s\hat{d}_{22}/\hat{m}_{22}$, $b_1 = T_s\alpha/\hat{m}_{11}$, and $b_2 = T_s\alpha/\hat{m}_{22}$. Both of these inputs satisfy the inequalities in Theorem 1 with the parameters in Table C.2, and thus, they can be used as conservative inputs defined in Definitions 1 and 2.

*Reinforcement Learning Algorithm and Reference Method:* We have combined our proposed safe exploration method (Algorithm 1) with the Deep Deterministic Policy Gradient (DDPG) algorithm [16], a representative RL algorithm applicable to (7) and (8), in each experimental setting with the immediate costs and conservative inputs described above. We also combined the safe exploration method given in the previous work [19] that does not take disturbance into account with the DDPG algorithm for the reference where we set $u_k^{stay} = 0$ as in the original paper. The network structure and hyperparameters we used throughout this evaluation are listed in Tables C.1 and C.2 in Appendix.

*Parameters for Safe Exploration:* We set the pre-specified probabilities in both problems to be $\eta = 0.95$. Other parameters for safe exploration are listed in Tables C.1 and C.2 in Appendix.

## 5.2    Simulation Results

We evaluated our method and the previous one with 100 episodes × 10 runs of the simulation (each episode consists of 100 timesteps). The source code is publicly available as described in Code Availability Statement. The computational resource and running time information for this evaluation is given in Appendix C.4.

Figure 3 shows the results of the cumulative costs at each episode and the relative frequencies of constraint satisfaction at each timestep. The lines shown in the left figures are the mean values of the cumulative cost at each episode calculated over the 10 runs, while the shaded areas show their 95% confidence intervals. We can see that both methods enabled to reduce their cumulative costs as the number of episode increases. However, as shown in the right figures, the previous method [19] (blue triangles) could not meet the chance constraint (6) (went below the green dashed lines that show the pre-specified probability $\eta$) at several timesteps. In contrast, our proposed method (red crosses) could make the relative frequencies of constraint satisfaction greater than or equal to $\eta$ for all timesteps. Both simulations support our theoretical results and show the effectiveness of the proposed method.

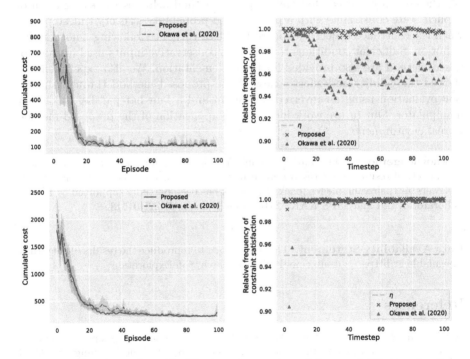

**Fig. 3.** Simulation results with (**Top**) an inverted-pendulum and (**Bottom**) a four-bar parallel link robot manipulator: (**Left**) Cumulative costs at each episode, (**Right**) Relative frequencies of constraint satisfaction at each timestep. Both the proposed method (red) and the previous one (blue) [19] enabled to reduce their cumulative costs; however, only the proposed method made the relative frequencies of constraint satisfaction greater than or equal to $\eta$ for all timesteps in both experimental settings.

## 6   Limitations

There are two main things we need to care about to use the proposed method. First, although it is relaxed compared to the previous work [19], the controlled object and disturbance should satisfy several conditions and we need partial prior knowledge about them as described in Assumptions 1 through 4. In addition, the proposed method requires calculations including matrices, vectors, nonlinear functions and probabilities. This additional computational cost may become a problem if the controller should be implemented as an embedded system.

## 7   Conclusion

In this study, we proposed a safe exploration method for RL to guarantee the safety during learning under the existence of disturbance. The proposed method uses partial prior knowledge of both the controlled object and disturbances. We theoretically proved that the proposed method achieves the satisfaction of explicit state constraints with a pre-specified probability at every timestep even when the controlled object is exposed to the disturbance following a normal distribution. Sufficient conditions to construct conservative inputs used in the proposed method are also provided for its implementation. We also experimentally showed the validity and effectiveness of the proposed method through simulation evaluation using an inverted pendulum and a four-bar parallel link robot manipulator. Our future work includes the application of the proposed method to real environments.

**Acknowledgements.** The authors thank Yusuke Kato for the fruitful discussions on theoretical results about the proposed method. The authors also thank anonymous reviewers for their valuable feedback. This work has been partially supported by Fujitsu Laboratories Ltd and JSPS KAKENHI Grant Number JP22H01513.

**Data Availability Statement.** The source code to reproduce the results of this study is available at https://github.com/FujitsuResearch/SafeExploration

## References

1. Achiam, J., Held, D., Tamar, A., Abbeel, P.: Constrained policy optimization. In: Proceedings of the 34th International Conference on Machine Learning, pp. 22–31 (2017)
2. Ames, A.D., Coogan, S., Egerstedt, M., Notomista, G., Sreenath, K., Tabuada, P.: Control barrier functions: theory and applications. In: Proceedings of the 18th European Control Conference, pp. 3420–3431 (2019)
3. Berkenkamp, F., Turchetta, M., Schoellig, A., Krause, A.: Safe model-based reinforcement learning with stability guarantees. In: Advances in Neural Information Processing Systems 30, pp. 908–919 (2017)

4. Biyik, E., Margoliash, J., Alimo, S.R., Sadigh, D.: Efficient and safe exploration in deterministic Markov decision processes with unknown transition models. In: Proceedings of the 2019 American Control Conference, pp. 1792–1799 (2019)
5. Boyd, S., Vandenberghe, L.: Convex Optimization. Cambridge University Press (2004)
6. Brockman, G., et al.: OpenAI Gym. arXiv preprint. arXiv:1606.01540 (2016). The code is available at https://github.com/openai/gym with the MIT License
7. Cheng, R., Orosz, G., Murray, R.M., Burdick, J.W.: End-to-end safe reinforcement learning through barrier functions for safety-critical continuous control tasks. In: Proceedings of the 33rd AAAI Conference on Artificial Intelligence, pp. 3387–3395 (2019)
8. Chow, Y., Nachum, O., Faust, A., Duenez-Guzman, E., Ghavamzadeh, M.: Lyapunov-based safe policy optimization for continuous control. arXiv preprint. arXiv:1901.10031 (2019)
9. Fan, D.D., Nguyen, J., Thakker, R., Alatur, N., Agha-mohammadi, A.A., Theodorou, E.A.: Bayesian learning-based adaptive control for safety critical systems. In: Proceedings of the 2020 IEEE International Conference on Robotics and Automation, pp. 4093–4099 (2020)
10. García, J., Fernández, F.: A comprehensive survey on safe reinforcement learning. J. Mach. Learn. Res. **16**, 1437–1480 (2015)
11. Ge, Y., Zhu, F., Ling, X., Liu, Q.: Safe Q-learning method based on constrained Markov decision processes. IEEE Access **7**, 165007–165017 (2019)
12. Glavic, M., Fonteneau, R., Ernst, D.: Reinforcement learning for electric power system decision and control: past considerations and perspectives. In: Proceedings of the 20th IFAC World Congress, pp. 6918–6927 (2017)
13. Khojasteh, M.J., Dhiman, V., Franceschetti, M., Atanasov, N.: Probabilistic safety constraints for learned high relative degree system dynamics. In: Proceedings of the 2nd Conference on Learning for Dynamics and Control, pp. 781–792 (2020)
14. Kiran, B.R., et al.: Deep reinforcement learning for autonomous driving: a survey. IEEE Transactions on Intelligent Transportation Systems (2021). (Early Access)
15. Koller, T., Berkenkamp, F., Turchetta, M., Boedecker, J., Krause, A.: Learning-based model predictive control for safe exploration and reinforcement learning. arXiv preprint. arXiv:1906.12189 (2019)
16. Lillicrap, T.P., et al.: Continuous control with deep reinforcement learning. arXiv preprint. arXiv:1509.02971 (2015)
17. Liu, Z., Zhou, H., Chen, B., Zhong, S., Hebert, M., Zhao, D.: Safe model-based reinforcement learning with robust cross-entropy method. ICLR 2021 Workshop on Security and Safety in Machine Learning Systems (2021)
18. Namerikawa, T., Matsumura, F., Fujita, M.: Robust trajectory following for an uncertain robot manipulator using $H_\infty$ synthesis. In: Proceedings of the 3rd European Control Conference, pp. 3474–3479 (1995)
19. Okawa, Y., Sasaki, T., Iwane, H.: Automatic exploration process adjustment for safe reinforcement learning with joint chance constraint satisfaction. In: Proceedings of the 21st IFAC World Congress, pp. 1588–1595 (2020)
20. Sutton, R.S., Barto, A.G.: Reinforcement Learning: An Introduction, 2nd edn. MIT Press (2018)
21. Thananjeyan, B., et al.: Recovery RL: safe reinforcement learning with learned recovery zones. IEEE Robot. Autom. Lett. **6**(3), 4915–4922 (2021)
22. Yang, T.Y., Rosca, J., Narasimhan, K., Ramadge, P.J.: Projection-based constrained policy optimization. In: Proceedings of the 8th International Conference on Learning Representations (2020)

# Model Selection in Reinforcement Learning with General Function Approximations

Avishek Ghosh[1]([⊠]) and Sayak Ray Chowdhury[2]

[1] Halıcıoğlu Data Science Institute (HDSI), UC San Diego, San Diego, USA
a2ghosh@ucsd.edu
[2] Boston University, Boston, USA
sayak@bu.edu

**Abstract.** We consider model selection for classic Reinforcement Learning (RL) environments – Multi Armed Bandits (MABs) and Markov Decision Processes (MDPs) – under general function approximations. In the model selection framework, we do not know the function classes, denoted by $\mathcal{F}$ and $\mathcal{M}$, where the true models – reward generating function for MABs and transition kernel for MDPs – lie, respectively. Instead, we are given $M$ nested function (hypothesis) classes such that true models are contained in at-least one such class. In this paper, we propose and analyze efficient model selection algorithms for MABs and MDPs, that *adapt* to the smallest function class (among the nested $M$ classes) containing the true underlying model. Under a separability assumption on the nested hypothesis classes, we show that the cumulative regret of our adaptive algorithms match to that of an oracle which knows the correct function classes (i.e., $\mathcal{F}$ and $\mathcal{M}$) a priori. Furthermore, for both the settings, we show that the cost of model selection is an additive term in the regret having weak (logarithmic) dependence on the learning horizon $T$.

**Keywords:** Model selection · Bandits · Reinforcement learning

## 1 Introduction

We study the problem of *model selection* for Reinforcement Learning problems, which refers to choosing the appropriate hypothesis class, to model the mapping from actions to expected rewards. We choose two particular frameworks—(a) Multi-Armaed Bandits (MAB) and (b) markov Decision Processes (MDP). Specifically, we are interested in studying the model selection problems for these frameworks without any function approximations (like linear, generalized linear etc.). Note that, the problem of model selection plays an important role

---

A. Ghosh and S. R. Chowdhury contributed equally.

**Supplementary Information** The online version contains supplementary material available at https://doi.org/10.1007/978-3-031-26412-2_10.

M.-R. Amini et al. (Eds.): ECML PKDD 2022, LNAI 13716, pp. 148–164, 2023.
https://doi.org/10.1007/978-3-031-26412-2_10

in applications such as personalized recommendations, autonomous driving, robotics as we explain in the sequel. Formally, a family of nested hypothesis classes $\mathcal{H}_f$, $f \in \mathcal{F}$ is specified, where each class posits a plausible model for mapping actions to expected rewards. Furthermore, the family $\mathcal{F}$ is totally ordered, i.e., if $f_1 \leq f_2$, then $\mathcal{H}_{f_1} \subseteq \mathcal{H}_{f_2}$. It is assumed that the true model is contained in at least one of these specified families. Model selection guarantees then refer to algorithms whose regret scales in the complexity of the *smallest hypothesis class containing the true model*, even though the algorithm was not aware apriori.

Multi-Armed Bandits (MAB) [7] and Markov decision processes (MDP) [25] are classical frameworks to model a reinforcement learning (RL) environment, where an agent interacts with the environment by taking successive decisions and observe rewards generated by those decisions. One of the objectives in RL is to maximize the total reward accumulated over multiple rounds, or equivalently minimize the *regret* in comparison with an optimal policy [7]. Regret minimization is useful in several sequential decision-making problems such as portfolio allocation and sequential investment, dynamic resource allocation in communication systems, recommendation systems, etc. In these settings, there is no separate budget to purely explore the unknown environment; rather, exploration and exploitation need to be carefully balanced.

Optimization over large domains under restricted feedback is an important problem and has found applications in dynamic pricing for economic markets [6], wireless communication [8] and recommendation platforms (such as Netflix, Amazon Prime). Furthermore, in many applications (e.g., robotics, autonomous driving), the number of actions and the observable number of states can be very large or even infinite, which makes RL challenging, particularly in generalizing learnt knowledge across unseen states and actions. For example, the game of Go has a state space with size $3^{361}$, and the state and action spaces of certain robotics applications can even be continuous. In recent years, we have witnessed an explosion in the RL literature to tackle this challenge, both in theory (see, e.g., [4,10,18,22,29]), and in practice (see, e.g., [21,31]).

In the first part of the paper, we focus on learning an unknown function $f^* \in \mathcal{F}$, supported over a compact domain, via online noisy observations. If the function class $\mathcal{F}$ is known, the optimistic algorithm of [26] learns $f^*$, yielding a regret that depends on *eluder dimension* (a complexity measure of function classes) of $\mathcal{F}$. However, in the applications mentioned earlier, it is not immediately clear how one estimates $\mathcal{F}$. Naive estimation techniques may yield an unnecessarily big $\mathcal{F}$, and as a consequence, the regret may suffer. On the other hand, if the estimated class, $\hat{\mathcal{F}}$ is such that $\hat{\mathcal{F}} \subset \mathcal{F}$, then the learning algorithm might yield a linear regret because of this infeasibility. Hence, it is important to estimate the function class properly, and here is where the question of model selection appears. The problem of model selection is formally stated as follows— we are given a family of $M$ hypothesis classes $\mathcal{F}_1 \subset \mathcal{F}_2 \subset \ldots \subset \mathcal{F}_M$, and the unknown function $f^*$ is assumed to be contained in the family of nested classes. In particular, we assume that $f^*$ lies in $\mathcal{F}_{m^*}$, where $m^*$ is unknown. Model selection guarantees refer to algorithms whose regret scales in the complexity of the *smallest model class containing the true function $f^*$*, i.e., $\mathcal{F}_{m^*}$, even though the algorithm is not aware of that a priori.

In the second part of the paper, we address the model selection problem for generic MDPs without funcction approximation. The most related work to ours is by [4], which proposes an algorithm, namely UCRL-VTR, for model-based RL without any structural assumptions, and it is based on the upper confidence RL and value-targeted regression principles. The regret of UCRL-VTR depends on the *eluder dimension* [26] and the *metric entropy* of the corresponding family of distributions $\mathcal{P}$ in which the unknown transition model $P^*$ lies. In most practical cases, however, the class $\mathcal{P}$ given to (or estimated by) the RL agent is quite pessimistic; meaning that $P^*$ actually lies in a small subset of $\mathcal{P}$ (e.g., in the game of Go, the learning is possible without the need for visiting all the states [27]). We are given a family of $M$ nested hypothesis classes $\mathcal{P}_1 \subset \mathcal{P}_2 \subset \ldots \subset \mathcal{P}_M$, where each class posits a plausible model class for the underlying RL problem. The true model $P^*$ lies in a model class $\mathcal{P}_{m*}$, where $m^*$ is unknown apriori. Similar to the functional bandits framework, we propose learning algorithms whose regret depends on the *smallest model class containing the true model $P^*$*.

The problem of model selection have received considerable attention in the last few years. Model selection is well studied in the contextual bandit setting. In this setting, minimax optimal regret guarantees can be obtained by exploiting the structure of the problem along with an eigenvalue assumption [9,12,15] We provide a comprehensive list of recent works on bandit model selection in Sect. 1.2. However, to the best of our knowledge, this is the first work to address the model selection question for generic (functional) MAB without imposing any assumptions on the reward structure.

In the RL framework, the question of model selection has received little attention. In a series of works, [23,24] consider the corralling framework of [2] for contextual bandits and reinforcement learning. While the corralling framework is versatile, the price for this is that the cost of model selection is multiplicative rather than additive. In particular, for the special case of linear bandits and linear reinforcement learning, the regret scales as $\sqrt{T}$ in time with an additional multiplicative factor of $\sqrt{M}$, while the regret scaling with time is strictly larger than $\sqrt{T}$ in the general contextual bandit. These papers treat all the hypothesis classes as bandit arms, and hence work in a (restricted) partial information setting, and as a consequence explore a lot, yielding worse regret. On the other hand, we consider all $M$ classes at once (full information setting) and do inference, and hence explore less and obtain lower regret.

Very recently, [20] study the problem of model selection in RL with function approximation. Similar to the *active-arm elimination* technique employed in standard multi-armed bandit (MAB) problems [11], the authors eliminate the model classes that are dubbed misspecified, and obtain a regret of $\mathcal{O}(T^{2/3})$. On the other hand, our framework is quite different in the sense that we consider model selection for RL with *general* transition structure. Moreover, our regret scales as $\mathcal{O}(\sqrt{T})$. Note that the model selection guarantees we obtain in the linear MDPs are partly influenced by [15], where model selection for linear contextual bandits are discussed. However, there are a couple of subtle differences: (a) for linear contextual framework, one can perform pure exploration, and [15] crucially leverages that and (b) the contexts in linear contextual framework is assumed to be i.i.d, whereas for linear MDPs, the contexts are implicit and depend on states, actions and transition probabilities.

## 1.1   Our Contributions

In this paper, our setup considers *any general* model class (for both MAB and MDP settings) that are totally bounded, i.e., for arbitrary precision, the metric entropy is bounded. Note that this encompasses a significantly larger class of environments compared to the problems with function approximation. Assuming nested families of reward function and transition kernels, respectively for MABs and MDPs, we propose adaptive algorithms, namely *Adaptive Bandit Learning* (ABL) and *Adaptive Reinforcement Learning* (ARL). Assuming the hypothesis classes are separated, both ABL and ARL construct a test statistic and thresholds it to identify the correct hypothesis class. We show that these *simple* schemes achieve the regret of $\tilde{\mathcal{O}}(d^* + \sqrt{d^* \mathbb{M}^* T})$ for MABs and $\tilde{\mathcal{O}}(d^* H^2 + \sqrt{d^* \mathbb{M}^* H^2 T})$ for MDPs (with episode length $H$), where $d_{\mathcal{E}}^*$ is the *eluder dimension* and $\mathbb{M}^*$ is the *metric entropy* corresponding to the smallest model classes containing true models ($f^*$ for MAB and $P^*$ for MDP). The regret bounds show that both ABL and ARL adapts to the true problem complexity, and the cost of model section is only $\mathcal{O}(\log T)$, which is minimal compared to the total regret.

**Notation.** For a positive integer $n$, we denote by $[n]$ the set of integers $\{1, 2, \ldots, n\}$. For a set $\mathcal{X}$ and functions $f, g : \mathcal{X} \to \mathbb{R}$, we denote $(f - g)(x) := f(x) - g(x)$ and $(f - g)^2(x) := (f(x) - g(x))^2$ for any $x \in \mathcal{X}$. For any $P : \mathcal{Z} \to \Delta(\mathcal{X})$, we denote $(Pf)(z) := \int_{\mathcal{X}} f(x) P(x|z) dx$ for any $z \in \mathcal{Z}$, where $\Delta(\mathcal{X})$ denotes the set of signed distributions over $\mathcal{X}$.

## 1.2   Related Work

**Model Selection in Online Learning:** Model selection for bandits are only recently being studied [9,13]. These works aim to identify whether a given problem instance comes from contextual or standard setting. For linear contextual bandits, with the dimension of the underlying parameter as a complexity measure, [12,15] propose efficient algorithms that adapts to the *true* dimension of the problem. While [12] obtains a regret of $\mathcal{O}(T^{2/3})$, [15] obtains a $\mathcal{O}(\sqrt{T})$ regret (however, the regret of [15] depends on several problem dependent quantities and hence not instance uniform). Later on, these guarantees are extended to the generic contextual bandit problems without linear structure [16,19], where $\mathcal{O}(\sqrt{T})$ regret guarantees are obtained. The algorithm Corral was proposed in [2], where the optimal algorithm for each model class is casted as an expert, and the forecaster obtains low regret with respect to the best expert (best model class). The generality of this framework has rendered it fruitful in a variety of different settings; see, for example [2,3].

**RL with Function Approximation:** Regret minimization in RL under function approximation is first considered in [22]. It makes explicit model-based assumptions and the regret bound depends on the eluder dimensions of the models. In contrast, [32] considers a low-rank linear transition model and propose a model-based algorithm with regret $\mathcal{O}(\sqrt{d^3 H^3 T})$. Another line of work parameterizes the $Q$-*functions* directly, using state-action feature maps, and develop model-free algorithms with regret $\mathcal{O}(\texttt{poly}(dH)\sqrt{T})$ bypassing the need for fully learning the transition model [17,30,35]. A recent line of work [29,33] generalize these approaches by designing algorithms that work with general and neural function approximations, respectively.

## 2   Model Selection in Functional Multi-armed Bandits

Consider the problem of sequentially maximizing an unknown function $f^* : \mathcal{X} \to \mathbb{R}$ over a compact domain $\mathcal{X} \subset \mathbb{R}^d$. For example, in a machine learning application, $f^*(x)$ can be the validation accuracy of a learning algorithm and $x \in \mathcal{X}$ is a fixed configuration of (tunable) hyper-parameters of the training algorithm. The objective is to find the hyper-parameter configuration that achieves the highest validation accuracy. An algorithm for this problem chooses, at each round $t$, an input (also called action or arm) $x_t \in \mathcal{X}$, and subsequently observes a function evaluation (also called reward) $y_t = f^*(x_t) + \epsilon_t$, which is a noisy version of the function value at $x_t$. The action $x_t$ is chosen causally depending upon the history $\{x_1, y_1, \ldots, x_{t-1}, y_{t-1}\}$ of arms and reward sequences available before round $t$.

**Assumption 1 (Sub-Gaussian noise).** *The noise sequence $\{\epsilon_t\}_{t\geq 1}$ is conditionally zero-mean, i.e., $\mathbb{E}[\epsilon_t|\mathcal{F}_{t-1}] = 0$ and $\sigma$-sub-Gaussian for known $\sigma$ ,i.e.,*

$$\forall t \geq 1, \ \forall \lambda \in \mathbb{R}, \ \mathbb{E}\left[\exp(\lambda \epsilon_t)|\mathcal{F}_{t-1}\right] \leq \exp\left(\frac{\lambda^2 \sigma^2}{2}\right)$$

*almost surely, where $\mathcal{F}_{t-1} := \sigma(x_1, y_1, \ldots, x_{t-1}, y_{t-1}, x_t)$ is the $\sigma$-field summarizing the information available just before $y_t$ is observed.*

This is a mild assumption on the noise (it holds, for instance, for distributions bounded in $[-\sigma, \sigma]$) and is standard in the literature [1,26,28].

**Regret:** The learner's goal is to maximize its (expected) cumulative reward $\sum_{t=1}^{t} f^*(x_t)$ over a time horizon $T$ (not necessarily known a priori) or, equivalently, minimize its cumulative *regret*

$$\mathcal{R}_T := \sum_{t=1}^{T} \left(f^*(x^*) - f^*(x_t)\right),$$

where $x^* \in \text{argmax}_{x \in \mathcal{X}} f(x)$ is a maximizer of $f$ (assuming the maximum is attained; not necessarily unique). A sublinear growth of $\mathcal{R}_T$ implies the time-average regret $\mathcal{R}_T/T \to 0$ as $T \to \infty$, implying the algorithm eventually chooses actions that attain function values close to the optimum most of the time.

## 2.1    Model Selection Objective

In the literature, it is assumed that $f^*$ belongs to a known class of functions $\mathcal{F}$. In this work, in contrast to the standard setting, we do not assume the knowledge of $\mathcal{F}$. Instead, we are given $M$ nested function classes $\mathcal{F}_1 \subset \mathcal{F}_2 \subset \ldots \subset \mathcal{F}_M$. Among the nested classes $\mathcal{F}_1, ..\mathcal{F}_M$, the ones containing $f^*$ is denoted as *realizable* classes, and the ones not containing $f^*$ are dubbed as *non-realizable* classes. The smallest such family where the unknown function $f^*$ lies is denoted by $\mathcal{F}_{m^*}$, where $m^* \in [M]$. However, we do not know the index $m^*$, and our goal is to propose adaptive algorithms such that the regret depends on the complexity of the function class $\mathcal{F}_{m^*}$. In order to achieve this, we need a separability condition on the nested models.

**Assumption 2 (Local Separability).** *There exist $\Delta > 0$ and $\eta > 0$ such that*

$$\inf_{f \in \mathcal{F}_{m^*-1}} \inf_{x_1 \neq x_2 : D^*(x_1, x_2) \leq \eta} |f(x_1) - f^*(x_2)| \geq \Delta,$$

*where*[1], $D^*(x_1, x_2) = |f^*(x_1) - f^*(x_2)|$.

The above assumption[2] ensures that for action pairs $(x_1, x_2)$, where the obtained (expected) rewards are close (since it is generated by $f^*$), there is a gap between the true function $f^*$ and the ones belonging to the function classes not containing $f^*$ (i.e., the non-realizable function classes). Note that we do not require this separability to hold for all actions – just the ones which are indistinguishable from observing the rewards. Note that separability is needed for model selection since we neither assume any structural assumption on $f^*$, nor on the set $\mathcal{X}$.

We emphasize that separability is quite standard and assumptions of similar nature appear in a wide range of model selection problems, specially in the setting of contextual bandits [16, 19]. It is also quite standard in statistics, specifically in the area of clustering and latent variable modelling [5, 14, 34].

*Separability for Lipschitz $f^*$:* If the true function $f^*$ is 1-Lipschitz. In that setting, the separability assumption takes the following form: for $\Delta > 0$ and $\eta > 0$,

$$\inf_{f \in \mathcal{F}_{m^*-1}} \inf_{x_1 \neq x_2 : \|x_1 - x_2\| \leq \eta} |f(x_1) - f^*(x_2)| \geq \Delta$$

However, note that the above assumption is quite strong – any (random) arbitrary algorithm can perform model selection (with the knowledge of $\eta$ and $\Delta$)[3] in the following way: first choose action $x_1$. Using $\|x_1 - x_2\| \leq \eta$, choose $x_2$. Pick any function $f$ belonging to some class $\mathcal{F}_m$ in the nested family and evaluate $|f(x_1) - y_t(x_2)|$, which is a good proxy for $|f(x_1) - f^*(x_2)|$. The algorithm continues to pick different $f \in \mathcal{F}_m$. With the knowledge of $\Delta$, depending on how

---

[1] Here the roles of $x_1$ and $x_2$ are interchangeable without loss of generality.

[2] We assume that the action set $\mathcal{X}$ is compact and continuous, and so such action pairs $(x_1, x_2)$ always exist, i.e., given any $x_1 \in \mathcal{X}$, an action $x_2$ such that $D^*(x_1, x_2) \leq \eta$ always exists.

[3] This can be found using standard trick like doubling.

big $\mathcal{F}_m$ is, the algorithm would be able to identify whether $\mathcal{F}_m$ is realizable or not. Continuing it for all hypothesis classes, it would identify the correct class $\mathcal{F}_{m^*}$. Hence, for structured $f^*$, the problem of model selection with separation is not interesting and we do not consider that setup in this paper.

*Separability for Linear $f^*$:* If $f^*$ is linear, the separability assumption is not necessary for model selection. In this setting, $f^*$ is parameterized by some properties of the parameter, such as sparsity and norm, denotes the nested function classes. [12,15] addresses the linear bandit model selection problem without the separability assumption.

## 2.2   Algorithm: Adaptive Bandit Learning (ABL)

In this section, we provide a novel model selection algorithm (Algorithm 2) that, over multiple epochs, successively refine the estimate of the true model class $\mathcal{F}_{m^*}$ where the unknown function $f^*$ lies. At each epoch, we run a fresh instance of a base bandit algorithm for the estimated function class, which we call Bandit Learning. Note that our model selection algorithm works with any provable bandit learning algorithm, and is agnostic to the particular choice of such base algorithm. In what follows, we present a generic description of the base algorithm and then specialize to a special case.

*The Base Algorithm.* Bandit Learning (Algorithm 1), in its general form, takes a function class $\mathcal{F}$ and a confidence level $\delta \in (0, 1]$ as its inputs. At each time $t$, it maintains a (high-probability) confidence set $\mathcal{C}_t(\mathcal{F}, \delta)$ for the unknown function $f^*$, and chooses the most optimistic action with respect to this confidence set,

$$x_t \in \underset{x \in \mathcal{X}}{\operatorname{argmax}} \ \max_{f \in \mathcal{C}_t(\mathcal{F}, \delta)} f(x) \,. \tag{1}$$

The confidence set $\mathcal{C}_t(\mathcal{F}, \delta)$ is constructed using all the data $\{x_s, y_s\}_{s<t}$ gathered in the past. First, a regularized least square estimate of $f^*$ is computed as $\hat{f}_t \in \operatorname{argmin}_{f \in \mathcal{F}} \mathcal{L}_{t-1}(f)$, where $\mathcal{L}_t(f) := \sum_{s=1}^{t} (y_s - f(x_s))^2$ is the cumulative squared prediction error. The confidence set $\mathcal{C}_t(\mathcal{F}, \delta)$ is then defined as the set of all functions $f \in \mathcal{F}$ satisfying

$$\sum_{s=1}^{t-1} \left( f(x_s) - \hat{f}_t(x_s) \right)^2 \leq \beta_t(\mathcal{F}, \delta) \,, \tag{2}$$

where $\beta_t(\mathcal{F}, \delta)$ is an appropriately chosen confidence parameter. We now specialize to the bandit learning algorithm of [26] by setting the confidence parameter

$$\beta_t(\mathcal{F}, \delta) := 8\sigma^2 \log\left(2\mathcal{N}\left(\mathcal{F}, 1/T, \|\cdot\|_\infty\right)/\delta\right) + 2\left(8 + \sqrt{8\sigma^2 \log\left(8t(t+1)/\delta\right)}\right),$$

where $\mathcal{N}(\mathcal{F}, \alpha, \|\cdot\|_\infty)$ is the $(\alpha, \|\cdot\|_\infty)$-covering number[4] of $\mathcal{F}$, one can ensure that $f^*$ lies in the confidence set $\mathcal{C}_t(\mathcal{F}, \delta)$ at all time instant $t \geq 1$ with probability at least $1 - \delta$. The theoretical guarantees presented in the paper are also for this particular choice of base algorithm.

---

[4] For any $\alpha > 0$, we call $\mathcal{F}^\alpha$ an $(\alpha, \|\cdot\|_\infty)$ cover of the function class $\mathcal{F}$ if for any $f \in \mathcal{F}$ there exists an $f'$ in $\mathcal{F}^\alpha$ such that $\|f' - f\|_\infty := \sup_{x \in \mathcal{X}} |f'(x) - f(x)| \leq \alpha$.

---

**Algorithm 1.** Bandit Learning

1: **Input:** Function class $\mathcal{F}$, confidence level $\delta \in (0, 1]$
2: **for** time $t = 1, 2, 3, \ldots$ **do**
3:     Compute an estimate $\widehat{f}_t$ of $f^*$
4:     Construct confidence set $\mathcal{C}_t(\mathcal{F}, \delta)$ using (2)
5:     Choose an action $x_t$ using (1)
6:     Observe reward $y_t$
7: **end for**

---

**Algorithm 2.** Adaptive Bandit Learning (ABL)

1: **Input:** Nested function classes $\mathcal{F}_1 \subset \mathcal{F}_2 \subset \ldots \subset \mathcal{F}_M$, confidence level $\delta \in (0, 1]$, threshold $\gamma_i > 0$
2: **for** epochs $i = 1, 2 \ldots$ **do**
3:     **Model Selection:**
4:     Compute elapsed time $\tau_{i-1} = \sum_{j=1}^{i-1} t_j$
5:     **for** function classes $m = 1, 2 \ldots, M$ **do**
6:         Compute the minimum average squared prediction error using (3)
7:     **end for**
8:     Choose index $m^{(i)} = \min\{m \in [M] : T_m^{(i)} \leq \gamma_i\}$
9:     **Model Learning:**
10:     Set epoch length $t_i = 2^i$, confidence level $\delta_i = \delta/2^i$
11:     Run Bandit Learning (Algorithm 1) over a time horizon $t_i$ with function class $\mathcal{F}_{m^{(i)}}$ and confidence level $\delta_i$ as its inputs
12: **end for**

---

*Our Approach–Adaptive Bandit Learning (*ABL*):* The description of our model selection algorithm is given in Algorithm 2. We consider doubling epochs – at each epoch $i \geq 1$, the base algorithm is run over time horizon $t_i = 2^i$. At the beginning of $i$-th epoch, using all the data of the previous epochs, we employ a model selection module as follows. First, we compute, for each class $\mathcal{F}_m$, the minimum average squared prediction error (via an offline regression oracle)

$$T_m^{(i)} = \min_{f \in \mathcal{F}_m} \frac{1}{\tau_{i-1}} \sum_{s=1}^{\tau_{i-1}} (y_s - f(x_s))^2 \,, \tag{3}$$

where $\tau_{i-1} := \sum_{j=1}^{i-1} t_j$ denotes the total time elapsed before epoch $i$. Finally, we compare $T_m^{(i)}$ to a pre-calculated threshold $\gamma$, and pick the function class for which $T_m^{(i)}$ falls below such threshold (with smallest $m$, see Algorithm 2). After selecting the function class, we run the base algorithm for this class with confidence level $\delta_i = \delta/2^i$. We call the complete procedure Adaptive Bandit Learning (ABL).

## 2.3     Performance Guarantee of ABL

We now provide model selection and regret guarantees of ABL (Algorithm 2), when the base algorithm is chosen as [26]. Though the results to be presented in this section are quite general, they do not apply to any arbitrary function classes. In what follows, we will make the following boundedness assumption.

**Assumption 3 (Bounded functions).** *We assume that $f(x) \in [0, 1] \; \forall \; x \in \mathcal{X}$ and $f \in \mathcal{F}_m$ ($\forall \; m \in [M]$).*[5]

It is worth noting that this same assumption is also required in the standard setting, i.e., when the true model class is known ($\mathcal{F}_{m^*} = \mathcal{F}$).

We denote by $\log \mathcal{N}(\mathcal{F}_m) = \log\left(\mathcal{N}(\mathcal{F}_m, 1/T, \|\cdot\|_\infty)\right)$ the metric entropy (with scale $1/T$) of the class $\mathcal{F}_m$. We have the following guarantee for ABL.

**Lemma 1 (Model selection of ABL).** *Fix a $\delta \in (0, 1]$ and $\lambda > 0$. Suppose, Assumptions 1, 2 and 3 hold and we set the threshold $\gamma_i = T_M^{(i)} + C_1$, for a sufficiently small constant $C_1$. Then, with probability at least $1 - O(M\delta)$, ABL identifies the correct model class $\mathcal{F}_{m^*}$ from epoch $i \geq i^*$ when the time elapsed before epoch $i^*$ satisfies*

$$\tau_{i^*-1} \geq C\sigma^4(\log T) \max\left\{\frac{\log(1/\delta)}{(\frac{\Delta^2}{2} - 4\eta)^2}, \log\left(\frac{\mathcal{N}(\mathcal{F}_M)}{\delta}\right)\right\},$$

*provided $\Delta \geq 2\sqrt{2\eta}$, where $C > 1$ is a sufficiently large universal constant.*

*Remark 1 (Dependence on the biggest class).* Note that we choose a threshold that depends on the epoch number and the test statistic of the biggest class. Here we crucially exploit the fact that the biggest class always contains the true model class and use this to design the threshold.

We characterize the complexity of each function class $\mathcal{F}_m$ by its *eluder dimension*, first introduced by [26] in the standard setting.

**Definition 1 (Eluder dimension).** *The $\varepsilon$-eluder dimension $\dim_{\mathcal{E}}(\mathcal{F}_m, \varepsilon)$ of a function class $\mathcal{F}$ is the length of the longest sequence $\{x_i\}_{i=1}^n \subseteq \mathcal{X}$ of input points such that for some $\varepsilon' \geq \varepsilon$ and for each $i \in \{2, \ldots, n\}$,*

$$\sup_{f_1, f_2 \in \mathcal{F}} \left\{(f_1 - f_2)(x_i) \mid \sqrt{\sum_{j=1}^{i-1}(f_1 - f_2)^2(x_i)} \leq \varepsilon'\right\} > \varepsilon'.$$

Define $\mathcal{F}^* = \mathcal{F}_{m^*}$. Denote by $d_{\mathcal{E}}(\mathcal{F}^*) = \dim_{\mathcal{E}}(\mathcal{F}^*, 1/T)$, the $(1/T)$-eluder dimension of the (realizable) function class $\mathcal{F}^*$, where $T$ is the time horizon. Then, armed with Lemma 1, we obtain the following regret bound for ABL.

---

[5] We can extent the range to $[0, c]$ without loss of generality.

**Theorem 1 (Cumulative regret of ABL).** *Suppose the condition of Lemma 1 holds. Then, for any $\delta \in (0,1]$, the regret of ABL for horizon $T$ is*

$$\mathcal{R}_T \leq \mathcal{O}\left(\sigma^4(\log T)\max\left\{\frac{\log(1/\delta)}{(\frac{\Delta^2}{2}-4\eta)^2}, \log\left(\frac{\mathcal{N}(\mathcal{F}_M)}{\delta}\right)\right\}\right)$$
$$+ \mathcal{O}\left(d_{\mathcal{E}}(\mathcal{F}^*)\log T + c\sqrt{Td_{\mathcal{E}}(\mathcal{F}^*)\log(\mathcal{N}(\mathcal{F}^*)/\delta)\log^2(T/\delta)}\right),$$

*with probability at least[6] $1 - O(M\delta)$.*

*Remark 2 (Cost of model selection).* We retain the regret bound of [26] in the standard setting, and the first term in the regret bound captures the cost of model selection – the cost suffered before accumulating enough samples to infer the correct model class (with high probability). It has weak (logarithmic) dependence on horizon $T$ and hence considered as a minor term, in the setting where $T$ is large. Hence, model selection is essentially *free* upto log factors. Let us now have a close look at this term. It depends on the metric entropy of the biggest model class $\mathcal{F}_M$. This stems from the fact that the thresholds $\{\gamma_i\}_{i \geq 1}$ depend on the test statistic of $\mathcal{F}_M$ (see Remark 1). We believe that, without additional assumptions, one can't get rid of this (minor) dependence on the complexity of the biggest class.

The second term is the major one ($\sqrt{T}$ dependence on total number of steps), which essentially is the cost of learning the true kernel $f^*$. Since in this phase, we basically run the base algorithm for the correct model class, our regret guarantee matches to that of an oracle with the apriori knowledge of the correct class. Note that if we simply run a non model-adaptive algorithm for this problem, the regret would be $\widetilde{\mathcal{O}}(H\sqrt{Td_{\mathcal{E}}(\mathcal{F}_M)\log\mathcal{N}(\mathcal{F}_M)})$, where $d_{\mathcal{E}}(\mathcal{F}_M)$ denotes the eluder dimension of the largest model class $\mathcal{F}_M$. In contrast, by successively testing and thresholding, our algorithm adapts to the complexity of the smallest function class containing the true model class.

*Remark 3 (Requires no knowledge of $(\Delta, \eta)$).* Our algorithm ABL doesn't require the knowledge of $\Delta$ and $\eta$. Rather, it automatically adapts to these parameters, and the dependence is reflected in the regret expression. The separation $\Delta$ implies how complex the job of model selection is. If the separation is small, it is difficult for ABL to separate out the model classes. Hence, it requires additional exploration, and as a result the regret increases. Another interesting fact of Theorem 1 is that it does not require any minimum separation across model classes. This is in sharp contrast with existing results in statistics (see, e.g. [5,34]). Even if $\Delta$ is quite small, Theorem 1 gives a model selection guarantee. Now, the cost of separation appears anyways in the minor term, and hence in the long run, it does not effect the overall performance of the algorithm.

---

[6] One can choose $\delta = 1/\mathrm{poly}(M)$ to obtain a high-probability bound which only adds an extra $\log M$ factor.

# 3  Model Selection in Markov Decision Processes

An (episodic) MDP is denoted by $\mathcal{M}(\mathcal{S}, \mathcal{A}, H, P^*, r)$, where $\mathcal{S}$ is the state space, $\mathcal{A}$ is the action space (both possibly infinite), $H$ is the length of each episode, $P^* : \mathcal{S} \times \mathcal{A} \to \Delta(\mathcal{S})$ is an (unknown) transition kernel (a function mapping state-action pairs to signed distribution over the state space) and $r : \mathcal{S} \times \mathcal{A} \to [0,1]$ is a (known) reward function. In episodic MDPs, a (deterministic) policy $\pi$ is given by a collection of $H$ functions $(\pi_1, \ldots, \pi_H)$, where each $\pi_h : \mathcal{S} \to \mathcal{A}$ maps a state $s$ to an action $a$. In each episode, an initial state $s_1$ is first picked by the environment (assumed to be fixed and history independent). Then, at each step $h \in [H]$, the agent observes the state $s_h$, picks an action $a_h$ according to $\pi_h$, receives a reward $r(s_h, a_h)$, and then transitions to the next state $s_{h+1}$, which is drawn from the conditional distribution $P^*(\cdot|s_h, a_h)$. The episode ends when the terminal state $s_{H+1}$ is reached. For each state-action pair $(s, a) \in \mathcal{S} \times \mathcal{A}$ and step $h \in [H]$, we define action values $Q_h^\pi(s, a)$ and and state values $V_h^\pi(s)$ corresponding to a policy $\pi$ as

$$Q_h^\pi(s,a) = r(s,a) + \mathbb{E}\left[\sum\nolimits_{h'=h+1}^{H} r(s_{h'}, \pi_{h'}(s_{h'}))|s_h = s, a_h = a\right], \quad V_h^\pi(s) = Q_h^\pi\big(s, \pi_h(s)\big),$$

where the expectation is with respect to the randomness of the transition distribution $P^*$. It is not hard to see that $Q_h^\pi$ and $V_h^\pi$ satisfy the Bellman equations:

$$Q_h^\pi(s,a) = r(s,a) + (P^* V_{h+1}^\pi)(s,a), \quad \forall h \in [H], \quad \text{with} V_{H+1}^\pi(s) = 0 \text{for all} s \in \mathcal{S}.$$

A policy $\pi^*$ is said to be optimal if it maximizes the value for all states $s$ and step $h$ simultaneously, and the corresponding optimal value function is denoted by $V_h^*(s) = \sup_\pi V_h^\pi(s)$ for all $h \in [H]$, where the supremum is over all (non-stationary) policies. The agent interacts with the environment for $K$ episodes to learn the unknown transition kernel $P^*$ and thus, in turn, the optimal policy $\pi^*$. At each episode $k \geq 1$, the agent chooses a policy $\pi^k := (\pi_1^k, \ldots, \pi_H^k)$ and a trajectory $(s_h^k, a_h^k, r(s_h^k, a_h^k), s_{h+1}^k)_{h \in [H]}$ is generated. The performance of the learning agent is measured by the cumulative (pseudo) regret accumulated over $K$ episodes, defined as

$$\mathcal{R}(T) := \sum\nolimits_{k=1}^{K} \left[V_1^*(s_1^k) - V_1^{\pi^k}(s_1^k)\right],$$

where $T = KH$ is total steps in $K$ episodes.

In this work, we consider general MDPs without any structural assumption on the unknown transition kernel $P^*$. In the standard setting [4], it is assumed that $P^*$ belongs to a known family of transition models $\mathcal{P}$. Here, in contrast to the standard setting, we do not have the knowledge of $\mathcal{P}$. Instead, we are given $M$ nested families of transition kernels $\mathcal{P}_1 \subset \mathcal{P}_2 \subset \ldots \subset \mathcal{P}_M$. The smallest such family where the true transition kernel $P^*$ lies is denoted by $\mathcal{P}_{m^*}$, where $m^* \in [M]$. However, we do not know the index $m^*$, and our goal is to propose adaptive algorithms such that the regret depends on the complexity of the family $\mathcal{P}_{m^*}$. We assume a similar separability condition on these nested model classes.

**Assumption 4 (Local Separability).** *There exist constants $\Delta > 0$ and $\eta > 0$ such that for any function $V : \mathcal{S} \to \mathbb{R}$,*

$$\inf_{P \in \mathcal{P}_{m^*-1}} \inf_{D*((s_1,a_1),(s_2,a_2)) \leq \eta} |PV(s_1,a_1) - P^*V(s_2,a_2)| \geq \Delta,$$

*where $(s_1,a_1) \neq (s_2,a_2)$ and $D^*((s_1,a_1),(s_2,a_2)) = |P^*V(s_1,a_1) - P^*V(s_2,a_2)|$.*

This assumption ensures that expected values under the true model is well-separated from those under models from non-realizable classes for two distinct state-action pairs for which values are close under true model. Once again, we need state and action spaces to be compact and continuous to guarantee such pairs always exist. Note that the assumption might appear to break down for any constant function $V$. However, we will be invoking this assumption with the value functions computed by the learning algorithm (see (4)). For reward functions that *vary sufficiently* across states and actions, and transition kernels that admit densities, the chance of getting hit by constant value functions is admissibly low. In case the rewards are constant, every policy would anyway incur zero regret rendering the learning problem trivial. The value functions appear in the separability assumption in the first place since we are interested in minimizing the regret. Instead, if one cares only about learning the true model, then separability of transition kernels under some suitable notion of distance (e.g., the KL-divergence) might suffice. Note that in [16, 19], the regret is defined in terms of the regression function and hence the separability is assumed on the regression function itself. Model selection without separability is kept as an interesting future work.

### 3.1  Algorithm: Adaptive Reinforcement Learning (ARL)

In this section, we provide a novel model selection algorithm ARL (Algorithm 2) that use successive refinements over epochs. We use UCRL-VTR algorithm of [4] as our base algorithm, and add a model selection module at the beginning of each epoch. In other words, over multiple epochs, we successively refine our estimates of the proper model class where the true transition kernel $P^*$ lies.

*The Base Algorithm:* UCRL-VTR, in its general form, takes a family of transition models $\mathcal{P}$ and a confidence level $\delta \in (0,1]$ as its input. At each episode $k$, it maintains a (high-probability) confidence set $\mathcal{B}_{k-1} \subset \mathcal{P}$ for the unknown model $P^*$ and use it for optimistic planning. First, it finds the transition kernel $P_k = \mathrm{argmax}_{P \in \mathcal{B}_{k-1}} V^*_{P,1}(s_1^k)$, where $V^*_{P,h}$ denote the optimal value function of an MDP with transition kernel $P$ at step $h$. UCRL-VTR then computes, at each step $h$, the optimal value function $V_h^k := V^*_{P_k,h}$ under the kernel $P_k$ using dynamic programming. Specifically, starting with $V_{H+1}^k(s,a) = 0$ for all pairs $(s,a)$, it defines for all steps $h = H$ down to 1,

$$Q_h^k(s,a) = r(s,a) + (P_k V_{h+1}^k)(s,a), \quad V_h^k(s) = \max_{a \in \mathcal{A}} Q_h^k(s,a). \qquad (4)$$

Then, at each step $h$, UCRL-VTR takes the action that maximizes the $Q$-function estimate, i,e. it chooses $a_h^k = \text{argmax}_{a \in \mathcal{A}} Q_h^k(s_h^k, a)$. Now, the confidence set is updated using all the data gathered in the episode. First, UCRL-VTR computes an estimate of $P^*$ by employing a non-linear value-targeted regression model with data $\left(s_h^j, a_h^j, V_{h+1}^j(s_{h+1}^j)\right)_{j \in [k], h \in [H]}$. Note that $\mathbb{E}[V_{h+1}^k(s_{h+1}^k)|\mathcal{G}_{h-1}^k] = (P^* V_{h+1}^k)(s_h^k, a_h^k)$, where $\mathcal{G}_{h-1}^k$ denotes the $\sigma$-field summarizing the information available just before $s_{h+1}^k$ is observed. This naturally leads to the estimate $\widehat{P}_k = \text{argmin}_{P \in \mathcal{P}} \mathcal{L}_k(P)$, where

$$\mathcal{L}_k(P) := \sum_{j=1}^{k} \sum_{h=1}^{H} \left(V_{h+1}^j(s_{h+1}^j) - (P V_{h+1}^j)(s_h^j, a_h^j)\right)^2. \tag{5}$$

The confidence set $\mathcal{B}_k$ is then updated by enumerating the set of all transition kernels $P \in \mathcal{P}$ satisfying $\sum_{j=1}^{k} \sum_{h=1}^{H} \left((P V_{h+1}^j)(s_h^j, a_h^j) - (\widehat{P}_k V_{h+1}^j)(s_h^j, a_h^j)\right)^2 \leq \beta_k(\delta)$ with the confidence width being defined as $\beta_k(\delta) := 8H^2 \log\left(\frac{2\mathcal{N}(\mathcal{P}, \frac{1}{kH}, \|\cdot\|_{\infty,1})}{\delta}\right) + 4H^2\left(2 + \sqrt{2\log\left(\frac{4kH(kH+1)}{\delta}\right)}\right)$, where $\mathcal{N}(\mathcal{P}, \cdot, \cdot)$ denotes the covering number of the family $\mathcal{P}$.[7] Then, one can show that $P^*$ lies in the confidence set $\mathcal{B}_k$ in all episodes $k$ with probability at least $1 - \delta$. Here, we consider a slight different expression of $\beta_k(\delta)$ as compared to [4], but the proof essentially follows the same technique. Please refer to Appendix ?? for further details.

*Our Approach:* We consider doubling epochs - at each epoch $i \geq 1$, UCRL-VTR is run for $k_i = 2^i$ episodes. At the beginning of $i$-th epoch, using all the data of previous epochs, we add a model selection module as follows. First, we compute, for each family $\mathcal{P}_m$, the transition kernel $\widehat{P}_m^{(i)}$, that minimizes the empirical loss $\mathcal{L}_{\tau_{i-1}}(P)$ over all $P \in \mathcal{P}_m$ (see (5)), where $\tau_{i-1} := \sum_{j=1}^{k-1} k_j$ denotes the total number of episodes completed before epoch $i$. Next, we compute the average empirical loss $T_m^{(i)} := \frac{1}{\tau_{i-1}H} \mathcal{L}_{\tau_{i-1}}(\widehat{P}_m^{(i)})$ for the model $\widehat{P}_m^{(i)}$. Finally, we compare $T_m^{(i)}$ to a pre-calculated threshold $\gamma_i$, and pick the transition family for which $T_m^{(i)}$ falls below such threshold (with smallest $m$, see Algorithm 3). After selecting the family, we run UCRL-VTR for this family with confidence level $\delta_i = \frac{\delta}{2^i}$, where $\delta \in (0,1]$ is a parameter of the algorithm.

### 3.2 Performance Guarantee of ARL

First, we present our main result which states that the model selection procedure of ARL (Algorithm 3) succeeds with high probability after a certain number of epochs. To this end, we denote by $\log \mathcal{N}(\mathcal{P}_m) = \log(\mathcal{N}(\mathcal{P}_m, 1/T, \|\cdot\|_{\infty,1}))$ the metric entropy (with scale $1/T$) of the family $\mathcal{P}_m$. We also use the shorthand notation $\mathcal{P}^* = \mathcal{P}_{m^*}$.

---

[7] For any $\alpha > 0$, $\mathcal{P}^\alpha$ is an $(\alpha, \|\cdot\|_{\infty,1})$ cover of $\mathcal{P}$ if for any $P \in \mathcal{P}$ there exists an $P'$ in $\mathcal{P}^\alpha$ such that $\|P' - P\|_{\infty,1} := \sup_{s,a} \int_{\mathcal{S}} |P'(s'|s,a) - P(s'|s,a)| ds' \leq \alpha$.

---

**Algorithm 3.** Adaptive Reinforcement Learning – ARL

---
1: **Input:** Parameter $\delta$, function classes $\mathcal{P}_1 \subset \mathcal{P}_2 \subset \ldots \subset \mathcal{P}_M$, thresholds $\{\gamma_i\}_{i \geq 1}$
2: **for** epochs $i = 1, 2 \ldots$ **do**
3:     Set $\tau_{i-1} = \sum_{j=1}^{i-1} k_j$
4:     **for** function classes $m = 1, 2 \ldots, M$ **do**
5:         Compute $\widehat{P}_m^{(i)} = \operatorname{argmin}_{P \in \mathcal{P}_m} \sum_{k=1}^{\tau_{i-1}} \sum_{h=1}^{H} \left( V_{h+1}^k(s_{h+1}^k) - (PV_{h+1}^k)(s_h^k, a_h^k) \right)^2$
6:         Compute $T_m^{(i)} = \frac{1}{\tau_{i-1}H} \sum_{k=1}^{\tau_{i-1}} \sum_{h=1}^{H} \left( V_{h+1}^k(s_{h+1}^k) - (\widehat{P}_m^{(i)} V_{h+1}^k)(s_h^k, a_h^k) \right)^2$
7:     **end for**
8:     Set $m^{(i)} = \min\{m \in [M] : T_m^{(i)} \leq \gamma_i\}$, $k_i = 2^i$ and $\delta_i = \delta/2^i$
9:     Run UCRL-VTR for the family $\mathcal{P}_{m^{(i)}}$ for $k_i$ episodes with confidence level $\delta_i$
10: **end for**

---

**Lemma 2 (Model selection of ARL).** *Fix a $\delta \in (0, 1]$ and suppose Assumption 2 holds. Suppose the thresholds are set as $\gamma_i = T_M^{(i)} + C_2$, for some sufficiently small constant $C_2$. Then, with probability at least $1 - O(M\delta)$, ARL identifies the correct model class $\mathcal{P}_{m^*}$ from epoch $i \geq i^*$, where epoch length of $i^*$ satisfies*

$$2^{i^*} \geq C' \log K \max \left\{ \frac{H^3}{(\frac{1}{2}\Delta^2 - 2H\eta)^2} \log(2/\delta), 4H \log\left( \frac{\mathcal{N}(\mathcal{P}_M)}{\delta} \right) \right\},$$

*provided $\Delta \geq 2\sqrt{H\eta}$, for a sufficiently large universal constant $C' > 1$.*

*Regret Bound:* In order to present our regret bound, we define, for each model class $\mathcal{P}_m$, a collection of functions $\mathcal{M}_m := \{f : \mathcal{S} \times \mathcal{A} \times \mathcal{V}_m \to \mathbb{R}\}$ such that any $f \in \mathcal{M}_m$ satisfies $f(s, a, V) = (PV)(s, a)$ for some $P \in \mathcal{P}_m$ and $V \in \mathcal{V}_m$, where $\mathcal{V}_m := \{V_{P,h}^* : P \in \mathcal{P}_m, h \in [H]\}$ denotes the set of optimal value functions under the transition family $\mathcal{P}_m$. By one-to-one correspondence, we have $\mathcal{M}_1 \subset \mathcal{M}_2 \subset \ldots \subset \mathcal{M}_M$, and the complexities of these function classes determine the learning complexity of the RL problem under consideration. We characterize the complexity of each function class $\mathcal{M}_m$ by its *eluder dimension*, which is defined similarly as Definition 1. (We take domain of function class $\mathcal{M}_m$ to be $\mathcal{S} \times \mathcal{A} \times \mathcal{V}_m$.)

We define $\mathcal{M}^* = \mathcal{M}_{m^*}$, and denote by $d_{\mathcal{E}}(\mathcal{M}^*) = \dim_{\mathcal{E}}(\mathcal{M}^*, 1/T)$, the $(1/T)$-eluder dimension of the (realizable) function class $\mathcal{M}^*$, where $T$ is the time horizon. Then, armed with Lemma 2, we obtain the following regret bound.

**Theorem 2 (Cumulative regret of ARL).** *Suppose the conditions of Lemma 2 hold. Then, for any $\delta \in (0, 1]$, running ARL for $K$ episodes yields a regret bound*

$$\mathcal{R}(T) = \mathcal{O}\left( \log K \max \left\{ \frac{H^4 \log(1/\delta)}{(\frac{\Delta^2}{2} - 2H\eta)^2}, H^2 \log\left( \frac{\mathcal{N}(\mathcal{P}_M)}{\delta} \right) \right\} \right)$$
$$+ \mathcal{O}\left( H^2 d_{\mathcal{E}}(\mathcal{M}^*) \log K + H\sqrt{T d_{\mathcal{E}}(\mathcal{M}^*) \log(\mathcal{N}(\mathcal{P}^*)/\delta)} \log K \log(T/\delta) \right).$$

*with probability at least $1 - O(M\delta)$.*

Similar to Theorem 1, the first term in the regret bound captures the cost of model selection, having weak (logarithmic) dependence on the number of episodes $K$ and hence considered as a minor term, in the setting where $K$ is large. Hence, model selection is essentially *free* upto log factors. The second term is the major one ($\sqrt{T}$ dependence on total number of steps), which essentially is the cost of learning the true kernel $P^*$. Since in this phase, we basically run UCRL-VTR for the correct model class, our regret guarantee matches to that of an oracle with the apriori knowledge of the correct class. ARL doesnot require the knowledge of $(\Delta, \eta)$ and it adapts to the complexity of the problem.

## 4   Conclusion

We address the problem of model selection for MAB and MDP and propose algorithms that obtains regret similar to an oracle who knows the true model class apriori. Our algorithms leverage the separability conditions crucially, and removing them is kept as a future work.

**Acknowledgements.** We thank anonymous reviewers for their useful comments. Moreover, we would like to thank Prof. Kannan Ramchandran (EECS, UC Berkeley) for insightful discussions regarding the topic of model selection. SRC is grateful to a CISE postdoctoral fellowship of Boston University.

## References

1. Abbasi-Yadkori, Y., Pál, D., Szepesvári, C.: Improved algorithms for linear stochastic bandits. In: Advances in Neural Information Processing Systems, pp. 2312–2320 (2011)
2. Agarwal, A., Luo, H., Neyshabur, B., Schapire, R.E.: Corralling a band of bandit algorithms. In: Conference on Learning Theory, pp. 12–38. PMLR (2017)
3. Arora, R., Marinov, T.V., Mohri, M.: Corralling stochastic bandit algorithms. In: International Conference on Artificial Intelligence and Statistics, pp. 2116–2124. PMLR (2021)
4. Ayoub, A., Jia, Z., Szepesvari, C., Wang, M., Yang, L.F.: Model-based reinforcement learning with value-targeted regression. arXiv preprint arXiv:2006.01107 (2020)
5. Balakrishnan, S., Wainwright, M.J., Yu, B., et al.: Statistical guarantees for the EM algorithm: from population to sample-based analysis. Ann. Stat. **45**(1), 77–120 (2017)
6. Besbes, O., Zeevi, A.: Dynamic pricing without knowing the demand function: risk bounds and near-optimal algorithms. Oper. Res. **57**(6), 1407–1420 (2009)
7. Cesa-Bianchi, N., Lugosi, G.: Prediction, Learning, and Games. Cambridge University Press, Cambridge (2006)
8. Chang, F., Lai, T.L.: Optimal stopping and dynamic allocation. Adv. Appl. Probabil. **19**(4), 829–853 (1987). http://www.jstor.org/stable/1427104
9. Chatterji, N.S., Muthukumar, V., Bartlett, P.L.: Osom: a simultaneously optimal algorithm for multi-armed and linear contextual bandits. arXiv preprint arXiv:1905.10040 (2019)

10. Chowdhury, S.R., Gopalan, A.: Online learning in kernelized Markov decision processes. In: The 22nd International Conference on Artificial Intelligence and Statistics, pp. 3197–3205 (2019)
11. Even-Dar, E., Mannor, S., Mansour, Y.: Action elimination and stopping conditions for the multi-armed bandit and reinforcement learning problems. J. Mach. Learn. Res. **7**(39), 1079–1105 (2006). http://jmlr.org/papers/v7/evendar06a.html
12. Foster, D.J., Krishnamurthy, A., Luo, H.: Model selection for contextual bandits. In: Advances in Neural Information Processing Systems, pp. 14714–14725 (2019)
13. Ghosh, A., Chowdhury, S.R., Gopalan, A.: Misspecified linear bandits. In: Proceedings of the AAAI Conference on Artificial Intelligence, vol. 31 (2017)
14. Ghosh, A., Pananjady, A., Guntuboyina, A., Ramchandran, K.: Max-affine regression: Provable, tractable, and near-optimal statistical estimation. arXiv preprint arXiv:1906.09255 (2019)
15. Ghosh, A., Sankararaman, A., Kannan, R.: Problem-complexity adaptive model selection for stochastic linear bandits. In: International Conference on Artificial Intelligence and Statistics, pp. 1396–1404. PMLR (2021)
16. Ghosh, A., Sankararaman, A., Ramchandran, K.: Model selection for generic contextual bandits. arXiv preprint arXiv:2107.03455 (2021)
17. Jin, C., Yang, Z., Wang, Z., Jordan, M.I.: Provably efficient reinforcement learning with linear function approximation. arXiv preprint arXiv:1907.05388 (2019)
18. Kakade, S., Krishnamurthy, A., Lowrey, K., Ohnishi, M., Sun, W.: Information theoretic regret bounds for online nonlinear control. arXiv preprint arXiv:2006.12466 (2020)
19. Krishnamurthy, S.K., Athey, S.: Optimal model selection in contextual bandits with many classes via offline oracles. arXiv preprint arXiv:2106.06483 (2021)
20. Lee, J.N., Pacchiano, A., Muthukumar, V., Kong, W., Brunskill, E.: Online model selection for reinforcement learning with function approximation. CoRR abs/2011.09750 (2020). https://arxiv.org/abs/2011.09750
21. Mnih, V., et al.: Playing Atari with deep reinforcement learning. arXiv preprint arXiv:1312.5602 (2013)
22. Osband, I., Van Roy, B.: Model-based reinforcement learning and the eluder dimension. In: Advances in Neural Information Processing Systems 27 (NIPS), pp. 1466–1474 (2014)
23. Pacchiano, A., Dann, C., Gentile, C., Bartlett, P.: Regret bound balancing and elimination for model selection in bandits and RL. arXiv preprint arXiv:2012.13045 (2020)
24. Pacchiano, A., et al.: Model selection in contextual stochastic bandit problems. In: Larochelle, H., Ranzato, M., Hadsell, R., Balcan, M.F., Lin, H. (eds.) Advances in Neural Information Processing Systems, vol. 33, pp. 10328–10337. Curran Associates, Inc. (2020). https://proceedings.neurips.cc/paper/2020/file/751d51528afe5e6f7fe95dece4ed32ba-Paper.pdf
25. Puterman, M.L.: Markov Decision Processes: Discrete Stochastic Dynamic Programming. Wiley, New York (2014)
26. Russo, D., Van Roy, B.: Eluder dimension and the sample complexity of optimistic exploration. In: Advances in Neural Information Processing Systems, pp. 2256–2264 (2013)
27. Silver, D., et al.: Mastering the game of go without human knowledge. Nature **550**(7676), 354–359 (2017)
28. Srinivas, N., Krause, A., Kakade, S.M., Seeger, M.W.: Information-theoretic regret bounds for gaussian process optimization in the bandit setting. IEEE Trans. Inf. Theory **58**(5), 3250–3265 (2012)

29. Wang, R., Salakhutdinov, R., Yang, L.F.: Provably efficient reinforcement learning with general value function approximation. arXiv preprint arXiv:2005.10804 (2020)
30. Wang, Y., Wang, R., Du, S.S., Krishnamurthy, A.: Optimism in reinforcement learning with generalized linear function approximation. arXiv preprint arXiv:1912.04136 (2019)
31. Williams, G., Aldrich, A., Theodorou, E.A.: Model predictive path integral control: From theory to parallel computation. J. Guid. Control. Dyn. **40**(2), 344–357 (2017)
32. Yang, L.F., Wang, M.: Reinforcement leaning in feature space: matrix bandit, kernels, and regret bound. arXiv preprint arXiv:1905.10389 (2019)
33. Yang, Z., Jin, C., Wang, Z., Wang, M., Jordan, M.: Provably efficient reinforcement learning with kernel and neural function approximations. In: Advances in Neural Information Processing Systems, vol. 33 (2020)
34. Yi, X., Caramanis, C., Sanghavi, S.: Solving a mixture of many random linear equations by tensor decomposition and alternating minimization. CoRR abs/1608.05749 (2016). http://arxiv.org/abs/1608.05749
35. Zanette, A., Brandfonbrener, D., Brunskill, E., Pirotta, M., Lazaric, A.: Frequentist regret bounds for randomized least-squares value iteration. In: International Conference on Artificial Intelligence and Statistics, pp. 1954–1964 (2020)

# Multi-agent Reinforcement Learning

# Heterogeneity Breaks the Game: Evaluating Cooperation-Competition with Multisets of Agents

Yue Zhao[1] and José Hernández-Orallo[2,3(✉)]

[1] School of Computer Science and Engineering, Northwestern Polytechnical University, Xi An, China
[2] Valencian Research Institute for Artifcial Intelligence (VRAIN), Universitat Politècnica de València, Valencia, Spain
`jorallo@upv.es`
[3] Leverhulme Centre for the Future of Intelligence, University of Cambridge, Cambridge, UK

**Abstract.** The value of an agent for a team can vary significantly depending on the heterogeneity of the team and the kind of game: cooperative, competitive, or both. Several evaluation approaches have been introduced in some of these scenarios, from homogeneous competitive multi-agent systems, using a simple average or sophisticated ranking protocols, to completely heterogeneous cooperative scenarios, using the Shapley value. However, we lack a general evaluation metric to address situations with both cooperation and (asymmetric) competition, and varying degrees of heterogeneity (from completely homogeneous teams to completely heterogeneous teams with no repeated agents) to better understand whether multi-agent learning agents can adapt to this diversity. In this paper, we extend the Shapley value to incorporate both repeated players and competition. Because of the combinatorial explosion of team multisets and opponents, we analyse several sampling strategies, which we evaluate empirically. We illustrate the new metric in a predator and prey game, where we show that the gain of some multi-agent reinforcement learning agents for homogeneous situations is lost when operating in heterogeneous teams.

**Keywords:** Multi-agent reinforcement learning · Cooperation-competition game · Evaluation

## 1 Introduction

The evaluation of how much a member contributes to a team is a key question in many disciplines, from economics to biology, and has been an important element of study in artificial intelligence, mostly in the area of multi-agent systems (MAS). When a homogeneous multi-agent system has to achieve a collaborative goal, evaluation can be based on measuring overall performance under several

---

**Supplementary Information** The online version contains supplementary material available at https://doi.org/10.1007/978-3-031-26412-2_11.

M.-R. Amini et al. (Eds.): ECML PKDD 2022, LNAI 13716, pp. 167–182, 2023.
https://doi.org/10.1007/978-3-031-26412-2_11

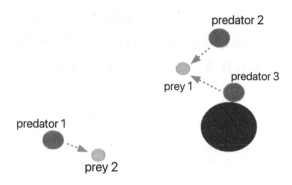

**Fig. 1.** Predator-prey game using the Multi-agent Particle Environment (MPE) [3,4] where we see 3 predators (in red), 2 preys (in green), and landmarks (in black). With $m = 5$ agents playing in total, and $l = 3$ different kinds of agents to choose from (MAD-DPG, DDPG and random), the combinations with repetitions of the team sizes configurations $(l_{pred}, l_{prey}) = (4,1), (3,2), (2,3), (1,4)$ make a total of 45+60+60+45= 210 experiments, and a larger number if we also consider experiments with $m < 5$. Determining which agent has the most contributions to the team considering all roles, and estimating this number with a small number of experiments is the goal of this paper.

agent configurations. However, a more general and realistic version of the problem is when teams are heterogeneous, with players behaving differently and reacting in various ways depending on their teammates. The Shapley value [1] is a well-known metric of the contribution of a player to a heterogeneous team taking into account different coalition formations.

Things become more sophisticated in situations where the players are learning agents [2]. Even if some of these agents use the same algorithm, they may end up having different behaviour after training, with important variations when the same episode is re-run. Despite this variability, they still should be considered as 'repeated' players, something that the original Shapley value does not account for well. Finally, and yet more generally, teams may compete against other teams in asymmetric games, and the contribution of each player will depend on the composition of its team but also on the composition of the opponent team, with the same algorithm possibly appearing once or more on one team or both. This is the general situation we address in this paper.

This situation suffers from poor stability in the payoffs when teams are composed of several learning agents: the same algorithm will lead to very different payoffs depending on the configuration of teams [5]. This requires many iterations in the evaluation protocols, which makes each value for a team configuration expensive to calculate. Consequently, it is even more difficult than in other uses of the Shapley value to collect all possible team configurations. As a result, approximations based on sampling become necessary to deal with the huge number of combinations [6].

Motivated by these issues, we present the following contributions. First, we extend the Shapley value to incorporate repeated players and opposing teams: more technically, the new Shapley value can be applied to cooperative-competitive scenarios when asymmetric teams are multisets of players. Second, we analyse several sampling strategies to approximate this new Shapley value, which we evaluate empirically. Third, we apply this extended Shapley value estimation to a popular asymmetric multi-team multi-agent reinforcement learning (MARL) scenario: predator and prey teams composed of three different kinds of algorithms, which accounts for the heterogeneity of the team. An example of the scenarios we want to evaluate is presented in Fig. 1. We show that some MARL algorithms that work well in homogeneous situations, such as MADDPG, degrade significantly in heterogeneous situations.

The rest of the paper is organised as follows. The following section overviews related work on the evaluation of multi-agent systems, and multi-agent reinforcement learning in particular. Section 3 builds on the original definition of the Shapley value to the extension for multisets and opposing teams (cooperative-competitive), also showing what original properties are preserved. Some sampling methods for approximating this extension are explained in Sect. 4. Section 5 discusses the MARL and single-agent reinforcement learning algorithms and defines the experimental setting. Section 6 covers the experimental results and Sect. 7 closes the paper.

## 2  Background

The evaluation of competitive and cooperative games is at the heart of game theory, pervading many other disciplines. Let us start the analysis with competitive (non-cooperative) games, for two-player games or in multi-agent games. Nash equilibrium [7] is the most common way to define the solution of a non-cooperative game and is invariant to redundant tasks and games, but discovering the Nash equilibrium is not always easy or possible in a multi-agent system [8]. Some new methods are based on the idea of playing a meta-game, that is, a pair-wise win-rate matrix between $N$ agents, as in [9] and the recently proposed $\alpha$-Rank method [10], which was shown to apply to general games. These methods are also inspired by early ranking systems used in (symmetric) competitive games, like the Elo score in chess [11], which estimates the strength of a player, based on the player's performance against *some* of the other opponents. With sparse match results and strongly non-transitive and stochastic players, the predictive power of Elo may be compromised, and this gets even worse in multi-agent games with more than two players per game. As a result, other rating systems such as Glicko [12], TrueSkill [13] and Harkness [14] have been proposed. However, these extensions still show problems of consistency [15,16], very sensitive to non-transitivity and high variability of results between matches.

On the other hand, in purely cooperative games, players are organised into a coalition, a group of players that need to cooperate for the same goal. When the team is homogeneous, the evaluation is easy, as $n$ equal copies of an algorithm or policy are evaluated each time. The best policy or algorithm can be selected just by averaging results. However, in heterogeneous teams, we need to determine the contribution of each specific player in a wide range of situations with complex interactions –the attribution problem. The Shapley value [1] has emerged as a key concept in multi-agent systems to determine each agent's contribution. Given all the coalitions and their payoffs, the Shapley value determines the final contribution of each player. Because of the combinatorial explosion in team formations, approximations are required, both to reduce the computational cost [17] but more importantly to reduce the number of experiments to be run or actual games to be played. Still, in cooperative game theory, the Shapley value provides a key tool for analysing situations with strong interdependence between players [18,19].

The general situation when both competition and cooperation need to be evaluated has been present in many disciplines for centuries, from economics to biology, from sports to sociology. It is also increasingly more prevalent in artificial intelligence, with areas such as reinforcement learning introducing better algorithms for cooperative games but also for cooperative-competitive environments. For instance, DDPG is a deep reinforcement learning agent based on the actor-critic framework, with each agent learning the policy independently without considering the influence of other agents. MADDPG [3] is also based on an actor-critic algorithm, but extends DDPG into a multi-agent policy gradient algorithm where each agent learns a centralised critic based on the observations and actions of all agents. This and other methods (e.g., [20]) are illustrated on some testbed tasks showing that they outperform the baseline algorithms. However, this comparison assumes a homogeneous situation (all the agents in the team use the same algorithm). It is unclear whether these algorithms can still operate in heterogeneous situations. In some cases, the algorithms do not work well when the exchange of information only happens for a subset of agents in the coalition, but in many other cases it is simply that the only available metric is an average reward and the problem of attribution reemerges [21].

Finally, things become really intricate when we consider both competition and cooperation, and we assume that teams can be heterogeneous. But this scenario is becoming increasingly more common as more algorithms could potentially be evaluated in mixed settings (cooperation and competition) [3,22–24]. It is generally believed that more collaboration always leads to better system performance, but usually because systems are evaluated in the homogeneous case. Are these 'better' agents robust when used in a mixed environment, when they can take different roles (in either team in a competitive game) and have to collaborate with different agents? This is fundamental for understanding how well AI systems perform in more realistic situations where agents have to collaborate with other different agents (including humans). This question remains unanswered because of several challenges: (1) No formalism exists to determine the contribution of each agent —its value— in these (possibly asymmetric) competition-cooperation situations with repeated agents (2) Heterogeneous situations are avoided because any robust estimation requires a combinatorially high number of experiments to evaluate all possible formations. These two challenges are what we address in this paper. We start by extending the Shapley value for competitive games and repeated agents next.

## 3    Extending the Shapley Value

A cooperative $game(N, v)$ is defined from a set of $n = |N|$ players, and a characteristic function $v : 2^N \to \mathbb{R}$. If $S \in 2^N$ is a coalition of players (a team), then $v(S)$ is the worth of coalition $S$, usually quantifying the benefits the members of $S$ can get from the cooperation. The Shapley value of player $i$ reflects its contribution to the overall goal by distributing benefits fairly among players, defined as follows:

$$\varphi_i(v) = \frac{1}{n} \sum_{S \subseteq N \setminus \{i\}} \binom{n-1}{|S|}^{-1} [v(S \cup \{i\}) - v(S)] \tag{1}$$

where $(v(S \cup \{i\}) - v(S))$ is the marginal contribution of $i$ to the coalition $S$, and $N \setminus \{i\}$ is the set of players excluding $i$. The combinatorial normalisation term divides by the number of coalitions of size $|S|$ excluding $i$.

Note that the above expression assumes that the size of the largest team, let us denote it by $m$, is equal to the number of players we have, $n$. However, in general, these two values may be different, with $m \leq n$, and a generalised version of the Shapley value is expressed as:

$$\varphi_i(v) = \frac{1}{m} \sum_{j=0}^{m-1} \binom{n-1}{j}^{-1} \sum_{S \subseteq N \setminus \{i\}:|S|=j} [v(S \cup \{i\}) - v(S)] \tag{2}$$

It is now explicit that the marginal contributions are grouped by the size of $S$, i.e., $|S| = j$. Also, we see that the number of 'marginal contributions' to compute for each $\varphi_i$ is $r_i = \sum_{j=0}^{m-1} \binom{n-1}{j}$, and for all $\varphi_i$ in total this is $r = \sum_{j=0}^{m} \binom{n}{j}$. This counts the sets with $\leq m$ elements including $\emptyset$, even if we assume $v(\emptyset) = 0$. For the special case of $n = m$ we have $r = |2^N| = 2^n$, i.e., we have to calculate as many experiments as the power set of $N$.

### 3.1    Multisets of Agents

One first limitation of the Shapley value is that coalitions are sets of players. If we have $n$ agents then the coalitions will have sizes up to $n$. However, a common situation in artificial intelligence is that we can replicate some agents as many times as we want. This decouples the number of agents from the size of the coalitions. For instance, with agents $\{a, b, c\}$ and coalitions up to $m = 4$ agents, we could have coalitions as multisets such as $S_1 = \{a, a, b\}$ or $S_2 = \{b, b, b, d\}$. A straightforward way of extending the Shapley value with multisets is to consider that, if there are $l$ different players and the coalitions are of size $m$, we can define $m$ 'copies' of each of the $l$ different agents into a new set $N=\{a_1,a_2,a_3,a_4,b_1,b_2,b_3,b_4,c_1,c_2,c_3,c_4\}$. With this we end up having $|N| = n = l \cdot m$ agents and no multisets. In the examples above, we would have $m=4$ and $l=3$, $n=12$, with $S_1=\{a_1,a_2,b_1\}$ or $S_2=\{b_1,b_2,b_3,d_1\}$ (actually there are several possible equivalent variants of each of them).

We can now use Eq. 2, but many results should be equivalent, e.g., $v(\{a_1,b_3\})=v(\{a_2,b_3\})$ with all possible variants. Suppose $R$ is the subset of $2^N$, where all redundant coalitions have been removed and only a canonical one has been kept. Then Eq. 2 can be simplified into:

$$\varphi_i(v) = \frac{1}{m} \sum_{j=0}^{m-1} \left(\binom{l}{j}\right)^{-1} \sum_{S \in R: |S|=j} [v(S \cup \{i\}) - v(S)] \tag{3}$$

where $\left(\binom{x}{y}\right)$ denotes the combinations of size $y$ of $x$ elements with repetitions. The derivation simply replaces the combinations of $j$ elements taken from $n$ by the combinations of $j$ elements taken from $l=\frac{n}{m}$ with repetitions. Note that $S$ can now contain $i$, and we have situations where the marginal contribution is calculated over a coalition $S$ that already has one or more instances of $i$ compared to $S$ with an extra instance of $i$. Now, the number of required values (or experiments) for each $\varphi_i(v)$ is $r_i = \sum_{j=0}^{m-1} \left(\binom{l}{j}\right)$, with a total of

$$r = |R| = \sum_{j=0}^{m} \left(\binom{l}{j}\right) \tag{4}$$

For instance, with $l=m=3$, we have $n=9$ and we have $r=1+3+6+10=20$ possible sets. With $l=3$ and $m=4$, this would be $r=1+3+6+10+15=35$. With $l=m=4$, this would be $r=1+4+10+20+35=70$. With $l=m=5$, this would be $r=1+5+15+35+70+126=252$.

## 3.2    Cooperation-Competition Games

The Shapley value was designed for cooperation, so there is only one team, with the same goal and share of the payoff for each agent. However, in situations where there are more than one team competing against each other, several instances of the same type of agents can be part of one or more teams. An agent cooperates with the members of the same team, while different teams compete against them. We extend the Shapley value for this situation. We will work with two opposing teams, but this can be extended to any number of teams.

Consider the two team roles $\{A, B\}$ in a competitive game, e.g., $A$ could be predators and $B$ could be preys. When considering the role $A$ we define the $game^A(v^A, N^A)$, where $B$ is the opponent. Similarly, for $game^B(v^B, N^B)$ the role is $B$ and $A$ is the opponent. Role $A$ can have teams up to $m^A$ players, and role $B$ can have teams up to $m^B$ agents. $N^A$ is the set of the $l^A$ different agents of role $A$, with this we end up having $n^A = l^A \cdot m^A$ agents, and similarly for $B$. The possible teams for role $A$, namely $R^A = \{T_1^A, T_2^A, ...\}$, are the same as we did for $R$ for cooperative games avoiding repetitions. Similarly, $R^B = \{T_1^B, T_2^B, ...\}$ for $B$. Then we now extend $v$ for competition by defining $v^A(T^A, T^B)$, as the value of team $T^A \in R^A$ in role $A$ against $T^B \in R^B$ as opponent in role $B$. Note that if we fix the opponent, e.g., $T^B$, from the point of the role $A$, we have its Shapley value from Eq. 3:

$$\varphi_i^A(v^A, T^B) = \frac{1}{m^A} \sum_{j=0}^{m^A-1} \left\{ \left( \binom{n^A/m^A}{j} \right)^{-1} \sum_{S \in R^A : |S|=j} dv^A(S, i, T^B) \right\} \quad (5)$$

where $dv^A(S, i, T^B) = \left[ v^A(S \cup \{i\}, T^B) - v^A(S, T^B) \right]$ is the marginal contribution of agent $i$ to coalition $S$ when the opponent team is $T^B$. Then, if we have all possible teams for role $B$, then we can define $\varphi_i^A(v^A)$:

$$\varphi_i^A(v^A) = \frac{1}{m^A |R^B|} \sum_{j=0}^{m^A-1} \left\{ \left( \binom{n^A/m^A}{j} \right)^{-1} \right.$$
$$\left. \cdot \sum_{S \in R^A : |S|=j} \sum_{T^B \in R^B} dv^A(S, i, T^B) \right\} \quad (6)$$

The amount that agent $i$ gets given a team $game_T^B(v^B, N^B)$ when playing against $T^A$ in role $A$ is $\varphi_i^B(v^B, T^A)$. And the Shapley value $\varphi_i^B(v^B)$ is defined symmetrically to Eq. 6.

The value of agent $i$ for all its possible participations in any team of any role is finally given by:

$$\varphi_i(v^A, v^B) = \frac{1}{2} \left[ \varphi_i^A(v^A) + \varphi_i^B(v^B) \right] \quad (7)$$

The above equation makes sense when $v^A$ and $v^B$ have commensurate values (e.g., through normalisation), otherwise one role will dominate over the other. A particular case where this equation is especially meaningful is for symmetric team games, where both roles have the same scoring system. Finally, the total required values (experiments) for all $\varphi_i^A$ is:

$$r^A = |R^A| \cdot |R^B| = \left[ \sum_{j=0}^{m^A} \left( \binom{n^A/m^A}{j} \right) \right] \cdot \left[ \sum_{j=0}^{m^B} \left( \binom{n^B/m^B}{j} \right) \right] \quad (8)$$

For instance, for $l^A = l^B = 3$ and $m^A = m^B = 4$, we have $r = 35^2 = 1225$ experiments (note that they are the same experiments for $\varphi_i^B$, so we do not have to double this). The huge numbers that derive from the above expression, also illustrated in Fig. 1 for a small example, means that calculating this extension of the Shapley value with repetitions and opposing teams exacerbates the combinatorial problem of computing the value of a huge number of coalitions. Consequently, we need to find ways of approximating the value, through sampling, as we see next.

## 3.3    Properties

In this work, we propose extending the Shapley value to calculate the benefits of each agent in the case of mixed settings (cooperation and competition games). The original Shapley value is characterised by the well-known properties of efficiency, symmetry, linearity, and null player. Let us analyse these properties for Eq. 6. We will see here that if $game^A(N^A, v^A)$ is defined from a set of $n^A = |N^A|$ players, we find that a special case of efficiency holds, the symmetry and the linearity property are met completely, while the null player property does not make sense in our case.

1. Efficiency. Efficiency in $game^A(v^A, N^A)$ requires that the sum of all the Shapley values of all agents is equal to the worth of grand coalition:

$$\sum_{i \in N^A} \varphi_i^A(v^A) = v^A(N^A)$$

For $m^A < n^A$, we cannot define the grand coalition if the maximum number of team members (in the lineup) is lower than the total number of agents. This is similar to many games such as football or basketball, where only a subset of players (11 and 5 respectively) can play at the same time. Accordingly, it is impossible to have a coalition with $n^A$ agents. For the very special case where $m^A = n^A$, then we have the simple case of only one kind of agent $l_A = 1$ and the property is not insightful any more.

2. Symmetry. Now we see that the symmetry property holds in full:

**Proposition 1.** *If for a pair $i, k \in N^A$, we have that $v^A(S \cup \{i\}, T^B) = v^A(S \cup \{k\}, T^B)$ for all the sets $S$ that contain neither $i$ nor $k$, then $\varphi_i^A(v^A) = \varphi_k^A(v^A)$.*

3. Linearity. The easiest one is linearlity as we only have composition of linear functions.

**Proposition 2.** *If $v^A(S, T^B)$ and $w^A(S, T^B)$ are the value functions describing the worth of coalition $S$, then the Shapley value should be represented by the sum of Shapley values of the player derived from $v^A$ and $v^B$: $\varphi_i^A(v^A + w^A) = \varphi_i(v^A) + \varphi_i(w^A)$. And for a, we have $\varphi_i^A(av^A) = a\varphi_i(v^A)$.*

4. Null player. A null player refers to a player who does not contribute to the coalition regardless of whether the player is in the coalition or not. For many team formations having both cooperation and competition, e.g., predator and prey game, even if a player is completely motionless, the other team members and the opponent team's members are affected by this agent, and it cannot have null effect. For instance, in the prey and predator game, if there is a collision, it will produce a reward to this player, which is not in line with the understanding of a null player. This property is not really important in our setting, as many factors affect the result to look for a normalised case where an absolute zero value is meaningful.

# 4 Approximating the Shapley Value

Applying the Shapley value requires the calculation of many $v^A$ and $v^B$ as per Eq. 6. In deep reinforcement learning, for instance, calculating $v^A$ for a pair of teams in simple games such as predator-prey with a reasonable number of steps and episodes may require enormous resources. We explore what kind of sampling is most appropriate for the new extension of the Shapley value taking into account the trade-off between the number of experiments to be run (e.g., number of different $v^A$ and $v^B$ that are calculated) while keeping a good approximations to the actual Shapley values (note that we need a value for each $i$ of the $l$ different players).

## 4.1 Algorithms for Sampling

Monte-Carlo is the common and practical approach approximating the Shapley value [25, 26]. Castro et al. [27] propose a sampling method to approximate the Shapley value by using a polynomial method. The stratified sampling method was first applied by Maleki [28]. These methods and many extensions have been successfully applied to approximate the Shapley value [29–31].

In what follows we present three methods for our setting. We have to sample from $R^A$ and $R^B$, but we will only discuss sampling for one role to simply notation.

*Simple Random Sampling.* This method simply chooses $k$ elements $\mathcal{S} = \{S_1, S_2, ..., S_k\}$ from $R$ with a uniform distribution and without replacement. Then, for each agent $i$, where $i = 1..l$, we compose all $S \cup \{i\}$ for each $S \in \mathcal{S}$, and check whether the new composed set is already in the sample. Then we calculate the approximation $\varphi_i^*$. Note that we sample on the population of experiments (the sets $S$ in $\mathcal{S}$) and not on the population of marginal contribution pairs $\{v(S \cup \{i\}), v(S)\}$. If we fix $s$, the number of sampled experiments, and then try to find or generate the case when $i$ is added, then the exact number of complete pairs will depend on the number of overlaps. If we want to get a particular value of pairs, we can sample elements from $R$ incrementally until we reach the desired value.

*Stratified Random Sampling.* The way the Shapley value is calculated by groups of coalitions of the same size (with $j$ going from 0 to $m-1$) suggests a better way of sampling that ensures a minimum of coalitions to calculate at least some marginal contribution pairs for each value of $j$. Stratified sampling divides $R$ into strata, which each stratum containing all the sets $S$ such that $|S|=j$. If the size of a stratum $\Gamma_j$ is lower than or equal to a specific value $\Gamma_{min}$, we will sample all the elements from the stratum. For all the other strata, we will pick the same number for each. For instance, with $l=m=4$ and $\Gamma_{min}=5$, and $s=14$ , we would do $s=1+4+3+3+3$ from the total of $r=1+4+10+20+35=70$. If $s=23$ we would do $s=1+4+6+6+6$. Once $\mathcal{S}$ is done, we proceed as in the simple random sampling when $\Gamma_j > 5$ : for each $i = 1..l$, we compose all $S \cup \{i\}$ for each $S \in \mathcal{R}$, and check whether the new composed set is already in the sample. Then we calculate the approximation $\varphi_i^*$.

*Information-driven Sampling.* While stratified sampling tries to get information from all sizes, when samples are small, we may end having very similar coalitions, e.g., $\{a, a, b\}$ and $\{a, b, b\}$. Information-driven sampling usually aims for a more diversified sampling procedure. In our case, we use the Levenshtein distance as a metric of similarity between the different samples (assuming the multisets are ordered). Our version of information-driven sampling is actually based on the stratified random sampling presented above, where similarity is used intra-stratum and the coalitions are ordered with the largest average Levenshtein distance from the previous ones in the stratum.

## 4.2 Analysis of Sampling Methods

To evaluate which sampling method is best, we need to be able to calculate the actual Shapley values for several values of $l$ and $m$. Doing this in a real scenario would be unfeasible, so we use synthetically generated data and explore different degrees of sampling for each method, to determine the method with best tradeoff between the approximation of the Shapley value and the number of experiments required.

The synthetic data is generated as follows. First, the worth $v$ of each player (singleton sets) is generated from a uniform distribution $v \sim U(0, 1)$. Second, the contribution of a coalition is the sum of separate player contributions, i.e., $v(\{a, b\}) = v(\{a\}) + v(\{b\})$. Third, we corrupt a number $\nu$ of these $v$ for multisets, also using a uniform distribution $\sim U(0, 1)$. With this procedure, we have created six datasets. Synthetic data 1 is a game with $m=l=4$. There are three variants with $\nu = 1, 5$ and 10 corrupted data, named 'test1', 'test2' and 'test3' respectively. Synthetic data 2 is a game with $m=l=5$. There are also three variants with $\nu = 1, 10$ and 30 corrupted data. We used $\Gamma_{min} = 5$.

With these six synthetic datasets, we now evaluate the three methods and compare the approximate Shapley value with the true Shapley value using all coalitions. The total number of different coalitions (range of the $x$-axis) for synthetic data 1, with $m=l=4$ ($n=16$) is 70, and synthetic data 2, with $m=l=5$ ($n=25$) is exactly 252, coming from Eq. 4. To achieve a stable and robust evaluation, we repeat sampling 50 times before corruption and create 50 repetitions in each case for the corruptions. Then, we have $50 \times 50$ repetitions in total.

We computed the Spearman correlation and Mean Square Error (MSE) between the true Shapley and the approximation value. Figure 2 shows the three sampling methods for increasing sampling size. The stratified and information-driven sampling methods only need a few coalitions (around 20 for $m=l=4$, and around 40 for $m=l=5$) to reach high Spearman correlation (0.98) and very low MSE. Since we do not see a clear difference between the stratified and information-driven methods, we will use the former in what follows.

## 5    Experimental Setting

Now we can explore how the new extensions are useful to determine the value of different algorithms in heterogeneous multi-agent systems with both competition

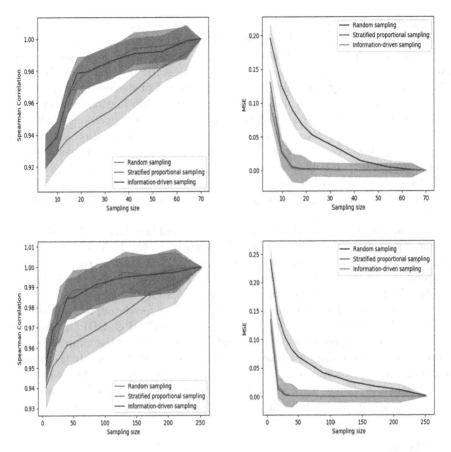

**Fig. 2.** Evolution of sampling methods (simple random, stratified proportional and information-driven) for $m = l = 4$ (top) and $m = l = 5$ (bottom). Left: Spearman correlation between the true Shapley values and the approximate values. Right: MSE.

and cooperation. In order to do this, we choose MPE (multi-agent particle environments), a simple multi-agent particle world [3, 4] that integrates the flexibility of considering several game configurations with different kinds of learning agents. In particular, MPE comes with a single-agent actor-critic algorithm, Deep Deterministic Policy Gradient (DDPG), in which the agent will learn directly from the observation spaces through the policy gradient method, and a multi-agent variation, Multi-Agent DDPG (MADDPG), where decentralised agents learn a centralised critic based on the observations and actions of all agents.

We will explore their behaviour in the predator-prey game, a common cooperative and competitive game, where several predators ($A$) have to coordinate to capture one or more preys ($B$). Preys can also coordinate as well to avoid being caught. In the MPE standard implementation of this game, preys are faster than predators. The arena is a rectangular space with continuous coordinates. Apart from the agents themselves, there are also some static obstacles, which agents must learn to avoid or take advantage of. Agents and obstacles are circles of different size, as represented in Fig. 1. The observation information for each agent combines data from the physical velocity, physical position, positions of all landmarks in the agent's reference frame, all the other agents' position, and all the other agents' velocity. The prey will increase the reward for increased distance from the adversary. If collision, the reward will be –10. Contrarily, the adversary will decrease the reward for increased distance from the prey. If collision, the reward will be +10. In addition, prey agents will be penalised for exiting the screen.

Several questions arise when trying to understand how MADDPG and DDPG perform in heterogeneous situations. In particular, (1) Is MADDPG robust when it has to cooperate with different agents? (2) Is this the case when non-cooperative agents, such as a random agent is included? (3) Are the results similar for the predator role as for the prey role? To answer these questions, we will explore a diversity of situations (roles as prey or predator) and three types of agents (in both teams, so $l^A=l^B=3$). These are MADDPG, DDPG, and a random walk agent, represented by M, D and R respectively in the team. The total number of training episodes in the experiments is 60,000. We variate the number of agents in our experiments with a maximum of $m^A=m^B=4$. The number of combinations is $35\times35=1225$, according to Eq. 8. We do stratified sampling with sizes ranging from 37 to 199, using $\Gamma_{min}=3$. We use the same sampling for prey.

## 6   Results

We report here a summary of results. Further results with all the code and data readily available at a git repository[1].

One of the main motivations for MADDPG was showing that when several agents of this kind cooperate they can achieve better results than their single-agent version, DDPG. In this homogeneous setting, [3] show that "MADDPG predators are far more successful at chasing DDPG prey (16.1 collisions/episode) than the converse (10.3 collisions/episode)". We analysed the same situation with homogeneous teams of predators of 2, 3 and 4 MADDPG agents against 13 variations of prey teams of size 1, 2, 3 and 4. We do the same experiments with DDPG predators with exactly the same preys. In Table 1 (first two rows) we show the average rewards of the 39 games each. As expected, the predator M teams scored better than those with only D agents. The values are consistent with the apparent superiority of MADDPG over DDPG.

[1] https://github.com/EvaluationResearch/ShapleyCompetitive.

**Table 1.** Average reward for 39 homogeneous predator teams composed of two to four agents (first row with Ms only and second row with Ds only) against a diversity of 13 prey coalitions of size 1 to 4 (the same in both cases). Average rewards for 22 heterogeneous predator teams of sizes between 2 and 4 (all including at least one random agent R) against a diversity of 11 prey coalitions of size one to four (the teams in the third row contain an agent M that is systematically replaced by an agent D in the fourth row)

| Teams | Predator | Prey |
|---|---|---|
| M (hom.: MM, MMM, MMMM) | 3788K | −3324K |
| D (hom.: DD, DDD, DDDD) | 3517K | −3375K |
| M (het.: MR, MR$X$, MR$XY$) | 3121K | −3437K |
| D (het.: DR, DR$X$, DR$XY$) | 3184K | −3380K |

Source: The teams composed of two to four agents (first row with Ms only and second row with Ds only, details in Table 2 (labeled with homogeneousM) and Table 3 (labeled with homogeneousD) in appendix) against a diversity of 13 prey coalitions of size 1 to 4 (the same in both cases). Average rewards for 22 heterogeneous predator teams of sizes between 2 and 4 (all including at least one random agent R) against a diversity of 11 prey coalitions of size one to four (the teams in the third row contain an agent M that is systematically replaced by an agent D in the fourth row, see Table 4 (labeled with RandomM) and Table 5 (labeled with RandomD) in appendix).

We can tentatively explore whether this advantage is preserved in heterogeneous teams. If we now build predator teams where apart from M or D agents we include other agents (and always a non-cooperative random agent R), we now get worse results (Table 1, last two rows) as expected, but interestingly we see that the average reward of the M teams is now worse than the D teams.

Because of the careful pairing of the experiments M vs D in Table 1, the average return are meaningful to illustrate the difference, but they do not really clarify whether the contributions of M and D are positive or negative (the average for predator will typically be positive as most results are positive, and the opposite for prey). This phenomenon is replicated when we calculate the average for all the experiments. In this predator-prey game, adding more preys (even if they are good) usually leads to more negative rewards, and hence the averages are negative. But could we still have a good agent, whose contribution is positive for the prey team? This is possible for the Shapley value, as the difference between two negative values can be positive, and does happen for some examples. Consequently, the Shapley values in Fig. 3 show a clearer picture of the actual contributions of each agent to the team (for either roles). While there are some fluctuations, the trends seem to stabilise around a sample size of 130, showing that the sampling method is effective beyond this level.

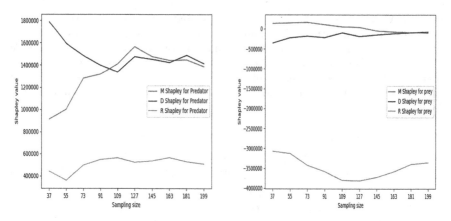

**Fig. 3.** Approximating Shapley for predator-prey environment with increasing sample size. Left: Predator. Right: Prey.

Looking at the sample size 199, as predators, the MADDPG agent has a value of 1387K while DDPG is at 1413K. The value of the random agent plummets to 507K, which makes sense. As preys, the DDPG agent is also the most valuable, with a Shapley value of –79K while MADDPG goes down to –102K. The random agent is further down, at –3361K. Comparing with the results of Table 1, the approximation of the Shapley value integrates both homogeneous and heterogeneous cases, and shows that the gains of M in the homogeneous situations are counteracted by the poorer performance in the heterogeneous situations. Overall, for both predator and prey, the results for M and D are very close. The take-away message in this particular game is that D or M should be chosen depending on the proportion of heterogeneous coalitions that are expected or desirable.

## 7    Conclusions

The Shapley value provides a direct way of calculating the value of an agent in a coalition, originally introduced in cooperative scenarios with no repeated agents (completely heterogeneous). For the first time, we introduce an extension that covers both cooperative and competitive scenarios and a range of situations from complete heterogeneity (all agents being different) to complete homogeneity (all agents in a team equal). These multisets, and the existence of two or more teams competing, increase the combinatorial explosion. To address this, we have analysed several sampling methods, with stratified sampling finding good approximations with a relatively small number of experiments. We have applied these approximations to a prey-predator game, showing that the benefits of a centralised RL agent (MADDPG) in the homogeneous case are counteracted by the loss of value in the heterogeneous case, being comparable overall to DDPG.

There are a few limitations of this extension. First, as we have seen in the asymmetric game of predator-prey, the Shapley value as predator is not commensurate with the Shapley value as prey, and these values should be normalised before being integrated into a single value for all roles in the game. Second, the extension does not take into account the diversity of the team, something that might be positive in some games (or some roles of a game). Third, the Shapley value does not consider that some coalitions may be more likely than others, something that could be addressed by including weights or probabilities over agents or teams in the formulation, or in the sampling method. These are all directions for future work.

**Acknowledgements.** This work has been partially supported by the EU (FEDER) and Spanish MINECO grant RTI2018-094403-B-C32 funded by MCIN/AEI/10.13039/ 501100011033 and by "ERDF A way of making Europe", Generalitat Valenciana under grant PROMETEO/2019/098, EU's Horizon 2020 research and innovation programme under grant agreement No. 952215 (TAILOR), the EU (FEDER) and Spanish grant AEI/PID2021-122830OB-C42 (Sfera) and China Scholarship Council (CSC) scholarship (No. 202006290201). We thank the anonymous reviewers for their comments and interaction during the discussion process. All authors declare no competing interests.

# References

1. Roth, A.E.: The Shapley value: essays in honor of Lloyd S. Shapley. Cambridge University Press (1988)
2. Li, S., Wu, Y., Cui, X., Dong, H., Fang, F., Russell, S.: Robust multi-agent reinforcement learning via minimax deep deterministic policy gradient. AAAI **33**, 4213–4220 (2019)
3. Lowe, R., Wu, Y., Tamar, A., Harb, J., Abbeel, P., Mordatch, I.: Multi-agent actor-critic for mixed cooperative-competitive environments. arXiv preprint arXiv:1706.02275 (2017)
4. Mordatch, I., Abbeel, P.: Emergence of grounded compositional language in multi-agent populations. In: Proceedings of the AAAI Conference on Artificial Intelligence, vol. 32, no. 1 (2018)
5. Sessa, P.G., Bogunovic, I., Kamgarpour, M., Krause, A.: No-regret learning in unknown games with correlated payoffs. In: NeurIPS (2019)
6. Aas, K., Jullum, M., Løland, A.: Explaining individual predictions when features are dependent: more accurate approximations to Shapley values. arXiv preprint arXiv:1903.10464 (2019)
7. Nash, J.J.F.: Equilibrium points in n-person games. Proceed. Nation. Acad. Sci. **36**(1), 48–49 (1950)
8. Aleksandrov, M., Walsh, T.: Pure nash equilibria in online fair division. In: IJCAI, pp. 42–48 (2017)
9. Balduzzi, D., Tuyls, K., Perolat, J., et al.: Re-evaluating evaluation. In: Advances in Neural Information Processing Systems 31 (2018)
10. Omidshafiei, S., Papadimitriou, C., Piliouras, G., et al.: $\alpha$-rank: multi-agent evaluation by evolution. Sci. Rep. **9**(1), 1–29 (2019)
11. Elo, A.E.: The rating of chess players, past and present. Acta Paediatrica **32**(3–4), 201–217 (1978)
12. Glickman, M.E., Jones, A.C.: Rating the chess rating system. Chance-Berlin then New york **12**, 21–28 (1999)
13. Minka, T., Cleven, R., Zaykov, Y.: Trueskill 2: an improved Bayesian skill rating system. Technical Report (2018)

14. Harkness, K.: Official chess hand- book. D. McKay Company (1967)
15. Kiourt, C., Kalles, D., Pavlidis, G.: Rating the skill of synthetic agents in competitive multi-agent environments. Knowl. Inf. Syst. **58**(1), 35–58 (2019)
16. Kiourt, C., Kalles, D., Pavlidis, G.: Rating the skill of synthetic agents in competitive multi-agent environments. Knowl. Inf. Syst. **58**(1), 35–58 (2019)
17. Fatima, S.S., Wooldridge, M., Jennings, N.R.: A linear approximation method for the Shapley value. Artif. Intell. **172**(14), 1673–1699 (2008)
18. Kotthoff, L., Fréchette, A., Michalak, T.P., et al.: Quantifying algorithmic improvements over time. In: IJCAI, pp. 5165–5171 (2018)
19. Li, J., Kuang, K., Wang, B., et al.: Shapley counterfactual credits for multi-agent reinforcement learning. In: Proceedings of the 27th ACM SIGKDD Conference on Knowledge Discovery & Data Mining, pp. 934–942 (2021)
20. Yu, C., Velu, A., Vinitsky, E., et al.: The surprising effectiveness of PPO in cooperative, multi-agent games. arXiv preprint arXiv:2103.01955 (2021)
21. Omidshafiei, S., Pazis, J., Amato, C., et al.: Deep decentralized multi-task multi-agent reinforcement learning under partial observability. In: International Conference on Machine Learning, pp. 2681–2690. PMLR (2017)
22. Bowyer, C., Greene, D., Ward, T., et al.: Reinforcement learning for mixed cooperative/competitive dynamic spectrum access. In: 2019 IEEE International Symposium on Dynamic Spectrum Access Networks (DySPAN), pp. 1–6. IEEE (2019)
23. Iqbal, S., Sha, F.: Actor-attention-critic for multi-agent reinforcement learning. In: International Conference on Machine Learning, pp. 2961–2970. PMLR (2019)
24. Ma, J., Lu, H., Xiao, J., et al.: Multi-robot target encirclement control with collision avoidance via deep reinforcement learning. J. Intell. Robotic Syst. **99**(2), 371–386 (2020)
25. Touati, S., Radjef, M.S., Lakhdar, S.: A Bayesian Monte Carlo method for computing the Shapley value: application to weighted voting and bin packing games. Comput. Oper. Res. **125**, 105094 (2021)
26. Ando, K., Takase, K.: Monte Carlo algorithm for calculating the Shapley values of minimum cost spanning tree games. J. Oper. Res. Soc. Japan **63**(1), 31–40 (2020)
27. Castro, J., Gómez, D., Tejada, J.: Polynomial calculation of the Shapley value based on sampling. Comput. Oper. Res. **36**(5), 1726–1730 (2009)
28. Maleki, S.: Addressing the computational issues of the Shapley value with applications in the smart grid. University of Southampton (2015)
29. Burgess, M.A., Chapman, A.C.: Approximating the shapley value using stratified empirical Bernstein sampling. In: International Joint Conferences on Artificial Intelligence Organization (2021)
30. Gnecco, G., Hadas, Y., Sanguineti, M.: Public transport transfers assessment via transferable utility games and Shapley value approximation. Transport. A Trans. Sci. **17**(4), 540–565 (2021)
31. Illés, F., Kerényi, P.: Estimation of the Shapley value by ergodic sampling. arXiv preprint arXiv:1906.05224 (2019)

# Constrained Multiagent Reinforcement Learning for Large Agent Population

Jiajing Ling[1(✉)], Arambam James Singh[2], Nguyen Duc Thien[1], and Akshat Kumar[1]

[1] School of Computing and Information Systems, Singapore Management University, Singapore, Singapore
{jjling.2018,akshatkumar}@smu.edu.sg
[2] School of Computing, National University of Singapore, Singapore, Singapore
jamesa@nus.edu.sg

**Abstract.** Learning control policies for a large number of agents in a decentralized setting is challenging due to partial observability, uncertainty in the environment, and scalability challenges. While several scalable multiagent RL (MARL) methods have been proposed, relatively few approaches exist for large scale *constrained* MARL settings. To address this, we first formulate the constrained MARL problem in a collective multiagent setting where interactions among agents are governed by the aggregate count and types of agents, and do not depend on agents' specific identities. Second, we show that standard Lagrangian relaxation methods, which are popular for single agent RL, do not perform well in constrained MARL settings due to the problem of credit assignment—how to identify and modify behavior of agents that contribute most to constraint violations (and also optimize primary objective alongside)? We develop a fictitious MARL method that addresses this key challenge. Finally, we evaluate our approach on two large-scale real-world applications: maritime traffic management and vehicular network routing. Empirical results show that our approach is highly scalable, can optimize the cumulative global reward and effectively minimize constraint violations, while also being significantly more sample efficient than previous best methods.

**Keywords:** Multi-agent systems · Multiagent reinforcement learning · Constraint optimization

## 1 Introduction

Sequential multiagent decision making allows multiple agents operating in an uncertain, partially observable environment to take coordinated decision towards a long term goal [4]. The decentralized partially observable MDP (Dec-POMDP) model [20] has emerged as a popular framework for cooperative multiagent control problems with several applications in multiagent robotics [2], packet rout-

**Supplementary Information** The online version contains supplementary material available at https://doi.org/10.1007/978-3-031-26412-2_12.

M.-R. Amini et al. (Eds.): ECML PKDD 2022, LNAI 13716, pp. 183–199, 2023.
https://doi.org/10.1007/978-3-031-26412-2_12

ing in networks [12], and vehicle fleet optimization [17,32]. However, solving Dec-POMDPs optimally is computationally intractable even for a small two-agent system [4]. When the planning model is not known, multiagent reinforcement learning (MARL) for Dec-POMDPs also suffers from scalability challenges. However, good progress has been made recently towards scalable MARL methods [21,24,34,36].

To address the complexity, various models have been explored where agent interactions are limited by design by enforcing various conditional and contextual independencies such as transition and observation independence among agents [16] and event driven interactions [3]. However, their impact remains limited due to narrow application scope. To address practical applications, recently introduced multiagent decision theoretic frameworks (and corresponding MARL algorithms) model the behavior of a population of nearly identical agents operating collaboratively in an *uncertain* and *partially observable* environment. The key enabling insight and related assumption is that in several urban environments (such as transportation, supply-demand matching) agent interactions are governed by the aggregate count and types of agents, and do not depend on the specific identities of individual agents. Several scalable methods have been developed for this setting such as mean field RL [26–28,36], collective Dec-POMDPs [17,18,35], anonymity based multiagent planning and learning [32,33] among others [10].

A key challenge in MARL is that of multiagent credit assignment, which enables different agents to deduce their individual contribution to the team's success, and is challenging in large multiagent systems [6,31]. Recently, there has been progress in addressing this issue for large scale MARL [8,19]. However, such previous methods address the credit assignment problem in a *constraint-free* setting. With the introduction of constraints, we need to perform credit assignment jointly both for primary objective and for the cost incurred by constraints, and deduce accurately the role of each agent in optimizing the primary objective, and lowering constraint violations. In our work, we develop novel techniques that address this issue for *constrained* collective MARL settings.

**Constrained RL.** Most existing works focus on single agent constrained RL and deal with cumulative constraints (discounted and mean valued). The most common approach to solve this problem is the Lagrangian relaxation (LR) [5]. The constrained RL problem is converted to an unconstrained one by adding Lagrangian multipliers, and both Lagrange multipliers and policy parameters are updated iteratively [30]. Methods such as CPO [1] extend the trust region optimization to the constrained RL setting and solve an approximate quadratically constrained problem for policy updates. IPO [13] algorithm uses a logarithm barrier function as the penalty to the original objective to force the constraint to be satisfied. Forming a max-min problem by constructing a lower bound for the objective is used in [9].

Although the above mentioned approaches can solve the single agent constrained RL well, extending these approaches such as LR to multiagent constrained RL directly is not trivial. Since credit assignment for costs also remains unsolved, searching for a policy that satisfies the constraints becomes challenging. There are few works aiming to solve multi-agent constrained RL. [7] used the LR method and proposed to learn a centralized policy critic and penalty critics to guide the update of policy parameters and Lagrangian multipliers. However, centralized critics can be noisy since contributions from each individual agents are not clear. [14] also proposed LR but in a setting where agents are allowed to communicate over a pre-defined communication network (in contrast, our method requires no communication during policy execution). The most recent work CMIX [12] combines the multi-objective programming and Q-mix framework [8]. However, scalability is still a big challenge since different Q-function approximators for each constraint and each agent are required. To summarize, LR is one of the most common approach to solve both single and multiagent constrained RL. However, how to decide the credit assignment with respect to constraint costs and how to scale to large-scale multiagent systems still remain challenging.

**Our Contributions.** First, we formulate the MARL problem for settings where agent interactions are primarily governed by the aggregate count and types of agents using the collective Dec-POMDP framework [17,18] augmented with constraints. Second, we develop a fictitious constrained MARL method which is also based on Lagrangian relaxation, but addresses the issue of credit assignment for both primary objective and constraints. Finally, we test on both real world and synthetic datasets for the maritime traffic management problem [25], and network routing problem [12]. We show that our method is significantly better in satisfying constraints than the standard LR method for MARL. Similarly, when compared against CMIX [12], our approach reduces both average and peak constraint violations to within the threshold using significantly lower number of samples, while achieving similar global objective.

## 2    Fictitious Constrained Reinforcement Learning

### 2.1    Collective CDec-POMDP

We consider the collective decentralized POMDP (CDec-POMDP) framework to model multi-agent systems (MAS) where the transition and the reward of each individual agent depends on the number (count values) of agents in different local states. CDec-POMDP MAS has a wide range of applications in many real world domains such as traffic control, transport management or resource allocation [17, 25,35]. Formally, a CDec-POMDP model is defined by:

- A finite planning horizon $H$.
- The number of agents $M$. An agent $m$ can be in one of the states in the state space $S$. We denote a single state as $i \in S$. We assume that different agents share the same state space $S$. Therefore, the joint state-space is $\boldsymbol{S} = S^M$.
- A set of actions $A$ for each agent $m$. We denote an individual action as $j \in A$.
- $\boldsymbol{s}_t, \boldsymbol{a}_t$ denote the joint state and joint action of agents at time $t$.
- Let $(s_{1:H}, a_{1:H})^m = (s_1^m, a_1^m, s_2^m \ldots, s_H^m, a_H^m)$ denote the complete state-action trajectory of an agent $m$. We denote the state and action of agent $m$ at time $t$ using random variables $s_t^m$, $a_t^m$. We use the individual indicator function $\mathbb{I}(s_t^m = i, a_t^m = j) \in \{0; 1\}$ to indicate whether the agent $m$ is in local state $i$ and taking action $j$ at time step $t$. Other indicators are defined similarly. Given different indicator functions, the count variables are defined as follows:

- $n_t(i,j,i') = \sum_{m=1}^{M} \mathbb{I}(s_t^m = i, a_t^m = j, s_{t+1}^m = i') \; \forall i, i' \in S, j \in A$
- $n_t(i,j) \quad = \sum_{m=1}^{M} \mathbb{I}(s_t^m = i, a_t^m = j) \qquad \forall i \in S, j \in A$
- $n_t(i) \qquad = \sum_{m=1}^{M} \mathbb{I}(s_t^m = i) \; \forall i \in S$

When states and actions are not specified, we denote the state count table as $n_t^s = (n_t(i) \; \forall i \in S)$, state-action count table as $n_t^{sa} = (n_t(i,j) \; \forall i \in S, j \in A)$ and transition count table as $n_t = (n_t(i,j,i') \; \forall i, i' \in S, j \in A)$. For a given subset $S' \subseteq S$, we define the count table for agents in $S'$ as $n_t(S') = (n_t(i,j,i') \; \forall i \in S', j \in A, i' \in S)$.

- The local transition function of an individual $m$ is $P(s_{t+1}^m | s_t^m, a_t^m, n_t^{sa})$. The transition function is the same for all the agents. Note that it is also affected by $n_t^{sa}$, which depends on the collective behavior of the agent population.
- Each agent $m$ has a policy $\pi_t^m(j|i, n_t^s)$ denoting the probability of agent $m$ taking action $j$ given its local state $i$ and the count table $n_t^s$. Note that when agent cannot fully observe the whole count table, we can model an observation function $o(i, n_t^s)$ as a non-trainable component of $\pi$. When agents have the same policy, we can ignore the index and denote the common policy with $\pi$.
- Initial state distribution, $b_o = (P(i) \forall i \in S)$, is the same for all agents.

We define a set of reward functions $r_l(n_t)$, $l = 1 : L$ and a set of cost functions $c_k(n_t)$, $k = 1 : K$ that depend on the count variables $n_t$. constrained program:

$$\max_{\pi_\theta} \quad V(\pi_\theta) = \mathbb{E}_{n_{1:H}} \left[ \sum_{t=1}^{H} \sum_l r_l(n_t) | \pi_\theta \right] \tag{1a}$$

$$\text{s.t} \quad \mathbb{E}_{n_{1:H}} \left[ \sum_{t=1}^{H} c_k(n_t) | \pi_\theta \right] \geq 0, \qquad \forall k \tag{1b}$$

**Agents with Types.** We can also associate different types with different agents to distinguish them (e.g., 4-seater taxi, 6-seater taxi). This can be done using a type-augmented state space as $S' = S \times T$, where $T$ is the set of possible agent types. The main benefit of the collective modeling is that we can exploit the aggregate nature of interactions among agents when the number of types is much smaller than the number of agents

**Simulator for MARL.** In the MARL setting, we do not have access to the transition and reward function. As shown in [18], a count based simulator provides the experience tuple for the centralized learner as $(n_t^s, n_t^{sa}, n_t, r_t)$. In other words, simulation and learning in the collective setting can be done at the abstraction of counts. This avoids the need to keep track of individual agents' state-action trajectories, and increases the computational scalability to large number of agents.

## 2.2   Individual Value Representation

Solving Problem(1) is difficult because the constraints are globally coupled with the joint counts $n_{1:H}$. In many domains in practice, the reward and cost functions only involve the count variables over a subset of states $S$. For example, in congestion domain, we have the penalty cost defined for a specific area/zone. We consider a general framework where we can define the subset $S_k \subseteq S$ that affects constraint $k$, and the subset $S_l \subseteq S$ that affects reward $r_l$. In extreme case where a function is non-decomposable, $S_k$ can be set to $S$. Let $n_t(S_l)$ denote count table that summarizes the distribution of agents in states $s \in S_l$ (as defined in Sect. 2.1). Let $|n_t(S_l)|$ denote the number of agents in $S_l$. We can re-write (1) as follows:

$$\max_{\pi_\theta} \quad V(\pi_\theta) = \mathbb{E}_{n_{1:H}}\left[\sum_{t=1}^{H}\sum_{l} r_l(n_t(S_l))|\pi_\theta\right] \tag{2a}$$

$$\text{s.t} \quad \mathbb{E}_{n_{1:H}}\left[\sum_{t=1}^{H} c_k(n_t(S_k))|\pi_\theta\right] \geq 0, \qquad \forall k \tag{2b}$$

Furthermore, we show that we can re-write the global constrained program in the form of an individual agent's constrained program. For a specific function $f_l$, which can be either cost $f_l = c_l(S_l)$ or reward function $f_l = r_l(S_l)$, we define an auxiliary individual function:

$$f_l^m\left(s_t^m, a_t^m, n_t(S_l)\right) = \begin{cases} \frac{f_l\left(n_t(S_l)\right)}{|n_t(S_l)|} & \text{if} s_t^m \in S_l \\ 0 & \text{otherwise} \end{cases}$$

We use $\mathbf{s}_{1:H}^m, \mathbf{a}_{1:H}^m$ to denote the state-action trajectory with length $H$ of agent $m$, and use $\mathbf{s}_{1:H}, \mathbf{a}_{1:H}$ to denote the join state-action trajectory of all $M$ agents in our system.

**Proposition 1.** *Consider any reward/cost component $f_l$. The global expected value of $f_l$ is equal to a factor of individual value function:*

$$\mathbb{E}_{n_{1:H}}\left[\sum_{t=1}^{H} f_l(n_t(S_l))|\pi_\theta\right] = M \times \mathbb{E}_{\mathbf{s}_{1:H},\mathbf{a}_{1:H}}\left[\sum_{t=1}^{H} f_l^m(s_t^m, a_t^m, n_t(S_l))\right] \tag{3}$$

*Proof.* By applying the exchangeability theorem from [17], we can derive the individual function for reward/cost component $f_l$ as:

$$\mathbb{E}_{\mathbf{s}_{1:H},\mathbf{a}_{1:H}}[\sum_{t=1}^{H} f_l^m(s_t^m, a_t^m, n_t(S_l))]$$

$$= \sum_{t=1}^{H} \mathbb{E}_{\mathbf{s}_{1:H},\mathbf{a}_{1:H}}[f_l^m(s_t^m, a_t^m, n_t(S_l))] \tag{4}$$

We replace the joint probability $P(\mathbf{s}_{1:H}, \mathbf{a}_{1:H})$ with $P(\mathbf{s}_{1:t}^m, \mathbf{a}_{1:t}^m, n_{1:t})$.

$$= \sum_{t=1}^{H} \sum_{\mathbf{s}_{1:t}^m, \mathbf{a}_{1:t}^m, n_{1:t}} P(\mathbf{s}_{1:t}^m, \mathbf{a}_{1:t}^m, n_{1:t}) f_l^m(s_t^m, a_t^m, n_t(S_l)) \tag{5}$$

$$= \sum_{t=1}^{H} \sum_{\mathbf{s}_{1:t}^m, \mathbf{a}_{1:t}^m, n_{1:t}} P(\mathbf{s}_{1:t}^m, \mathbf{a}_{1:t}^m, n_{1:t}) \sum_{s' \in S_l} \mathbb{I}(s_t^m = s') \frac{f_l(n_t(S_l))}{|n_t(S_l)|} \tag{6}$$

We now apply the exchangeability of agents with respect to the count variables n [17]:

$$= \sum_{t=1}^{H} \sum_{n_{1:t}} P(n_{1:t}) \sum_{s' \in S_l} \frac{n_t(s')}{M} \frac{f_l(n_t(S_l))}{|n_t(S_l|} \tag{7}$$

$$= \frac{1}{M} \mathbb{E}_{n_{1:H}} \left[ \sum_{t=1}^{H} f_l(n_t(S_l) | \pi_\theta \right] \tag{8}$$

□

By applying Proposition 1 to reward and cost functions in (2), we have the following lemma.

**Lemma 1.** *Solving a collective constrained reinforcement learning problem defined in (1) is equivalent to solve the individual constrained reinforcement learning problem defined as follows:*

$$\max_{\pi_\theta} \quad V^m(\pi_\theta) = \mathbb{E}_{\mathbf{s}_{1:H}, \mathbf{a}_{1:H}} \left[ \sum_{t=1}^{H} \sum_l r_l^m(s_t^m, a_t^m, n_t(S_l)) | \pi_\theta \right] \tag{9a}$$

$$s.t \quad \mathbb{E}_{\mathbf{s}_{1:H}, \mathbf{a}_{1:H}} \left[ \sum_{t=1}^{H} c_k^m(s_t^m, a_t^m, n_t(S_k)) \right] \geq 0, \qquad \forall k \tag{9b}$$

To solve Problem (9), we apply fictitious-play [15] based constrained optimization (FICO) in which at each iteration, agent tries to optimize its own policy given the joint state-action samples and ignore the effect of its policy change on other agents. Amongst popular methods to solve constrained RL, in this work we apply the Lagrange relaxation method to solve FICO.

We also highlight that Problem (9) is the key to performing the credit assignment for primary objective and constraints. This problem clearly separates out the contribution of each agent $m$ to the value function (or $V^m$) and each constraint $k$ (or $c_k^m$). Therefore, the FICO method enables effective credit assignment for both primary objective and constraints.

## 2.3   Fictitious Collective Lagrangian Relaxation

We consider applying Lagrangian relaxation to solve FICO (9). The Lagrange dual problem is given as follows.

$$\min_{\lambda \geq 0} \max_{\pi_\theta} \mathbb{E}_{\mathbf{s}_{1:H}, \mathbf{a}_{1:H}} \left[ \sum_{t=1}^{H} \sum_l r_l^m(s_t^m, a_t^m, n_t(S_l)) + \sum_k \lambda_k c_k^m(s_t^m, a_t^m, n_t(S_k)) | \pi_\theta \right] \tag{10}$$

To solve this dual Problem (10), we apply stochastic gradient ascent-descent to alternatively update parameters $\theta$ of the policy and the Lagrange multiplier $\lambda$ following the two-time scale approximation [30].

**Individual Policy Update.** To optimize $\pi_\theta$, we first compute the modified reward as follows.

$$R(s_t^m, a_i^m, \mathrm{n}_t) = \sum_l r_l^m(s_t^m, a_t^m, \mathrm{n}_t(S_l)) + \sum_k \lambda_k c_k^m(s_t^m, a_t^m, \mathrm{n}_t(S_k))$$

Given the fixed Lagrange multipliers, parameters $\theta$ are optimized by solving the following problem,

$$\max_{\pi_\theta} \mathbb{E}_{\mathbf{s}_{1:H}, \mathbf{a}_{1:H}} \Big[ \sum_{t=1}^{H} R(s_t^m, a_i^m, \mathrm{n}_t) | \pi_\theta \Big] \tag{11}$$

The benefit of the above representation is that now we can apply various techniques developed for collective Dec-POMDPs to optimize (11) using stochastic gradient ascent. To optimize (11), we consider a fictitious play approach to compute policy gradient for policy $\pi_\theta$ of agent $m$ over all possible local state $i$ and individual action $j$. Using the standard policy gradient [29] with respect to an individual agent $m$, we can perform the update for $\theta$ as follows:

$$\theta' = \theta + \alpha_\theta \sum_t \sum_{i,j,\mathrm{n}_t} \mathbb{E}_{\mathbf{s}_{t:H}, \mathbf{a}_{t:H}} \Big[ \mathbb{I}(s_t^m = i, a_t^m = j) \mathbb{I}(\mathrm{n}_t \sim \mathbf{s}_t, \mathbf{a}_t, \mathbf{s}_{t+1})$$

$$\times \sum_{T=t}^{H} R(s_T^m, a_T^m, \mathrm{n}_T) | \pi_\theta \Big] \nabla_\theta \log \pi_\theta(j|i, \mathrm{n}_t) \tag{12}$$

where $\mathbb{I}(\mathrm{n}_t \sim \mathbf{s}_t, \mathbf{a}_t, \mathbf{s}_{t+1})$ is an indicator function for whether the count table of the joint transition $\mathbf{s}_t, \mathbf{a}_t, \mathbf{s}'_t$ is identical to $\mathrm{n}_t$. Applying results from [17] for collective Dec-POMDPs, we can sample the counts (using the current policy) and use these counts to compute the gradient term in (12) as:

$$\mathbb{E}_{\mathbf{s}_{t:H}, \mathbf{a}_{t:H}} \Big[ \mathbb{I}(s_t^m = i, a_t^m = j) \mathbb{I}(\mathrm{n}_t \sim \mathbf{s}_t, \mathbf{a}_t, \mathbf{s}_{t+1}) \sum_{T=t}^{H} R(s_T^m, a_T^m, \mathrm{n}_T) | \pi_\theta \Big]$$

$$= \sum_{\mathrm{n}'_{1:H}} P(\mathrm{n}'_{1:H}) \mathbb{I}(\mathrm{n}'_t = \mathrm{n}_t) \frac{n'_t(i,j)}{M} \sum_{T=t}^{H} \sum_{s_T^m, a_T^m} \frac{n'_t(s_T^m, a_T^m)}{M} R(s_T^m, a_T^m, \mathrm{n}'_T) \tag{13}$$

The above expected value can be estimated by a Monte-Carlo approximation $\hat{Q}_t^m(i, j, \mathrm{n}_t)$ with samples $\xi = 1, \ldots, K$ of counts [17]. For a given count sample $\mathrm{n}_{1:H}^\xi$:

$$V_H^\xi(i,j) = R_H(i, j, n_H^\xi(i)) \tag{14}$$

$$V_t^\xi(i,j) = R_t(i, j, n_t^\xi(i)) + \sum_{j'} \frac{n_t^\xi(i, j', i')}{n_t^\xi(i)} V_{t+1}^\xi(i', j') \tag{15}$$

$$Q_t^\xi(i,j) = \frac{n_t^\xi(i,j)}{M} \times V_t^\xi(i,j) \tag{16}$$

$$\hat{Q}_t^m(i, j, \mathrm{n}_t) = \frac{1}{K} \sum_{\xi | \mathrm{n}^\xi = \mathrm{n}_t} Q_t^\xi(i,j) \tag{17}$$

**Continuous Actions.** We highlight that even though we have formulated FICO for discrete action spaces, our method works for continuous action space also as long as the policy gradient analogue of (12) is available for continuous actions. Empirically, we do test on the maritime traffic control problem where action space is continuous, and policy gradient is derived in [25].

**Lagrange Multiplier Update.** Given a fixed policy $\pi_\theta$, the Lagrange multiplier $\lambda_k$ for each constraint $k$ is optimized by solving the following problem.

$$\min_{\lambda_k} \mathbb{E}_{\mathbf{s}_{1:H}, \mathbf{a}_{1:H}} \Big[ \sum_{t=1}^{H} \lambda_k c_k^m(s_t^m, a_t^m, n_t(S_k)) | \pi_\theta \Big] \tag{18}$$

We re-write the objective in the collective way as follows.

$$= \lambda_k \sum_{s' \in S_k} \sum_{t=1}^{H} \sum_{\mathbf{s}_{1:t}^m, \mathbf{a}_{1:t}^m, n_{1:t}} \mathbb{I}(s_t^m = s') P(\mathbf{s}_{1:t}^m, \mathbf{a}_{1:t}^m, n_{1:t}) \frac{c_k(n_t(S_k))}{|n_t(S_k)|}$$

$$= \lambda_k \frac{1}{M} \mathbb{E}_{n_{1:H}} \Big[ \sum_{t=1}^{H} c_k(n_t(S_k)) | \pi_\theta \Big] \tag{19}$$

By applying gradient descent, we have $\lambda_k' = \lambda_k - \alpha_k \frac{1}{M} \mathbb{E}_{n_{1:H}} \Big[ \sum_{t=1}^{H} c_k(n_t(S_k)) | \pi_\theta \Big]$ where $\alpha_k$ is the learning rate for the update of $\lambda_k$.

# 3    Experiments

In the section, we evaluate our proposed approach FICO on two real-world tasks: Maritime traffic management (MTM) and vehicular network routing problem with a large scale of agent population. For the MTM problem, we compare FICO with two baseline approaches LR-MACPO and LR-MACPO+. LR-MACPO is a Lagrangian based approach without any credit assignment. The policy in this case is trained with global reward and global cost signal, similar to RCPO algorithm [30]. LR-MACPO+ is also a standard Lagrangian based approach with credit assignment only for the reward signal but not for the cost function in constraint. The detailed problem formulations for LR-MACPO and LR-MACPO+ are provided in the supplementary. We compare FICO with CMIX [12] in the vehicular network routing problem. Since CMIX only deals with discrete action space, we did not evaluate CMIX in the MTM problem where the action space is continuous. Our code is publicly available (link in supplementary).

## 3.1    Maritime Traffic Management

The main objective in the MTM problem is to minimize the travel delay incurred by vessels while transiting busy port waters and also to reduce the congestion developed due to uncoordinated movement of vessels. The previous formulations of the MTM problem in [23, 25] involved unconstrained policy optimization—the objective is a weighted combination of the delay and congestion costs.

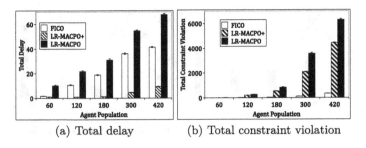

(a) Total delay                    (b) Total constraint violation

**Fig. 1.** Results on the map with 23 zones and different agent population. The lower is the better for both metrics.

This requires an additional tuning of the weight parameters in the objective. We propose a new MTM formulation using constrained MARL as in (1). In our formulation, we introduce constraints that the cumulative congestion cost should be within a threshold and minimize the travel delay as the main objective. The new formulation with constraint is more interpretable, and avoids the intensive search over weight parameters to formulate a single objective as in previous models [25]. Additional description of the formulation is in supplementary.

We evaluate our constrained based approach of the MTM problem in both synthetic and real-data instances. In synthetic data experiments we test the scalability and robustness of our proposed algorithm. Real-data instances are used to measure the effectiveness of the approach in a real-world problem.

**Synthetic Data Instances.** For synthetic data experiments, we first randomly generate directed graphs (provided in supplementary) similar to the procedure described in [23]. The edge of the graph represents a zone, vessels move from left to right through the zones. Each zone has some capacity i.e. the maximum number of vessels the zone can accommodate at any time. Each zone is also associated with a minimum and maximum travel time to cross the zone. Vessels arrive at the source zone following an arrival distribution, and its next heading zone is sampled from a pre-determined distribution. More details on the experimental settings are provided in supplementary.

We first evaluate the scalability of our approach with varying agent population size from 60 agents to 420 agents on the map with 23 zones. We show the results on total delay and total constraint violation respectively as in Fig. 1(a) and Fig. 1(b). Delay is computed as the difference between actual travel time and minimum travel time in the zone. Total violation computes the total constraint violations over all zones. X-axis denotes the agent population size in both the subfigures, and y-axis denotes the total delay and total constraint violation in (a) and (b) respectively. We observe LR-MACPO baseline perform poorly than other approaches in terms both the metric of delay and violation. This is because LR-MACPO is trained with global system reward and global cost function, which is without any credit assignment technique. LR-MACPO+'s performance on delay metric is superior than our approach FICO, but it suffers severely on

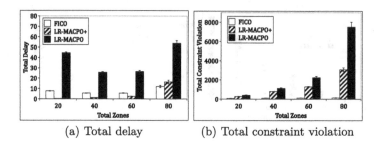

(a) Total delay                    (b) Total constraint violation

**Fig. 2.** Results on agent population 300 and different maps

violation metric in Fig. 1(b). This is an expected result because in LR-MACPO+ credit assignment is provided only for the reward signal not for the cost signal. This makes each agent's credit for the cost component to be noisy resulting in ineffective handling of the constraints. Also, low delay benefits from the high constraint violation in LR-MACPO+.

We next evaluate the robustness of our approach with different maps with varying number of zones (from 20 zones to 80 zones). As in Fig. 2, x-axis denotes different maps with varying number of zones in both the subfigures, and y-axis denoted the total delay and total constraint violation in (a) and (b) respectively. In this experiment the complexity of the problem increases with the increasing number of zones. We observe that our approach FICO is able to reduce the violation consistently for all the settings as shown in Fig. 2(b). In settings with less than 80 zones, LR-MACPO+ beats our approach in terms of total delay. However, it fails to satisfy the constraints poorly. We see that at the most difficult setting with 80 zones, our approach performs better than LR-MACPO+ in terms of both total delay and constraint violation.

Finally, we evaluate the robustness of our approach with different constraint thresholds on the map with 23 zones and 420 agents. The constraint threshold specifies the upper bound of cumulative resource violation over the horizons. The constraint threshold is defined as a percentage of the total resource violations over the horizons when agents are moving with the fastest speed. As shown in Fig. 3, x-axis denotes different constraint thresholds in both the subfigures, and y-axis denotes the total delay and total constraint violation in (a) and (b) respectively. We observe that FICO performs better than LR-MACPO baseline in terms of both the metric of delay and constraint violation. In Fig. 3(b), we see that FICO performs better than other baselines consistently over different constraint thresholds. With the increase of constraint threshold, the total constraint violation is decreasing. FICO is almost able to make the constraint satisfied with the loosest constraint threshold (30%). The constraint violation comes from the zone in the middle of the map which is the busiest zone.

**Real Data Instances.** We also evaluated our proposed approach on real-world data instances from Singapore strait. The strait is considered to be one of the busiest in the world. It connects the maritime traffic of South China Sea and

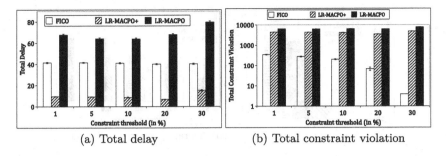

(a) Total delay                    (b) Total constraint violation

**Fig. 3.** Results on agent population 420 and 23 zones with different constraint threshold

Indian Ocean. We use 6 months (2017-July - 2017-December) of historical AIS data of vessel movement in Singapore strait. Each AIS record consists of vital navigation information such as lat-long, speed over ground, heading etc., and is logged every 15 s. In our evaluation we mainly focus on tankers and cargo vessel types because majority of traffic belongs to these two types.

Figure 6 shows the electronic navigation chart of Singapore strait. Vessels enter and leave the strait through one-way sea lanes called traffic separation scheme (TSS). It is created for easy transit of vessels in the strait and helps in minimizing any collision risks. TSS is further sub-divided into smaller zones for better management of the traffic. From the total datasets of 6 months (180 days), 150 days are used for training and 30 days for testing.

**Training.** From the historical data, we first estimate the problem instance parameters such as capacity of each zones, minimum and maximum travel time in each zone. The simulator that we use is the same as in [25], and is publicly available at [22].

The capacity of a zone is computed as 60% of the maximum number of vessels present at any time in the zone overall all days. Each zone can have a different capacity value. We treat the physical sea space in a zone as a resource. Each vessel occupies 1 unit of resource of that zone. The constraint for each zone is expressed as the cumulative resource violation over time should be within a threshold. There are also other problem parameters which are specific to a particular day such as vessels' arrival time on the strait and initial count distribution of vessels present at the strait in beginning of the day. For each training day we estimate the two parameters. Our constrained based policy FICO is trained on varying scenarios of training days. From historical data we observe that there are peak hour periods of traffic intensity during 3rd - 7th hour of the day. Therefore, in our evaluation we focus on optimizing the peak hour periods.

**Testing.** We test our trained policy on separate 30 testing days. Figure 4(a) shows the results of average travel time of vessels crossing the strait averaged over 30 test days. We observe all the three baselines achieve better travel time than the historical data baseline Hist-Data. LR-MACPO performs poorly among

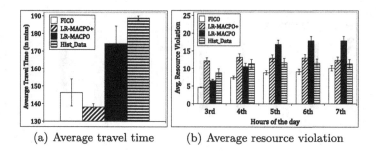

(a) Average travel time          (b) Average resource violation

**Fig. 4.** Results on maritime real-data over 30 testing days. (lower is better for both metrics).

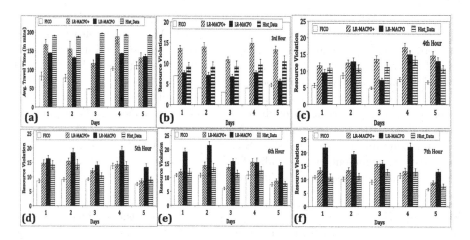

**Fig. 5.** (a) Average travel time(lower is better) (b-f) Resource violation for peak hours 3rd-7th (lower is better)

the three. This is because LR-MACPO is trained with the system reward and cost signals which are without any credit assignment techniques. We also observe that LR-MACPO+ baseline is able to further reduce the travel time slightly better than our approach FICO but at the expense of higher resource violation as seen in Fig. 4(b).

Results in Fig. 4(b) show the average violation of resource over 30 testing days. X-axis denotes the peak hour periods. During the peak hour period, FICO achieves reduced violation of resource among all the baselines. Since LR-MACPO+ lacks the credit assignment signal on the cost function it performs sub-optimally than FICO. The results in Fig. 4 validate the benefit of providing efficient credit assignment technique to both reward and cost function.

In Fig. 5 we show the results of top 5 busiest testing days and results for remaining 25 days are provided in supplementary. Figure 5(a) shows the results

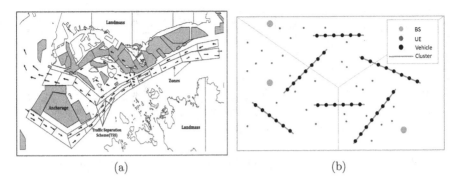

**Fig. 6.** (a) Electronic navigation chart(ENC) of Singapore Strait; (b)Vehicular network model (adapted from [11])

for travel time, x-axis denotes days and y-axis denotes average travel time for crossing the strait. Figure 5(b-f) show the results of resource violation during peak hour periods (3rd-7th hour), x-axis denotes the days and y-axis denotes the resource violation. In all 5 days and during the peak hour periods, our approach achieves an improved reduction in violation of resource while also reducing the travel time better than other baselines.

### 3.2   Vehicular Network Routing Problem

We compare our approach with CMIX [12] in their cooperative vehicular networking problem, which is their largest tested domain. Figure 6 shows a network model with three cells and six clusters. There are two types of cluster - inter cluster (between two cells) and intra cluster. Each cluster contains well connected vehicles that can communicate with high throughput via V2V (Vehicle-to-Vehicle) links. The base station (BS) cell is shared by other mobile user equipments (UEs) and can communicate with vehicles via the direct V2I (Vehicle-to-Infrastructure) links. In this paper, we consider the problem of downlink data transmission where the data are transmitted from BSs to vehicles in the clusters. The objective for all vehicles/agents here is to find the network routes such that the total throughput is maximized (i.e., delivering high volumes of data to destination vehicles), while satisfying both the peak and average latency constraints. Peak constraint means that the latency due to the execution of an agent's action at any time step should be bounded. In CMIX, each agent requires an individual policy to perform action selection. In contrast, agents that belong to the same cluster share the same policy in our collective method. Time limit is set to 180 mins. Further details on hyperparams and neural network structure are in supplementary.

We first follow the same experimental settings as in the CMIX paper to evaluate the performance. There are total three cells and six clusters that are randomly distributed over cells. The number of vehicles in each cluster is randomly generated between a range $[5, 10]$ so that there will be total $30 \sim 60$ vehicles.

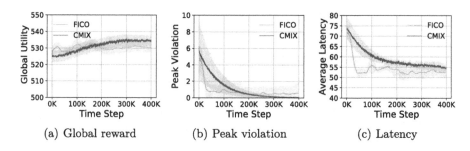

**Fig. 7.** Convergence results over time steps with total 30 ~ 60 agents

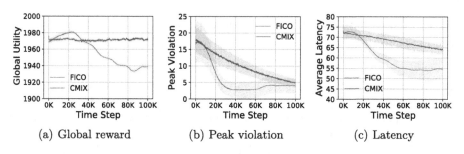

**Fig. 8.** Convergence results over time steps with total 150 ~ 180 agents

The throughput and latency of these V2V and V2I links are also randomly generated. Figure 7 shows the learning curves of global reward, peak violation and latency over time steps. The average latency over time steps in one episode is bounded by 60. The threshold for peak constraint is also 60. From Fig. 7(c), we can see that the average constraint is satisfied when convergence occurs in both approaches. However, our approach converged much faster and is more sample efficient than the CMIX. In Fig. 7(b), the peak constraint is almost satisfied in our approach. Only one agent's latency is greater than 60 in average. The reason is that we use shared policy for agents in the same cluster. And it is challenging for the policy to consider each agent's peak constraint. In Fig. 7(a), the global rewards in both approaches are increasing (higher is better), and converged to almost the same values (530 in our approach v.s. 535 in CMIX). It shows that our approach is also able to maximize the delivered volumes of data.

We next evaluate the scalability of our approach. We increase agents in each cluster to [5, 10]; total number of agents are between 150 and 180. CMIX is only able to train around 100K steps within the limit and our approach can finish 400K steps. Therefore, we show the learning process over 100K steps. Figures 8(b) and (c) show that peak violation and average latency are decreasing in both two approaches. However, the average latency and peak violation in FICO are decreasing with a much faster speed than CMIX. Also, our approach is able to find a policy to satisfy the latency constraint within 100K steps, confirming the effectiveness of our method for large scale problems. The average latency in CMIX is still greater than the threshold 60. The global rewards over time steps are almost unchanged in CMIX, and decreased slightly from 1970 to 1940 in our approach as our approach results in lower constraint violations versus CMIX.

## 4    Conclusion

We presented a new approach for solving constrained MARL for large agent population. We formulate the constrained MARL problem in a collective multiagent setting then propose to use the fictitious collective Lagrangian relaxation to solve the constrained problem. We developed a credit assignment scheme for both the reward and cost signals under the fictitious play framework. We evaluate our proposed approach on two real-world problems: maritime traffic management and vehicular network routing. Experimental results show that our approach is able to scale up to large agent population and can optimize the cumulative global reward while minimizing the constraint violations.

**Acknowledgement.** This research/project is supported by the National Research Foundation Singapore and DSO National Laboratories under the AI Singapore Programme (AISG Award No: AISG2-RP-2020-016). Any opinions, findings and conclusions or recommendations expressed in this material are those of the author(s) and do not reflect the views of National Research Foundation, Singapore and DSO National Laboratories.

## References

1. Achiam, J., Held, D., Tamar, A., Abbeel, P.: Constrained policy optimization. In: International Conference on Machine Learning, pp. 22–31 (2017)
2. Amato, C., Konidaris, G.D., Kaelbling, L.P., How, J.P.: Modeling and planning with macro-actions in decentralized POMDPs. JAIR **64**, 817–859 (2019)
3. Becker, R., Zilberstein, S., Lesser, V.: Decentralized Markov decision processes with event-driven interactions. In: AAMAS, pp. 302–309 (2004)
4. Bernstein, D.S., Givan, R., Immerman, N., Zilberstein, S.: The complexity of decentralized control of Markov decision processes. Math. Oper. Res. **27**(4), 637–842 (2002)
5. Bertsekas, D.P.: Nonlinear programming. Athena Scientific (1999)
6. Chang, Y.H., Ho, T., Kaelbling, L.P.: All learning is local: multi-agent learning in global reward games. In: NeurIPS, pp. 807–814 (2004)
7. Diddigi, R.B., Danda, S.K.R., Bhatnagar, S., et al.: Actor-critic algorithms for constrained multi-agent reinforcement learning. arXiv preprint:1905.02907 (2019)
8. Foerster, J.N., Farquhar, G., Afouras, T., Nardelli, N., Whiteson, S.: Counterfactual multi-agent policy gradients. In: AAAI Conference on Artificial Intelligence (2018)
9. Gattami, A., Bai, Q., Agarwal, V.: Reinforcement learning for multi-objective and constrained Markov decision processes. arXiv preprint arXiv:1901.08978 (2019)
10. Hüttenrauch, M., Šošić, A., Neumann, G.: Deep reinforcement learning for swarm systems. IN: JMLR (2018)

11. Kassir, S., de Veciana, G., Wang, N., Wang, X., Palacharla, P.: Enhancing cellular performance via vehicular-based opportunistic relaying and load balancing. In: INFOCOM IEEE Conference on Computer Communications, pp. 91–99 (2019)
12. Liu, C., Geng, N., Aggarwal, V., Lan, T., Yang, Y., Xu, M.: CMIX: deep multi-agent reinforcement learning with peak and average constraints. In: ECML PKDD (2021)
13. Liu, Y., Ding, J., Liu, X.: IPO: interior-point policy optimization under constraints. In: AAAI (2020)
14. Lu, S., Zhang, K., Chen, T., Basar, T., Horesh, L.: Decentralized policy gradient descent ascent for safe multi-agent reinforcement learning. In: AAAI (2021)
15. Meyers, C.A., Schulz, A.S.: The complexity of congestion games. Networks (2012)
16. Nair, R., Varakantham, P., Tambe, M., Yokoo, M.: Networked distributed POMDPs: a synthesis of distributed constraint optimization and POMDPs. In: AAAI Conference on Artificial Intelligence, pp. 133–139 (2005)
17. Nguyen, D.T., Kumar, A., Lau, H.C.: Collective multiagent sequential decision making under uncertainty. In: AAAI (2017)
18. Nguyen, D.T., Kumar, A., Lau, H.C.: Policy gradient with value function approximation for collective multiagent planning. In: NeurIPS, pp. 4322–4332 (2017)
19. Nguyen, D.T., Kumar, A., Lau, H.C.: Credit assignment for collective multiagent RL with global rewards. In: NeurIPS, pp. 8113–8124 (2018)
20. Oliehoek, F.A., Amato, C.: A Concise Introduction to Decentralized POMDPs. SpringerBriefs in Intelligent Systems, Springer (2016). https://doi.org/10.1007/978-3-319-28929-8
21. Rashid, T., Samvelyan, M., de Witt, C.S., Farquhar, G., Foerster, J.N., Whiteson, S.: Monotonic value function factorisation for deep multi-agent reinforcement learning. JMLR **21**, 1–51 (2020)
22. Singh, A.J.: Multiagent decision making for maritime traffic management. https://github.com/rlr-smu/camarl/tree/main/PG_MTM (2019)
23. Singh, A.J., Kumar, A., Lau, H.C.: Hierarchical multiagent reinforcement learning for maritime traffic management. In: Proceedings of the 19th AAMAS (2020)
24. Singh, A.J., Kumar, A., Lau, H.C.: Learning and exploiting shaped reward models for large scale multiagent RL. In: ICAPS (2021)
25. Singh, A.J., Nguyen, D.T., Kumar, A., Lau, H.C.: Multiagent decision making for maritime traffic management. In: AAAI (2019)
26. Subramanian, J., Mahajan, A.: Reinforcement learning in stationary mean-field games. In: AAMAS, pp. 251–259 (2019)
27. Subramanian, S.G., Poupart, P., Taylor, M.E., Hegde, N.: Multi type mean field reinforcement learning. In: AAMAS (2020)
28. Subramanian, S.G., Taylor, M.E., Crowley, M., Poupart, P.: Partially observable mean field reinforcement learning. In: AAMAS, pp. 537–545 (2021)
29. Sutton, R.S., McAllester, D., Singh, S., Mansour, Y.: Policy gradient methods for reinforcement learning with function approximation. In: NeurIPS (1999)
30. Tessler, C., Mankowitz, D.J., Mannor, S.: Reward constrained policy optimization. In: International Conference on Learning Representations (2018)
31. Tumer, K., Agogino, A.: Distributed agent-based air traffic flow management. In: AAMAS, pp. 1–8 (2007)
32. Varakantham, P., Adulyasak, Y., Jaillet, P.: Decentralized stochastic planning with anonymity in interactions. In: AAAI, pp. 2505–2511 (2014)
33. Verma, T., Varakantham, P., Lau, H.C.: Entropy based independent learning in anonymous multi-agent settings. In: ICAPS, pp. 655–663 (2019)

34. Wang, J., Ren, Z., Liu, T., Yu, Y., Zhang, C.: QPLEX: duplex dueling multi-agent q-learning. In: ICLR (2021)
35. Wang, W., Wu, G., Wu, W., Jiang, Y., An, B.: Online collective multiagent planning by offline policy reuse with applications to city-scale mobility-on-demand systems. In: AAMAS (2022)
36. Yang, Y., Luo, R., Li, M., Zhou, M., Zhang, W., Wang, J.: Mean field multi-agent reinforcement learning. In: ICML, vol. 80, pp. 5567–5576 (2018)

# Reinforcement Learning for Multi-Agent Stochastic Resource Collection

Niklas Strauss[(✉)] , David Winkel , Max Berrendorf ,
and Matthias Schubert

LMU Munich, Munich, Germany
{strauss,winkel,berrendorf,schubert}@dbs.ifi.lmu.de

**Abstract.** Stochastic Resource Collection (SRC) describes tasks where an agent tries to collect a maximal amount of dynamic resources while navigating through a road network. An instance of SRC is the traveling officer problem (TOP), where a parking officer tries to maximize the number of fined parking violations. In contrast to vehicular routing problems, in SRC tasks, resources might appear and disappear by an unknown stochastic process, and thus, the task is inherently more dynamic. In most applications of SRC, such as TOP, covering realistic scenarios requires more than one agent. However, directly applying multi-agent approaches to SRC yields challenges considering temporal abstractions and inter-agent coordination. In this paper, we propose a novel multi-agent reinforcement learning method for the task of Multi-Agent Stochastic Resource Collection (MASRC). To this end, we formalize MASRC as a Semi-Markov Game which allows the use of temporal abstraction and asynchronous actions by various agents. In addition, we propose a novel architecture trained with independent learning, which integrates the information about collaborating agents and allows us to take advantage of temporal abstractions. Our agents are evaluated on the multiple traveling officer problem, an instance of MASRC where multiple officers try to maximize the number of fined parking violations. Our simulation environment is based on real-world sensor data. Results demonstrate that our proposed agent can beat various state-of-the-art approaches.

**Keywords:** Multi-Agent RL · Navigation · Deep RL

## 1 Introduction

In many sequential planning tasks, agents travel on a transportation network, like road or public transportation networks, to reach certain points of interest (POIs) to earn rewards. One way to differentiate these tasks is according to the time intervals for which POIs grant rewards and whether these intervals are known to the agents. For example, for the traveling salesman and the basic vehicular routing problem (VRP), reaching POIs grants rewards regardless of the time they are visited. In

**Supplementary Information** The online version contains supplementary material available at https://doi.org/10.1007/978-3-031-26412-2_13.

more sophisticated tasks such as windowed VRPs [11], POIs only grant rewards during given time windows that are known to the agent. In contrast, in applications like taxi dispatching and ride-sharing, the agent does not know in advance at which time intervals rewards can be earned. Thus, policies try to guide the agents into areas where collecting rewards is more likely, i.e., passengers might show up.

The task of Stochastic Resource Collection (SRC) [21] assumes that resources have fixed locations and change their availability based on an unknown random process. Thus, the agent observes currently collectible resources and can try to reach these before the resources are not collectible anymore. An instance of the SRC task is the TOP [22] in which a parking officer is guided to fine a maximal amount of parking offenders. The setting is based on the assumption that information about parking sensors is available from sensors registering the duration of parking events. As offenders might leave before the officer arrives, not all resources remain collectible, and thus, agents have to consider the chance of reaching resources in time. [21] model SRCs as Semi-Markov Decision Processes (SMDP) and propose an action space that lets the agent travel to any resource location on a pre-computed shortest path. To find effective policies maximizing the number of collected resources in a given time interval, a reinforcement learning (RL) algorithm based on deep Q-Networks (DQN) is proposed. Though the proposed method learns successful policies for single agents, it often requires more than one agent to handle sufficiently large areas. Thus, [18] propose a multi-agent heuristics for guiding multiple officers in a larger area. As RL methods already showed better performance than known heuristic methods in the single-agent case, it makes sense to examine multi agent reinforcement learning (MARL) methods to improve policies. However, known MARL approaches usually are not designed for Semi-Markov models where agents' actions require varying amounts of time. In addition, they often require mechanisms that counter the problem of the size of the joint action space, which grows exponentially with the number of agents, and the credit assignment problem when using joint rewards. Though there are several methods to counter each of these problems, most of them do not consider the properties of the MASRC environments with asynchronous agent actions in a Semi-Markov environment.

In this paper, we formalize MASRC as a selfish Semi-Markov Game (SMG). We adapt the action space of [21] to let each agent target any resource in the network. Thus, agents generally terminate their actions in varying time steps. We propose a selfish formulation where each agent optimizes its own individual rewards. We argue that a group of independent agents still optimizes the sum of collected resources sufficiently well as the agents learn that evading other agents decreases the chances of another agent collecting close-by resources. We empirically verify our reward design by comparing it to joint rewards. To approximate Q-values, we propose a neural network architecture that processes information about resources, agents, actions, and the relation between them. To combine these types of information, we employ attention and graph neural network mechanisms. This way, our agent can estimate the likelihood of reaching a collectible resource before it becomes uncollectible or another agent reaches the resource first. Furthermore, our resource embedding considers the spatial closeness of additional collectible resources to make actions moving the agent into a region with multiple collectible resources more attractive. To evaluate our new approach, we developed a multi-agent simulation based on real-world parking data from the city of Melbourne. Our experiments demonstrate superior performance compared to several baselines [18] and (adaptions of) state-of-the-art approaches [1,9,21]. We compare our methods with heuristic methods proposed in [18], an adaption of the single-agent SRC method from [21] and an architecture proposed for dynamic multi-agent VRP [1] which is based on the well-known single-agent architecture [9]. We evaluated the last benchmark to demonstrates that state-of-the-art solutions for the dynamic VRP do not sufficiently cope with the additional stochasticity of MASRC problems. To further justify the design choices in our architecture, we provide ablations studies. To conclude, we summarize the contributions of our paper as:

– A formulation of the MASRC as a Semi-Markov Game building a solid theoretical foundation for the development of MARL approaches
– A novel architecture for learning rich state representations for MARL
– A scalable simulation environment for the multi-agent traveling officer problem (MTOP) problem based on real-world data

## 2   Related Work

In this section, we review work on related tasks routing an agent through spatial environments to collect rewards. In addition, we will discuss general multi agent reinforcement learning approaches.

### 2.1   Stochastic Resource Collection

One of the most recognized routing tasks in the AI community is the *vehicular routing problem (VRP)* where a group of agents needs to visit a set of customer locations in an efficient way. There exist various variations of the VRP [4] and some of them include the appearance of new customers during the day [1]. In contrast to SRC, the setting does not include customers disappearing after an unknown time interval. This is a decisive difference as it makes the reward of an action uncertain. In recent years, several approaches have been developed to solve the vehicular routing problem or some of its variations using DRL [1,9,16,17]. MARDAM [1] is an actor-critic RL-agent - based on [9] - designed to solve VRP with multiple agents using attention mechanisms. While state and action spaces of dynamic VRP and MASRC can be considered as very similar, the behavior of the environment is not. To demonstrate these differences, we compare to an agent using the architecture of [1] in our experiments.

There exist various papers on *multi-agent taxi dispatching* [8,12,13,28,32] which can be formulated as a MASRC task. However, in most settings there are significant differences to MASRC as the resources are usually not claimed at arrival. Instead, customers are assigned to close-by taxis the moment the guest publishes a request to the dispatcher. Thus, reaching the guest in time is usually not considered. Furthermore, to the best of our knowledge, only a single approach works directly on the road network [8]. All other approaches work on grid abstractions which are too coarse for MASRC. Finally, taxi dispatching tasks usually involve large and time variant sets of agents. To conclude, known solutions to taxi dispatching are not applicable to solve MASRC.

The *traveling officer problem (TOP)*, first described by [22] is an instance of SRC. In [21], the authors propose an Semi-Markov RL-based agent to solve the single-agent TOP task and name other tasks that can be formulated as SRC. Later on, the authors of [18] study the MTOP. They propose a population-based encoding, which can be solved using various heuristics for optimization problems like cuckoo search or genetic algorithms. Additionally, they propose a simple greedy baseline that assigns idling officers to the resource in violation using "first-come-first-serve". Competition between officers is handled by assigning a collectible resource to the officer with the highest probability that the resource is still in violation when the officer arrives.

## 2.2   Multi-Agent Reinforcement Learning

After reviewing solutions to similar tasks, we will now discuss general multi agent reinforcement learning (MARL) approaches w.r.t. their suitability for training on MASRC environments. In MARL, a group of agents shares the same environment they interact with. There are various challenges in MARL: the non-stationarity of the environment from the perspective of an individual agent, the exponentially increasing joint action space, the coordination between agents, and the credit assignment problem. A plethora of different approaches to tackle these challenges exists [6] and we will give a brief overview of the most important MARL approaches in the following.

Joint action learners reduce the multi-agent problem to a single-agent problem by utilizing a single centralized controller that directly selects a joint action. While joint action learners can naturally handle coordination and avoid the non-stationarity, in practice, these approaches are often infeasible because of the exponential growth of the joint action space w.r.t. the number of agents [7].

On the opposite site, we can use multiple independent learners [26]. The agents interact in parallel in a shared environment using a single-agent RL algorithm. In many cases, it has been shown that independent learners can yield strong performance while allowing for efficient training. However, in some settings, independent learners can suffer from the non-stationarity of the environment induced by simultaneously learning and exploring agents.

In recent years, approaches have been developed that utilize centralized training and decentralized execution (CLDE). In [24], the authors presented VDN that decomposes the joint action-value function as a sum of the individual agents' Q-function values obtained solely from the agents' local observation. The authors of [19] propose QMIX, a method that extends VDN by learning a non-linear monotonic combination of the individual Q-functions, which allows representing a larger class of problems. The authors of [3] propose a counterfactual multi-agent actor-critic method (COMA) that uses a centralized critic that allows estimating how the action of a single agent affects the global reward in order to address the credit assignment problem.

Another way to tackle the problem of coordination between agents is to facilitate communication between the agents. CommNet [23] is a prominent approach that learns a differentiable communication model between the agents. Both CLDE and communication-based approaches suffer from the credit-assignment problem, which we mitigate through our individual reward design.

A drawback of the named approaches when applied to MASRC is that these algorithms do not consider *temporal abstractions*, i.e., actions with varying duration. The application of temporal abstraction to CLDE requires the modification of the problem in a way that the decision epochs are synchronized or experience needs to be trimmed [27]. This way of training is inefficient as it exponentially increases the number of decision epochs with respect to the number of agents. The authors of [2,5,14,20,27] investigate temporal abstraction in multi-agent settings. The authors of [20] first introduce different termination schemes for actions with different temporal abstractions that are executed in parallel. [14] propose independent learners to efficiently handle the asynchronous termination setting, [27] adapt CommNet and QMIX to a setting with temporal abstraction, while [2] propose a version of COMA in decentralized settings with temporal abstractions. Let us note that some of these approaches, like COMA, QMIX, or CommNet, can be adapted to train our function approximation and thus, can be applied to MASRC. We experimented with these approaches but could not observe any convincing benefit for solving MASRC. In addition, the use of those methods tries to learn complex coordination schemes between agents. However, in MASRC agents basically cannot directly support each other as the only action impacting other agents is collecting resources.

## 3    Problem Formulation

We consider the problem of MASRC, where $n$ agents try to maximize the collection of resources in a road network $G = (V, E, C)$, where $V$ is a set of nodes, $E$ denotes a set of edges and $C : E \rightarrow \mathbb{R}^+$ are the corresponding travel costs. Each resource $p \in P$ is located on an edge $e \in E$ in the road network. Whether a resource $p$ is collectible can be observed by the agents but might change over time. The state changes of resources follow an unknown stochastic process. Whenever an agent passes a collectible resource, the resource is collected by the agent.

Formally, we model the MASRC problem as a Semi-Markov Game (SMG) $\langle I, S, \mathbf{A}, P, R, \gamma \rangle$, where $I$ is a set of agents indexed by $1, \ldots, n$, $S$ is the set of states, $\mathbf{A}$ denotes the joint action space, $P$ is the transition probability functions, $R$ denotes the reward functions of the individual agents, and $\gamma$ is the discount factor.

**Agent:** A set of $n$ agents moving in road network and collecting resources.

**State:** $\mathbf{s}_t \in S$ denotes the global state of the environment at time $t$. The exact information included in the state depends on the actual instantiation, e.g., TOP.

Nonetheless, all MASRC tasks share a common structure that can be decomposed into resources, agents, and environment:

- *Resources* characterized by the current status, e.g., availability and position.
- *Agents* defined by their position and ID.
- *Environment* with features such as the time of the day or an indication of holidays.

**Action:** $\mathbf{a}_t \in \mathbf{A} = A_1 \times \ldots \times A_n$ : is the joint action at time $t$. Following the single-agent formulation of [21], we define the individual action space $A_i$ of an agent to correspond to the set of edges $E$, i.e., the agent will travel on the shortest path to the corresponding edge. This allows to focus on the MASRC task itself rather than solving the routing problem, where high-performance deterministic algorithms are available. Therefore, the individual actions have varying duration, depending on the agent's position and target location. As a result, agents may have to asynchronously select actions at different decision times. Between those decision times, agents continue to their target. Formally, we can reduce this to a synchronous setting, and thus the given joint action space, by introducing a special "continue" action, as described in [14].

**Reward:** Each agent $i$ has an independent reward function $R_i \in R$, where $R : (S \times \mathbf{A}) \to \mathbb{R}$. Each agent $i$ independently tries to maximize its own expected discounted return $\mathbb{E}\left[\sum_{j=0}^{\infty} \gamma^j r_{i,t+j}\right]$. Each agent's individual reward function corresponds to the resources collected by the agent itself. The reward is incremented by 1 for each collected resource. A resource is collected when an agent passes a collectible resource.

**State Transition Probability:** With

$$(\mathbf{s}_{t+1}, \tau \mid \mathbf{s}_t, \mathbf{a}_t) : (S \times \mathbb{R}^+ \times \mathbf{A} \times S) \to [0, 1] \subset \mathbb{R} \tag{1}$$

we denote the probability of transitioning to the state $\mathbf{s}_{t+1}$ from the current state $\mathbf{s}_t$ by taking the joint action $\mathbf{a}_t$. Although, some effects of an action are deterministic (e.g., the positions of the agents), the state changes of resources are uncertain and the exact dynamics are unknown. Unlike in a Markov Game, in a SMG, we additionally sample the number of elapsed time-steps $\tau$ of the action $\mathbf{a}_t$. The smallest feasible temporal abstraction is the greatest common divisor of all edge travel times. The duration is determined by the individual action $a_i \in \mathbf{a}$ with the shortest duration, which is a multiple of the smallest feasible duration. As a result, an agent receives a time-discounted reward $\zeta = \sum_{j=0}^{\tau-1} \gamma^j r_{t+j+1}$.

## 4    Method

In this section, we introduce our novel multi-agent RL agent. At first, we provide some insight into our reward design. Secondly, we present our training procedure that is based on independent DQN [25]. After that, we describe the inputs to our architecture and name the particular features for our evaluation on the MTOP task. Finally, we introduce our novel function approximator for the MASRC problem that facilitates coordination between the agents.

## 4.1  Individual Rewards

In general, the goal of MASRC is to maximize the expected joint reward $\mathbb{E}\left[\sum_{i=0}^{|I|}\sum_{j=0}^{\infty}\gamma^j r_{i,t+j}\right]$. However, we decided to use individual rewards to avert the credit assignment problem, which leads to a Markov Game where agents act selfishly. In literature, the impact of such selfish behavior is commonly denoted as the "price of anarchy" [10]. In the context of MASRC, we argue that the price of anarchy is likely to be very low and outweighed by the benefits of having a reward function that allows the agents to assess the impact of their actions more directly and thus mitigates the credit assignment problem. This is because in MASRC helping other agents directly is not possible. Therefore, coordination boils down to not getting in the way of other agents. There might be cases where a joint reward might lead to policies where particular agents would target far-off resources decreasing their own but increasing the sum of collected resources. However, we observed in our experiments that these cases are rare. We provide an empirical evaluation of our reward design choice compared to joint rewards in Sect. 6.3.

## 4.2  Training

Independent DQN [25] combines independent learners [26] and DQN [15]. To speed up learning, we share the network parameters between agents and distinguish them by their IDs [31]. Independent learning provides a natural way to handle settings with asynchronous termination [14]. In independent learning, each agent treats the other agents as part of the environment. However, this may lead to sub-optimal coordination between the agents. To mitigate this problem, we introduce an architecture that allows each agent to efficiently reason about the intents of other agents. We utilize a DoubleDQN [29] adapted to the Semi-Markov setting. We update the network parameters with respect to a batch of transitions collected from all agents by minimizing the following loss function:

$$\mathcal{L}(\Theta) = \mathbb{E}_{\mathbf{s}_t, a_t, \tau, r_{t:t+\tau-1}, \mathbf{s}_{t+\tau}} \left[ loss(y_t, Q(\mathbf{s}_t, a_t; \Theta)) \right] \qquad (2)$$

where $y_t = \sum_{j=0}^{\tau-1}\gamma^j r_{t+j} + \gamma^\tau Q(s_{t+\tau}, a'_{t+\tau}; \Theta')$. The action $a'_{t+\tau}$ is the optimal action w.r.t to $\Theta$, i.e., $a'_{t+\tau} = \text{argmax}_{a_{t+\tau} \in A(s_{t+\tau})} Q(s_{t+\tau}, a_{t+\tau}; \Theta)$. $\Theta$ denotes the parameters of the behavior $Q$ network and $\Theta'$ denotes the parameters of the frozen target $Q$ network which are periodically copied from $\Theta$. To improve clarity, we omitted the indices indicating the individual agents. We use a smooth L1 loss.[1].

## 4.3  Input Views

In the following, we will briefly describe the inputs to our function approximation and name the particular features for our evaluation on the MTOP task.

---

[1] cf. https://pytorch.org/docs/stable/generated/torch.nn.SmoothL1Loss.html.

**Resource Features.** We encode each resource from the perspective of each individual agent separately. To this end, we add features describing the relation of the agent to the resource, e.g., the distance or arrival time. This results in $n$ times different views of each resource. The resource view for the MTOP contains a one-hot-encoding of the resource's current status, i.e., free, occupied, in violation, or fined. Additionally, we provide a flag that indicates whether a parked car would be in violation if it remains parked and the officer would directly go there. Finally, we add the current time of the day, walking time, agent arrival time, and distance to the resource. All these features are normalized. We add a real-valued number between -1 and 2, indicating how long a car is still allowed to occupy the resource and how long it is in violation, respectively. A score greater than zero indicates a violation. Finally, we add the normalized coordinates of the resource's position.

**Agent Features.** For MTOP, it consists of a one-hot encoding of the agents' ID, the normalized coordinates of its current position and target, as well as the normalized walking time and distance to its target.

**Spatial Relation.** To capture the spatial interaction between the resources, we create a distance matrix for each agent. There is one row in the matrix for each action consisting of the network distance of the action target to each resource, the distance between the agent, and the action target to each row.

## 4.4   Architecture

An effective policy in MASRC requires an agent to consider the complex interaction between resources, actions, and other agents to estimate the likelihood of reaching a *collectible* resource. To capture those dependencies, our novel architecture first encodes the action-level intents of each agent using the resources and their spatial relationship solely from the perspective of each individual agent, i.e., ignoring the other agents. We call this module the *Shared Action Encoder*. After that we continue by combining the perspective of the current agent with the action-level intents of the other agents using multi-head attention in the *Intent Combination Module*. This allows an agent to asses the likelihood that another agent catches collectible resources first. While the inputs are different, the parameters of all networks are shared between all agents.

**Shared Action Encoder.** In the context of SRC, the value of an action, i.e., the likelihood of reaching resources in time, depends largely on the state of resources near the target [21]. We argue that in a multi-agent setting, the simple distance weighting from [21] is not expressive enough to capture the complex dependency of an action's value on, e.g., the uncertainty of reaching the resource in time. Thus, we propose an extended *Shared Action Encoder* to calculate agent-specific action embeddings, based upon the agent's features, and resources' features, as well as the distances. We provide the pseudo-code roughly following PyTorch

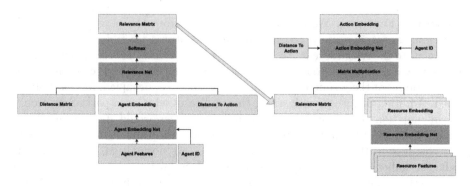

**Fig. 1.** Conceptual overview of the *Shared Action Encoder*. This module creates a rich representation for each action based on the resource states from the perspective of an individual agent. The module captures the spatial relationship of resources around each action's target using a graph neural network mechanism. Networks and operations are colored yellow, the output of the module and crucial intermediate representations are purple, while blue denotes input features.

```
def sea(feat_ag, feat_res, i_ag, dist, dist_ag2ac):        1
    """Shared action encoder for a single agent."""        2
    x_ag        = mlp1(feat_ag)                             3
    # shape: (dim_ag,)                                      4
    rel_act_res = mlp2(cat(broadcast([                      5
        x_ag[None],                                         6
        dist_ag2ac[None],                                   7
        dist,                                               8
    ]), dim=-1))                                            9
    # shape: (n_action, n_res)                              10
    rel_act_res = softmax(rel_act_res, dim=-1)              11
    # shape: (n_action, n_res)                              12
    x_res       = mlp3(feat_res)                            13
    # shape: (n_res, dim_res)                               14
    x_act       = rel_act_res @ x_res                       15
    # shape: (n_act, dim_res)                               16
    return mlp4(cat(x_act, i_ag, dist_ag2ac))               17
```

**Fig. 2.** Pseudocode for the *Shared Action Encoder* following PyTorch style. `feat_ag` denotes the agent's features, `feat_res` the (agent-specific) resource features, `i_ag` the agent's ID, and `dist` the action-resource distance matrix, and `dist_ag2ac` the distance from the current agent to all actions (which are target edges). `mlp1` to `mlp4` are separate MLPs.

style in Fig. 2, and show an overview in Fig. 1. We begin by transforming the agent's features with an MLP (cf. line 3). Next, we calculate unnormalized agent-specific resource to action relevance scores combining information from the agent representation, the distance from the agent to the action (i.e., edge), and the

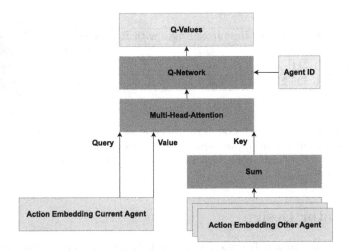

**Fig. 3.** In the *Intent Combination Module*, we enrich the action embedding of a single agent with information about the other agents' actions using multi-head attention. Afterwards, we reduce the enriched action representations to a $Q$-value for every action with an MLP. Networks and operations are colored yellow, inputs coming from the *Shared Action Encoder* are purple, and blue denotes input features.

action-to-resource distance matrix using another MLP (cf. lines 5–9). These relevance scores are subsequently normalized using the softmax operator (cf. line 11). The resource features are first transformed by an MLP, before we use the previously computed relevance scores for aggregating them per action (cf. line 13–15). A final MLP combines this information with distance to the agent as well as the agent ID (cf. line 17). In the following, we denote the result of this component as $\mathbf{E}$.

**Intent Combination Module.** Information about the other agents' intents is crucial in multi-agent settings - thus, we propose an attention-based mechanism to update an agent's action representations by considering the ones of other agents. Let $\mathbf{E}_i$ denote the output of the shared action encoder for agent $i$. For scalability, we first aggregate the latent actions of all other agents $\overline{\mathbf{E}}_i := \sum_{j \in I \setminus \{i\}} \mathbf{E}_j$. Next, apply a multi-head attention mechanism [30] with $\mathbf{E}_i$ as query and value and $\overline{\mathbf{E}}_i$ as key. Finally, we reduce every row of the result, corresponding to the co-agents-aware action representation, to a single $Q$-value using an MLP. This MLP also receives the agent ID as an additional input to allow diversification of the agents.

| Area | Nodes | Edges | Resources | Edges with Resources |
|------|-------|-------|-----------|----------------------|
| Docklands | 1,435 | 4,307 | 487 | 166 |
| Queensberry | 1,711 | 5,356 | 639 | 177 |
| Downtown | 6,806 | 21,369 | 1,481 | 493 |

**Fig. 4.** Description and illustration of the different areas used in our evaluation: Docklands (blue), Downtown (red), and Queensberry (green). Notice that typically only a small fraction of edges contains resources and there can be more than one resource per edge. (Color figure online)

## 5  Simulator Design

To the best of our knowledge, there is no publicly available simulation for MTOP. To enable effective training of reinforcement learning agents, we implement a simulator that can replay real-world sensor data and parking restrictions, which allows us to simulate as close as possible to the real world. The walking graph, i.e., road network, is extracted from OpenStreetMap[2]. We assign parking spots to the closest edge in the graph. When an agent passes a resource in violation, it will be fined. The time for fining a violation is set to zero in our simulation. The agent collects a reward of +1 for every fined resource. All agents start at the same place every day. They work for 12 h from 7 am to 7 pm. Each agent has a walking speed of 5km/h.

For our evaluation, we use openly available on-street parking sensor data and parking restrictions from the city of Melbourne in 2019[3]. We divide Melbourne into three areas to study different graph structures and hyperparameter transferability. Details regarding the areas can be found in Fig. 4. Each run was trained using a single GPU on a cluster consisting of RTX A6000 (48GB) and A100 (40GB) GPUs. The code of our simulation and agents is publicly available.[4]

## 6  Experimental Evaluation

We split the parking event dataset into a training, validation, and test set. Parking follows weekly patterns. To avoid biases introduced through weekdays, we split the dataset as follows: If the remainder of the day in the year divided by 13 is 0, we add the day to the test set. In case the remainder is 1, we add the day to the validation set. The remaining 308 d are added to the training set. An episode is equivalent to a working day. The order of the training days is shuffled. To speed up training, the agent interacts with eight environments in parallel.

The transferability of the hyperparameters across different regions and numbers of agents is important. We tuned the hyperparameters in a single area

---

[2] https://www.openstreetmap.org.

[3] https://data.melbourne.vic.gov.au/browse?tags=parking.

[4] https://github.com/niklasdbs/masrc.

(Docklands) with two agents. Agents were trained using early stopping. The test results reported are with respect to the best validation results. The full hyperparameter setting can be found in the supplement.

### 6.1 Baselines

**Greedy.** We modify the greedy baseline from [18] for better performance: Instead of assigning agents to the resource with the earliest violation time, we directly use the catching probability from the the the tie-breaking mechanism.

**LERK.** The authors of [18] propose to solve the MTOP by representing it using leader-based-random-key encoding (LERK) and then solve it using various classical heuristic solvers developed for combinatorial issues. One of these heuristics that yielded the best performance was the genetic algorithm, which we have implemented.

**MARDAM.** [1] is an actor-critic RL-agent - based on [9] - designed to solve dynamic-VRP with multiple agents using attention mechanisms. While they propose a method to transform the underlying Markov game into a sequential MDP, this transformation is not possible for MASRC tasks. Therefor, we train their architecture using independent actor-critic. Due to the dynamic state of resources in MASRC tasks, we need to calculate the customer-embeddings (i.e., resource), in every step using a Transformer which is computationally expensive and memory intense. As a result, we are not able to train the agent on full episodes and need to rely on bootstrapping.

**SASRC.** We train the architecture of [21] that has been proposed for the SASRC using independent learning with shared independent learners. We add the agent-id to the final network so that agents can differentiate their behavior. Additionally, we add information about the targets of all agents to the resources, which allows the agent to incorporate information about other agents and thus benefits learning [31].

### 6.2 Results

As the evaluation metric, we use the average number of violations fined per day. We evaluate in three different areas using two, four, and eight agents. The results in Table 1 show that our proposed approach can surpass the other approaches and baselines in various regions and across different numbers of agents. We can beat MARDAM, a state-of-the-art algorithm designed for multi-agent dynamic-VRP, by a large margin. This underlines that approaches for SRC tasks need to be able to handle the increased stochasticity. Moreover, MARDAM requires a massive amount of GPU memory due to the use of the transformer encoder in large settings like Downtown with eight agents. For this setting, our approach

**Table 1.** Average number of violations fined per day in Docklands, Queensberry, and Downtown for 2, 4, and 8 agents on the validation and test set.

| Area | Algorithm | 2 Agents | | 4 Agents | | 8 Agents | |
|---|---|---|---|---|---|---|---|
| | | Validation | Test | Validation | Test | Validation | Test |
| Docklands | Greedy | 186.93 | 192.67 | 304.32 | 300.22 | 442.57 | 439.33 |
| | LERK | 244.78 | 245.11 | 328.61 | 330.56 | 424.32 | 418.19 |
| | MARDAM | 339.79 | 336.37 | 418.96 | 416.44 | 482.57 | 479.78 |
| | SASRC | 343.82 | 304.19 | 476.07 | 465.52 | 551.86 | 544.63 |
| | **OURS** | **388.32** | **379.59** | **527.21** | **518.52** | **588.04** | **580.00** |
| Queensberry | Greedy | 180.46 | 189.15 | 240.5 | 250.07 | 277.54 | 286.56 |
| | LERK | 192.18 | 198.78 | 233.86 | 245.07 | 260.79 | 271.67 |
| | MARDAM | 225.18 | 229.85 | 247.29 | 255.41 | 257.04 | 267.81 |
| | SASRC | 222.61 | 231.41 | 257.11 | 266.00 | 263.79 | 273.15 |
| | **OURS** | **244.43** | **255.41** | **271.39** | **281.59** | **284.75** | **294.37** |
| Downtown | Greedy | 138.79 | 144.63 | 256.14 | 257.33 | 429.93 | 435.22 |
| | LERK | 213.57 | 219.19 | 298.32 | 305.40 | 430.18 | 428.19 |
| | MARDAM | 340.54 | 342.37 | 469.18 | 471.26 | 255.82 | 260.93 |
| | SASRC | 425.68 | 418.48 | 657.61 | 658.04 | 815.68 | 815.41 |
| | **OURS** | **495.07** | **494.70** | **710.75** | **713.3** | **866.04** | **867.93** |

uses approximately 16 times less GPU memory during training. As a result, the batch size needs to be reduced for those settings, which may impact performance. Additionally, the episode length in MASRC tasks is much longer than in a typical VRP, which makes learning on whole episodes impossible. Furthermore, the experiments show that our approach yields considerably better results than existing heuristic solvers designed for the MTOP, such as LERK, which require intensive computational resources at inference time. While our approach requires several days of training it only needs a few milliseconds at inference time.The authors of LERK [18] state a runtime of 4.67 min for making a single decision with seven officers using their fastest approach. This makes the application of their algorithm in real-world settings infeasible.

### 6.3   Ablation Studies

To assess the individual components' impact on the final performance, we provide several ablation studies. We conduct the ablations with two agents on the validation set in the Docklands area and report the average number of violations fined per day. The results of the ablations can be found in Table 2, where they are sorted decreasingly by performance, i.e., the highest impact is on the right.

We observe that not using an *action embedding network* has the strongest impact on the performance resulting in an 8.3% reduction in performance. Since the reduction is less severe when removing inputs of this network, the effect can be primarily attributed to the additional non-linear transformation after resource aggregation. Switching from individual to joint rewards is next in terms of relevance. We observe that using *joint rewards* performs considerably worse, leading

**Table 2.** Ablations performed in the area of Docklands with two agents. We report the average number of caught violations per day. The second row shows the relative performance compared to the base configuration. The values are sorted decreasingly, i.e., the highest impact is on the right.

|  | OURS | With Other Agents' Target | Without Agent ID | Without Resource Position | Without Agent Embedding Network | Without Distance To Action | Joint Reward | Without Action Embedding Network |
|---|---|---|---|---|---|---|---|---|
| Absolute | **388.32** | 378.54 | 375.39 | 370.36 | 365.25 | 360.96 | 359.71 | 356.18 |
| Relative | 100.0% | 97.5% | 96.7% | 95.4% | 94.1% | 93.0% | 92.6% | 91.7 |

to a 7.4% reduction in performance,[5] which we attribute to the credit assignment problem. Ignoring the *distance to the action* leads to a reduction of 7.0%. Without this information the agent lacks input to assess the inherent reward uncertainty in far actions. The *agent embedding network* is the next crucial component, with a reduction of 5.9%. Without it, the model cannot utilize the agent features, such as its position. Not having access to the *agent ID* aggravates diversification of agent policies and leads to a performance decrease of 3.3%. Finally, *adding other agents' target information* to the agent-specific views of the resources leads to slightly worse performance of around 2.5%, despite yielding improvements in the SASRC baseline. This indicates that our architecture can already sufficiently incorporate the intents of other agents for effective coordination.

## 7    Conclusion

In this work, we have formalized Multi-Agent Stochastic Resource Collection (MASRC) as a Semi-Markov Game, providing a solid theoretical framework for the development of new approaches. We further proposed a novel architecture to solve MASRC tasks featuring an innovative intent combination model which permits re-assessment of action representations based on the other agents' action representations. To enable evaluation, we introduced an efficient agent-based simulation for the MTOP task, for which we publish the source code to support the community in future research. Using the simulation, we could demonstrate that our approach is able to beat existing heuristic baselines, adaptions of state-of-the-art single-agent SRC solutions, and approaches for the multi-agent dynamic-VRP in terms of fined violations. On a more fundamental level, our results indicate that existing approaches for multi-agent dynamic-VRP struggle to handle the increased dynamics in MASRC tasks, and thus MASRC requires specialized solutions. In future work, we want to include dynamic travel times. Furthermore, we want to investigate the transfer of trained policies between different areas and numbers of agents. Finally, we will research further scaling our approach to very large graphs.

---

[5] Notice though that even with joint rewards, our approach is able to beat baselines trained with individual rewards.

**Acknowledgments.** We thank the City of Melbourne, Australia, for providing the parking datasets used in this paper and Oliver Schrüfer for contributing the implementation of LERK. This work has been funded by the German Federal Ministry of Education and Research (BMBF) under Grant No. 01IS18036A. The authors of this work take full responsibilities for its content.

# References

1. Bono, G., Dibangoye, J.S., Simonin, O., Matignon, L., Pereyron, F.: Solving multi-agent routing problems using deep attention mechanisms. IEEE Trans. Intell. Transp. Syst. **22**(12), 7804–7813 (2020)
2. Chakravorty, J., et al.: Option-critic in cooperative multi-agent systems. arXiv preprint arXiv:1911.12825 (2019)
3. Foerster, J., Farquhar, G., Afouras, T., Nardelli, N., Whiteson, S.: Counterfactual multi-agent policy gradients. In: AAAI, vol. 32 (2018)
4. Gendreau, M., Laporte, G., Séguin, R.: Stochastic vehicle routing. Eur. J. Oper. Res. **88**(1), 3–12 (1996)
5. Han, D., Böhmer, W., Wooldridge, M., Rogers, A.: Multi-agent hierarchical reinforcement learning with dynamic termination. In: Nayak, A.C., Sharma, A. (eds.) PRICAI 2019. LNCS (LNAI), vol. 11671, pp. 80–92. Springer, Cham (2019). https://doi.org/10.1007/978-3-030-29911-8_7
6. Hernandez-Leal, P., Kartal, B., Taylor, M.E.: A survey and critique of multi-agent deep reinforcement learning. Auton. Agents Multi-Agent Syst. **33**(6), 750–797 (2019). https://doi.org/10.1007/s10458-019-09421-1
7. Hu, J., Wellman, M.P., et al.: Multiagent reinforcement learning: theoretical framework and an algorithm. In: ICML, vol. 98, pp. 242–250. Citeseer (1998)
8. Kim, J., Kim, K.: Optimizing large-scale fleet management on a road network using multi-agent deep reinforcement learning with graph neural network. In: ITSC, pp. 990–995. IEEE (2021)
9. Kool, W., Van Hoof, H., Welling, M.: Attention, learn to solve routing problems! arXiv preprint arXiv:1803.08475 (2018)
10. Koutsoupias, E., Papadimitriou, C.: Worst-case equilibria. In: Meinel, C., Tison, S. (eds.) STACS 1999. LNCS, vol. 1563, pp. 404–413. Springer, Heidelberg (1999). https://doi.org/10.1007/3-540-49116-3_38
11. Kumar, S.N., Panneerselvam, R.: A survey on the vehicle routing problem and its variants (2012)
12. Li, M., et al.: Efficient ridesharing order dispatching with mean field multi-agent reinforcement learning. In: The world wide web conference, pp. 983–994 (2019)
13. Liu, Z., Li, J., Wu, K.: Context-aware taxi dispatching at city-scale using deep reinforcement learning. IEEE Trans. Intell. Transp. Syst. **99**, 1–14 (2020)
14. Makar, R., Mahadevan, S., Ghavamzadeh, M.: Hierarchical multi-agent reinforcement learning. In: Proceedings of the fifth International Conference on Autonomous agents, pp. 246–253 (2001)
15. Mnih, V., et al.: Playing Atari with deep reinforcement learning. arXiv preprint arXiv:1312.5602 (2013)
16. Nazari, M., Oroojlooy, A., Snyder, L., Takác, M.: Playing Atari with deep reinforcement learning. In: Advance Neural Information Processing System, vol. 31 (2018)

17. Peng, B., Wang, J., Zhang, Z.: A deep reinforcement learning algorithm using dynamic attention model for vehicle routing problems. In: Li, K., Li, W., Wang, H., Liu, Y. (eds.) ISICA 2019. CCIS, vol. 1205, pp. 636–650. Springer, Singapore (2020). https://doi.org/10.1007/978-981-15-5577-0_51

18. Qin, K.K., Shao, W., Ren, Y., Chan, J., Salim, F.D.: Solving multiple travelling officers problem with population-based optimization algorithms. Neural Comput. Appl. **32**(16), 12033–12059 (2020)

19. Rashid, T., Samvelyan, M., Schroeder, C., Farquhar, G., Foerster, J., Whiteson, S.: QMIX: monotonic value function factorisation for deep multi-agent reinforcement learning. In: ICML, pp. 4295–4304. PMLR (2018)

20. Rohanimanesh, K., Mahadevan, S.: Learning to take concurrent actions. In: Advance Neural Information Processing System, vol. 15 (2002)

21. Schmoll, S., Schubert, M.: Semi-markov reinforcement learning for stochastic resource collection. In: IJCAI, pp. 3349–3355 (2021)

22. Shao, W., Salim, F.D., Gu, T., Dinh, N.T., Chan, J.: Traveling officer problem: managing car parking violations efficiently using sensor data. IEEE Internet Things J. **5**(2), 802–810 (2017)

23. Sukhbaatar, S., Fergus, R., et al.: Learning multiagent communication with backpropagation. In: Advance Neural Information Processing System, vol. 29 (2016)

24. Sunehag, P., et al.: Value-decomposition networks for cooperative multi-agent learning. arXiv preprint arXiv:1706.05296 (2017)

25. Tampuu, A., Matiisen, T., Kodelja, D., Kuzovkin, I., Korjus, K., Aru, J., Aru, J., Vicente, R.: Multiagent cooperation and competition with deep reinforcement learning. PLoS ONE **12**(4), e0172395 (2017)

26. Tan, M.: Multi-agent reinforcement learning: Independent vs. cooperative agents. In: ICML, pp. 330–337 (1993)

27. Tang, H., et al.: Hierarchical deep multiagent reinforcement learning with temporal abstraction. arXiv preprint arXiv:1809.09332 (2018)

28. Tang, X., et al.: A deep value-network based approach for multi-driver order dispatching. In: Proceedings of the 25th ACM SIGKDD, pp. 1780–1790 (2019)

29. Van Hasselt, H., Guez, A., Silver, D.: Deep reinforcement learning with double q-learning. In: AAAI, vol. 30 (2016)

30. Vaswani, A., et al.: Attention is all you need. In: Advance Neural Information Processing System, vol. 30 (2017)

31. Zheng, L., et al.: Magent: a many-agent reinforcement learning platform for artificial collective intelligence. In: AAAI, vol. 32 (2018)

32. Zhou, M., et al.: Multi-agent reinforcement learning for order-dispatching via order-vehicle distribution matching. In: Proceedings of the 28th ACM Int'l Conf on Information and Knowledge Management, pp. 2645–2653 (2019)

# Team-Imitate-Synchronize for Solving Dec-POMDPs

Eliran Abdoo, Ronen I. Brafman[(✉)], Guy Shani, and Nitsan Soffair

Ben-Gurion University of the Negev, Beersheba, Israel
{eliranab,brafman,shanigu,soffair}@bgu.ac.il

**Abstract.** Multi-agent collaboration under partial observability is a difficult task. Multi-agent reinforcement learning (MARL) algorithms that do not leverage a model of the environment struggle with tasks that require sequences of collaborative actions, while Dec-POMDP algorithms that use such models to compute near-optimal policies, scale poorly. In this paper, we suggest the Team-Imitate-Synchronize (TIS) approach, a heuristic, model-based method for solving such problems. Our approach begins by solving the joint team problem, assuming that observations are shared. Then, for each agent we solve a single agent problem designed to imitate its behavior within the team plan. Finally, we adjust the single agent policies for better synchronization. Our experiments demonstrate that our method provides comparable solutions to Dec-POMDP solvers over small problems, while scaling to much larger problems, and provides collaborative plans that MARL algorithms are unable to identify.

## 1 Introduction

Problems that require collaborative effort by several agents, operating under partial observability, are extremely challenging. Such problems can be tackled by a centralized planning algorithm that creates a policy for each agent. Then, each agent executes its policy in a distributed manner, restricting communication with other agents to explicit actions dictated by the policy.

Recently, cooperative multi agent problems are often tackled by deep multi-agent reinforcement learning (MARL), often under the term *centralized learning for decentralized execution*, showing impressive improvements [16,18]. RL is able to learn a policy directly, without requiring access to a model of the environment's dynamics. In the single-agent case, model-free RL methods are sometimes employed even when a model exists, because in many problems, a specification of a policy can be orders of magnitude smaller than a model of the environment.

However, in many MA domains, a sequence of collaborations, conditioned on appropriate observations, is needed to complete a task and earn a reward. MARL algorithms must explore the policy space, blindly at first, to identify such beneficial behaviors. As we show in this paper, current MARL algorithms have significant difficulty discovering such sequences by pure exploration.

Supported by ISF Grants 1651/19 and 1210/18, Ministry of Science and Technology's Grant #3-15626 and the Lynn and William Frankel Center for Computer Science.

M.-R. Amini et al. (Eds.): ECML PKDD 2022, LNAI 13716, pp. 216–232, 2023.
https://doi.org/10.1007/978-3-031-26412-2_14

On the other side of the spectrum lie algorithms that rely on a complete specification of the environment, typically as a Dec-POMDP model [3]. Given such a specification the agents can identify good behaviors more easily. Some Dec-POMDP solvers can compute solutions with optimality guarantees or error bounds. But these solvers have difficulty scaling up beyond very small problems – not surprising given the NEXP-Time hardness of this problem [3].

In this paper we suggest a new approach for solving MA problems *given* a model. Like Deep MARL methods, our approach does not provide optimality guarantees, yet scales significantly better than existing Dec-POMDP algorithms. Unlike MARL methods, we can use the world model to better guide the agents towards complex beneficial behaviors. This allows us to solve problems that require a sequence of collaborative actions, where MARL methods utterly fail.

Our approach, Team-Imitate-Synchronize (TIS) works in 3 phases. First, we solve a *team POMDP* in which every agent's observations are implicitly available to the other agents. Hence, all agents share the same belief state. The solution to the team POMDP is typically not executable because it may condition an agent's actions on observations made by other agents. Hence, in the next step, TIS tries to produce a policy for each agent that *imitates* that agent's behavior within the team policy. The resulting policies are executable by the agents, as they depend on their own observations only. However, they are not well synchronized. The last step improves the synchronization of the timing of action execution by different agents, while still relying on information available to that agent only. In the Dec-POMDPs with a *no-op* action we consider here, this can be done by delaying the execution of particular parts of the policy.

TIS is a general approach for solving Dec-POMDPs—there are different ways of instantiating the Imitate and Synchronize steps, and we offer here a simple, heuristic instantiation. We create a specific imitation-POMDP for each agent in which it receives reward for behaving similarly to its behavior in the team policy. The synchronization step employs a heuristic approach that analyzes the agents' policy trees to improve synchronization. That being said, our chosen methods for these steps enable us to handle many MA problems that cannot be currently solved by any other method. TIS does not provide optimality guarantees because the Team step solves a relaxed version of the original multi-agent problem. We know that an optimal solution to a relaxed problem may not be refinable to an optimal solution of the original problem (e.g., see the case of hierarchical planning and RL [7]). Yet, relaxation-based methods offer a very practical approach in many areas.

We experiment on 3 problems, comparing our approach to exact and approximate Dec-POMDP solution methods, as well as MARL algorithms. We demonstrate that TIS scales significantly better than all other methods, especially in domains that require non-trivial coordination of actions. Such collaborations include both the ability to order actions properly, so that one agent's actions help set up conditions needed for the success of other agents' actions, and the ability to perform appropriate actions concurrently. Code and domain encodings are available at https://github.com/neuronymous/decpomdp-solver.

## 2   Background

*POMDPs:* A POMDP models single-agent sequential decision making under uncertainty and partial observability. It is a tuple $P = \langle S, A, T, R, \Omega, O, \gamma, h, b_0 \rangle$. $S$ is the set of states. $A$ is the set of actions. $T(s, a, s')$ is the probability of transitioning to $s'$ when applying $a$ in $s$. $R(s, a)$ is the immediate reward for action $a$ in state $s$. $\Omega$ is the set of observations. $O(a, s', o)$ is the probability of observing $o \in \Omega$ when performing $a$ and *reaching* $s'$. $\gamma \in (0, 1)$ is the discount factor. $h$ is the planning horizon. A *belief state* is a distribution over $S$, with $b_0 \in \prod(S)$ denoting the initial belief state.

We focus on factored models where each state is an assignment to some set of variables $X_1, \ldots, X_k$, and each observation $\Omega$ is an assignment to observation variables $W_1, \ldots, W_d$. Thus, $S = Dom(X_1) \times \cdots \times Dom(X_k)$ and $\Omega = Dom(W_1) \times \cdots \times Dom(W_d)$. In that case, $\tau$, $O$, and $R$ can be represented compactly by, e.g., a dynamic Bayesian network [4].

For ease of representation, we assume that actions are either sensing actions or non-sensing actions. Sensing actions do not modify the state of the world, and may result in different observations in different states. An agent that applies a non-sensing action always receives the *null-obs* observation. We assume that every action has one effect on the world that we consider as its successful outcome, while all other effects are considered failures. Both assumptions are realistic in many domains. Both can be removed, at the cost of a more complex algorithm.

A solution to a POMDP is a *policy* that assigns an action to every history of actions and observations (*AO-history*). It is often represented using a *policy tree* or *graph* (a.k.a. finite-state controller). Each vertex is associated with an action, and each edge is associated with an observation.

A *trace* $T$ is a sequence of quintuplets $e_i = (s_i, a_i, s'_i, o_i, r_i)$, where $s_i$ is the state in step $i$, $a_i$ is the action taken in step $i$; $s'_i = s_{i+1}$ is the resulting state, and $o_i$ and $r_i$ are the observation and reward received after taking action $a_i$ in $s_i$ and reaching $s'_i$. For brevity, in our description, we typically ignore $s'_i$ and $r_i$.

*Dec-POMDPs:* Dec-POMDPs model problems where $n > 1$ fully cooperative agents seek to maximize the expected sum of rewards received by the team. The agents act in a distributed manner and obtain different observations, so their information states may differ. Formally, a Dec-POMDP for $n$ agents is a tuple $P = \langle S, A = \times_{i=1}^{n} \{A_i\}, T, R, \Omega = \times i = 1^n \{\Omega_i\}, O, \gamma, h, b_0 \rangle$. The components are similar to those of a POMDP with the following differences: each agent $i$ has its own set of actions $A_i$ and its own set of observations $\Omega_i$. These sets define the *joint-action* set $A = A_1 \times A_2 \times .. \times A_n$ and the *joint-observation* set $\Omega = \Omega_1 \times \Omega_2 \times \ldots \times \Omega_n$. All other elements are defined identically as in a POMDP w.r.t. the set of joint-actions and joint-observations. We assume $A_i$ *always* contains a *no-op* action that does not modify the state of the world nor generates any meaningful observation. (This essentially implies that there are no exogenous processes.) We also use the *no-ops* for *reward-calibration*: a joint action consisting of *no-ops* only has a reward of 0. The agents share the initial belief state, $b_0$. However, during execution, agent $i$ receives only its component

$\omega_i$ of the joint observation $\omega = (\omega_1, \ldots, \omega_n)$. A solution to a Dec-POMDP is a *set* of policies $\rho_i$ (as defined for POMDP), one for each agent.

Dec-POMDPs can use a factored specification [15], although most work to date uses the flat-state representation. An important element of a factored specification is a compact formalism for specifying joint-actions. If each agent has $k$ actions, then, in principle, there are $O(k^n)$ possible joint actions. Yet, in many problems of interest most actions do not interact with each other. If $a \in A_i$ is an action of agent $i$, we identify $a$ with the joint action (no-op, ..., a, ..., no-op). Actions $a_i \in A_i$, $a_j \in A_j$ are said to be non interacting, if their effect distribution, when applied jointly (i.e., (no-op, ..., $a_i$, ..., $a_j$ ..., no-op)), is identical to their effect distribution when applied sequentially. Thus, our specification language focuses on specifying the effects of single-agent actions and specific combinations of single-agent actions that interact with each other, which we call *collaborative* actions [2]. As above, we identify the collaborative action of a group of agents with the joint action in which all other agents perform a *no-op*. Then, we can decompose every joint action into some combination of non-interacting single-agent and collaborative actions, defining its dynamics.

*Example 1.* Our running example consists of a 2-cell box-pushing domain, with cells $L$(left) and $R$(right), two agents, and two boxes. $B_1$ is light, and $B_2$ is heavy (Fig. 1). The state is composed of 4 state variables: the location of each box – $(X_{B1}, X_{B2})$ – and the location of each agent – $(X_{A1}, X_{A2})$. In addition, there are two observation variables for each agent $(\omega_1^i, \omega_2^i)$. $\omega_j^i$, indicates to $Agent_i$ whether it is co-located with $B_j$. Initially, $A_1$ and $B_1$ are at $L$ and $A_2$ and $B_2$ are at $R$. The goal is to swap the boxes, i.e. $(X_{B1} = R, X_{B2} = L)$. Agents can move, push a box, or sense their current cell for a box. Move and Push can be done in any direction. Push actions fail with some probability, and a single-agent cannot succeed pushing the heavy box. The action in which both agents push a heavy box is modeled as a *collaborative-push* action.

*Public, Private and Relevant Variables:* A state variable $X_i$ is *affected* by $a$ if there is some state $s$ for which there is a non zero probability that the value of $X_i$ changes following $a$. We denote the variables affected by $a$ by $\mathit{eff}(a)$. $X_i$ is *affected* by agent $j$, if $X_i \in \mathit{eff}(a)$ for some $a \in A_j$. $X_i$ is called *public* if it is affected by two or more agents, that is, there exist $j \neq k$ and actions $a \in A_j$, $a' \in A_k$ such that $X_i \in \mathit{eff}(a) \cap \mathit{eff}(a')$. An action $a$ is called *public* if one of its effects is public. Otherwise, $a$ is *private*. Thus, collaborative actions are always public. Sensing actions are private, by nature. Here, we also assume they are non-collaborative, i.e., they affect one agent's observation variables only.

A state variable $X_i$ is an *influencer* of $a$ if $a$ behaves differently given different values of $X_i$. That is, if there are two states $s_1, s_2$ that differ only in the value of $X_i$ such that $R(s_1, a, s') \neq R(s_2, a, s')$, or $T(s_1, a, s') \neq T(s_2, a, s')$ for some state $s'$, or in the case of sensing actions, $O(a, s_1, o) \neq O(a, s_2, o)$ for some observation $o$. We denote influencers of $a$ by $\mathit{inf}(a)$. We refer to the union of the influencers and effects of $a$ as the *relevant* variables of $a$, denoted $\mathit{rel}(a)$.

*Example 2.* In our running example, $X_{B1}$ and $X_{B2}$ are the public variables, as they are the effects of both agents' push actions. $X_{A1}$, $X_{A2}$ are private variables of agent 1 and agent 2, respectively, as they are the effect of a single agent's move action. $\omega_j^i$ is private to $Agent_i$, being the effect of its sensing actions. Actions move and sense are private, while the push actions are public.

## 3    TIS – Team-Imitate-Synchronize

We begin with a high level description of the major components of TIS. Then, we explain each step in more depth.

**1. Team Problem:** Given a Dec-POMDP model $P$ as input, this step outputs a near-optimal policy $\pi_{team}$ for the *team* POMDP $P_{team}$. $P_{team}$ is identical to $P$ but ignores the underlying multi-agent structure. That is, actions and observations in $P_{team}$ are the joint actions and the joint observations of $P$ with the same transition and reward function. $P_{team}$ models a single agent that controls all agents and receives the joint observations. We obtain $\pi_{team}$ by solving $P_{team}$ using a POMDP solver. This can be a model-based solver or an RL algorithm that can handle POMDPs, such as DRQN [8].

**2. Generate Tree or Traces:** Some POMDP solvers output a policy tree. If this tree is very large, we approximate the set of path in it by simulating $\pi_{team}$ on $P_{team}$, obtaining a set T of execution traces.[1]

**3. Imitate:** Given the Dec-POMDP model $P$ and the traces T as input, in this step every agent tries to imitate its behavior in the team policy. This is a non-standard imitation learning problem. First, each agent has access to less information than the expert (=team). Second, the agent can sometimes obtain additional information by applying sensing actions that are not part of the team policy. Third, how good one agent's imitation policy depends on how other agents imitate their part of the team policy. While this can be a very interesting imitation learning problem, instead, in this paper we use our model to construct an imitation POMDP $P_i$, for each agent $i$. $P_i$'s dynamics are similar to $P_{team}$, ignoring certain variables and actions that are not relevant for agent $i$, and the observations available to other agents. $P_i$ rewards the agent when its action choice is similar to that which appears in a comparable trace of $\pi_{team}$. Its solution, $\pi_i'$, is the basis for $i$'s policy.

**4. Synchronize:** Given the agents' policies, $\{\pi_i'\}_{i=1}^n$, generate a policy graph for each agent and compute modified single-agent policies $\{\pi_i\}_{i=1}^n$ aiming at (probabilistically) better coordination between agents. This is done by inserting additional *no-op* actions to agents' policies to affect the timing of action execution. Notice that in Dec-POMDPs, one can always insert a *no-op* action. Whether this helps or not depends on what other agents do at the same time. Specifically, we focus on improving the probability that individual parts of a collaborative

---

[1] Our implementation uses the simulation function of the SARSOP solver. We precompute sample size based on concentration bounds that ensure that distribution over initial state will match the true belief state.

**Table 1.** Two example traces. $P, S, M, CoP$ abbreviate *Push, Sense, Move, Collaborative-Push*

|      | $X_{A1}$ | $X_{A2}$ | $X_{B1}$ | $X_{B2}$ | $a_1$ | $a_2$ | $\omega_1^1$ | $\omega_2^1$ | $\omega_1^2$ | $\omega_2^2$ |
|------|----------|----------|----------|----------|-------|-------|--------------|--------------|--------------|--------------|
| 1.1  | L | L | R | R | $PushRight(A_1, B_1)$ | $no\text{-}op$ | $\phi$ | $\phi$ | $\phi$ | $\phi$ |
| 1.2  | L | R | R | R | $SenseBox(A_1, B_1)$ | $no\text{-}op$ | no | $\phi$ | $\phi$ | $\phi$ |
| 1.3  | L | R | R | R | $MoveRight(A_1)$ | $no\text{-}op$ | $\phi$ | $\phi$ | $\phi$ | $\phi$ |
| 1.4  | R | R | R | R | $CPushLeft(A_1, B_2)$ | $CPushLeft(A_2, B_2)$ | $\phi$ | $\phi$ | $\phi$ | $\phi$ |
| 1.5  | R | R | R | L | $no\text{-}op$ | $SenseBox(A_2, B_2)$ | $\phi$ | $\phi$ | $\phi$ | no |
| 2.1  | L | L | R | R | $PushRight(A_1, B_1)$ | $no\text{-}op$ | $\phi$ | $\phi$ | $\phi$ | $\phi$ |
| 2.2  | L | R | R | R | $SenseBox(A_1, B_1)$ | $no\text{-}op$ | no | $\phi$ | $\phi$ | $\phi$ |
| 2.3  | L | R | R | R | $MoveRight(A_1)$ | $no\text{-}op$ | $\phi$ | $\phi$ | $\phi$ | $\phi$ |
| 2.4  | R | R | R | R | $CPushLeft(A_1, B_2)$ | $CPushLeft(A_2, B_2)$ | $\phi$ | $\phi$ | $\phi$ | $\phi$ |
| 2.5  | R | R | R | R | $no\text{-}op$ | $SenseBox(A_2, B_2)$ | $\phi$ | $\phi$ | $\phi$ | yes |
| 2.6  | R | R | R | R | $CPushLeft(A_1, B_2)$ | $CPushLeft(A_2, B_2)$ | $\phi$ | $\phi$ | $\phi$ | $\phi$ |
| 2.7  | R | R | R | L | $no\text{-}op$ | $SenseBox(A_2, B_2)$ | $\phi$ | $\phi$ | $\phi$ | no |

action will be synchronized and that the action order in $\pi_{team}$ between agents is maintained. $\{\pi_i\}_{i=1}^n$ is the final output of the entire algorithm.

Steps 1 and 2 are straightforward. Below, we detail Steps 3 and 4.

*Example 3.* A possible team policy for our example is shown in Fig. 1a. Edges are labeled by the joint observations of the team. $A_1$ begins by pushing $B_1$ to the right, then senses whether it succeeded. It then moves right to assist $A_2$ to push the heavy box, $B_2$, to the left. $A_2$ then senses for success. As observations are shared in $P_{team}$, $A_1$ is also aware of the resulting observation. Two example traces are shown in Table 1.

## 3.1 Generating the Individual Agent POMDPs

We now aim to generate agent policies that, combined, will behave similarly to the team policy. This can be achieved in several ways. For example, we could try to imitate the behavior based on the agents' individual belief state. Here, we suggest a heuristic approach, motivated by a simple intuition. We design for each agent $i$ a POMDP $P_i$, in which the world dynamics remains the same, but agent $i$ is rewarded whenever it imitates its role in the team policy. That is, $i$ is rewarded when executing an action similarly to the public plan.

We focus on imitating public actions as they influence other agents. We wish to reward an agent when it executes a public action in a context in which it was applied in the collected traces T. Hence, we define a context $c$ for an action $a$, and reward $i$ only when it applies $a$ in $c$. Public actions not encountered in T are not relevant for imitation, and thus we remove them from the imitation POMDPs, $P_i$. For private actions, we maintain the same reward as in $P$.

Defining the *context* in which a public action $a$ is applied in a trace to be the state it was applied in, is too narrow. We must generalize the context to capture only the relevant variables within the state. To better generalize from the states at which $a$ was executed in T, we remove irrelevant variables from the context.

**Definition 1.** *The* context $c$ *of action* $a$ *of agent* $i$ *in state* $s$ *is the projection of* $s$ *to the set of variables consisting of all public variables and any private variable of* $i$ *relevant to* $a$. *The pair* $\langle c, a \rangle$ *is called a* contexted action (CA).

$CA_i$, *the set of contexted actions for agent* $i$, *is the union of all contexted actions* $\langle c, a \rangle$ *for all state,action pairs for agent* $i$ *appearing in any trace in* T.

*Example 4.* The public actions of $A_1$ in the trace elements shown in Table 1 are $PushRight(A_1, B_1)$ in 1.1,2.1, and $CPushLeft(A_1, B_2)$ in 1.4,2.4,2.6. These actions appear multiple times in identical contexts. Context of $PushRight(A_1, B_1)$ contains the public variable $X_{B1}$, $X_{B2}$, and $X_{A1}$ which is the only private relevant variable of $PushRight(A_1, B_1)$. The context of $CPushLeft(A_1, B_2)$ for $A_1$ is identical. Thus: $CA_1 = \{\langle\langle X_{A1} = L, X_{B1} = L, X_{B2} = R\rangle, PushRight(A_1, B_1)\rangle, \langle\langle X_{A1} = R, X_{B1} = R, X_{B2} = R\rangle, CPushLeft(A_1, B_2)\rangle\}$. Also, $CA_2 = \{\langle\langle X_{A2} = R, X_{B1} = R, X_{B2} = R, \rangle, CPushLeft(A_2, B_2)\rangle\}$.

Encouraging the execution of a public action in an appropriate context is insufficient. We must also discourage execution of public actions outside their context. Public actions modify public variables' values, which may cause future actions by other agents to have undesirable outcomes that differ from the team plan. Hence, we associate negative reward with out-of-context public actions.

This must be done carefully. $P_i$ contains the non-sensing actions of all agents. This helps the synchronizing agent $i$'s policy with those of other agents. That is, the agent has to simulate in its policy actions of other agents that are needed for its own reward. Thus, it must time its actions appropriately w.r.t. other agents' actions (simulated by it), which leads to better coordination. However, $P_i$ does not contain the sensing actions of other agents. Therefore, we should not penalize it for performing actions when the value of a variable it cannot gain information on is "wrong". For this reason, we define a relaxed context.

**Definition 2.** *Let* $X^i_{obs}$ *be the set of variables that agent* $i$ *can learn about through, possibly noisy, observations. The* relaxed context $c'$ *of a contexted action* $\langle c, a \rangle$ *is the projection of* $c$ *to the set of public variables and* $X^i_{obs}$.

That is, $c'$ is obtained from $c$ by ignoring some context variables. Therefore, $\neg c' \to \neg c$, and fewer states are associated with this penalty than had we applied it to any state *not* satisfying $c$. This leads to the following definition of the factored single-agent POMDP $P_i$, solved by agent $i$.

**Actions:** $P_i$ contains all private non-sensing actions of all agents, all public non-sensing actions that appeared in some trace of the team plan, and all the sensing actions of agent $i$.

**Transitions:** the transition function of public actions and agent $i$'s private actions is identical to that of the original problem definition. Private actions of other agents are determinized, leveraging our assumption about a desired effect for actions. The deterministic version of $a$ always achieves its desired effect. This relaxation is not essential, but reduces the difficulty in solving $P_i$.

**State and Observation Variables:** Observations of other agents do not appear in $P_i$. State variables that correspond only to the removed observations are also ignored, as they do not affect the transition functions of the actions in $P_i$. All other variables appear in $P_i$.

**Rewards:** (1) The reward for a private action is identical to its reward in the original Dec-POMDP and in $P_{team}$. (2) The reward $R(s,a)$ for a public action $a$ in state $s$ is defined as: ($i$) if $s \models c$ for some context $c$ such that $\langle c, a \rangle \in CA_i$ $R(s,a)$ is positive. ($ii$) If $s \not\models c'$ for any relaxed context $c'$ of some contexted action $\langle c, a \rangle \in CA_i$ then $R(s,a)$ is negative.

We use the following method for defining reward values:

1. Reward for an in-context action. We use the following steps:
   (a) Associate a value $R_\tau$ with each trace, reflecting the utility of the trace. Let $\tau$ be a trace, $R_\tau^+$ and $R_\tau^-$ the sum of positive and negative rewards, respectively, in $\tau$.
   (b) Distribute $R_\tau^+$ among the agents based on their relative contribution to attaining this utility, yielding $R_\tau^i$ for each $i$. Let $R_\tau^{-,i}$ be the total negative reward in $\tau$ associated with agent $i$'s actions only (including collaborative actions). Define $R_\tau^i = R_\tau^+ \cdot \frac{R_\tau^{-,i}}{R_\tau^-}$. This is the relative portion of reward we want to allocate to agent $i$'s actions in the trace.
   (c) For each agent and trace, distribute $R_\tau^i$ to each instance $e$ of the agent's public actions in this trace, yielding $R_{\tau,e}^i$. Let $e_j = (s_j, a_j, s'_j, o_j, r_j)$ be the $j^{th}$ step in trace $\tau$. We distinguish between contributing and non-contributing steps (defined below). If $e_j$ is non-contributing then $R_{\tau,e}^i = 0$. Otherwise, $R_{\tau,e}^i$ is defined using the following process:
       i. Associate a cost $c_e$ with $e$ (defined below).
       ii. Compute the relative weight, $w_e$, of step $e$ in $\tau$: $w_e = \frac{c_e}{R_\tau^{-,i}}$.
       iii. Define $R_{\tau,e}^i = w_e \cdot R_\tau^i$.
   (d) Associate with $\langle c, a \rangle$ the average value $r_{c,a}$ of $R_{\tau,e}^i$ over all traces $\tau \in T$, and all steps $e \in \tau$ such that $e$ involves the execution of $a$ in a state satisfying $c$. The reward assigned to a CA in the model is the average reward $r_{CA}$ of steps in which it appears in the traces. Formally:

$$r_{CA} = \frac{\sum_{\tau \in T} \sum_{e \in \tau \wedge proj_i(e) = CA} R_{\tau,e}^i}{|\{e | e \in \tau, \tau \in T, proj_i(e) = CA\}|} \qquad (1)$$

$proj_i(e)$ is the contexted action obtained when projecting the state and action of $e$ w.r.t. agent $i$.

2. Penalty for an out-of relaxed-context action. We associate with the execution of an action in a state that does not satisfy any of its *relaxed* contexts, a negative reward $- \max_{ca \in CA_i} r_{ca} \cdot |CA_i|$, an upper bound on the sum of rewards that can be achieved from applying contexted actions.

To complete the description above, step $e_j = (s_j, a_j, s'_j, o_j, r_j)$ is a *contributing* step of agent $i$, if $a_j$ contains an action of agent $i$, and either (i) $s_j$ did not appear earlier in the trace, or (ii) $r_j > 0$, i.e., a positive reward was gained. To define $c_e$, we iterate over the steps in $\tau$, accumulating the step costs (negative rewards). When encountering a contributing step of agent $i$, we assign it with the accumulated cost, and reset the accumulation.

*Example 5.* We now construct $A_1$'s single-agent problem. We denote the CAs from the previous example by $ca_1, ca_2$, and their reward with $r_{ca_1}, r_{ca_2}$. We follow the projection stages one by one: (1) Push actions are the only public actions and we leave only the ones observed in traces: $PushRight(A_1, B_1)$, $CPushLeft(A_1, B_2)$, $CPushLeft(A_2, B_2)$. All other Push actions are removed. Notice that we keep $A_2$'s $CPushLeft(A_2, B_2)$ as $A_1$ might need to simulate it. (2) We remove the two sensing actions of $A_2$ and its observation variables $\omega_1^2, \omega_2^2$. (3) We leave all private actions of both agents. They are deterministic to start with, so no change is needed. (4) We set rewards $r_{ca_1}, r_{ca_2}$ to $ca_1$ and $ca_2$ respectively. (5) We set a penalty of $-2 \cdot max(r_{ca_1}, r_{ca_2})$ to the remaining public actions, applied in any context except for the CA's *relaxed* contexts. The relaxed context for $A_1$'s CAs is: $\{\langle\langle X_{A1} = L\rangle, PushRight(A_1, B_1)\rangle, \langle\langle X_{A1} = R\rangle, CPushLeft(A_1, B_2)\rangle\}$. (6) The reward for pushing the boxes to the target cells is set to 0, as we reward the agent only for doing its public actions in context.

## 3.2  Policy Adjustment and Alignment

We now solve the agent specific POMDPs $P_i$ and obtain agent specific policies $\pi_i'$, in the form of a policy graph for each agent. These policy graphs may contain private actions of other agents, and are likely not well synchronized. For example, there may be a low probability that collaborative actions are executed jointly at the same time. We now adjust the policies to obtain better synchronization.

First, we remove the private actions of all other agents from $i$'s policy graph, introduced to"simulate" the behavior of other agents in $P_i$. Next, we attempt to align policies to increase the probability that actions of different agents occur in the same order as in the team plan. An action $a_1$ of agent 1 that sets the context value of a variable in the context of action $a_2$ of agent 2 should be executed before $a_2$. Collaborative actions should be executed at the same time by all participating agents. As each policy is local to an agent, and action effects are stochastic, one cannot guarantee perfect synchronization. However, using a few heuristics, we attempt to increase the probability of synchronization.

For each public action $a$ in an agent's policy graph we select all simple paths from the root to $a$, and map them to their *identifiers*, where an *identifier* of a path is the sequence of public actions along it. Our goal is to equalize execution time of public actions occurrences with a shared identifier. For a public action $a$ with a shared identifier in multiple agents' graphs: let $l$ be the length of the longest simple path to the action in all relevant policy graphs, including private actions. In any graph where the length is less than $l$, we add *no-op* actions prior to $a$ to delay its execution. We use an iterative process—we begin with the action with the shortest identifier (breaking ties arbitrarily), and delay its execution where needed using *no-ops*. Then, we move to the next action, and so

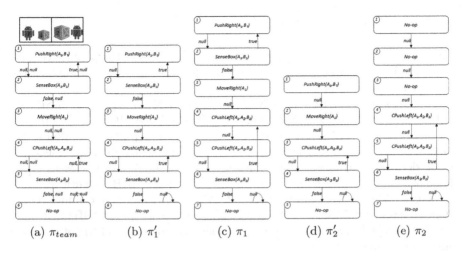

**Fig. 1.** A 2-cell box pushing problem, and the resulting plan graphs. The agents must switch between the boxes. Box $B_1$ is light, $B_2$ is heavy and must be pushed by the two agents jointly.

forth. After the alignment, we replace in each agent's aligned policy graph all actions of other agents with a *no-op*.

Finally, we handle the problem of a potential "livelock" between *collaborative* actions. Consider a scenario where two agents need to perform a non-deterministic collaborative action whose effect can be directly sensed. Each agent executes its part of the action and then senses whether it succeeded. Due to the stochastic nature of previous actions, one agent may execute the collaborative action one step before the other. In that case, it will sense its failure and repeat, with the second agent acting identically, one step later. To handle this, given a collaborative action with $n$ collaborating agents, we modify the graph so that every collaborative action that is part of a cycle is repeated by every agent for $n$ times instead of just once, preventing this livelock. This may be non-beneficial if a cost is associated with a collaborative action. To decide whether to apply it, during the reward definition stage, for each CA that contains a collaborative action, we calculate the expected reward of repeating the action, and sum it over all CAs. If the sum is positive, we enable the method.

*Example 6.* Figures 1b, 1d show plan graphs generated from the agent POMDPs. Here, edges are labeled by single-agent observations. In $\pi'_1$, the sensing action allocated to $A_2$ in $\pi_{team}$ is replaced by a sensing action of $A_1$ (node 5). In $\pi'_2$ appear the simulated actions of $A_1$ (nodes 1 and 2). Figures 1c, 1e show the policy graphs after the alignment and adjustments procedure. $\pi_1$ shows the repeated collaborative push action for live-lock avoidance (nodes 4 and 5). In $\pi_2$ the simulated actions of $A_1$ are replaced by *no-op* (nodes 1 and 2), and another *no-op* is added for alignment with $\pi_1$ (node 3).

## 4    Empirical Evaluation

We evaluate our algorithm on 3 domains: two variations of the popular cooperative Box Pushing, Dec-Tiger and Rock Sample. TIS uses SARSOP [9] as the underlying POMDP solver.

We compare TIS, with 3 Dec-POMDP solvers, GMAA*-ICE [13], JESP [10], and DICEPS [12] using MADP-tools [14]. The GMAA*-ICE solver provides optimality guarantees. JESP searches in policy space, performing alternating maximizations using dynamic programming. DICEPS is an approximate method, which does not excel on smaller domains, but presumably can scale to larger domains.[2] We also compare TIS to 3 state-of-the-art MARL algorithms: two versions of WQMix [16], and QTRAN [18] All experiments were conducted on a PC with Intel-core i7-5500U CPU @ 2.40GHz with 4 cores and 7.7GB of memory. TIS, GMMA-ICE, and JESP were run under Windows-11 and the others were run under Ubuntu 20.04.1.

GMAA*-ICE and DP-JESP require an horizon specification, specified under column $H$. TIS computes a policy for an unbounded horizon and its $H$-value specifies the average number of steps until reaching the goal state. DICEPS uses restarts, and repeatedly returned an overflow error. To generate comparable running times to TIS, we rerun it multiple time, and used the maximal score over these runs, which is equivalent to simply letting it run longer.

For GMAA*-ICE and DP-JESP we report the computed policy value. For TIS the *value* column provides the average discounted accumulated reward over 1000 simulations truncated at the relevant horizon. The *avg* column denotes average number of steps for all agents to reach the goal state. For MARL algorithms, we measure maximum average discounted reward over a number of test runs. The discount factor was set to $\gamma = 0.99$.

Planners were given 1 h to solve each ⟨*configuration, horizon*⟩ pair. MARL algorithms were given longer times. We report running times in seconds.

### 4.1    Domains

**Box Pushing:** agents on a grid must push boxes to their destinations [5,6]. Light boxes can be pushed by a single agent. Heavy boxes must be pushed collaboratively by two agents. Agents can sense for a box at the present location. The initial box locations are uncertain. We also consider a variant of this domain in which we add a penalty when a push action is performed in a cell with no box. Problem names are composed of 5 elements, specifying width, height, number of agents, number of light boxes, and number of heavy boxes.

**Dec-Tiger:** Agents must open a door avoiding a tiger [11]. The tiger's location resets following a door opening. Collaboratively opening a tiger door results in a lower penalty compared to single agent opening actions. Agents have noisy

---

[2] DICEPS, while not new, was recommended to us, independently, by two senior researchers as still being a state-of-the-art approximate solver.

**Table 2.** Results for Collaborative Dec-Tigeron state-of-the-art Dec-POMDP and MARL solvers and TIS. Best overall value for each problem in bold.

| Collaborative Dec-Tiger, $|S| = 8, |A| = 5, I = \langle (0,1,0),(0,1,1) \rangle$ | | | | | | | | | | | |
|---|---|---|---|---|---|---|---|---|---|---|---|
| H | DP-JESP | | GMAA*-ICE | | DICESP | | OW QMix | | QTran | | TIS | |
| | Time | Value | Time | Value | Time | Value | 300s | 3600 s s | 300s | 3600 s s | Time | Value |
| 3 | 0.42 | 1.96 | 0.24 | **11.15** | 121.54 | 0 | −7.84 | 7.44 | 0 | 2.55 | 6.03 | 11.10 |
| 4 | 33.15 | 10.05 | 841.67 | 11.03 | 155.98 | 3.2 | −7.87 | **13.602** | 0 | 0.958 | ″ | 9.12 |
| 5 | 1792.22 | 5.78 | × | − | 206.56 | 4.03 | 1.68 | **14.454** | 0 | 4.08 | ″ | 8.56 |
| 6 | × | − | × | − | 243.07 | 5 | 2.61 | **15.473** | 0 | 1.96 | ″ | 9.05 |
| 10 | × | − | × | − | x | x | −17.79 | **18.136** | 0 | 0.758 | ″ | 10.19 |
| 20 | × | − | × | − | x | x | −13.91 | **20.058** | 0 | 1.79 | ″ | 16.42 |
| 30 | × | − | × | − | x | x | −19.43 | 18.954 | 0.58 | **54.907** | ″ | 20.10 |
| 40 | × | − | × | − | x | x | −11.702 | **26.128** | 0 | 10.49 | ″ | 24.62 |

observations on the tiger's location. We use a larger Dec-Tiger version that requires agents to move on a grid [1].

**Decentralized Rock-Sample:** We suggest a new variant of Mars exploration problem [17], where two rovers move on a grid that contains rocks that must be sampled . The grid is divided into overlapping control areas, one per rover. The rovers can move around the grid, sense the rocks' quality, which is initially unknown, and sample them. Agents are rewarded for sampling all the good rocks in their control area, but penalized directly for sampling a bad quality rock. Once a good quality rock is sampled, it turns bad. Rovers have a long range sensor, whose quality deteriorates with increased distance from the rock. Problem names are composed of the grid size and the number of rocks.

These problems call for solutions of different types. In Collaborative Dec-Tiger, the problem resets after one of the doors is opened and so the planning horizon can be short, there are no interdependent actions (i.e., ones setting up conditions for the execution of others), but tight coordination in the execution of door opening actions is desirable. Collaborative Box-Pushing rewards the agents for pushing a box to the target tile, requiring aggregating a chain of properly-directed pushes of a box to a single reward. With light boxes, a simple plan can use a single agent. Yet, efficient plans make use of the different agent positions to reduce (costly) move actions. With heavy boxes, agents must push the box at the same time, requiring even better coordination. The second box-pushing variant calls for trading off the cost of additional sensing with the penalty for moving a non-existent box. Decentralized Rock-Sampling rewards an agent only for achieving all of its objectives, which makes large-horizon planning ability crucial, giving no reward at all to partial solutions.

### 4.2 Results

Table 2 describes the results for Dec-Tiger. GMAA*-ICE, which is an optimal solver, produces the best policy for the shortest horizon, but cannot scale beyond

**Table 3.** TIS vs. Dec-POMDP solvers and vs. MARL solvers on Box-Pushing (BP) and Rock-Sample (DRS). Best value per problem in bold.

Collaborative Box-Pushing

| Problem | | | DP-JESP | | | GMAA*-ICE | | | DICEPS | | | TIS | | |
|---|---|---|---|---|---|---|---|---|---|---|---|---|---|---|
| Name | $|S|$ | $|A|$ | H | Time | Value | H | Time | Value | H | Time | Value | Avg | Time | Value |
| 3,1,2,1,1 | 81 | 16 | 4 | 1861.30 | 279 | 4 | 30.23 | 330 | 4 | 221.71 | 0 | 15 | 6.07 | 613 |
| 2,2,2,0,2 | 256 | 225 | 3 | 267.24 | 271 | 3 | 160.18 | 320 | 3 | 348.56 | 0 | 9 | 7.08 | **348** |
| 2,2,2,0,3 | 1024 | 400 | 2 | 59.06 | 0 | 2 | 1053.27 | 414 | 2 | 514.09 | 0 | 17 | 125.50 | **514** |
| 1-Penalty | 81 | 16 | 3 | 25.95 | 0 | 4 | 79.28 | 265 | – | – | – | 15 | 8.54 | 587 |
| 2-Penalty | 256 | 225 | 3 | 495.61 | 135 | 3 | 446.03 | 214 | – | – | – | 9 | 11.66 | **354** |
| 3-Penalty | 1024 | 400 | 2 | 38.70 | 0 | 2 | 1054.5 | 327 | – | – | – | 15 | 31.09 | **510** |

| Problem | | | OW QMIX | | | CW QMIX | | | QTRAN | | | TIS | | |
|---|---|---|---|---|---|---|---|---|---|---|---|---|---|---|
| | | | H | 300s | 7200 s s | H | 300s | 7200 s s | H | 300s | 7200 s s | Avg | Time | Value |
| 3,1,2,1,1 | 81 | 16 | 20 | 120.27 | 1165 | 20 | -8.91 | **1177** | 20 | 345.17 | 348 | 15 | 6.07 | **614** |
| 2,2,2,0,2 | 256 | 225 | 15 | -3.57 | -1.98 | 15 | 0 | 0 | 15 | 0 | -0.79 | 9 | 7.08 | **349** |
| 2,2,2,0,3 | 1024 | 400 | 20 | -2 | 0 | 20 | 0 | -2 | 20 | -0.8 | 0 | 17 | 125.50 | **514** |
| 1-Penalty | 81 | 16 | 20 | -36.32 | 1144 | 20 | 76.97 | **1191** | 20 | -77.94 | 372 | 15 | 8.54 | **587** |
| 2-Penalty | 256 | 225 | 15 | 0 | -3.9 | 15 | 0 | 0 | 15 | 0 | 0 | 9 | 11.66 | **354** |
| 3-Penalty | 1024 | 400 | 20 | 0 | 0 | 20 | 0 | 0 | 20 | -35.62 | -29 | 15 | 31.09 | **510** |
| 3,2,3,0,2 | 7776 | 3375 | 20 | 0 | 0 | 20 | 0 | – | 20 | 0 | 0 | 12 | 89.83 | **276** |
| 3,2,3,0,3 | 46656 | 8000 | 20 | 0 | 0 | 20 | 0 | 0 | 20 | 0 | 0 | 18 | 2210.28 | **406** |
| 3,3,2,2,1 | 59049 | 324 | 20 | -4.37 | -2 | 20 | 0 | 0 | 20 | -72.76 | 0 | 39 | 3014.04 | **322** |

Decentralized Rock-Sampling

| Problem | | | DP-JESP | | | GMAA*-ICE | | | DICEPS | | | TIS | | |
|---|---|---|---|---|---|---|---|---|---|---|---|---|---|---|
| 12,3 | 512 | 90 | 3 | 314.35 | 224 | 3 | 82.86 | 739 | 3 | 2018 | 755 | 18 | 1725.80 | **1028** |
| 12,4 | 1024 | 100 | 3 | 1081.41 | 518 | 3 | 2867.06 | 508 | 3 | 2763 | 495 | 20 | 2438.82 | **1048** |
| 20,4 | 2304 | 100 | 3 | 1838.84 | 111 | 3 | – | – | 3 | 2868 | 514 | 16 | 2434.88 | **1158** |

| Problem | | | OW QMIX | | | CW QMIX | | | QTRAN | | | TIS | | |
|---|---|---|---|---|---|---|---|---|---|---|---|---|---|---|
| | | | H | 300s | 7200 s s | H | 300s | 7200 s s | H | 300s | 7200 s s | Avg | Time | Value |
| 12,3 | 512 | 90 | 25 | -56.51 | 474 | 25 | -54.42 | 509 | 25 | -34.28 | 609 | 18 | 1725.80 | **1028** |
| 12,4 | 1024 | 100 | 25 | -45.6 | 198 | 25 | -12.00 | 203 | 25 | 0 | 617 | 20 | 2438.82 | **1048** |
| 20,4 | 2304 | 100 | 25 | -55.20 | 452 | 25 | -45.63 | 219 | 25 | -40.05 | 327 | 16 | 2434.88 | **1158** |
| 20,6 | 9216 | 143 | 25 | -19.34 | 135 | 25 | -43.85 | 20 | 25 | -21.83 | 408 | 21 | 2537.10 | **1121** |
| 28,6 | 16384 | 143 | 25 | -29.23 | 47 | 25 | 91.78 | 24 | 25 | -1.99 | 0 | 23 | 3310.95 | **1046** |

4 steps. JESP scales to horizon 5, and DICEPS to horizon 6, but they produce policies of much lower quality than TIS, which can handle horizon 40 in about 6 s. DICESP, which should be able to scale well, terminated with an error after roughly 10–20 s for horizons 10 and higher. For the MARL algorithms we show results for two running times: 300 and 3600 s. In this domain, OW-QMix was able to perform better than TIS, but only when given 3600 s, as opposed to 6 for TIS. In the horizon 30 case, QTran was able to perform much better than TIS, but performed far worse on other horizons. Except for this case, TIS performed comparatively to, or better than other MARL algorithms but required much less time. As can be seen from the results for 300 s, MARL algorithms cannot compete with TIS as far a policy quality with shorter running times, even when

**Table 4.** Comparing to optimal on small box pushing with $|S| = 8$ and $|A| = 16$. *Max* row is for maximal value for any horizon. ArgMax Horizon in parenthesis.

| H | GMAA*-ICE | | TIS | |
|---|---|---|---|---|
| | Time | Value | Time | Value |
| 4 | 1.15 | **426.91** | 3.40 | 306.62 |
| 5 | 2.09 | **438.34** | " | 301.56 |
| 6 | 6.97 | **448.19** | " | 357.44 |
| 7 | 8.98 | **450.97** | " | 412.54 |
| Max | 17.1 | **454.70** (25) | 3.40 | 412.54 (7) |

given 50X running time. CW-QMix was always dominated by OW-QMix and is therefore omitted.

Table 3 shows results for Box Pushing and Rock Sample. Both DP-JESP and GMAA*-ICE could not handle horizons larger than 3, while TIS can consider much longer action sequences, and, as such, collect much higher rewards. As noted above, Decentralized Rock-Sampling requires much lengthier horizons to achieve any reward, and TIS's advantage here is pretty clear. Among the MARL algorithms, QMix was able to produce much better results on one problem instance (again, requiring orders of magnitude more time), but MARL solvers failed on harder Box Pushing domains that contain more heavy boxes that require a collaborative push action. Similar results were obtained with the alternative reward function, punishing attempts to push a non-existent box. TIS is consistently better and faster than the MARL solvers on Rock Sample, again due to MARL difficulty in learning policies that require more complex coordination.

In the Box Pushing and Rock Sample we can also test the scalability of TIS. These are domains that current Dec-POMDP solvers cannot handle, and hence we only provide a comparison with MARL algorithms. As can be seen, TIS can consider very long horizons even in these significantly larger problems. The size of the hardest configuration, BP-33221, approaches the maximal problems that the single agent POMDP solver SARSOP can solve, indicating that TIS could scale to even larger problems given a stronger POMDP solver. We are not aware of any other Dec-POMDP solver that is able to approach state sizes even close to these on these domains: over 59000 in Box Pushing and over 16000 in Rock Sample. The MARL solvers were unable to generate reasonable solution within 7200 s for these larger problems. (For this reason, we do not provide results for shorter running times). To complete the picture, we note that in the hardest Box-Pushing and Rock-Sample instances, TIS was able to achieve the goal in 100% of all trials for Box Pushing and in 97% of all trials for Rock Sample.

TIS contains many heuristic choices that may cause it to obtain sub-optimal policies. To measure this, Table 4 focuses on a small box pushing problem that GMAA*-ICE can solve optimally. We see that TIS manages to produce reasonable results that are about 10% worse on the larger horizons. Furthermore, in Dec-Tiger, which calls for strong agent synchronization, TIS does virtually the same as GMAA\*-ICE on horizon 3. On medium size horizons TIS performs worse, probably because there is more opportunity for unsynchronized actions.

### 4.3 Discussion

As we observed above, over the 3 domains that we experiment with, TIS is substantially better than all other solvers. It produces policies which are close to optimal on smaller problems with shorter horizons, and scales many orders of magnitude beyond existing Dec-POMDP solvers.

In comparison to state-of-the-art MARL algorithms, TIS is much faster, and often returns much better policies. Of course, MARL algorithms do not receive the model as input and must compensate for it by running many trials. Nevertheless, we provide this comparison for two reasons. First, there is a perception that RL algorithms are a magic bullet and are the state-of-the-art for stochastic sequential decision making. Second, model-free RL algorithms are often suggested as an alternative to model-based solvers on domains with large state spaces, as they do not need to maintain and query an explicit model, which in larger problems can be difficult to represent. While this is certainly true, the time and resources required to compensate for the lack of a model can be substantial. As we have shown here, domains that require better coordination are still challenging for state-of-the-art MARL solvers. In longer (non-exhaustive) experiments conducted on QMIX, the results on these domains did not improve even given over 10 h. Thus, we think that it can be safely concluded that when a model is available and a plan is needed quickly, TIS is currently the best option.

Finally, we briefly discuss the MCEM Dec-POMDP solver [19]. This algorithm is an example of an approximate solver that can scale to huge problems. Indeed, MCEM was able to solve a grid-based traffic problem with 2500 agents and roughly $10^{100}$ states, which is certainly remarkable. MCEM excels on this domain because it is very loosely coupled, and a simple local controller per agent can generate good behavior. MCEM exploits these domain properties to truly factor the problem (aided by a hand-coded MDP policy for policy exploration).

TIS cannot handle such a domain because the team problem would be too large to describe formally, and no POMDP solver can scale to such problem sizes. On the other hand, [19] also tested MCEM on standard Dec-POMDP benchmarks that do not enjoy such weak coupling, and in which stronger implicit coordination between the agents' actions is required. In these domains, MCEM was unable to scale to the problems sizes TIS handles. For example, the largest box pushing problem solved by [19] had 100 states, and the largest Mars Rover problem solved (which is quite similar to rock-sample) had 256 states compared with 59K and 16K states, respectively, for TIS, limited only by the capabilities of the underlying POMDP solver.

## 5    Conclusion

TIS is a general approach for solving Dec-POMDPs. We described a particular implementation that solves Dec-POMDP by solving multiple POMDPs. First, we solve a team POMDP – a single-agent relaxation of the Dec-POMDP. Then, we solve an imitation POMDP for each agent that seeks to generate a policy that imitates that agent's behavior in the team policy. Finally, policy graphs are aligned to improve synchronization. We report promising empirical results. Our implementation of TIS solves significantly larger problems and horizons than existing approximate Dec-POMDP solvers on standard benchmarks in which the problem is not very loosely coupled. It compares well to near-optimal solvers on the problems they can solve, and is much better than MARL algorithms on domains that require some agent coordination.

While the high level flow of TIS is attractive, the particular implementation of the steps is often very specific, not unlike many RL/MARL/DL approaches. In particular, our current synchronization step relies heavily on no-op insertion. The ability to use no-ops implies that the agents are the sole cause of change in the environment. Yet, one exciting aspect of the TIS schema is its generality, and the many exciting opportunities it offers for instantiating each element, in more general and more effective ways.

# References

1. Amato, C., Bernstein, D.S., Zilberstein, S.: Optimizing fixed-size stochastic controllers for pomdps and decentralized pomdps. JAAMAS **21**(3), 293–320 (2010)
2. Bazinin, S., Shani, G.: Iterative planning for deterministic qdec-pomdps. In: GCAI-2018, 4th Global Conference on Artificial Intelligence, vol. 55, pp. 15–28 (2018)
3. Bernstein, D.S., Givan, R., Immerman, N., Zilberstein, S.: The complexity of decentralized control of Markov decision processes. Math. Oper. Res. **27**(4), 819–840 (2002)
4. Boutilier, C., Dean, T., Hanks, S.: Decision-theoretic planning: structural assumptions and computational leverage. J. Artif. Int. Res. **11**(1), 1–94 (1999)
5. Brafman, R.I., Shani, G., Zilberstein, S.: Qualitative planning under partial observability in multi-agent domains. In: AAAI 2013 (2013)
6. Carlin, A., Zilberstein, S.: Value-based observation compression for dec-pomdps. In: Proceedings of the 7th International Joint Conference on Autonomous Agents and Multiagent Systems, vol. 1, pp. 501–508 (2008)
7. Dietterich, T.G.: Hierarchical reinforcement learning with the maxq value function decomposition. J. AI Res. **13**, 227–303 (2000)
8. Hausknecht, M.J., Stone, P.: Deep recurrent q-learning for partially observable mdps. In: 2015 AAAI Fall Symposium, pp. 29–37. AAAI Press (2015)
9. Kurniawati, H., Hsu, D., Lee, W.S.: Sarsop: efficient point-based pomdp planning by approximating optimally reachable belief spaces. In: Proceedings Robotics: Science and Systems (2008)
10. Nair, R., Tambe, M., Yokoo, M., Pynadath, D., Marsella, S.: Taming decentralized pomdps: towards efficient policy computation for multiagent settings. In: IJCAI 2003, pp. 705–711 (2003)
11. Nair, R., Tambe, M., Yokoo, M., Pynadath, D., Marsella, S.: Taming decentralized pomdps: towards efficient policy computation for multiagent settings. In: IJCAI, vol. 3, pp. 705–711 (2003)
12. Oliehoek, F., Kooij, J., Vlassis, N.: The cross-entropy method for policy search in decentralized pomdps. Informatica **32**, 341–357 (2008)

13. Oliehoek, F.A., Spaan, M.T.J., Amato, C., Whiteson, S.: Incremental clustering and expansion for faster optimal planning in decentralized POMDPs. JAIR **46**, 449–509 (2013)
14. Oliehoek, F.A., Spaan, M.T.J., Terwijn, B., Robbel, P., Messias, J.A.V.: The madp toolbox: an open source library for planning and learning in (multi-)agent systems. J. Mach. Learn. Res. **18**(1), 3112–3116 (2017)
15. Oliehoek, F.A., Spaan, M.T.J., Whiteson, S., Vlassis, N.: Exploiting locality of interaction in factored dec-pomdps. In: AAMAS, pp. 517–524 (2008)
16. Rashid, T., Farquhar, G., Peng, B., Whiteson, S.: Weighted QMIX: expanding monotonic value function factorisation for deep multi-agent reinforcement learning. Adv. Neural Inf. Process. Syst. **33**, 10199–10210 (2020)
17. Smith, T., Simmons, R.: Point-based pomdp algorithms: improved analysis and implementation. In: UAI, pp. 542–549 (2005)
18. Son, K., Kim, D., Kang, W.J., Hostallero, D., Yi, Y.: QTRAN: learning to factorize with transformation for cooperative multi-agent reinforcement learning. In: ICML, pp. 5887–5896 (2019)
19. Wu, F., Zilberstein, S., Jennings, N.R.: Monte-carlo expectation maximization for decentralized pomdps. In: IJCAI, pp. 397–403 (2013)

# DistSPECTRL: Distributing Specifications in Multi-Agent Reinforcement Learning Systems

Joe Eappen[✉] and Suresh Jagannathan

Purdue University, West Lafayette, IN 47907, USA
jeappen@purdue.edu, suresh@cs.purdue.edu

**Abstract.** While notable progress has been made in specifying and learning objectives for general cyber-physical systems, applying these methods to distributed multi-agent systems still pose significant challenges. Among these are the need to (a) craft specification primitives that allow expression and interplay of both local and global objectives, (b) tame explosion in the state and action spaces to enable effective learning, and (c) minimize coordination frequency and the set of engaged participants for global objectives. To address these challenges, we propose a novel specification framework that allows natural composition of local and global objectives used to guide training of a multi-agent system. Our technique enables learning expressive policies that allow agents to operate in a coordination-free manner for local objectives, while using a decentralized communication protocol for enforcing global ones. Experimental results support our claim that sophisticated multi-agent distributed planning problems can be effectively realized using specification-guided learning. Code is provided at https://github.com/yokian/distspectrl.

**Keywords:** Multi-agent reinforcement learning · Specification-guided learning

## 1 Introduction

Reinforcement Learning (RL) can be used to learn complex behaviors in many different problem settings. A main component of RL is providing feedback to an agent via a reward signal. This signal should encourage desired behaviors, and penalize undesirable ones, enabling the agent to eventually proceed through a sequence of tasks and is designed by the programmer beforehand. A commonly used technique to encode tasks in a reward signal is the sparse method of providing zero reward until a task is completed upon which a non-zero reward is given to the agent. Because this procedure has the significant shortcoming of delaying generating a useful feedback signal for a large portion of the agent-environment interaction process, a number of alternative techniques have been proposed [1,14].

**Supplementary Information** The online version contains supplementary material available at https://doi.org/10.1007/978-3-031-26412-2_15.

Formulating a reward signal that reduces the sparsity of this feedback is known as *reward shaping*. Often, this is done manually but a more general, robust method would be to automatically shape a reward given a specification of desired behavior. SPECTRL [5] proposes a reward shaping mechanism for a set of temporal logic specifications on a single-agent task that uses a compiled a finite-state automaton called a task monitor. Reward machines [2,18,19] are another objective-specifying method for RL problems that also define a finite automaton akin to the ones used in SPECTRL, with some subtle differences such as the lack of registers (used by the task monitor for memory).

Distributed multi-agent applications however, introduce new challenges in automating this reward shaping process. Agents have their own respective goals to fulfill as well as coordinated goals that must be performed in cooperation with other agents. While the expressiveness of the language in SPECTRL lends itself, with minor extensions, to specifying these kinds of goals, we require new compilation and execution algorithms to tackle inherent difficulties in multi-agent reinforcement learning (MARL); these include credit assignment of global objectives and the presence of large state and action spaces that grow as the number of agents increases. Learning algorithms for multi-agent problems have often encouraged distribution as a means of scaling in the presence of state and action space explosion. This is because purely centralized approaches have the disadvantages of not only requiring global knowledge of the system at all times but also induce frequent and costly synchronized agent control.

To address these issues, we develop a new specification-guided distributed multi-agent reinforcement learning framework. Our approach has four main features. First, we introduce two classes of predicates (*viz.* local and global) to capture tasks in a multi-agent world (Sect. 4). Second, we develop a new procedure for generating composite task monitors using these predicates and devise new techniques to distribute these monitors over all agents to address scalability and decentralization concerns (Sect. 5). Third, we efficiently solve the introduced problem of subtask synchronization (Sect. 6) among agents via synchronization states in the task monitors. Lastly, we describe a wide class of specification structures (Sect. 7) amenable to scaling in the number of agents and provide a means to perform such a scaling (Sect. 8).

By using these components in tandem, we provide the first solution to composing specifications and distributing them among agents in a scalable fashion within a multi-agent learning scenario supporting continuous state and action spaces. Before presenting details of our approach, we first provide necessary background information (Sect. 2) and formalize the problem (Sect. 3).

## 2   Background

**Markov Decision Processes.** Reinforcement learning is a tool to solve Markov Decision Processes (MDPs). MDPs are tuples of the form $\langle S, D, A, P, R, T \rangle$ where $S \in \mathbb{R}^n$ is the state space, $D$ is the initial state distribution, $A \in \mathbb{R}^m$ is the action space, $P : S \times A \times S \to [0, 1]$ is the transition function, and $T$ is the time horizon. A rollout $\zeta \in Z$ of length $T$ is a sequence of states and actions $\zeta = (s_0, a_0, ..., a_{T-1}, s_T)$ where $s_i \in \mathcal{S}$ and $a_i \in A$ are such that $s_{i+1} \sim P(s_i, a_i)$. $R : Z \to \mathbb{R}$ is a reward function used to score a rollout $\zeta$.

**Multi-agent Reinforcement Learning.** A Markov game with $\mathcal{N} = \{1, \cdots, N\}$ denoting the set of $N$ agents is a tuple $\mathcal{M}_g = \langle \mathcal{N}, \{\mathcal{S}^i\}_{i \in \mathcal{N}}, D, \{A^i\}_{i \in \mathcal{N}}, P, \{R^i\}_{i \in \mathcal{N}}, T \rangle$ where $A^i$, $R^i$ define their agent-specific action spaces and reward functions. They are a direct generalization of MDPs to the multi-agent scenario. Let $\mathcal{S}_m = \{\mathcal{S}^i\}_{i \in \mathcal{N}}$ and $A_m = \{A^i\}_{i \in \mathcal{N}}$, then $P : \mathcal{S}_m \times A_m \times \mathcal{S}_m \to [0, 1]$ is the transition function. A rollout $\zeta_m \in Z_m$ here corresponds to $\zeta_m = (\bar{s}_0, \bar{a}_0, ..., \bar{a}_{T-1}, \bar{s}_T)$ where $\bar{s} \in \mathcal{S}_m$ and $\bar{a} \in A_m$. We also define an agent specific rollout $\zeta_m^i \in Z_m^i$, $\zeta_m^i = (s_0^i, a_0^i, ..., a_{T-1}^i, s_T^i)$ where $s^i \in \mathcal{S}^i$ and $a^i \in A^i$. $D$ is the initial state distribution over $\mathcal{S}_m$.

Agents attempt to learn a policy $\pi^i : \mathcal{S}^i \to \Delta(A^i)$ such that $\mathbb{E}\left[\sum_t R_t^i | \pi^i, \pi^{-i}\right]$ is maximized, where $\Delta(A^i)$ is a probability distribution over $A^i$ and $\pi^{-i}$ is the set of all policies apart from $\pi^i$. We use $\Pi = \{\pi^i\}_{i \in \mathcal{N}}$ to denote the set of all agent policies. For simplicity, we restrict our formulation to a *homogeneous* set of agents which operate over the same state ($\mathcal{S}^i = \mathcal{S}_A$) and action space ($A^i = A_A$).

**SPECTRL.** Jothimurugan *et. al* [5] introduce a specification language for reinforcement learning problems built using temporal logic constraints and predicates. It is shown to be adept at handling complex compositions of task specifications through the use of a *task monitor* and well-defined monitor transition rules. Notably, one can encode Non-Markovian tasks into the MDP using the additional states of the automaton (task monitor) compiled from the given specification.

The atomic elements of this language are Boolean predicates $b$ defined as functions of a state $\mathcal{S}$ with output $[\![b]\!] : \mathcal{S} \to \mathbb{B}$. These elements have quantitative semantics $[\![b]\!]_q$ with the relation being $[\![b]\!](s) = \texttt{True} \iff [\![b]\!]_q(s) > 0$. Specifications $\phi$ are Boolean functions of the state trajectory $\zeta = (s_1, s_2, ..., s_T)$. The specification language also includes composition functions for a specification $\phi$ and Boolean predicate $b$, with the language defined as

$$\phi :: = \texttt{achieve } b \mid \phi \texttt{ ensuring } b \mid \phi_1;\ \phi_2 \mid \phi_1 \texttt{ or } \phi_2$$

The description of these functions is as follows. $\texttt{achieve } b$ is true when the trajectory satisfies $b$ at least once. $\phi \texttt{ ensuring } b$ is true when $b$ is satisfied at all timesteps in the trajectory. $\phi_1;\ \phi_2$ is a sequential operator that is true when, in a given trajectory $\zeta = (s_1, s_2, ..., s_T)$, $\exists\, k > 1$ such that $\phi_1(s_1, ..., s_k)$ is true and $\phi_2(s_{k+1}, ..., s_T)$ is true. In other words, $\phi_1;\ \phi_2$ represents an ordered sequential completion of specification $\phi_1$ followed by $\phi_2$. Lastly, $\phi_1 \texttt{ or } \phi_2$ is true when a trajectory satisfies either $\phi_1$ or $\phi_2$.

Given a specification $\phi$ on a Markov Decision Process $\langle S, D, A, P, T \rangle$ (MDP) defined using SPECTRL, a task monitor $\langle Q, X, \Sigma, U, \Delta, q_0, v_0, F, \rho \rangle$ (a finite state automaton [20]) is compiled to record the completion status of tasks with monitor states $Q$; final monitor states $F$ denote a satisfied trajectory. This is used to create an augmented version of the MDP $\langle \tilde{S}, \tilde{s}_0, \tilde{A}, \tilde{P}, \tilde{R}, \tilde{T} \rangle$ with an expanded state, action space and modified reward function . The task monitor provides a scoring function for trajectories in the augmented MDP to guide policy behavior.

While SPECTRL has been shown to work with trajectory-based algorithms for reinforcement learning [12], it is not immediately evident how to translate it to common RL algorithms such as DDPG [10] and PPO [16]. A simple solution would be to keep the episodic format with a trajectory $\zeta = (s_0, \cdots, s_T)$ and assign the trajectory value of SPECTRL (a function of $\zeta$) to the final state transition in the trajectory $s_{T-1} \to s_T$ and zero for all other states. Importantly, this maintains the trajectory ordering properties of SPECTRL in the episodic return $(\sum_{t=0}^{T} r_t)$.

## 3    Problem Statement

Directly appropriating SPECTRL for our use case of imposing specifications on multi-agent problems poses significant scalability issues. Consider the case

$$\phi_a = \texttt{achieve}(\texttt{reach}(P)); \texttt{achieve}(\texttt{reach}(Q))$$

where $[\![\texttt{reach}(P)]\!] = \texttt{True}$ when an agent reaches state $P$. To ease the illustration of our framework, we assume that all agents are homogeneous, *i.e.* $S^i = S_A, \forall i \in \mathcal{N}$. Now, the state space of the entire multi-agent system is $S = (S_A)^N$ for $N$ agents (we omit $m$ for perspicuity).

If the predicate $\texttt{reach}$ was defined on the entire state $S$, it would yield a specification forcing synchronization between agents. On the other hand, if $\texttt{reach}$ was defined on the agent state $S_A$, then it would create a *localized* specification where synchronization is not required. This would be akin to allowing individual agents to act independently of other agent behaviors.

However, using a centralized task monitor for the localized predicate would cause the number of monitor states to exponentially increase with the number of agents $N$ and subtasks $K$ since the possible stages of task completion would be $\mathcal{O}(K^N)$.

To address this scalability issue, the benefits of task monitor distribution are apparent. In the case of $\phi_a$ above, assume $\texttt{reach}$ is defined on the local state space $S_A$. If each agent had a separate task monitor stored locally to keep track of the task completion stages, the new number of monitor states is now reduced to $\mathcal{O}(NK)$.

Consider an example of robots in a warehouse. A few times a day, all robots must gather at a common point for damage inspection at the same time (akin to a global reach) to minimize the frequency of inspection (an associated cost). To ensure satisfaction of the entire specification, the reward given to an RL agent learning this objective should capture both the global and local tasks. For example, if the global reach task for the routine inspection is made local instead, the cost incurred may be larger than if it was a synchronized global objective.

**Main Objective.** Given a specification $\phi$ on a system of $N$ agents, we wish to find policies $\Pi = \{\pi^1, \cdots, \pi^N\}$ to maximize the probability of satisfying $\phi$ for all agents. Formally, we seek

$$\Pi^* \in \underset{\pi^1, \cdots, \pi^N}{\arg\max} \; \underset{\zeta_m \sim D_\Pi}{\Pr} \; [\![\phi(\zeta_m)]\!] = \texttt{True}]$$

where $D_\Pi$ is the distribution of all system rollouts when all agents collectively follow policy set $\Pi$. We emphasize that $\phi$ acts on the entire rollout, $\phi : Z_m \rightarrow \{0, 1\}$ and not in an agent-specific manner, $\phi' : Z_m^i \rightarrow \{0, 1\}$. This discourages agents from attempting to simply satisfy their local objectives while preventing the system from achieving necessary global ones.

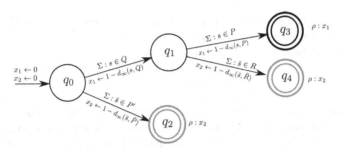

**Fig. 1.** Example Composite Task Monitor for specification $\phi_{ex}$ (Sect. 4) with 4 task goals denoted by Q,P,P' and R where the agent starts at $q_0$. Double circles represent *final* states while green circles represent *global* states. The diagram removes a state between $q_1$ and $(q_3, q_4)$ as well as self-loops for ease of explanation.

# 4   SPECTRL in a Multi-Agent World

Unlike the single agent case, multi-agent problems have two major classes of objectives. Agents have individual goals to fulfill as well as collective goals that require coordination and/or global system knowledge. These individual goals are often only dependent on the agent-specific state $s^i$ while collective goals require full system knowledge $\bar{s}$.

Consequently, for a multi-agent problem, we see the need for two types of predicates *viz.* local and global. Local predicates are of the form $p_{lo} : \mathcal{S}_A \rightarrow \mathbb{B}$ whereas global predicates have the form $p_{gl} : \mathcal{S} \rightarrow \mathbb{B}$ where $\mathbb{B}$ is the Boolean space. We introduce two simple extensions of **reach** [5] to demonstrate the capabilities of this distinction.

Local predicates are defined with respect to each agent and represent our individual goals. As an example, closely related to the problems observed in SPECTRL, we introduce the following local predicates for a state $s_a \in \mathcal{S}_A$,

$$[\![\texttt{reach}_{lo}x]\!](s_a) = (||s_a - x||_\infty < 1)$$

which represents reaching near location $x$ in terms of the $L_\infty$ norm. Now to enforce global restrictions, we introduce counterparts to these predicates that act on a global state $\bar{s} \in \mathcal{S}$.

$$[\![\mathbf{reach}_{gl}\bar{x}]\!](\bar{s}) = (||\bar{s} - \bar{x}||_\infty < 1)$$

where we now have a set of locations $\bar{x} \in \mathcal{S}$.

As in SPECTRL, each of these predicates $b$ require quantitative semantics $[\![b]\!]_q$ to facilitate our reward shaping procedure. We define these semantics as follows:

- $\mathbf{reach}_{lo}$ has the same semantics as $\mathbf{reach}$ in [5] yet is defined on space $\mathcal{S}_A$.

$$[\![\mathbf{reach}_{lo}x]\!]_q(s_a) = 1 - d_\infty(s_a, x)$$

where $d_\infty(a, b)$ represents the $L_\infty$ distance between $a$ and $b$ with the usual extension to the case where $b$ is a set.

- $\mathbf{reach}_{gl}$ is defined on the state space $\mathcal{S}$ as

$$[\![\mathbf{reach}_{gl}\bar{x}]\!]_q(\bar{s}) = 1 - d_\infty(\bar{s}, \bar{x})$$

We observe that the same composition rules can apply to these predicates and we thus attempt to solve RL systems described with these compositions. As shown in Sect. 3, using a centralized SPECTRL compilation algorithm on the entire state space, even for simple sequences of tasks, leads to an explosion in monitor states. We, therefore, distribute task monitors over agents to handle scalability. Furthermore, we also need to change SPECTRL's compilation rules to handle mixed objective compositions such as[1]

$$\phi = \mathbf{reach}_{lo}(P); \mathbf{reach}_{gl}(Q); \mathbf{reach}_{gl}(R)$$

To compile these specifications into a usable format, we utilize a *composite task monitor* as described in Sect. 5 and develop a new algorithm to achieve our goal. As an example, see Fig. 1 depicting a task monitor whose specification is:

$$\phi_{ex} = \mathbf{reach}_{gl}(P') \text{ or } \mathbf{reach}_{lo}(Q); [\mathbf{reach}_{lo}(P) \text{ or } \mathbf{reach}_{gl}(R)]$$

Here, we have 4 task goals denoted by $P, Q, R$ and $P'$. The agents all start at the root node $q_0$. States $q_2, q_3$ and $q_4$ are all final states in the task monitor while $q_2$ and $q_4$ are global monitor states. As shown in Sect. 6, $q_0$ and $q_1$ are a *synchronization states*. While it may seem that agents only require coordination at global states, it is also necessary for the agents to have the same task transition at these synchronization states as well.

---

[1] We omit $\mathbf{achieve}$ in $\mathbf{achieve}(\mathbf{reach}_{lo}(P))$ and $\mathbf{achieve}(\mathbf{reach}_{gl}(P))$ from here on to reduce clutter; this specification is implied when we compose $\mathbf{reach}(P)$ with ; and or.

# 5   Compilation Steps

Given a specification $\phi$ and the Markov game $\mathcal{M}_g$, we create a task monitor $M$ that is distributed among agents by making agent-specific copies. This is used to create an augmented Markov game $\mathcal{M}'_g = \langle \mathcal{N}, \{\tilde{\mathcal{S}}_A\}_{i \in \mathcal{N}}, \tilde{D}, \{\tilde{A}_A\}_{i \in \mathcal{N}}, \tilde{P}, \{\tilde{R}^i\}_{i \in \mathcal{N}}, T \rangle$ on which the individual agent policies are trained.

**Create Composite Task Monitor.** When the types of specifications are divided into two based on the domain, the solution can be modeled with a *composite task monitor* $M_\phi = \langle Q, \tilde{X}, \tilde{\Sigma}, \tilde{U}, \tilde{\Delta}, q_0, v_0, F, \rho \rangle$. As in SPECTRL, $Q$ is a finite set of monitor states. $\tilde{X} = X_l \cup X_g$ is a finite set of registers that are partitioned into $X_l$ for local predicates and $X_g$ for global predicates. These registers are used to keep track of the degree of completion of the task at the current monitor state for local and global tasks respectively.

We describe below how to use the compiled composite task monitor to create an augmented Markov game $\mathcal{M}'_g$. Each $\tilde{\mathcal{S}}_A$ in $\mathcal{M}'_g$ is an augmented state space with an augmented state being a tuple $(s_A, q, v) \in \mathcal{S}_A \times Q \times V$ where $V \in \mathbb{R}^X$ and $v \in V$ is a vector describing the register values. $\tilde{\Delta} = \Delta_l \cup \Delta_g$ houses the transitions of our task monitor. We require that: i) different transitions are allowed only under certain conditions defined by our states and register values; and, ii) furthermore, they must also provide rules on how to update the register values during each transition. To define these conditions for transition availability, we use $\tilde{\Sigma} = \Sigma_l \cup \Sigma_g$ where $\Sigma_l$ is a set of predicates over $\mathcal{S}_A \times V$ and $\Sigma_g$ is a set of predicates over $\mathcal{S} \times V$. Similarly, $\tilde{U} = U_l \cup U_g$ where $U_l$ is a set of functions $u_l : \mathcal{S}_A \times V \to V$ and $U_g$ is a set of functions $u_g : \mathcal{S} \times V \to V$. Now, we can define $\tilde{\Delta} \subseteq Q \times \tilde{\Sigma} \times \tilde{U} \times Q$ to be a finite set of transitions that are non-deterministic. Transition $(q, \sigma, u, q') \in \tilde{\Delta}$ is an augmented transition either representing $(s^i, q, v) \xrightarrow{a^i | \Pi_{-i}} ((s^i)', q', u_l(s^i, v))$ or the form $(\bar{s}, q, v) \xrightarrow{a^i | \Pi_{-i}} (\bar{s}', q', u_g(s, v))$ depending on whether $\sigma \in \Sigma_l$ or $\sigma \in \Sigma_g$ respectively. Let $\delta_l \in \Delta_l$ represent the former (localized) and $\delta_g \in \Delta_g$ the latter (global) transition types. Here $\Pi_{-i}$ denotes the policy set of all agents except agent $i$. Lastly, $q_0$ is the initial monitor state and $v_0$ is the initial register value (for all agents), $F \subseteq Q$ is the set of final monitor states, and $\rho : \mathcal{S} \times F \times V \to \mathbb{R}$ is the reward function.

Copies of these composite task monitors $M$ are distributed over agents $\mathcal{N}$ to form the set $\{M^i\}_{i \in \mathcal{N}}$. These individually stored task monitors are used to let each agent $i \in \mathcal{N}$ keep track of its subtasks and the degree of completion of those subtasks by means of monitor state $q^i$ and register value $v^i$.

**Create Augmented Markov Game.** From our specification $\phi$ we create the augmented Markov game $\mathcal{M}'_g = \langle \mathcal{N}, \{\tilde{\mathcal{S}}_A\}_{i \in \mathcal{N}}, \tilde{D}, \{\tilde{A}_A\}_{i \in \mathcal{N}}, \tilde{P}, \{\tilde{R}^i\}_{i \in \mathcal{N}}, T \rangle$ using the compiled composite task monitor $M$ . A set of policies $\tilde{\Pi}^*$ that maximizes rewards in $\mathcal{M}'_g$ should maximize the chance of the specification $\phi$ being satisfied.

Each $\tilde{\mathcal{S}}_A = \mathcal{S}_A \times Q \times V$ and $\tilde{D} = (\{s_0\}_{i \in \mathcal{N}}, q_0, v_0)$. We use $\Delta$ to augment the transitions of $P$ with monitor transition information. Since $\Delta$ may contain non-deterministic transitions, we require the policies $\tilde{\Pi}$ to decide which transition to choose. Thus $\tilde{A}_A = A_A \times A_\phi$ where $A_\phi = \Delta$ chooses among the set of available transitions at a monitor state $q$. Since monitors are distributed among all agents in $\mathcal{N}$, we denote the set of current monitor states as $\bar{q} = \{q^i\}_{i \in \mathcal{N}}$ and the set of register values as $\bar{v} = \{v^i\}_{i \in \mathcal{N}}$. Now, each agent policy must output an *augmented action* $(a, \delta) \in \tilde{A}_A$ with the condition that $\delta_l = (q, \sigma_l, u_l, q')$ is possible in local augmented state $\tilde{s}_a = (s_a, q, v)$ if $\sigma_l(s_a, v)$ is True and $\delta_g = (q, \sigma_g, u_g, q')$ is possible in global augmented state $\tilde{s} = (\bar{s}, \bar{q}, \bar{v})$ if $\sigma_g(\bar{s}, v)$ is True. We can write the augmented transition probability $\tilde{P}$ as,

$$\tilde{P}((\bar{s}, q, v), (a, (q, \sigma, u, q')), (\bar{s}', q', u(\bar{s}, v)))) = P(\bar{s}, a, \bar{s}')$$

for transitions $\delta_g \in \Delta_g$ with $(\sigma, u) = (\sigma_g, u_g)$ and transitions $\delta_l \in \Delta_l$ with $(\sigma, u) = (\sigma_l, u_l)$. Here, we let $u_l(\bar{s}, v) = u_l(s^i, v)$ for agent $i$ since $s^i$ is included in $\bar{s}$. An *augmented rollout* $\tilde{\zeta}_m$ where

$$\tilde{\zeta}_m = ((\bar{s}_0, \bar{q}_0, \bar{v}_0), \bar{a}_0, ..., \bar{a}_{T-1}, (\bar{s}_T, \bar{q}_T, \bar{v}_T))$$

is formed by these augmented transitions. To translate this trajectory back into the Markov game $\mathcal{M}_g$ we can perform projection $\texttt{proj}(\tilde{\zeta}_m) = (\bar{s}_0, \bar{a}_0, ..., \bar{a}_{T-1}, \bar{s}_T, )$.

**Determine Shaped Rewards.** Now that we have the augmented Markov game $\mathcal{M}_g'$ and compiled our composite task monitor, we proceed to form our reward function that encourages the set of policies $\Pi$ to satisfy our specification $\phi$. We can perform shaping in a manner similar to SPECTRL's single-agent case on our distributed task monitor. Crucially, since reward shaping is done during the centralized training phase, we can assume we have access to the entire augmented rollout namely $\tilde{s}_t = (\bar{s}_t, \bar{q}_t, \bar{v}_t)$ at any given $t \in [0, T]$. From the monitor reward function $\rho$, we can determine the weighting for a complete augmented rollout as

$$\tilde{R}^i(\tilde{\zeta}_m) = \begin{cases} \rho(\bar{s}_T, q_T^i, v_T^i), & \text{if } q_T^i \in F \\ -\infty & \text{otherwise} \end{cases}$$

**Theorem 1.** *(Proof in Appendix Sec. F.) For any Markov game $\mathcal{M}_g$, specification $\phi$ and rollout $\zeta_m$ of $\mathcal{M}_g$, $\zeta_m$ satisfies $\phi$ if and only if there exists an augmented rollout $\tilde{\zeta}_m$ such that i) $\tilde{R}^i(\tilde{\zeta}_m) > 0 \ \forall \ i \in \mathcal{N}$ and ii) $\texttt{proj}(\tilde{\zeta}_m) = \zeta_m$.*

The $\tilde{R}^i$ specified is $-\infty$ unless a trajectory reached a final state of the composite task monitor. To reduce the sparsity of this reward signal, we transform this into a shaped reward $\tilde{R}_s^i$ that gives partial credit to completing subtasks in the composite task monitor.

Define for a non-final monitor state $q \in Q \setminus F$, function $\alpha : \mathcal{S} \times Q \times V \to \mathbb{R}$.

$$\alpha(\bar{s}, q, v) = \max_{(q, \sigma, u, q') \in \Delta, q \neq q'} [\![\sigma]\!]_q(\bar{s}, v)$$

This represents how close an augmented state $\tilde{s} = (\bar{s}, q, v)$ is to transition to another state $\tilde{s}'$ with a different monitor state. Intuitively, the larger $\alpha$ is, the higher the chance of moving deeper into the task monitor. In order to use this definition on all $\sigma$, we overload $\sigma_l$ to also act on elements $\bar{s} = \{s^i\}_{i \in \mathcal{N}} \in \mathcal{S}$ by yielding for agent $i$, the value $\sigma_l(\bar{s}) = \sigma_l(s^i)$.

Let $C_l$ be a lower bound on the final reward at a final monitor state, and $C_u$ being an upper bound on the absolute value of $\alpha$ over non-final monitor states. Also for $q \in Q$, let $d_q$ be length of the longest path from $q_0$ to $q$ in the graph $M_\phi$ (ignoring the self-loops in $\Delta$) and $D = \max_{q \in Q} d_q$. For an augmented rollout $\tilde{\zeta}_m$ let $\tilde{s}_k = (\bar{s}_k, q_k^i, \bar{v})$ be the first augmented state in $\tilde{\zeta}_m$ such that $q_k^i = q_{k+1}^i = \cdots = q_T^i$. Then we have the shaped reward,

$$\tilde{R}_s^i(\tilde{\zeta}_m) = \begin{cases} \max_{k \le j < T} \alpha(\bar{s}_j, q_T^i, v_j) + 2C_u \cdot (d_{q_T^i} - D) + C_l & \text{if } q_T^i \notin F \\ \tilde{R}^i(\tilde{\zeta}_m) & \text{otherwise} \end{cases} \quad (1)$$

**Theorem 2.** *(Proof in Appendix Sec. F.) For two augmented rollouts $\tilde{\zeta}_m, \tilde{\zeta}'_m$, (i) if $\tilde{R}^i(\tilde{\zeta}_m) > \tilde{R}^i(\tilde{\zeta}'_m)$, then $\tilde{R}_s^i(\tilde{\zeta}_m) > \tilde{R}_s^i(\tilde{\zeta}'_m)$, and (ii) if $\tilde{\zeta}_m$ and $\tilde{\zeta}'_m$ end in distinct non-final monitor states $q_T^i$ and $(q_T^i)'$ such that $d_{q_T^i} > d_{(q_T^i)'}$, then $\tilde{R}_s^i(\tilde{\zeta}_m) \ge \tilde{R}_s^i(\tilde{\zeta}'_m)$.*

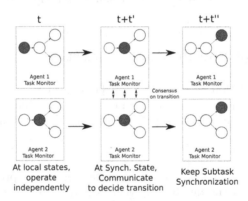

**Fig. 2.** Overview of the DistSPECTRLprocess for task synchronization. Branching in the task monitor diagram denotes potential non-deterministic choices between future tasks (such as in $\phi_{\text{ex}}$). Left to right represents the order of policy actions over a trajectory. Green states represent the current monitor state of that agent.

## 6    Sub-task Synchronization

**Importance of Task Synchronization.** Consider the following example specification:

$$\phi'_{1a} = \text{reach}_{lo}(P) \text{ or } \text{reach}_{gl}(Q)$$

where $P, Q$ are some goals. To ensure flexibility with respect to the possible acceptable rollouts within $\phi'_{1a}$, the individual agent policies $\pi^i$ are learnable and the task transition chosen is dependent on the agent-specific observations. This flexibility between agents however, adds an additional possible failure method in achieving a global specification - if even a single agent attempts to fulfill the global objective while the others decide to follow their local objectives, the specification would never be satisfied.

**Identifying Synchronization States.** As emphasized above, task synchronization is an important aspect of deploying these composite task monitors in the Markov game $\mathcal{M}_g$ with specification $\phi$. We show the existence of a subset of monitor states $\texttt{Sync} \in Q$ where in order to maintain task synchronization, agents simply require a consensus on which monitor transition $\delta = (q, \sigma, u, q')$ to take. If we use $Q_g$ to symbolize the set of global monitor states, viz. all $q \in Q$ such that $\exists (q, \sigma_g, u_g, q') \in \Delta_g$, then we see that $Q_g \subseteq \texttt{Sync}$. A valid choice for $q \in \texttt{Sync}$ with $q \notin Q_g$ is all branching states in the graph of $M_\phi$ with a set refinement presented in the Appendix (Sec. D).

During training, we enforce the condition that when an agent $i$ has monitor state $q_t^i \in \texttt{Sync}$, it must wait for time $t_1 > t$ such that $q_{t_1}^j = q_t^i \ \forall j \in \mathcal{N}$ and then choose a common transition as the other agents. This is done during the centralized training phase by sharing the same transition between agents based on a majority vote.

# 7    Multi-agent Specification Properties

Consider a specification $\phi$ and let $\mathcal{N} = \{1, \ldots, N\}$ be the set of all agents with $\zeta_m$ being a trajectory sampled from the environment. $\phi(\zeta_m, n)$ is used to denote that the specification is satisfied on $\zeta_m$ for the set of agents $n \subseteq \mathcal{N}$ (i.e. $[\![\phi(\zeta_m, n)]\!] == \texttt{True}$).

**MA-Distributive.** Many specifications pertaining to MA problems can be satisfied independent of the number of agents. At its core, we have the condition that a specification being satisfied with respect to a union of two disjoint sets of agents implies that it can be satisfied on both sets independently. Namely if $n_1, n_2 \subset \mathcal{N}$ with $n_1 \cap n_2 = \emptyset$ then an MA-Distributive specification satisifies the following condition:

$$\phi(\zeta_m, n_1 \cup n_2) \implies \phi(\zeta_m, n_1) \wedge \phi(\zeta_m, n_2)$$

**MA-Decomposable.** Certain specifications satisfy a decomposibility property particular to multi-agent problems that can help in scaling with respect to the number of agents.

Say $\exists\, k \in \{1, \ldots, N-1\}$ such that

$$\phi(\zeta_m, \mathcal{N}) \implies \phi_k(\zeta_m, \mathcal{N}) = \bigwedge_{j \in \{1, \ldots, J\}} \phi(\zeta_m, n_j)$$

where

$$n_j \subset \mathcal{N} \,,\, k \le |n_j| < N \,,\, J = \lfloor \frac{N}{k} \rfloor \,,\, \bigcap_j n_j = \emptyset \,,\, \bigcup_j n_j = \mathcal{N}$$

with $\lfloor\rfloor$ representing the floor function. Each $n_j$ is a set of at least $k$ unique agents and $\{n_j\}_j$ forms a partition over $\mathcal{N}$.

We then call the specification $\phi$ *MA-Decomposable* with decomposibility factor $k$. Here $\phi_k$ can be thought of as a means to approximate the specification $\phi$ to smaller groups of agents within the set of agents $\mathcal{N}$. Provided we find a value of $k$, we can then use this as the basis of our MA-Dec scaling method to significantly improve training times for larger numbers of agents.

**Theorem 3.** *(Proof in Appendix Sec. F.)  All MA-Distributive specifications are also MA-Decomposable with decomposability factors $k \in \mathbb{Z}^+, 1 \le k < N$.*

Notably all compositions of $\text{reach}_{gl}$ and $\text{reach}_{lo}$ within our language are MA-Distributive and are thus MA-Decomposable with factor $k = 2$. [2] This is far from a general property however, as one can define specifications on $N$ robots such as $\text{achieve}(\text{"collect } x \text{ fruits"})$ where each robot can carry at most $x/N$ fruits . In this case, no single subset of agents can satisfy the specification as the total capacity of fruits would be less than $x$ and the specification is neither MA-Distributive nor MA-Decomposable.

## 8   Algorithm

**Training.** Agents learn $\pi^i(s^i, v^i, q^i) = (a^i, \delta^i)$ on the augmented Markov game $\mathcal{M}'_g$ where $s^i, v^i, q^i$ are agent-specific state, register value and task monitor state respectively. Since training is centralized, all agent task monitors receive the same global state. Based on our discussion in Sect. 6, if an agent is in any given global monitor state, we wait for other agents to enter the same state, then do the $\arg\max$ task transition for all agents in the same state. In addition, at the synchronization states (Sect. 6), we perform a similar process to select the task transition. These trained *augmented policies* are then projected into policies that can act in the original $\mathcal{M}_g$.

---

[2]  It is satisfied with $k = 1$ as well but this is the trivial case where $\text{reach}_{gl}$ and $\text{reach}_{lo}$ are equivalent.

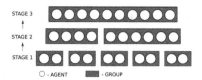

**Fig. 3.** Example MA-Dec Scaling Process with $N = 10, k = 2, f = 2$ on an MA-Decomposable spec $\phi$ with decomposability factor 2. At Stage 1, $g_1 = 2$ to start with 5 groups. Next $g_2 = fg_1 = 4$ which forms 2 groups. Finally at Stage 3, $g_3 = fg_2 = 8$ which forms one group ($\mathcal{N}$).

**Scaling MA-Decomposable Specifications.** Our algorithm for scaling based off the MA-Decomposable property is shown in Algorithm 1 (refer Appendix) and we name it MA-Dec scaling. Essentially, we approximate the spec. $\phi$ by first independently considering smaller groups within the larger set of agents $\mathcal{N}$ and try to obtain a policy satisfying $\phi$ on these smaller groups. By progressively making the group sizes larger over stages and repeating the policy training process while continuing from the previous training stage's policy parameters , we form a curriculum that eases solving the original problem $\phi$ on all agents $\mathcal{N}$.

In Fig. 3 we demonstrate MA-Dec scaling for $N = 10$ agents on a spec. $\phi$ which is MA-Decomposable with decomposability factor 2. For this example we set the scaling parameters $k = 2$ and $f = 2$. Initially we have a min. group size $g_1 = 2$ and this is changed to $g_2 = 4$ and $g_3 = 8$ from setting the scaling factor. We increment the stage number every time all the groups of a stage have satisfied the entire specification $\phi$ w.r.t. their group. While separating training into stages, agents must be encouraged to move from stage $i$ to stage $i + 1$. To ensure this, we need to scale rewards based on the stage. We chose a simple linear scaling where for stage number $i$ and time step $t$, each agent receives reward $r_{i,t} = ic_k + CTM_{i,t}$ where $CTM_{i,t}$ is the original composite task monitor reward at stage $i$ and $c_k \in \mathbb{R}$ is a constant. By bounding the reward terms such that rewards across stages are monotonically increasing ($r_{i,t} < r_{i+1,t'}$) we can find a suitable $c_k$ to be $(2D+1)C_u$ (refer Appendix Sec. B) where the terms are the same as in Eq. 1.

From setting the initial min. group size $g_1$ and scaling factor $f$, we get the total number of learning stages ($T_s$) as $T_s = \lfloor \log_f(N) - \log_f(k) \rfloor = \mathcal{O}(\log_f(N))$. We build the intuition behind why MA-Dec scaling is effective in the Appendix (Sec. B), by describing it as a form of curriculum learning.

**Deployment.** Policies are constructed to proceed with only local information $(s^i, v^i, q^i)$. Since we cannot share the whole system state with the agent policies during deployment yet our composite task monitor requires access to this state at all times, we allow the following relaxations: 1) Global predicates $\sigma_g(\bar{s}, v)$ enabling task monitor transitions need global state and access it during deployment. 2) Global register updates $u_g(\bar{s}, v)$ are also a function of global state and access it during deployment.

In order to maintain task synchronization, agents use a consensus based communication method to decide task monitor transitions at global and synchronization states. If agents choose different task transitions at these monitor states, the majority vote is used as done during training.

## 9   Experiment Setup

Our experiments aim to  validate that the use of a distributed task monitor can achieve synchronization during the deployment of multiple agents on a range of specifications.

In addition, to emphasize the need for distribution of task monitors to alleviate the state space explosion caused by mixing local and global specifications, we include experiments with SPECTRL applied to a centralized controller.

Lastly, we provide results showing the efficacy of the MA-Dec scaling approach for larger numbers of agents when presented with a specification that satisfies the MA-Decomposability property (Sect. 7).

As a baseline comparison, we also choose to run our algorithm without giving policies access to the monitor state (**no_mon**). These are trained with the same shaped reward as DistSPECTRL. We also provide a Reward Machine baseline (**RM**) for $\phi_1$ with continuous rewards since $\phi_1$ is similar to the 'Rendezvous' specification in [13].

**Environment.** Our first set of experiments are done on a 2D Navigation problem with $N = 3$ agents. The observations ($\mathcal{S} \in \mathbb{R}^2$) used are coordinates within the space with the action space ($\mathcal{A} \in \mathbb{R}^2$) providing the velocity of the agent.

The second set of experiments towards higher dimension 3D benchmarks, represent particle motion in a 3D space. We train multiple agents ($N = 3$) in the 3D space ($\mathcal{S} \in \mathbb{R}^3$) with a 3D action space ($\mathcal{A} \in \mathbb{R}^3$) to show the scaling potential of our framework.

The final set of experiments were on a modern discrete-action MARL benchmark built in Starcraft 2 [15] with $N = 8$ agents (the "8m" map). Each agent has 14 discrete actions with a state space $\mathcal{S} \in \mathbb{R}^{80}$ representing a partial view of allies and enemies.

**Algorithm Choices.** For the scaling experiments (Fig. 6) we used the 2D Navigation problem with horizon $T = 500$ and the scaling parameters[3] $k = 2$ and $f = 2$. We also choose a version of PPO with a centralized Critic to train the augmented Markov Game noting that our framework is agnostic to the choice of training algorithm. The current stage is passed to the agents as an extra integer dimension. For other experiments we chose PPO with independent critics as our learning algorithm. Experiments were implemented using the RLLib toolkit [9].

---

[3] While we could start with $k = 1$, we set $k = 2$ to reduce the number of learning stages.

**Specifications (2D Navigation).** The evaluated specifications are a mix of local and global objectives. The reach predicates have an error tolerance of 1 (the $L_\infty$ distance from the goal).

(i) $\phi_1 = \texttt{reach}_{\texttt{gl}}(5,0); \texttt{reach}_{\texttt{gl}}(0,0)$ , (ii) $\phi_2 = \phi_1; \texttt{reach}_{\texttt{gl}}(3,0)$

(iii) $\phi_3 = \texttt{reach}_{\texttt{lo}}(5,0); \texttt{reach}_{\texttt{gl}}(0,0); \texttt{reach}_{\texttt{gl}}(3,0)$

(iv) $\phi_4 = [\texttt{reach}_{\texttt{lo}}(3,0)$ or $\texttt{reach}_{\texttt{lo}}(5,10)]; \phi_3$

**(SC2)** $\phi_{sc}$ represents 'kiting' behaviour and is explained further in the Appendix (Sec. E). $\phi_{sc} = \phi_{sc_a}; \phi_{sc_a}; \phi_{sc_a}$ where $\phi_{sc_a} = \texttt{away\_from\_enemy}_{gl}$; $\texttt{shooting\_range}_{lo}$;

**(3D Environment )** $\phi_a = \texttt{reach}_{\texttt{lo}}(5,0,0); \texttt{reach}_{\texttt{gl}}(0,0,0); \texttt{reach}_{\texttt{gl}}(3,0,0)$ is the specification considered within X-Y-Z coordinates.

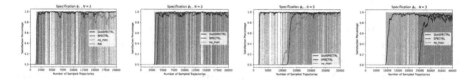

**Fig. 4.** Satisfaction percentages on specifications $\phi_1, \phi_2, \phi_3$ and $\phi_4$ with $N = 3$ agents. The shaded regions show the maximum and minimum achieved over 5 separate evaluation runs

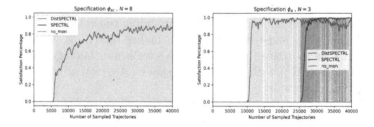

**Fig. 5.** Specification satisfaction percentages (left) for the StarCraft 2 specification $\phi_{sc}$ with $N = 8$ agents and (right) for the 3D Navigation experiments on specification $\phi_a$.

## 10    Results

**Handling Expressive Specifications.** The experiments in Fig. 4 demonstrate execution when the task monitor predicates have access to the entire system state. This provides agents with information sufficient to calculate global predicates for task monitor transitions. The overall satisfaction percentage is reported

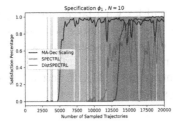 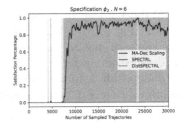

**Fig. 6.** Specification satisfaction percentages for $N = 10$ agents on $\phi_1$ (left) and $N = 6$ agents on $\phi_3$ (right) comparing the MA-Dec scaling (red) to centralized SPECTRL (blue) and vanilla DistSPECTRL(green) i.e. without scaling enhancements. (Color figure online)

**Table 1.** Specification satisfaction percentages on convergence for Fig. 4,5

| Spec | DistSPECTRL | no_mon | SPECTRL |
|---|---|---|---|
| $\phi_1$ | 99.62 | 91.17 | 100.00 |
| $\phi_2$ | 99.05 | 00.00 | 97.38 |
| $\phi_3$ | 97.59 | 94.77 | 96.81 |
| $\phi_4$ | 97.31 | 00.00 | 90.78 |
| $\phi_a$ | 98.49 | 00.00 | 99.60 |
| $\phi_{sc}$ | 86.79 | 00.00 | 00.00 |

**Table 2.** Specification satisfaction percentages on convergence for Fig. 6, (Scaling to more Agents)

| Spec./# Agents | MA-Dec | DistSPECTRL | SPECTRL |
|---|---|---|---|
| $\phi_3/N =6$ | 94.09 | 0.00 | 0.00 |
| $\phi_1/N =6$ | 97.83 | 80.67 | 98.96 |
| $\phi_1/N =10$ | 97.03 | 72.30 | 99.28 |

with the value 0 being an incomplete task to 1.0 being the entire specification satisfied.

While SPECTRL has often been shown to be more effective [5,6] than many existing methods (e.g. **RM** case) for task specification, the further utility of the monitor state in enhancing coordination between agents is clearly evident in a distributed setting. The task monitor state is essential for coordination as our baseline **no_mon** is often unable to complete the entire task (even by exhaustively going through possible transitions) and global task completion requires enhanced levels of synchronization between agents.

From Table 1 we see that upon convergence of the learning algorithm, the agent is able to maintain a nearly 100% task completion rate for our tested specifications, a significant improvement in comparison to the **no_mon** case, showing the importance of the task monitor as part of a multi-agent policy.

**Benefits of Distribution over Centralization.** The centralized SPECTRL graphs (blue curves in Figs. 4, 5, 6) show that while distribution may not be necessary for certain specifications with few local portions (e.g. $\phi_2$), concatenating them will quickly lead to learning difficulties with larger number of agents (Fig. 5, $\phi_{sc}$ and Fig. 6, $\phi_3$). This difficulty is due in large part to state space explosion of the task monitor in these cases as is apparent by the significantly better performance of our distributed algorithm. We also remind the reader that a centralized algorithm is further disadvantageous in MARL settings due to the added synchronization cost between agents during deployment.

**Scaling to Larger State Spaces.** The results in Fig. 5 show promise that the DistSPECTRLframework can be scaled up to larger dimension tasks as well. The 3D environment results exhibits similar behavior to the 2D case with the **no_mon** showing difficulty in progressing beyond the local tasks in the larger state space with sparser predicates. The $\phi_{sc}$ results also show promise in defining relevant predicates and achieving general specifications for modern MARL benchmarks.

**Scaling to More Agents.** Table 2 and Fig. 6 demonstrate the benefits of MA-Dec scaling for larger $N$ when presented with an MA-Decomposable specification. At smaller ranges of $N$ as well as less complex combinations of mixed and global objectives, the effect of MA-Dec scaling is less pronounced. We observe that the stage based learning is crucial for a even simple mixed specification like $\phi_3$ with as little as $N = 6$ agents.

## 11   Related Work

Multi-agent imitation learning [7,17,23] uses demonstrations of a task to specify desired behavior. However in many cases, directly being able to encode a specification by means of our framework is more straightforward and removes the need to have demonstrations beforehand. Given demonstrations, one may be able to infer the specification [21] and make refinements or compositions for use in our framework.

TLTL [8] is another scheme to incorporate temporal logic constraints into learning enabled controllers, although its insufficiency in handling non-Markovian specifications led us to choose SPECTRL as the basis for our methodology. Reward Machines (RMs) [2,18,19] are an automaton-based framework to encode different tasks into an MDP. While RMs can handle many non-Markovian reward structures, a major difference is that SPECTRL starts with a logical temporal logic specification and includes with the automaton the presence of memory (in the form of registers capable of storing real-valued information). Recent work [6] shows the relative advantages SPECTRL -based solutions may have over a range of continuous benchmarks.

Concurrent work has introduced the benefits of a temporal logic based approach to reward specification [4]. While experimental results are not yet displayed, the convergence guarantees of the given algorithm are promising. Since we use complex non-linear function approximators (neural networks) in our work, such guarantees are harder to provide. Reward Machines have also been explored as a means of specifying behavior in multi-agent systems [13] albeit in discrete state-action systems that lend themselves to applying tabular RL methods such as Q-learning. One may extend this framework to continuous systems by means of function approximation but to the best of our knowledge, this has not been attempted yet. Similar to our synchronization state, the authors use a defined local event set to sync tasks between multiple agents and requires being aware of shared events visible to the other agents.

In the same spirit as our stage-based approach, transferring learning from smaller groups of agents to larger ones has also been explored [22]. Lastly, while we chose PPO to train the individual agents for its simplicity, our framework is agnostic to the RL algorithm used and can be made to work with other modern multi-agent RL setups [3,11] for greater coordination capabilities.

## 12   Conclusion

We have introduced a new specification language to help detail MARL tasks and describe how it can be used to compile a desired description of a distributed execution in order to achieve specified objectives. Our framework makes task synchronization realizable among agents through the use of: 1) Global predicates providing checks for task completion that are easily computed, well-defined and tractable; 2) A monitor state to keep track of task completion; and 3) Synchronization states to prevent objectives from diverging among agents.

**Acknowledgements.** This work was supported in part by C-BRIC, one of six centers in JUMP, a Semiconductor Research Corporation (SRC) program sponsored by DARPA.

## References

1. Andrychowicz, M., et al.: Hindsight experience replay. In: NIPS (2017)
2. Camacho, A., Toro Icarte, R., Klassen, T.Q., Valenzano, R., McIlraith, S.A.: LTL and beyond: formal languages for reward function specification in RL. In: IJCAI (2019)
3. Foerster, J., Farquhar, G., Afouras, T., Nardelli, N., Whiteson, S.: Counterfactual multi-agent policy gradients. In: AAAI (2018)
4. Hammond, L., Abate, A., Gutierrez, J., Wooldridge, M.: Multi-agent reinforcement learning with temporal logic specifications. In: AAMAS (2021)
5. Jothimurugan, K., Alur, R., Bastani, O.: A composable specification language for reinforcement learning tasks. In: NeurIPS (2019)
6. Jothimurugan, K., Bansal, S., Bastani, O., Alur, R.: Compositional reinforcement learning from logical specifications. In: NeurIPS (2021)
7. Le, H.M., Yue, Y., Carr, P., Lucey, P.: Coordinated MA imitation learning. In: ICML (2017)
8. Li, X., Vasile, C., Belta, C.: RL with temporal logic rewards. In: IROS (2017)
9. Liang, E., et al.: RLlib: abstractions for distributed reinforcement learning. In: ICML (2018)

10. Lillicrap, T.P., et al.: Continuous control with deep reinforcement learning. In: ICLR (2016)
11. Lowe, R., Wu, Y., Tamar, A., Harb, J., Abbeel, P., Mordatch, I.: Multi-agent actor-critic for mixed cooperative-competitive environments. In: NIPS (2017)
12. Mania, H., Guy, A., Recht, B.: Simple random search of static linear policies is competitive for reinforcement learning. In: NeurIPS (2018)
13. Neary, C., Xu, Z., Wu, B., Topcu, U.: Reward machines for cooperative multi-agent reinforcement learning. In: AAMAS (2021)
14. Ng, A.Y., Harada, D., Russell, S.: Policy invariance under reward transformations: theory and application to reward shaping. In: ICML (1999)
15. Samvelyan, M., et al.: Starcraft challenge. In: AAMAS (2019)
16. Schulman, J., Wolski, F., Dhariwal, P., Radford, A., Klimov, O.: Proximal policy optimization algorithms. CoRR (2017)
17. Song, J., Ren, H., Sadigh, D., Ermon, S.: Multi-agent generative adversarial imitation learning. In: NeurIPS (2018)
18. Toro Icarte, R., Klassen, T.Q., Valenzano, R., McIlraith, S.A.: Using reward machines for high-level task specification and decomposition in reinforcement learning. In: ICML (2018)
19. Toro Icarte, R., Klassen, T.Q., Valenzano, R., McIlraith, S.A.: Reward machines: exploiting reward function structure in reinforcement learning. arXiv preprint arXiv:2010.03950 (2020)
20. Vardi, M., Wolper, P.: Reasoning about infinite computations. Inf. Comp. **115**, 1–37 (1994)
21. Vazquez-Chanlatte, M., Jha, S., Tiwari, A., Ho, M.K., Seshia, S.: Learning task specifications from demonstrations. In: NeurIPS (2018)
22. Wang, W., et al.: From few to more: large-scale dynamic multiagent curriculum learning. In: AAAI, no. 05 (2020)
23. Yu, L., Song, J., Ermon, S.: Multi-agent adversarial inverse RL. In: ICML (2019)

# MAVIPER: Learning Decision Tree Policies for Interpretable Multi-agent Reinforcement Learning

Stephanie Milani[1(✉)], Zhicheng Zhang[1,2], Nicholay Topin[1],
Zheyuan Ryan Shi[1], Charles Kamhoua[3], Evangelos E. Papalexakis[4],
and Fei Fang[1]

[1] Carnegie Mellon University, Pittsburgh, USA
smilani@andrew.cmu.edu
[2] Shanghai Jiao Tong University, Shanghai, China
[3] Army Research Lab, Adelphi, USA
[4] University of California, Riverside, USA

**Abstract.** Many recent breakthroughs in multi-agent reinforcement learning (MARL) require the use of deep neural networks, which are challenging for human experts to interpret and understand. On the other hand, existing work on interpretable reinforcement learning (RL) has shown promise in extracting more interpretable decision tree-based policies from neural networks, but only in the single-agent setting. To fill this gap, we propose the first set of algorithms that extract interpretable decision-tree policies from neural networks trained with MARL. The first algorithm, IVIPER, extends VIPER, a recent method for single-agent interpretable RL, to the multi-agent setting. We demonstrate that IVIPER learns high-quality decision-tree policies for each agent. To better capture coordination between agents, we propose a novel centralized decision-tree training algorithm, MAVIPER. MAVIPER jointly grows the trees of each agent by predicting the behavior of the other agents using their anticipated trees, and uses resampling to focus on states that are critical for its interactions with other agents. We show that both algorithms generally outperform the baselines and that MAVIPER-trained agents achieve better-coordinated performance than IVIPER-trained agents on three different multi-agent particle-world environments.

**Keywords:** Interpretability · Explainability · Multi-agent reinforcement learning

## 1 Introduction

Multi-agent reinforcement learning (MARL) is a promising technique for solving challenging problems, such as air traffic control [5], train scheduling [27], cyber

---

S. Milani and Z. Zhang—Equal contribution.

---

**Supplementary Information** The online version contains supplementary material available at https://doi.org/10.1007/978-3-031-26412-2_16.

M.-R. Amini et al. (Eds.): ECML PKDD 2022, LNAI 13716, pp. 251–266, 2023.
https://doi.org/10.1007/978-3-031-26412-2_16

defense [22], and autonomous driving [4]. In many of these scenarios, we want to train a *team* of cooperating agents. Other settings, like cyber defense, involve an adversary or set of adversaries with goals that may be at odds with the team of defenders. To obtain high-performing agents, most of the recent breakthroughs in MARL rely on neural networks (NNs) [10,35], which have thousands to millions of parameters and are challenging for a person to interpret and verify. Real-world risks necessitate learning *interpretable* policies that people can inspect and verify before deployment, while still performing well at the specified task and being robust to a variety of attackers (if applicable).

Decision trees [34] (DTs) are generally considered to be an *intrinsically* interpretable model family [28]: sufficiently small trees can be contemplated by a person at once (simulatability), have subparts that can be intuitively explained (decomposability), and are verifiable (algorithmic transparency) [18]. In the RL setting, DT-like models have been successfully used to model transition functions [40], reward functions [8], value functions [33,43], and policies [24]. Although learning DT policies for interpretability has been investigated in the single-agent RL setting [24,32,37], it has yet to be explored in the multi-agent setting.

To address this gap, we propose two algorithms, IVIPER and MAVIPER, which combine ideas from model compression and imitation learning to learn DT policies in the multi-agent setting. Both algorithms extend VIPER [2], which extracts DT policies for single-agent RL. IVIPER and MAVIPER work with most existing NN-based MARL algorithms: the policies generated by these algorithms serve as "expert policies" and guide the training of a set of DT policies.

The main contributions of this work are as follows. First, we introduce the IVIPER algorithm as a novel extension of the single-agent VIPER algorithm to multi-agent settings. Indeed, IVIPER trains DT policies that achieve high individual performance in the multi-agent setting. Second, to better capture coordination between agents, we propose a novel centralized DT training algorithm, MAVIPER. MAVIPER jointly grows the trees of each agent by predicting the behavior of the other agents using their anticipated trees. To train each agent's policy, MAVIPER uses a novel resampling scheme to find states that are considered critical for its interactions with other agents. We show that MAVIPER-trained agents achieve better coordinated performance than IVIPER-trained agents on three different multi-agent particle-world environments.

## 2    Background and Preliminaries

We focus on the problem of learning interpretable DT policies in the multi-agent setting. We first describe the formalism of our multi-agent setting, then discuss DT policies and review the single-agent version of VIPER.

### 2.1    Markov Games and MARL Algorithms

In MARL, agents act in an environment defined by a Markov game [19,38]. A Markov game for $N$ agents consists of a set of states $\mathcal{S}$ describing all possible configurations for all agents, the initial state distribution $\rho : \mathcal{S} \to [0,1]$, and the set of actions $\mathcal{A}_1, ..., \mathcal{A}_N$ and observations $\mathcal{O}_1, ..., \mathcal{O}_N$ for each agent $i \in [N]$. Each agent aims to maximize its own total expected return $R_i = \sum_{t=0}^{\infty} \gamma^t r_i^t$, where $\gamma$ is the discount factor that weights the relative importance of future rewards. To do so, each agent selects actions using a policy $\pi_{\theta_i} : \mathcal{O}_i \to \mathcal{A}_i$. After the agents simultaneously execute their actions $\overrightarrow{a}$ in the environment, the environment produces the next state according to the state transition function $P : \mathcal{S} \times \mathcal{A}_1 \times ... \times \mathcal{A}_N \to \mathcal{S}$. Each agent $i$ receives reward according to a reward function $r_i : \mathcal{S} \times \mathcal{A}_i \to \mathbb{R}$ and a private observation, consisting of a vector of *features*, correlated with the state $o_i : \mathcal{S} \to \mathcal{O}_i$.

Given a policy profile $\pi = (\pi_1, ..., \pi_N)$, agent $i$'s value function is defined as: $V_i^\pi(s) = r_i + \gamma \sum_{s' \in \mathcal{S}} P(s, \pi_1(o_1), ..., \pi_N(o_N), s') V_i^\pi(s')$ and state-action value function is: $Q_i^\pi(s, a_1, ..., a_N) = r_i + \gamma \sum_{s' \in \mathcal{S}} P(s, a_1, ..., a_N, s') V_i^\pi(s')$. We refer to a policy profile excluding agent $i$ as $\pi_{-i}$.

MARL algorithms fall into two categories: value-based [35,39,41] and actor-critic [11,16,20,48]. Value-based methods often approximate $Q$-functions for individual agents in the form of $Q_i^\pi(o_i, a_i)$ and derive the policies $\pi_i$ by taking actions with the maximum Q-values. In contrast, actor-critic methods often follow the centralized training and decentralized execution (CTDE) paradigm [30]. They train agents in a centralized manner, enabling agents to leverage informa-

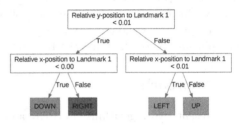

**Fig. 1.** A decision tree of depth two that MAVIPER learns in the Cooperative Navigation environment. The learned decision tree captures the expert's behavior of going to one of the landmarks.

tion beyond their private observation during training; however, agents must behave in a decentralized manner during execution. Each agent $i$ uses a centralized critic network $Q_i^\pi$, which takes as input some state information $x$ (including the observations of all agents) and the actions of all agents. This assumption addresses the stationarity issue in MARL training: without access to the actions of other agents, the environment appears non-stationary from the perspective of any one agent. Each agent $i$ also has a policy network $\pi_i$ that takes as input its observation $o_i$.

## 2.2   Decision Tree Policies

DTs are tree-like models that recursively partition the input space along a specific feature using a cutoff value. These models produce axis-parallel partitions: internal nodes are the intermediate partitions, and leaf nodes are the final partitions. When used to represent policies, the internal nodes represent the features and values of the input state that the agent uses to choose its action, and the leaf nodes correspond to chosen actions given some input state. For an example of a DT policy, see Fig. 1.

## 2.3  VIPER

VIPER [2] is a popular algorithm [7,21,25] that extracts DT policies for a finite-horizon Markov decision process given an *expert* policy trained using any single-agent RL algorithm. It combines ideas from model compression [6,13] and imitation learning [1]—specifically, a variation of the DAGGER algorithm [36]. It uses a high-performing deep NN that approximates the state-action value function to guide the training of a DT policy.

VIPER trains a DT policy $\hat{\pi}^m$ in each iteration $m$; the final output is the best policy among all iterations. More concretely, in iteration $m$, it samples $K$ trajectories $\{(s, \hat{\pi}^{m-1}(s)) \sim d^{\hat{\pi}^{m-1}}\}$ following the DT policy trained at the previous iteration. Then, it uses the expert policy $\pi^*$ to suggest actions for each visited state, leading to the dataset $D^m = \{(s, \pi^*(s)) \sim d^{\hat{\pi}^{m-1}}\}$ (Line 4, Alg. 3). VIPER adds these relabeled experiences to a dataset $\mathcal{D}$ consisting of experiences from previous iterations. Let $V^{\pi^*}$ and $Q^{\pi^*}$ be the state value function and state-action value function given the expert policy $\pi^*$. VIPER resamples points $(s, a) \in \mathcal{D}$ according to weights: $\tilde{l}(s) = V^{\pi^*}(s) - \min_{a \in A} Q^{\pi^*}(s, a)$. See Algorithm 3 in Appendix A for the full VIPER algorithm.

# 3  Approach

We present two algorithms: IVIPER and MAVIPER. Both are general policy extraction algorithms for the multi-agent setting inspired by the single-agent VIPER algorithm. At a high level, given an expert policy profile $\pi^* = (\pi_1^*, ... \pi_N^*)$ with associated state-action value functions $Q^{\pi^*} = (Q_1^{\pi^*}, ..., Q_N^{\pi^*})$ trained by an existing MARL algorithm, both algorithms produce a DT policy $\hat{\pi}_i$ for each agent $i$. These algorithms work with various state-of-art MARL algorithms, including value-based and multi-agent actor-critic methods. We first discuss IVIPER, the basic version of our multi-agent DT learning algorithm. We then introduce additional changes that form the full MAVIPER algorithm.

## 3.1  IVIPER

Motivated by the practical success of single-agent RL algorithms in the MARL setting [3,23], we extend single-agent VIPER to the multi-agent setting by independently applying the single-agent algorithm to each agent, with a few critical changes described below. Algorithm 1 shows the full IVIPER pseudocode.

First, we ensure that each agent has sufficient information for training its DT policy. Each agent has its own dataset $\mathcal{D}_i$ of training tuples. When using VIPER with multi-agent actor-critic methods that leverage a per-agent centralized critic network $Q_i^\pi$, we ensure that each agent's dataset $\mathcal{D}_i$ has not only its observation and actions, but also the complete state information $x$—which consists of the observations of all of the agents—and the expert-labeled actions of all of the other agents $\pi_j^*(o_j) \forall j \neq i$. By providing each agent with the information about all other agents, we avoid the stationarity issue that arises when the policies of all agents are changing throughout the training process (like in MARL).

---

**Algorithm 1.** IVIPER in Multi-Agent Setting

---

**Input:** $(X, A, P, R)$, $\pi^*$, $Q^{\pi^*} = (Q_1^{\pi^*}, ..., Q_N^{\pi^*})$, $K$, $M$
**Output:** $\hat{\pi}_1, ..., \hat{\pi}_N$

1: **for** i=1 to N **do**
2:     Initialize dataset $\mathcal{D}_i \leftarrow \emptyset$ and policy $\hat{\pi}_i^0 \leftarrow \pi_i^*$
3:     **for** $m = 1$ to $M$ **do**
4:         Sample $K$ trajectories: $\mathcal{D}_i^m \leftarrow \{(x, \pi_1^*(o_1), ..., \pi_N^*(o_N)) \sim d^{\hat{\pi}_i^{m-1}, \pi_{-i}^*}\}$
5:         Aggregate dataset $\mathcal{D}_i \leftarrow \mathcal{D}_i \cup \mathcal{D}_i^m$
6:         Resample dataset according to loss:
            $\mathcal{D}_i' \leftarrow \{(x, \overrightarrow{a}) \sim p((x, \overrightarrow{a})) \propto \tilde{l}_i(x) \mathbb{I}[(x, \overrightarrow{a}) \in \mathcal{D}_i]\}$
7:         Train decision tree $\hat{\pi}_i^m \leftarrow \text{TrainDecisionTree}(\mathcal{D}_i')$
8:     Get best policy $\hat{\pi}_i \leftarrow \text{BestPolicy}(\hat{\pi}_i^1, ..., \hat{\pi}_i^M, \pi_{-i}^*)$
9: **return** Best policies for each agent $\hat{\pi} = (\hat{\pi}_1, ..., \hat{\pi}_N)$

---

Second, we account for important changes that emerge from moving to a multi-agent formalism. When we sample and relabel trajectories for training each agent's DT policy, we sample from the distribution $d^{\hat{\pi}_i^{m-1}, \pi_{-i}^*}$ induced by agent $i$'s policy at the previous iteration $\hat{\pi}_i^{m-1}$ and the expert policies of all other agents $\pi_{-i}^*$. We only relabel the action for agent $i$ because the other agents choose their actions according to $\pi^*$. It is equivalent to treating all other expert agents as part of the environment and only using DT policy for agent $i$.

Third, we incorporate the actions of all agents when resampling the dataset to construct a new, weighted dataset (Line 6, Algorithm 1). If the MARL algorithm uses a centralized critic $Q(s, \overrightarrow{a})$, we resample points according to:

$$p((x, a_1, ..., a_N)) \propto \tilde{l}_i(x) \mathbb{I}[(x, a_1, ..., a_N) \in \mathcal{D}_i], \tag{1}$$

where,

$$\tilde{l}_i(x) = V_i^{\pi^*}(x) - \min_{a_i \in \mathcal{A}_i} Q_i^{\pi^*}(x, a_i, \overrightarrow{a}_{-i})|_{\overrightarrow{a}_{-i} = \pi_j^*(o_j) \forall j \neq i}. \tag{2}$$

Crucially, we include the actions of all other agents in Eq. (2) to select agent $i$'s minimum Q-value from its centralized state-action value function.

When applied to value-based methods, IVIPER is more similar to single-agent VIPER. In particular, in Line 4, Algorithm 1, it is sufficient to only store $o_i$ and $\pi_i^*(o_i)$ in the dataset $\mathcal{D}_i^m$, although we still must sample trajectories according to $\hat{\pi}_i^{m-1}$ and $\pi_{-i}^*$. In Line 6, we use $\tilde{l}(x) = V_i^{\pi^*}(s) - \min_{a_i \in \mathcal{A}_i} Q_i^{\pi^*}(o_i, a_i)$ from single-agent VIPER, removing the reliance of the loss on a centralized critic.

Taken together, these algorithmic changes form the basis of the IVIPER algorithm. This algorithm can be viewed as transforming the multi-agent learning problem to a single-agent one, in which other agents are folded into the environment. This approach works well if i) we only want an interpretable policy for a single agent in a multi-agent setting or ii) agents do not need to *coordinate* with each other. When coordination is needed, this algorithm does not reliably capture coordinated behaviors, as each DT is trained independently without consideration for what the other agent's resulting DT policy will learn. This issue is particularly apparent when trees are constrained to have a small maximum depth, as is desired for interpretability.

## 3.2  MAVIPER

---

**Algorithm 2.** MAVIPER (Joint Training)

---

**Input:** $(\mathcal{X}, A, P, R)$, $\pi^*$, $Q^{\pi^*} = (Q_1^{\pi^*}, \ldots, Q_N^{\pi^*})$, $K$, $M$
**Output:** $(\hat{\pi}_1, \ldots, \hat{\pi}_N)$

1: Initialize dataset $\mathcal{D} \leftarrow \emptyset$ and policy for each agent $\hat{\pi}_i^0 \leftarrow \pi_i^* \ \forall i \in N$
2: **for** $m = 1$ to $M$ **do**
3:     Sample $K$ trajectories: $\mathcal{D}^m \leftarrow \{(x, \pi_1^*(o_1), \ldots, \pi_N^*(o_N)) \sim d^{(\hat{\pi}_1^{m-1}, \ldots, \hat{\pi}_N^{m-1})}\}$
4:     Aggregate dataset $\mathcal{D} \leftarrow \mathcal{D} \cup \mathcal{D}^m$
5:     For each agent $i$, resample $\mathcal{D}_i$ according to loss:
          $\mathcal{D}_i \leftarrow \{(x, a) \sim p((x, a)) \propto \tilde{l}_i(x)\mathbb{I}[(x, a) \in \mathcal{D}]\}\forall i \in N$
6:     Jointly train DTs: $(\hat{\pi}_1^m, \ldots, \hat{\pi}_N^m) \leftarrow$ TrainJointTrees$(\mathcal{D}_1, \ldots, \mathcal{D}_N)$
7: **return** Best set of agents $\hat{\pi} = (\hat{\pi}_1, \ldots, \hat{\pi}_N) \in \{(\hat{\pi}_1^1, \ldots, \hat{\pi}_N^1), \ldots, (\hat{\pi}_1^M, \ldots, \hat{\pi}_N^M)\}$

8: **function** TRAINJOINTTREES$(\mathcal{D}_1, \ldots, \mathcal{D}_N)$
9:     Initialize decision trees $\hat{\pi}_1^m, \ldots, \hat{\pi}_N^m$.
10:    **repeat**
11:        Grow one more level for agent $i$'s tree $\hat{\pi}_i^m \leftarrow$ Build$(\hat{\pi}_1^m, \ldots, \hat{\pi}_N^m, \mathcal{D}_i)$
12:        Move to the next agent: $i \leftarrow (i+1)\%N$
13:    **until** all trees have grown to the maximum depth allowed
14:    **return** decision trees $\hat{\pi}_1^m, \ldots, \hat{\pi}_N^m$

15: **function** BUILD$(\hat{\pi}_1^m, \ldots, \hat{\pi}_N^m, \mathcal{D}_i)$
16:    **for** each data point $(x, a) \in \mathcal{D}_i$ **do**
17:        // Will agent $j$'s (projected) final DT predict its action correctly?
18:        $v_j \leftarrow \mathbb{I}\left[\text{Predict}(\hat{\pi}_j^m, x) = a_j\right] \forall j \in [1, N]$
19:        // This data point is useful only if many agents' final DTs predict correctly.
20:        **if** $\sum_{j=1}^N v_j <$ threshold **then** Remove $d$ from dataset: $\mathcal{D}_i \leftarrow \mathcal{D}_i \setminus \{(x, a)\}$
21:    $\hat{\pi}_i^m \leftarrow$ Calculate best next feature split for DT $\hat{\pi}_i^m$ using $\mathcal{D}_i$.
22:    **return** $\hat{\pi}_i^m$

23: **function** PREDICT$(\hat{\pi}_j^m, x)$
24:    Use $x$ to traverse $\hat{\pi}_j^m$ until leaf node $l(x)$
25:    Train a projected final DT $\hat{\pi}_j' \leftarrow$ TrainDecisionTree$(\mathcal{D}_j)$
26:    **return** $\pi$.predict$(x)$

---

To address the issue of coordination, we propose MAVIPER, our novel algorithm for centralized training of coordinated multi-agent DT policies. For expository purpose, we describe MAVIPER in a fully cooperative setting, then explain how to use MAVIPER for mixed cooperative-competitive settings. At a high-level, MAVIPER trains all of the DT policies, one for each agent, in a centralized manner. It jointly grows the trees of each agent by predicting the behavior of the other agents in the environment using their anticipated trees. To train each DT policy, MAVIPER employs a new resampling technique to find states that are critical for its interactions with other agents. Algorithm 2 shows the full MAVIPER algorithm. Specifically, MAVIPER is built upon the following extensions to IVIPER that aim at addressing the issue of coordination.

First, MAVIPER does not calculate the probability $p(x)$ of a joint observation $x$ by viewing the other agents as stationary experts. Instead, MAVIPER focuses on the critical states where a good joint action can make a difference. Specifically, MAVIPER aims to measure how much worse off agent $i$ would be, taking expectation over all possible joint actions of the other agents, if it acts in the worst way possible compared with when it acts in the same way as the expert agent. So, we define $l_i(x)$, as in Eq. (2), as:

$$\tilde{l}_i(x) = \mathbb{E}_{\overrightarrow{a}_{-i}} \left[ Q_i^{\pi^*} \left( x, \pi_i^*(x), \overrightarrow{a}_{-i} \right) - \min_{a_i \in \mathcal{A}_i} Q_i^{\pi^*} \left( x, a_i, \overrightarrow{a}_{-i} \right) \right]. \tag{3}$$

MAVIPER uses the DT policies $(\hat{\pi}_1^{m-1}, \ldots, \hat{\pi}_N^{m-1})$ from the last iteration to perform rollouts and collect new data.

Second, we add a prediction module to the DT training process to increase the *joint* accuracy, as shown in the Predict function. The goal of the prediction module is to predict the actions that the other DTs $\{\hat{\pi}_j\}_{j \neq i}$ might make, given their partial observations. To make the most of the prediction module, MAVIPER grows the trees evenly using a breadth-first ordering to avoid biasing towards the result of any specific tree. Since the trees are not complete at the time of prediction, we use the output of another DT trained with the full dataset associated with that node for the prediction. Following the intuition that the correct prediction of one agent alone may not yield much benefit if the other agents are wrong, we use this prediction module to remove all data points whose proportion of correct predictions is lower than a predefined threshold. We then calculate the splitting criteria based on this modified dataset and continue iteratively growing the tree.

In some mixed cooperative-competitive settings, agents in a team share goals and need to coordinate with each other, but they face other agents or other teams whose goals are not fully aligned with theirs. In these settings, MAVIPER follows a similar procedure to jointly train policies for agents in the same team to ensure coordination. More specifically, for a team $Z$, the Build and Predict function is constrained to only make predictions for the agents in the same team. Equation (3) now takes the expectation over the joint actions for agents outside the team and becomes:

$$\tilde{l}_i(x) = \mathbb{E}_{\overrightarrow{a}_{-Z}} \left[ Q_i^{\pi^*} \left( x, \pi_Z^*(x), \overrightarrow{a}_{-Z} \right) - \min_{a_i \in \mathcal{A}_i, i \in Z} Q_i^{\pi^*} \left( x, \overrightarrow{a}_Z, \overrightarrow{a}_{-Z} \right) \right]. \tag{4}$$

Taken together, these changes comprise the MAVIPER algorithm. Because we explicitly account for the anticipated behavior of other agents in both the predictions and the resampling probability, we hypothesize that MAVIPER will better capture coordinated behavior.

## 4   Experiments

We now investigate how well MAVIPER and IVIPER agents perform in a variety of environments. Because the goal is to learn high-performing yet interpretable policies, we evaluate the quality of the trained policies in three multi-agent environments: two mixed competitive-cooperative environments and one fully cooperative environment. We measure how well the DT policies perform in the environment because our goal is to *deploy* these policies, not the expert ones.

Since small DTs are considered interpretable, we constrain the maximum tree depth to be at most 6. The expert policies used to guide the DT training are generated by MADDPG [20][1]. We compare to two baselines:

1. Fitted Q-Iteration. We iteratively approximate the Q-function with a regression DT [9]. We discretize states to account for continuous state values. More details in Appendix B.2. We derive the policy by taking the action associated with the highest estimated Q-value for that input state.
2. Imitation DT. Each DT policy is directly trained using a dataset collected by running the expert policies for multiple episodes. No resampling is performed. The observations for an agent are the features, and the actions for that agent are the labels.

We detail the hyperparameters and the hyperparameter-selection process in Appendix B.3. We train a high-performing MADDPG expert, then run each DT-learning algorithm 10 times with different random seeds. We evaluate all policies by running 100 episodes. Error bars correspond to the 95% confidence interval. Our code is available through our project website: https://stephmilani. github.io/maviper/.

### 4.1   Environments

We evaluate our algorithms on three multi-agent particle world environments [20], described below. Episodes terminate when the maximum number of timesteps $T = 25$ is reached. We choose the primary performance metric based on the environment (detailed below), and we also provide results using expected return as the performance metric in Appendix C.

*Physical Deception.* In this environment, a team of $N$ defenders must protect $N$ targets from one adversary. One of the targets is the true target, which is known to the defenders but not to the adversary. For our experiments, $N = 2$. Defenders succeed during an episode if they split up to cover all of the targets simultaneously; the adversary succeeds if it reaches the true target during the episode. Covering and reaching targets is defined as being $\epsilon$-close to a target for at least one timestep during the episode. We use the defenders' and the adversary's success rate as the primary performance metric in this environment.

---

[1] We use the Pytorch [31] implementation https://github.com/shariqiqbal2810/maddpg-pytorch.

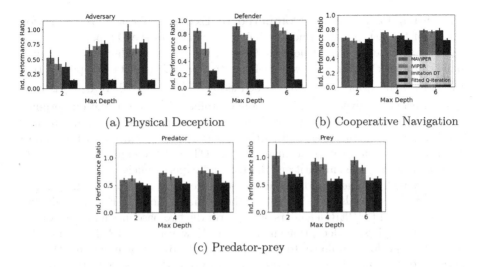

(a) Physical Deception          (b) Cooperative Navigation

(c) Predator-prey

**Fig. 2.** Individual performance ratio: Relative performance when only one agent adopts DT policy and all other agents use expert policy.

*Cooperative Navigation.* This environment consists of a team of $N$ agents, who must learn to cover all $N$ targets while avoiding collisions with each other. For our experiments, $N = 3$. Agents succeed during an episode if they split up to cover all of the targets without colliding. Our primary performance metric is the summation of the distance of the closest agent to each target, for all targets. Low values of the metric indicate that the agents correctly learn to split up.

*Predator-Prey.* This variant involves a team of $K$ slower, cooperating predators that chase $M$ faster prey. There are $L = 2$ landmarks impeding the way. We choose $K = M = 2$. We assume that each agent has a restricted observation space mostly consisting of binarized relative positions and velocity (if applicable) of the landmarks and other agents in the environment. See Appendix B.1 for full details. Our primary performance metric is the number of collisions between predators and prey. For prey, lower is better; for predators, higher is better.

## 4.2 Results

For each environment, we compare the DT policies generated by different methods and check if IVIPER and MAVIPER agents achieve better performance ratio than the baselines overall. We also investigate whether MAVIPER learns better coordinated behavior than IVIPER. Furthermore, we investigate which algorithms are the most robust to different types of opponents. We conclude with an ablation study to determine which components of the MAVIPER algorithm contribute most to its success.

**Individual Performance Compared to Experts.** We analyze the performance of the DT policies when only one agent adopts the DT policy while all other agents use the expert policies. Given a DT policy profile $\hat{\pi}$ and the expert policy profile $\pi^*$, if agent $i$ who belongs to team $Z$ uses its DT policy, then the individual performance ratio is defined as: $\frac{U_Z(\hat{\pi}_i, \pi^*_{-i})}{U_Z(\pi^*)}$, where $U_Z(\cdot)$ is team $Z$'s performance given the agents' policy profile (since we define our primary performance metrics at the team level). A performance ratio of 1 means that the DT policies perform as well as the expert ones. We can get a ratio above 1, since we compare the performance of the DT and the expert policies in the environment, not the similarity of the DT and expert policies.

We report the mean individual performance ratio for each team in Fig. 2, averaged over all trials and all agents in the team. As shown in Fig. 2a, individual MAVIPER and IVIPER defenders outperform the two baselines for all maximum depths in the physical deception environment. However, MAVIPER and IVIPER adversaries perform similarly to the Imitation DT adversary, indicating that the correct strategy may be simple enough to capture with a less-sophisticated algorithm. Agents also perform similarly on the cooperative navigation environment (Fig. 2b). As mentioned in the original MADDPG paper [20], this environment has a less stark contrast between success and failure, so these results are not unexpected.

In predator-prey, we see the most notable performance difference when comparing the predator. When the maximum depth is 2, only MAVIPER achieves near-expert performance. When the maximum depths are 4 and 6, MAVIPER and IVIPER agents achieve similar performance and significantly outperform the baselines. The preys achieve similar performance across all algorithms. We suspect that the complexity of this environment makes it challenging to replace even a single prey's policy with a DT.

Furthermore, MAVIPER achieves a performance ratio above 0.75 in all environments with a maximum depth of 6. The same is true for IVIPER, except for the adversaries in physical deception. That means DT policies generated by IVIPER and MAVIPER lead to a performance degradation of less than or around 20% compared to the less interpretable NN-based expert policies. These results show that IVIPER and MAVIPER generate reasonable DT policies and outperform the baselines overall when adopted by a single agent.

**Joint Performance Compared to Experts.** A crucial aspect in multi-agent environments is agent coordination, especially when agents are on the same team with shared goals. To ensure that the DT policies capture this coordination, we analyze the performance of the DT policies when all agents in a team adopt DT policies, while other agents use expert policies. We define the joint performance ratio as: $\frac{U_Z(\hat{\pi}_Z, \pi^*_{-Z})}{U_Z(\pi^*)}$, where $U_Z(\hat{\pi}_Z, \pi^*_{-Z})$ is the utility of team $Z$ when using their DT policies against the expert policies of the other agents $-Z$. Figure 3 shows the mean joint performance ratio for each team, averaged over all trials.

Figure 3a shows that MAVIPER defenders outperform IVIPER and the baselines, indicating that it better captures the coordinated behavior necessary to

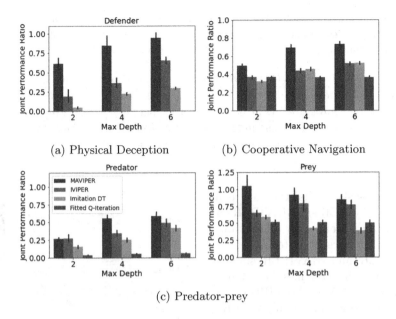

(a) Physical Deception          (b) Cooperative Navigation

(c) Predator-prey

**Fig. 3.** Joint performance ratio: Relative performance when all agents in a team adopt DT policy and other agents use expert policy.

succeed in this environment. Fitted Q-Iteration struggles to achieve coordinated behavior, despite obtaining non-zero success for individual agents. This algorithm cannot capture the coordinated behavior, which we suspect is due to poor Q-value estimates. We hypothesize that the superior performance of MAVIPER is partially due to the defender agents correctly splitting their "attention" to the two targets to induce the correct behavior of covering both targets. To investigate this, we inspect the normalized average feature importances of the DT policies of depth 4 for both IVIPER and MAVIPER over 5 of the trials, as shown in Fig. 4. Each of the MAVIPER defenders (top) most commonly focuses on the attributes associated with one of the targets. More specifically, defender 1 focuses on target 2 and defender 2 focuses on target 1. In contrast, both IVIPER defenders (bottom) mostly focus on the attributes associated with the goal target. Not only does this overlap in feature space mean that defenders are unlikely to capture the correct covering behavior, but it also leaves them more vulnerable to an adversary, as it is easier to infer the correct target.

Figure 3b shows that MAVIPER agents significantly outperform all other algorithms in the cooperative navigation environment for all maximum depths. IVIPER agents significantly outperform the baselines for a maximum depth of 2 but achieve similar performance to the Imitation DT for the other maximum depths (where both algorithms significantly outperform the Fitted Q-Iteration baseline). MAVIPER better captures coordinated behavior, even as we increase the complexity of the problem by introducing another cooperating agent.

**Table 1.** Robustness results. We report mean team performance and standard deviation of DT policies for each team, averaged across a variety of opponent policies. The best-performing algorithm for each agent type is shown in **bold**.

| Environment | Team | MAVIPER | IVIPER | Imitation DT | Fitted Q-Iteration |
|---|---|---|---|---|---|
| Physical | Defender | **.77** (.01) | .33 (.01) | .24 (.03) | .004 (.00) |
| Deception | Adversary | **.42** (.03) | **.41** (.03) | **.42** (.03) | .07 (.01) |
| Predator- | Predator | **2.51** (0.72) | **1.98** (0.58) | 1.14 (0.28) | 0.26 (0.11) |
| prey | Prey | 1.76 (0.80) | **2.16** (1.24) | **2.36** (1.90) | 1.11 (0.82) |

Figure 3c shows that the prey teams trained by IVIPER and MAVIPER outperform the baselines for all maximum depths. The predator teams trained by IVIPER and MAVIPER similarly outperform the baselines for all maximum depths. Also, MAVIPER leads to better performance than IVIPER in two of the settings (prey with depth 2 and predator with depth 4) while having no statistically significant advantage in other settings. Taken together, these results indicate that IVIPER and MAVIPER better capture the coordinated behavior necessary for a team to succeed in different environments, with MAVIPER significantly outperforming IVIPER in several environments.

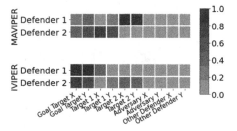

**Fig. 4.** Features used by the two defenders in the physical deception environment. Actual features are the relative positions of that agent and the labeled feature. Darker squares correspond to higher feature importance. MAVIPER defenders most commonly split importance across the two targets. (Color figure online)

**Robustness to Different Opponents.** We investigate the robustness of the DT policies when a team using DT policies plays against a variety of opponents in the mixed competitive-cooperative environments. For this set of experiments, we choose a maximum depth of 4. Given a DT policy profile $\hat{\pi}$, a team $Z$'s performance against an alternative policy file $\pi'$ used by the opponents is: $U_Z(\hat{\pi}_Z, \pi'_{-Z})$. We consider a broad set of opponent policies $\pi'$, including the policies generated by MAVIPER, IVIPER, Imitation DT, Fitted Q-Iteration, and MADDPG. We report the mean team performance averaged over all opponent policies in Table 1. See Tables 3 and 4 in Appendix C for the full results.

For physical deception, MAVIPER defenders outperform all other algorithms, with a gap of 0.44 between its performance and the next-best algorithm, IVIPER. This result indicates that MAVIPER learns coordinated defender policies that perform well against various adversaries. MAVIPER, IVIPER, and Imitation DT adversaries perform similarly on average, with a similar standard deviation, which supports the idea that the adversary's desired behavior is simple enough

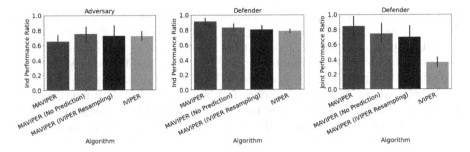

**Fig. 5.** Ablation study for MAVIPER for a maximum depth of 4. MAVIPER (No Prediction) does not utilize the predicted behavior of the anticipated DTs of the other agents to grow each agent's tree. MAVIPER (IVIPER Resampling) uses the same resampling method as IVIPER.

to capture with a less-sophisticated algorithm. For predator-prey, MAVIPER predators and prey outperform all other algorithms. The standard deviation of the performance of all algorithms is high due to this environment's complexity.

**Ablation Study.** As discussed in Sect. 3.2, MAVIPER improves upon IVIPER with a few critical changes. First, we utilize the predicted behavior of the anticipated DTs of the other agents to grow each agent's tree. Second, we alter the resampling probability to incorporate the average Q-values over all actions for the other agents. To investigate the contribution of these changes to the performance, we run an ablation study with a maximum depth of 4 on the physical deception environment. We report both the mean independent and joint performance ratios for the defender team in Fig. 5, comparing MAVIPER and IVIPER to two variants of MAVIPER without one of the two critical changes. Results show that both changes contributed to the improvement of MAVIPER over IVIPER, especially in the joint performance ratio.

## 5  Related Work

Most work on interpretable RL is in the single-agent setting [26]. We first discuss techniques that directly learn DT policies. CUSTARD [42] extends the action space of the MDP to contain actions for constructing a DT, i.e., choosing a feature to branch a node. Training an agent in the augmented MDP yields a DT policy for the original MDP while still enabling training using any function approximator, like NNs, during training. By redefining the MDP, the learning problem becomes more complex, which is problematic in multi-agent settings where the presence of other agents already complicates the learning problem. A few other works directly learn DT policies [9,24,44] for single-agent RL but not for the purpose of interpretability. Further, these works have custom learning algorithms and cannot utilize a high-performing NN policy to guide training.

VIPER [2] is considered to be a post-hoc DT-learning method [2]; however, we use it to produce *intrinsically* interpretable policies for deployment. MOET [45] extends VIPER by learning a mixture of DT policies trained on different regions of the state space. The resulting policy is a linear combination of multiple trees with non-axis-parallel partitions of the state. We find that the performance difference between VIPER and MOET is not significant enough to increase the complexity of the policy structure, which would sacrifice interpretability.

Despite increased interest in interpretable single-agent RL, interpretable MARL is less commonly explored. One line of work generates explanations from non-interpretable policies. Some work uses attention [14,17,29] to select and focus on critical factors that impact agents in the training process. Other work generates explanations as verbal explanations with predefined rules [47] or Shapley values [12]. The most similar line of work to ours [15] approximates non-interpretable MARL policies to interpretable ones using the framework of abstract argumentation. This work constructs argument preference graphs given manually-provided arguments. In contrast, our work does not need these manually-provided arguments for interpretability. Instead, we generate DT policies.

# 6   Discussion and Conclusion

We proposed IVIPER and MAVIPER, the first algorithms, to our knowledge, that train interpretable DT policies for MARL. We evaluated these algorithms on both cooperative and mixed competitive-cooperative environments. We showed that they can achieve individual performance of at least 75% of expert performance in most environment settings and over 90% in some of them, given a maximum tree depth of 6. We also empirically validated that MAVIPER effectively captures coordinated behavior by showing that teams of MAVIPER-trained agents outperform the agents trained by IVIPER and several baselines. We further showed that MAVIPER generally produces more robust agents than the other DT-learning algorithms.

Future work includes learning these high-quality DT policies from fewer samples, e.g., by using dataset distillation [46]. We also note that our algorithms can work in some environments where the experts and DTs are trained on different sets of features. Since DTs can be easier to learn with a simpler set of features, future work includes augmenting our algorithm with an automatic feature selection component that constructs simplified yet still interpretable features for training the DT policies.

**Acknowledgements.** This material is based upon work supported by the Department of Defense (DoD) through the National Defense Science & Engineering Graduate (NDSEG) Fellowship Program. This research was sponsored by the U.S. Army Combat Capabilities Development Command Army Research Laboratory and was accomplished under Cooperative Agreement Number W911NF-13-2-0045 (ARL Cyber

Security CRA). Any opinions, findings and conclusions or recommendations expressed in this material are those of the author(s) and do not reflect the views of the funding agencies or government agencies. The U.S. Government is authorized to reproduce and distribute reprints for Government purposes notwithstanding any copyright notation here on.

# References

1. Abbeel, P., Ng, A.: Apprenticeship learning via inverse reinforcement learning. In: ICML (2004)
2. Bastani, O., et al.: Verifiable reinforcement learning via policy extraction. In: NeurIPS (2018)
3. Berner, C., et al.: Dota 2 with large scale deep reinforcement learning. arXiv preprint 1912.06680 (2019)
4. Bhalla, S., et al.: Deep multi agent reinforcement learning for autonomous driving. In: Canadian Conference Artificial Intelligent (2020)
5. Brittain, M., Wei, P.: Autonomous air traffic controller: a deep multi-agent reinforcement learning approach. arXiv preprint arXiv:1905.01303 (2019)
6. Buciluǎ, C., et al.: Model compression. In: KDD (2006)
7. Chen, Z., et al.: Relace: Reinforcement learning agent for counterfactual explanations of arbitrary predictive models. arXiv preprint arXiv:2110.11960 (2021)
8. Degris, T., et al.: Learning the structure of factored Markov decision processes in reinforcement learning problems. In: ICML (2006)
9. Ernst, D., et al.: Tree-based batch mode reinforcement learning. JMLR 6 (2005)
10. Foerster, J., et al.: Stabilising experience replay for deep multi-agent reinforcement learning. In: ICML (2017)
11. Foerster, J., et al.: Counterfactual multi-agent policy gradients. In: AAAI (2018)
12. Heuillet, A., et al.: Collective explainable ai: explaining cooperative strategies and agent contribution in multiagent reinforcement learning with shapley values. IEEE Comput. Intell. Mag. 17, 59–71 (2022)
13. Hinton, G., et al.: Distilling the knowledge in a neural network. arXiv preprint arXiv:1503.02531 (2015)
14. Iqbal, S., Sha, F.: Actor-attention-critic for multi-agent reinforcement learning. In: ICML (2019)
15. Kazhdan, D., et al.: Marleme: a multi-agent reinforcement learning model extraction library. In: IJCNN (2020)
16. Li, S., et al.: Robust multi-agent reinforcement learning via minimax deep deterministic policy gradient. In: AAAI (2019)
17. Li, W., et al.: Sparsemaac: sparse attention for multi-agent reinforcement learning. In: International Conference on Database Systems for Advanced Applications (2019)
18. Lipton, Z.: The mythos of model interpretability. ACM Queue 16(3) (2018)
19. Littman, M.: Markov games as a framework for multi-agent reinforcement learning. In: Mach. Learning (1994)
20. Lowe, R., et al.: Multi-agent actor-critic for mixed cooperative-competitive environments. arXiv preprint arXiv:1706.02275 (2017)
21. Luss, R., et al.: Local explanations for reinforcement learning. arXiv preprint arXiv:2202.03597 (2022)
22. Malialis, K., Kudenko, D.: Distributed response to network intrusions using multiagent reinforcement learning. Eng. Appl. Artif. Intell. 40, 270–284 (2015)

23. Matignon, L., et al.: Independent reinforcement learners in cooperative Markov games: a survey regarding coordination problems. Knowl. Eng. Rev. **27**(1), 1–31 (2012)
24. McCallum, R.: Reinforcement learning with selective perception and hidden state. Ph.D. thesis, Univ. Rochester, Dept. of Comp. Sci. (1997)
25. Meng, Z., et al.: Interpreting deep learning-based networking systems. In: Proceedings of the Annual Conference of the ACM Special Interest Group on Data Communication on the Applications, Technologies, Architectures, and Protocols for Computer Communication (2020)
26. Milani, S., et al.: A survey of explainable reinforcement learning. arXiv preprint arXiv:2202.08434 (2022)
27. Mohanty, S., et al.: Flatland-rl: multi-agent reinforcement learning on trains. arXiv preprint arXiv:2012.05893 (2020)
28. Molnar, C.: Interpretable Machine Learning (2019)
29. Motokawa, Y., Sugawara, T.: MAT-DQN: toward interpretable multi-agent deep reinforcement learning for coordinated activities. In: ICANN (2021)
30. Oliehoek, F., et al.: Optimal and approximate q-value functions for decentralized pomdps. JAIR **32**, 289–353 (2008)
31. Paszke, A., et al.: Automatic differentiation in pytorch (2017)
32. Pyeatt, L.: Reinforcement learning with decision trees. In: Appl. Informatics (2003)
33. Pyeatt, L., Howe, A.: Decision tree function approximation in reinforcement learning. In: Int. Symp. on Adaptive Syst.: Evol. Comput. and Prob. Graphical Models (2001)
34. Quinlan, J.: Induction of decision trees. Mach. Learn. **1**, 81–106 (1986)
35. Rashid, T., et al.: Qmix: monotonic value function factorisation for deep multi-agent reinforcement learning. In: ICML (2018)
36. Ross, S., et al.: A reduction of imitation learning and structured prediction to no-regret online learning. In: AISTATS (2011)
37. Roth, A., et al.: Conservative q-improvement: reinforcement learning for an interpretable decision-tree policy. arXiv preprint arXiv:1907.01180 (2019)
38. Shapley, L.: Stochastic games. PNAS **39**(10), 1095–1100 (1953)
39. Son, K., et al.: Qtran: Learning to factorize with transformation for cooperative multi-agent reinforcement learning. arXiv preprint arXiv:1905.05408 (2019)
40. Strehl, A., et al.: Efficient structure learning in factored-state mdps. In: AAAI (2007)
41. Sunehag, P., et al.: Value-decomposition networks for cooperative multi-agent learning. arXiv preprint arXiv:1706.05296 (2017)
42. Topin, N., et al.: Iterative bounding mdps: learning interpretable policies via non-interpretable methods. In: AAAI (2021)
43. Tuyls, K., et al.: Reinforcement learning in large state spaces. In: Robot Soccer World Cup (2002)
44. Uther, W., Veloso, M.: The lumberjack algorithm for learning linked decision forests. In: International Symposium on Abstraction, Reformulation, and Approximation (2000)
45. Vasic, M., et al.: Moët: Interpretable and verifiable reinforcement learning via mixture of expert trees. arXiv preprint arXiv:1906.06717 (2019)
46. Wang, T., et al.: Dataset distillation. arXiv preprint arXiv:1811.10959 (2018)
47. Wang, X., et al.: Explanation of reinforcement learning model in dynamic multi-agent system. arXiv preprint arXiv:2008.01508 (2020)
48. Yu, C., et al.: The surprising effectiveness of mappo in cooperative, multi-agent games. arXiv preprint arXiv:2103.01955 (2021)

# Bandits and Online Learning

# Hierarchical Unimodal Bandits

Tianchi Zhao$^{(\boxtimes)}$, Chicheng Zhang, and Ming Li

University of Arizona, Tucson, AZ 85721, USA
{tzhao7,chichengz,lim}@email.arizona.edu

**Abstract.** We study a multi-armed bandit problem with clustered arms and a unimodal reward structure, which has applications in millimeter wave (mmWave) communication, road navigation, etc. More specifically, a set of $N$ arms are grouped together to form $C$ clusters, and the expected reward of arms belonging to the same cluster forms a Unimodal function (a function is Unimodal if it has only one peak, e.g. parabola). First, in the setting when $C = 1$, we propose an algorithm, SGSD (Stochastic Golden Search for Discrete Arm), that has better guarantees than the prior Unimodal bandit algorithm [Yu and Mannor 2011]. Second, in the setting when $C \geq 2$, we develop HUUCB (Hierarchical Unimodal Upper Confidence Bound (UCB) algorithm), an algorithm that utilizes the clustering structure of the arms and the Unimodal structure of the rewards. We show that the regret upper bound of our algorithm grows as $O(\sqrt{CT \log(T)})$, which can be significantly smaller than UCB's $O(\sqrt{NT \log(T)})$ regret guarantee. We perform a multi-channel mmWave communication simulation to evaluate our algorithm. Our simulation results confirm the advantage of our algorithm over the UCB algorithm [Auer et al. 2002] and a two-level policy (TLP) proposed in prior works [Pandey et al. 2007].

## 1 Introduction

### 1.1 Motivation

The multi-armed bandit (MAB) problem [Thompson 1933] models many real-world scenarios where a decision maker learns to take a sequence of action (arms) to maximize reward. Here, the decision maker is given access to an arm set, and chooses an arm from the arm set resulting in a reward drawn from an unknown distribution. The objective of the decision maker is to maximize its expected cumulative reward over a time horizon of $T$. To this end, it faces a tradeoff between exploration and exploitation.

In this work, we consider a multi-armed bandit problem with clustered arms, where the arm set can be partitioned into $C$ clusters, and each cluster's rewards exhibits a unimodal structure. This arises naturally in various decision problems, as shown in the following two examples:

---

This work was partly supported by ONR YIP grant N00014-16-1-2650. The authors would like to thank Zhiwu Guo for his help on drawing figures, and the anonymous reviewers for their helpful comments.

M.-R. Amini et al. (Eds.): ECML PKDD 2022, LNAI 13716, pp. 269–283, 2023.
https://doi.org/10.1007/978-3-031-26412-2_17

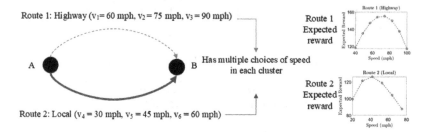

**Fig. 1.** Road navigation example: each route represents a cluster. The arms in each cluster are represented by different speeds. For cluster (route) 1, it contains $v_1 = 60$ mph, $v_2 = 75$ mph, $v_3 = 90$ mph). The safety indices for speeds in route 1 are $p_1 = 0.9, p_2 = 0.8, p_3 = 0.5$. The expected reward values in cluster 1 are $r_1 = 150$, $r_2 = 155$ and $r_3 = 140$. For cluster (route) 2, it contains $v_4 = 30$ mph, $v_5 = 45$ mph, $v_6 = 60$ mph). The safety indices for speeds in route 2 are $p_4 = 0.9, p_5 = 0.8, p_6 = 0.5$. The expected reward values in route 2 are $r_4 = 120$, $r_5 = 125$ and $r_6 = 110$. We can see that each cluster's expected reward function has only one peak, which satisfies the Unimodal property.

*Example 1: Road navigation.* A person driving from A to B has the option to choose two routes: highway and local way. After choosing a route, she needs to further choose a speed. In this example, a (route, speed) combination corresponds to an arm, and a route corresponds to a cluster. The expected reward (Utility) is defined as follows: $r_j = v_j + 10 \times p_j$, where $v_j$ denotes velocity for arm $j$ and $p_j$ denotes safety [Sun et al. 2018] for arm $j$. Note that, if velocity increases, safety will decrease, and thus, each cluster's reward structure is oftentimes Unimodal. See Fig. 1 for a numerical example.

*Example 2: Multi-channel mmWave communication.* Let us consider optimal antenna beam selection for a mmWave communication link with multiple frequency channels. Theoretical analysis [Wu et al. 2019] and experimental results [Hashemi et al. 2018] indicate that the received signal strength (RSS) function over the beam space in the channel with a single path (or a dominant line-of-sight path) can be characterized by a Unimodal function. Our goal is to select the best channel and beam combination to maximize the link RSS. In this example, the arm is the combination of frequency channel and beam, and the reward is the signal strength. We regard the beams under each channel as a cluster. Our goal is to select the optimal channel and beam for communication in an online manner. In Fig. 2, we provide an illustration of the multi-channel mmWave communication example.

## 1.2  Related Work

**Bandits with Hierarchical Structures.** Hierarchical bandit problem, where the arm space is partitioned into multiple clusters, has been studied in Nguyen and Lauw [2014], Jedor et al. [2019], Bouneffouf et al. [2019], Carlsson et al.

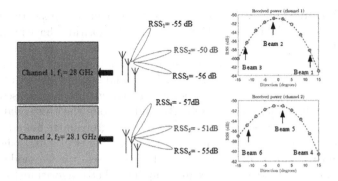

**Fig. 2.** Multi-channel mmWave communication example. There are two channels: $f_1 = 28\,\text{GHz}$, $f_2 = 28\,\text{GHz} + 100\,\text{MHz}$ (These two frequencies are based on 3GPP TS 38.101-1/2, 38.104-1/2 [Lopez et al. 2019]). For each channel, the algorithm can select three beams. Experimental results in Hashemi et al. [2018] show that the RSS function over the beam space in a fixed frequency is a Unimodal function.

2021]. These papers give regret bounds under different assumptions on the clustering. Specifically, Pandey et al. [2007] proposed a Two-level Policy (TLP) algorithm. It divided the arms into multiple clusters. However, their work does not provide a theoretical analysis of the algorithm. Zhao et al. [2019] proposed a novel Hierarchical Thompson Sampling (HTS) algorithm to solve this problem. The beams under the same chosen channel can be regarded as a cluster of arms in MAB. However, it does not utilize the Unimodal property in each cluster. Bouneffouf et al. [2019] considered a two-level UCB scheme that the arm set is pre-clustered, and the reward distributions of the arms within each cluster are similar. However, they did not consider the Unimodal property in each cluster. Jedor et al. [2019] introduced a MAB setting where arms are grouped in one of three types of categories. Each type has a different ordering between clusters, and our work does not have such assumption among the clusters. Yang et al. [2022] considered a problem of online clustering: a set of arms can be partitioned into various groups that are unknown. Note that the partition of cluster is time-varying, and we study a different setting where the clusters are pre-specified. Kumar et al. [2019] addressed the problem of hidden population sampling problem in online social platforms. They proposed a hierarchical Multi-Arm Bandit algorithm (Decision-Tree Thompson Sampling (DT-TMP)) that uses a decision tree coupled with a reinforcement learning search strategy to query the combinatorial search space. However, their algorithm is based on Thompson Sampling, and no theoretical analysis of its regret is given. Singh et al. [2020] studies a multi-armed bandit problem with dependent arms. When an agent pulls arm $i$, it not only reveals information about its own reward distribution but also reveal all those arms that belong to the same cluster with arm $i$, which is not the case in our problem. Carlsson et al. [2021] proposed a Thompson Sampling based algorithm with clustered arms, and give a regret bound which depends on the number of clusters. However, they do not utilize the Unimodal property as well.

**Unimodal Bandit.** In a Unimodal bandit problem, the expected reward of arms forms a Unimodal function. Here, specialized algorithms have been designed to exploit the Unimodality structure, to achieve faster convergence rate (compared to standard bandit algorithms such as UCB). Yu and Mannor [2011] is the first work to propose an algorithm for Unimodal bandits for both continuous arm and discrete arm settings. Combes and Proutiere [2014] proposed Optimal Sampling for Unimodal Bandits (OSUB), and exploits the Unimodal structure under the discrete arm setting. They provided a regret upper bound for OSUB which does not depend on the number of arms. Zhang et al. [2021] showed that the effective throughputs of mmWave codebooks possess the Unimodal property and proposed a Unimodal Thompson Sampling (UTS) algorithm to deal with mmWave codebook selection. However, both papers only consider Unimodal property without clustered arms. Blinn et al. [2021] proposed Hierarchical Optimal Sampling of Unimodal Bandits. The difference with our work is that they use the OSUB algorithm to select an arm in each cluster, and they did not provide a theoretical regret analysis.

### 1.3   Main Contributions

Our main contributions can be summarized as follows:

1. In the single-cluster setting ($C = 1$), we propose a new Unimodal bandit algorithm, called Stochastic Golden Search with Discrete arm (SGSD), that improves over an existing Unimodal bandit algorithm [Yu and Mannor, 2011], in that it simultaneously achieves gap-dependent and gap-independent regret bounds. In addition, its regret bounds are competitive with UCB, and can sometimes be much better.
2. In the multi-cluster setting ($C \geq 2$), built on the SGSD, we present a UCB-based, hierarchical Unimodal bandit algorithm, called HUUCB, to solve the MAB with Clustered arms and a Unimodal reward structure (MAB-CU) problem. The key insight is a new setting of reward UCB for each cluster, taking into account the regret incurred for each cluster. We prove a gap-independent regret bound for this algorithm, and show that they can be better compared with the baseline strategy of UCB on the "flattened" arm set.
3. We evaluate our algorithms experimentally in both the single-cluster setting and the multi-cluster setting, using two different datasets (synthetic/simulated).
   (a) In the single-cluster setting, our SGSD algorithm outperforms UCB.
   (b) In the multi-cluster setting, our HUUCB algorithm outperforms UCB with flatten arms, and TLP [Pandey et al. 2007].

## 2   Hierarchical Unimodal Bandits: Problem Setup

The problem statement is as follows: There are $N$ arms available and each arm $j$'s reward comes from a particular distribution (supported on $[0, 1]$) with an

---

**Algorithm 1.** Stochastic Golden Search for discrete arm (SGSD)

---
1: Parameters: $\epsilon_1, .... > 0$:
2: Initialize $x_A = 0, x_B = \frac{1}{\phi^2}, x_c = 1$ ($\phi = \frac{1+\sqrt{5}}{2}$)
3: **for** each stage $s = 1, 2, ...S$, **do**
4:    **if** there has more than one discrete arms $j/N$ in $[x_A, x_C]$ **then**
5:       Let

$$x'_B = \begin{cases} x_B - \frac{1}{\phi^2}(x_B - x_A) & x_B - x_A > x_C - x_B \\ x_B + \frac{1}{\phi^2}(x_C - x_B) & \text{otherwise,} \end{cases}$$

6:       Obtain the reward of each continuous point $\{x_A, x_B, x'_B, x_C\}$ according to Algorithm 2, each point for $\frac{2}{\epsilon_s^2} \log(8T)$ times, and let $\hat{x}$ be the point with highest empirical mean in this stage
7:       If $\hat{x} \in \{x_A, x_B\}$ then eliminate interval $(x'_B, x_C]$ and let $x_C = x'_B$,
8:       else eliminate interval $[x_A, x_B)$ and let $x_A = x_B$
9:    **else**
10:       Break
11:    **end if**
12:    Keep pulling the only discrete arm $j/N$ in $[x_A, x_C]$
13: **end for**

---

unknown mean $\mu_j$. The arms are partitioned to $C$ clusters, where we denote Cluster$_i$ as the $i$-th cluster. In each cluster $i$, we assume that the expected rewards of arm $j \in$ Cluster$_i$ form a Unimodal function (a function is Unimodal if the function has only one local maximum, e.g. a negative parabola). We assume that every cluster have the same number of arms $B$, therefore, $N = CB$.

The Multi-armed bandit (MAB) model focuses on the essential issue of trade-off between exploration and exploitation [Auer et al. 2002]. At each time step, the algorithm selects one arm $j_t$. Then a reward of this arm is independently drawn, and observed by the algorithm. The objective of the algorithm is to gather as much cumulative reward as possible. The expected cumulative regret can be expressed as (Bubeck and Cesa-Bianchi [2012]):

$$E[R(T)] = \sum_{t=1}^{T}(\mu_{j^*} - \mu_{j_t}) \tag{1}$$

where $j^* = \arg\max_{j \in \{1,...,N\}} \mu_j$ is the optimal arm, $T$ is the total number of time steps. Note that the algorithm only observes the reward for the selected arm, also known as the bandit feedback setting.

## 3    Algorithm for the Single-Cluster Setting

We first study the single-cluster setting ($C = 1$), where the problem degenerates to a Unimodal bandit problem [Yu and Mannor 2011, Combes and

---

**Algorithm 2.** Reward sampling algorithm for an arbitrary continuous point $x'$

---

1: Input: $x'$
2: Output: a stochastic reward of conditional mean $f(x')$ (Eq. (2))
3: $j = \lfloor Nx' \rfloor$
4: set

$$l = \begin{cases} j & \text{with probability} \quad j + 1 - Nx' \\ j + 1 & \text{otherwise,} \end{cases}$$

5: $r \leftarrow$ reward of pulling arm $l$
6: **return** $r$

---

Proutiere 2014]. One drawback of prior works is that their guarantees have limited adaptivity: achieving gap-dependent and gap-independent regret bounds require setting parameters differently. In this work, we provide an algorithm that simultaneously enjoys gap-dependent and gap-independent regret guarantees, which is useful for practical deployment. Our algorithm is built on the SGS algorithm [Yu and Mannor 2011], and we call it SGS for discrete arm setting algorithm (SGSD), namely, Algorithm 1.

The high level idea of SGSD is to reduce the discrete-arm Unimodal bandits problem to a continuous-arm Unimodal bandits, and use the SGS algorithm in the continuous arm setting. Specifically, given a discrete-arm Unimodal bandit problem $\mu_1, \ldots, \mu_N$, we associate every arm $j$ to a point $j/N$ in the $[0, 1]$ interval and perform linear interpolation, inducing a function

$$f(x) = \mu_j \cdot (j + 1 - Nx) + \mu_{j+1} \cdot (Nx - j), \quad x \in [j/N, (j+1)/N] \quad (2)$$

over the continuous interval $[0, 1]$, and use SGS to optimize it. Observe that $f$ has minimum at $x^* = j^*/N$, and for $x \in [j/N, (j+1)/N)$, bandit feedback of $f(x)$ can be simulated by pulling arms randomly from $\{j, j+1\}$ (Algorithm 2; see subsequent paragraphs for more details). To this end, it narrows down the sampling interval, maintaining the invariant that with high probability, $j^*/N \in [x_A, x_C]$.

The SGSD algorithm proceeds as follows: first, the algorithm initialize parameters $x_A = 0, x_B = \frac{1}{\phi^2}$, $x_c = 1$ (line 2 in Algorithm 1, where $\phi = \frac{1+\sqrt{5}}{2}$). In line 4, the algorithm checks the number of discrete arms in the range $[x_A, x_C]$; if only one arm $j/N$ is in the range $[x_A, x_C]$, with high probability, it must be the case that $j = j^*$, i.e. we have identified the optimal arm – in this case, the algorithm breaks the loop and keep pulling that arm (line 12). Then, given three points $x_A < x_B < x_C$ where the distance of $x_B$ to the other two points satisfy the golden ratio. The reason we choose three point is to ensure the elimination of a constant fraction of the sample interval that does not contain $j^*/N$ in each iteration. Note that $x_B$ may be closer to $x_A$ or to $x_C$ depending on the past updating value of the SGSD algorithm. The point $x'_B$ is set in the larger interval between $x_B - x_A$ and $x_C - x_B$ (The updating procedure for $x'_B$ is in Algorithm 1's line 5). If we set $x_C - x_A = \ell$, the following equalities hold at any step of

SGS algorithm: $x_B - x_A = \frac{\ell}{\phi^2}, x'_B - x_B = \frac{\ell}{\phi^3}, x_C - x'_B = \frac{\ell}{\phi^2}$. Then, we eliminate $[x_A, x_B)$ or $(x'_B, x_C)$, depending on whether the smallest empirical mean value is found in set $\{x_A, x_B\}$ or $\{x'_B, x_C\}$ (Shown in Algorithm 1's line 7 and 8). Algorithm 1 gives the detail of the algorithm.

Note that we convert the expected rewards of discrete arms into a continuous function, we need to simulate noisy values of $f$ on $\{x_A, x_B, x'_B, x_C\}$ via queries to the discrete arms $\{1, \ldots, N\}$. We use Algorithm 2 to calculate such "virtual" instantaneous rewards. Given input arm $x' \in [0, 1]$, we determine the interval $[j/N, (j+1)/N)$ that $x'$ belongs to (Algorithm 2's line 3). In each iteration, we obtain its reward by probabilistic sampling of the two discrete arms in $x$'s neighborhood (where the sampling probability of each neighboring arm is shown in line 4), such that the output reward has expectation $f(x')$ (Shown in Algorithm 2's line 4 -line 5).

To analyze Algorithm 1, we make the following assumptions similar to Yu and Mannor [2011]:

**Assumption 1.** (1)$\mu$ is strongly Unimodal: there exists a unique maximizer $j^*$ of $\mu_1, \ldots, \mu_N$[1].

(2) There exist positive constants $D_L$ and $D_H > 0$ such that $|\mu_j - \mu_{j+1}| \leq D_H$, and $|\mu_j - \mu_{j+1}| \geq D_L$ for all $j \in \{1, \ldots, N\}$.

Assumption 1.(1) ensures that the continuous function has one peak value. The valid domain of assumption 1.(2) is on both $[0, v_{j^*}]$ and $[v_{j^*}, 1]$. Note that each neighbor is connected by linear interpolation. So, our new continuous function has the lowest slope value which is determined by linear interpolation and $D_L$. Then, we have the following regret bound.

**Theorem 1.** Under Assumption 1, the expected regret of Algorithm 1, with $\epsilon_s = ND_L\phi^{-(s+3)}$, is:

$$E[R(T)] \leq O\left( \min\left\{ \frac{D_H}{D_L} \log(8T)\sqrt{T}, \frac{D_H}{(D_L)^2} \log(8T) \right\} \right). \tag{3}$$

The proof of the first bound in Theorem 1 is inspired by the analysis of SGS in Yu and Mannor [2011] after linear interpolation to reduce the discrete-arm setting to a continuous-arm setting. The second bound is inspired by the proof of Theorem IV.4 in Yu and Mannor [2011]. From Theorem 1, we can see that the upper bound is independent of the number of arms. However, it depends on the problem-dependent constants $(D_L, D_H)$.

We now compare this regret bound with that of the UCB algorithm Auer et al. [2002]. UCB has a gap-independent regret bound of $O(\sqrt{TN \log(T)})$, and gap-dependent regret bound of $O(\sum_{j \neq j^*} \frac{\log(T)}{\Delta_j})$ (where $\Delta_j = \mu_{j^*} - \mu_j$). Then, we examine UCB's gap-dependent bound in terms of $D_H$. Note that, the function is a Unimodal function, and the number of arms on either the left or the right side of the optimal arm $j^*$ must be greater than $\frac{N}{2}$. Then, the gap-dependent regret

---

[1] Strong Unimodality means that it only has one optimal arm among the arm set.

---

**Algorithm 3.** Hierarchical Unimodal UCB Algorithm

---

1: Input: $D_H, D_L$
2: For each cluster $i = 1, \ldots, C$: $\hat{\nu}_i(0) = 0$, $M_i(0) = 0$, initialize $\mathcal{A}_i$, a copy of Alg. 1.
3: For each arm $j = 1, \ldots, N$: $\hat{\mu}_j(0) = 0$, $m_j(0) = 0$
4: **for** $t = 1, 2, \ldots N$, **do**
5:     Play arm $j = t$, and update corresponding $\hat{\mu}_j(t)$, $m_j(t) = 1$ ,
6: **end for**
7: **for** each cluster i **do**
8:     $M_i(t) = \sum_{j \in \text{Cluster}_j} m_j(t)$
9:     $\hat{\nu}_i(t) = \sum_{j \in \text{Cluster}_i} \frac{m_j}{M_i} \hat{\mu}_j(t)$
10: **end for**
11: **for** stage $t = N + 1, N + 2, \ldots$, **do**
12:     Choose the cluster

$$
i_t := \arg\max \left\{ \hat{\nu}_i(t) + \sqrt{\frac{2 \log(t)}{M_i(t)}} + \frac{D_H}{D_L} \sqrt{\frac{\log(t)}{M_i(t)}} \right\}, \tag{4}
$$

13:     Resume $\mathcal{A}_{i_t}$ and run it for one time step, select an arm $j_t \in \text{Cluster}_{i_t}$, and obtain the reward of selected arm $r_{j_t}(t)$ at stage $t$
14:     Update empirical mean rewards and counts for all clusters:

$$
(\hat{\nu}_i(t), M_i(t)) = \begin{cases} \left( \frac{\hat{\nu}_i(t-1) \cdot M_i(t-1) + r_{j_t}(t)}{M_i(t-1)+1}, M_i(t-1) + 1 \right), & i = i_t, \\ (\hat{\nu}_i(t-1), M_i(t-1)), & i \neq i_t. \end{cases}
$$

15: **end for**

---

bound of UCB must be larger than $\sum_{j=1}^{N/2} \frac{\log(T)}{j D_H} = \frac{\log(T)}{D_H} \sum_{j=1}^{N/2} \frac{1}{j} = \Omega(\log(\frac{N}{2}) \cdot \frac{\log(T)}{D_H})$. We therefore see that the regret bound of the UCB algorithm depends on the number of arms in both gap-independent and gap-dependent bounds, which does not apply to SGSD.

## 4   Hierarchical Unimodal UCB Algorithm

We now turn to study the more challenging multi-cluster setting ($C \geq 2$). Existing works such as Two-Level Policy (TLP, Pandey et al. [2007]) approaches this problem using the following strategy: treat each cluster as a "virtual arm", and view the cluster selection problem (which we call *inter-cluster selection*) as a stationary MAB problem. In each step, the TLP algorithm chooses a virtual arm first using UCB, and then an actual arm within the selected cluster using some *intra-cluster arm selection* algorithm. However, due to the nonstationary nature of the rewards within a cluster (as the intra-cluster arm selection algorithm may gradually converge to pulling the cluster's optimal arm), TLP do not have theoretical guarantees.

In contrast, in this section, we propose a Hierarchical Unimodal UCB Algorithm (HUUCB) (Algorithm 3) that has a provable regret guarantee. Our algorithm design follows the "optimism in the face of uncertainty" principle: clusters are chosen according to their optimistic upper confidence bounds on their maximum expected rewards $\nu_i = \max_{j \in \text{Cluster}_i} \mu_j$'s, a property not satisfied by TLP. This ensures a sublinear regret for the cluster selection task. The algorithm proceeds as follows: it first takes into $D_H, D_L$ as inputs, which are the reward gap parameters specified in Assumption 1. Then, the initialization phase (lines 4 to 10) begins by selecting each arm at least once to ensure $M_i(t)$ and $\hat{\nu}_i(t)$ are updated. $M_i(t)$ is number of times that cluster $i$ has been selected and $\hat{\nu}_i(t)$ is the empirical mean value for the cluster $i$. Once the initialization is completed, the algorithm selects the cluster that maximizes our designed UCB (Equation 4). From the equation, we can see that the UCB for cluster $i$,

$$\hat{\nu}_i(t) + \sqrt{\frac{2\log(t)}{M_i(t)}} + \frac{D_H}{D_L}\sqrt{\frac{\log(t)}{M_i(t)}}$$

is the sum of three terms. The first term is the empirical mean value of the $M_i(t)$ rewards obtained by pulling the arms in the cluster $i$. The second term accounts for the concentration between the sum of the noisy rewards and the sum of their corresponding expected rewards. The third term is new and unique to HUUCB – it accounts for the suboptimality of the arm selection in cluster $i$ by SGSD so far, calculated by dividing SGSD's regret $O\left(\frac{D_H}{D_L}\sqrt{M_i(t)}\right)$ by $M_i(t)$. The three terms jointly ensures that the UCB is indeed a high-probability upper bound of $\nu_i$. In line 13, Algorithm 3 selects an arm $j_t \in \text{Cluster}_{i_t}$ using $\mathcal{A}_{i_t}$ after selecting a cluster $i_t$ and obtaining the reward $r_{j_t}$ ($\mathcal{A}_{i_t}$ is a copy of Algorithm 1 for cluster $i_t$). Last, in line 14, the algorithm updates the chosen cluster $i_t$'s statistics, empirical reward mean $\hat{\nu}_{i_t}(t)$ and count $M_{i_t}(t)$. Other clusters' statistics remain the same as time step $t-1$.

We have the following regret guarantee of Algorithm 3:

**Theorem 2.** *If each cluster satisfies Assumption 1, the regret of Hierarchical Unimodal UCB is upper bounded by,*

$$E[R(T)] \leq O\left(\frac{D_H}{D_L}\sqrt{2CT\log(T)}\right), \tag{5}$$

*where $C$ is the number of clusters.*

**Outline of the proof for Theorem 2**: First, we define the event

$$E = \left\{ |\hat{\nu}_i(t) - \nu_i| \leq \sqrt{\frac{2\log(T)}{M_i(t)}} + \frac{D_H}{D_L}\sqrt{\frac{\log(T)}{M_i(t)}}, \forall i, t \right\}.$$

Without loss of generality, assume that Cluster$_1$ contains the globally optimal arm. The high-level idea of the proof is as follows:

(1) We bound the regret incurred when the algorithm chooses cluster $i \neq 1$ when the event $\boldsymbol{E}$ holds.

(2) We bound the probability of the event $\boldsymbol{E}$ does not happen using Azuma's inequality.

(3) Lastly, we bound the regret incurred when the algorithm chooses optimal cluster Cluster$_1$ but selects a sub-optimal arm using Theorem 1. The detailed proof is in Appendix.

**Remark:** Theorem 2 shows that the regret bound depends on the number of clusters (instead of the number of arms) because we incorporate the SGSD algorithm. Compared to the "flattened" UCB algorithm with a total of $N = CB$ arms ($B$ is the number of arms in each cluster), whose regret is $O(\sqrt{TCB\log(T)})$, when $\frac{D_H}{D_L} \ll \sqrt{B}$, HUUCB has a much better regret.

Alternatively, we can also apply a general bandit model selection algorithm over SGSD for the MAB-CU problem. Specifically, we regard each cluster $i$'s algorithm as a base algorithm defined in [Abbasi-Yadkori et al. 2020, Cutkosky et al. 2021]. In our problem, $C$ is the number of the base algorithm. Then, we define $R_i(T)$ as the regret upper bound for cluster $i$, represented as

$$R_i(T) \leq O\left(\frac{D_H}{D_L}\log(T)\sqrt{T}\right) \leq \text{const}_1 d_i \log(T)\sqrt{T}, \tag{6}$$

where const$_1$ is a positive constant independent of $T$ and $i$, $d_i = \frac{D_H}{D_L}$. According to Theorem 2.1 in [Abbasi-Yadkori et al. 2020], the regret is upper bounded by,

$$E[R(T)] \leq C \max_i R_i(T) \leq C\text{const}_1 d_i \log(T)\sqrt{T}, \tag{7}$$

Comparing (5) and (7), we can see that our result is better than their result (Our result's $C$ term (number of cluster) is in the square root). According to Theorem 1 in [Cutkosky et al. 2021], the regret is upper bounded by,

$$E[R(T)] \leq O\left(\sqrt{CT} + (C^{\frac{1}{2}}(\frac{D_H}{D_L})^2 + \frac{D_H}{D_L} + C^{\frac{1}{2}})\log(T)T^{\frac{1}{2}}\right), \tag{8}$$

Comparing (5) and (8), we can see that our result is better than their result (Our result's $\log(T)$ term is in the square root).

## 5    Experiments

We aim to answer the following questions through experiments:

1. Can SGSD outperform other algorithms in Unimodal bandit environments?
2. Can HUUCB outperform other hierarchical bandit algorithms (such as TLP) in MAB-CU environments? Meanwhile, we intend to validate whether the simulation result conforms to our theoretical analysis – specifically, does HUUCB's new $\frac{D_H}{D_L}\sqrt{\frac{\text{const}}{M_i(t)}}$ bonus term help in cluster selection?

To answer these questions, we consider two sets of experiments:

1. Learning in a synthetic Unimodal bandit setting, taken from Combes and Proutiere [2014]. First, we consider $N = 17$ arms with Bernoulli rewards which $\mu = [0.1, 0.2...0.9, 0.8...0.1]$ and the rewards are Unimodal. Then, we consider $N = 129$, and the expected rewards form a triangular shape as in the previous example $N = 17$ ($\mu$ is between $[0.1, 0.9]$). We evaluate three algorithms: our SGSD algorithm, UCB [Auer et al. 2002], and OSUB [Combes and Proutiere 2014].

2. Bandit learning in the MAB-CU setting. We use a simulated environment of multi-channel mmWave communication. We perform our simulations using MATLAB. Recall from Sect. 1.1 that in this application, an arm is a combination of channel and beam (chosen by the transmitter), and the reward is the received signal strength (RSS) at the receiver. We regard the beams under the same channel as a cluster. We fix the transmitter (i.e., base station) at location [0,0], and we randomly generate four receiver locations from a disk area with a radius of 10 m. The base station is equipped with a uniform linear array (ULA) with four antennas, which are separated by a half wavelength. For the wireless channel model, we assume that there either exists only one line-of-sight (LOS) path or one non-line-of-sight (NLOS) path if the LOS path is blocked. We obtain the RSS under channel $i$ and beam $j$ in each time step using Monte-Carlo simulations, following the free-space signal propagation model: $RSS_{ij} = \alpha_i P^{TX} G_j^{RX} G_j^{TX} (\frac{\lambda_i}{4\pi d})^2$ [Molisch 2012], where $\alpha_i$ is the random path fading amplitude under channel $i$ (since there's a dominant LoS path, $\alpha_i$ is assumed to follow the Rician distribution [Samimi et al. 2016]), $G_j^{RX}$ and $G_j^{TX}$ are the gains of the receive and transmit antennas for beam $j$ (in the directions of angle-of-arrival (AoA) and angle-of-departure (AoD)), respectively, $\lambda_i$ is the wavelength for channel $i$ ($j \in \text{Cluster}_i$), $d$ is the distance between transmitter and receiver, and $P^{TX}$ is the transmit power. Note that, the AoA, AoD, distance $d$, and fading $\alpha_i$ are all unknown to the transmitter during the bandit algorithm execution. We denote $RSS_{ij}$ as the reward for beam $j$ under channel $i$. The system is assumed to operate at 28 GHz center carrier frequency (based on the 3GPP TS 38.101-1/2 standard Lopez et al. [2019]), has a bandwidth of 100 MHz, and uses 16-QAM modulation. We consider two scenarios: 1) two channels (clusters): [28–28.1, 28.1–28.2] GHz, 2) five channels (clusters): [27.8–27.9, 27.9–28, 28–28.1, 28.1–28.2, 28.2–28.3] GHz. For each channel, there are a total of 16 beams and we only consider TX beam selection (each beam's width is 5 °C and the step between adjacent beams' angles is 10 °C.).

We evaluate the following algorithms: (1) our HUUCB algorithm; (2) UCB algorithm [Auer et al. 2002] and (3) Instantiations of the Two-level Policy framework (TLP) of Pandey et al. [2007] using different base algorithms for intra-cluster arm selection. In each step, the TLP algorithm chooses a cluster first, and then an actual arm within the cluster is selected. The key difference between algorithms under TLP framework and our HUUCB algorithm is that, TLP uses an aggressive confidence bound for selecting clusters, which

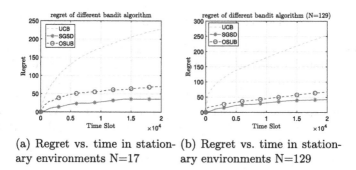

(a) Regret vs. time in station-   (b) Regret vs. time in stationary environments N=17            ary environments N=129

**Fig. 3.** Comparison between UCB and SGSD algorithm under Unimodal setting

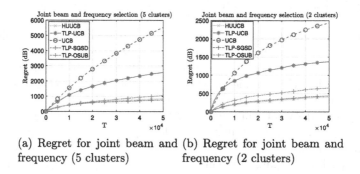

(a) Regret for joint beam and   (b) Regret for joint beam and
frequency (5 clusters)          frequency (2 clusters)

**Fig. 4.** Comparison of cumulative regret among HUUCB and existing algorithms

does not follow the "optimism in the face of uncertainty" principle, and does
not have theoretical guarantees. We consider TLP composed with three base
algorithms: first, UCB, which does not utilize the Unimodal property in each
cluster; second, SGSD, our Algorithm 1; third, OSUB [Combes and Proutiere
2014] – we call the resulting algorithms TLP-UCB, TLP-SGSD, and TLP-
OSUB respectively.

## 5.1    Simulation Result

Figure 3 shows the cumulative regret of our SGSD algorithm in the above-
mentioned synthetic Unimodal bandit setting. Regrets are calculated averaging
over 100 independent runs. SGSD significantly outperforms the UCB algorithm.
This is because the UCB algorithm does not utilize the Unimodal property.
Meanwhile, the SGSD algorithm has better performance than the OSUB algorithm. We speculate that SGSD's improved performance is due to its use of $D_H$
and $D_L$, in contrast to OSUB.

Figure 4(a) shows the cumulative regret of the joint beam and frequency
selection with 5 clusters, and Fig. 4(b) shows the same result with 2 clusters.

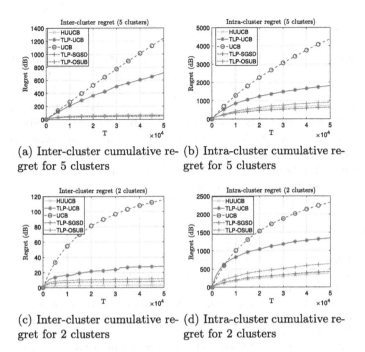

(a) Inter-cluster cumulative regret for 5 clusters

(b) Intra-cluster cumulative regret for 5 clusters

(c) Inter-cluster cumulative regret for 2 clusters

(d) Intra-cluster cumulative regret for 2 clusters

**Fig. 5.** Comparison of Intra-cluster and inter-cluster cumulative regret

Regrets are calculated averaging over 20 independent runs. From Fig. 4(a) and 4(b), we can see that HUUCB has lower regret than the UCB and TLP-UCB algorithm. This result is consistent with our expectation since the Unimodal property in each cluster can help the algorithm converge faster. Meanwhile, we can see that our HUUCB algorithm has a similar performance as TLP-SGSD and TLP-OSUB algorithms, and has the best performance in the 2-cluster setting.

To further examine the advantage of our proposed algorithm over the baseline, we analyze the *inter-cluster* and *intra-cluster* cumulative regret of all algorithms. Intra-cluster cumulative regret is the regret that the algorithm chooses an arm that is not the optimal arm in the currently chosen cluster; formally, $R_{\text{intra}}(T) = \sum_{t=1}^{T}(\nu_{i_t} - \mu_{j_t})$. Inter-cluster cumulative regret is the regret that the algorithm chooses a suboptimal cluster, i.e. the cluster that does not contain the optimal arm; formally $R_{\text{inter}}(T) = \sum_{t=1}^{T}(\mu_{j^*} - \nu_{i_t})$. It can be seen that the regret can be decomposed as: $R(T) = R_{\text{intra}}(T) + R_{\text{inter}}(T)$. From Fig. 5, we can see that UCB and TLP-UCB algorithms incur both large inter-cluster cumulative regret and intra-cluster cumulative regret. This is because both algorithms do not fully utilize Unimodal and Hierarchical properties. Meanwhile, we can see that HUUCB has comparable performance to TLP-SGSD. In the 2-cluster setting, HUUCB has better inter-cluster regret than TLP-SGSD - this may be due to the setting of the extra bonus term in HUUCB.

## 6    Conclusion

In this paper, we combine the ideas of the Hierarchical bandit and Unimodal bandit algorithms and propose a novel Hierarchical Unimodal UCB algorithm. First, we adapt the Stochastic Golden Search (SGS) algorithm into discrete arm settings (called SGSD), and we derive a regret bound for SGSD. Then, we propose a novel HUUCB algorithm that is based on the SGSD algorithm. Simulation result shows that our HUUCB algorithm outperforms TLP-UCB, using one benchmark dataset. For future work, we plan to derive a gap-dependent $\log(T)$-style regret bound for the HUUCB algorithm and validate the regret bound in various simulation scenarios.

## References

Yu, J.Y., Mannor, S.: Unimodal bandits (2011)

Auer, P., Cesa-Bianchi, N., Fischer, P.: Finite-time analysis of the multiarmed bandit problem. Mach. Learn. **47**(2–3), 235–256 (2002)

Pandey, S., Chakrabarti, D., Agarwal, D.: Multi-armed bandit problems with dependent arms. In: Proceedings of the 24th international conference on Machine learning, pp. 721–728 (2007)

Thompson, W.R.: On the likelihood that one unknown probability exceeds another in view of the evidence of two samples. Biometrika **25**(3–4), 285–294 (1933)

Sun, M., Li, M., Gerdes, R.: Truth-aware optimal decision-making framework with driver preferences for v2v communications. In: 2018 IEEE Conference on Communications and Network Security (CNS), pp. 1–9. IEEE (2018)

Wen, W., Cheng, N., Zhang, N., Yang, P., Zhuang, W., Shen, X.: Fast mmwave beam alignment via correlated bandit learning. IEEE Trans. Wirel. Commun. **18**(12), 5894–5908 (2019)

Hashemi, M., Sabharwal, A., Koksal, C.E., Shroff, N.B.: Efficient beam alignment in millimeter wave systems using contextual bandits. In: IEEE INFOCOM 2018, pp. 2393–2401. IEEE (2018)

Lopez, A.V., Chervyakov, A., Chance, G., Verma, S., Tang, Y.: Opportunities and challenges of mmwave nr. IEEE Wirel. Commun. **26**(2), 4–6 (2019)

Nguyen, T.T., Lauw, H.W.: Dynamic clustering of contextual multi-armed bandits. In: Proceedings of the 23rd ACM International Conference on Conference on Information and Knowledge Management, pp. 1959–1962 (2014)

Jedor, M., Perchet, V., Louedec, J.: Categorized bandits. Adv. Neural Inf. Process. Syst. **32**, 1–11 (2019)

Bouneffouf, D., Parthasarathy, S., Samulowitz, H., Wistub, M.: Optimal exploitation of clustering and history information in multi-armed bandit. arXiv preprint arXiv:1906.03979 (2019)

Carlsson, E., Dubhashi, D., Johansson, F.D.: Thompson sampling for bandits with clustered arms. arXiv preprint arXiv:2109.01656 (2021)

Zhao, T., Li, M., Poloczek, M.: Fast reconfigurable antenna state selection with hierarchical thompson sampling. In: ICC 2019–2019 IEEE International Conference on Communications (ICC), pp. 1–6. IEEE (2019)

Yang, J., Zhong, Z., Tan, V.Y.F.: Optimal clustering with bandit feedback. arXiv preprint arXiv:2202.04294 (2022)

Kumar, S., Gao, H., Wang, C., Chang, K.C.C., Sundaram, H.: Hierarchical multi-armed bandits for discovering hidden populations. In: Proceedings of the 2019 IEEE/ACM International Conference on Advances in Social Networks Analysis and Mining, pp. 145–153 (2019)

Singh, R., Liu, F., Sun, Y., Shroff, N.: Multi-armed bandits with dependent arms. arXiv preprint arXiv:2010.09478 (2020)

Combes, R., Proutiere, A.: Unimodal bandits: regret lower bounds and optimal algorithms. In: International Conference on Machine Learning, pp. 521–529. PMLR (2014)

Zhang, Y., Basu, S., Shakkottai, S., Heath Jr, R.W.: Mmwave codebook selection in rapidly-varying channels via multinomial thompson sampling. In: Proceedings of the Twenty-second International Symposium on Theory, Algorithmic Foundations, and Protocol Design for Mobile Networks and Mobile Computing, pp. 151–160 (2021)

Blinn, N., Boerger, J., Bloch, M.: mmwave beam steering with hierarchical optimal sampling for unimodal bandits. In: ICC 2021-IEEE International Conference on Communications, pp. 1–6. IEEE (2021)

Bubeck, S.,Cesa-Bianchi, N.: Regret analysis of stochastic and nonstochastic multi-armed bandit problems. arXiv preprint arXiv:1204.5721 (2012)

Abbasi-Yadkori, Y., Pacchiano, A., Phan, M.: Regret balancing for bandit and rl model selection. arXiv preprint arXiv:2006.05491 (2020)

Cutkosky, A., Dann, C., Das, A., Gentile, C., Pacchiano, A., Purohit, M.: Dynamic balancing for model selection in bandits and rl. In: International Conference on Machine Learning, pp. 2276–2285. PMLR (2021)

Molisch, A.F.: Wireless Communications, vol. 34. John Wiley & Sons, Hoboken (2012)

Samimi, M.K., MacCartney, G.R., Sun, S., Rappaport, T.S.: 28 ghz millimeter-wave ultrawideband small-scale fading models in wireless channels. In: 2016 IEEE 83rd Vehicular Technology Conference (VTC Spring), pp. 1–6. IEEE (2016)

# Hypothesis Transfer in Bandits
# by Weighted Models

Steven Bilaj$^{(\boxtimes)}$, Sofien Dhouib, and Setareh Maghsudi

Eberhard Karls University of Tübingen, Tübingen, Germany
{steven.bilaj,sofiane.dhouib,setareh.maghsudi}@uni-tuebingen.de

**Abstract.** We consider the problem of contextual multi-armed bandits
in the setting of hypothesis transfer learning. That is, we assume having
access to a previously learned model on an unobserved set of contexts,
and we leverage it in order to accelerate exploration on a new bandit
problem. Our transfer strategy is based on a re-weighting scheme for
which we show a reduction in the regret over the classic Linear UCB
when transfer is desired, while recovering the classic regret rate when
the two tasks are unrelated. We further extend this method to an arbitrary amount of source models, where the algorithm decides which model
is preferred at each time step. Additionally we discuss an approach where
a dynamic convex combination of source models is given in terms of a
biased regularization term in the classic LinUCB algorithm. The algorithms and the theoretical analysis of our proposed methods substanti-
ated by empirical evaluations on simulated and real-world data.

**Keywords:** Multi-armed bandits · Linear reward models ·
Recommender systems · Transfer learning

## 1 Introduction

The *multi-armed bandit* problem (MAB) [7,22,27] revolves about maximizing the
reward collected by playing actions from a predefined set, with uncertainty and
limited information about the observed payoff. At each round, the bandit player
chooses an arm according to some rule that balances the exploitation of the currently available knowledge and the exploration of new actions that might have
been overlooked while being more rewarding. This is known as the exploration-
exploitation trade-off. MAB's find applications in several areas [6], notably in
recommender systems [13,16,18,33]. In these applications, the number of actions
to choose from can grow very large, and it becomes provably detrimental to the
algorithm's performance to ignore any side information provided when playing
an action or dependence between the arms [4]. Considering such information
defines the *Stochastic Contextual Bandits* [1,8,14,16] setting, where playing an
action outputs a context-dependent reward, where a context can correspond to

---

**Supplementary Information** The online version contains supplementary material
available at https://doi.org/10.1007/978-3-031-26412-2_18.

M.-R. Amini et al. (Eds.): ECML PKDD 2022, LNAI 13716, pp. 284–299, 2023.
https://doi.org/10.1007/978-3-031-26412-2_18

a user's profile and/or the item to recommend in recommender system applications. Hence, less exploration is required as arms with correlating context vectors share information, thus further reducing uncertainty in the reward estimation. This ultimately led to lower regret bounds and improved performance [1].

While the stochastic contextual bandit problem solves the aforementioned issues, it disregards the possibility of learning from previously trained bandits. For instance, assume a company deploys its services in a new region. Then it would waste the information it has already learned from its previous recommending experience if it is not leveraged to accelerate the recognition of the new users' preferences. Such scenarios have motivated transfer learning for bandits [13,18,24,26], which rely on the availability of contexts of the previously learned tasks to the current learner. However, regarding a setup where context vectors correspond to items which have been selected by a user, privacy issues are encountered in healthcare applications [21,25] for instance, the aim being to recommend a treatment based on a patient's health state. Indeed, accessing the contexts of the previous tasks entails the history of users' previous activities. Moreover, in engineering applications such as scheduling of radio resources [2], storage issues [17,19,29] might arise when needing access to the context history of previous tasks. These problems would render algorithms depending on previous tasks' contexts inapplicable.

In this work, we aim to reduce exploration by exploiting knowledge from a previously trained contextual bandit accessible only through its parameters, thus accelerating learning if such model is related to the one at hand, and ultimately decreasing the regret. We extend this idea by including an arbitrary amount of models increasing the likelihood of including useful knowledge. To summarize our contributions, we propose a variation of the Linear Upper Confidence Bound (*LinUCB*) algorithm, which has access to previously trained models called source models. The knowledge transfer takes place by using an evolving convex combination of sources models and a LinUCB model, called a target model, estimated with the collected data. The combination's weights are updated according to two different weighting update strategies which minimize the required exploration factor and consecutively the upper regret bound, while also taking a lack of information into consideration. Our regret bound is at least as good as the classic LinUCB one [1], where the improvement depends on the quality of the source models. Moreover, we prove that if the source model used for transfer is not related to our problem, then it will be discarded early on and we recover the LinUCB regret rate. In other words, our algorithm is immune against negative transfer. We test our algorithm on synthetic and real data sets and show experimentally how the overall regret improves on the classic model.

The rest of the paper is organized as follows. We discuss related work in Sect. 2 and formulate our problem in Sect. 3, then we provide and analyse our weighting solution in Sect. 4. This is followed by an extension to the case where one has access to more than one trained model in Sect. 5. Finally, the performance of our algorithm is assessed in Sect. 6.

## 2    Related Work

We hereby discuss two families of contributions related to ours, namely transfer for multi-armed bandits, and hypothesis transfer learning.

**Transfer for MAB's.** To the best of our knowledge, tUCB [5] is the first algorithm to tackle transfer in an MAB setting. Given a sequence of bandit problems picked from a finite set, it uses a tensor power method to estimate their parameters in order to transfer knowledge to the task at hand, leading to a substantial improvement over UCB. Regarding the richer contextual MAB setting, MT-LinUCB [24] reduces the confidence set of the reward estimator by using knowledge from previous episodes. More recently, transfer for MAB's has been applied to recommender systems [13,18], motivated by the cold start problem where a lack of initial information requires more exploration at the cost of higher regret. The TCB algorithm [18] assumes access to correspondence knowledge between the source and target tasks, in addition to contexts, and achieves a regret of $O(d\sqrt{n \log n})$ as in the classic LinUCB case, with empirical improvement. The same regret rate holds for the T-LinUCB algorithm [13], which exploits prior observations to initialize the set of arms, in order to accelerate the training process. The main difference of our formulation with respect to the previous ones is that we assume having access only the preference vectors of the previously learnt tasks, without their associated contexts, which goes in line with the Hypothesis Transfer Learning setting. Even with such a restriction, we keep the LinUCB regret rate and we show that the regret is lower in the case source parameters that are close to those of the task at hand.

**Hypothesis Transfer Learning.** Using previously learned models in order to improve learning on a new task defines the hypothesis transfer learning scenario, also known as model reuse or learning from auxiliary classifiers. Some lines of work consider building the predictor of the task at hand as the sum of a source one (possibly a weighted combination of different models) and the one learned from the available data points [10,28,30]. Such models were thoroughly analyzed in [11,12,20] by providing performance guarantees. The previously mentioned additive form of the learned model was further studied and generalized to a large family of transformation functions in [9]. In online learning, the pioneering work of [31] relies on a convex combination instead of a sum, with adaptive weights. More recently, the Condor algorithm [32] was proposed and theoretically analyzed to handle the concept drift scenario, relying on biased regularization w.r.t. a convex combination of source models. Our online setting involves transfer with decisions over a large set of alternatives at each time step, thus it becomes crucial to leverage transfer to improve exploration. To this end, we use a weighting scheme inspired by [31] but that relies on exploration terms rather than on how the models approximate the rewards.

## 3    Problem Formulation

We consider a contextual bandit setting in which at each time $k$, playing an action $a$ from a set $\mathcal{A}$ results in observing a context vector $\mathbf{x}_{a_k} \in \mathbb{R}^d$ assumed to satisfy $\|\mathbf{x}_{a_k}\| \leq 1$ , in addition to a reward $r(k)$. We further define the matrix induced norm: $\|\mathbf{x}\|_{\mathbf{A}} := \sqrt{\mathbf{x}^T \mathbf{A} \mathbf{x}}$ for any vector $\mathbf{x} \in \mathbb{R}^d$ and any matrix $\mathbf{A} \in \mathbb{R}^{d \times d}$. The classical case aims to find an estimation $\hat{\boldsymbol{\theta}}$ of an optimal bandit parameter $\boldsymbol{\theta}^* \in \mathbb{R}^d$ which determines the rewards $r$ of each arm with context vector $\mathbf{x}_a$ in a linear fashion $r = \mathbf{x}_a^T \boldsymbol{\theta}^* + \epsilon$ up to some $\sigma$-subgaussian noise $\epsilon$. The decision at time $k$ is made according to an upper confidence bound (UCB) associated to $\hat{\boldsymbol{\theta}}(k)$:

$$a_k = \arg\max_{a \in \mathcal{A}} \mathbf{x}_a^T \hat{\boldsymbol{\theta}}(k) + \gamma \sqrt{\mathbf{x}_a^T \mathbf{A}^{-1}(k) \mathbf{x}_a}, \tag{1}$$

where $\gamma > 0$ is a hyperparameter estimated through the derivation of the UCB later and $\mathbf{A}(k) := \lambda \mathbf{I}_d + \sum_{k'=1}^{k} \mathbf{x}_{a_{k'}} \mathbf{x}_{a_{k'}}^T$. The latter term in the sum (1) represents the exploration term which decreases the more arms are explored. $\hat{\boldsymbol{\theta}}(k)$ is computed through regularized least-squares regression with regularization parameter $\lambda > 0$: $\hat{\boldsymbol{\theta}}(k) = \mathbf{A}^{-1}(k) \mathbf{D}^T(k) \mathbf{y}(k)$, with $\mathbf{D}(k) = [\mathbf{x}_{a_i}^T]_{i \in \{1,...,k\}}$ and $\mathbf{y}(k) = [r(i)]_{i \in \{1,...,k\}}$ as the concatenation of selected arms' context vectors and corresponding rewards respectively. We alter this decision making approach with the additional use of a previously trained source bandit. Inspired by [31], we transfer knowledge from one linear bandit model to another by a weighting approach. We denote the parameters of the source bandit by $\boldsymbol{\theta}_S \in \mathbb{R}^d$. The bandit at hand's parameters are then estimated as:

$$\hat{\boldsymbol{\theta}} = \alpha_S \boldsymbol{\theta}_S + \alpha_T \hat{\boldsymbol{\theta}}_T(k), \tag{2}$$

with weights $\alpha_S, \alpha_T \geq 0$ satisfying $\alpha_S + \alpha_T = 1$. More important is how the exploration term changes and how it affects the classic regret bound. From [1] we know that the upper bound of the immediate regret in a linear bandit algorithm directly depends on the exploration term of the UCB. We aim to reduce the required exploration with the use of the source bandits knowledge, in order to accelerate the learning process as well as reducing the upper regret bound. For the analysis we consider the pseudo-regret [3] defined as:

$$R(n) = n \max_{a \in \mathcal{A}} \mathbf{x}_a^T \theta^* - \sum_{k=1}^{n} \mathbf{x}_{a_k}^T \theta^*. \tag{3}$$

Our goal is to prove that this quantity is reduced if the source bandit is related to the one at hand, whereas its rate is not worsened in the opposite case.

## 4    Weighted Linear Bandits

The model we use features dynamic weights, thus at time $k$, we use the following model for our algorithm:

$$\hat{\boldsymbol{\theta}}(k) = \alpha_S(k)\boldsymbol{\theta}_S + \alpha_T(k)\hat{\boldsymbol{\theta}}_T(k), \tag{4}$$

with $\hat{\boldsymbol{\theta}}_T(k)$ being updated like in the classic LinUCB case [1] and $\boldsymbol{\theta}_S$ remaining constant. To devise an update rules of the weights, we first re-write the new UCB expression as:

$$\text{UCB}(a) = \mathbf{x}_a^T \left( \alpha_S(k)\boldsymbol{\theta}_S + \alpha_T(k)\hat{\boldsymbol{\theta}}_T(k) \right) + (\alpha_S(k)\gamma_S + \alpha_T(k)\gamma_T) \|\mathbf{x}_a\|_{\mathbf{A}^{-1}}, \tag{5}$$

with $\gamma_S \geq \|\boldsymbol{\theta}^* - \boldsymbol{\theta}_S\|_{\mathbf{A}(k)}$ and $\gamma_T \geq \|\boldsymbol{\theta}^* - \hat{\boldsymbol{\theta}}_T(k)\|_{\mathbf{A}(k)}$ as confidence set bounds for the source bandit and target bandit respectively. We retrieve the classic case by setting $\alpha_S(k)$ to zero *i.e.* erasing all influence from the source. The confidence set bound $\gamma_T$ has already been determined in [1].

As mentioned in Sect. 3 we aim to reduce the required exploration in order to reduce the upper regret bound. Thus we select the weights such that the exploration term in (5) is minimized.

### 4.1 Weighting Update Strategies

We want to determine the weights after each time step such that:

$$\alpha_S, \alpha_T = \operatorname*{arg\,min}_{\substack{\alpha_S', \alpha_T' \geq 0 \\ \alpha_S' + \alpha_T' = 1}} \alpha_S' \gamma_S + \alpha_T' \gamma_T. \tag{6}$$

The above minimization problem is solved for:

$$\alpha_S = \mathbb{1}_{\gamma_S \leq \gamma_T}, \quad \alpha_T = 1 - \alpha_S. \tag{7}$$

This strategy would guarantee an upper regret bound at least as good as the LinUCB bound in [1] as will be shown in the analysis section later. However, without any knowledge of the relation between source and target tasks, our upper bound on the confidence set of the source bandit is rather loose:

$$\|\boldsymbol{\theta}^* - \boldsymbol{\theta}_S\|_{\mathbf{A}(k)} = \sqrt{\lambda U^2 + \|\mathbf{D}(k)(\boldsymbol{\theta}^* - \boldsymbol{\theta}_S)\|_2^2} \leq \sqrt{4\lambda + \|\overline{\mathbf{y}}(k) - \mathbf{y}_S(k)\|_2^2},$$

with $\|\theta^* - \theta_S\|_2 = U$, $\mathbf{y}_S$ as the concatenation of the source estimated rewards and $\overline{\mathbf{y}}$ as the concatenation of the observed mean rewards for each arm. Naturally after every time step, each entry in $\overline{\mathbf{y}}$ corresponding to the latest pulled arm needs to be updated to their mean value. The mean values are taken in order to cancel out the noise term in the observations. Also, we have $U \leq 2$ in case the vectors show in opposing directions and we additionally assume that $\|\theta^*\|, \|\theta_S\| \leq 1$. An upper bound on the confidence set $\gamma_T$ of the target bandit has been determined in [1]:

$$\gamma_T = \sqrt{d \log\left(1 + \frac{k}{d\lambda}\right) + \log\left(\frac{1}{\delta^2}\right)}. \tag{8}$$

As such, $\gamma_T$ grows with $\sqrt{\log(k)}$ and later on in the analysis we show if $\boldsymbol{\theta}_S \neq \boldsymbol{\theta}^*$ then an upper bound on $\gamma_S$ grows with at least $\sqrt{k}$. Consequently, in theory there is some point in time where $\gamma_S$ will outgrow $\gamma_T$, meaning that the source bandit will be discarded. As already mentioned, our estimation of $\gamma_S$ can be loose due to our lack of information on the euclidean distance term $U$, thus we potentially waste a good source bandit with this strategy. Additionally we would only use one bandit at a time this way instead of the span of two bandits for example. Alternatively we can adjust the strategy in (6) by adding a regularization term in the form of KL-divergence. By substituting $\alpha_T = 1 - \alpha_S$ we get:

$$\alpha_S(k+1) = \arg\min_{\alpha_S \in [0,1]} \left\langle \begin{pmatrix} \alpha_S \\ 1 - \alpha_S \end{pmatrix}, \begin{pmatrix} \gamma_S \\ \gamma_T \end{pmatrix} \right\rangle + \frac{\mathrm{KL}(\boldsymbol{\alpha}\|\boldsymbol{\alpha}(k))}{\beta}, \qquad (9)$$

with $\boldsymbol{\alpha} := (\alpha_S, 1 - \alpha_S)^T$ being a vector containing both weights. The addition of the KL divergence term forces both weights to stay close to their previous value, where $\beta > 0$ is a hyper parameter controlling the importance of the regularization. Problem (9) is solved for:

$$\alpha_S(k+1) = \frac{1}{1 + \frac{1-\alpha_S(k)}{\alpha_S(k)} \exp(\beta(\gamma_S - \gamma_T))}, \qquad (10)$$

which is a softened version of our solution in (7), but in this case the source bandit will not be immediately discarded if the upper bound on its confidence set becomes larger than the target bandit's.

## 4.2   Analysis

We are going to analyse how the upper regret bound changes, within our model in comparison to [1]. All proofs are given in the appendix. First we bound the regret for the hard update approach, not including the KL-divergence term in (7):

**Theorem 1.** *Let $\{\mathbf{x}_{a_k}\}_{k=1}^{N}$ be sequence in $\mathbb{R}^d$, $U := \|\boldsymbol{\theta}_S - \boldsymbol{\theta}^*\|$ and $R_T$ be the classic regret bound of the linear model [1]. Let $m := \min(\kappa, n)$ and $\delta \leq \exp(-2\lambda)$. Then, with a probability at least $1 - \delta$, the regret of the hard update approach for the weighted LinUCB algorithm is bounded as follows:*

$$R(n) \leq U\sqrt{8md\log\left(1 + \frac{m}{d\lambda}\right)(\lambda + m)} + R_T(n) - R_T(m) \leq R_T(n) \qquad (11)$$

*with $\kappa$ satisfying:*

$$\kappa = \left\lfloor 2\left[d\left(\frac{1}{U^2} - \lambda\right) + \lambda\left(\frac{2}{U^2} - \frac{1}{2}\right)\right]\right\rfloor. \qquad (12)$$

The value for $\kappa$ essentially gives a threshold such that we have $\gamma_S < \gamma_T$ for every $k < \kappa$. As expected, for better sources *i.e.* low values $U$, $\kappa$ increases meaning the source is viable for more time steps. Also notable is how we see an increasing value for $\kappa$ at high dimensional spaces. This is most likely due to the fact, that at higher dimensions the classic algorithm requires more time steps, in order to find a suitable estimation, thus having a larger confidence set bound. In these instances a trained source bandit would be viable early on. The regret is reduced for lower values of $U$ and the time $\kappa$ at which a source is discarded is extended. For source bandits satisfying $\|\boldsymbol{\theta}_S - \boldsymbol{\theta}^*\|_2 = 2$, we would retrieve the classic regret bound, preventing negative transfer.

Next we show what happens in case of a negative transfer for the softmax update strategy, *i.e.* the source does not provide any useful information at all and worsens the regret rate with $\gamma_S > \gamma_T$ at all time steps.

**Theorem 2.** *Let $\{\mathbf{x}_{a_k}\}_{k=1}^N$ be sequence in $\mathbb{R}^d$ and the minimal difference between confidence set bounds given as $\Delta_{\min} = \min_{k \in \{0,\dots,N\}}(\gamma_S(k) - \gamma_T(k))$, with $\gamma_S > \gamma_T$ for all time steps and the initial target weight denoted by $\alpha_T(0)$. Then with probability of at least $1 - \delta$ an upper regret bound $R(n)$ in case of a negative transfer scenario is given by:*

$$R(n) \leq \frac{(1 - \alpha_T(0))}{e\beta\alpha_T(0)(1 - \exp(-\beta\Delta_{\min}))} + R_T(n) \tag{13}$$

Theorem 2 shows that in case of a negative transfer, the upper regret bound is increased by at most a constant term and vanishes in the case of $\beta \to \infty$ retrieving the hard update rule.

## 5   Weighted Linear Bandits with Multiple Sources

Up until now we only used a single source bandit, but our model can easily be extended to an arbitrary amount of different sources. Assuming we have $M$ source bandits $\{\theta_{S,j}\}_{j=1}^M$, we define $\hat{\theta}$ as:

$$\hat{\boldsymbol{\theta}} = \sum_{j=1}^M \alpha_{S,j}\boldsymbol{\theta}_{S,j} + \alpha_T\hat{\boldsymbol{\theta}}_T, \tag{14}$$

with $\alpha_{S,j}, \alpha_T \geq 0\ \forall 1 \leq j \leq M$ and $\alpha_T + \sum_{j=1}^M \alpha_{S,j} = 1$. With this each source bandit yields its own confidence set bound $\gamma_{S,j}$. Similarly to (5) we retrieve for the UCB with multiple sources:

$$\mathrm{UCB}(a) = \mathbf{x}_a^T\left(\sum_{j=1}^M \alpha_{S,j}(k)\boldsymbol{\theta}_{S,j} + \alpha_T(k)\hat{\boldsymbol{\theta}}_T(k)\right) + \boldsymbol{\alpha}^T(k)\boldsymbol{\gamma}\|\mathbf{x}_a\|_{\mathbf{A}^{-1}(k)}, \tag{15}$$

with $\boldsymbol{\alpha}(k) = (\alpha_{S,1}(k), ..., \alpha_{S,M}(k), \alpha_T(k))^T$ and $\boldsymbol{\gamma} = (\gamma_{S,1}, ..., \gamma_{S,M}, \gamma_T)^T$. As for the weight updates the same single source strategies apply *i.e.* the minimization of the exploration term in the UCB function:

$$\boldsymbol{\alpha}(k+1) = \underset{\alpha \in \mathfrak{P}_{M+1}}{\arg\min} \, \boldsymbol{\alpha}^T(k)\boldsymbol{\gamma} + \frac{1}{\beta}\mathrm{KL}(\boldsymbol{\alpha}||\boldsymbol{\alpha}(k)), \tag{16}$$

where $\mathfrak{P}_{M+1}$ is the $(M+1)$−dimensional probability simplex. The solution of the previous problem is:

$$\alpha_{S,m}(k+1) = \frac{\alpha_{S,m}(k) \exp{(-\beta\gamma_{S,m})}}{\sum_{j=1}^{M} \alpha_{S,j}(k) \exp(-\beta\gamma_{S,j}) + \alpha_T(k) \exp(-\beta\gamma_T)}. \tag{17}$$

This is basically the solution of (10) generalized to multiple sources. In the decisions making it favours the bandit with the lowest upper bound $\gamma$ of their confidence set. When we take the limit $\beta \to \infty$ in (16) the KL-divergence term vanishes and we retrieve the hard case:

$$\alpha_{S,j} = \mathbb{1}_{\gamma_{S,j}=\min(\min_i \gamma_{S,i}, \gamma_T)} \tag{18}$$

which forces the weights to satisfy $\alpha_{S,m}, \alpha_T \in \{0,1\}$ for every source index and for all time steps. Thus decision making is done by selecting one single bandit in each round with the lowest value of their respective confidence set bound $\gamma$. The regret of hard update strategy for multiple sources is given by the following theorem:

**Theorem 3.** *Let $\{\mathbf{x}_{a_k}\}_{k=1}^N$ be sequence in $\mathbb{R}^d$ and $\min_m \|\boldsymbol{\theta}_{S,m} - \boldsymbol{\theta}^*\| = U_{\min}$ and the classic regret bound of the linear model up to time step n given by $R_T(n)$ [1]. Let $m := \min(\kappa, n)$ and $\delta \leq \exp(-2\lambda)$. Then with probability of at least $1 - \delta$ the regret of the hard update approach for the weighted LinUCB algorithm with multiple sources is bounded by:*

$$R(n) \leq 4U_{\min}\sqrt{\kappa d \log(1 + \kappa/(d\lambda))(\lambda + \kappa)} - R_T(m) + R_T(n) \leq R_T(n), \tag{19}$$

*with $\kappa$ as:*

$$\kappa = \left\lfloor 2\left[d\left(\frac{1}{U_{\min}^2} - \lambda\right) + \lambda\left(\frac{2}{U_{\min}^2} - \frac{1}{2}\right)\right]\right\rfloor.$$

depending on $U_{\min}$ the multiple source approach benefits from the additional information as the upper bound corresponds to the best source overall. In case of the softmax update strategy, we need to show how the regret changes in case of a negative transfer scenario, *i.e.* the confidence set bounds of any source bandit is larger than the target bound at any time.

**Theorem 4.** *Let* $\{\mathbf{x}_{a_k}\}_{k=1}^{N}$ *be sequence in* $\mathbb{R}^d$, *a total of* $M$ *source bandits being available indexed by* $j$ *and the minimal difference between confidence set bounds set as* $\Delta_{\min,j} = \min_{k \in \{0,...,N\}}(\gamma_{S,j}(k) - \gamma_T(k))$ *for every source* $j$ *with* $\gamma_{S,j} > \gamma_T$ $\forall j$ *at every time step. Additionally the initial target weight is denoted by* $\alpha_T(0)$. *Then with probability* $1 - \delta$ *an upper regret bound* $R(n)$ *in case of a negative transfer scenario is given by:*

$$R(n) \leq \frac{(1 - \alpha_T(0))}{e\beta M\alpha_T(0)} \sum_{j=1}^{M} \frac{1}{(1 - \exp(-\beta\Delta_{\min,j}))} + R_T \qquad (20)$$

In comparison to the single source result, the additional constant is averaged over all sources. Depending on the quality, it can be beneficial to include more source bandits as potentially bad sources would be mitigated.

---

**Algorithm 1:** Weighted LinUCB

---

Initialize: $\hat{\boldsymbol{\theta}}_T(0)$ from $\mathcal{U}([0,1]^d)$, $\alpha_{S,j}(0) = (1 - \alpha_T(0))/M = \frac{1}{2M}$, $U_j > 0$
$\gamma_{S,j} > 0 \ \forall j \in \{1,...,M\}$, $\delta \in [0,1]$, $\gamma_T > 0$, $\lambda > 0$, $\beta > 0$, $\mathbf{A}(0) = \lambda\mathbf{I}$,
$\mathbf{b}(0) = \mathbf{0}$;
**for** $k = 0...N$ **do**

> Pull arm $a_k = \arg\max_a \text{UCB}(a)$ taken from (15);
> Receive estimated rewards from sources and real rewards:
> $r_{S,j}(k)|_{j \in \{0,...,M\}}, r(k)$;
> $\mathbf{A}(k+1) = \mathbf{A}(k) + \mathbf{x}_{a_k}\mathbf{x}_{a_k}^T$;
> $\mathbf{b}(k+1) = \mathbf{b}(k) + r(k)\mathbf{x}_{a_k}$;
> $\hat{\boldsymbol{\theta}}_T(k+1) = \mathbf{A}^{-1}(k+1)\mathbf{b}(k+1)$;
> Store rewards $r_{S,j}(k)|_{j \in \{0,...,M\}}, r(k)$ in vectors $\mathbf{y}_{S,j}(k)|_{j \in \{0,...,M\}}, \mathbf{y}(k)$
> respectively;
> Calculate $\overline{\mathbf{y}}(k)$ from $\mathbf{y}(k)$ such that each entry $r$ corresponding to the
> latest arm $a_k$ pulled is updated to the mean reward $\overline{r}$ of the
> respective arm;
> Update $U_j = \max_{i \in \{0,...,k\}} \frac{|\overline{r}(i) - r_{S,j}(i)|}{\|\mathbf{x}_{a_i}\|}$ for every $j$;
> $\gamma_{S,j} = \sqrt{\lambda U_j + \|\mathbf{y}_{S,j}(k) - \overline{\mathbf{y}}(k)\|}$;
> $\gamma_T = \sqrt{\lambda} + \sqrt{\log \frac{\|\mathbf{A}(k)\|}{\lambda^d \delta^2}}$;
> update source weights $\alpha_{S,j}(k+1)$ according either to softmax rule in
> (17):
> or to the hard update rule in (18);
> update target weight as:
> $\alpha_T(k+1) = 1 - \sum_{j=1}^{M} \alpha_{S,j}(k+1)$;

---

For the practical implementation we use $\gamma_T = \sqrt{\lambda} + \sqrt{\log \frac{\|\mathbf{A}(k)\|}{\lambda^d \delta^2}}$ which is also taken from [1] and gives a tighter confidence set bound on the target estimator. Also we give an estimation for $U_j$ by taking the maximum value of the lower bound induced by the Cauchy-Schwartz inequality $U_j = \|\boldsymbol{\theta}_{S,j} - \boldsymbol{\theta}^*\| \geq \max_{i \in \{0,...,k\}} \frac{|\overline{r}(i) - r_{S,j}(i)|}{\|\mathbf{x}_{a_i}\|}$ at each time step.

## 5.1    Biased Regularization

In [32] a similar approach of model reuse was used in a concept drift scenario for linear classifiers via biased regularization. In [12] the risk generalization analysis for this approach was delivered in a supervised offline learning setting. Their mathematical formulation is stated as following: A classifier is about to be trained given a target training set $(\mathbf{D}, \mathbf{y})$ and a source hypothesis $\boldsymbol{\theta}_{src}$, which is specifically used for a biased regularization term. In contrast to our approach the weighting is only applied the source model, giving an alternate solution to the target classifier. Adapted to a linear bandit model, the optimization problem can be formulated as:

$$\hat{\boldsymbol{\theta}} = \arg\min_{\boldsymbol{\theta}} \|\mathbf{D}\boldsymbol{\theta} - \mathbf{y}\|^2 + \lambda\|\boldsymbol{\theta} - \boldsymbol{\theta}_{src}\|^2. \tag{21}$$

$\boldsymbol{\theta}_{src}$ is a convex combination of an arbitrary amount of given source models $\{\boldsymbol{\theta}_j\}_{j \in \{1,...,M\}}$:

$$\boldsymbol{\theta}_{src} = \sum_{j=1}^{M} \alpha_j \boldsymbol{\theta}_j, \tag{22}$$

As in our model, these weights are not static and are updated after each time step. The update strategy is not chosen to minimize the upper regret bound but can be chosen such that the convex combination is as close as possible to the optimal bandit parameter. The UCB function is then simply given by:

$$\text{UCB}(a) = \mathbf{x}_a^T \hat{\boldsymbol{\theta}} + \gamma\|\mathbf{x}_a\|_{\mathbf{A}^{-1}(k)}, \tag{23}$$

with $\gamma = \sqrt{d\log\left(1 + \frac{k}{d\lambda}\right) + \log\left(\frac{1}{\delta^2}\right)} + \sqrt{\lambda}\|\boldsymbol{\theta}_{src} - \boldsymbol{\theta}^*\|_2$ and the solution to (21):

$$\hat{\boldsymbol{\theta}} = \mathbf{A}^{-1}\mathbf{D}^T\mathbf{y} - (\mathbf{A}^{-1}\mathbf{D}^T\mathbf{D} - \mathbf{I})\boldsymbol{\theta}_{src}. \tag{24}$$

At some point in time we expect the weights to converge to a single source bandit closest to the optimal bandit. But contrary to our original model it is not possible for the model to discard all sources once the target estimation yield better upper bounds for their confidence sets. The upper regret bound is similar to the classic bound with the difference being in one term.

**Theorem 5.** *Let $\{\mathbf{x}_{a_k}\}_{k=1}^N$ be sequence in $\mathbb{R}^d$ and the upper bound of the biggest euclidean distance between any of the $M$ source bandit indexed by $m$ and optimal bandit parameter given by $\max_m \|\boldsymbol{\theta}_{S,m} - \boldsymbol{\theta}^*\| \leq U_{\max}$, then with probability of at least $1 - \delta$ the regret of the biased LinUCB algorithm with multiple sources is upper bounded by:*

$$R(n) \leq \sqrt{8nd\log(\lambda + n/d)}\left(\sqrt{d\log\left(1 + \frac{n}{d\lambda}\right) + \log\left(\frac{1}{\delta^2}\right)} + \sqrt{\lambda}U_{\max}\right) \tag{25}$$

Since we are looking for an upper bound, $U$ is dominated by the largest euclidean distance between the optimal bandit parameter and all given source bandits. Theorem 5 differs from the classic case in the regularization related parameters where we have $\sqrt{\lambda}U_{\max}$ instead of $\sqrt{\lambda}\|\boldsymbol{\theta}^*\|$. For sources with low values of $U$, we improve the overall regret.

## 6    Experimental Results

We test the presented algorithms, *i.e.* the weighted model algorithm as well as the biased regularization algorithm, for single source and multiple source transfers on synthetic and real data sets. The plots include the results from the classical LinUCB approach as well as the EXP4 approach from [15] with target and source models acting as expert, for comparison purposes. Additionally to the regret plots we also showcase the mean of the target weight as a function of time to see how the relevancy of the target estimation evolved.

### 6.1    Synthetic Data Experiments

Our synthetic experiments follow a similar approach to [18]. The target context feature vectors $\mathbf{x}_a$ are drawn from a multivariate Gaussian with variances sampled from a uniform distribution. We chose the number of dimensions $d = 20$ and the number of arms to be 1000. Our optimal target bandit parameter is sampled from a uniform distribution and scaled such that $\|\boldsymbol{\theta}^*\| \leq 1$, thus the rewards are implicitly initialized as well with $r = \mathbf{x}_a^T \boldsymbol{\theta}^* + \epsilon$, with some Gaussian noise $\epsilon \sim \mathcal{N}(0, \sigma^2)$ and $\sigma = 1/\sqrt{2\pi}$. The source bandit parameters $\boldsymbol{\theta}_{S,m}$ are initialized by adding a random noise vector $\boldsymbol{\eta}_m$ to the optimal target bandit parameters for every source bandit to be generated $\boldsymbol{\theta}_{S,m} = \boldsymbol{\theta}^* + \boldsymbol{\eta}_m$. This way we ensure that there is actual information of the target domain in the source bandit parameter. We could also scale $\boldsymbol{\eta}_m$ to determine how much information the respective source yields about the target domain. The regularization parameter was constantly chosen to be $\lambda = 1$ and the initial weights are equally distributed among all available bandit parameters: $\alpha_T = \alpha_{S,m} = \frac{1}{M+1}$. The shown results are the averaged values over 20 runs (Fig. 1).

As we showed in Sect. 4 the upper regret bound is lower for $\beta \to \infty$ *i.e.* the hard update rule which ignores the KL-divergence in the optimization, but we see overall better results than in the classic case with the softmax update strategy as well. The inclusion of eight more source bandits in Fig. 2 improves the sources slightly, though it should be mentioned that all sources generated were similar in quality. Thus we would expect higher improvements in the regret when including significantly better sources. The EXP4 algorithm on the other hand does not perform as well when increasing the number of experts.

### 6.2    Real Data Experiments

The real data sets used for our purposes are taken from the MovieLens sets. Their data include an assemble of thousands of users and corresponding traits

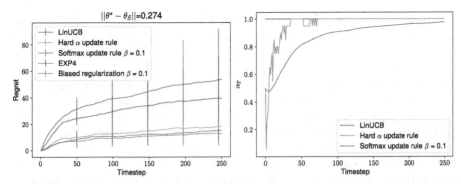

(a) Regret evolution plot labeled by confidence set bound.

(b) Evolution of the target weight $\alpha_T$.

**Fig. 1.** Regret and weight evolution for single source transfer scenario on synthetic data sets. The blue lines showcase the classic LinUCB results. The vertical lines indicate the standard deviation. (Color figure online)

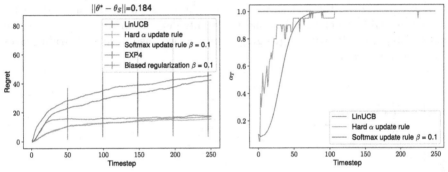

(a) Regret evolution plot labeled by the lowest confidence set bound of all available sources.

(b) Evolution of the target weight $\alpha_T$. Since multiple sources are present, the initial weight is reduced

**Fig. 2.** Regret and weight evolution for multiple source transfer scenario (9 sources) on synthetic data sets. The blue lines showcase the classic LinUCB results. The vertical lines indicate the standard deviation. (Color figure online)

such as age, gender and profession as well as thousands of movies and their genres. Every user has a rating from 1 to 5 given to at least 20 different movies. The movies, rated by a user, function as the available arms for that particular user. The information of the movies apart from the title itself are solely given by their genres. Each movie may have up to three different genres and there are 18 different genres in total. Arms, which are linked to the movies, have context vectors depending on the movies genre only. We design 18-dimensional context vector with each dimension representing a genre. If the movie is associated with a particular genre, the respective dimensional feature is set as $x_i = \frac{1}{\sqrt{S}}$ with $S$ as the total number of genres the movie is associated with. This way we guarantee that every context vector is bounded by 1. the reward of an arm in our bandit setting is simply given by the user rating.

For our purposes we require source bandits for the transfer learning to take place. Therefore we pretrained a bandit for every single user, given all of the movie information, with the classic LinUCB algorithm and stored the respective parameters. This way every single user can function as a potential source for a different user. With all of the users available we grouped them according their age, gender and profession. We enforce every user to only act as source to other users with similar traits. This stems from a general assumption that people with matching traits may also have similar interests. This is a very general assumption made but given all of the information, it is the easiest way to find likely useful sources for every user. In Fig. 3 the results for two individuals of two different groups of users respectively are showcased. Instead of only using one source, we used the multiple source strategy and made use of every user of the same group the individuals are located in, since this way we have a higher chance to find good sources. Even though the real data is far from guaranteed to have a linear reward structure, as well as the fact that important information on the arms' contexts are not available, since ratings usually not only depend on the movie genre, we find satisfying results with converging regrets as well as improved learning rates when including sources.

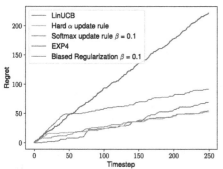 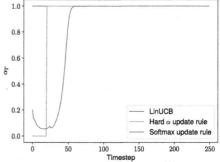

(a) Regret evolution plot with user data taken from the group of 35 to 44 years old female lawyers.

(b) Target weight evolution plot with the respective algorithms labeled with user data taken from the group of 35 to 44 years old female lawyer.

**Fig. 3.** Regret evolution for multiple source transfer scenario on real data sets taken from Movielens data. A group of users are shown with one bandit trained for a random user of each group, while the rest of the users act as source to the respective user. The blue lines showcase the classic LinUCB results. (Color figure online)

# 7    Discussion and Outlook

This work shows that our approach to make use of information from different tasks, without having actually access to concrete data points, is efficient, given the improved regrets. We have proven an upper regret bound of our weighted LinUCB algorithm with the hard update strategy at least as good as the classic LinUCB bound with a regret rate of $O(d\sqrt{n \log n})$, and a converging sublinear negative-transfer term when using the softmax update strategy. Further argument for the utility of our model was given with synthetic and real data experiments. The synthetic data sets showed promising results especially with the softmax update strategy, even without having a guaranteed improved regret bound. The softmax approach uses a convex combination of models, which might be more practical than using one model at a time especially when it comes to high quality sources. This further raises the question whether different weighting update rules, which yield solutions consisting of a span of source models, might be more efficient for transfer. The inclusion of multiple sources further improved the results, indicating that using information from multiple different tasks is more effective then just one, which aligns with our theoretical result in Theorem 3. The real-world data experiments showed improvements as well, even when considering that the rewards did not necessarily follow a linear model and that the available features for the context vector were rather sparse, the transfer of information from similar users almost always led to lower regrets.

In upcoming projects we intend to adapt our approach to non-linear models such as kernelized bandits, since the convex weighting is not limited to just linear models, as well as give a proper regret bound for the softmax update strategy. There is potential in using our transfer model to non stationary bandits, such that each prior estimation of the bandit parameter may act as source for the current setting, thus making use of the information collected in prior instances of the bandit setting. In this case we would need to make assumptions of the change rate of the tasks after a certain amount of time steps. Previous algorithms on non-stationary bandits [23] perform weighting on data points and discard them after some time steps, without evaluating the benefit of the data beforehand. In our setting, previously trained bandit parameters would be used according to their performance.

**Acknowledgements.** This work was supported by Grant 01IS20051 from the German Federal Ministry of Education and Research (BMBF). S. Maghsudi is a member of the Machine Learning Cluster of Excellence, EXC number 2064/1 - Project number 390727645. The authors thank the International Max Planck Research School for Intelligent Systems (IMPRS-IS) for supporting Steven Bilaj.

# References

1. Abbasi-Yadkori, Y., Pál, D., Szepesvári, C.: Improved algorithms for linear stochastic bandits. In: Advances in Neural Information Processing Systems (2011)
2. Amrallah, A., Mohamed, E., Tran, G., Sakaguchi, K.: Radio resource management aided multi-armed bandits for disaster surveillance system. In: Proceedings of the 2020 International Conference On Emerging Technologies For Communications (ICETC2020), Virtual, K1–4 (2020)
3. Audibert, J., Munos, R., Szepesvári, C.: Exploration-exploitation tradeoff using variance estimates in multi-armed bandits. Theoret. Comput. Sci. (2009)
4. Auer, P., Cesa-Bianchi, N., Freund, Y., Schapire, R.: The nonstochastic multiarmed bandit problem. SIAM J. Comput. **32**, 48–77 (2002)
5. Azar, M., Lazaric, A., Brunskill, E.: Sequential transfer in multi-armed bandit with finite set of models. In: Advances In Neural Information Processing Systems (2013)
6. Bouneffouf, D., Rish, I., Aggarwal, C.: Survey on applications of multi-armed and contextual bandits. In: 2020 IEEE Congress On Evolutionary Computation (CEC) (2020)
7. Bush, R., Mosteller, F.: A stochastic model with applications to learning. Ann. Math. Stat. **24**, 559–585 (1953)
8. Chu, W., Li, L., Reyzin, L., Schapire, R.: Contextual bandits with linear payoff functions. In: AISTATS (2011)
9. Du, S., Koushik, J., Singh, A., Póczos, B.: Hypothesis transfer learning via transformation functions. In: Advances In Neural Information Processing Systems (2017)
10. Duan, L., Tsang, I., Xu, D., Chua, T.: Domain adaptation from multiple sources via auxiliary classifiers. In: Proceedings of the 26th Annual International Conference On Machine Learning (2009)
11. Kuzborskij, I., Orabona, F.: Stability and hypothesis transfer learning. In: International Conference On Machine Learning (2013)
12. Kuzborskij, I., Orabona, F.: Fast rates by transferring from auxiliary hypotheses. Mach. Learn. **106**, 171–195 (2017)
13. Labille, K., Huang, W., Wu, X.: Transferable contextual bandits with prior observations. In: Pacific-Asia Conference On Knowledge Discovery And Data Mining. (2021)
14. Langford, J., Zhang, T.: The epoch-greedy algorithm for multi-armed bandits with side information. In: Advances In Neural Information Processing Systems (2007)
15. Lattimore, T., Szepesvári, C.: Bandit Algorithms (2020)
16. Li, L., Chu, W., Langford, J., Schapire, R.: A contextual-bandit approach to personalized news article recommendation. In: Proceedings of the 19th International Conference on World Wide Web (2010)
17. Liau, D., Song, Z., Price, E., Yang, G.: Stochastic multi-armed bandits in constant space. In: International Conference on Artificial Intelligence and Statistics (2018)
18. Liu, B., Wei, Y., Zhang, Y., Yan, Z., Yang, Q. Transferable contextual bandit for cross-domain recommendation. In: Proceedings of the AAAI Conference on Artificial Intelligence (2018)
19. Maiti, A., Patil, V., Khan, A.: Multi-armed bandits with bounded arm-memory: near-optimal guarantees for best-arm identification and regret minimization. In: Advances in Neural Information Processing Systems, vol. 34 (2021)
20. Perrot, M., Habrard, A.: A theoretical analysis of metric hypothesis transfer learning. In: International Conference on Machine Learning (2015)

21. Ras, Z., Wieczorkowska, A., Tsumoto, S.: Recommender systems for medicine and music (2021)
22. Robbins, H. Some aspects of the sequential design of experiments. Bulletin of the American Mathematical Society (1952)
23. Russac, Y., Vernade, C., Cappé, O.: Weighted linear bandits for non-stationary environments. In: Advances In Neural Information Processing Systems (2019)
24. Soare, M., Alsharif, O., Lazaric, A., Pineau, J.: Multi-task linear bandits. In: NIPS2014 Workshop On Transfer and Multi-task Learning: Theory Meets Practice (2014)
25. Stark, B., Knahl, C., Aydin, M., Elish, K.: A literature review on medicine recommender systems. Int. J. Adv. Comput. Sci. Appl. **10**, 6–13 (2019)
26. Suk, J., Kpotufe, S.: Self-tuning bandits over unknown covariate-shifts. In: Algorithmic Learning Theory (2021)
27. Thompson, W.: On the likelihood that one unknown probability exceeds another in view of the evidence of two samples. Biometrika (1933)
28. Tommasi, T., Orabona, F., Caputo, B.: Learning categories from few examples with multi model knowledge transfer. IEEE Trans. Pattern Anal. Mach. Intell. **36**, 928–941 (2014)
29. Xu, X., Zhao, Q.: Memory-constrained no-regret learning in adversarial multi-armed bandits. IEEE Trans. Signal Process. **69**, 2371–2382 (2021)
30. Yang, J., Yan, R., Hauptmann, A.: Cross-domain video concept detection using adaptive SVMs. In: Proceedings of the 15th ACM International Conference on Multimedia (2007)
31. Zhao, P., Hoi, S., Wang, J., Li, B.: Online transfer learning. Artif. Intell. **213**, 76–102 (2014)
32. Zhao, P., Cai, L., Zhou, Z.: Handling concept drift via model reuse. Mach. Learn. **109**, 533–568 (2020)
33. Zhou, Q., Zhang, X., Xu, J., Liang, B.: Large-scale bandit approaches for recommender systems. In: International Conference On Neural Information Processing (2017)

# Multi-agent Heterogeneous Stochastic Linear Bandits

Avishek Ghosh[1(✉)], Abishek Sankararaman[2], and Kannan Ramchandran[3]

[1] Halıcıoğlu Data Science Institute (HDSI), UC San Diego, San Diego, USA
a2ghosh@ucsd.edu
[2] AWS AI, Palo Alto, USA
abisanka@amazon.com
[3] Electrical Engineering and Computer Sciences, UC Berkeley, Berkeley, USA
kannanr@eecs.berkeley.edu

**Abstract.** It has been empirically observed in several recommendation systems, that their performance improve as more people join the system by learning *across heterogeneous users*. In this paper, we seek to theoretically understand this phenomenon by studying the problem of minimizing regret in an $N$ users heterogeneous stochastic linear bandits framework. We study this problem under two models of heterogeneity; *(i)* a personalization framework where no two users are necessarily identical, but are all similar, and *(ii)* a clustering framework where users are partitioned into groups with users in the same group being identical, but different across groups. In the personalization framework, we introduce a natural algorithm where, the personal bandit instances are initialized with the estimates of the global average model and show that, any agent $i$ whose parameter deviates from the population average by $\epsilon_i$, attains a regret scaling of $\widetilde{O}(\epsilon_i \sqrt{T})$. In the clustered users' setup, we propose a successive refinement algorithm, which for any agent, achieves regret scaling as $\mathcal{O}(\sqrt{T/N})$, if the agent is in a 'well separated' cluster, or scales as $\mathcal{O}(T^{\frac{1}{2}+\varepsilon}/(N)^{\frac{1}{2}-\varepsilon})$ if its cluster is not well separated, where $\varepsilon$ is positive and arbitrarily close to 0. Our algorithms enjoy several attractive features of being *problem complexity adaptive and parameter free*—if there is structure such as well separated clusters, or all users are similar to each other, then the regret of every agent goes down with $N$ (collaborative gain). On the other hand, in the worst case, the regret of any user is no worse than that of having individual algorithms per user that does not leverage collaborations.

**Keywords:** Linear bandits · Personalization · Clustering

## 1 Introduction

Large scale web recommendation systems have become ubiquitous in the modern day, due to a myriad of applications that use them including online shopping

---

A. Ghosh and A. Sankararaman—Contributed equally.

---

**Supplementary Information** The online version contains supplementary material available at https://doi.org/10.1007/978-3-031-26412-2_19.

services, video streaming services, news and article recommendations, restaurant recommendations etc., each of which are used by thousands, if not more users, across the world. For each user, these systems make repeated decisions under uncertainty, in order to better learn the preference of each individual user and serve them. A unique feature these large platforms have is that of *collaborative learning*—namely applying the learning from one user to improve the performance on another [26]. However, the sequential online setting renders this complex, as two users are seldom identical [39].

We study the problem of multi-user contextual bandits [6], and quantify the gains obtained by collaborative learning under user heterogeneity. We propose two models of user-heterogeneity: (a) personalization framework where no two users are necessarily identical, but are close to the population average, and (b) clustering framework where only users in the same group are identical. Both these models are widely used in practical systems involving a large number of users (ex. [28, 32, 39, 42]). The personalization framework in these systems is natural in many neural network models, wherein users represented by learnt embedding vectors are not identical; nevertheless similar users are embedded nearby [37, 38, 45, 48]. Moreover, user clustering in such systems can be induced from a variety of factors such as affinity to similar interests, age-groups etc. [33, 38, 43].

Formally, our model consists of $N$ users, all part of a common platform. The interaction between the agents and platform proceeds in a sequence of rounds. Each round begins with the platform receiving $K$ contexts corresponding to $K$ items from the environment. The platform then recommends an item to each user and receives feedback from them about the item. We posit that associated with user $i$, is an preference vector $\theta_i^*$, initially unknown to the platform. In any round, the average reward (the feedback) received by agent $i$ for a recommendation of item, is the inner product of $\theta_i^*$ with the context vector of the recommended item. The goal of the platform is to maximize the reward collected over a time-horizon of $T$ rounds. Following standard terminology, we henceforth refer to an "arm" and item interchangeably, and thus "recommending item $k$" is synonymous to "playing arm $k$". We also use agents and users interchangeably.

**Example Application:** Our setting is motivated through a caricature of a news recommendation system serving $N$ users and $K$ publishers [27]. Each day, each of the $K$ publishers, publishes a news article, which corresponds to the context vector in our contextual bandit framework. In practice, one can use standard tools to embed articles in vector spaces, where the dimensions correspond to topics such as politics, religion, sports etc. [44]. The user preference indicates the interest of a user, and the reward, being computed as an inner product of the context vector and the user preference, models the observation that the more aligned an article is to a user's interest, the higher the reward.

For both frameworks, we propose *adaptive* algorithms; in the personalization framework, our proposed algorithm, namely Personalized Multi-agent Linear Bandits (PMLB) adapts to the level of common representation across users. In particular, if an agents' preference vector is close to the population average, PMLB exploits that and incurs low regret for this agent due to collaboration. On the other hand if an agent's preference vector is far from the population average,

PMLB yields a regret similar to that of OFUL [6] or Linear Bandit algorithms [1] that do not benefit from multi-agent collaboration. In the clustering setup, we propose Successive Clustering of Linear Bandits (SCLB), which is agnostic to the number of clusters, the gap between clusters and the cluster size. Yet SCLB yields regret that depends on these parameters, and is thus adaptive.

## 2  Main Contributions

### 2.1  Algorithmic: Problem Complexity Adaptive and (almost) Parameter-Free

We propose adaptive and parameter free algorithms. Roughly speaking, an algorithm is parameter-free and adaptive, if does not need input about the difficulty of the problem, yet has regret guarantees that scale with the inherent complexity. We show in the two frameworks that, if there is structure, then the regret attained by our algorithms is much lower as they learn across users. Simultaneously, in the worst case, the regret guarantee is no worse than if every agent had its own algorithm without collaborations.

**In the personalization framework,** we give PMLB, a parameter free algorithm, whose regret adapts to an appropriately defined problem complexity – if the users are similar, then the regret is low due to collaborative learning while, in the worst case, the regret is no worse than that of individual learning. Formally, we define the complexity as the *factor of common representation*, which for agent $i$ is $\epsilon_i := \|\theta_i^* - \frac{1}{N}\sum_{l=1}^{N}\theta_l^*\|$, where $\theta_i^* \in \mathbb{R}^d$ is agent $i$'s representation, and $\frac{1}{N}\sum_{l=1}^{N}\theta_l^*$ is the average representation of $N$ agents. PMLB adapts to $\epsilon_i$ gracefully (without knowing it apriori) and yields a regret of $\mathcal{O}(\epsilon_i\sqrt{dT})$. Hence, if the agents share representations, i.e., $\epsilon_i$ is small, then PMLB obtains low regret. On the other hand, if $\epsilon_i$ is large, say $\mathcal{O}(1)$, the agents do not share a common representation, the regret of PMLB is $\mathcal{O}(\sqrt{dT})$, which matches that obtained by each agent playing OFUL, independently of other agents. Thus, PMLB benefits from collaborative learning and obtains small regret, if the problem structure admits, else the regret matches the baseline strategy of every agent running an independent bandit instance.

**The clustering framework** considers the scenario when not all users are identical or near identical. In this framework, the large number of users belong to a few types, with users of the same type having identical parameters, but users across types have different parameters. Assuming that all users are near identical in this setting will not lead to good performance as all users can be far from the average. We give a multi-phase, successive refinement based algorithm, SCLB, which is parameter free—specifically no knowledge of cluster separation and number of clusters is needed. SCLB *automatically* identifies whether a given problem instance is 'hard' or 'easy' and adapts to the corresponding regret. Concretely, SCLB attains per-agent regret $\mathcal{O}(\sqrt{T/N})$, if the agent is in a 'well separated' (i.e. 'easy') cluster, or $\mathcal{O}(T^{\frac{1}{2}+\varepsilon}/(N)^{\frac{1}{2}-\varepsilon})$ if the agent's cluster is not well separated (i.e., 'hard'), where $\varepsilon$ is positive and arbitrarily close to 0. *This* *result holds true, even in the limit when the cluster separation approaches* 0. This shows that when the underlying instance gets harder to cluster, the regret is increased. Nevertheless, despite the clustering being hard to accomplish, every user still experiences collaborative gain of $N^{1/2-\varepsilon}$ and regret sub-linear in $T$. Moreover, if clustering is easy i.e., well-separated, then the regret rate *matches that of an oracle that knows the cluster identities.*

**Empirical Validation:** We empirically verify the theoretical insights on both synthetic and Last.FM real data. We compare with three benchmarks—CLUB [18], SCLUB [29], and a simple baseline where every agent runs an independent bandit model, i.e., no collaboration. We observe that our algorithms have superior performance compared to the benchmarks in a variety of settings.

## 2.2   Theoretical: Improved Bounds for Clustering

It is worth pointing out that SCLB works for *all* ranges of separation, which is starkly different from standard algorithms in bandit clustering [17,18,23] and statistics [3,24]. We now compare our results to CLUB [18], that can be modified to be applicable to our setting (c.f. Sect. 7) (note that we make *identical assumptions* to that of CLUB). First, CLUB is non-adaptive and its regret guarantees hold only when the clusters are separated. Second, even in the separated setting, the separation (gap) cannot be lower than $\mathcal{O}(1/T^{1/4})$ for CLUB, while it can be as low as $\mathcal{O}(1/T^{\alpha})$, where $\alpha < 1/2$ for SCLB. Moreover, in simulations (Sect. 7) we observe that SCLB outperforms CLUB in a variety of synthetic and a real data setting.

## 2.3   Technical Novelty

The key innovations we introduce in the analysis are that of *'shifted OFUL'* and *'perturbed OFUL'* algorithms in the personalization and clustering setup respectively. In the personalization setup, our algorithm first estimates the mean vector $\bar{\theta}^* := \frac{1}{N}\sum_{i=1}^{N}\theta_i^*$ of the population. Subsequently, the algorithm subtracts the effect of the mean and only learns the component $\theta_i^* - \bar{\theta}^*$ by compensating the rewards. Our technical innovation is to show that with high probability, shifting the rewards by any fixed vector can only increase overall regret. In the clustering setup, our algorithm first runs individual OFUL instances per agent, estimates the parameter, then clusters the agents and treats all agents of a single cluster as one entity. In order to prove that this works even when the cluster separation is small, we need to analyze the behaviour of OFUL where the rewards come from a slightly perturbed model.

# 3   Related Work

Collaborative gains in multi-user recommendation systems have long been studied in Information retrieval and recommendation systems (ex. [26,28,32,42]). The focus has been in developing effective ideas to help practitioners deploy large scale systems. Empirical studies of recommendation system has seen renewed interest lately due to the integration of deep learning techniques with classical ideas (ex. [9,34,36,37,47,49]). Motivated by the empirical success, we undertake a theoretical approach to quantify collaborative gains achievable in a contextual bandit setting. Contextual bandits has proven to be fruitful in modeling sequential decision making in many applications [5,18,27].

The framework of personalized learning has been exploited in a great detail in representation learning and meta-learning. While [11,21,25,40,41] learn common representation across agents in Reinforcement Learning, [2] uses it for imitation learning. We remark that representation learning is also closely connected to meta-learning [10,15,22], where close but a common initialization is learnt from leveraging non identical but similar representations. Furthermore, in Federated learning, the problem of personalization is a well studied problem [12,13,35].

The paper of [18] is closest to our clustering setup, where in each round, the platform plays an arm for a single randomly chosen user. This model was then subsequently improved by [30] and [29] which all exploit the fact that the users' unknown vectors are clustered. As outlined before, our algorithm obtains a superior performance, both in theory and empirically. For personalization, the recent papers of [46] and [4] are the closest, which posits all users's parameters to be in a common low dimensional subspace. [46] proposes a learning algorithm under this assumption. In contrast, we make no parametric assumptions, and demonstrate an algorithm that achieves collaboration gain, if there is structure, while degrading gracefully to the simple baseline of independent bandit algorithms in the absence of structure.

# 4   Problem Setup

**Users and Arms:** Our system consists of $N$ users, interacting with a centralized system (termed as 'center' henceforth) repeatedly over $T$ rounds. At the beginning of each round, environment provides the center with $K$ context vectors corresponding to $K$ arms, and for each user, the center recommends one of the $K$ arms to play. At the end of the round, every user receives a reward for the arm played, which is observed by the center. The $K$ context vectors in round $t$ are denoted by $\beta_t = [\beta_{1,t}, \ldots, \beta_{K,t}] \in \mathbb{R}^{d \times K}$.

**User Heterogeneity:** Each user $i$, is associated with a preference vector $\theta_i^* \in \mathbb{R}^d$, and the reward user $i$ obtains from playing arm $j$ at time $t$ is is given by $\langle \beta_{j,t}, \theta^* \rangle + \xi_t$. Thus, the structure of the set of user representations $(\theta_i^*)_{i=1}^N$ govern how much benefit from collaboration can be expected. In the rest of the paper, we consider two instantiations of the setup - a clustering framework and the personalization framework.

**Stochastic Assumptions:** We follow the framework of [1,6] and assume that $(\xi_t)_{t \geq 1}$ and $(\beta_t)_{t \geq 1}$ are random variables. We denote by $\mathcal{F}_{t-1}$, as the sigma algebra generated by all noise random variables upto and including time $t-1$. We denote by $\mathbb{E}_{t-1}(.)$ and $\mathbb{V}_{t-1}(.)$ as the conditional expectation and conditional variance operators respectively with respect to $\mathcal{F}_{t-1}$. We assume that the $(\xi_t)_{t \geq 1}$

are conditionally sub-Gaussian noise with known parameter $\sigma$, conditioned on all the arm choices and realized rewards in the system upto and including time $t-1$. Without loss of generality, we assume $\sigma = 1$ throughout. The contexts $\beta_{i,t}$ are assumed to be drawn from a (coordinate-wise)[1] bounded distribution (i.e., in any distribution supported on $[-c, c]^{\otimes d}$ for some constant $c$) independent of both the past and $\{\beta_{j,t}\}_{j \neq i}$, satisfying

$$\mathbb{E}_{t-1}[\beta_{i,t}] = 0 \qquad \mathbb{E}_{t-1}[\beta_{i,t}\,\beta_{i,t}^{\top}] \succeq \rho_{\min} I. \tag{1}$$

Moreover, for any fixed $z \in \mathbb{R}^d$, of unity norm, the random variable $(z^{\top}\beta_{i,t})^2$ is conditionally sub-Gaussian, for all $i$, with $\mathbb{V}_{t-1}[(z^{\top}\beta_{i,t})^2)] \leq 4\rho_{\min}$. This means that the conditional mean of the covariance matrix is zero and the conditional covariance matrix is positive definite with minimum eigenvalue at least $\rho_{\min}$.

Furthermore, the conditional variance assumption is crucially required to apply (1) for contexts of (random) bandit arms selected by our learning algorithm (see [18, Lemma 1]). Note this set of assumptions is not new and the exact set of assumptions were used in [6,18][2] for online clustering and binary model selection respectively. Furthermore, [16] uses similar assumptions for stochastic linear bandits and [19] uses it for model selection in Reinforcement learning problems with function approximation.

**Example of Contexts:** Contexts, $\beta_{i,t}$, drawn iid from $\mathrm{Unif}[-1/\sqrt{d}, 1/\sqrt{d}]^{\otimes d}$ satisfy the above conditions, with $\rho_{\min} = c_0/d$ ($c_0$ : constant). The $1/\sqrt{d}$ scaling ensures that the norm is $\mathcal{O}(1)$. Observe that our stochastic assumption also includes the setting where the distribution of contexts over time follows a random process independent of the actions and rewards from the learning algorithm.

**Performance Metric:** At time $t$, we denote by $B_{i,t} \in [K]$ to be the arm played by any agent $i$ with preference vector $\theta_i^*$. The corresponding regret, over a time horizon of $T$ is given by $R_i(T) = \sum_{t=1}^{T} \mathbb{E}\max_{j \in [K]}\langle \theta_i^*, \beta_{j,t} - \beta_{B_{i,t},t}\rangle$.

Throughout, OFUL refers to the linear bandit algorithm of [1], which we use as a blackbox. In particular we use a variant of the OFUL as prescribed in [6][3].

## 5   Personalization

In this section, we assume that the users' representations $\{\theta_i^*\}_{i=1}^N$ are similar but not necessarily identical. Of course, without any structural similarity among $\{\theta_i^*\}_{i=1}^N$, the only way-out is to learn the parameters separately for each user. In the setup of personalized learning, it is typically assumed that (see [8,14,31,46] and the references therein) that the parameters $\{\theta_i^*\}_{i=1}^N$ share some commonality, and the job is to learn the shared components or representations of $\{\theta_i^*\}_{i=1}^N$

---

[1] In the clustering framework, we were able to remove this coordinate-wise bounded assumption. We only assume boundedness in $\ell_2$ norm.

[2] The conditional variance assumption is implicitly used in [6].

[3] We use OFUL as used in the OSOM algorithm of [6] without bias for the linear contextual setting.

---

**Algorithm 1:** Personalized Multi-agent Linear Bandits (PMLB)

---

1: **Input:** Agents $N$, Horizon $T$

  **Common representation learning : Estimate** $\bar{\theta}^* = \frac{1}{N}\sum_{i=1}^{N}\theta_i^*$

2: Initialize a single instance of OFUL($\delta$), called common OFUL

3: **for** times $t \in \{1,\cdots,\sqrt{T}\}$ **do**

4:    All agents play the action given by the common OFUL

5:    Common OFUL's state updated by average of observed rewards
    at all agents

6: **end for**

7: $\widehat{\theta}^* \leftarrow$ the parameter estimate of Common OFUL at the end of round $\sqrt{T}$

  **Personal Learning**

8: **for** agents $i \in \{1,\ldots,N\}$ **in parallel do**

9:    Initialize modified ALB-Norm($\delta$) of [20] instance per agent (reproduced in
    Supplementary Material)

10:    **for** times $t \in \{\sqrt{T}+1,\ldots,T\}$ **do**

11:      Agents play arm output by their personal copy of ALB-Norm (denoted
      as $\beta_{b_t^{(i)},t}$) and receive reward $y_t$

12:      Every agent updates their ALB-Norm state with corrected reward
      $\tilde{y}_i^{(t)} = y_i^{(t)} - \langle \beta_{b_t^{(i)},t}, \hat{\theta}^* \rangle$

13:    **end for**

14: **end for**

---

collaboratively. After learning the common part, the individual representations
can be learnt locally at each agent.

We assume, that the contexts are drawn iid from $\mathsf{Unif}[-1/\sqrt{d}, 1/\sqrt{d}]^{\otimes d}$. This
is for clarity of exposition and concreteness and without loss of generality, our
analysis can be extended to any distribution supported on $[-c, c]^{\otimes d}$. Moreover,
we relax this assumption in Sect. 6. We now define the notion of common repre-
sentation across users. Let $\|\theta_l^*\| \leq 1$ for all $l \in [N]$. We define $\bar{\theta}^* = \frac{1}{N}\sum_{l=1}^{N}\theta_l^*$
as the average parameter.

**Definition 1.** *($\epsilon$ common representation) An agent $i$ has $\epsilon_i$ common represen-
tation across $N$ agents if $\|\theta_i^* - \bar{\theta}^*\| \leq \epsilon_i$, where $\epsilon_i$ is defined as the common
representation factor.*

The above definition characterizes how far the representation of agent $i$ is from
the average representation $\bar{\theta}^*$. Note that since $\|\theta_l^*\| \leq 1$ for all $l$, we have $\epsilon_i \leq 2$.
Furthermore, if $\epsilon_i$ is small, one can hope to exploit the common representation
across users. On the other hand, if $\epsilon_i$ is large (say $\mathcal{O}(1)$), there is no hope to
leverage collaboration across agents.

### 5.1 The PMLB Algorithm

Algorithm 1 has *(i)* a common learning and *(ii)* a personal fine-tuning phase.

**Common Representation Learning:** In the first phase, PMLB learns the
average representation $\bar{\theta}^*$ by recommending the same arm to all users and averaging the obtained rewards. At the end of this phase, the center has the estimate

$\hat{\theta}^*$ of the average representation $\bar{\theta}^*$. Since the algorithm aggregates the reward from all $N$ agents, it turns out that the common representation learning phase can be restricted to $\sqrt{T}$ steps.

**Personal Fine-Tuning:** In the personal learning phase, the center learns the vector $\theta_i^* - \hat{\theta}^*$, *independently* for every agent. For learning $\theta_i^* - \hat{\theta}^*$, we employ the Adaptive Linear Bandits-norm (ALB-norm) algorithm of [20][4]. ALB-norm is adaptive, yielding a norm dependent regret, i.e., depends on $\|\theta_i^* - \hat{\theta}^*\|$. The idea here is to exploit the fact that in the common learning phase we have a good estimate of $\bar{\theta}^*$. Hence, if the common representation factor $\epsilon_i$ is small, then $\|\theta_i^* - \hat{\theta}^*\|$ is small, and it reflects in the regret expression. In order to estimate the difference, the center *shifts* the reward by the inner product of the estimate $\hat{\theta}^*$. By exploiting the anti-concentration property of Chi-squared distribution along with some standard results from optimization, we show that the regret of the shifted system is worse than the regret of agent $i$ (both in expectation and in high probability)[5].

Without loss of generality, in what follows, we focus on an arbitrary agent belonging to cluster $i$ and characterize the regret. We assume

$$T \geq C \frac{1}{N} \left[ \frac{\tau_{\min}(\delta)\rho_{\min}}{d\log(1/\delta)} \right]^{\frac{1}{2\alpha}}, \tau_{\min}(\delta) = \left[ \frac{16}{\rho_{\min}^2} + \frac{8}{3\rho_{\min}} \right] \log(\frac{2dT}{\delta}) \quad (2)$$

## 5.2 Regret Guarantee for PMLB

**Theorem 1.** *Playing Algorithm 1 with $T$ time and $\delta$, where $T \geq \tau_{\min}^2(\delta)$ (defined in Eq. (2)) and $d \geq C\log(K^2T)$, then the regret of agent $i$ satisfies*

$$R_i(T) \leq \tilde{\mathcal{O}}(\epsilon_i \sqrt{dT} + T^{1/4} \sqrt{\frac{d^2}{\rho_{\min}N}}) \log^2(1/\delta),$$

*with probability at least $1 - c\delta - \frac{1}{\text{poly}(T)}$.*

*Remark 1.* The leading term in regret is $\tilde{\mathcal{O}}(\epsilon_i\sqrt{dT})$. If the common representation factor $\epsilon_i$ is small, PMLB exploits that across agents and as a result the regret is small as well.

*Remark 2.* Moreover, if $\epsilon_i$ is big enough, say $\mathcal{O}(1)$, this implies that there is no common representation across users, and hence collaborative learning is meaningless. In this case, the agents learn individually (by running OFUL), and obtain a regret of $\tilde{\mathcal{O}}(\sqrt{dT})$ with high probability. Note that this is being reflected in Theorem 1, as the regret is $\tilde{\mathcal{O}}(\sqrt{dT})$, when $\epsilon_i = \mathcal{O}(1)$.

---

[4] In the supp. material, we indeed modify ALB-Norm. For parameter $\theta$, the original ALB-Norm yields a regret of $\mathcal{O}[(\|\theta\|+1)d\sqrt{T}]$, while our modified algorithm obtains $\mathcal{O}(\|\theta\|d\sqrt{T})$.

[5] This is intuitive since, otherwise one can find *appropriate shifts* to reduce the regret of OFUL, which contradicts the optimality of OFUL.

The above remarks imply the adaptivity of PMLB. Without knowing the common representation factor $\epsilon_i$, PMLB indeed adapts to it—meaning that yields a regret that depends on $\epsilon_i$. If $\epsilon_i$ is small, PMLB leverages common representation learning across agents, otherwise when $\epsilon_i$ is large, it yields a performance equivalent to the individual learning. Note that this is intuitive since with high $\epsilon_i$, the agents share no common representation, and so we do not get a regret improvement in this case by exploiting the actions of other agents.

*Remark 3.* (Lower Bound) When $\epsilon_i = 0$, i.e., in the case when all agents have the identical vectors $\theta_i^*$, then Theorem 1 gives a regret scaling as $R_i(T) \leq \tilde{\mathcal{O}}(T^{1/4} d \sqrt{\frac{1}{\rho_{\min} N}})$. When the contexts are adversarily generated, [7] obtain a lower bound (in expectation) of $\Omega(\sqrt{dT})$. However, in the presence of stochastic context, a lower bound on the contextual bandit problem is unknown to the best of our knowledge.

The requirement on $d$ in Theorem 1 can be removed for expected regret.

**Corollary 1.** *(Expected Regret) Suppose $T \geq \tau_{\min}^2(\delta)$ for $\delta > 0$. The expected regret of the $i$-th agent after running Algorithm 1 for $T$ time steps is given by*

$$\mathbb{E}[R_i(T)] \leq \tilde{\mathcal{O}}(\epsilon_i \sqrt{dT} + T^{1/4} \sqrt{\frac{d^2}{\rho_{\min} N}}).$$

## 6   Clustering

We now propose the clustering framework. Here, we assume that instead of being coordinate-wise bounded, the contexts, $\beta_{i,t} \in \mathbb{B}^d(1)$. The users' vectors $\{\theta_u^*\}_{u=1}^N$ are clustered into $L$ groups, with $p_i \in (0, 1]$ denoting the fraction of users in cluster $i$. All users in the same cluster have the same the preference vector–denoted by $\theta_i^*$ for cluster $i \in [L]$. We define *separation parameter*, or SNR (signal to noise ratio) of cluster $i$ as $\Delta_i := \min_{j \in [L] \setminus \{i\}} \|\theta_i^* - \theta_j^*\|$, smallest distance to another cluster.

**Learning Algorithm:** We propose the Successive Clustering of Linear Bandits (SCLB) algorithm in Algorithm 2. SCLB does not need any knowledge of the gap $\{\Delta_i\}_{i=1}^L$, the number of clusters $L$ or the cluster size fractions $\{p_i\}_{i=1}^L$.

---

**Algorithm 2:** Successive Clustering of Linear Bandits (SCLB)

---

1: **Input:** No. of users $N$, horizon $T$, parameter $\alpha < 1/2$, constant $C$, high probability bound $\delta$

2: **for** phases $1 \leq j \leq \log_2(T)$ **do**

3:     Play CMLB ($\gamma = 3/(N2^j)^\alpha$, horizon $T = 2^j$, high probability $\delta/2^j$, cluster-size $p^* = j^{-2}$)

4: **end for**

---

---

**Algorithm 3:** Clustered Multi-Agent Bandits (CMLB)

---

1: **Input:** No. of users $N$, horizon $T$, parameter $\alpha < 1/2$, constant $C$, high
   probability bound $\delta$, threshold $\gamma$, cluster-size parameter $p^*$

   **Individual Learning Phase**

2: $T_{\text{Explore}} \leftarrow C^{(2)} d(NT)^{2\alpha} \log(1/\delta)$

3: All agents play OFUL($\delta$) independently for $T_{\text{explore}}$ rounds

4: $\{\hat{\theta}^{(u)}\}_{u=1}^N \leftarrow$ All agents' estimates at the end of round $T_{\text{explore}}$.

   **Cluster the Users**

5: User-Clusters $\leftarrow$ MAXIMAL-CLUSTER($\{\hat{\theta}^{(u)}\}_{u=1}^N, \gamma, \ p^*$)

   **Collaborative Learning Phase**

6: Initialize one OFUL($\delta$) instance per-cluster

7: **for** clusters $\ell \in \{1, \ldots, |\text{User-Clusters}|\}$ **in parallel do**

8:     **for** times $t \in \{T_{\text{explore}} + 1, \cdots, T\}$ **do**

9:         All users in the $\ell$-th cluster play the arm given by the OFUL algorithm
   of cluster $l$.

10:        Average of the observed rewards of all users of cluster $l$ is used to
   update the OFUL($\delta$) state of cluster $l$

11:     **end for**

12: **end for**

---

Nevertheless, SCLB adapts to the problem SNR and yields regret accordingly. One attractive feature of Algorithm 2 is that it works uniformly *for all* ranges of the gap $\{\Delta_i\}_{i=1}^L$. This is in sharp contrast with the existing algorithms [18] which is only guaranteed to give good performance when the gap $\{\Delta_i\}_{i=1}^L$ are large enough. Furthermore, our uniform guarantees are in contrast with the works in standard clustering algorithms, where theoretical guarantees are only given for a sufficiently large separation [3,24].

SCLB is a multi-phase algorithm, invoking Clustered Multi-agent Linear Bandits (CMLB) (Algorithm 3) repeatedly, by decreasing the size parameter, namely $p^*$ polynomially and high probability parameter $\delta_j$ exponentially. Algorithm 2 proceeds in phases of exponentially growing phase length with phase $j \in \mathbb{N}$ lasting for $2^j$ rounds. In each phase, a fresh instance of CMLB is instantiated with high probability parameter $\delta/2^j$ and the minimum size parameter $j^{-2}$. As the phase length grows, the size parameter sent as input to Algorithm 3 decays. This simple strategy suffices to show that the size parameter converges to $p_i$, and we obtain collaborative gains without knowledge of $p_i$.

**CMLB (Algorithm 3):** CMLB works in the three phases: (a) (Individual Learning) the $N$ users play an independent linear bandit algorithm to (roughly) learn their preference; (b) (Clustering) users are clustered based on their estimates using MAXIMAL CLUSTER (Algorithm 4); and (c) (Collaborative Learning) one Linear Bandit instance per cluster is initialized and all users of a cluster play the same arm. The average reward over all users in the cluster is used to update the per-cluster bandit instance. When clustered correctly, the learning is faster, as the noise variance is reduced due to averaging across users. Note that MAXIMAL CLUSTER algorithm requires a size parameter $p^*$.

## 6.1    Regret Guarantee of SCLB

As mentioned earlier, SCLB is an adaptive algorithm that yields provable regret for *all ranges* of $\{\Delta_i\}_{i=1}^L$. When $\{\Delta_i\}_{i=1}^L$ are large, SCLB can cluster the agents perfectly, and thereafter exploit the collaborative gains across users in same cluster. On the other hand, if $\{\Delta_i\}_{i=1}^L$ are small, SCLB still adapts to the gap, and yields a non-trivial (but sub-optimal) regret. As a special case, we show that if all the clusters are very close to one another, then with high probability, SCLB identifies treats all agents as *one big* cluster, yielding highest collaborative gain.

**Definition 2 ($\alpha$-Separable Cluster).** *For a fixed $\alpha < 1/2$, cluster $i \in [L]$ is termed $\alpha$-separable if $\Delta_i \geq \frac{5}{(NT)^\alpha}$. Otherwise, it is termed as $\alpha$-inseparable.*

**Lemma 1.** *If CMLB is run with parameters $\gamma = 3/(NT)^\alpha$ and $p^* \leq p_i$ and $\alpha < \frac{1}{2}$, then with probability at least $1 - 2\binom{N}{2}\delta$, any cluster $i$ that is $\alpha$-separable is clustered correctly. Furthermore, the regret of any user in the $\alpha$-separated cluster $i$ satisfies,*

$$R_i(T) \leq C_1 \left[ \frac{d}{\rho_{\min}}(NT)^\alpha + \sqrt{\frac{d}{\rho_{\min}}} \left( \sqrt{\frac{T - \frac{d(NT)^{2\alpha}}{\rho_{\min}}\log(1/\delta)}{p_i N}} \right) \right] \log(1/\delta),$$

*with probability exceeding $1 - 4\binom{N}{2}\delta$.*

We now present the regret of SCLB for the setting with separable cluster

**Theorem 2.** *If Algorithm 2 is run for $T$ steps with parameter $\alpha < \frac{1}{2}$, then the regret of any agent in a cluster $i$ that is $\alpha$-separated satisfies*

$$R_i(T) \leq 4 \left( 2^{\frac{1}{\sqrt{p_i}}} \right) + C_2 \left[ \frac{d}{\rho_{\min}}(NT)^\alpha + \sqrt{\frac{dT}{\rho_{\min} N}} \right] \log^2(T) \log(1/\delta),$$

*with probability at-least $1 - cN^2\delta$. Moreover, if $\alpha \leq \frac{1}{2}(\frac{\log\left[\frac{\rho_{\min} T}{dp_i N}\right]}{\log(NT)})$, we have $R_i(T) \leq \tilde{O}[2^{\frac{1}{\sqrt{p_i}}} + \sqrt{\frac{d}{\rho_{\min}}} \sqrt{\frac{T}{N}}] \log(1/\delta)$.*

*Remark 4.* Note that we obtain the regret scaling of $\tilde{O}(\sqrt{T/N})$, which is optimal, i.e., the regret rate matches an oracle that knows cluster membership. The cost of successive clustering is $O(2^{\frac{1}{\sqrt{p_i}}})$, which is a $T$-independent (problem dependent) constant.

*Remark 5.* Note that the separation we need is only $5/(NT)^\alpha$. This is a weak condition since in a collaborative system with large $N$ and $T$, this quantity is sufficiently small.

*Remark 6.* Observe that $R_i(T)$ is a decreasing function of $N$. Hence, more users in the system ensures that the regret decreases. This is collaborative gain.

---

**Algorithm 4: MAXIMAL-CLUSTER**

---

1: **Input:** All estimates $\{\hat{\theta}^{(i)}\}_{i=1}^{N}$, size parameter $p^* > 0$, threshold $\gamma \geq 0$.
2: Construct an undirected Graph $G$ on $N$ vertices as follows:
   $\|\hat{\theta}_i^* - \hat{\theta}_j^*\| \leq \gamma \Leftrightarrow i \sim_G j$
3: $\mathcal{C} \leftarrow \{C_1, \cdots, C_k\}$ all the connected components of $G$
4: $\mathcal{S}(p^*) \leftarrow \{C_j : |C_j| < p^*N\}$ {All Components smaller than $p^*N$}
5: $C^{(p)} \leftarrow \cup_{C \in \mathcal{S}(p^*)} C$ {Collapse all small components into one}
6: **Return :** $\mathcal{C} \setminus \mathcal{S}(p^*) \bigcup C^{(p)}$ {Each connected component larger than $p^*N$ is a cluster, and all small components are a single cluster}

---

*Remark 7.* (Comparison with [18]) Note that in a setup where clusters are separated, [18] also yields a regret of $\tilde{\mathcal{O}}(\sqrt{T/N})$. However, the separation between the parameters (gap) for [18] cannot be lower than $\mathcal{O}(1/T^{1/4})$, in order to maintain order-wise optimal regret. On the other hand, we can handle separations of the order $\mathcal{O}(1/T^{\alpha})$, and since $\alpha < 1/2$, this is a strict improvement over [18].

*Remark 8.* The constant term $\mathcal{O}(2^{\frac{1}{\sqrt{p_i}}})$ can be removed if we have an estimate of the $p_i$. Here, instead of SCLB, we simply run CMLB with the estimate of $p_i$ and obtain the regret of Lemma 1, without the term $\mathcal{O}(2^{\frac{1}{\sqrt{p_i}}})$.

We now present our results when cluster $i$ is $\alpha$-inseparable.

**Lemma 2.** *If CMLB is run with input* $\gamma = 3/(NT)^{\alpha}$ *and* $p^* \leq p_i$ *and* $\alpha < \frac{1}{2}$, *then any user in a cluster $i$ that is $\alpha$-inseparable satisfies*

$$R(T) \leq C_1 L(\frac{T^{1-\alpha}}{N^{\alpha}}) + C_2 \sqrt{\frac{d}{\rho_{\min}}} \left[ \sqrt{\frac{T - \frac{d(NT)^{2\alpha}}{\rho_{\min}} \log(1/\delta)}{p^*N}} \right] \log(1/\delta),$$

*with probability at least* $1 - 4\binom{N}{2}\delta$.

**Theorem 3.** *If Algorithm 2 is run for $T$ steps with parameter $\alpha < \frac{1}{2}$, then the regret of any agent in a cluster $i$ that is $\alpha$-inseparable satisfies*

$$R_i(T) \leq 4(2^{\frac{1}{\sqrt{p_i}}}) + C L(\frac{T^{1-\alpha}}{N^{\alpha}}) \log(T) + C_1 \sqrt{\frac{dT}{N\rho_{\min}}} \log(1/\delta) \log^2(T),$$

*with probability at-least* $1 - cN^2\delta$. *Moreover, if If* $\alpha = \frac{1}{2} - \varepsilon$, *where $\varepsilon$ is a positive constant arbitrarily close to 0,* $R(T) \leq \tilde{\mathcal{O}}\left[ 2^{\frac{1}{\sqrt{p_i}}} + L(\frac{T^{\frac{1}{2}+\varepsilon}}{N^{\frac{1}{2}-\varepsilon}}) + \sqrt{\frac{d}{\rho_{\min}}} (\sqrt{\frac{T}{N}}) \log(1/\delta) \right]$.

*Remark 9.* As $\varepsilon > 0$, the regret scaling of $\tilde{\mathcal{O}}(\frac{T^{\frac{1}{2}+\varepsilon}}{N^{\frac{1}{2}-\varepsilon}})$ is strictly worse than the optimal rate of $\tilde{\mathcal{O}}(\sqrt{T/N})$. This can be attributed to the fact that the gap (or SNR) can be arbitrarily close to 0, and inseparability of the clusters makes the problem harder to address.

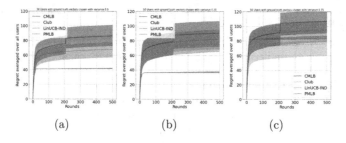

**Fig. 1.** Synthetic simulations of PMLB.

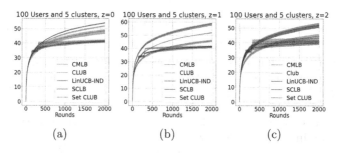

**Fig. 2.** Synthetic data simulations for clustering.

*Remark 10.* In this setting of low gap (or SNR), where the clusters are insepara-
ble, most existing algorithms (for example [18]) are not applicable. However, we
still manage to obtain sub-optimal but non-trivial regret with high probability.

*Special case of all clusters being close* If $\max_{i \neq j} \|\theta_i^* - \theta_j^*\| \leq 1/(NT)^\alpha$, CMLB
puts all the users in one big cluster. The collaborative gain in this setting is the
largest. Here the regret guarantee of SCLB will be similar to that of Theorem 3
with $p_i = 1$. We defer to the Appendix for a detailed analysis.

*Remark 11.* Observe that if all agents are identical $\max_{i \neq j} \|\theta_i^* - \theta_j^*\| = 0$ our
regret bound does not match that of an *oracle* which knows such information.
The oracle guarantee would be $\mathcal{O}(\sqrt{T/N})$, whereas our guarantee is strictly
worse. The additional regret stems from the universality of our algorithm as it
works for all ranges of $\Delta_i$.

## 7   Simulations

**Personalization Setting:** In Fig. 1, we consider a system where the $N$ ground-
truth $\theta^*$ vectors are sampled independently from $\mathcal{N}(\mu, \sigma\mathbb{I})$. We choose $\mu$ from
the standard normal distribution in each experiment and test performance for
different values of $\sigma$. Observe that for small $\sigma$, all the ground-truth vectors will be
close-by (high structure) and when $\sigma$ is large, the ground-truth vectors are more
spread out. We observe in Fig. 1 that PMLB adapts to the available structure.

With small $\sigma$ where all users are close to the average, PMLB has much lower regret compared to the baselines. On the other hand, at large $\sigma$ when there is no structure to exploit, PMLB is comparable to the baselines. This demonstrates empirically that PMLB *adapts* to the problem structure and exploits it whenever present, while not being wore off in the worst case.

**Clustering Setting:** For each plot of Figs. 2, users are clustered such that the frequency of cluster $i$ is proportional to $i^{-z}$ (identical to that done in [18]), where $z$ is mentioned in the figures. Thus for $z = 0$, all clusters are balanced, and for larger $z$, the clusters become imbalanced. For each cluster, the unknown parameter vector $\theta^*$ is chosen uniformly at random from the unit sphere. We compare SCLB (ALgorithm 2), CMLB (Algorithm 3) with CLUB [18], Set CLUB [29] and LinUCB-Ind the baseline where every agent has an independent copy of OFUL, i.e., no collaboration. (Details in Appendix). We observe that our algorithm is competitive with respect to CLUB and Set CLUB, and is superior compared to the baseline where each agent is playing an independent copy of OFUL. In particular, we observe either as the clusters become more imbalanced, or as the number of users increases, SCLB and CMLB have a superior performance compared to CLUB and Set CLUB. Furthermore, since SCLB only clusters users logarithmically many number of times, its run-time is faster compared to CLUB.

# 8   Conclusion

We consider the problem of leveraging user heterogeneity in a multi-agent stochastic bandit problem under (i) a personalization and, (ii) a clustering framework. In both cases, we give novel adaptive algorithms that, without any knowledge of the underlying instance, provides sub-linear regret guarantees. A natural avenue for future work will be to combine the two frameworks, where users are all not necessarily identical, but at the same time, their preferences are spread out in space (for example the preference vectors are sampled from a Gaussian mixture model). Natural algorithms here will involve first performing a clustering on the population, followed by algorithms such as PMLB. Characterizing performance and demonstrating adaptivity in such settings is left to future work.

# References

1. Abbasi-yadkori, Y., Pál, D., Szepesvári, C.: Improved algorithms for linear stochastic bandits. In: Shawe-Taylor, J., Zemel, R., Bartlett, P., Pereira, F., Weinberger, K.Q. (eds.) Advances in Neural Information Processing Systems, vol. 24, pp. 2312–2320. Curran Associates, Inc. (2011)
2. Arora, S., Du, S., Kakade, S., Luo, Y., Saunshi, N.: Provable representation learning for imitation learning via bi-level optimization. In: International Conference on Machine Learning, pp. 367–376. PMLR (2020)

3. Balakrishnan, S., Wainwright, M.J., Yu, B., et al.: Statistical guarantees for the EM algorithm: from population to sample-based analysis. Ann. Stat. **45**(1), 77–120 (2017)
4. Ban, Y., He, J.: Local clustering in contextual multi-armed bandits. In: Proceedings of the Web Conference 2021, pp. 2335–2346 (2021)
5. Cesa-Bianchi, N., Gentile, C., Zappella, G.: A gang of bandits. arXiv preprint arXiv:1306.0811 (2013)
6. Chatterji, N., Muthukumar, V., Bartlett, P.: OSOM: a simultaneously optimal algorithm for multi-armed and linear contextual bandits. In: International Conference on Artificial Intelligence and Statistics, pp. 1844–1854. PMLR (2020)
7. Chu, W., Li, L., Reyzin, L., Schapire, R.: Contextual bandits with linear payoff functions. In: Proceedings of the Fourteenth International Conference on Artificial Intelligence and Statistics, pp. 208–214. JMLR Workshop and Conference Proceedings (2011)
8. Collins, L., Hassani, H., Mokhtari, A., Shakkottai, S.: Exploiting shared representations for personalized federated learning. arXiv preprint arXiv:2102.07078 (2021)
9. Covington, P., Adams, J., Sargin, E.: Deep neural networks for Youtube recommendations. In: Proceedings of the 10th ACM Conference on Recommender Systems, pp. 191–198 (2016)
10. Denevi, G., Ciliberto, C., Grazzi, R., Pontil, M.: Learning-to-learn stochastic gradient descent with biased regularization. In: Chaudhuri, K., Salakhutdinov, R. (eds.) Proceedings of the 36th International Conference on Machine Learning. Proceedings of Machine Learning Research, vol. 97, pp. 1566–1575. PMLR, 09–15 June 2019. http://proceedings.mlr.press/v97/denevi19a.html
11. D'Eramo, C., Tateo, D., Bonarini, A., Restelli, M., Peters, J.: Sharing knowledge in multi-task deep reinforcement learning. In: International Conference on Learning Representations (2019)
12. Fallah, A., Mokhtari, A., Ozdaglar, A.: On the convergence theory of gradient-based model-agnostic meta-learning algorithms. In: International Conference on Artificial Intelligence and Statistics, pp. 1082–1092. PMLR (2020)
13. Fallah, A., Mokhtari, A., Ozdaglar, A.: Personalized federated learning: a meta-learning approach. arXiv preprint arXiv:2002.07948 (2020)
14. Fallah, A., Mokhtari, A., Ozdaglar, A.: Personalized federated learning with theoretical guarantees: a model-agnostic meta-learning approach. In: Larochelle, H., Ranzato, M., Hadsell, R., Balcan, M.F., Lin, H. (eds.) Advances in Neural Information Processing Systems, vol. 33, pp. 3557–3568. Curran Associates, Inc. (2020). https://proceedings.neurips.cc/paper/2020/file/24389bfe4fe2eba8bf9aa9203a44cdad-Paper.pdf
15. Finn, C., Rajeswaran, A., Kakade, S., Levine, S.: Online meta-learning. In: International Conference on Machine Learning, pp. 1920–1930. PMLR (2019)
16. Foster, D.J., Krishnamurthy, A., Luo, H.: Model selection for contextual bandits (2019)
17. Gentile, C., Li, S., Kar, P., Karatzoglou, A., Zappella, G., Etrue, E.: On context-dependent clustering of bandits. In: International Conference on Machine Learning, pp. 1253–1262. PMLR (2017)
18. Gentile, C., Li, S., Zappella, G.: Online clustering of bandits. In: International Conference on Machine Learning, pp. 757–765. PMLR (2014)
19. Ghosh, A., Chowdhury, S.R., Ramchandran, K.: Model selection with near optimal rates for reinforcement learning with general model classes. arXiv preprint arXiv:2107.05849 (2021)

20. Ghosh, A., Sankararaman, A., Kannan, R.: Problem-complexity adaptive model selection for stochastic linear bandits. In: Banerjee, A., Fukumizu, K. (eds.) Proceedings of The 24th International Conference on Artificial Intelligence and Statistics. Proceedings of Machine Learning Research, vol. 130, pp. 1396–1404. PMLR, 13–15 April 2021. http://proceedings.mlr.press/v130/ghosh21a.html

21. Higgins, I., et al.: Darla: improving zero-shot transfer in reinforcement learning. In: International Conference on Machine Learning, pp. 1480–1490. PMLR (2017)

22. Khodak, M., Balcan, M.F., Talwalkar, A.: Adaptive gradient-based meta-learning methods. arXiv preprint arXiv:1906.02717 (2019)

23. Korda, N., Szorenyi, B., Li, S.: Distributed clustering of linear bandits in peer to peer networks. In: International Conference on Machine Learning, pp. 1301–1309. PMLR (2016)

24. Kwon, J., Caramanis, C.: The EM algorithm gives sample-optimality for learning mixtures of well-separated gaussians. In: Conference on Learning Theory, pp. 2425–2487. PMLR (2020)

25. Lazaric, A., Restelli, M.: Transfer from multiple mdps. In: Shawe-Taylor, J., Zemel, R., Bartlett, P., Pereira, F., Weinberger, K.Q. (eds.) Advances in Neural Information Processing Systems, vol. 24. Curran Associates, Inc. (2011). https://proceedings.neurips.cc/paper/2011/file/fe7ee8fc1959cc7214fa21c4840dff0a-Paper.pdf

26. Lee, W.S.: Collaborative learning for recommender systems. In: ICML, vol. 1, pp. 314–321. Citeseer (2001)

27. Li, L., Chu, W., Langford, J., Schapire, R.E.: A contextual-bandit approach to personalized news article recommendation. In: Proceedings of the 19th International Conference on World Wide Web, pp. 661–670 (2010)

28. Li, Q., Kim, B.M.: Clustering approach for hybrid recommender system. In: Proceedings IEEE/WIC International Conference on Web Intelligence (WI 2003), pp. 33–38. IEEE (2003)

29. Li, S., Chen, W., Leung, K.S.: Improved algorithm on online clustering of bandits. arXiv preprint arXiv:1902.09162 (2019)

30. Li, S., Karatzoglou, A., Gentile, C.: Collaborative filtering bandits. In: Proceedings of the 39th International ACM SIGIR conference on Research and Development in Information Retrieval, pp. 539–548 (2016)

31. Li, T., Hu, S., Beirami, A., Smith, V.: Federated multi-task learning for competing constraints. CoRR abs/2012.04221 (2020). https://arxiv.org/abs/2012.04221

32. Linden, G., Smith, B., York, J.: Amazon. com recommendations: item-to-item collaborative filtering. IEEE Internet Comput. 7(1), 76–80 (2003)

33. Liu, Y., Liu, Z., Chua, T.S., Sun, M.: Topical word embeddings. In: Proceedings of the AAAI Conference on Artificial Intelligence, vol. 29 (2015)

34. Ma, Y., Narayanaswamy, B., Lin, H., Ding, H.: Temporal-contextual recommendation in real-time. In: Proceedings of the 26th ACM SIGKDD International Conference on Knowledge Discovery and Data Mining, pp. 2291–2299 (2020)

35. Mansour, Y., Mohri, M., Ro, J., Suresh, A.T.: Three approaches for personalization with applications to federated learning. CoRR abs/2002.10619 (2020). https://arxiv.org/abs/2002.10619

36. Naumov, M., et al.: Deep learning recommendation model for personalization and recommendation systems. arXiv preprint arXiv:1906.00091 (2019)

37. Okura, S., Tagami, Y., Ono, S., Tajima, A.: Embedding-based news recommendation for millions of users. In: Proceedings of the 23rd ACM SIGKDD International Conference on Knowledge Discovery and Data Mining, pp. 1933–1942 (2017)

38. Ozsoy, M.G.: From word embeddings to item recommendation. arXiv preprint arXiv:1601.01356 (2016)
39. Pal, A., Eksombatchai, C., Zhou, Y., Zhao, B., Rosenberg, C., Leskovec, J.: Pinnersage: multi-modal user embedding framework for recommendations at Pinterest. In: Proceedings of the 26th ACM SIGKDD International Conference on Knowledge Discovery and Data Mining, pp. 2311–2320 (2020)
40. Parisotto, E., Ba, J.L., Salakhutdinov, R.: Actor-mimic: deep multitask and transfer reinforcement learning. arXiv preprint arXiv:1511.06342 (2015)
41. Rusu, A.A., et al.: Policy distillation. arXiv preprint arXiv:1511.06295 (2015)
42. Sarwar, B.M., Karypis, G., Konstan, J., Riedl, J.: Recommender systems for large-scale e-commerce: scalable neighborhood formation using clustering. In: Proceedings of the Fifth International Conference on Computer and Information Technology, vol. 1, pp. 291–324. Citeseer (2002)
43. Saveski, M., Mantrach, A.: Item cold-start recommendations: learning local collective embeddings. In: Proceedings of the 8th ACM Conference on Recommender systems, pp. 89–96 (2014)
44. Wang, S., Tang, J., Aggarwal, C., Liu, H.: Linked document embedding for classification. In: Proceedings of the 25th ACM International on Conference on Information and Knowledge Management, pp. 115–124 (2016)
45. Xue, H.J., Dai, X., Zhang, J., Huang, S., Chen, J.: Deep matrix factorization models for recommender systems. In: IJCAI, vol. 17, pp. 3203–3209. Melbourne, Australia (2017)
46. Yang, J., Hu, W., Lee, J.D., Du, S.S.: Impact of representation learning in linear bandits. In: International Conference on Learning Representations (2021). https://openreview.net/forum?id=edJ_HipawCa
47. Yao, T., et al.: Self-supervised learning for deep models in recommendations. arXiv preprint arXiv:2007.12865 (2020)
48. Zhao, H., Ding, Z., Fu, Y.: Multi-view clustering via deep matrix factorization. In: Proceedings of the AAAI Conference on Artificial Intelligence, vol. 31 (2017)
49. Zhao, Z., et al.: Recommending what video to watch next: a multitask ranking system. In: Proceedings of the 13th ACM Conference on Recommender Systems, pp. 43–51 (2019)

# On the Complexity of All $\varepsilon$-Best Arms Identification

Aymen al Marjani[1($\boxtimes$)], Tomas Kocak[2], and Aurélien Garivier[1]

[1] UMPA, ENS Lyon, Lyon, France
{aymen.al_marjani,aurelien.garivier}@ens-lyon.fr
[2] University of Potsdam, Potsdam, Germany

**Abstract.** We consider the question introduced by [16] of identifying all the $\varepsilon$-optimal arms in a finite stochastic multi-armed bandit with Gaussian rewards. We give two lower bounds on the sample complexity of any algorithm solving the problem with a confidence at least $1 - \delta$. The first, unimprovable in the asymptotic regime, motivates the design of a Track-and-Stop strategy whose average sample complexity is asymptotically optimal when the risk $\delta$ goes to zero. Notably, we provide an efficient numerical method to solve the convex max-min program that appears in the lower bound. Our method is based on a complete characterization of the alternative bandit instances that the optimal sampling strategy needs to rule out, thus making our bound tighter than the one provided by [16]. The second lower bound deals with the regime of high and moderate values of the risk $\delta$, and characterizes the behavior of any algorithm in the initial phase. It emphasizes the linear dependency of the sample complexity in the number of arms. Finally, we report on numerical simulations demonstrating our algorithm's advantage over state-of-the-art methods, even for moderate risks.

**Keywords:** Multi-armed bandits · Best-arm identification · Pure exploration

## 1 Introduction

The problem of finding all the $\varepsilon$-good arms was recently introduced by [16]. For a finite family of distributions $(\nu_a)_{a \in [K]}$ with vector of mean rewards $\boldsymbol{\mu} = (\mu_a)_{a \in [K]}$, the goal is to return $G_\varepsilon(\boldsymbol{\mu}) \triangleq \{a \in [K] : \mu_a \geq \max_i \mu_i - \varepsilon\}$ in the additive case and $G_\varepsilon(\boldsymbol{\mu}) \triangleq \{a \in [K] : \mu_a \geq (1 - \varepsilon) \max_i \mu_i\}$ in the multiplicative case. This problem is closely related to two other pure-exploration problems in the multi-armed bandit literature, namely the TOP$-k$ arms selection and the THRESHOLD bandits. The former aims to find the $k$ arms with the highest means, while the latter seeks to identify all arms with means larger than a given threshold $s$. As argued by [16], finding all the $\varepsilon$-good arms is a more robust objective than the TOP-K and THRESHOLD problems, which require some prior knowledge of the distributions

**Supplementary Information** The online version contains supplementary material available at https://doi.org/10.1007/978-3-031-26412-2_20.

in order to return a relevant set of solutions. Take for example drug discovery applications, where the goal is to perform an initial selection of potential drugs through *in vitro* essays before conducting more expensive clinical trials: setting the number of arms $k$ too high or the threshold $s$ too low may result into poorly performing solutions. Conversely, if we set $k$ to a small number or the threshold $s$ too high we might miss promising drugs that will prove to be more efficient under careful examination. The All-$\varepsilon$ objective circumvents this issues by requiring to return all drugs whose efficiency lies within a certain range from the best. In this paper, we want to identify $G_\varepsilon(\mu)$ in a PAC learning framework with fixed confidence: for a risk level $\delta$, the algorithm samples arms $a \in [K]$ in a sequential manner to gather information about the distribution means $(\mu_a)_{a \in [K]}$ and returns an estimate $\widehat{G}_\varepsilon$ such that $\mathbb{P}_\mu(\widehat{G}_\varepsilon \neq G_\varepsilon(\mu)) \leq \delta$. Such an algorithm is called $\delta$-PAC and its performance is measured by the expected number of samples $\mathbb{E}[\tau_\delta]$, also called the *sample complexity*, needed to return a good answer with high probability. [16] provided two lower bounds on the sample complexity: fhe first bound is based on a classical change-of-measure argument and exhibits the behavior of sample complexity in the low confidence regime ($\delta \to 0$). The second bound resorts to the Simulator technique [18] combined with an algorithmic reduction to Best Arm Identification and shows the dependency of the sample complexity on the number of arms $K$ for moderate values of $\delta$. They also proposed FAREAST, an algorithm matching the first lower bound, up to some numerical constants and log factors, in the asymptotic regime $\delta \to 0$. Our contributions can be summarized as follows:

- Usual lower bounds on the sample complexity write as $f(\nu) \log(1/\delta) + g(\nu)$ for an instance $\nu$. We derive a tight bound in terms of the first-order term which writes as $T_\varepsilon^*(\mu) \log(1/\delta)$, where the characteristic time $T_\varepsilon^*(\mu)$ is the value of a concave max-min optimization program. Our bound is tight in the sense that any lower bound of the form $f(\nu) \log(1/\delta)$ that holds for all $\delta \in (0,1)$ is such that $f(\nu) \leq T_\varepsilon^*(\mu)$. To do so, we investigate all the possible alternative instances $\boldsymbol{\lambda}$ that one can obtain from the original problem $\boldsymbol{\mu}$ by a change-of-measure, including (but not only) the ones that were considered by [16].
- We derive a second lower bound that writes as $g(\nu)$ in Theorem 2. $g(\nu)$ shows an additional linear dependency on the number of arms which is negligible when $\delta \to 0$ but can be dominant for moderate values of the risk. This result generalizes Theorem 4.1 in [16], since it also includes cases where there can be several arms with means close to the top arm. The proof of this result relies on a personal rewriting of the Simulator method of [18] which was proposed for the Best Arm Identification and TOP-k problems. As we explain in Sect. 3.3, our proof can be adapted to derive lower bounds for other pure exploration problems, *without resorting to algorithmic reduction of these problems to Best Arm Identification*. Therefore, we believe that the proof itself constitutes a significant contribution.
- We present two efficient methods to solve the minimization sub-problem (resp. the entire max-min program) that defines the characteristic time. These methods are used respectively in the stopping and sampling rule of our Track-and-

Stop algorithm, whose sample complexity matches the lower bound when $\delta$ tends to 0.

– Finally, to corroborate our asymptotic results, we conduct numerical experiments for a wide range of the confidence parameters and number of arms. Empirical evaluation shows that Track-and-Stop is optimal either for a small number of arms $K$ or when $\delta$ goes to 0, and excellent in practice for much larger values of $K$ and $\delta$. We believe these are significant improvements in performance to be of interest for ML practitioners seeking solutions for this kind of problem.

In Sect. 2 we introduce the setting and the notation. Section 3 is devoted to our lower bounds on the sample complexity of identifying the set of ε-good arms and the pseudo-code of our algorithm, along with the theoretical guarantees on its sample complexity. In Sects. 4 and 5, we present our method for solving the optimization program that defines the characteristic time, which is at the heart of the sampling and stopping rules of our algorithm.

## 2    Setting and Notation

The stochastic multi-armed bandit is a sequential learning framework where a learner faces a set of unknown probability distributions $(\nu_a)_{a\in[K]}$ with means $(\mu_a)_{a\in[K]}$, traditionally called *arms*. The learner collects information on the distributions by, at each time step $t$, choosing an arm based on past observations, and receiving an independent sample of this arm. The goal of *fixed-confidence pure exploration* is to answer some question about this set of distributions while using a minimum number of samples. In our case, we define the set of ε-good arms as $G_\varepsilon(\mu) \triangleq \{a \in [K] : \mu_a \geq \max_i \mu_i - \varepsilon\}$; we wish to devise an algorithm that will collect samples and stop as soon as it can produce an estimate of $G_\varepsilon(\mu)$ that is certified to be correct with a prescribed probability $1 - \delta$. This algorithm has three components. The *sampling rule* is $\{\pi_t\}_{t\geq 1}$, where $\pi_t(a| \ a_1, r_1, \ldots, a_{t-1}, r_{t-1})$ denotes the probability of choosing arm $a$ at step $t$ after a sequence of choices $(a_1, \ldots, a_{t-1})$ and the corresponding observations $(r_1, \ldots, r_t)$. The *stopping rule* $\tau_\delta$ is a stopping time w.r.t the filtration of sigma-algebras $\mathcal{F}_t = \sigma(a_1, r_1, \ldots, a_{t-1}, r_{t-1})$ generated by the observations up to time $t$. Finally, the *recommendation rule* $\widehat{G}_\varepsilon$ is measurable w.r.t. $\mathcal{F}_{\tau_\delta}$ and should satisfy $\mathbb{P}_{\nu,\mathcal{A}}(\widehat{G}_\varepsilon = G_\varepsilon(\mu)) \geq 1 - \delta$. Algorithms obeying this inequality are called *δ-correct*, and among all of them we aim to find one with a minimal expected stopping time $\mathbb{E}_{\nu,\mathcal{A}}[\tau_\delta]$. In this work, like in [16], we restrict our attention to Gaussian arms with variance one. Even though this assumption is not mandatory, it considerably simplifies the presentation of the results[1].

---

[1] For $\sigma^2$-subgaussian distributions, we only need to multiply our bounds by $\sigma^2$. For bandits coming from another single-parameter exponential family, we lose the closed-form expression of the best response oracle that we have in the Gaussian case, but one can use binary search to solve the best response problem.

# 3 Lower Bounds and Asymptotically Matching Algorithm

We start by proving a lower bound on the sample complexity of any $\delta$-correct algorithm. This lower bound will later motivate the design of our algorithm.

## 3.1 First Lower Bound

Let $\Delta_K$ denote the $K$-dimensional simplex and $\mathrm{kl}(p, q)$ be the KL-divergence between two Bernoulli distributions with parameters $p$ and $q$. Finally, define the set of *alternative* bandit problems $\mathrm{Alt}(\boldsymbol{\mu}) = \{\boldsymbol{\lambda} \in \mathbb{R}^K : G_\varepsilon(\boldsymbol{\mu}) \neq G_\varepsilon(\boldsymbol{\lambda})\}$. Using change-of-measure arguments introduced by [13] , we derive the following lower bound on the sample complexity in our special setting.

**Proposition 1.** *For any $\delta$-correct strategy $\mathcal{A}$ and any bandit instance $\boldsymbol{\mu}$, the expected stopping time $\tau_\delta$ can be lower-bounded as*

$$\mathbb{E}_{\nu, \mathcal{A}}[\tau_\delta] \geq T_\varepsilon^*(\boldsymbol{\mu}) \log(1/2.4\delta)$$

*where*

$$T_\varepsilon^*(\boldsymbol{\mu})^{-1} \triangleq \sup_{\boldsymbol{\omega} \in \Delta_K} T_\varepsilon(\boldsymbol{\mu}, \boldsymbol{\omega})^{-1} \qquad and \tag{1}$$

$$T_\varepsilon(\boldsymbol{\mu}, \boldsymbol{\omega})^{-1} \triangleq \inf_{\boldsymbol{\lambda} \in \mathrm{Alt}(\boldsymbol{\mu})} \sum_{a \in [K]} \omega_a \frac{(\mu_a - \lambda_a)^2}{2} . \tag{2}$$

The characteristic time $T_\varepsilon^*(\boldsymbol{\mu})$ above is an instance-specific quantity that determines the difficulty of our problem. The optimization problem in the definition of $T_\varepsilon^*(\boldsymbol{\mu})$ can be seen as a two-player game between an algorithm which samples each arm $a$ proportionally to $\omega_a$ and an adversary who chooses an alternative instance $\boldsymbol{\lambda}$ that is difficult to distinguish from $\boldsymbol{\mu}$ under the algorithm's sampling scheme. This suggests that an optimal strategy should play the optimal allocation $\boldsymbol{\omega}^*$ that maximizes the optimization problem (1) and, as a consequence, rules out all alternative instances as fast as possible. This motivates our algorithm, presented in Sect. 3.2.

## 3.2 Algorithm

We propose a simple Track-and-Stop strategy similar to the one proposed by [8] for the problem of Best-Arm Identification. It starts by sampling once from every arm $a \in [K]$ and constructs an initial estimate $\widehat{\boldsymbol{\mu}}_K$ of the vector of mean rewards $\boldsymbol{\mu}$. After this burn-in phase, the algorithm enters a loop where at every iteration it plays arms according to the estimated optimal sampling rule (3) and updates its estimate $\widehat{\boldsymbol{\mu}}_t$ of the arms' expectations. Finally, the algorithm checks if the stopping rule (4) is satisfied, in which case it stops and returns the set of empirically $\varepsilon$-good arms.

*Sampling rule:* our sampling rule performs so-called C-tracking: first, we compute $\widetilde{\omega}(\widehat{\mu}_t)$, an allocation vector which is $\frac{1}{\sqrt{t}}$-optimal in the lower-problem (1) for the instance $\widehat{\mu}_t$. Then we project $\widetilde{\omega}(\widehat{\mu}_t)$ on the set $\Delta_K^{\eta_t} = \Delta_K \cap [\eta_t, 1]^K$. Given the projected vector $\widetilde{\omega}^{\eta_t}(\widehat{\mu}_t)$, the next arm to sample from is defined by:

$$a_{t+1} = \arg\min_a N_a(t) - \sum_{s=1}^{t} \widetilde{\omega}_a^{\eta_t}(\widehat{\mu}_s) \tag{3}$$

where $N_a(t)$ is the number of times arm $a$ has been pulled up to time $t$. In other words, we sample the arm whose number of visits is farther behind its corresponding sum of empirical optimal allocations. In the long run, as our estimate $\widehat{\mu}_t$ tends to the true value $\mu$, the sampling frequency $N_a(t)/t$ of every arm $a$ will converge to the oracle optimal allocation $\omega_a^*(\mu)$. The projection on $\Delta_K^{\eta_t}$ ensures exploration at minimal rate $\eta_t = \frac{1}{2\sqrt{(K^2+t)}}$ so that no arm is left-behind because of bad initial estimates.

*Stopping rule:* To be sample-efficient, the algorithm should should stop as soon as the collected samples are sufficiently informative to declare that $G_\varepsilon(\widehat{\mu}_t) = G_\varepsilon(\mu)$ with probability larger than $1-\delta$. For this purpose we use the Generalized Likelihood Ratio (GLR) test [3]. We define the $Z$-statistic:

$$Z(t) = t \times T_\varepsilon\left(\widehat{\mu}_t, \frac{N(t)}{t}\right)^{-1}$$

where $N(t) = \left(N_a(t)\right)_{a\in[K]}$. As shown in [6,8], the Z-statistic is equal to the ratio of the likelihood of observations under the most likely model where $G_\varepsilon(\widehat{\mu}_t)$ is the correct answer, i.e. $\widehat{\mu}_t$, to the likelihood of observations under the most likely model where $G_\varepsilon(\widehat{\mu}_t)$ is not the set of $\varepsilon$-good arms. The algorithm rejects the hypothesis $G_\varepsilon(\widehat{\mu}_t) \neq G_\varepsilon(\mu)$ and stops as soon as this ratio of likelihoods becomes larger than a certain threshold $\beta(\delta, t)$, properly tuned to ensure that the algorithm is $\delta$-PAC. The stopping rule is defined as:

$$\tau_\delta = \inf\left\{t \in \mathbb{N} \ : \ Z(t) > \beta(t, \delta)\right\} \tag{4}$$

One can find many suitable thresholds from the bandit literature [7,12,15], all of which are of the order $\beta(\delta, t) \approx \log(1/\delta) + \frac{K}{2}\log(\log(t/\delta))$ is enough to ensure that $\mathbb{P}\left(G_\varepsilon(\widehat{\mu}_{\tau_\delta}) \neq G_\varepsilon(\mu)\right) \leq \delta$, i.e. that the algorithm is $\delta$-correct.

Now we state our sample complexity result which we adapted from Theorem 14 in [8]. Notably, while their Track-and-Stop strategy relies on tracking the exact optimal weights to prove that the expected stopping time matches the lower bound when $\delta$ tends to zero, our proof shows that it is enough to track some slightly sub-optimal weights with a decreasing gap in the order of $\frac{1}{\sqrt{t}}$ to enjoy the same sample complexity guarantees.

**Theorem 1.** *For all $\delta \in (0,1)$, Track-and-Stop terminates almost-surely and its stopping time $\tau_\delta$ satisfies:*

$$\limsup_{\delta\to 0} \frac{\mathbb{E}[\tau_\delta]}{\log(1/\delta)} \leq T_\varepsilon^*(\mu).$$

---

**Algorithm 1:** Track and Stop

---

**Input:** Confidence level $\delta$, accuracy parameter $\varepsilon$.

1 Pull each arm once and observe rewards $(r_a)_{a \in [K]}$.
2 Set initial estimate $\widehat{\boldsymbol{\mu}}_K = (r_1, \ldots, r_K)^T$.
3 Set $t \leftarrow K$ and $N_a(t) \leftarrow 1$ for all arms $a$.
4 **while** *Stopping condition (4) is not satisfied* **do**
5     Compute $\tilde{\omega}(\widehat{\boldsymbol{\mu}}_t)$, a $\frac{1}{\sqrt{t}}$-optimal vector for (1) using mirror-ascent.
6     Pull next arm $a_{t+1}$ given by (3) and observe reward $r_t$.
7     Update $\widehat{\boldsymbol{\mu}}_t$ according to $r_t$.
8     Set $t \leftarrow t + 1$ and update $\big(N_a(t)\big)_{a \in [K]}$.
9 **end**

**Output:** Empirical set of $\varepsilon$-good arms: $G_\varepsilon(\widehat{\boldsymbol{\mu}}_{\tau_\delta})$

---

**Remark 1.** Suppose that the arms are ordered decreasingly $\mu_1 \geq \mu_2 \geq \cdots \geq \mu_K$. [16] define the upper margin $\alpha_\varepsilon = \min_{k \in G_\varepsilon(\mu)} \mu_k - (\mu_1 - \varepsilon)$ and provide a lower bound of the form $f(\nu) \log(1/\delta)$ where:

$$f(\nu) \triangleq 2 \sum_{a=1}^{K} \max \left( \frac{1}{(\mu_1 - \varepsilon - \mu_i)^2}, \frac{1}{(\mu_1 + \alpha_\varepsilon - \mu_a)^2} \right).$$

It can be seen directly (or deduced from Theorem 1) that $f(\nu) \leq T_\varepsilon^*(\boldsymbol{\mu})$. In a second step, they proposed FAREAST, an algorithm whose sample complexity in the asymptotic regime $\delta \to 0$ matches their bound up to some universal constant $c$ that does not depend on the instance $\nu$. From Proposition 1, we deduce that $T_\varepsilon^*(\boldsymbol{\mu}) \leq cf(\nu)$, which can be seen directly from the particular changes of measure considered in that paper. The sample complexity of our algorithm improves upon previous work by multiplicative constants that can possibly be large, as illustrated in Sect. 6.

## 3.3  Lower Bound for Moderate Confidence Regime

The lower bound in Proposition 1 and the upper bound in Theorem 1 show that in the asymptotic regime $\delta \to 0$ the optimal sample complexity scales as $T_\varepsilon^*(\boldsymbol{\mu}) \log(1/\delta)$. However, one may wonder whether this bound catches all important aspects of the complexity, especially for large or moderate values of the risk $\delta$. Towards answering this question, we present the following lower bound which shows that there is an additional cost, linear in the number of arms, that any $\delta$-PAC algorithm must pay in order to learn the set of All-$\varepsilon$ good arms. Before stating our result, let us introduce some notation. We denote by $\mathbf{S}_K$ the group of permutations over $[K]$. For a bandit instance $\nu = (\nu_1, \ldots, \nu_K)$ we define the *permuted instance* $\pi(\nu) = (\nu_{\pi(1)}, \ldots, \nu_{\pi(K)})$. $\mathbf{S}_K(\nu) = \{\pi(\nu), \ \pi \in \mathbf{S}_K\}$ refers to the set of all permuted instances of $\nu$. Finally, we will write $\pi \sim \mathbf{S}_K$ to indicate that a permutation is drawn uniformly at random from $\mathbf{S}_K$. These results are much inspired from [16], but come with quite different proofs that we hope can be useful to the community.

**Theorem 2.** *Fix $\delta \leq 1/10$ and $\varepsilon > 0$. Consider an instance $\nu$ such that there exists at least one bad arm: $G_\varepsilon(\boldsymbol{\mu}) \neq [K]$. Without loss of generality, suppose the arms are ordered decreasingly $\mu_1 \geq \mu_2 \geq \cdots \geq \mu_K$ and define the lower margin $\beta_\varepsilon = \min\limits_{k \notin G_\varepsilon(\boldsymbol{\mu})} \mu_1 - \varepsilon - \mu_k$. Then any $\delta$-PAC algorithm has an average sample complexity over all permuted instances satisfying*

$$\mathbb{E}_{\pi \sim \mathbf{S}_K} \mathbb{E}_{\pi(\nu)}[\tau_\delta] \geq \frac{1}{12|G_{\beta_\varepsilon}(\boldsymbol{\mu})|^3} \sum_{b=1}^{K} \frac{1}{(\mu_1 - \mu_b + \beta_\varepsilon)^2},$$

The proof of the lower bound can be found in Appendix C. In the special case where $|G_{2\beta_\varepsilon}| = 1$, then $|G_{\beta_\varepsilon}| = 1$ also (since $\{1\} \subset G_{\beta_\varepsilon} \subset G_{2\beta_\varepsilon}$) and we recover the bound in Theorem 4.1 of [16]. The lower bound above informs us that we must pay a linear cost in $K$, *even when there are several arms close to the top one*, provided that their cardinal does not scale with the total number of arms, i.e. $|G_{\beta_\varepsilon}| = \mathcal{O}(1)$.

**The bound of Thm 2 can be arbitrarily large compared to** $T_\varepsilon^*(\boldsymbol{\mu}) \log(1/\delta)$. Fix $\delta = 0.1$ and let $\varepsilon, \beta > 0$ with $\beta \ll \varepsilon$ and consider the instance such that $\mu_1 = \beta, \mu_K = -\varepsilon$ and $\mu_a = -\beta$ for $a \in [\![2, K-1]\!]$. Then we show in Appendix C that $T_\varepsilon^*(\boldsymbol{\mu}) \log(1/\delta) = \mathcal{O}(1/\beta^2 + K/\varepsilon^2)$. In contrast the lower bound above scales as $\Omega(K/\beta^2)$. Since $\beta \ll \varepsilon$, the second bound exhibits a better scaling w.r.t the number of arms.

**The intuition behind this result** comes from the following observations: first, note that arms in $G_{\beta_\varepsilon}(\boldsymbol{\mu})$ must be sampled at least $\Omega(1/\beta_\varepsilon^2)$ times, because otherwise we might underestimate their means and misclassify the arms in $\arg\min_{k \notin G_\varepsilon(\boldsymbol{\mu})} \mu_1 - \varepsilon - \mu_k$ as good arms. Second, in the initial phase the algorithm does not know which arms belong to $G_{\beta_\varepsilon}(\boldsymbol{\mu})$ and we need at least $\Omega(1/(\mu_1 - \mu_b)^2)$ samples to distinguish any arm $b$ from arms in $G_{\beta_\varepsilon}(\boldsymbol{\mu})$. Together, these observations tell us that we must pay a cost of $\Omega(\min(1/\beta_\varepsilon^2, 1/(\mu_1 - \mu_b)^2))$ samples to either declare that $b$ is not in $G_{\beta_\varepsilon}(\boldsymbol{\mu})$ or learn its mean up to $\mathcal{O}(\beta_\varepsilon)$ precision. More generally, consider a pure exploration problem with a unique answer, where some particular arm $i^{\star 2}$ needs to be estimated up to some precision $\eta > 0$ in order to return the correct answer. In this case, one can adapt our proof, *without using any algorithmic reduction to Best Arm Identification*, to show that every arm $a$ must be played at least $\Omega(1/(|\mu_{i^\star} - \mu_a| + \eta)^2)$ times. For example, consider the problem of testing whether the minimum mean of a multi-armed bandit is above or below some threshold $\gamma$. Let $\nu$ be an instance such that $\{a \in [K] : \mu_a < \gamma\} = \{i^\star\}$ and define $\eta \triangleq \gamma - \mu_{i^\star} > 0$. Then our proof can be

---

[2] or a subset of arms, as in our case.

adapted in a straightforward fashion to prove that any $\delta$-PAC algorithm for this task has a sample complexity of at least $\Omega\left(\sum_{a=1}^{K} \frac{1}{(\mu_a - \mu_{i^*} + \eta)^2}\right)$.[3]

## 4    Solving the Min Problem: Best Response Oracle

Note that Algorithm 1 requires to solve the best response problem, i.e. the minimization problem in (2), in order to be able to compute the $Z$-statistic of the stopping rule, and also to solve the entire lower bound problem in (1) to compute the optimal weights for the sampling rule. The rest of the paper is dedicated to presenting the tools necessary to solve these two problems. For a given vector $\boldsymbol{\omega}$, we want to compute the best response

$$\boldsymbol{\lambda}^*_{\varepsilon,\mu}(\boldsymbol{\omega}) \triangleq \underset{\boldsymbol{\lambda} \in \text{Alt}(\boldsymbol{\mu})}{\arg\min} \sum_{a \in [K]} \omega_a \frac{(\mu_a - \lambda_a)^2}{2}. \tag{5}$$

For the simplicity of the presentation, we assume that the arms are ordered decreasingly $\mu_1 \geq \mu_2 \geq \cdots \geq \mu_K$ and start by presenting the additive case (i.e. $G_\varepsilon(\boldsymbol{\mu}) \triangleq \{a \in [K] : \mu_a \geq \max_i \mu_i - \varepsilon\}$). The multiplicative case can be treated in the same fashion and is deferred to appendix A. Finally, we denote by $B_\varepsilon(\boldsymbol{\mu}) \triangleq [K] \setminus G_\varepsilon(\boldsymbol{\mu})$ the set of bad arms.

Since an alternative problem $\boldsymbol{\lambda} \in \text{Alt}(\boldsymbol{\mu})$ must have a different set of $\varepsilon$-optimal arms than the original problem $\boldsymbol{\mu}$, we can obtain it from $\boldsymbol{\mu}$ by changing the expected reward of some arms. We have two options to create an alternative problem $\boldsymbol{\lambda}$:

- **Making one of the $\varepsilon$-optimal arms bad.** We can achieve it by decreasing the expectation of some $\varepsilon$-optimal arm $k$ while increasing the expectation of some other arm $\ell$ to the point where $k$ is no more $\varepsilon$-optimal. This is illustrated in Fig. 1.
- **Making one of the $\varepsilon$-sub-optimal arms good.** We can achieve it by increasing the expectation of some sub-optimal arm $k$ while decreasing the expectations of the arms with the largest means -as many as it takes- to the point where $k$ becomes $\varepsilon$-optimal. This is illustrated in Fig. 1.

In the following, we solve both cases separately.

**Case 1: Making one of the $\varepsilon$-optimal arms bad.** Let $k \in G_\varepsilon(\boldsymbol{\mu})$ be one of the $\varepsilon$-optimal arms. In order to make arm $k$ sub-optimal, we need to set the expectation of arm $k$ to some value $\lambda_k = t$ and the maximum expectation over

---

[3]   The phenomenon discussed above is essentially already discussed in [16], a very rich study of the problem. However, we do not fully understand the proof of Theorem 4.1. Define a sub-instance to be a bandit $\tilde{\nu}$ with fewer arms $m \leq K$ such that $\{\tilde{\nu}_1, \ldots, \tilde{\nu}_m\} \subset \{\nu_1, \ldots, \nu_K\}$. Lemma D.5 in [16] actually shows that there exists some sub-instance of $\nu$ on which the algorithm must pay $\Omega(\sum_{b=2}^{m} 1/(\mu_1 - \mu_b)^2)$ samples. But this does not imply that such cost must be paid for the instance of interest $\nu$ instead of some sub-instance with very few arms.

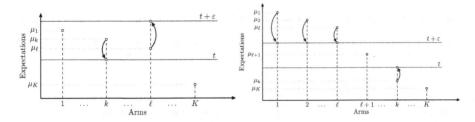

**Fig. 1.** Left: Making One of the ε-Optimal Arms Bad. Right: Making One of the ε-Sub-Optimal Arms Good.

all arms to $\max_a \lambda_a = t + \varepsilon$. Note that the index of the arm $\ell$ with maximum expectation can be chosen in $G_\varepsilon(\boldsymbol{\mu})$. Indeed, if we choose some arm from $B_\varepsilon(\boldsymbol{\mu})$ to become the arm with maximum expectation in $\boldsymbol{\lambda}$ then we would make an ε-suboptimal arm good which is covered in the other case below. The expectations of all the other arms should stay the same as in the instance $\boldsymbol{\mu}$, since changing their values would only increase the value of the objective. Now given indices $k$ and $\ell$, computing the optimal value of $t$ is rather straightforward since the objective function simplifies to

$$\omega_k \frac{(\mu_k - t)^2}{2} + \omega_\ell \frac{(\mu_\ell - t - \varepsilon)^2}{2}$$

for which the optimal value of $t$ is:

$$t = \overline{\mu}_\varepsilon^{k,\ell}(\boldsymbol{\omega}) \triangleq \frac{\omega_k \mu_k + \omega_\ell(\mu_\ell - \varepsilon)}{\omega_k + \omega_\ell}.$$

and the corresponding alternative bandit is:

$$\boldsymbol{\lambda}_\varepsilon^{k,\ell}(\boldsymbol{\omega}) \triangleq (\mu_1, \ldots, \underbrace{\overline{\mu}_\varepsilon^{k,\ell}(\boldsymbol{\omega})}_{\text{index } k}, \ldots, \underbrace{\overline{\mu}_\varepsilon^{k,\ell}(\boldsymbol{\omega}) + \varepsilon}_{\text{index } \ell}, \ldots, \mu_K)^\mathsf{T}.$$

The last step is taking the pair of indices $(k, \ell) \in G_\varepsilon(\boldsymbol{\mu}) \times (G_\varepsilon(\boldsymbol{\mu}) \setminus \{k\})$ with the minimal value in the objective (2).

**Case 2: Making one of the sub-optimal arms good.** Let $k \in B_\varepsilon(\boldsymbol{\mu})$ be a sub-optimal arm, if such arm exists, and denote by $t$ the value of its expectation in $\boldsymbol{\lambda}$. In order to make this arm ε-optimal, we need to decrease the expectations of all the arms that are above the threshold $t + \varepsilon$. We pay a cost of $\frac{1}{2}\omega_k(t - \mu_k)^2$ for moving arm $k$ and of $\frac{1}{2}\omega_i(t + \varepsilon - \mu_i)^2$ for every arm $i$ such that $\mu_i > t + \varepsilon$. Consider the functions:

$$f_k(t) = \frac{1}{2}\omega_k(t - \mu_k)^2$$

and for $i \in [K] \setminus \{k\}$

$$f_i(t) = \begin{cases} \frac{1}{2}\omega_i(t + \varepsilon - \mu_i)^2 & \text{for } t < \mu_i - \varepsilon, \\ 0 & \text{for } t \geq \mu_i - \varepsilon. \end{cases}$$

Each of these functions is convex. Therefore the function $f(t) = \sum_{i=1}^{K} f_i(t)$ is convex and has a unique minimizer $t^*$. One can easily check that $f'(\mu_k) \leq 0$ and $f'(\mu_1 - \varepsilon) \geq 0$, implying that $\mu_k - \varepsilon < \mu_k \leq t^* \leq \mu_1 - \varepsilon$. Therefore:

$$\ell = \min\{i \geq 1 \ : \ t^* > \mu_i - \varepsilon\} - 1$$

is well defined and satisfies $\ell \in [|1, k-1|]$. Note that by definition $\mu_{\ell+1} - \varepsilon < t^*$ and $t^* \leq \mu_a - \varepsilon$ for all $a \leq \ell$, hence:

$$0 = f'(t^*) = \omega_k(t^* - \mu_k) + \sum_{a=1}^{\ell} \omega_a(t^* + \varepsilon - \mu_a).$$

Implying that[4]:

$$t^* = \overline{\mu}_\varepsilon^{k,\ell}(\omega) \triangleq \frac{\omega_k \mu_k + \sum_{a=1}^{\ell} \omega_a(\mu_a - \varepsilon)}{\omega_k + \sum_{a=1}^{\ell} \omega_a}$$

and the alternative bandit in this case writes as:

$$\lambda_\varepsilon^{k,\ell}(\omega) \triangleq (\underbrace{\overline{\mu}_\varepsilon^{k,\ell}(\omega) + \varepsilon, \mu_{\ell+1}, \ldots, \overline{\mu}_\varepsilon^{k,\ell}(\omega)}_{\text{indices 1 to } \ell}, \ldots, \underbrace{\mu_K}_{\text{index } k})^\mathsf{T}.$$

Observe that since $\ell$ depends on $t^*$, we can't directly compute $t^*$ from the expression above. Instead, we use the fact that $\ell$ is unique by definition. Therefore, to determine $t^*$ one can compute $\overline{\mu}_\varepsilon^{k,\ell}(\omega)$ for all values of $\ell \in [|1, k-1|]$ and search for the index $\ell$ satisfying $\mu_{\ell+1} - \varepsilon < \overline{\mu}_\varepsilon^{k,\ell}(\omega) \leq \mu_\ell - \varepsilon$ and with minimum value in the objective (2).

As a summary, we have reduced the minimization problem over the infinite set $\mathrm{Alt}(\mu)$ to a combinatorial search over a finite number of alternative bandit instances whose analytical expression is given in the next definition.

**Definition 1.** *Let $\lambda_\varepsilon^{k,\ell}(\omega)$ be a vector created form $\mu$ by replacing elements on positions $k$ and $\ell$ (resp. 1 to $\ell$), defined as:*

$$\lambda_\varepsilon^{k,\ell}(\omega) \triangleq (\mu_1, \ldots, \underbrace{\overline{\mu}_\varepsilon^{k,\ell}(\omega)}_{\text{index } k}, \ldots, \underbrace{\overline{\mu}_\varepsilon^{k,\ell}(\omega) + \varepsilon}_{\text{index } \ell}, \ldots, \mu_K)^\mathsf{T}$$

*for $k \in G_\varepsilon(\mu)$ and*

$$\lambda_\varepsilon^{k,\ell}(\omega) \triangleq (\underbrace{\overline{\mu}_\varepsilon^{k,\ell}(\omega) + \varepsilon, \mu_{\ell+1}, \ldots, \overline{\mu}_\varepsilon^{k,\ell}(\omega)}_{\text{indices 1 to } \ell}, \ldots, \underbrace{\mu_K}_{\text{index } k})^\mathsf{T}$$

*for $k \in B_\varepsilon(\mu)$ where $\overline{\mu}_\varepsilon^{k,\ell}(\omega)$ is a weighted average of elements on positions $k$ and $\ell$ (resp. 1 to $\ell$) defined as:*

$$\overline{\mu}_\varepsilon^{k,\ell}(\omega) \triangleq \frac{\omega_k \mu_k + \omega_\ell(\mu_\ell - \varepsilon)}{\omega_k + \omega_\ell}$$

---

[4] $\overline{\mu}_\varepsilon^{k,\ell}(\omega)$ has a different definition depending on $k$ being a good or a bad arm.

*for $k \in G_\varepsilon(\boldsymbol{\mu})$ and*

$$\overline{\mu}_\varepsilon^{k,\ell}(\boldsymbol{\omega}) \triangleq \frac{\omega_k \mu_k + \sum_{a=1}^{\ell} \omega_a (\mu_a - \varepsilon)}{\omega_k + \sum_{a=1}^{\ell} \omega_a}$$

*for $k \in B_\varepsilon(\boldsymbol{\mu})$.*

The next lemma then states that the best response oracle belongs to the finite set of $(\boldsymbol{\lambda}_\varepsilon^{k,\ell}(\boldsymbol{\omega}))_{k,\ell}$.

**Lemma 1.** *Using the previous definition, $\boldsymbol{\lambda}_{\varepsilon,\boldsymbol{\mu}}^*(\boldsymbol{\omega})$ can be computed as*

$$\boldsymbol{\lambda}_{\varepsilon,\boldsymbol{\mu}}^*(\boldsymbol{\omega}) = \underset{\boldsymbol{\lambda} \in \Lambda_G \cup \Lambda_B}{\arg\min} \sum_{a \in [K]} \omega_a \frac{(\mu_a - \lambda_a)^2}{2}$$

*where*

$$\Lambda_G = \{\boldsymbol{\lambda}_\varepsilon^{k,\ell}(\boldsymbol{\omega}) : k \in G_\varepsilon(\boldsymbol{\mu}), \ell \in G_\varepsilon(\boldsymbol{\mu})/\{k\}\}$$

*and*

$$\Lambda_B = \{\boldsymbol{\lambda}_\varepsilon^{k,\ell}(\boldsymbol{\omega}) : k \in B_\varepsilon(\boldsymbol{\mu}), \ell \in [|1, k-1|]$$
$$\text{s.t. } \mu_\ell \geq \overline{\mu}_\varepsilon^{k,\ell}(\boldsymbol{\omega}) + \varepsilon > \mu_{\ell+1}\}.$$

## 5  Solving the Max-Min Problem: Optimal Weights

First observe that we can rewrite $T_\varepsilon(\boldsymbol{\mu}, .)^{-1}$ as a minimum of linear functions:

$$T_\varepsilon(\boldsymbol{\mu}, \boldsymbol{\omega})^{-1} = \inf_{\boldsymbol{d} \in \mathcal{D}_{\varepsilon,\boldsymbol{\mu}}} \boldsymbol{\omega}^\mathsf{T} \boldsymbol{d} \tag{6}$$

where

$$\mathcal{D}_{\varepsilon,\boldsymbol{\mu}} \triangleq \left\{ \left( \frac{(\lambda_a - \mu_a)^2}{2} \right)_{a \in [K]}^\mathsf{T} \mid \boldsymbol{\lambda} \in \mathrm{Alt}(\boldsymbol{\mu}) \right\}.$$

Note that by using $\mathcal{D}_{\varepsilon,\boldsymbol{\mu}}$ instead of $\mathrm{Alt}(\boldsymbol{\mu})$, the optimization function becomes simpler for the price of more complex domain (see Fig. 2 for an example). As a result, $T_\varepsilon(\boldsymbol{\mu}, .)^{-1}$ is concave and we can compute its supergradients thanks to Danskin's Theorem [4] which we recall in the lemma below.

**Lemma 2.** *(Danskin's Theorem) Let $\boldsymbol{\lambda}^*(\boldsymbol{\omega})$ be a best response to $\boldsymbol{\omega}$ and define $\boldsymbol{d}^*(\boldsymbol{\omega}) \triangleq \left( \frac{(\boldsymbol{\lambda}^*(\boldsymbol{\omega})_a - \mu_a)^2}{2} \right)_{a \in [K]}^\mathsf{T}$. Then $\boldsymbol{d}^*(\boldsymbol{\omega})$ is a supergradient of $T_\varepsilon(\boldsymbol{\mu}, .)^{-1}$ at $\boldsymbol{\omega}$.*

Next we prove that $T_\varepsilon(\boldsymbol{\mu}, .)^{-1}$ is Liptschiz.

**Lemma 3.** *The function $\boldsymbol{\omega} \mapsto T_\varepsilon(\boldsymbol{\mu}, \boldsymbol{\omega})^{-1}$ is L-Lipschitz with respect to $\|\cdot\|_1$ for any*

$$L \geq \max_{a,b \in [K]} \frac{(\mu_a - \mu_b + \varepsilon)^2}{2}.$$

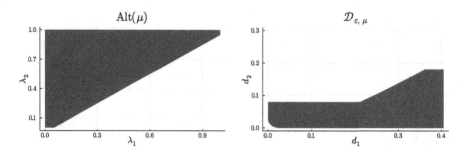

**Fig. 2.** Comparison of $\mathrm{Alt}(\boldsymbol{\mu})$ with Simple Linear Boundaries (First Figure) and $\mathcal{D}_{\varepsilon,\boldsymbol{\mu}}$ with Non-Linear Boundaries (Second Figure) for $\boldsymbol{\mu} = [0.9,\ 0.6]$ and $\varepsilon = 0.05$.

*Proof.* As we showed in Lemma 1, the best response $\boldsymbol{\lambda}^*_{\varepsilon,\boldsymbol{\mu}}(\boldsymbol{\omega})$ to $\boldsymbol{\omega}$ is created from $\boldsymbol{\mu}$ by replacing some of the elements by $\overline{\boldsymbol{\mu}}^{k,\ell}_\varepsilon(\boldsymbol{\omega})$ or $\overline{\boldsymbol{\mu}}^{k,\ell}_\varepsilon(\boldsymbol{\omega}) + \varepsilon$. We also know that $\overline{\boldsymbol{\mu}}^{k,\ell}_\varepsilon(\boldsymbol{\omega})$ is a weighted average of an element of $\boldsymbol{\mu}$ with one or more elements of $\boldsymbol{\mu}$ decreased by $\varepsilon$. This means that:

$$\max_{a\in[K]} \mu_a \geq \overline{\boldsymbol{\mu}}^{k,\ell}_\varepsilon(\boldsymbol{\omega}) \geq \min_{a\in[K]} \mu_a - \varepsilon$$

and, as a consequence, we have:

$$|\mu_i - \boldsymbol{\lambda}^*_{\varepsilon,\boldsymbol{\mu}}(\boldsymbol{\omega})_i| \leq \max_{a,b\in[K]} (\mu_a - \mu_b + \varepsilon)$$

for any $i \in [K]$. Let $f(\boldsymbol{\omega}) \triangleq T_\varepsilon(\boldsymbol{\mu},\boldsymbol{\omega})^{-1}$. Using the last inequality and the definition of $\boldsymbol{d}^*(\boldsymbol{\omega})$, we can obtain:

$$\begin{aligned} f(\boldsymbol{\omega}) - f(\boldsymbol{\omega}') &\leq (\boldsymbol{\omega} - \boldsymbol{\omega}')^\mathsf{T} \boldsymbol{d}^*(\boldsymbol{\omega}') \\ &\leq \|\boldsymbol{\omega} - \boldsymbol{\omega}'\|_1 \|\boldsymbol{d}^*(\boldsymbol{\omega}')\|_\infty \\ &\leq \|\boldsymbol{\omega} - \boldsymbol{\omega}'\|_1 \max_{a,b\in[K]} \frac{(\mu_a - \mu_b + \varepsilon)^2}{2} \end{aligned}$$

for any $\boldsymbol{\omega}, \boldsymbol{\omega}' \in \Delta_K$.

As a summary $T_\varepsilon(\boldsymbol{\mu},.)^{-1}$ is concave, Lipschitz and we have a simple expression to compute its supergradients through the best response oracle. Therefore we have all the necessary ingredients to apply a gradient-based algorithm in order to find the optimal weights and therefore, the value of $T^*_\varepsilon(\boldsymbol{\mu})$. The algorithm of our choice is the mirror ascent algorithm which provides the following guarantees:

**Proposition 2.** *[2] Let* $\boldsymbol{\omega}_1 = (\frac{1}{K}, \ldots, \frac{1}{K})^\mathsf{T}$ *and learning rate* $\alpha_n = \frac{1}{L}\sqrt{\frac{2\log K}{n}}$. *Then using mirror ascent algorithm to maximize a $L$-Lipschitz function $f$, with respect to $\|\cdot\|_1$, defined on $\Delta_K$ with generalized negative entropy $\Phi(\boldsymbol{\omega}) = \sum_{a\in[K]} \omega_a \log(\omega_a)$ as the mirror map enjoys the following guarantees:*

$$f(\boldsymbol{\omega}^*) - f\left(\frac{1}{N}\sum_{n=1}^{N} \boldsymbol{\omega}_n\right) \leq L\sqrt{\frac{2\log K}{N}}.$$

**Computational Complexity of Our Algorithm.** To simplify the presentation and analysis, we chose to focus on the vanilla version of Track and Stop. However, in practice this requires solving the optimization program that appears in the lower bound at every time step, which can result in large run times. Nonetheless, we note that there are many possible adaptations of Track and Stop that reduce the computational complexity, while retaining the guarantees of asymptotic optimality in terms of the sample complexity (and with a demonstrated small performance loss experimentally). A first solution is to use Franke-Wolfe style algorithms [17,19], which only perform a gradient step of the optimization program at every step. Once can also apply the Gaming approach initiated by [5] which only needs to solve the best response problem, and runs a no-regret learner such as AdaHedge to determine the weights to be tracked at each step. This approach was used for example by [10] in a similar setting of Pure Exploration with semi-bandit feedback. Another adaptation is the Lazy Track-and-Stop [9], which updates the weights that are tracked by the algorithms every once in a while. We chose the latter solution in our implementation, where we updated the weights every $100K$ steps.

# 6    Experiments

We conducted three experiments to compare Track-and-Stop with state-of-the-art algorithms, mainly $(\text{ST})^2$ and FAREAST from [16]. In the first experiment, we simulate a multi-armed bandit with Gaussian rewards of means $\boldsymbol{\mu} = [1, 1, 1, 1, 0.05]$, variance one and a parameter $\varepsilon = 0.9$. We chose this particular instance $\boldsymbol{\mu}$ because its difficulty is two-fold: First, the last arm $\mu_5$ is very close to the threshold $\max_a \mu_a - \varepsilon$. Second, the argmax is realized by more than one arm, which implies that any algorithm must estimate all the means to high precision to produce a confident guess of $G_\varepsilon(\boldsymbol{\mu})$. Indeed, a small underestimation error of $\max_a \mu_a$ would mean wrongly classifying $\mu_5$ as a good arm. We run the three algorithms for several values of $\delta$ ranging from $\delta = 0.1$ to $\delta = 10^{-10}$, with $N = 100$ Monte-Carlo simulations for each risk level. Figure 3 shows the expected stopping time along with the 10% and 90% quantiles (shaded area) for each algorithm. Track-and-Stop consistently outperforms $(\text{ST})^2$ and FAREAST, even for moderate values of $\delta$. Also note that, as we pointed out in Remark 1, the sample complexity of Track-and-Stop is within some multiplicative constant of $(\text{ST})^2$.

Next, we examine the performance of the algorithms w.r.t the number of arms. For any given $K$, we consider a bandit problem $\boldsymbol{\mu}$ similar to the previous instance: $\forall a \in [\![1, K-1]\!]$, $\mu_a = 1$ and $\mu_K = 0.05$. We fix $\varepsilon = 0.9$ and $\delta = 0.1$ and run $N = 30$ Monte-Carlo simulations for each $K$. Figure 4 shows, in log-scale, the ratio of the sample complexities of $(\text{ST})^2$ and FAREAST w.r.t to the sample complexity of Track-and-Stop. We see that Track-and-Stop performs better than $(\text{ST})^2$ (resp. FAREAST) for small values of $K$. However when the number of arms grows larger than $K = 40$ (resp. $K = 60$), $(\text{ST})^2$ (resp. FAREAST) have a smaller sample complexity.

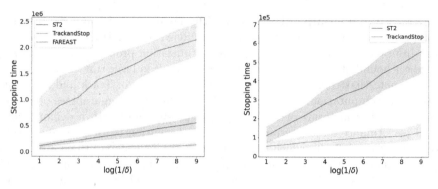

**Fig. 3.** Expected Stopping Time on $\mu = [1,1,1,1,0.05]$. Left: All three Algorithms. Right: Track-and-Stop vs FAREAST.

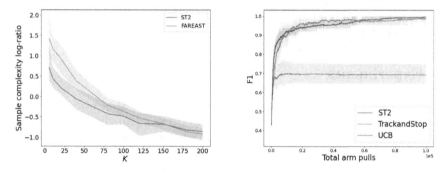

**Fig. 4.** Left: $\log_{10}\left(\mathbb{E}_{\text{Alg}}[\tau_\delta]/\mathbb{E}_{\text{TaS}}[\tau_\delta]\right)$ for Alg $\in \{(\text{ST})^2, \text{FAREAST}\}$ and TaS = Track-and-Stop, $K_{\min} = 5$ arms. Right: F1 scores for Cancer Drug discovery.

Finally, we rerun the Cancer Drug Discovery experiment from [16]. Note that this experiment is more adapted to a *fixed budget setting* where we fix a sampling budget and the algorithm stops once it has reached this limit, which is different from the *fixed confidence* setting that our algorithm was designed for. The goal is to find, among a list of 189 chemical compounds, potential inhibitors to **ACRVL1**, a Kinaze that researchers [1] have linked to several forms of cancer. We use the same dataset as [16], where for each compound a percent control[5] is reported. We fix a budget of samples $N = 10^5$ and try to find all the $\varepsilon$-good compounds in the multiplicative case with $\varepsilon = 0.8$. For each algorithm, we compute the F1-score[6] of its current estimate $\widehat{G}_\varepsilon = \{i : \widehat{\mu}_i \geq (1-\varepsilon)\max_a \widehat{\mu}_a\}$ after every iteration. The F1-score in this fixed-budget setting

---

[5] percent control is a metric expressing the efficiency of the compound as an inhibitor against the target Kinaze.

[6] F1 score is the harmonic mean of precision (the proportion of arms in $\widehat{G}$ that are actually good) and recall (the proportion of arms in $G_\varepsilon(\mu)$ that were correctly returned in $\widehat{G}$).

reflects how good is the sampling scheme of an algorithm, independently of its stopping condition. In Fig. 4 we plot the average F1-score along with the 10% and 90% quantiles (shaded area). We see that $(ST)^2$ and Track-and-Stop have comparable performance and that both outperform UCB's sampling scheme.

## 7   Conclusion

We shed a new light on the sample complexity of finding all the $\varepsilon$-good arms in a multi-armed bandit with Gaussian rewards. We derived two lower bounds, identifying the characteristic time that reflects the true hardness of the problem in the asymptotic regime. Moreover, we proved a second bound highlighting an additional cost that is linear in the number of arms and can be arbitrarily larger than the first bound for moderate values of the risk. Then, capitalizing on an algorithm solving the optimization program that defines the characteristic time, we proposed an efficient Track-and-Stop strategy whose sample complexity matches the lower bound for small values of the risk level. Finally, we showed through numerical simulations that our algorithm outperforms state-of-the-art methods for bandits with small to moderate number of arms. Several directions are worth investigating in the future. Notably, we observe that while Track-and-Stop performs better in the fixed-$K$-small-$\delta$ regime, the elimination based algorithms $(ST)^2$ and FAREAST become more efficient in the large-$K$-fixed-$\delta$ regime. It would be interesting to understand the underlying tradeoff between the number of arms and confidence parameter. This will help design pure exploration strategies having best of both worlds guarantees.

**Acknowledgements.** The authors acknowledge the support of the Chaire SeqALO (ANR-20-CHIA-0020-01) and of Project IDEXLYON of the University of Lyon, in the framework of the Programme Investissements d'Avenir (ANR-16-IDEX-0005).

## References

1. Bocci, M., et al.: Activin receptor-like kinase 1 is associated with immune cell infiltration and regulates *CLEC14A* transcription in cancer. Angiogenesis **22**(1), 117–131 (2018). https://doi.org/10.1007/s10456-018-9642-5
2. Bubeck, S.: Convex optimization: algorithms and complexity. Foundations and Trends in Machine Learning (2015)
3. Chernoff, H.: Sequential design of experiments. Ann. Math. Stat. **30**(3), 755–770 (1959)
4. Danskin, J.M.: The theory of max-min, with applications. SIAM J. Appl. Math. **14**, 641–664 (1966)
5. Degenne, R., Koolen, W.M., Ménard, P.: Non-asymptotic pure exploration by solving games. In: Wallach, H., Larochelle, H., Beygelzimer, A., d'Alché-Buc, F., Fox, E., Garnett, R. (eds.) Advances in Neural Information Processing Systems, vol. 32. Curran Associates, Inc. (2019). https://proceedings.neurips.cc/paper/2019/file/8d1de7457fa769ece8d93a13a59c8552-Paper.pdf

6. Garivier, A., Kaufmann, E.: Non-asymptotic sequential tests for overlapping hypotheses and application to near optimal arm identification in bandit models. Sequential Anal. **40**, 61–96 (2021)
7. Garivier, A.: Informational confidence bounds for self-normalized averages and applications. In: 2013 IEEE Information Theory Workshop (ITW) (Sep 2013). https://doi.org/10.1109/itw.2013.6691311
8. Garivier, A., Kaufmann, E.: Optimal best arm identification with fixed confidence. In: Proceedings of the 29th Conference On Learning Theory, pp. 998–1027 (2016)
9. Jedra, Y., Proutiere, A.: Optimal best-arm identification in linear bandits. In: Larochelle, H., Ranzato, M., Hadsell, R., Balcan, M.F., Lin, H. (eds.) Advances in Neural Information Processing Systems, vol. 33, pp. 10007–10017. Curran Associates, Inc. (2020). https://proceedings.neurips.cc/paper/2020/file/7212a6567c8a6c513f33b858d868ff80-Paper.pdf
10. Jourdan, M., Mutn'y, M., Kirschner, J., Krause, A.: Efficient pure exploration for combinatorial bandits with semi-bandit feedback. In: ALT (2021)
11. Kaufmann, E., Cappé, O., Garivier, A.: On the complexity of best arm identification in multi-armed bandit models. J. Mach. Learn. Res. (2015)
12. Kaufmann, E., Koolen, W.M.: Mixture martingales revisited with applications to sequential tests and confidence intervals. arXiv preprint arXiv:1811.11419 (2018)
13. Lai, T., Robbins, H.: Asymptotically efficient adaptive allocation rules. Adv. Appl. Math. **6**(1), 4–22 (1985)
14. Lattimore, T., Szepesvári, C.: Bandit Algorithms. Cambridge University Press, Cambridge (2019)
15. Magureanu, S., Combes, R., Proutiere, A.: Lipschitz bandits: regret lower bounds and optimal algorithms. In: Conference on Learning Theory (2014)
16. Mason, B., Jain, L., Tripathy, A., Nowak, R.: Finding all $\varepsilon$-good arms in stochastic bandits. In: Larochelle, H., Ranzato, M., Hadsell, R., Balcan, M.F., Lin, H. (eds.) Advances in Neural Information Processing Systems, vol. 33, pp. 20707–20718. Curran Associates, Inc. (2020). https://proceedings.neurips.cc/paper/2020/file/edf0320adc8658b25ca26be5351b6c4a-Paper.pdf
17. Ménard, P.: Gradient ascent for active exploration in bandit problems. arXiv e-prints p. arXiv:1905.08165 (May 2019)
18. Simchowitz, M., Jamieson, K., Recht, B.: The simulator: understanding adaptive sampling in the moderate-confidence regime. In: Kale, S., Shamir, O. (eds.) Proceedings of the 2017 Conference on Learning Theory. Proceedings of Machine Learning Research, vol. 65, pp. 1794–1834. PMLR, Amsterdam, Netherlands (07–10 Jul 2017), http://proceedings.mlr.press/v65/simchowitz17a.html
19. Wang, P.A., Tzeng, R.C., Proutiere, A.: Fast pure exploration via frank-wolfe. In: Advances in Neural Information Processing Systems, vol. 34 (2021)

# Improved Regret Bounds for Online Kernel Selection Under Bandit Feedback

Junfan Li[ID] and Shizhong Liao[(✉)][ID]

College of Intelligence and Computing, Tianjin University, Tianjin 300350, China
{junfli,szliao}@tju.edu.cn

**Abstract.** In this paper, we improve the regret bound for online kernel selection under bandit feedback. Previous algorithm enjoys a $O((\|f\|^2_{\mathcal{H}_i} + 1)K^{\frac{1}{3}}T^{\frac{2}{3}})$ expected bound for Lipschitz loss functions. We prove two types of regret bounds improving the previous bound. For smooth loss functions, we propose an algorithm with a $O(U^{\frac{2}{3}}K^{-\frac{1}{3}}(\sum_{i=1}^{K} L_T(f_i^*))^{\frac{2}{3}})$ expected bound where $L_T(f_i^*)$ is the cumulative losses of optimal hypothesis in $\mathbb{H}_i = \{f \in \mathcal{H}_i : \|f\|_{\mathcal{H}_i} \leq U\}$. The data-dependent bound keeps the previous worst-case bound and is smaller if most of candidate kernels match well with the data. For Lipschitz loss functions, we propose an algorithm with a $O(U\sqrt{KT}\ln^{\frac{2}{3}}T)$ expected bound asymptotically improving the previous bound. We apply the two algorithms to online kernel selection with time constraint and prove new regret bounds matching or improving the previous $O(\sqrt{T\ln K} + \|f\|^2_{\mathcal{H}_i} \max\{\sqrt{T}, \frac{T}{\sqrt{\mathcal{R}}}\})$ expected bound where $\mathcal{R}$ is the time budget. Finally, we empirically verify our algorithms on online regression and classification tasks.

**Keywords:** Model selection · Online learning · Bandit · Kernel method

## 1 Introduction

Selecting a suitable kernel function is critical for online kernel learning algorithms, and is more challenge than offline kernel selection since the data are provided sequentially and may not be i.i.d.. Such kernel selection problems are named online kernel selection [22]. To address those challenges, many online kernel selection algorithms reduce it to a sequential decision problem, and then randomly select a kernel function or use a convex combination of multiple kernel functions on the fly [7,11,16,19,22]. Let $\mathcal{K} = \{\kappa_i\}_{i=1}^K$ be predefined base kernels. An adversary sequentially sends the learner instances $\{\mathbf{x}_t\}_{t=1}^T$. The learner will choose a sequence of hypotheses $\{f_t\}_{t=1}^T$ from the $K$ reproducing kernel Hilbert

This work was supported in part by the National Natural Science Foundation of China under grants No. 62076181.

**Supplementary Information** The online version contains supplementary material available at https://doi.org/10.1007/978-3-031-26412-2_21.

spaces (RKHSs) induced by kernels in $\mathcal{K}$. At each round $t$, the learner suffers a prediction loss $\ell(f_t(\mathbf{x}_t), y_t)$. The goal is to minimize the regret defined as follows,

$$\forall \kappa_i \in \mathcal{K}, \forall f \in \mathcal{H}_i, \quad \mathrm{Reg}_T(f) = \sum_{t=1}^{T} \ell(f_t(\mathbf{x}_t), y_t) - \sum_{t=1}^{T} \ell(f(\mathbf{x}_t), y_t). \quad (1)$$

Effective online kernel selection algorithms must keep sublinear regret bounds w.r.t. the unknown optimal RKHS $\mathcal{H}_{i^*}$ induced by $\kappa_{i^*} \in \mathcal{K}$.

Previous work reduces online kernel selection to a sequential decision problem, including (i) prediction with expert advice [4], (ii) $K$-armed bandit problem [3], (iii) prediction with limited advice [15]. The online multi-kernel learning algorithms [5,14] which reduce the problem to prediction with expert advice, use a convex combination of $K$ hypotheses and enjoy a $O(\mathrm{poly}(\|f\|_{\mathcal{H}_i})\sqrt{T \ln K})$ regret bound. Combining $K$ hypotheses induces a $O(Kt)$ per-round time complexity which is linear with $K$. To reduce the time complexity, the OKS algorithm (Online Kernel Selection) [19] reduces the problem to an adversarial $K$-armed bandit problem. OKS randomly selects a hypothesis per-round and only provides a $O(\mathrm{poly}(\|f\|_{\mathcal{H}_i})K^{\frac{1}{3}}T^{\frac{2}{3}}))$ [1] expected bound. The per-round time complexity of OKS is $O(t)$. The B(AO)$_2$KS algorithm [11] reduces the problem to predict with limited advice and randomly selects two hypotheses per-round. B(AO)$_2$KS can provide a $\tilde{O}(\mathrm{poly}(\|f\|_{\mathcal{H}_i})\sqrt{KT})$ high-probability bound and suffers a $O(t/K)$ per-round time complexity. From the perspective of algorithm design, an important question arises: does there exist some algorithm only selecting a hypothesis (or under bandit feedback) improving the $O(\mathrm{poly}(\|f\|_{\mathcal{H}_i})K^{\frac{1}{3}}T^{\frac{2}{3}}))$ expected bound? The significances of answering the question include (i) explaining the information-theoretic cost induced by only selecting a hypothesis (or observing a loss); (ii) designing better algorithms for online kernel selection with time constraint. In this paper, we will answer the question affirmatively.

We consider Lipschitz loss functions and smooth loss functions (Assumption 1). For Lipschitz loss functions, we propose an algorithm whose expected regret bound is $O(U\sqrt{KT} \ln^{\frac{2}{3}} T)$ asymptotically improving the $O(\mathrm{poly}(\|f\|_{\mathcal{H}_i})K^{\frac{1}{3}}T^{\frac{2}{3}}))$ expected bound. Our regret bound proves that selecting a or multiple hypotheses will not induce significant variation on the worst-case regret bound. For smooth loss functions, we propose an adaptive parameter tuning scheme for OKS and prove a $O(U^{\frac{2}{3}}K^{-\frac{1}{3}}(\sum_{j=1}^{K} L_T(f_j^*))^{\frac{2}{3}})$ expected bound where $L_T(f_j^*) = \min_{f \in \mathbb{H}_j} \sum_{t \in [T]} \ell(f(\mathbf{x}_t), y_t)$. If most of base kernels in $\mathcal{K}$ match well with the data, i.e., $L_T(f_j^*) \ll T$, then the data-dependent regret bound significantly improves the previous worst-case bound. In the worst case, i.e., $L_T(f_j^*) = O(T)$, the data-dependent bound is still same with the previous bound. Our new regret bounds answer the above question. We summary the results in Table 1.

We apply the two algorithms to online kernel selection with time constraint where the time of kernel selection and online prediction is limited to $\mathcal{R}$ quanta [9]. It was proved that any budgeted algorithm must suffer an expected regret of order $\Omega(\|f_i^*\|_{\mathcal{H}_i} \max\{\sqrt{T}, \frac{T}{\sqrt{\mathcal{R}}}\})$ and the LKMBooks algorithm enjoys

---

[1] $\mathrm{poly}(\|f\|_{\mathcal{H}_i}) = \|f\|_{\mathcal{H}_i}^2 + 1$. The original paper shows a $O((\|f\|_{\mathcal{H}_i}^2 + 1)\sqrt{KT})$ expected regret bound. We will clarify the difference in Sect. 2.

**Table 1.** Expected regret bounds for online kernel selection under bandit feedback. $\mathcal{R}$ is the time budget. $\bar{L}_T = \sum_{j=1}^{K} L_T(f_j^*)$. $\nu$ is a parameter in the definition of smooth loss (see Assumption 1). There is no algorithm under bandit feedback in the case of a time budget. Thus we report the result produced under the expert advice model [9].

| $\mathcal{R}$ | Loss function | Previous results | Our results |
|---|---|---|---|
| No | Lipschitz loss | $O\left(\text{poly}(\|f\|_{\mathcal{H}_i})K^{\frac{1}{3}}T^{\frac{2}{3}}\right)$ [19] | $O(U\sqrt{KT}\ln^{\frac{2}{3}}T)$ |
|  | Smooth loss $\nu = 1$ |  | $O(U^{\frac{2}{3}}K^{-\frac{1}{3}}\bar{L}_T^{\frac{2}{3}})$ |
|  | Smooth loss $\nu = 2$ |  | $O(U^{\frac{2}{3}}K^{-\frac{1}{3}}\bar{L}_T^{\frac{2}{3}})$ |
| Yes | Lipschitz loss | $O\left(\|f\|_{\mathcal{H}_i}^2 \max\{\sqrt{T}, \frac{T}{\sqrt{\mathcal{R}}}\}\right)$ [9] | $O(U\sqrt{KT}\ln^{\frac{2}{3}}T + \frac{UT\sqrt{\ln T}}{\sqrt{\mathcal{R}}})$ |
|  | Smooth loss $\nu = 1$ |  | $\tilde{O}(U^{\frac{2}{3}}K^{-\frac{1}{3}}\bar{L}_T^{\frac{2}{3}} + \frac{UL_T(f_i^*)}{\sqrt{\mathcal{R}}})$ |
|  | Smooth loss $\nu = 2$ |  | $\tilde{O}(U^{\frac{2}{3}}K^{-\frac{1}{3}}\bar{L}_T^{\frac{2}{3}} + \frac{U\sqrt{TL_T(f_i^*)}}{\sqrt{\mathcal{R}}})$ |

a $O(\sqrt{T \ln K} + \|f\|_{\mathcal{H}_i}^2 \max\{\sqrt{T}, \frac{T}{\sqrt{\mathcal{R}}}\})$ expected bound [9]. LKMBooks uses convex combination to aggregate $K$ hypotheses. Raker uses random features to approximate kernel functions and also aggregates $K$ hypotheses [16]. Raker enjoys a $\tilde{O}((\sqrt{\ln K} + \|f\|_1^2)\sqrt{T} + \|f\|_1 \frac{T}{\sqrt{\mathcal{R}}})$ bound where $f = \sum_{t=1}^{T} \alpha_t \kappa_i(\mathbf{x}_t, \cdot)$ and $\|f\|_1 = \|\boldsymbol{\alpha}\|_1$ [16]. The two algorithms reduce the problem to prediction with expert advice, while our algorithms just use bandit feedback.

We also use random features and make a mild assumption that reduces the time budget $\mathcal{R}$ to the number of features. For smooth loss functions, we prove two data-dependent regret bounds which can improve the previous worst-case bounds [9,16] if there is a good kernel in $\mathcal{K}$ that matches well with the data. For Lipschitz loss functions, our algorithm enjoys a similar upper bound with LKMBooks. We also summary the results in Table 1.

## 2   Problem Setting

Denote by $\{(\mathbf{x}_t, y_t)\}_{t \in [T]}$ a sequence of examples, where $\mathbf{x}_t \in \mathcal{X} \subseteq \mathbb{R}^d, y \in [-1, 1]$ and $[T] = \{1, 2, \ldots, T\}$. Let $\kappa(\cdot, \cdot) : \mathcal{X} \times \mathcal{X} \to \mathbb{R}$ be a positive definite kernel and $\mathcal{K} = \{\kappa_1, \ldots, \kappa_K\}$. For each $\kappa_i \in \mathcal{K}$, let $\mathcal{H}_i = \{f | f : \mathcal{X} \to \mathbb{R}\}$ be the associated RKHS satisfying $\langle f, \kappa_i(\mathbf{x}, \cdot)\rangle_{\mathcal{H}_i} = f(\mathbf{x}), \forall f \in \mathcal{H}_i$. Let $\|f\|_{\mathcal{H}_i}^2 = \langle f, f\rangle_{\mathcal{H}_i}$. We assume that $\kappa_i(\mathbf{x}, \mathbf{x}) \leq 1, \forall \kappa_i \in \mathcal{K}$. Let $\ell(\cdot, \cdot) : \mathbb{R} \times \mathbb{R} \to \mathbb{R}$ be the loss function.

### 2.1   Online Kernel Selection Under Bandit Feedback

We formulate online kernel selection as a sequential decision problem. At any round $t \in [T]$, an adversary gives an instance $\mathbf{x}_t$. The learner maintains $K$ hypotheses $\{f_{t,i} \in \mathcal{H}_i\}_{i=1}^{K}$ and selects $f_t \in \text{span}(f_{t,i} : i \in [K])$, and outputs $f_t(\mathbf{x}_t)$. Then the adversary gives $y_t$. The learner suffers a prediction loss $\ell(f_t(\mathbf{x}_t), y_t)$. The learner aims to minimize the regret w.r.t. any $f \in \cup_{i=1}^{K}\mathcal{H}_i$ which is defined in (1). If the learner only computes a loss $\ell(f_{t,I_t}(\mathbf{x}_t), y_t), I_t \in [K]$, then we call it *bandit feedback setting*. The learner can also compute $N \in \{2, \ldots, K\}$ losses, i.e., $\{\ell(f_{t,i_j}(\mathbf{x}_t), y_t)\}_{j=1}^{N}, i_j \in [K]$. The OKS algorithm [19] follows the bandit feedback setting. The online multi-kernel learning algorithms

---

**Algorithm 1.** OKS
___
**Input:** $\mathcal{K} = \{\kappa_1, \ldots, \kappa_K\}$, $\delta \in (0,1)$, $\eta$, $\lambda$
**Initialization:** $\{f_{1,i} = 0, w_{1,i} = 1\}_{i=1}^{K}$, $\mathbf{p}_1 = \frac{1}{K}\mathbf{1}_K$
1: **for** t=1,...,T **do**
2:     Receive $\mathbf{x}_t$;
3:     Sample a kernel $\kappa_{I_t}$ where $I_t \sim \mathbf{p}_t$;
4:     Update $w_{t+1,I_t} = w_{t,I_t} \exp(-\eta \frac{\ell(f_{t,I_t}(\mathbf{x}_t),y_t)}{p_{t,I_t}})$;
5:     Update $f_{t+1,I_t} = f_{t,I_t} - \lambda \frac{\nabla_{f_{t,I_t}} \ell(f_{t,I_t}(\mathbf{x}_t),y_t)}{p_{t,I_t}}$;
6:     Update $\mathbf{q}_{t+1} = \frac{\mathbf{w}_{t+1}}{\sum_{j=1}^{K} w_{t+1,j}}$ and set $\mathbf{p}_{t+1} = (1-\delta)\mathbf{q}_{t+1} + \frac{\delta}{K}\mathbf{1}_K$;
7: **end for**
8: Output: $\mathbf{q}_T$.

---

[5,14,16] correspond to $N = K$. The B(AO)$_2$KS algorithm [11] corresponds to $N = 2$. From the perspective of computation, the per-round time complexity of computing $N$ losses is $N$ times larger than the bandit feedback setting. From the perspective of regret bound, we aim to reveal the information-theoretic cost induced by observing a loss (or bandit feedback) not multiple losses (or $N \geq 2$).

## 2.2  Regret Bound of OKS

We first prove that the regret bound of OKS [19] is $O((\|f\|_{\mathcal{H}_i}^2 + 1)K^{\frac{1}{3}}T^{\frac{2}{3}})$, and then explain the technical weakness of OKS.

The pseudo-code of OKS is shown Algorithm 1. Let $\Delta_K$ be the $(K-1)$-dimensional simplex. At any round $t$, OKS maintains $\mathbf{p}_t, \mathbf{q}_t \in \Delta_K$. OKS samples $f_{t,I_t}$ where $I_t \sim \mathbf{p}_t$, and outputs $f_{t,I_t}(\mathbf{x}_t)$. For simplicity, we define two notations,

$$L_T(f) := \sum_{t=1}^{T} \ell(f(\mathbf{x}_t), y_t), \quad \bar{L}_{\mathbf{q}_{1:T}} := \sum_{t=1}^{T}\sum_{i=1}^{K} q_{t,i}\ell(f_{t,i}(\mathbf{x}_t), y_t).$$

**Theorem 1 ([19]).** *Assuming that $\ell(f_{t,i}(\mathbf{x}),y) \in [0, \ell_{\max}]$, $\forall i \in [K]$, $t \in [T]$, and $\|\nabla_f \ell(f(\mathbf{x}),y)\|_{\mathcal{H}_i} \leq G$, $\forall f \in \mathcal{H}_i$. The expected regret of OKS satisfies*

$$\forall i \in [K], f \in \mathcal{H}_i, \quad \mathbb{E}[\bar{L}_{\mathbf{q}_{1:T}}] \leq L_T(f) + \frac{\|f\|_{\mathcal{H}_i}^2}{2\lambda} + \frac{\lambda KTG^2}{2\delta} + \frac{\eta KT\ell_{\max}^2}{2(1-\delta)} + \frac{\ln K}{\eta}.$$

*In particular, let $\delta \in (0,1)$ be a constant and $\eta, \lambda = \Theta((KT)^{-\frac{1}{2}})$, then the expected regret bound is $O((\|f\|_{\mathcal{H}_i}^2 + 1)\sqrt{KT})$.*

*Remark 1.* Since $I_t \sim \mathbf{p}_t$, the expected cumulative losses of OKS should be $\mathbb{E}[\bar{L}_{\mathbf{p}_{1:T}}]$ which is different from $\mathbb{E}[\bar{L}_{\mathbf{q}_{1:T}}]$ as stated in Theorem 1. Since $\mathbf{p}_t = (1-\delta)\mathbf{q}_t + \frac{\delta}{K}\mathbf{1}_K$, the expected regret of OKS should be redefined as follows

$$\forall i \in [K], f \in \mathcal{H}_i, \quad \mathbb{E}[\bar{L}_{\mathbf{p}_{1:T}}] - L_T(f)$$

$$\leq \delta\mathbb{E}[\bar{L}_{\frac{1}{K}\mathbf{1}}] - \delta L_T(f) + \frac{\|f\|_{\mathcal{H}_i}^2}{2\lambda} + \frac{\lambda KTG^2}{2\delta} + \frac{\eta KT\ell_{\max}^2}{2(1-\delta)} + \frac{\ln K}{\eta}$$

$$\leq \delta T\ell_{\max} + \frac{\|f\|_{\mathcal{H}_i}^2}{2\lambda} + \frac{\lambda KTG^2}{2\delta} + \frac{\eta KT\ell_{\max}^2}{2(1-\delta)} + \frac{\ln K}{\eta}.$$

To minimize the upper bound, let $\delta = (G/\ell_{\max})^{\frac{2}{3}} K^{\frac{1}{3}} T^{-\frac{1}{3}}$, $\lambda = \sqrt{\delta/(KTG^2)}$ and $\eta = \sqrt{2(1-\delta)\ln K}/\sqrt{KT\ell_{\max}^2}$. The upper bound is $O((\|f\|_{\mathcal{H}_i}^2 + 1)K^{\frac{1}{3}}T^{\frac{2}{3}})$.

*Remark 2.* OKS is essentially an offline kernel selection algorithm, since it aims to output a hypothesis following $\mathbf{q}_T$ for test datasets (see line 8 in Algorithm 1). Thus Theorem 1 defines the expected regret using $\{\mathbf{q}_1, \ldots, \mathbf{q}_T\}$, and the $O((\|f\|_{\mathcal{H}_i}^2 + 1)\sqrt{KT})$ bound is reasonable. For online kernel selection, we focus on the online prediction performance. Since OKS selects $f_{t,I_t}$ following $\mathbf{p}_t$, the expected regret should be defined using $\{\mathbf{p}_1, \ldots, \mathbf{p}_T\}$.

We find that the dependence on $O(K^{\frac{1}{3}}T^{\frac{2}{3}})$ comes from the term $\frac{\lambda KTG^2}{2\delta}$ which upper bounds the cumulative variance of gradient estimators, i.e.,

$$\frac{\lambda}{2}\mathbb{E}\left[\sum_{t=1}^{T}\|\tilde{\nabla}_{t,i}\|_{\mathcal{H}_i}^2\right] \leq \frac{\lambda KTG^2}{2\delta}, \quad \tilde{\nabla}_{t,i} = \frac{\nabla_{t,i}}{p_{t,i}}\mathbb{I}_{i=I_t}, \nabla_{t,i} = \nabla_{f_{t,i}}\ell(f_{t,i}(\mathbf{x}_t), y_t).$$

Next we give a simple analysis. To start with, it can be verified that

$$\mathbb{E}\left[\|\tilde{\nabla}_{t,i}\|_{\mathcal{H}_i}^2\right] = \mathbb{E}\left[p_{t,i}\frac{\|\nabla_{t,i}\|_{\mathcal{H}_i}^2}{p_{t,i}^2} + (1-p_{t,i})\cdot 0\right] \leq \mathbb{E}\left[\max_{t=1,\ldots,T}\left(\frac{1}{p_{t,i}}\right)\|\nabla_{t,i}\|_{\mathcal{H}_i}^2\right].$$

Recalling that $p_{t,i} \geq \frac{\delta}{K}$, $\forall i \in [K]$, $t \in [T]$. Summing over $t = 1, \ldots, T$ yields

$$\sum_{t=1}^{T}\mathbb{E}\left[\|\tilde{\nabla}_{t,i}\|_{\mathcal{H}_i}^2\right] \leq \frac{K}{\delta}\sum_{t=1}^{T}\mathbb{E}\left[\|\nabla_{t,i}\|_{\mathcal{H}_i}^2\right] \leq \frac{KTG^2}{\delta}.$$

The regret bound of online gradient descent (this can be found in our supplementary materials) depends on $\frac{\lambda}{2}\mathbb{E}\left[\sum_{t=1}^{T}\|\tilde{\nabla}_{t,i}\|_{\mathcal{H}_i}^2\right] \leq \frac{\lambda KTG^2}{2\delta}$. Thus it is the high variance of $\tilde{\nabla}_{t,i}$ that causes the $O(K^{\frac{1}{3}}T^{\frac{2}{3}})$ regret bound.

OKS selects a hypothesis per-round, reduces the time complexity to $O(t)$ but damages the regret bound. It was proved selecting two hypotheses can improve the regret bound to $\tilde{O}((\|f\|_{\mathcal{H}_i}^2 + 1)\sqrt{KT})$ [11]. A natural question arises: will selecting a hypothesis induce worse regret bound than selecting two hypotheses? From the perspective of algorithm design, we concentrate on the question:

- does there exist some algorithm selecting a hypothesis (or under bandit feedback) that can improve the $O((\|f\|_{\mathcal{H}_i}^2 + 1)K^{\frac{1}{3}}T^{\frac{2}{3}}))$ bound?

## 3   Improved Regret Bounds for Smooth Loss Functions

In this section, we propose the OKS++ algorithm using an adaptive parameter tuning scheme for OKS. Specifically, we reset the value of $\delta, \eta$ and $\lambda$ in Theorem 1 and prove data-dependent regret bounds for smooth loss functions. Such regret bounds can improve the previous worst-case bound if most of candidate kernel functions match well with the data. Although OKS++ just resets the value of parameters, deriving the new regret bounds requires novel and non-trivial analysis. To start with, we define the smooth loss functions.

**Assumption 1 (Smoothness condition)** $\ell(\cdot,\cdot)$ *is convex w.r.t. the first parameter. Denote by* $\ell'(a,b) = \frac{\mathrm{d}\,\ell(a,b)}{\mathrm{d}\,a}$. *For any* $f(\mathbf{x})$ *and* $y$, *there is a constant* $C_0 > 0$ *such that*

$$|\ell'(f(\mathbf{x}),y)|^\nu \leq C_0 \ell(f(\mathbf{x}),y), \quad \nu \in \{1,2\}.$$

Zhang et al. [21] considered online kernel learning under smooth loss functions with $\nu = 1$. The logistic loss $\ell(f(\mathbf{x}),y) = \ln(1 + \exp(-yf(\mathbf{x})))$ satisfies Assumption 1 with $\nu = 1$ and $C_0 = 1$. The square loss $\ell(f(\mathbf{x}),y) = (f(\mathbf{x}) - y)^2$ and the squared hinge loss $\ell(f(\mathbf{x}),y) = (\max\{0, 1 - yf(\mathbf{x})\})^2$ satisfy Assumption 1 with $\nu = 2$ and $C_0 = 4$.

Let $U > 0$ be a constant. We define $K$ restricted hypothesis spaces. $\forall i \in [K]$, let $\mathbb{H}_i = \{f \in \mathcal{H}_i : \|f\|_{\mathcal{H}_i} \leq U\}$. Then it is natural to derive Assumption 2.

**Assumption 2** $\forall \kappa_i \in \mathcal{K}$ *and* $\forall f \in \mathbb{H}_i$, *there exists a constant* $G > 0$ *such that* $\max_{t \in [T]} |\ell'(f(\mathbf{x}_t), y_t)| \leq G$.

It can be verified that many loss functions satisfy the assumption and $G$ may depend on $U$. For instance, if $\ell$ is the square loss, then $G \leq 2(U + 1)$. For simplicity, denote by $c_{t,i} = \ell(f_{t,i}(\mathbf{x}_t), y_t)$ for all $i \in [K]$ and $t \in [T]$. It can be verified that $\max_{t,i} c_{t,i}$ is bounded and depends on $U$. Then our algorithm updates $\mathbf{q}_t$ using $c_t$ (see line 4 and line 6 in Algorithm 1). Since we use restricted hypothesis spaces, our algorithm changes line 5 in Algorithm 1 as follows

$$f_{t+1,I_t} = \arg\min_{f \in \mathbb{H}_{I_t}} \left\| f - \left( f_{t,I_t} - \lambda_{t,I_t} \frac{\nabla_{f_{t,I_t}} \ell(f_{t,I_t}(\mathbf{x}_t), y_t)}{p_{t,I_t}} \right) \right\|_{\mathcal{H}_{I_t}}^2 . \tag{2}$$

Except for $\{\lambda_{t,i}\}_{i=1}^K$, our algorithm also uses time-variant $\delta_t$ and $\eta_t$. We omit the pseudo-code of OKS++ since it is similar with Algorithm 1.

Next we show the regret bound. For simplicity, let $\tilde{C}_{t,K} = \sum_{\tau=1}^t \sum_{i=1}^K \tilde{c}_{\tau,i}$ where $\tilde{c}_{\tau,i} = \frac{c_{\tau,i}}{p_{\tau,i}} \mathbb{I}_{I_\tau = i}$, and $\bar{L}_T = \sum_{j=1}^K L_T(f_j^*)$ where $L_T(f_j^*) = \min_{f \in \mathbb{H}_j} L_T(f)$.

**Theorem 2.** *Let* $\ell$ *satisfy Assumption 1 with* $\nu = 1$ *and Assumption 2. Let*

$$\delta_t = \frac{(GC_0)^{\frac{1}{3}}(UK)^{\frac{2}{3}}}{2\max\left\{(GC_0)^{\frac{1}{3}}(UK)^{\frac{2}{3}}, 2\tilde{C}_{t,K}^{\frac{1}{3}}\right\}}, \eta_t = \frac{\sqrt{2\ln K}}{\sqrt{1 + \sum_{\tau=1}^t \sum_{i=1}^K q_{\tau,i}\tilde{c}_{\tau,i}^2}},$$

$$\forall i \in [K], \lambda_{t,i} = \frac{U^{\frac{4}{3}}(\max\{GC_0U^2K^2, 8\tilde{C}_{t,K}\})^{-\frac{1}{6}}}{\sqrt{4/3}K^{\frac{1}{6}}(GC_0)^{\frac{1}{3}}\sqrt{1 + \Delta_{t,i}}}, \Delta_{t,i} = \sum_{\tau=1}^t \frac{\ell(f_{\tau,i}(\mathbf{x}_\tau), y_\tau)}{p_{\tau,i}} \mathbb{I}_{I_\tau = i}.$$

*Then the expected regret of OKS++ satisfies,* $\forall i \in [K]$,

$$\mathbb{E}\left[\bar{L}_{\mathbf{p}_{1:T}}\right] - L_T(f_i^*) = O\left(U^{\frac{2}{3}}(GC_0)^{\frac{1}{3}}K^{-\frac{1}{3}}\bar{L}_T^{\frac{2}{3}} + U^{\frac{2}{3}}(GC_0)^{\frac{1}{3}}K^{\frac{1}{6}}\bar{L}_T^{\frac{1}{6}}L_T^{\frac{1}{2}}(f_i^*)\right).$$

*Let* $\ell$ *satisfy Assumption 1 with* $\nu = 2$. *Let* $G = 1$ *in* $\delta_t$ *and* $\lambda_{t,i}$. $\eta_t$ *keeps unchanged. Then the expected regret of OKS++ satisfies*

$$\forall i \in [K], \mathbb{E}\left[\bar{L}_{\mathbf{p}_{1:T}}\right] - L_T(f_i^*) = O\left(U^{\frac{2}{3}}C_0^{\frac{1}{3}}K^{-\frac{1}{3}}\bar{L}_T^{\frac{2}{3}} + U^{\frac{2}{3}}C_0^{\frac{1}{3}}K^{\frac{1}{6}}\bar{L}_T^{\frac{1}{6}}L_T^{\frac{1}{2}}(f_i^*)\right).$$

The values of $\lambda_{t,i}$, $\delta_t$ and $\eta_t$ which depend on the observed losses, are important to obtain the data-dependent bounds. Beside, it is necessary to set different $\lambda_{t,i}$ for each $i \in [K]$. OKS sets a same $\lambda$. Thus the changes on the values of $\delta$, $\eta$ and $\lambda$ are non-trivial. Our analysis is also non-trivial. OKS++ sets time-variant parameters and does not require prior knowledge of the nature of the data.

Now we compare our results with the regret bound in Theorem 1. The main difference is that we replace $KT$ with a data-dependent complexity $\bar{L}_T$. In the worst case, $\bar{L}_T = O(KT)$ and our regret bound is $O(K^{\frac{1}{3}}T^{\frac{2}{3}})$ which is same with the result in Theorem 1. In some benign environments, we expect that $\bar{L}_T \ll KT$ and our regret bound would be smaller. For instance, if $L_T(f_i^*) = o(T)$ for all $i \in [K]$, then our regret is $o(T^{\frac{2}{3}})$ improving the result in Theorem 1. If there are only $M < K$ hypothesis spaces such that $L_T(f_i^*) = O(T)$, where $M$ is independent of $K$, then our regret bound is $O((MT)^{\frac{2}{3}}K^{-\frac{1}{3}})$. Such a result still improves the dependence on $K$. A more interesting result is that, if $L_T(f_i^*) = O(T^{\frac{1}{4}})$ for all $i \in [K]$, then OKS++ achieves a $O(K^{\frac{1}{3}}\sqrt{T})$ regret bound which is better than the $\tilde{O}(\text{poly}(\|f\|_{\mathcal{H}_i})\sqrt{KT})$ bound achieved by B(AO)$_2$KS [11].

# 4    Improved Regret Bound for Lipschitz Loss Functions

In this section, we consider Lipschitz loss functions and propose a new algorithm with improved worst-case regret bound.

## 4.1    Algorithm

For the sake of clarity, we decompose OKS into two levels. At the outer level, it uses a procedure similar with Exp3 [3] to update $\mathbf{p}_t$ and $\mathbf{q}_t$. At the inner level, it updates $f_{t,I_t}$ using online gradient descent. Exp3 can be derived from online mirror descent framework with negative entropy regularizer [1], i.e.,

$$\nabla_{\mathbf{q}'_{t+1}}\psi_t(\mathbf{q}'_{t+1}) = \nabla_{\mathbf{q}_t}\psi_t(\mathbf{q}_t) - \tilde{c}_t, \qquad \mathbf{q}_{t+1} = \arg\min_{\mathbf{q}\in\Delta_K}\mathcal{D}_{\psi_t}(\mathbf{q}, \mathbf{q}'_{t+1}), \qquad (3)$$

where $\psi_t(\mathbf{p}) = \sum_{i=1}^{K}\frac{1}{\eta}p_i\ln p_i$ is the negative entropy and $\mathcal{D}_{\psi_t}(\mathbf{p},\mathbf{q}) = \psi_t(\mathbf{p}) - \psi_t(\mathbf{q}) - \langle\nabla\psi_t(\mathbf{q}),\mathbf{p}-\mathbf{q}\rangle$ is the Bregman divergence. Different regularizer yields different algorithm. We will use $\psi_t(\mathbf{p}) = \sum_{i=1}^{K}\frac{-\alpha}{\eta_{t,i}}p_i^{\frac{1}{\alpha}}$, $\alpha > 1$, which slightly modifies the $\alpha$-Tsallis entropy [17,23]. We also use the increasing learning rate scheme in [1], that is $\eta_{t,i}$ is increasing. The reason is that if $\eta_{t,i}$ is increasing, then there will be a negative term in the regret bound which can be used to control the large variance of gradient estimator, i.e., $\mathbb{E}\left[\sum_{t=1}^{T}\|\tilde{\nabla}_{t,i}\|_{\mathcal{H}_i}^2\right]$ (see Sect. 2.2). If we use the log-barrier [1] or $\alpha$-Tsallis entropy with $\alpha = 2$ [2,23], then the regret bound will increase a $O(\ln T)$ factor. This factor can be reduced to $O(\ln^{\frac{2}{3}} T)$ for $\alpha \geq 3$. We choose $\alpha = 8$ for achieving a small regret bound.

At the beginning of round $t$, our algorithm first samples $I_t \sim \mathbf{p}_t$ and outputs the prediction $f_{t,I_t}(\mathbf{x}_t)$ or $\text{sign}(f_{t,I_t}(\mathbf{x}_t))$. Next our algorithm updates $f_{t,I_t}$

---

**Algorithm 2.** IOKS

---

**Input:** $\mathcal{K} = \{\kappa_1, \ldots, \kappa_K\}$, $\alpha = 8$, $\upsilon = e^{\frac{2}{3\ln T}}$, $\eta$
**Initialization:** $\{f_{1,i} = 0, \eta_{1,i} = \eta\}_{i=1}^K$, $\mathbf{q}_1 = \mathbf{p}_1 = \frac{1}{K}\mathbf{1}_K$
1:  **for** $t = 1, \ldots, T$ **do**
2:      Receive $\mathbf{x}_t$
3:      Sample a kernel $\kappa_{I_t}$ where $I_t \sim \mathbf{p}_t$
4:      Output $\hat{y}_t = f_{t,I_t}(\mathbf{x}_t)$ or $\text{sign}(\hat{y}_t)$
5:      Compute $f_{t+1,I_t}$ according to (2)
6:      Compute $\tilde{c}_{t,I_t}$ according to (4)
7:      $\forall i \in [K]$, compute $q_{t+1,i}$ according to (5)
8:      Compute $\mathbf{p}_{t+1} = (1-\delta)\mathbf{q}_{t+1} + \frac{\delta}{K}\mathbf{1}_K$
9:      **for** $i = 1, \ldots, K$ **do**
10:         **if** $\frac{1}{p_{t+1,i}} > \rho_{t,i}$ **then**
11:             $\rho_{t+1,i} = \frac{2}{p_{t+1,i}}$, $\eta_{t+1,i} = \upsilon\eta_{t,i}$
12:         **else**
13:             $\rho_{t+1,i} = \rho_{t,i}$, $\eta_{t+1,i} = \eta_{t,i}$
14:         **end if**
15:     **end for**
16: **end for**

---

following (2). $\forall i \in [K]$, let $c_{t,i} = \ell(f_{t,i}(\mathbf{x}_t), y_t)/\ell_{\max} \in [0, 1]$. We redefine $\tilde{c}_t$ by

$$\text{if } p_{t,I_t} \geq \max_i \eta_{t,i}, \text{ then } \tilde{c}_{t,i} = \frac{c_{t,i}}{p_{t,i}}\mathbb{I}_{i=I_t}, \text{ otherwise } \tilde{c}_{t,i} = \frac{c_{t,i} \cdot \mathbb{I}_{i=I_t}}{p_{t,i} + \max_i \eta_{t,i}}. \quad (4)$$

It is worth mentioning that $\tilde{c}_t$ is essentially different from that in OKS, and aims to ensure that (3) has a computationally efficient solution as follows

$$\forall i \in [K], \quad q_{t+1,i} = \left(q_{t,i}^{-\frac{7}{8}} + \eta_{t,i}(\tilde{c}_{t,i} - \mu^*)\right)^{-\frac{8}{7}}, \quad (5)$$

where $\mu^*$ can be solved using binary search. We show more details in the supplementary materials. We name this algorithm IOKS (Improved OKS).

### 4.2  Regret Bound

**Assumption 3 (Lipschitz condition)** $\ell(\cdot, \cdot)$ *is convex w.r.t. the first parameter. There is a constant $G_1$ such that $\forall \kappa_i \in \mathcal{K}$, $f \in \mathbb{H}_i$, $\|\nabla_f \ell(f(\mathbf{x}), y)\|_{\mathcal{H}_i} \leq G_1$.*

**Theorem 3.** *Let $\ell$ satisfy Assumption 3. Let $\delta = T^{-\frac{3}{4}}$,*

$$\eta = \frac{3\ell_{\max}K^{\frac{3}{8}}}{2UG_1\sqrt{T\ln T}}, \quad \forall i \in [K], \ \lambda_{t,i} = \frac{U}{\sqrt{2}\sqrt{1 + \sum_{\tau=1}^t \|\tilde{\nabla}_{\tau,i}\|_{\mathcal{H}_i}^2}}. \quad (6)$$

*Let $T \geq 40$. Then the expected regret of IOKS satisfies,*

$$\forall i \in [K], f \in \mathbb{H}_i, \ \mathbb{E}\left[\bar{L}_{\mathbf{p}_{1:T}}\right] - L_T(f) = O\left(UG_1\sqrt{KT}\ln^{\frac{2}{3}}T + \frac{\ell_{\max}^3 K^{\frac{11}{4}}}{U^2G_1^2\ln T}\right).$$

$\ell_{\max}$ is a normalizing constant and can be computed given the loss function, such as $\ell_{\max} \leq U + 1$ in the case of absolute loss. Next we compare our regret bound with previous results. On the positive side, IOKS gives a $O(U\sqrt{KT}\ln^{\frac{2}{3}} T)$ bound which asymptotically improves the $O(K^{\frac{1}{3}}T^{\frac{2}{3}})$ bound achieved by OKS. On the negative side, if $T$ is small, then $\sqrt{KT}\ln^{\frac{2}{3}} T > K^{\frac{1}{3}}T^{\frac{2}{3}}$ and thus IOKS is slightly worse than OKS. B(AO)$_2$KS [11] which selects two hypotheses per-round, can provide a $\tilde{O}(\text{poly}(\|f\|_{\mathcal{H}_i})\sqrt{KT})$ bound which is same with our result.

We further compare the implementation of IOKS and OKS. It is obvious that OKS is easier than IOKS, since IOKS uses binary search to compute $\mathbf{q}_{t+1}$ (see (5)). The computational cost of binary search can be omitted since the main computational cost comes from the computing of $f_{t,I_t}(\mathbf{x}_t)$ which is $O(t)$.

## 5 Application to Online Kernel Selection with Time Constraint

In practice, online algorithms must face time constraint. We assume that there is a time budget of $\mathcal{R}$ quanta. Both OKS++ and IOKS suffer a $O(t)$ per-round time complexity, and do not satisfy the time constraint. In this section, we will use random features [12] to approximate kernel functions and apply our two algorithms to online kernel selection with time constraint [9].

We consider kernel function $\kappa(\mathbf{x}, \mathbf{v})$ that can be decomposed as follows

$$\kappa(\mathbf{x}, \mathbf{v}) = \int_\Omega \phi_\kappa(\mathbf{x}, \omega)\phi_\kappa(\mathbf{v}, \omega) \mathrm{d}\,\mu_\kappa(\omega), \ \forall \mathbf{x}, \mathbf{v} \in \mathcal{X} \qquad (7)$$

where $\phi_\kappa : \mathcal{X} \times \Omega \to \mathbb{R}$ is the eigenfunctions and $\mu_\kappa(\cdot)$ is a distribution function on $\Omega$. Let $p_\kappa(\cdot)$ be the density function of $\mu_\kappa(\cdot)$. We can approximate the integral via Monte-Carlo sampling. We sample $\{\omega_j\}_{j=1}^D \sim p_\kappa(\omega)$ independently and compute $\tilde{\kappa}(\mathbf{x}, \mathbf{v}) = \frac{1}{D}\sum_{j=1}^D \phi_\kappa(\mathbf{x}, \omega_j)\phi_\kappa(\mathbf{v}, \omega_j)$. For any $f \in \mathcal{H}_\kappa$, let $f(\mathbf{x}) = \int_\Omega \alpha(\omega)\phi_\kappa(\mathbf{x}, \omega)p_\kappa(\omega)\mathrm{d}\omega$. We can approximate $f(\mathbf{x})$ by $\hat{f}(\mathbf{x}) = \frac{1}{D}\sum_{j=1}^D \alpha(\omega_j)\phi_\kappa(\mathbf{x}, \omega_j)$. It can be verified that $\mathbb{E}[\hat{f}(\mathbf{x})] = f(\mathbf{x})$. Such an approximation scheme also defines an explicit feature mapping denoted by $z(\mathbf{x}) = \frac{1}{\sqrt{D}}(\phi_\kappa(\mathbf{x}, \omega_1), \ldots, \phi_\kappa(\mathbf{x}, \omega_D))$. The approximation scheme is the so called random features [12]. $\forall \kappa_i \in \mathcal{K}$, we define two hypothesis spaces [13] as follows

$$\mathbb{H}_i = \left\{ f(\mathbf{x}) = \int_\Omega \alpha(\omega)\phi_{\kappa_i}(\mathbf{x}, \omega)p_{\kappa_i}(\omega)\mathrm{d}\omega \,\big|\, |\alpha(\omega)| \leq U \right\},$$

$$\mathcal{F}_i = \left\{ \hat{f}(\mathbf{x}) = \sum_{j=1}^{D_i} \alpha_j\phi_{\kappa_i}(\mathbf{x}, \omega_j) \,\Big|\, |\alpha_j| \leq \frac{U}{D_i} \right\}.$$

We can rewrite $\hat{f}(\mathbf{x}) = \mathbf{w}^\top z_i(\mathbf{x})$, where $\mathbf{w} = \sqrt{D_i}(\alpha_1, \ldots, \alpha_{D_i}) \in \mathbb{R}^{D_i}$. Let $\mathcal{W}_i = \{\mathbf{w} \in \mathbb{R}^{D_i} \,|\, \|\mathbf{w}\|_\infty \leq \frac{U}{\sqrt{D_i}}\}$. It can be verified that $\|\mathbf{w}\|_2^2 \leq U^2$. For all $\kappa_i$ satisfying (7), there is a constant $B_i$ such that $|\phi_{\kappa_i}(\mathbf{x}, \omega_j)| \leq B_i$ for all $\omega_j \in \Omega$ and $\mathbf{x} \in \mathcal{X}$ [10]. Thus we have $|f(\mathbf{x})| \leq UB_i$ for any $f \in \mathbb{H}_i$ and $f \in \mathcal{F}_i$.

Next we define the time budget $\mathcal{R}$ and then present an assumption that establishes a reduction from $\mathcal{R}$ to $D_i$.

**Definition 1 (Time Budget [9]).** *Let the interval of arrival time between* $\mathbf{x}_t$ *and* $\mathbf{x}_{t+1}, t = 1, \ldots, T$ *be less than* $\mathcal{R}$ *quanta. A time budget of* $\mathcal{R}$ *quanta is the maximal time interval that any online kernel selection algorithm outputs the prediction of* $\mathbf{x}_t$ *and* $\mathbf{x}_{t+1}$.

**Assumption 4** *For each* $\kappa_i \in \mathcal{K}$ *satisfying (7), there exist online leaning algorithms that can run in some* $\mathcal{F}_i$ *whose maximal dimension is* $D_i = \beta_{\kappa_i}\mathcal{R}$ *within a time budget of* $\mathcal{R}$ *quanta, where* $\beta_{\kappa_i} > 0$ *is a constant depending on* $\kappa_i$.

The online gradient descent algorithm (OGD) satisfies Assumption 4. The main time cost of OGD comes from computing the feature mapping. For shift-invariant kernels, it requires $O(D_i d)$ time complexity [12]. For the Gaussian kernel, it requires $O(D_i \log(d))$ time complexity [8,20]. Thus the per-round time complexity of OGD is linear with $D_i$. Since the running time of algorithm is linear with the time complexity, it natural to assume that $\mathcal{R} = \Theta(D_i)$.

## 5.1   Algorithm

At any round $t$, our algorithm evaluates a hypothesis and avoids allocating the time budget. Thus we can construct $\mathcal{F}_i$ satisfying $D_i = \beta_{\kappa_i}\mathcal{R}$. Our algorithm is extremely simple, that is, we just need to run OKS++ or IOKS in $\{\mathcal{F}_i\}_{i=1}^{K}$. It is worth mentioning that, learning $\{\hat{f}_{t,i} \in \mathcal{F}_i\}_{t=1}^{T}$ is equivalent to learn $\{\mathbf{w}_t^i \in \mathcal{W}_i\}_{t=1}^{T}$, where $\hat{f}_{t,i}(\mathbf{x}_t) = (\mathbf{w}_t^i)^{\top} z_i(\mathbf{x}_t)$. We replace the update (2) with (8),

$$
\begin{aligned}
\tilde{\mathbf{w}}_{t+1}^i &= \mathbf{w}_t^i - \lambda_{t,i} \nabla_{\mathbf{w}_t^i} \ell\left(\hat{f}_{t,i}(\mathbf{x}_t), y_t\right) \frac{1}{p_{t,i}} \mathbb{I}_{i=I_t}, \\
\mathbf{w}_{t+1}^i &= \underset{\mathbf{w} \in \mathcal{W}_i}{\arg\min} \left\| \mathbf{w} - \tilde{\mathbf{w}}_{t+1}^i \right\|_2^2 .
\end{aligned}
\tag{8}
$$

The solution of the projection operation in (8) is as follows,

$$
\forall j = 1, \ldots, D_i, \ w_{t+1,j}^i = \min\left\{ 1, \frac{U}{|\tilde{w}_{t+1,j}^i|\sqrt{D_i}} \right\} \tilde{w}_{t+1,j}^i.
$$

The time complexity of projection is $O(D_i)$ and thus can be omitted relative to the time complexity of computing feature mapping. We separately name the two algorithms RF-OKS++ (Random Features for OKS++) and RF-IOKS (Random Features for IOKS). We show the pseudo-codes in the supplementary materials due to the space limit. The pseudo-codes are similar with OKS++ and IOKS.

*Remark 3.* The application of random features to online kernel algorithms is not a new idea [6,16,18]. Previous algorithms did not restrict hypothesis spaces, while our algorithms consider restricted hypothesis spaces, i.e., $\mathbb{H}_i$ and $\mathcal{F}_i$. This is one of the differences between our algorithms and previous algorithms. The restriction on the hypothesis spaces is necessary since we must require $\|\mathbf{w}_t^i\|_2 \leq U$ for any $i \in [K]$ and $t \in [T]$.

## 5.2  Regret Bound

**Theorem 4.** *Let $\ell$ satisfy Assumption 1 with $\nu = 1$ and Assumption 2. Let $\delta_t$, $\eta_t$ and $\{\lambda_{t,i}\}_{i=1}^K$ follow Theorem 2. For a fixed $\delta \in (0,1)$, let $\mathcal{R}$ satisfy $D_i > \frac{32}{9} C_0^2 U^2 B_i^2 \ln \frac{1}{\delta}$, $\forall i \in [K]$. Under Assumption 4, with probability at least $1 - \delta$, the expected regret of RF-OKS++ satisfies*

$$\forall i \in [K],\ \mathbb{E}\left[\bar{L}_{\mathbf{p}_{1:T}}\right] - L_T(f_i^*) = O\left(\frac{C_0 U B_i}{\sqrt{\beta_{\kappa_i} \mathcal{R}}} L_T(f_i^*) \sqrt{\ln \frac{KT}{\delta}}\right.$$
$$\left. + U^{\frac{2}{3}}(GC_0)^{\frac{1}{3}} K^{-\frac{1}{3}} \bar{L}_T^{\frac{2}{3}} + U^{\frac{2}{3}}(GC_0)^{\frac{1}{3}} K^{\frac{1}{6}} \bar{L}_T^{\frac{1}{6}} L_T^{\frac{1}{2}}(f_i^*)\right).$$

*Let $\ell$ satisfy Assumption 1 with $\nu = 2$. Let $G = 1$ in $\delta_t$ and $\lambda_{t,i}$. $\eta_t$ keeps unchanged. For a fixed $\delta \in (0,1)$, with probability at least $1 - \delta$, the expected regret of RF-OKS++ satisfies*

$$\forall i \in [K],\ \mathbb{E}\left[\bar{L}_{\mathbf{p}_{1:T}}\right] - L_T(f_i^*) = O\left(U B_i \frac{\sqrt{C_0 T L_T(f_i^*)}}{\sqrt{\beta_{\kappa_i} \mathcal{R}}} \sqrt{\ln \frac{KT}{\delta}}\right.$$
$$\left. + \frac{C_0 U^2 B_i^2 T}{\beta_{\kappa_i} \mathcal{R}} \ln \frac{KT}{\delta} + U^{\frac{2}{3}} C_0^{\frac{1}{3}} K^{-\frac{1}{3}} \bar{L}_T^{\frac{2}{3}} + U^{\frac{2}{3}} C_0^{\frac{1}{3}} K^{\frac{1}{6}} \bar{L}_T^{\frac{1}{6}} L_T^{\frac{1}{2}}(f_i^*)\right).$$

The regret bounds depend on $\frac{L_T(f_i^*)}{\sqrt{\mathcal{R}}}$ or $\frac{1}{\sqrt{\mathcal{R}}}\sqrt{T L_T(f_i^*)} + \frac{T}{\mathcal{R}}$. The larger the time budget is, the smaller the regret bound will be, which proves a trade-off between regret bound and time constraint. If $L_T(f_i^*) \ll T$, then RF-OKS++ can achieve a sublinear regret bound under a small time budget.

**Theorem 5.** *Let $\ell$ satisfy Assumption 2 and Assumption 3. Let $\{\lambda_{t,i}\}_{i=1}^K$, $\eta$ and $\delta$ follow Theorem 3. Under Assumption 4, with probability at least $1 - \delta$, the expected regret of RF-IOKS satisfies, $\forall i \in [K], \forall f \in \mathbb{H}_i$,*

$$\mathbb{E}\left[\bar{L}_{\mathbf{p}_{1:T}}\right] - L_T(f) = O\left(U G_1 \sqrt{KT} \ln^{\frac{2}{3}} T + \frac{\ell_{\max}^3 K^{\frac{11}{4}}}{U^2 G_1^2 \sqrt{\ln T}} + \frac{G B_i U T}{\sqrt{\beta_{\kappa_i} \mathcal{R}}} \sqrt{\ln \frac{KT}{\delta}}\right).$$

The regret bound depends on $\frac{T}{\sqrt{\mathcal{R}}}$ which also proves a trade-off between regret bound and time constraint. Achieving a $\tilde{O}(T^\alpha)$ bound requires $\mathcal{R} = \Omega(T^{2(1-\alpha)})$, $\alpha \in [\frac{1}{2}, 1)$. The regret bounds in Theorem 4 depend on $L_T(f_i^*)$, while the regret bound in Theorem 5 depends on $T$. Under a same time budget $\mathcal{R}$, if $L_T(f_i^*) \ll T$, then RF-OKS++ enjoys better regret bounds than RF-IOKS.

## 5.3  Comparison with Previous Results

For online kernel selection with time constraint, if the loss function is Lipschitz continuous, then there is a $\Omega(\|f_i^*\|_{\mathcal{H}_i} \max\{\sqrt{T}, \frac{T}{\sqrt{\mathcal{R}}}\})$ lower bound on expected regret [9]. Theorem 5 gives a nearly optimal upper bound. LKMBooks [9] gives a $O(\sqrt{T \ln K} + \|f\|_{\mathcal{H}_i}^2 \max\{\sqrt{T}, \frac{T}{\sqrt{\mathcal{R}}}\})$ bound in the case of $K \leq d$, and thus is slightly better than RF-IOKS. LKMBooks selects $K$ hypotheses per-round. RF-IOKS just selects a hypothesis per-round and is suitable for $K > d$.

For smooth loss functions, the dominated terms in Theorem 4 are $O(\frac{L_T(f_i^*)}{\sqrt{\mathcal{R}}})$ and $O(\frac{1}{\sqrt{\mathcal{R}}}\sqrt{TL_T(f_i^*)} + \frac{T}{\mathcal{R}})$. If the optimal kernel $\kappa_{i*}$ matches well with the data, that is, $L_T(f_{i*}^*) \ll T$, then $O(\frac{L_T(f_{i*}^*)}{\sqrt{\mathcal{R}}})$ and $O(\frac{1}{\sqrt{\mathcal{R}}}\sqrt{TL_T(f_{i*}^*)})$ are much smaller than $O(\frac{T}{\sqrt{\mathcal{R}}})$. To be specific, in the case of $L_T(f_{i*}^*) = o(T)$, RF-OKS++ is better than LKMBooks within a same time budget $\mathcal{R}$.

Our algorithms are similar with Raker [16] which also adopts random features. Raker selects $K$ hypotheses and provides a $\tilde{O}((\sqrt{\ln K} + \|f\|_1^2)\sqrt{T} + \|f\|_1 \frac{T}{\sqrt{\mathcal{R}}})$ bound, where $f = \sum_{t=1}^{T} \alpha_t \kappa_i(\mathbf{x}_t, \cdot)$ and $\|f\|_1 = \|\boldsymbol{\alpha}\|_1$. The regret bounds of RF-OKS++ are better, since (i) they depend on $L_T(f_i^*)$ and $\sum_{j=1}^{K} L_T(f_j^*)$ while the regret bound of Raker depends on $T$; (ii) they depend on $U$, while the regret bound of Raker depends on $\|f\|_1$ which is hard to bound and explain.

# 6    Experiments

We adopt the Gaussian kernel $\kappa(\mathbf{x}, \mathbf{v}) = \exp(-\frac{\|\mathbf{x}-\mathbf{v}\|_2^2}{2\sigma^2})$ and select 6 kernel widths $\sigma = 2^{-2:1:3}$. We choose four classification datasets (*magic04:19,020, phishing:11,055, a9a:32,561, SUSY:20,000*) and four regression datasets (*bank:8,192, elevators:16,599, ailerons:13,750, Hardware:28,179*). The datasets are downloaded from UCI [2], LIBSVM website [3] and WEKA. The features of all datasets are rescaled to fit in $[-1, 1]$. The target variables are rescaled in $[0, 1]$ for regression and $\{-1, 1\}$ for classification. We randomly permutate the instances in the datasets 10 times and report the average results. All algorithms are implemented with R on a Windows machine with 2.8 GHz Core(TM) i7-1165G7 CPU [4]. We separately consider online kernel selection without and with time constraint.

## 6.1    Online Kernel Selection Without Time Constraint

We compare OKS++, IOKS with OKS and aim to verify Theorem 2 and Theorem 3. We consider three loss functions: (i) the logistic loss satisfying Assumption 1 with $\nu = 1$ and $C_0 = 1$; (ii) the square loss satisfying Assumption 1 with $\nu = 2$ and $C_0 = 4$; (iii) the absolute loss which is Lipschitz continuous. We do not compare with B(AO)$_2$KS [11], since it is only used for the hinge loss. If $\ell$ is logistic loss, then we use classification datasets and measure the average mistake rate, i.e., AMR $:= \frac{1}{T}\sum_{t=1}^{T} \mathbb{I}_{\hat{y}_t \neq y_t}$, and set $U = 15$. Otherwise, we use regression datasets and measure the average loss, i.e., AL $:= \frac{1}{T}\sum_{t=1}^{T} \ell(f_{t,I_t}(\mathbf{x}_t), y_t)$, and set $U = 1$. The parameters of OKS++ and IOKS follow Theorem 2 and Theorem 3 where we change $\eta = \frac{8\ell_{\max} K^{3/8}}{UG_1 \sqrt{T} \ln T}$ in Theorem 3 and set $\ell_{\max} = 1$. For OKS, we

[2] http://archive.ics.uci.edu/ml/datasets.php.

[3] https://www.csie.ntu.edu.tw/~cjlin/libsvmtools/datasets/.

[4] The codes are available at https://github.com/JunfLi-TJU/OKS-Bandit.

**Table 2.** Online kernel selection without time constraint in the regime of logistic loss

| Algorithm | Phishing | | a9a | |
| --- | --- | --- | --- | --- |
| | AMR (%) | Time (s) | AMR (%) | Time (s) |
| OKS | 13.80 ± 0.34 | 17.34 ± 1.48 | 19.65 ± 0.12 | 208.84 ± 31.16 |
| IOKS | 13.25 ± 0.28 | 6.58 ± 0.18 | 17.46 ± 0.12 | 103.91 ± 13.89 |
| OKS++ | **7.80 ± 0.49** | 32.31 ± 3.98 | **16.57 ± 0.31** | 474.65 ± 117.43 |
| Algorithm | magic04 | | SUSY | |
| | AMR (%) | Time (s) | AMR (%) | Time (s) |
| OKS | 22.23 ± 0.22 | 6.31 ± 0.95 | 32.98 ± 0.66 | 9.97 ± 1.85 |
| IOKS | 21.50 ± 0.18 | 4.02 ± 0.11 | 31.75 ± 0.30 | 6.68 ± 0.15 |
| OKS++ | **17.88 ± 0.57** | 11.06 ± 3.08 | **27.84 ± 0.70** | 19.88 ± 5.28 |

**Table 3.** Online kernel selection without time constraint in the regime of square loss

| Algorithm | Elevators | | Bank | |
| --- | --- | --- | --- | --- |
| | AL | Time (s) | AL | Time (s) |
| OKS | 0.0068 ± 0.0001 | 3.23 ± 0.25 | 0.0240 ± 0.0002 | 1.51 ± 0.17 |
| IOKS | 0.0077 ± 0.0001 | 4.08 ± 0.05 | 0.0252 ± 0.0002 | 1.57 ± 0.11 |
| OKS++ | **0.0046 ± 0.0001** | 12.75 ± 3.12 | **0.0205 ± 0.0006** | 4.24 ± 0.76 |
| Algorithm | ailerons | | Hardware | |
| | AL | Time (s) | AL | Time (s) |
| OKS | **0.0176 ± 0.0060** | 6.94 ± 0.82 | 0.0012 ± 0.0000 | 53.84 ± 1.80 |
| IOKS | 0.0351 ± 0.0003 | 5.59 ± 0.08 | 0.0010 ± 0.0001 | 49.36 ± 1.14 |
| OKS++ | **0.0166 ± 0.0006** | 22.79 ± 3.41 | **0.0008 ± 0.0001** | 114.47 ± 23.42 |

**Table 4.** Online kernel selection without time constraint in the regime of absolute loss

| Algorithm | Elevators | | Bank | |
| --- | --- | --- | --- | --- |
| | AL | Time (s) | AL | Time (s) |
| OKS | 0.0507 ± 0.0001 | 4.76 ± 0.17 | 0.0961 ± 0.0009 | 1.55 ± 0.13 |
| IOKS | **0.0492 ± 0.0004** | 5.20 ± 0.54 | 0.0961 ± 0.0008 | 1.64 ± 0.20 |
| Algorithm | ailerons | | Hardware | |
| | AL | Time (s) | AL | Time (s) |
| OKS | **0.0723 ± 0.0005** | 8.20 ± 0.19 | **0.0105 ± 0.0001** | 56.14 ± 1.07 |
| IOKS | 0.0771 ± 0.0007 | 9.86 ± 0.68 | 0.0155 ± 0.0002 | 52.01 ± 3.72 |

set $\delta, \lambda$ and $\eta$ according to Remark 1, where $\lambda \in \{1, 5, 10, 25\} \cdot \sqrt{\delta/(KT)}$ and $\ell_{\max} = G = 1$. The results are shown in Table 2, Table 3 and Table 4.

Table 2 and Table 3 prove that OKS++ performs better than OKS and IOKS for smooth loss functions. The reason is that OKS++ adaptively tunes the parameters using the observed losses, while OKS and IOKS do not use this information to tune the parameters. The experimental results coincide with Theorem 2. Besides IOKS performs similar with OKS, since IOKS is only asymptotically better than OKS. If $T$ is small, then the regret bound of OKS is smaller. The theoretical significance of IOKS is that it proves that selecting a hypothesis does not produce high information-theoretic cost in the worst case.

**Table 5.** Online kernel selection with time constraint in the regime of logistic loss

| Algorithm | B-D | Phishing | | B-D | a9a | |
|---|---|---|---|---|---|---|
| | | AMR (%) | $t_p * 10^5(s)$ | | AMR (%) | $t_p * 10^5(s)$ |
| RF-OKS | 500 | 14.61 ± 0.65 | 9.63 | 450 | 21.25 ± 0.12 | 11.61 |
| LKMBooks | 250 | 12.50 ± 1.03 | 9.46 | 220 | 20.06 ± 0.54 | 11.53 |
| Raker | 70 | 13.60 ± 1.00 | 9.35 | 90 | 24.08 ± 0.00 | 11.30 |
| RF-IOKS | 380 | 15.59 ± 0.39 | 9.66 | 380 | 22.99 ± 0.20 | 11.95 |
| RF-OKS++ | 400 | **9.15 ± 0.56** | 9.20 | 400 | **17.28 ± 0.29** | 11.19 |

**Table 6.** Online kernel selection with time constraint in the regime of square loss

| Algorithm | B-D | Elevators | | B-D | Hardware | |
|---|---|---|---|---|---|---|
| | | AL * $10^2$ | $t_p * 10^5(s)$ | | AL * $10^2$ | $t_p * 10^5(s)$ |
| RF-OKS | 450 | 0.72 ± 0.02 | 6.47 | 420 | 0.13 ± 0.00 | 10.48 |
| LKMBooks | 220 | 0.90 ± 0.04 | 6.72 | 200 | 0.21 ± 0.01 | 10.76 |
| Raker | 40 | 0.70 ± 0.04 | 6.57 | 80 | 0.20 ± 0.00 | 10.25 |
| RF-IOKS | 380 | 0.89 ± 0.01 | 6.83 | 400 | 0.12 ± 0.01 | 10.20 |
| RF-OKS++ | 400 | **0.51 ± 0.02** | 6.45 | 400 | **0.09 ± 0.01** | 10.31 |

**Table 7.** Online kernel selection with time constraint in the regime of absolute loss

| Algorithm | B-D | Elevators | | B-D | Hardware | |
|---|---|---|---|---|---|---|
| | | AL | $t_p * 10^5$ | | AL | $t_p * 10^5$ |
| RF-OKS | 530 | **0.0515 ± 0.0004** | 7.13 | 400 | **0.0108 ± 0.0001** | 10.39 |
| LKMBooks | 230 | 0.0550 ± 0.0014 | 7.35 | 200 | 0.0203 ± 0.0020 | 10.41 |
| Raker | 50 | 0.0550 ± 0.0012 | 7.41 | 80 | 0.0154 ± 0.0001 | 10.37 |
| RF-IOKS | 400 | **0.0515 ± 0.0007** | 7.63 | 400 | 0.0164 ± 0.0002 | 10.97 |

## 6.2   Online Kernel Selection with Time Constraint

We compare RF-OKS++, RF-IOKS with Raker [16], LKMBooks [9] and RF-OKS [19]. We construct RF-OKS by combining random features with OKS. The parameter setting of Raker and LKMBooks follows original paper, except that the learning rate of Raker is chosen from $\{1, 5, 10, 25\} \cdot 1/\sqrt{T}$. The parameter setting of RF-OKS++, RF-IOKS and RF-OKS is same with that of OKS++, IOKS and OKS, respectively. We limit time budget $\mathcal{R}$ by fixing the number of random features. To be specific, we choose RF-OKS++ as the baseline and set $D_i = D = 400$ for all $i \in [K]$ satisfying the condition in Theorem 4. Let the average per-round running time of RF-OKS++ be $t_p$. We tune $D$ or $B$ of other algorithms for ensuring the same running time with $t_p$. The results are shown in Table 5, Table 6 and Table 7. In Table 7, we use RF-IOKS as the baseline.

For smooth loss functions, RF-OKS++ still performs best under a same time budget. The reason is also that RF-OKS++ adaptively tunes the parameters using the observed losses, while the other algorithms do not use the observed losses. For the square loss function, Theorem 4 shows the regret bound depends on $O(\frac{1}{\sqrt{\mathcal{R}}}\sqrt{T L_T(f_i^*)})$ which becomes $O(\frac{T}{\sqrt{\mathcal{R}}})$ in the worst case and thus is same with previous results. To explain the contradiction, we recorded the cumulative square losses of RF-OKS++, i.e., $\sum_{t=1}^{T}(f_{t,I_t}(\mathbf{x}_t) - y_t)^2$ and use it as a proxy for $L_T(f_i^*)$. In our experiments, $L_T(f_i^*) \approx 88.6$ on the *elevators* dataset and $L_T(f_i^*) \approx 23.8$ on the *Hardware* dataset. Thus $L_T(f_i^*) \ll T$ and $O(\frac{1}{\sqrt{\mathcal{R}}}\sqrt{TL_T(f_i^*)})$ is actually smaller than $O(\frac{T}{\sqrt{\mathcal{R}}})$. The above results coincide with Theorem 4.

RF-IOKS shows similar performance with the baseline algorithms, which is consistent with Theorem 5. The regret bound of RF-IOKS is slightly worse than that of LKMBooks and Raker, and is only asymptotically better than RF-OKS. All of the baseline algorithms tune the stepsize in hindsight, which is impossible in practice since the data can only be predicted once. RF-IOKS also proves that selecting a hypothesis does not damage the regret bound much in the worst case. More experiments are shown in the supplementary materials.

## 7 Conclusion

In this paper, we have proposed two algorithms for online kernel selection under bandit feedback and improved the previous worst-case regret bound. OKS++ which is applied for smooth loss functions, adaptively tunes parameters of OKS and achieves data-dependent regret bounds depending on the minimal cumulative losses. IOKS which is applied for Lipschitz loss functions, achieves a worst-case regret bound asymptotically better than previous result. We further apply the two algorithms to online kernel selection with time constraint and obtain better or similar regret bounds.

From the perspective of algorithm design, there is a trade-off between regret bound and the amount of observed information. IOKS proves that selecting a hypothesis or multiple hypotheses per-round will not induce significant variation on the worst-case regret bound. OKS++ which performs well both in theory and practice, implies that there may be differences in terms of data-dependent regret bounds. This question is left to future work.

## References

1. Agarwal, A., Luo, H., Neyshabur, B., Schapire, R.E.: Corralling a band of bandit algorithms. In: Proceedings of the 30th Conference on Learning Theory, pp. 12–38 (2017)
2. Audibert, J., Bubeck, S.: Minimax policies for adversarial and stochastic bandits. In: Proceedings of the 22nd Annual Conference on Learning Theory, pp. 217–226 (2009)
3. Auer, P., Cesa-Bianchi, N., Freund, Y., Schapire, R.E.: The non-stochastic multi-armed bandit problem. SIAM J. Comput. **32**(1), 48–77 (2002)
4. Cesa-Bianchi, N., Lugosi, G.: Prediction, Learning, and Games. Cambridge University Press, Cambridge (2006)
5. Foster, D.J., Kale, S., Mohri, M., Sridharan, K.: Parameter-free online learning via model selection. Adv. Neural. Inf. Process. Syst. **30**, 6022–6032 (2017)

6.  Ghari, P.M., Shen, Y.: Online multi-kernel learning with graph-structured feedback. In: Proceedings of the 37th International Conference on Machine Learning, pp. 3474–3483 (2020)
7.  Hoi, S.C.H., Jin, R., Zhao, P., Yang, T.: Online multiple kernel classification. Mach. Learn. **90**(2), 289–316 (2013)
8.  Le, Q.V., Sarlós, T., Smola, A.J.: Fastfood-computing hilbert space expansions in loglinear time. In: Proceedings of the 30th International Conference on Machine Learning, pp. 244–252 (2013)
9.  Li, J., Liao, S.: Worst-case regret analysis of computationally budgeted online kernel selection. Mach. Learn. **111**(3), 937–976 (2022)
10. Li, Z., Ton, J.F., Oglic, D., Sejdinovic, D.: Towards a unified analysis of random Fourier features. In: Proceedings of the 36th International Conference on Machine Learning, pp. 3905–3914 (2019)
11. Liao, S., Li, J.: High-probability kernel alignment regret bounds for online kernel selection. In: Proceedings of the European Conference on Machine Learning and Knowledge Discovery in Databases, pp. 67–83 (2021)
12. Rahimi, A., Recht, B.: Random features for large-scale kernel machines. Adv. Neural. Inf. Process. Syst. **20**, 1177–1184 (2007)
13. Rahimi, A., Recht, B.: Weighted sums of random kitchen sinks: replacing minimization with randomization in learning. Adv. Neural. Inf. Process. Syst. **21**, 1313–1320 (2008)
14. Sahoo, D., Hoi, S.C.H., Li, B.: Online multiple kernel regression. In: Proceedings of the 20th ACM SIGKDD International Conference on Knowledge Discovery and Data Mining, KDD, pp. 293–302 (2014)
15. Seldin, Y., Bartlett, P.L., Crammer, K., Abbasi-Yadkori, Y.: Prediction with limited advice and multiarmed bandits with paid observations. In: Proceedings of the 31st International Conference on Machine Learning, pp. 280–287 (2014)
16. Shen, Y., Chen, T., Giannakis, G.B.: Random feature-based online multi-kernel learning in environments with unknown dynamics. J. Mach. Learn. Res. **20**(22), 1–36 (2019)
17. Tsallis, C.: Possible generalization of boltzmann-gibbs statistics. J. Stat. Phys. **52**(1), 479–487 (1988)
18. Wang, J., Hoi, S.C.H., Zhao, P., Zhuang, J., Liu, Z.: Large scale online kernel classification. In: Proceedings of the 23rd International Joint Conference on Artificial Intelligence, pp. 1750–1756 (2013)
19. Yang, T., Mahdavi, M., Jin, R., Yi, J., Hoi, S.C.H.: Online kernel selection: algorithms and evaluations. In: Proceedings of the Twenty-Sixth AAAI Conference on Artificial Intelligence, pp. 1197–1202 (2012)
20. Yu, F.X., Suresh, A.T., Choromanski, K.M., Holtmann-Rice, D.N., Kumar, S.: Orthogonal random features. Adv. Neural. Inf. Process. Syst. **29**, 1975–1983 (2016)
21. Zhang, L., Yi, J., Jin, R., Lin, M., He, X.: Online kernel learning with a near optimal sparsity bound. In: Proceedings of the 30th International Conference on Machine Learning, pp. 621–629 (2013)
22. Zhang, X., Liao, S.: Online kernel selection via incremental sketched kernel alignment. In: Proceedings of the Twenty-Seventh International Joint Conference on Artificial Intelligence, pp. 3118–3124 (2018)
23. Zimmert, J., Seldin, Y.: An optimal algorithm for stochastic and adversarial bandits. In: Proceedings of the 22nd International Conference on Artificial Intelligence and Statistics, pp. 467–475 (2019)

# Online Learning of Convex Sets on Graphs

Maximilian Thiessen$^{(\boxtimes)}$ and Thomas Gärtner

Research Unit of Machine Learning, TU Wien, Vienna, Austria
{maximilian.thiessen,thomas.gaertner}@tuwien.ac.at

**Abstract.** We study online learning of general convex sets and halfspaces on graphs. While online learning of halfspaces in Euclidean space is a classical learning problem, the corresponding problem on graphs is understudied. In this context, a set of vertices is convex if it contains all connecting shortest paths and a halfspace is a convex set whose complement is also convex. We discuss mistake bounds based on the Halving algorithm and shortest path covers. Halving achieves near-optimal bounds but is inefficient in general. The shortest path cover based algorithm is efficient but provides optimal bounds only for restricted graph families such as trees. To mitigate the weaknesses of both approaches, we propose a novel polynomial time algorithm which achieves near-optimal bounds on graphs that are $K_{2,k}$ minor-free for some constant $k \in \mathbb{N}$. In contrast to previous mistake bounds on graphs, which typically depend on the induced cut of the labelling, our bounds only depend on the graph itself. Finally, we discuss the agnostic version of this problem and introduce an adaptive variant of Halving for $k$-intersections of halfspaces.

**Keywords:** Online learning · Graph convexity · Node classification

## 1 Introduction

We study online learning of halfspaces and general convex sets on graphs. While most previous mistake bounds in online learning on graphs are based on the *cut-size*, that is, the number of edges with differently labelled endpoints, we focus on *label-independent* bounds, which can be computed directly from the graph itself. Our approach makes small mistake bounds possible even if the cut-size is large. To achieve that, we assume that the vertices with positive labels are convex in the graph. Here convex means that the vertex set is connected and belongs to some intersection-closed hypothesis space. We will focus on the *geodesic* convexity defined by shortest paths. In the special case of halfspaces, where both the positively and negatively labelled vertex sets are convex, we prove a strong bound given solely by the *diameter* of the input graph and the size of a the largest complete bipartite graph $K_{2,m}$ that is a minor of the input graph.

While the problem of online learning halfspaces in Euclidean space is a classical machine learning problem [32,34], the graph variant of this problem has not yet been studied. Convexity on graphs is a well-studied topic outside of the field of machine learning [14,15,33]. However, only recently it was used in

© The Author(s), under exclusive license to Springer Nature Switzerland AG 2023
M.-R. Amini et al. (Eds.): ECML PKDD 2022, LNAI 13716, pp. 349–364, 2023.
https://doi.org/10.1007/978-3-031-26412-2_22

learning problems. Seiffarth et al. [35] initiated the study of node classification under the assumption that both classes are geodesically convex, hence halfspaces. Stadtländer et al. [36] studied a more relaxed version of geodesic convexity allowing multiple disconnected convex regions, called *weak convexity*. In a previous work, we study the active learning version of this problem and have shown near-tight bounds on the query complexity of learning halfspaces [37]. Bressan et al. [6] developed bounds for the same problem under additional margin assumptions and stronger query oracles. While active learning of general convex sets on graphs is not possible with a sub-linear query bound, as it requires to query the whole graph already in the case of a single path, non-vacuous mistake bounds in the online setting can still be achieved, as we will show. Previously, the special case of learning monotone classes on partially ordered sets has been studied [18,31]. Monotone classes can be seen as halfspaces under the *order convexity* [38] on directed acyclic graphs, hence a special case of graph convexity spaces.

## 2  Background

We introduce necessary concepts in online learning and convexity spaces.

### 2.1  Online Learning

We will focus on the *realisable online-learning* setting [28]. Given a set $X$ and a hypothesis space $\mathcal{H} \subseteq \{h(\cdot) \mid h : X \to \{-1, 1\}\}$, our learner $A_t$ knows $\mathcal{H}$ and its strategy might change in each round $t \in \mathbb{N}$. It plays the following iterative game against a potentially adversarial opponent. Any round $t$ has the following steps:

1. opponent picks $x_t \in X$;
2. learner predicts $A_t(x_t) = \hat{y}_t \in \{-1, 1\}$ as the label of $x_t$;
3. opponent reveals the correct label $y_t \in \{-1, 1\}$ to the learner;
4. if $\hat{y}_t \neq y_t$ the learner makes a mistake;
5. $A_t$ potentially updates its strategy.

The opponent is forced to play realisable, that is, there is an $h \in \mathcal{H}$ such that $y_t = h(x_t)$ for all $t \in \mathbb{N}$. The learner's predictions $\hat{y}_t$ are allowed to be *improper*, that is, there is not necessarily a hypothesis in $\mathcal{H}$ determining $\hat{y}_t$. Let $M_A(h)$ denote the worst-case number of mistakes an algorithm $A$ makes on any sequence of points labelled by $h \in \mathcal{H}$. The goal is to minimise the worst-case number of mistakes over all hypotheses $M_A(\mathcal{H}) = \max_{h \in \mathcal{H}} M_A(h)$.

For any given hypothesis space $\mathcal{H}$, a lower bound on the number of mistakes for any online learning algorithm is given by the *Littlestone dimension* $\mathrm{Ldim}(\mathcal{H})$, which is the size of the largest *mistake tree* [28]. It is a combinatorial quantity similar to the VC dimension $\mathrm{VC}(\mathcal{H})$. The *Standard Optimal Algorithm* (SOA) [28] achieves the optimal mistake bound $\mathrm{Ldim}(\mathcal{H})$ for any finite hypothesis space. In general, it is intractable as it requires computing $\mathrm{Ldim}(\mathcal{H}')$ for multiple $\mathcal{H}' \subseteq \mathcal{H}$ in each step, which is known to be hard [16].

## 2.2  Convexity Spaces

For a thorough introduction on convexity theory and graph convexity theory we refer the reader to [33,38].

For a set $X$ and a family $\mathcal{C} \subseteq 2^X$ of subsets, the pair $(X,\mathcal{C})$ is a *convexity space* if *(i)* $\emptyset, X \in \mathcal{C}$, *(ii)* $\mathcal{C}$ is closed under intersection, and *(iii)* $\mathcal{C}$ is closed under unions of sets totally ordered by inclusion. For finite set systems, property *(iii)* always holds. The sets in $\mathcal{C}$ are called *convex*. If a set $C$ and its complement $X \setminus C$ are convex, both are called *halfspaces*. We denote by $\mathcal{C}_H \subseteq \mathcal{C}$ the set of halfspaces of the convexity space $(X,\mathcal{C})$. Note that in general $\mathcal{C}_H$ is not intersection-closed. Two disjoint sets $A, B$ are *halfspace separable* if there exists a halfspace $C$ such that $A \subseteq C$ and $B \subseteq X \setminus C$. *Separation axioms* characterise the ability of a convexity space to separate sets via halfspaces.

**Definition 1 (Separation axioms [38]).** *A convexity space $(X,\mathcal{C})$ is:*

$S_1$ *if each singleton $x \in X$ is convex.*
$S_2$ *if each pair of distinct points $x, y \in X$ is halfspace separable.*
$S_3$ *if each convex set $C$ and points $x \in X \setminus C$ are halfspace separable.*
$S_4$ *if any two disjoint convex sets are halfspace separable.*

If $S_1$ holds the remaining axioms are increasingly stronger, that is, $S_2 \Leftarrow S_3 \Leftarrow S_4$. A mapping $\sigma : 2^X \to 2^X$ is a *convex hull* (or *closure*) *operator* if for all $A, B \subseteq X$ with $A \subseteq B$ *(i)* $\sigma(\emptyset) = \emptyset$, *(ii)* $\sigma(A) \subseteq \sigma(B)$, *(iii)* $A \subseteq \sigma(A)$, and *(iv)* $\sigma(\sigma(A)) = \sigma(A)$. Any convexity space $(X,\mathcal{C})$ induces a convex hull operator by $\sigma(A) = \bigcap\{C \mid A \subseteq C \in \mathcal{C}\}$. A set $A \subseteq X$ is convex, that is $A \in \mathcal{C}$, if and only if it is equal to its convex hull, $A = \sigma(A)$. A set $H \subseteq X$ is a *hull set* if its convex hull is the whole space, $\sigma(H) = X$. For $A, B \subseteq X$, the set $A/B = \{x \in X \mid A \cap \sigma(B \cup \{x\}) \neq \emptyset\}$ is the *extension* of $A$ away from $B$. For $a, b \in X$, the extension $\{a\}/\{b\}$ is also called a *ray $a/b$*. Two disjoint sets $A_1, A_2$ form a partition of $A \subseteq X$ if $A_1 \cup A_2 = A$. The partition $A_1, A_2$ of $A$ is a *Radon partition* if $\sigma(A_1) \cap \sigma(A_2) \neq \emptyset$. The *Radon number* is the minimum number $r(\mathcal{C})$ such that any subset of $X$ of size $r(\mathcal{C})$ has a Radon partition.

A particular type of convexity is *interval convexity*. It is given by an *interval mapping* $I : X \times X \to 2^X$ such that for all $x, y \in X$, *(i)* $x, y \in I(x,y)$ and *(ii)* $I(x,y) = I(y,x)$. $I(x,y)$ is the *interval* between $x$ and $y$. We denote $I(A) = \bigcup_{a,b \in A} I(a,b)$. A set $C$ in an interval convexity space is convex if and only if $C = I(C)$. The convex hull is given by $\sigma(A) = \bigcup_{k=1}^{\infty} I^k(A)$, where $I^1(\cdot) = I(\cdot)$ and $I^{k+1}(\cdot) = I(I^k(\cdot))$. Well-known interval convexity spaces are *metric spaces* $(X,\rho)$. There, the interval contains all the points for which the triangle inequality holds with equality: $I_\rho(x,y) = \{z \in X \mid \rho(x,y) = \rho(x,z) + \rho(z,y)\}$. In Euclidean space this corresponds to all points on a line segment and leads to the standard notion of convex sets.

We study convexity spaces induced by graphs. For a graph $G = (V,E)$, a convexity space $(V,\mathcal{C})$ is a *graph convexity space* if all $C \in \mathcal{C}$ are connected in the graph $G$. Typically, convex sets in graphs are defined through a set of paths $\mathcal{P}$ in the graph $G$. The set $\mathcal{P}$ could for example consists of all shortest or induced paths in $G$, or all paths up to a certain length. Then one can define the interval mapping $I_{\mathcal{P}}(x,y) = \bigcup\{V(P) \mid P \in \mathcal{P}$ has endpoints $x$ and $y\}$, where $V(\cdot)$ denotes the vertex set of the corresponding graph. The most commonly studied convexity on graphs is the *geodesic convexity* (or shortest path convexity) where $\mathcal{P}$ is the set of shortest paths in $G$. For a connected graph $G = (V, E)$ it is given by the interval mapping $I_d$, where $d : V^2 \to \mathbb{R}$ is the *shortest path distance* in $G$. Let $x, y \in V$. For unweighted graphs $d(x,y)$ is the minimum number of edges on any $x$-$y$-path and for graphs with edge weights, $w : E \to \mathbb{R}_{>0}$, it is the minimum total edge weight of any $x$-$y$-path. A set of vertices $C \subseteq V$ is, thus, geodesically convex if and only if $C$ contains every vertex on every shortest path joining vertices in $C$, corresponding again to the Euclidean case.

We denote the size of the geodesic minimum hull set in $G$ as $h(G)$ and the induced subgraph given by a vertex set $X \subseteq V(G)$ as $G[X]$. The *diameter* $d(G)$ of a weighted or unweighted graph $G$ is the maximum number of edges in any shortest path in $G$. We denote the treewidth of a graph, which is a measure of *tree-likeness*, as $\mathrm{tw}(G)$ [3]. Let $\mathrm{cbm}(G)$ be the largest integer $m$ such that the complete bipartite graph $K_{2,m}$ is a minor of $G$. For $n \in \mathbb{N}$, we let $[n] = \{1, \ldots, n\}$.

## 3    Learning Halfspaces

We start with the online learning of halfspaces, corresponding to the special case where the positive class and its complement, the negative class are convex. We start by discussing near-optimal bounds based on the Halving algorithm, which most likely has no polynomial runtime. After that we show how to use *shortest path covers* to get an efficient algorithm achieving near-optimal mistake bounds only on restricted graph families. We mitigate the weak points of both approaches by a novel polynomial-time algorithm that achieves near-optimal bounds on graphs with bounded $\mathrm{cbm}(G)$. See Table 1 for an overview on our resulting bounds.

**Table 1.** Overview on mistake bounds

|  | halfspaces | $k$-intersection | convex sets |
|---|---|---|---|
| Halving | $\mathcal{O}(r(G) \log |V(G)|)$ | $\mathcal{O}(k \, r(G) \log |V(G)|)$ | $\mathcal{O}(\mathrm{VC}(\mathcal{C}) \log |V(G)|)$ |
| tree-based | $\mathcal{O}(\mathrm{cbm}(G)^2 \log d(G))$ | / | / |
| shortest path cover | $\mathcal{O}(|\mathcal{S}^*| \log d(G))$ | $\mathcal{O}(|\mathcal{S}^*| \log d(G))$ | $\mathcal{O}(|\mathcal{S}^*| \log d(G))$ |

### 3.1    Halving

The well-known Halving algorithm is a very simple yet near-optimal approach to online learning. Let $\mathbb{1}_{\{\cdot\}}$ be the indicator function and $\mathrm{VS}_t = \{h \in \mathcal{H} \mid \forall n \in [t-1] : h(x_n) = y_n\}$ be the *version space* at round $t$. The idea is to predict each

$\hat{y}_t$ using the majority vote $\mathbb{1}_{\{\bar{h}(x_t)\geq 0\}}$, where $\bar{h} = \sum_{h\in \mathrm{VS}_t} h(x_t)$. That way, on any mistake, half of the hypotheses in the version space can be discarded. If $\mathcal{H}$ is finite, we can bound the number of Halving's mistakes $M_{1/2}(\mathcal{H})$ as follows.

**Proposition 2 (Angluin [1] and Littlestone [28]).** *For any hypothesis space $\mathcal{H}$ it holds that*

$$\mathrm{VC}(\mathcal{H}) \leq \mathrm{Ldim}(\mathcal{H}) \leq M_{1/2}(\mathcal{H}) \leq \log|\mathcal{H}| \leq 2\,\mathrm{VC}(\mathcal{H})\log|X|.$$

The last inequality follows from the Sauer-Shela lemma. Note that Halving achieves the optimal mistake bound $\mathrm{Ldim}(\mathcal{H})$ up to the $\log|X|$ factor. For the set of halfspaces $\mathcal{C}_H$ of a graph convexity space $(V, \mathcal{C})$, it holds that [37]

$$\mathrm{VC}(\mathcal{C}_H) \leq r(\mathcal{C}) \leq 2\,\mathrm{tw}(G) + 1 \leq 3\,\mathrm{tw}(G), \tag{1}$$

and thus we additionally get the following proposition.

**Proposition 3.** *For the set of halfspaces $\mathcal{C}_H$ of any convexity space $(X, \mathcal{C})$ it holds that*

$$M_{1/2}(\mathcal{C}_H) \leq 2\,\mathrm{VC}(\mathcal{C}_H)\log|V| \leq 2r(\mathcal{C})\log|V| \leq 6\,\mathrm{tw}(G)\log|V|.$$

While Halving achieves a near-optimal mistake bound, it is unclear whether it is possible to run Halving in polynomial time. In particular, checking whether there exists any consistent geodesically convex halfspace $h \in \mathcal{C}_H$ for the given partially labelled graph is NP-hard [35]. It is well-known that for many hypothesis spaces on graphs, it is hard to compute Halving's predictions, that is, decide whether $\bar{h} \geq 0$ [8,21]. This makes an exact polynomial time implementation of Halving unlikely. However, in the discussion section we will mention possible directions based on sampling to potentially overcome this problem. If the VC dimension of the set of halfspaces is bounded, we can run Halving in polynomial time by enumerating the version space, as long as we can efficiently compute convex hulls.

**Theorem 4.** *For any finite $S_4$ convexity space $(X, \mathcal{C})$, Halving on $\mathcal{C}_H$ can be implemented in time $\mathcal{O}(|X|^{\mathrm{VC}(\mathcal{C}_H)+2}\sigma_T)$ per step, where $\sigma_T$ is the time complexity to compute convex hulls in $(X, \mathcal{C})$.*

*Proof.* By the Sauer-Shelah lemma, the hypothesis space and hence any version space has size $\mathcal{O}(|X|^{\mathrm{VC}(\mathcal{C}_H)})$. For any given partially labelled $S_4$ convexity space, the question whether there exists a consistent hypothesis reduces to the question whether the convex hulls of the positively and negatively labelled points overlap. This follows directly from the definition of $S_4$ spaces, as in this case, we can find a halfspaces separating the two convex hulls.

A naive enumeration of the version space would result in $2^{|X|}$ such checks. To achieve an enumeration in time $\mathcal{O}(|X|^{\mathrm{VC}(\mathcal{C}_H)+2}\sigma_T)$ we have to be more careful. We first compute the *region of disagreement* $D$, which consists of all $x \in X$ such that there exist consistent $h, h' \in \mathcal{C}_H$ with $h(x) \neq h'(x)$ in $\mathcal{O}(|X|\sigma_T)$ time.

We will perform the following recursive enumeration. Let $x \in D$. We know that there exists hypotheses $h, h'$ consistent with the remaining labelled points and $h(x) \neq h'(x)$. So, we branch on $x$ by setting its label either to 1 or to $-1$. Recursively we recompute the disagreement region for this new set of points and continue. Any leaf in this recursion tree corresponds to a unique hypothesis consistent with the original labelled points. Also, as $D$ shrinks in each branching step by at least one element, the path from root to leaf in the recursion tree has length at most $|X|$. In total this gives $\mathcal{O}\left(|X||X|^{\mathrm{VC}(\mathcal{C}_H)}\right)$ many recursion steps, as the number of leaves corresponding to unique hypotheses is bounded by the size of the hypothesis space. As each branching step takes $\mathcal{O}(|X|\sigma_T)$ we achieve the stated overall runtime.    $\square$

As in the geodesic convexity, convex hulls can be computed in time $\sigma_T = \mathcal{O}(|V(G)|^3)$ [37], we achieve a polynomial runtime for $S_4$ graphs of bounded treewidth, because $\mathrm{VC}(\mathcal{C}_H) \leq 3\,\mathrm{tw}(G)$ by Eq. (1). For non $S_4$ graphs we can still achieve polynomial time if we can enumerate the version space in polynomial time. For example, we can enumerate all consistent hypotheses of bipartite or planar graphs, which are in general not $S_4$ and have unbounded treewidth, in polynomial time [19] leading to the next result.

**Proposition 5.** *Let $G$ be a planar or bipartite graph and $(V(G), \mathcal{C})$ the geodesic convexity on $G$. Halving can be implemented in polynomial time for the hypothesis space of geodesically convex halfspaces $\mathcal{C}_H$ on $G$.*

## 3.2   Shortest Path Cover Based Approach

In contrast to the inefficient Halving algorithm, we discuss now a simple and efficient algorithm achieving optimal bounds only on specific graph families.

To derive a simple upper bound, we note that one immediate consequence of the halfspace assumption is that any shortest path $P$ can have at most one *cut edge*, that is, an edge with differently labelled endpoints. We can follow Gärtner and Garriga [18] to perform online binary search on a path $P$ if we already know the labels of its endpoints. In this case, we can predict for any $x_t \in V(P)$ the label of the closer endpoint. That way if we make a mistake we can deduce at least half of the path's labels. That means that we will make at most $\lceil \log d(G) \rceil$ mistakes, as the length of $P$ is at most the diameter $|V(P)| - 1 \leq d(G)$. Here, log is the base 2 logarithm. If we do not know the endpoints' labels, we can apply the following simple strategy. First we make at most one mistake on the first point $a \in V(P)$. Then we will predict on any $b \in V(P)$ the same label as $a$, so that on mistake we would have two different labelled points on $P$. By the halfspace assumption we can infer the labels of all vertices but the $a$-$b$ sub-path and hence we are back in the previous case with endpoints with known labels.

**Lemma 6.** *Given any shortest path $P$ in a graph $G$, there exists a prediction strategy making at most $2 + \log(|V(P)| - 1)$ mistakes on $P$.*

We can generalise this approach to the whole graph using *shortest path covers* [37], which is a set $\mathcal{S}$ of shortest paths whose vertices cover the graph:

$\bigcup_{P \in \mathcal{S}} V(P) = V(G)$. Performing binary search on each path in $\mathcal{S}$ gives our next mistake upper bound. We call this approach the SPC algorithm SPC($\mathcal{S}$) based on a shortest path cover $\mathcal{S}$.

**Theorem 7.** *Let $(V, \mathcal{C})$ be the geodesic convexity space on a graph $G = (V, E)$ and $\mathcal{S}$ a shortest path cover of $G$. The mistake bound for the SPC algorithm using a shortest path cover $\mathcal{S}$ is*

$$M_{\mathrm{SPC}(\mathcal{S})}(\mathcal{C}_H) \leq |\mathcal{S}|(2 + \log d(G)).$$

As we can compute an $\mathcal{O}(\log d(G))$-approximation $\mathcal{S}$ to the minimum shortest path cover $\mathcal{S}^*$ in polynomial time [37] we get the following result.

**Theorem 8.** *Let $(V, \mathcal{C})$ be the geodesic convexity space on a graph $G = (V, E)$ and $\mathcal{S}^*$ a minimum shortest path cover of $G$. There exists a polynomial-time online learning algorithm, which computes a shortest path cover $\mathcal{S}$ such that*

$$M_{\mathrm{SPC}(\mathcal{S})}(\mathcal{C}_H) \leq \mathcal{O}(|\mathcal{S}^*|(\log d(G))^2).$$

There exist edge-weighted graph families where the bound of Theorem 7 is asymptotically tight. For example, we can use the same construction as in [37].

**Proposition 9.** *There exists a family of edge-weighted graphs $G_{k,\ell}$ with $k, \ell \in \mathbb{N}$, such that $G_{k,\ell}$ has a shortest path cover of size $k$ and diameter $\ell$ and for any online algorithm $A$ applied to the geodesically convex halfspaces $\mathcal{C}_H$ in $G_{k,\ell}$ the mistake lower bound $M_A(\mathcal{C}_H) \geq k \log \ell$ holds.*

### 3.3    Efficient Algorithms for Graphs with Bounded cbm($G$)

While the previously discussed SPC algorithm is tight on specific graphs, there exists graphs, where it is arbitrarily bad. For example, on the star graph, which is a tree $T$ with $V(T) - 1$ leaves, the minimum shortest path cover $\mathcal{S}^*$ has size $|\mathcal{S}^*| \geq \frac{V(T)-1}{2}$ resulting in a mistake bound linear in $V(T)$, whereas the optimal strategy makes at most two mistakes. In this section, we mitigate the weakness of the SPC algorithm and achieve a polynomial time algorithm with a near-optimal mistake bound on graphs with bounded cbm($G$).

---

**Algorithm 1:** Tree-based online halfspace learning on graphs

---

**Input:** unweighted graph $G$, with $n = |V(G)|$
1  compute a Dijkstra shortest path tree $T$ rooted at $x_1$
2  predict $\hat{y}_1 = 1$ and receive a mistake if $\hat{y}_1 \neq y_1$
3  **for** $t \in [n] \setminus \{1\}$ **do**
4      Let $P$ be the root-leaf path in $T$ containing $x_t$
5      Predict $\hat{y}_t$ with binary search on $P$.

---

**Theorem 10.** *Let $G$ be an unweighted graph. Algorithm 1 has a $\mathcal{O}(|V|^2)$ per step runtime and achieves a mistake bound of $\mathcal{O}(\mathrm{cbm}(G)^2 \log d(G))$ for the class of geodesically convex halfspaces in $G$.*

*Proof.* Without loss of generality we can assume that $x_1$ is positive. Correctness follows immediately by the fact that we are only applying the binary search strategy to each root-leaf shortest path in $T$.

By Lemma 6 we will make at most $\mathcal{O}(\log d(G))$ mistakes on each such path. Note however, that we only make mistakes on such a path if its leaf in $T$ is negative. Otherwise we will just predict positive all the time and do not make any mistake. We will bound the number of possible leaves in $T$ that can be negative by $\mathcal{O}(\mathrm{cbm}(G)^2)$ and hence show that only $\mathcal{O}(\mathrm{cbm}(G)^2)$ binary searches are required, while on all other paths we will not make any mistakes at all.

Let $H_- \subseteq \mathcal{C}_H$ be any negatively labelled geodesically convex halfspace with $x_1 \notin H_-$ and let $R_+ \subseteq T \setminus H_-$ be the set of vertices that are positive and have a neighbour in $H_-$. As $H_-$ is convex, and hence connected, we can contract all edges in $H_-$ such that only a single vertex $h_-$ remains. Additionally, we contract all edges leading to the $R_+$ vertices from $x_1$ but the ones with endpoints in $R_+$. This constructions shows that $G$ has a complete bipartite minor $K_{2,|R_+|}$ with $x_1$ and $h_-$ on the one side and $R_+$ on the other. Hence, $|R_+| \leq \mathrm{cbm}(G)$.

We inspect again the original non-contracted $G$. Let $r \in R_+$ and let $L_- \subseteq H_-$ be the children of $r$ in $T$. As $r$ is positive and all vertices in $L_-$ are negative, $L_-$ must be a clique in $G$. Otherwise, the halfspace assumption would be violated as the shortest path between $a, b \in L_-$ would go over $r$. For that, we use the fact that $G$ is unweighted. Any clique of size $f \geq 3$ contains a $K_{2,f-2}$ and hence $f \leq \mathrm{cbm}(G) + 2$. All together this gives $\mathcal{O}(\mathrm{cbm}(G)^2)$ cut edges on $T$. Note that there might be more cut edges in $G$, but the halfspaces is determined by the cut edges in $T$. The runtime is given by one run of Dijkstra and the repeated binary searches on each path. □

Note that for many graphs, the tree-based approach gives a significantly better bound than the shortest path cover based algorithm. On any tree the SPC mistake bound will be linear in the number of leaves, while the tree-based approach will just perform one binary search resulting in $\mathcal{O}(\log d(G))$. Also, for *outerplanar* graphs a constant number of binary searches suffice, as $\mathrm{cbm}(G) \leq 2$ [13].

### 3.4   Lower Bounds

We will use the separation axioms to discuss general mistake lower bounds for arbitrary graph convexity spaces. Let $(V, \mathcal{C})$ be a graph convexity space. Without any further assumptions, we have the VC dimension $\mathrm{VC}(\mathcal{C}_H)$ as a lower bound on the optimal number of mistakes $\mathrm{Ldim}(\mathcal{C}_H)$ as already discussed. For $S_4$ convexity spaces we have $r(\mathcal{C}) - 1 = \mathrm{VC}(\mathcal{C}_H)$ [37] and hence in this case, also the Radon number is a lower bound on the optimal number of mistakes, $r(\mathcal{C}) - 1 \leq \mathrm{Ldim}(\mathcal{C})$. Interestingly the minimum hull set size $h(G)$ is not a lower bound in general even though it is a lower bound in the active setting for $S_3$ convexity spaces [37]. For example, the star graph $T$ has $h(T) = |V(T)| - 1$ but $\text{Ldim}(\mathcal{C}_H) = 1$. Finally, in geodesic $S_2$ graph convexity spaces, we can place a cut edge arbitrarily on any shortest path, hence $\text{Ldim}(\mathcal{C}_H) \geq \log d(G)$. Compared to specific worst-case graphs, as in Proposition 9, these lower bounds hold in general for any graph, as long as the graph convexity space satisfies the corresponding separation axiom.

## 4  Learning General Convex Sets

Having discussed three different approaches to learn halfspaces on graphs, we now turn to general convex sets. We discuss Halving and an adapted shortest path cover based approach for this setting. In the special case of $k$-intersections of halfspaces, we discuss an adaptive strategy that does not require to know $k$. Let us start with a standard result on intersection-closed hypothesis spaces, adapted to our graph setting.

**Proposition 11 (Horváth and Turán [24]).** *For any graph convexity space* $(V(G), \mathcal{C})$ *on a graph* $G$ *it holds that*

$$\text{VC}(\mathcal{C}) = \max_{C \in \mathcal{C}} h(G[C]).$$

This immediately shows that the minimum hull set size $h(G)$ is a lower bound on $\text{VC}(\mathcal{C})$ and hence also on $\text{Ldim}(\mathcal{C})$. Also, $r(\mathcal{C}) - 1$ is a lower bound on $\text{VC}(\mathcal{C})$, as any set without a Radon partition can be shattered. It is unclear how to compute $\max_{C \in \mathcal{C}} h(G[C])$ efficiently, as already computing $h(G)$ is APX-hard [11]. We provide an efficiently computable upper bound. The VC dimension $\text{VC}(\mathcal{C})$ of any convexity space on a graph can be bounded by the VC dimension of the set of all connected sets $\mathcal{H}_{\text{con}} \supseteq \mathcal{C}$ of $G$. The quantity $\text{VC}(\mathcal{H}_{\text{con}})$ is bounded by the maximum number of leaves $\ell(G)$ in any spanning tree of $G$, $\ell(G) \leq \text{VC}(\mathcal{H}_{\text{con}}) \leq \ell(G) + 1$ [27]. Hence, we achieve the following proposition.

**Proposition 12.** *For any graph convexity space* $(V(G), \mathcal{C})$ *on a graph* $G$ *it holds that*

$$\text{VC}(\mathcal{C}) \leq \text{VC}(\mathcal{H}_{con}) \leq \ell(G) + 1.$$

Computing $\ell(G)$ is also APX-hard, yet, a near-linear time 3-approximation algorithm exists [30]. The first inequality of Proposition 12 is tight for specific convexity spaces: We can take a maximum vertex set $A \subseteq V(G)$ shatterable by connected sets and define $\mathcal{C}_A = 2^A \cup V(G)$. The resulting convexity space $(V(G), \mathcal{C}_A)$ satisfies $\ell(G) \leq \text{VC}(\mathcal{C}_A) = \text{VC}(\mathcal{H}_{\text{con}}) \leq \ell(G) + 1$.

By Proposition 2 we directly get that Halving achieves the mistake bound $\mathcal{O}(\text{VC}(\mathcal{C}) \log |V(G)|) = \mathcal{O}(\ell(G) \log |V(G)|)$. The next theorem shows that we can run Halving in polynomial-time if the VC dimension is a constant and convex hull computations are efficiently possible in $(X, \mathcal{C})$.

**Theorem 13.** *For any finite convexity space* $(X, \mathcal{C})$ *Halving can be implemented in time* $\mathcal{O}(|X|^{\text{VC}(\mathcal{C})+1} \sigma_T)$ *per step where* $\sigma_T$ *is the time complexity to compute convex hulls on* $(X, \mathcal{C})$.

To achieve the runtime in Theorem 13, we can use the algorithm of Boley et al. [5] to enumerate the whole version space in each step. Interestingly, we can adapt our SPC algorithm to be able to handle convex sets with the same asymptotic mistake bound.

**Theorem 14.** *Let $(V, \mathcal{C})$ be the geodesic convexity space on a graph $G = (V, E)$ and $\mathcal{S}$ a shortest path cover of $G$. The mistake bound of the SPC algorithm using an SPC $\mathcal{S}$ is*

$$M_{\mathrm{SPC}(\mathcal{S})}(\mathcal{C}) \leq \mathcal{O}(|\mathcal{S}| \log d(G)).$$

One can achieve the bound by performing two instead of one binary searches as soon as one point is known on any particular path. The idea is based on the same strategy on path covers [18]. We additionally remark that as on any fixed shortest path we cannot shatter three points, we get an upper bound on the VC dimension based on shortest path covers.

**Proposition 15.** *The VC dimension of geodesically convex sets in a graph is upper bounded by the size of the minimum shortest path cover $2|\mathcal{S}^*|$.*

Note that $|\mathcal{S}^*| \leq \ell(G)$ as any (Dijkstra-based) shortest-path tree with $k$ leaves is a spanning tree and can be covered with $k$ shortest paths.

### 4.1   Learning $k$-intersections of Halfspaces

In Euclidean space, any convex set can be represented as an intersection of a set of halfspaces. As general convex sets in Euclidean space have infinite VC dimension [25], a common way to bound the complexity is to only look at convex sets that can be represented as the intersection of $k$ halfspaces. The parameter $k$ linearly determines the VC dimension [25].

In general, not all convexity spaces have the property that all convex sets are intersections of halfspaces. Take for example the geodesically convex halfspaces in the complete bipartite graph $K_{2,3}$. The graph only has the two halfspaces $(\emptyset, V(K_{2,3}))$, while each vertex and edge on its own is convex. This property is actually exactly captured by the $S_3$ separation axiom.

**Proposition 16 (van de Vel [38]).** *Convex sets in a convexity space $(X, \mathcal{C})$ can be represented as an intersection of a set of halfspaces if and only if $(X, \mathcal{C})$ is an $S_3$ convexity space.*

For any hypothesis space $\mathcal{H}$ define $\mathcal{H}^{k\cap} = \{h_1 \cap \cdots \cap h_k \mid h_1, \ldots, h_k \in \mathcal{H}\}$ for $k \in \mathbb{N}$. Let $(X, \mathcal{C})$ be a convexity space and denote its halfspaces as $\mathcal{C}_H$. As $X \in \mathcal{C}_H$, we have $\mathcal{C}_H^{k'\cap} \subseteq \mathcal{C}_H^{k\cap}$ for all $k' \leq k$. Note that, Proposition 16 implies that for finite $S_3$ convexity spaces there is some $k \in \mathbb{N}$ such that $\mathcal{C} = \mathcal{C}_H^{k\cap}$. For any hypothesis space $\mathcal{H}$, the VC dimension of $\mathcal{H}^{k\cap}$ is bounded by $\mathcal{O}(k \log k \, \mathrm{VC}(\mathcal{H}))$ [12]. Additionally for finite $X$ we can again use the Sauer-Shelah Lemma and get

$$\mathrm{VC}(\mathcal{H}^{k\cap}) \leq \log|\mathcal{H}^{k\cap}| = k \log(|\mathcal{H}|) = \mathcal{O}(k \log(|X|^{\mathrm{VC}(\mathcal{H})})) = \mathcal{O}(k \, \mathrm{VC}(\mathcal{H}) \log|X|).$$

Hence, applying Halving results in the following bound.

**Proposition 17**

$$M_{1/2}\left(\mathcal{C}_H^{k\cap}\right) = \mathcal{O}(k\,\mathrm{VC}(\mathcal{C}_H)\log\min\{|X|,k\}\log|X|)$$

Applying this to $k$-intersections of halfspaces in graphs gives:

**Theorem 18.** *Let $G = (V,E)$ be a graph and $(V,\mathcal{C})$ a graph convexity space on $G$. Halving achieves the following mistake bound:*

$$M_{1/2}\left(\mathcal{C}_H^{k\cap}\right) = \mathcal{O}(k\,\mathrm{tw}(G)\log\min\{|X|,k\}\log|X|)\,.$$

On $S_3$ graphs we additionally get:

**Corollary 19.** *Let $G = (V,E)$ be a graph and $(V,\mathcal{C})$ an $S_3$ graph convexity space on $G$. Let $k \in \mathbb{N}$ such that $\mathcal{C} = \mathcal{C}_H^{k\cap}$. Halving achieves the following mistake bound:*

$$M_{1/2}(\mathcal{C}) = \mathcal{O}(k\,\mathrm{tw}(G)\log\min\{|X|,k\}\log|X|)\,.$$

The two previous bounds for Halving are difficult to use as to the best of our knowledge there is no obvious way to compute or upper bound $k$ for a given $S_3$ graph convexity space. Also, the minimum $k$ required to achieve $\mathcal{C} = \mathcal{C}_H^{k\cap}$ is non-trivial to compute. In this last paragraph of the section, we discuss how to make Halving *adaptive*, in the sense that if the target hypothesis $C^*$ is in $\mathcal{C}_H^{k'\cap}$ for some $k' \in \mathbb{N}$ which we do not know, we still get a bound linear in $k'$ instead of the globally required $k$. This can be achieved using the standard *doubling trick* [10]. It works by assuming that $k$ belongs to $\{2^{i-1},\dots,2^i\}$ for $i \in [\lceil \log k \rceil]$ and iteratively applying Halving to the hypothesis space $\mathcal{H}_{2^i}$. Each time the whole hypothesis space $\mathcal{H}_{2^i}$ is not consistent anymore with the labels seen so far, the $i$ is increased by one. We call this approach ADA-$1/2$, for *adaptive Halving*, and it achieves the following mistake bound.

**Proposition 20.** *Let $\mathcal{H}$ be a hypothesis space such that $\emptyset = \mathcal{H}_0 \subseteq \mathcal{H}_1 \subseteq \mathcal{H}_2 \subseteq \cdots \subseteq \mathcal{H}$. ADA-$1/2$ achieves the following bound if the target hypothesis $H$ is in $\mathcal{H}_k$ for some unknown $k \in \mathbb{N}$:*

$$M_{\mathrm{ADA}\text{-}1/2}(\mathcal{H}) \le \sum_{i\in[\lceil\log k\rceil]} \log\left|\mathcal{H}_{2^i}\setminus\mathcal{H}_{2^{i-1}}\right|\,.$$

Applied to $k$-intersections of halfspaces in graphs we achieve:

**Corollary 21**

$$M_{\mathrm{ADA}\text{-}1/2}(\mathcal{H}^{k\cap}) = \mathcal{O}(k\,\mathrm{tw}(G)\log\min\{|X|,k\}\log|X|)\,.$$

The additional constant factor to achieve the adaptive variant is negligible compared to standard Halving on the set of all convex sets $\mathcal{C}$, where the required number of halfspaces $k$ could be even linear in $|X|$, for example, on a star graph.

## 5   Discussion

We discuss efficiency aspects of Halving and how to generalise it to the agnostic case. We compare our results to previous bounds in online learning on graphs and show how we can use the *closure algorithm* in our setting.

**Efficient Halving by Sampling.** All discussed Halving based algorithms in this paper are in general not efficient. In particular, computing the weighted majority vote is in many cases hard. One possible way around this issue is to use the randomised version of Halving. It samples a consistent hypothesis uniformly at random and uses it for prediction of the current point $x_t$. This simple strategy RAND-$^1\!/_2$ is already enough to achieve the Halving bound in expectation.

**Proposition 22 (Littlestone and Warmuth [29])**

$$\mathbb{E}[M_{\text{RAND-}^1\!/_2}(\mathcal{H})] \leq \ln |\mathcal{H}|.$$

Thus, if we can sample uniformly at random from the version space we achieve Halving's bound in expectation. As a simple example, let us compare Halving and Rand-Halving on the simple learning problem of halfspaces on a path $P$. Standard Halving enumerates the whole version space of size $\mathcal{O}(|V(P)|)$, while Rand-Halving only needs to sample a number in $[|V(P)|]$, which can be achieved with $\mathcal{O}(\log |V(P)|)$ random binary draws; an exponential increase. However in general, sampling uniformly at random from a version space is a non-trivial task. Boley et al. [4] and Ganter [17] discuss sampling general convex sets in the context of *frequent pattern mining* and *formal concept analysis*. Nevertheless, their results are also applicable in our context. [4] shows that in general it is hard to sample a convex set uniformly at random, which corresponds to sampling a consistent convex hypothesis. Under additional assumptions they construct a Markov chain with polynomial mixing time and also discuss various practically efficient heuristics. Applying these techniques in our context is future work. A potential way to overcome the hardness might be to approximate sample, that is, only close to uniform, which still would provide an $\mathcal{O}(\log |\mathcal{H}|)$ bound.

**Agnostic Online Learning.** In the *agnostic* version of the problem, we drop the realisability assumption, hence the opponent is allowed to use arbitrary labels $y_t$ for $t \in [T]$ for some $T \in \mathbb{N}$. In this more general online learning model it is essentially hopeless to bound the number of mistakes, as the opponent can always set $y_t \neq \hat{y}_t$. Because of that, typically the *regret* is studied instead in agnostic online learning. The regret for any particular sequence $x_1, x_2, \ldots$ of an randomised algorithm $A$ with predictions $A_t(x_t) = \hat{y}_t$ is

$$R_A(\mathcal{H}) = \mathbb{E}\left[ \max_{x_1, \ldots, x_T} \sum_{t \in [T]} \mathbb{1}_{\{\hat{y}_t \neq y_t\}} - \min_{h \in \mathcal{H}} \sum_{t \in [T]} \mathbb{1}_{\{h(x_t) \neq y_t\}} \right],$$

where the expectation is taken over the random predictions of the algorithm $A$. Hence, we compare the performance of algorithm $A$ with the best fixed hypothesis from a given hypothesis space $\mathcal{H}$ in hind-sight.

Ben-David and Pál [2] have proven that there exists an algorithm $A$ achieving the optimal regret $R_A(\mathcal{H}) \leq \sqrt{1/2 \ln(|\mathcal{H}|)T}$. This means that if we let $\{x_1, \ldots, x_T\} = V$ for some graph $G = (V, E)$ we get the regret bound $R_A(\mathcal{H}) \leq \sqrt{1/2 \ln(|\mathcal{H}|)|V|}$ if each vertex $x_i$ appears only once. Note that in the realisable case we achieved bounds $\mathcal{O}(\log|V|)$ (not considering other parameters), while here in the agnostic case we get $\mathcal{O}(\sqrt{|V|})$. If we can expect that there is some hypothesis in $\mathcal{H}$ that performs rather well, say

$$\min_{h \in \mathcal{H}} \sum_{t \in [T]} \mathbb{1}_{\{h(x_t) \neq y_t\}} \leq M^\star$$

for some known $M^\star \in \mathbb{N}$, we can significantly improve the bound to

$$R_A(\mathcal{H}) \leq \sqrt{2M^\star \operatorname{Ldim}(\mathcal{H})} + \operatorname{Ldim}(\mathcal{H}).$$

For small $M^\star$ this asymptotically matches the realisable bound. By using the standard *doubling trick* [10] we can achieve almost the same bound without knowing $M^\star$. This bound allows learning in the following special case. Assume the target hypothesis $h^*$ is a vertex set that is a positive convex set but with at most $M^\star$ labels flipped to negative. That is, there exists $h \in \mathcal{C}$ which predicts everywhere positive where $h^*$ predicts positive, but can additionally predict at up to $M^\star$ many points positive, where $h^*$ is negative. Let $\mathcal{C}_{M^\star}$ be this space of *noisy* convex sets containing each set $C \in \mathcal{C}$ with all possible at most $M^\star$ label flips. That is $|\mathcal{C}_{M^\star}| = \mathcal{O}(|\mathcal{C}||X|^{M^\star})$. Applying adaptive Halving achieves a mistake bound similar to the the regret-based analysis:

$$M_{\text{ADA-}1/2}(\mathcal{C}_{M^\star}) = \mathcal{O}((\operatorname{VC}(\mathcal{C}) + M^\star)\log|X|).$$

## 5.1   Comparison to Cut-Based Learning

Common bounds in online learning on graphs do not make any hypothesis-space-based assumption and instead depend on the *cut-size* $\Phi_G(y) = \sum_{v,w \in E(G)} \mathbb{1}_{\{y(v) \neq y(w)\}}$ of the labelling $y$. Let $\tilde{\mathcal{H}}_c$ be the hypothesis space of labellings with bounded cut-size $\Phi_G(y) \leq c$. The VC dimension of $\tilde{\mathcal{H}}_c$ can be bounded linearly by $c$, $\operatorname{VC}(\tilde{\mathcal{H}}_c) \leq 2c+1$ [26]. By applying Halving to this hypothesis space we get the bound $\mathcal{O}(c \log|V|)$ [22]. Again, we can use the doubling trick to get the same bound without knowing the correct value of $c$.

**Proposition 23**

$$M_{\text{ADA-}1/2}(\tilde{\mathcal{H}}_c) = \mathcal{O}(c \log|V|).$$

Herbster et al. [21] proved that the majority vote in $\tilde{\mathcal{H}}_c$ is NP-hard, based on the fact counting label-consistent min-cuts is #P-hard. This makes the existance of an efficient and exact Halving algorithm for $\tilde{\mathcal{H}}_c$ unlikely.

Comparing with our bounds, we see that the cut-size $c$ now has the same role as previously $r(G)$ or $\text{tw}(G)$ in our bounds. Note however, that our bounds are label-independent, that is, under the halfspace or convexity assumption they hold for any labelling. Cut-size based bounds are complementary and depend on the actual labelling.

The problem of online learning on graphs was introduced by Herbster et al. [23]. They bound the number of mistakes as $4\Phi_G(y)d(G)\,\text{bal}(y)$, where $\text{bal}(y) = (1 - 1/|V(G)|\,|\sum y_i|)^{-2}$ is a *balancedness* term. The efficient Pounce algorithm [20] achieves the mistake bound $\mathcal{O}(\Phi_G(y)(\log|V(G)|)^4)$ for unweighted graphs, almost matching the near-optimal bound of Halving. In parallel to these works, Cesa-Bianchi et al. [7] first developed an efficient and optimal algorithm for online learning on trees and showed that Halving on trees actually also asymptotically achieves the optimal bound, which can be much smaller than $\mathcal{O}(c\log|V(G)|)$. The authors then generalised these ideas to general graphs [9] and achieved under mild assumptions an efficient algorithm that is optimal up to a $\log|V(G)|$ factor. We refer to [9,21] for an overview and in-depth discussion.

The convexity or the halfspace assumption are orthogonal to the standard assumption of small cut-size. For example, on a $2 \times k$ grid, we can have halfspaces corresponding to the two $1 \times k$ halves, that have a cut of size $k$. That means that the convexity or halfspace assumption can lead to strong bound in situations where the cut of the labelling might be large. However, assuming a small cut can also improve our bounds. For example, the shortest path cover based bound can be changed to $\mathcal{O}(\min\{|\mathcal{S}|, \Phi_G(y)\}\log d(G))$, as we only have to do at most $\min\{|\mathcal{S}|, \Phi_G(y)\}$ binary searches.

# 6   Conclusion

In this paper, we have studied online learning of halfspaces and general convex sets on graphs. On the one hand, we discussed that Littlestone's Halving algorithm achieves near-optimal bounds in general convexity spaces, yet is inefficient in general in its standard form. On the other hand, we have used shortest path covers to achieve a simple and efficient algorithm, which is however not optimal in many cases. For the special case of geodesic halfspaces on graphs with bounded $\text{cbm}(G)$, we proposed an algorithm with near-optimal mistake bound and quadratic runtime. We have discussed general lower bounds and specific worst-case examples. In the case of halfspaces we argued that general, increasingly stronger lower bounds are achieved through the separation axioms $S_1, \ldots, S_4$. We looked at the special case of $k$-intersections of halfspaces and discussed an adaptive version of Halving using the well-known doubling trick. Finally, we compared our bounds to previous label-dependent mistake bounds and discussed potential extensions to the agnostic case. As future work, we are looking into more general efficient and near-optimal algorithms, more relaxed assumptions on the labels, and multi-class online learning on graphs.

# References

1. Angluin, D.: Queries and concept learning. Mach. Learn. **2**(4), 319–342 (1988)
2. Ben-David, S., Pál, D., Shalev-Shwartz, S.: Agnostic online learning. In: COLT (2009)
3. Bodlaender, H.L.: A linear-time algorithm for finding tree-decompositions of small treewidth. SIAM J. Comput. **25**(6), 1305–1317 (1996)
4. Boley, M., Gärtner, T., Grosskreutz, H.: Formal concept sampling for counting and threshold-free local pattern mining. In: ICDM (2010)
5. Boley, M., Horváth, T., Poigné, A., Wrobel, S.: Listing closed sets of strongly accessible set systems with applications to data mining. Theoret. Comput. Sci. **411**(3), 691–700 (2010)
6. Bressan, M., Cesa-Bianchi, N., Lattanzi, S., Paudice, A.: Exact recovery of clusters in finite metric spaces using oracle queries. In: COLT (2021)
7. Cesa-Bianchi, N., Gentile, C., Vitale, F.: Fast and optimal prediction on a labeled tree. In: COLT (2009)
8. Cesa-Bianchi, N., Gentile, C., Vitale, F., Zappella, G.: A correlation clustering approach to link classification in signed networks. In: COLT (2012)
9. Cesa-Bianchi, N., Gentile, C., Vitale, F., Zappella, G.: Random spanning trees and the prediction of weighted graphs. JMLR **14**(1), 1251–1284 (2013)
10. Cesa-Bianchi, N., Lugosi, G.: Prediction, learning, and games (2006)
11. Coelho, E.M.M., Dourado, M.C., Sampaio, R.M.: Inapproximability results for graph convexity parameters. Theoret. Comput. Sci. **600**, 49–58 (2015)
12. Csikós, M., Mustafa, N.H., Kupavskii, A.: Tight lower bounds on the VC-dimension of geometric set systems. JMLR **20**(1), 2991–2998 (2019)
13. Diestel, R.: Graph Theory, 5th edn. Springer, Heidelberg (2017). https://doi.org/10.1007/978-3-662-53622-3
14. Duchet, P.: Convex sets in graphs, II. Minimal path convexity. J. Combinat. Theor. Ser. B **44**(3), 307–316 (1988)
15. Farber, M., Jamison, R.E.: Convexity in Graphs and Hypergraphs. SIAM J. Algeb. Discrete Methods **7**(3), 433–444 (1986)
16. Frances, M., Litman, A.: Optimal mistake bound learning is hard. Inf. Comput. **144**(1), 66–82 (1998)
17. Ganter, B.: Random extents and random closure systems. In: CLA, vol. 959, pp. 309–318 (2011)
18. Gärtner, T., Garriga, G.C.: The cost of learning directed cuts. In: ECMLPKDD (2007)
19. Glantz, R., Meyerhenke, H.: On finding convex cuts in general, bipartite and plane graphs. Theoret. Comput. Sci. **695**, 54–73 (2017)
20. Herbster, M., Lever, G., Pontil, M.: Online prediction on large diameter graphs. In: NIPS (2008)
21. Herbster, M., Pasteris, S., Ghosh, S.: Online prediction at the limit of zero temperature. In: NIPS (2015)
22. Herbster, M., Pontil, M.: Prediction on a graph with a perceptron. In: NIPS (2006)
23. Herbster, M., Pontil, M., Wainer, L.: Online learning over graphs. In: ICML (2005)
24. Horváth, T., Turán, G.: Learning logic programs with structured background knowledge. Artif. Intell. **128**(1–2), 31–97 (2001)
25. Kearns, M.J., Vazirani, U.: An introduction to computational learning theory (1994)
26. Kleinberg, J.: Detecting a network failure. Int. Math. **1**(1), 37–55 (2004)

27. Kranakis, E., Krizanc, D., Ruf, B., Urrutia, J., Woeginger, G.: The VC-dimension of set systems defined by graphs. Discret. Appl. Math. **77**(3), 237–257 (1997)
28. Littlestone, N.: Learning quickly when irrelevant attributes abound: a new linear-threshold algorithm. Mach. Learn. **2**(4), 285–318 (1988)
29. Littlestone, N., Warmuth, M.K.: The weighted majority algorithm. Inf. Comput. **108**(2), 212–261 (1994)
30. Lu, H.I., Ravi, R.: Approximating maximum leaf spanning trees in almost linear time. J. Algorithms **29**(1), 132–141 (1998)
31. Missura, O., Gärtner, T.: Predicting dynamic difficulty. In: NIPS (2011)
32. Novikoff, A.B.: On convergence proofs on perceptrons. In: Symposium on the Mathematical Theory of Automata (1962)
33. Pelayo, I.M.: Geodesic convexity in graphs. Springer, New York (2013). https://doi.org/10.1007/978-1-4614-8699-2
34. Rosenblatt, F.: The Perceptron: a probabilistic model for information storage and organization in the brain. Psychol. Rev. **65**(6), 386 (1958)
35. Seiffarth, F., Horváth, T., Wrobel, S.: Maximal closed set and half-space separations in finite closure systems. In: ECMLPKDD (2019)
36. Stadtländer, E., Horváth, T., Wrobel, S.: Learning weakly convex sets in metric spaces. In: ECMLPKDD (2021)
37. Thiessen, M., Gärtner, T.: Active learning of convex halfspaces on graphs. In: NeurIPS (2021)
38. van de Vel, M.L.: Theory of convex structures (1993)

# Active and Semi-supervised Learning

# Exploring Latent Sparse Graph
# for Large-Scale Semi-supervised Learning

Zitong Wang[1], Li Wang[2,3]($\boxtimes$), Raymond Chan[4], and Tieyong Zeng[5]

[1] Department of Industrial Engineering and Operations Research,
Columbia University in the City of New York, New York, USA
zw2690@columbia.edu
[2] Department of Computer Science and Engineering, University of Texas at
Arlington, Arlington, TX 76019-0408, USA
li.wang@uta.edu
[3] Department of Mathematics, University of Texas at Arlington, Arlington,
TX 76019-0408, USA
[4] Department of Mathematics, City University of Hong Kong, Kowloon, Hong Kong
rchan.sci@cityu.edu.hk
[5] Department of Mathematics, The Chinese University of Hong Kong,
Hong Kong, China
zeng@math.cuhk.edu.hk

**Abstract.** We focus on developing a novel scalable graph-based semi-supervised learning (SSL) method for input data consisting of a small amount of labeled data and a large amount of unlabeled data. Due to the lack of labeled data and the availability of large-scale unlabeled data, existing SSL methods usually either encounter suboptimal performance because of an improper graph constructed from input data or are impractical due to the high-computational complexity of solving large-scale optimization problems. In this paper, we propose to address both problems by constructing a novel graph of input data for graph-based SSL methods. A density-based approach is proposed to learn a latent graph from input data. Based on the latent graph, a novel graph construction approach is proposed to construct the graph of input data by an efficient formula. With this formula, two transductive graph-based SSL methods are devised with the computational complexity linear in the number of input data points. Extensive experiments on synthetic data and real datasets demonstrate that the proposed methods not only are scalable for large-scale data, but also achieve good classification performance, especially for an extremely small number of labeled data.

**Keywords:** Graph structure learning · Graph-based semi-supervised learning · Large-scale learning

---

**Supplementary Information** The online version contains supplementary material available at https://doi.org/10.1007/978-3-031-26412-2_23.

# 1  Introduction

Semi-supervised learning (SSL) is an important learning paradigm for the situations where a large amount of data are easily obtained, but only a few labeled data points are available due to the laborious or expensive annotation process [4]. A variety of SSL methods have been proposed over the past decades. Among them, graph-based SSL methods attract wide attention due to their superior performance including manifold regularization [1] and label propagation [34]. However, these methods usually suffer from the high computational complexity of computing the kernel matrix or graph Laplacian matrix and the optimization problem with a large number of optimized variables. Moreover, the quality of the input graph becomes critically important for graph-based SSL methods.

Many graph structure learning methods often provide a reliable similarity matrix (or graph) to characterize the underlying structure of the input data, but their performance can be significantly affected by graphs constructed from the input data with varying density, such as LLE-type graphs [18, 26, 29, 32] and $K$-NN graphs. Moreover, they are not scalable for large-scale data by learning a full similarity matrix. Graph neural networks are also exploited to infer graph structures from input data [10, 17], but it is not easy to control the sparsity of the graph weights. Some graph construction methods can recover a full graph from a small set of variables such as anchor points of the bipartite graph in [18]. However, these methods usually neglect the importance of the similarities among the small set of variables, e.g., the similarities of anchor points are not explored.

In this paper, we aim to design a novel graph construction approach by taking into account the latent sparse graph learning for high scalability of graph-based SSL. The proposed graph construction approach learns the latent sparse graph and the assignment probabilities to construct the graph of the input data in an efficient form so that graph-based SSL methods can be scalable for large-scale data without the need of explicitly computing the graph of the input data. The main contributions of this paper are summarized as follows:

- A density-based model is proposed to simultaneously learn a sparse latent graph and assignment probabilities from the input data. We further uncover the connection of our density-based model to reversed graph embedding [19] from the perspective of density estimation.
- A novel graph construction approach is proposed to take advantage of both the latent graph and the assignment probabilities learned by the proposed density-based model. We show the spectral properties of our constructed graph via the convergence property of a matrix series. We prove that the graph construction approach used in [18] is a special case of our approach.
- We demonstrate that the graph constructed by our approach can be efficiently integrated into two variants of graph-based SSL methods. We show that both methods have linear computation complexity in the number of data points.
- Extensive experiments on synthetic data and various real data sets are conducted. Results show that our methods not only achieve competitive performance to baselines but also are more efficient for large-scale data.

## 2    Related Work

### 2.1    Graph Construction and Graph Learning

Graphs can be constructed via heuristic approach or learned from input data. Manually crafted graphs are often used. Examples include: a dense matrix from a prefixed kernel function [34], a sparse matrix from a neighborhood graph [1], a pre-constructed graph using labeled data and side information [33], or a transformed graph from an initial one using graph filtering [17]. The dense matrix is computationally impractical for large-scale data due to high storage requirement, and the neighborhood graph is less robust for data with varying density regions [8]. Graph structure learning has also shown great successes in SSL. Learning sparse graphs from input data based on locally linear embedding (LLE) [22] has been widely studied to improve graph-based SSL methods. Linear neighborhood propagation [26] learns a sparse graph via LLE, which is then used in label propagation for SSL. For large-scale data, an anchor graph is constructed by local anchor embedding (LAE) [18]. The joint learning of an LLE-type graph and SSL model has also been explored [29,32]. The graphs obtained by the above methods highly rely on LLE, so they may not work well in cases where the LLE assumption fails [5]. In addition, various other strategies are also studied. The coefficients from the low-rank representation [35] or matrix completion based on the nuclear norm [25] are used to construct a graph for SSL. Metric learning is used to learn the weights of a graph with the fixed connectivities [27]. Graph neural networks [10,17] are also used to infer graph structures from input data. These methods update weights of graphs instead of learning a sparse representation, so it is not easy to control the sparsity of the graph weights.

### 2.2    Scalability Consideration for Large-Scale Data

Various methods have been proposed to solve the scalability issue by concentrating on either learning an efficient representation of a graph or developing scalable optimization methods. Graph construction approaches have been proposed to reduce the computation cost of graph-based SSL. The Nystrom method is used to approximate the graph adjacency matrix in [23,30], but the approximated graph Laplacian matrix is not guaranteed to be positive semi-definite. Numerical approximations to the eigenvectors of the normalized graph Laplacian are used to easily propagate labels through huge collections of images [9]. As pointed out by the authors of [9], the approximations are accurate only when the solution of the label propagation algorithm [34] is a linear combination of the single-coordinate eigenfunctions. This condition can be strong in general. Anchor graph regularization (AGR) [18] constructs the graph of the input data based on a small set of anchor points. Another approach is to design fast optimization algorithms for solving graph-based SSL problems. The primal problem of the Laplacian SVM was solved by the preconditioned conjugate descent [20] for fast approximation solutions. Distributed approaches have been explored by decomposing a large-scale problem into smaller ones [3].

# 3   Latent Sparse Graph Learning for SSL

We propose a new graph construction approach built on a density-based method and latent graph learning to construct a reliable graph of input data for large-scale graph-based SSL.

## 3.1   High-Density Points Learning

Given input data $\{\mathbf{x}_i\}_{i=1}^n$ with $\mathbf{x}_i \in \mathbb{R}^d$, we seek a small number of latent points called high-density points denoted by $\{\mathbf{c}_s\}_{s=1}^k$ that can best represent the high-density regions of the input data. Our goal here is to formulate a novel objective function for learning these latent points and their relationships with input data.

To model the density of the input data, we employ kernel density estimation (KDE) [6] on $\{\mathbf{c}_s\}_{s=1}^k$ to approximate the true distribution of data by assuming that the observed data $\{\mathbf{x}_i\}_{i=1}^n$ is sampled from the true distribution. The basic idea of KDE involves smoothing each point $\mathbf{c}_s$ by a kernel function. A typical choice of the kernel function is Gaussian. The density function of $\{\mathbf{x}_i\}_{i=1}^n$ becomes

$$p(\mathbf{x}_i | \{\mathbf{c}_s\}_{s=1}^k) = \frac{(2\pi)^{-\frac{d}{2}}}{k\sigma^d} \sum_{s=1}^k \exp(-\frac{1}{2\sigma^2}||\mathbf{x}_i - \mathbf{c}_s||^2), \tag{1}$$

where $\sigma$ is the bandwidth of the Gaussian kernel function. To obtain the optimal latent points $\{\mathbf{c}_s\}_{s=1}^k$, we can do the maximum log-likelihood estimation:

$$\max_{\{\mathbf{c}_s\}_{s=1}^k} f(C) := \sum_{i=1}^n \log \sum_{s=1}^k \exp(-\frac{1}{2\sigma^2}||\mathbf{x}_i - \mathbf{c}_s||^2), \tag{2}$$

where terms independent of $C = [\mathbf{c}_1, \ldots, \mathbf{c}_k] \in \mathbb{R}^{d \times k}$ are ignored. Maximizing (2) is equivalent to finding $k$ peaks of the density function (1). Each peak governs some local high-density region of the density function comparing with other peaks. Let $\{\mathbf{c}_s^*\}_{s=1}^k$ be the optimal solution of problem (2) and denote $C^* = [\mathbf{c}_1^*, \ldots, \mathbf{c}_k^*]$. The first order optimality condition of problem (2) is

$$\frac{\partial f(C^*)}{\partial \mathbf{c}_s} = \sum_{i=1}^n Z_{i,s}(\mathbf{x}_i - \mathbf{c}_s^*) = 0 \Rightarrow \mathbf{c}_s^* = \sum_{i=1}^n \frac{Z_{i,s}}{\sum_{i=1}^n Z_{i,s}}\mathbf{x}_i, \forall s = 1, \ldots, k. \tag{3}$$

where the assignment probability of $\mathbf{x}_i$ to high-density point $\mathbf{c}_k$ is

$$Z_{i,s} = \exp(-\frac{1}{2\sigma^2}||\mathbf{x}_i - \mathbf{c}_s^*||^2) \Big/ \sum_{s=1}^k \exp(-\frac{1}{2\sigma^2}||\mathbf{x}_i - \mathbf{c}_s^*||^2), \forall i, s. \tag{4}$$

We notice that our high-density points learning approach has close relations to probabilistic c-means (PCM) [15] and Gaussian mixture model (GMM) [2]. The key difference is that our unconstrained smooth objective function (2) can facilitate the joint optimization with other objectives as shown in Subsect. 3.2.

## 3.2   Joint Learning of High-Density Points and Latent Graph

We formulate a joint optimization problem for simultaneously learning a latent graph over high-density points and the probabilities of assigning each input

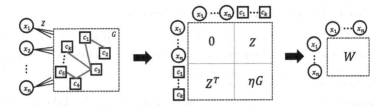

**Fig. 1.** The construction process of graph similarity matrix $W$ based on $Z$ and $G$. A larger graph by stacking the input points and the high-density points as vertexes is first constructed, and then $W$ is derived based on the random walk on the larger graph.

data point to these high-density points. The latent graph consists of the high-density points as vertexes and the similarities among these high-density points as edge weights. The latent graph learning aims to find optimal vertexes and edge weights.

We particularly concentrate on spanning trees since they are naturally connected and sparse. Let $\mathcal{T}$ be the set of all spanning trees over $\{c_i\}_{i=1}^k$, and define by $G \in \{0,1\}^{k \times k}$ the adjacency matrix for edge weights, where $G_{i,j} = 1$ means $c_i$ and $c_j$ are connected, and $G_{i,j} = 0$ otherwise. As each vertex has its associated high-density point as the node feature vector, the dissimilarity of two vertexes can be simply defined as the Euclidean distance between two high-density points.

Our goal is to find an adjacency matrix $G$ with minimum total cost from all feasible spanning trees. By combining the latent graph learning with the high-density points learning, we propose a joint optimization problem as

$$\max_{C, G \in \mathcal{T}} f(C) - \frac{\lambda_1}{4} \sum_{r=1}^k \sum_{s=1}^k G_{r,s} ||\mathbf{c}_r - \mathbf{c}_s||^2, \tag{5}$$

where $\lambda_1$ is a parameter to balance the two objectives.

Suppose $G$ is given. Similarly to (3), we have the following optimality condition

$$\left[ \sum_{i=1}^n Z_{i,1}(\mathbf{x}_i - \mathbf{c}_1), \ldots, \sum_{i=1}^n Z_{i,k}(\mathbf{x}_i - \mathbf{c}_k) \right] - \lambda_1 C L = 0, \tag{6}$$

where $L = \mathbf{diag}(G\mathbf{1}_k) - G$ is the graph Laplacian matrix over $G$. Accordingly, we have optimal $Z$ in (4) and the closed-form solution

$$C = XZ(\mathbf{diag}(Z^T\mathbf{1}_n) + \lambda_1 L)^{-1}. \tag{7}$$

Given $C$, problem (5) with respect to $G$ can be efficiently solved by Kruskal's algorithm [16]. Hence, the alternating method can be used to solve (5).

## 3.3   A Novel Graph Construction Approach

By solving (5) in Sect. 3.2, we can obtain $C$, $Z$ and $G$. It is worth noting that $G$ characterizing the relationships among high-density points is unique comparing with existing methods such as AGR. Below, we will show how $G$ can be leveraged to build a better graph of input data.

We propose to construct an $n \times n$ affinity matrix $W$ by taking advantage of both $Z$ and $G$ through the proposed process illustrated in Fig. 1. During the graph construction process, we first build a larger graph matrix of $(n+k) \times (n+k)$ with similarities formed by $Z$ and $G$, and then derive the similarity graph matrix of $n \times n$ based on random walks in order to satisfy certain criterion for SSL. Motivated by the stationary Markov random walks, we propose to construct the affinity matrix $W \in \mathbb{R}^{n \times n}$ by the following equation

$$\begin{bmatrix} W & A_1 \\ A_2 & A_3 \end{bmatrix} = P^2(I_{n+k} - \alpha P)^{-1}, \tag{8}$$

where $\alpha \in (0,1)$, and

$$P = \mathbf{diag}\left( \begin{bmatrix} \mathbf{0}_{n \times n} & Z \\ Z^T & \eta G \end{bmatrix} \mathbf{1}_{n+k} \right)^{-1} \begin{bmatrix} \mathbf{0}_{n \times n} & Z \\ Z^T & \eta G \end{bmatrix} = \begin{bmatrix} \mathbf{0}_{n \times n} & Z \\ P_{21} & P_{22} \end{bmatrix}. \tag{9}$$

$\mathbf{0}_{n \times n}$ is the $n \times n$ zero matrix, and $A_1, A_2, A_3, P_{21}$, and $P_{22}$ are sub-blocks in the partition. Here, $\eta$ is a positive parameter to balance the scale difference between $Z$ and $G$, and $Z$ is a positive matrix with $Z_{i,s} > 0$, $\forall i, s$ as defined in (4), and $G$ is a 0-1 matrix. The matrix inverse in (9) always exists. Later, we will show that $P$ is a stochastic matrix and possesses the stationary property.

Figure 2 demonstrates three key differences of our graph construction approach from LGC and AGR on the synthetic three-moon data: 1) the graph matrix $W$ over all input data is implicitly represented by both $Z$ and $G$; 2) the high-density points characterize the high-density regions of the input data, much better than the simple centroids obtained by the $k$-means method; 3) the tree structure can effectively model the relationships among these high-density points, while AGR does not have this property. Details of this experiment on synthetic data can be found in Subsect. 4.4.

Below, we conduct the theoretical analysis to justify the proposed graph construction approach. The proofs of propositions are given in the supplementary material. For convenience of analysis, we denote

$$Q = \begin{bmatrix} \mathbf{0}_{n \times n} & Z \\ Z^T & \eta G \end{bmatrix}, E = \mathbf{diag}(Z^T \mathbf{1}_n + \eta G \mathbf{1}_k), \Gamma = \mathbf{diag}(Q \mathbf{1}_{n+k}) = \begin{bmatrix} I_n & 0 \\ 0 & E \end{bmatrix}. \tag{10}$$

Then $P = \Gamma^{-1}Q$ and $P\mathbf{1}_{n+k} = \mathbf{1}_{n+k}$, satisfying the probability property over each row. We denote $M \geq 0$ if all elements in matrix $M$ are nonnegative.

Firstly, we show in Proposition 1 that the matrix $W$ is symmetric and nonnegative. This result is important since $W$ will be used as the weighted graph in Sect. 3.4 to compute a graph Laplacian for graph-based SSL.

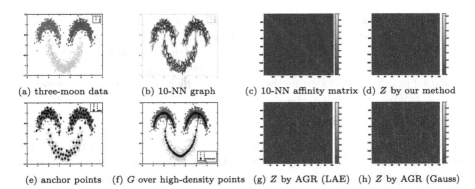

(a) three-moon data    (b) 10-NN graph    (c) 10-NN affinity matrix (d) $Z$ by our method

(e) anchor points    (f) $G$ over high-density points    (g) $Z$ by AGR (LAE)    (h) $Z$ by AGR (Gauss)

**Fig. 2.** The graph construction by three methods (LGC, AGR and our proposed method) on three-moon data. (a) the three-moon data points in 2-D space using the first two features. (b)-(c) the 10-NN graph and its affinity matrix used in LGC. (d) the $Z$ matrix obtained by our proposed method. (e) the anchor points obtained by the $k$-means method with 100 centroids. (f) the optimized high-density points and the learned tree structure. (g)-(h) the $Z$ matrices obtained by LAE and the Nadaraya-Watson kernel regression with Gaussian kernel function in AGR, respectively.

**Proposition 1.** *For $\alpha \in (0,1)$, $W$ defined in (8) is symmetric and nonnegative.*

Secondly, we would like to show that the anchor graph defined in AGR is a special case of our proposed formulation (8).

**Proposition 2.** *Suppose anchors in AGR are equal to $Z$ defined in (4). If either $\eta = 0$ or $G = 0$, and $\alpha = 0$, then $W$ defined in (8) is the same as anchor graph.*

Thirdly, we demonstrate that the matrix $W$ in (8) can be written in an explicit formula as shown in Proposition 3.

**Proposition 3.** *$W$ in (8) has an explicit formula:*

$$W = Z(I_k - \alpha\eta E^{-1}G - \alpha^2 E^{-1}Z^T Z)^{-1}E^{-1}Z^T. \tag{11}$$

Fourthly, let us consider the affinity $W$ defined in (11). Since $\widetilde{P} = \alpha\eta E^{-1}G + \alpha^2 E^{-1}Z^T Z$ has spectrum in $(-1,1)$, we have

$$(I_k - \alpha\eta E^{-1}G - \alpha^2 E^{-1}Z^T Z)^{-1} = \sum_{t=0}^{\infty} \widetilde{P}^t. \tag{12}$$

The series (12) motivate us to take the second-order approximation to the exact $W$ in (11) for cheaper computation. If we only keep the first two terms, i.e., $t = 0$ and $t = 1$, then we have an approximation of $W$ in (11) as

$$\widetilde{W} = ZE^{-1}Z^T + \alpha\eta ZE^{-1}GE^{-1}Z^T + \alpha^2 ZE^{-1}Z^T ZE^{-1}Z^T. \tag{13}$$

Obviously, matrix $\widetilde{W}$ is symmetric and nonnegative.

Finally, we can verify that much less storage requirement is needed to represent the full graph. Rather than storing the $n \times n$ graph matrices $W$ and $\widetilde{W}$, we only need to store the $n \times k$ matrix $Z$, $k \times k$ diagonal matrix $E$ and $k \times k$ matrix $Z^T Z$. We can easily use $Z, E, Z^T Z$ to compute (11) and (13). Hence, our proposed graph construction methods are very efficient for large $n$ but small $k$.

## 3.4   Graph-Based SSL

We apply $W$ derived in Sect. 3.3 as the learned graph of input data to two types of SSL methods: LGC-based approach and AGR-based approach. We will show that both approaches can become much more computationally efficient by using the proposed formulas (11) and (13) for large-scale data sets.

Let $F = [F_l; F_u] \in \mathbb{R}^{n \times c}$ be the label matrix of one hot representation $Y = [Y_l; Y_u]$ of class labels that maps $n$ sample data points in $X = [\mathbf{x}_1, \ldots, \mathbf{x}_n] \in \mathbb{R}^{d \times n}$ to $c$ labels, where $F_l$ is the submatrix corresponding to the $l$ samples with known labels and $F_u$ corresponds to $n - l$ unlabeled samples. We would like to infer $F_u$ by label propagation. Specifically, denote by $\mathbf{L}$ the graph Laplacian operator of $W$, i.e., $\mathbf{L}(W) = \mathbf{diag}(W \mathbf{1}_n) - W$, where $W$ is either (11) or (13).

**LGC-Based Approach.** By following the objective function of LGC [34], given a graph matrix $W$, we obtain $F_u$ by solving the following optimization problem:

$$\min_{F_u \in \mathbb{R}^{(n-l) \times c}} \mathbf{trace}(F^T \mathbf{L}(W) F) + \frac{\lambda_2}{2} \|F_u - Y_u\|_{\text{fro}}^2, \tag{14}$$

where $\lambda_2 > 0$ is a regularization parameter. The optimal $F_u$ is obtained by solving $c$ linear systems of equations:

$$(2L_3(W) + \lambda_2 I_{n-l}) F_u^i = b^i, \quad \forall i = 1, \cdots, c. \tag{15}$$

where $F_u = [F_u^1, \cdots, F_u^c]$, $\lambda_2 Y_u - 2L_2^T(W) F_l = [b^1, \cdots, b^c]$, and $L_2(W) \in \mathbb{R}^{l \times (n-l)}, L_3(W) \in \mathbb{R}^{(n-l) \times (n-l)}$ are sub-blocks of $\mathbf{L}(W)$. Taking the special form of $W$, (15) can be solved efficiently by conjugate gradient method [21].

**AGR-Based Approach.** Given a graph matrix $W$, we also study the label propagation model used in AGR [18] by learning a linear decision function, denoted by $F = ZA$, where $A \in \mathbb{R}^{k \times c}$ represents the linear coefficients for $c$ classes. Let $Z = [Z_l; Z_u]$. We solve the following optimization problem

$$\min_A \mathbf{trace}(A^T Z^T \mathbf{L}(W) Z A) + \frac{\lambda_2}{2} \|ZA - Y\|_{\text{fro}}^2. \tag{16}$$

The optimal $A^*$ has a closed-form expression

$$A^* = \lambda_2 (2Z^T \mathbf{L}(W) Z + \lambda_2 Z^T Z)^{-1} (Z^T Y). \tag{17}$$

Note that $Z^T \mathbf{L}(W) Z$ can be computed very efficiently. As in (17), the inverse is on a $k \times k$ matrix, so we simply calculate the matrix inversion since $k$ is small.

---

**Algorithm 1:** High-density graph learning (HiDeGL)

---

**Input:** $X, F_l = Y_l, k, \sigma, \lambda_1, \lambda_2, \alpha \in (0,1), \eta$
**Output:** $F, Z, C, G$

1  Initialization: $Y_u = 0$, $k$-means for $C$, $Z$ by (4)
2  **while** *not convergent* **do**
3  $\quad$ Solve $G$ using the minimal spanning tree algorithm;
4  $\quad$ $L = \mathbf{diag}(G\mathbf{1}_k) - G$, $\Xi = \mathbf{diag}(Z^T\mathbf{1}_n)$;
5  $\quad$ $C \leftarrow XZ(\Xi + \lambda_1 L)^{-1}$;
6  $\quad$ $Z_{i,s} \leftarrow \dfrac{\exp\left(-\|\mathbf{x}_i - \mathbf{c}_s\|^2/\sigma\right)}{\sum_{s=1}^{k}\exp\left(-\|\mathbf{x}_i - \mathbf{c}_s\|^2/\sigma\right)}, \forall i = 1, ..., n, s = 1, ..., k.$
7  Construct the graph $W$ using either (11) or (13)
8  Update $F_u$ by solving (15) using CG or (16) with $A^*$ in (17).
9  $y_i = \arg\max\limits_{j\in\{1,\cdots,c\}} \{(F_u)_{i,j}\}, \forall i = l+1, \ldots, n.$

---

### 3.5 Optimization Algorithm and Complexity Analysis

The proposed method is summarized in Algorithm 1 with two graph construction approaches and two inference approaches for the unlabeled data. The complexity of Algorithm 1 is determined by two individual subproblems. First, the complexity of finding the high-density points and the tree structure has the complexities of the following three components: 1) the complexity of Kruskal's algorithm requires $O(k^2d)$ for computing the fully connected graph and $O(k^2 \log k)$ for finding the spanning tree $G$; 2) computing the soft-assignment matrix $Z$ requires $O(nkd)$; 3) computing the inverse of a $k$ by $k$ matrix $(\Xi + \lambda_1 L)^{-1}$ requires $O(k^3)$ and doing the matrix multiplication to get $C$ takes $O(nkd + dk^2)$ flops. Therefore, the total complexity of each iteration is $O(k^3 + nkd + dk^2)$. The second subproblem is the inference of the unlabeled data. The computation complexity of each CG iteration requires $O(nk + k^2)$ for (11) and $O(nk + k^2 + k^3)$ for (13). The complexity of computing (17) needs $O(k^3 + nk(k+c))$. Hence, the complexity of Algorithm 1 is linear with $n$, no matter which inference method is used.

## 4 Experiments

### 4.1 Data Sets

One simulated three-moon data set is generated as follows: 500 points in two-dimensional space are first randomly generated on a lower half circle centered at (1.5,0.4) with radius 1.5; and then, another 500 points in two-dimensional space are randomly generated on two upper half unit circles centered at (0,0) and (3,0), respectively; finally, the 1500 points in total are expanded to have dimension 100 by filling up the bottom 98 entries of each point with noise following normal distribution with mean 0 and standard deviation 0.14 to each of the 100 dimensions. USPS-2 is popularly used as benchmark for evaluating the performance of SSL

**Table 1.** Average accuracies with standard deviations of nine methods over 10 randomly drawn labeled data with varied sizes on three-moon data. Best results are in bold.

| Method | $l=3$ | $l=10$ | $l=25$ | $l=50$ | $l=100$ |
|---|---|---|---|---|---|
| LGC | $94.19 \pm 6.69$ | $98.96 \pm 0.49$ | $99.02 \pm 0.30$ | $99.23 \pm 0.13$ | $99.40 \pm 0.12$ |
| TVRF(1) | $90.49 \pm 4.80$ | $97.48 \pm 1.15$ | $99.53 \pm 0.03$ | $99.52 \pm 0.05$ | $99.56 \pm 0.06$ |
| TVRF(2) | $99.52 \pm 0.07$ | $99.47 \pm 0.09$ | $99.46 \pm 0.11$ | $99.53 \pm 0.03$ | $99.56 \pm 0.06$ |
| AGR(Gauss) | $99.36 \pm 0.32$ | $99.46 \pm 0.20$ | $99.51 \pm 0.25$ | $99.65 \pm 0.08$ | $99.61 \pm 0.17$ |
| AGR(LAE) | $97.74 \pm 1.41$ | $98.68 \pm 0.31$ | $98.66 \pm 0.39$ | $98.83 \pm 0.29$ | $98.76 \pm 0.30$ |
| GCN | $94.58 \pm 4.58$ | $96.54 \pm 3.10$ | $98.60 \pm 0.15$ | $98.67 \pm 0.22$ | $98.74 \pm 0.19$ |
| IGCN(RNM) | $98.28 \pm 0.32$ | $98.99 \pm 0.09$ | $99.13 \pm 0.07$ | $99.15 \pm 0.08$ | $99.19 \pm 0.13$ |
| IGCN(AR) | $98.30 \pm 0.29$ | $99.01 \pm 0.09$ | $99.17 \pm 0.08$ | $99.17 \pm 0.10$ | $99.22 \pm 0.13$ |
| GLP(RNM) | $97.74 \pm 0.84$ | $98.19 \pm 0.36$ | $98.58 \pm 0.22$ | $98.68 \pm 0.13$ | $98.61 \pm 0.09$ |
| GLP(AR) | $95.04 \pm 4.51$ | $97.49 \pm 1.35$ | $98.28 \pm 0.26$ | $98.19 \pm 0.27$ | $98.22 \pm 0.20$ |
| KernelLP | $89.63 \pm 2.67$ | $92.90 \pm 4.22$ | $97.33 \pm 1.67$ | $97.89 \pm 1.19$ | $98.27 \pm 0.86$ |
| SSLRR | $87.28 \pm 4.00$ | $88.47 \pm 4.39$ | $96.81 \pm 0.29$ | $96.81 \pm 0.29$ | $97.04 \pm 0.25$ |
| HiDeGL(L-approx) | $99.85 \pm 0.06$ | $\mathbf{99.86 \pm 0.07}$ | $\mathbf{99.88 \pm 0.06}$ | $\mathbf{99.88 \pm 0.06}$ | $\mathbf{99.90 \pm 0.05}$ |
| HiDeGL(L-accurate) | $99.85 \pm 0.05$ | $99.85 \pm 0.06$ | $\mathbf{99.88 \pm 0.05}$ | $\mathbf{99.88 \pm 0.05}$ | $\mathbf{99.90 \pm 0.05}$ |
| HiDeGL(A-approx) | $\mathbf{99.87 \pm 0.05}$ | $\mathbf{99.86 \pm 0.07}$ | $\mathbf{99.88 \pm 0.05}$ | $\mathbf{99.88 \pm 0.06}$ | $99.89 \pm 0.06$ |
| HiDeGL(A-accurate) | $99.85 \pm 0.09$ | $\mathbf{99.86 \pm 0.06}$ | $99.87 \pm 0.05$ | $\mathbf{99.88 \pm 0.05}$ | $\mathbf{99.90 \pm 0.05}$ |

methods. COIL20 is used to show the performance of multi-class classification with a large number of classes. To demonstrate the capability of HiDeGL for medium-size data, we conduct experiments on Pendigits and MNIST. EMNIST-Digits [7] and Extended MNIST [13] are used for large-scale evaluation. The statistics of real datasets are respectively shown in Tables 2–4.

### 4.2    Compared Methods

For ease of references, we name our four proposed methods including the LGC-based approach with (11) and (13) as HiDeGL(L-accurate) and HiDeGL(L-approx), respectively, and the AGR-based approach with (11) and (13) as HiDeGL(A-accurate) and HiDeGL(A-approx), respectively. The comparing methods include LGC [34], AGR with Gaussian kernel regression (AGR-Gauss) and AGR with LAE (AGR-LAE) [18], $K$-NN classifier ($K$-NN), spectral graph transduction (SGT) [12], Laplacian regularized least squares (LapRLS) [1], $\mathcal{P}_{SQ}$ solved using SQ-Loss-1 [24], measure propagation (MP) [24]. TVRF with one edge (TVRF(1)) or two edges (TVRF(2)) [28], GCN [14], GLP and IGCN [17], KernelLP [31] and SSLRR [35]. Some methods cannot work for medium-large-size data sets such as KernelLP and SSLRR due to their high computational complexity, so their results on data sets with $n > 9000$ will not be reported.

### 4.3    Experimental Setting

Our experiments follow the work in [24]. For most graph-based SSL methods, there are some hyperparameters to tune. As the labeled data is very

**Table 2.** Average accuracies with standard deviations of compared methods over 10 randomly drawn labeled data with varied sizes on two datasets. Best results are in bold.

| Method | $l = 10$ | $l = 50$ | $l = 100$ | $l = 150$ |
|---|---|---|---|---|
| USPS-2 ($n = 1500, c = 2, d = 24$) | | | | |
| k-NN | 80.0 | 90.7 | 93.6 | 94.9 |
| SGT | 86.2 | 94.0 | 96.0 | **97.0** |
| LapRLS | 83.9 | 93.7 | 95.4 | 95.9 |
| SQ-Loss-I | 81.4 | 93.6 | 95.2 | 95.2 |
| MP | 88.1 | 93.9 | 96.2 | 96.8 |
| LGC | 85.21 ± 5.54 | 92.94 ± 3.36 | 95.94 ± 0.63 | 96.73 ± 0.28 |
| TVRF(1) | 82.00 ± 7.47 | 88.11 ± 2.85 | 92.47 ± 3.04 | 94.25 ± 1.80 |
| TVRF(2) | 73.66 ± 8.15 | 87.45 ± 4.19 | 92.86 ± 1.67 | 94.67 ± 1.05 |
| AGR(Gauss) | 75.01 ± 6.55 | 88.88 ± 2.65 | 91.92 ± 1.86 | 93.04 ± 1.04 |
| AGR(LAE) | 74.02 ± 8.60 | 88.01 ± 2.15 | 91.44 ± 1.39 | 92.33 ± 1.01 |
| GCN | 69.52 ± 10.97 | 88.01 ± 4.31 | 92.74 ± 2.32 | 94.59 ± 1.70 |
| IGCN(RNM) | 68.05 ± 10.06 | 88.96 ± 4.28 | 93.21 ± 1.77 | 94.35 ± 1.76 |
| IGCN_AR | 68.26 ± 9.61 | 88.17 ± 3.99 | 91.62 ± 1.68 | 94.22 ± 1.38 |
| GLP(RNM) | 71.78 ± 9.78 | 87.10 ± 5.98 | 91.36 ± 3.09 | 93.61 ± 2.67 |
| GLP(AR) | 69.65 ± 10.01 | 86.35 ± 5.98 | 90.45 ± 1.66 | 93.32 ± 1.82 |
| KernelLP | 72.22 ± 6.81 | 88.77 ± 2.50 | 92.66 ± 1.49 | 93.97 ± 1.09 |
| SSLRR | 64.36 ± 4.49 | 67.78 ± 2.25 | 68.59 ± 3.82 | 68.77 ± 3.58 |
| HiDeGL(L-approx) | 90.01 ± 3.94 | **95.88 ± 0.50** | **96.23 ± 0.43** | 96.77 ± 0.39 |
| HiDeGL(L-accurate) | 89.41 ± 1.64 | **95.88 ± 0.50** | **96.36 ± 0.71** | 96.95 ± 0.25 |
| HiDeGL(A-approx) | 91.93 ± 3.69 | 95.30 ± 0.79 | 95.68 ± 0.81 | 96.16 ± 0.53 |
| HiDeGL(A-accurate) | **91.94 ± 3.68** | 95.30 ± 0.79 | 95.68 ± 0.81 | 96.16 ± 0.53 |
| Method | $l = 40$ | $l = 80$ | $l = 100$ | $l = 160$ |
| COIL20 ($n = 1440, c = 20, d = 1024$) | | | | |
| LGC | 87.39 ± 1.43 | 90.88 ± 1.53 | 93.43 ± 1.22 | 95.66 ± 1.13 |
| TVRF(1) | 89.31 ± 2.13 | 92.65 ± 0.92 | 94.24 ± 1.47 | 95.20 ± 1.06 |
| TVRF(2) | 87.19 ± 2.23 | 90.32 ± 2.33 | 92.42 ± 1.44 | 95.04 ± 0.74 |
| AGR(Gauss) | 84.16 ± 3.55 | 93.81 ± 2.20 | 94.09 ± 1.84 | 95.70 ± 1.51 |
| AGR(LAE) | 89.55 ± 3.22 | 97.19 ± 1.67 | 96.91 ± 1.73 | 98.24 ± 0.78 |
| GCN | 72.42 ± 2.16 | 79.11 ± 2.04 | 82.14 ± 1.35 | 87.37 ± 1.41 |
| IGCN(RNM) | 74.44 ± 2.65 | 80.93 ± 1.97 | 83.12 ± 1.55 | 88.18 ± 0.79 |
| IGCN(AR) | 75.75 ± 1.73 | 81.14 ± 2.27 | 84.09 ± 1.43 | 88.88 ± 0.88 |
| GLP(RNM) | 73.81 ± 2.19 | 80.26 ± 1.94 | 82.77 ± 1.56 | 87.68 ± 1.23 |
| GLP(AR) | 76.18 ± 1.80 | 80.96 ± 1.89 | 83.24 ± 1.18 | 88.06 ± 1.14 |
| KernelLP | 71.76 ± 2.70 | 80.60 ± 1.60 | 83.19 ± 1.10 | 87.82 ± 1.47 |
| SSLRR | 62.96 ± 1.61 | 64.78 ± 2.25 | 65.66 ± 2.54 | 68.21 ± 2.90 |
| HiDeGL(L-approx) | 92.95 ± 1.55 | 96.23 ± 0.88 | 96.37 ± 1.38 | 97.16 ± 1.70 |
| HiDeGL(L-accurate) | 91.20 ± 1.65 | 95.45 ± 1.30 | 96.37 ± 1.41 | 97.45 ± 0.77 |
| HiDeGL(A-approx) | **96.75 ± 1.51** | **97.88 ± 0.44** | **98.16 ± 0.94** | 98.58 ± 0.73 |
| HiDeGL(A-accurate) | 96.74 ± 1.43 | 98.04 ± 0.98 | 98.09 ± 0.74 | **98.66 ± 0.52** |

**Table 3.** Average accuracies with standard deviations of compared methods over 10 randomly drawn labeled data with varied sizes on two datasets. Best results are in bold.

| Method | $l = 10$ | $l = 50$ | $l = 100$ | $l = 150$ |
|---|---|---|---|---|
| MNIST ($n = 70000, c = 10, d = 784$) | | | | |
| LGC | 66.66 ± 5.52 | 83.76 ± 2.33 | 87.84 ± 1.11 | 89.41 ± 0.88 |
| TVRF(1) | 53.44 ± 6.73 | 74.35 ± 1.64 | 78.50 ± 1.70 | 81.27 ± 1.38 |
| TVRF(2) | 61.73 ± 6.12 | 78.05 ± 2.58 | 84.70 ± 1.20 | 86.19 ± 0.95 |
| AGR (Gauss) | 51.97 ± 4.15 | 76.05 ± 4.37 | 79.26 ± 0.68 | 80.32 ± 1.41 |
| AGR (LAE) | 52.29 ± 3.92 | 76.97 ± 4.37 | 80.33 ± 0.93 | 81.30 ± 1.45 |
| GCN | 31.97 ± 6.63 | 57.59 ± 3.44 | 64.97 ± 2.21 | 69.09 ± 1.77 |
| IGCN(RNM) | 42.93 ± 5.53 | 64.67 ± 4.82 | 76.64 ± 2.39 | 81.06 ± 2.71 |
| IGCN(AR) | 42.26 ± 6.69 | 68.73 ± 2.81 | 79.66 ± 1.35 | 83.60 ± 1.90 |
| GLP(RNM) | 44.59 ± 4.49 | 69.30 ± 4.11 | 79.21 ± 4.26 | 83.04 ± 4.04 |
| GLP(AR) | 46.60 ± 5.14 | 70.79 ± 4.35 | 79.59 ± 5.13 | 83.27 ± 4.71 |
| HiDeGL(L-approx) | 83.38 ± 4.37 | 88.23 ± 1.87 | **90.36 ± 1.33** | 91.51 ± 0.78 |
| HiDeGL(L-accurate) | 83.38 ± 4.37 | 88.23 ± 1.87 | **90.36 ± 1.33** | 91.51 ± 0.78 |
| HiDeGL(A-approx) | **83.59 ± 4.19** | 88.22 ± 2.00 | 90.14 ± 1.15 | **91.28 ± 0.84** |
| HiDeGL(A-accurate) | **83.59 ± 4.19** | 90.73 ± 1.46 | 90.14 ± 1.15 | **91.28 ± 0.84** |
| Pendigits ($n = 10992, c = 10, d = 16$) | | | | |
| LGC | 80.97 ± 7.41 | 93.21 ± 1.99 | 94.44 ± 1.39 | 95.89 ± 1.02 |
| TVRF(1) | 43.57 ± 4.20 | 59.52 ± 2.11 | 66.23 ± 2.57 | 74.69 ± 1.76 |
| TVRF(2) | 52.50 ± 4.05 | 83.39 ± 2.86 | 89.54 ± 2.80 | 92.99 ± 1.62 |
| AGR(Gauss) | 52.56 ± 6.85 | 91.73 ± 1.95 | 95.01 ± 1.03 | 96.43 ± 0.85 |
| AGR(LAE) | 52.52 ± 6.67 | 91.60 ± 1.88 | 94.59 ± 1.24 | 96.18 ± 1.21 |
| GCN | 64.87 ± 5.56 | 83.90 ± 2.01 | 90.10 ± 1.66 | 92.72 ± 1.18 |
| IGCN(RNM) | 66.74 ± 4.33 | 83.19 ± 2.01 | 90.74 ± 1.18 | 94.00 ± 1.44 |
| IGCN(AR) | 71.90 ± 5.83 | 85.48 ± 2.41 | 91.41 ± 0.98 | 94.16 ± 1.34 |
| GLP(RNM) | 67.73 ± 5.80 | 84.46 ± 2.38 | 89.46 ± 1.45 | 92.25 ± 0.98 |
| GLP(AR) | 67.99 ± 3.63 | 85.74 ± 2.22 | 89.58 ± 1.70 | 92.33 ± 1.14 |
| HiDeGL(L-approx) | 85.26 ± 4.09 | 93.36 ± 1.80 | 95.54 ± 1.00 | **96.44 ± 1.06** |
| HiDeGL(L-accurate) | **85.72 ± 4.08** | 93.24 ± 1.77 | **95.56 ± 0.91** | 96.36 ± 1.13 |
| HiDeGL(A-approx) | 85.37 ± 4.61 | **93.67 ± 2.00** | 95.44 ± 1.72 | 96.13 ± 0.86 |
| HiDeGL(A-accurate) | 85.37 ± 4.61 | **93.67 ± 2.00** | 95.44 ± 1.74 | 96.14 ± 0.87 |

limited in our experimental setting, the commonly used cross-validation approach is not applicable [4]. To alleviate the difficulty of tuning hyperparameters, we choose to tune all hyperparameters in terms of the mean accuracies over the 10 random experiments for fair comparisons. For the two benchmark datasets, we directly take the results from [24] under the same setting of the number of labeled data $l \in \{10, 50, 100, 150\}$. In the experiments, we tune the parameters $k \in \{200, 500, 750, 1500\}$, $\sigma \in [0.01, 0.5]$, $\lambda_1 \in \{0.1, 1, 10, 100\}$, $\lambda_2 \in \{0.001, 0.01, 0.05\}$, $\eta \in \{0.01, 0.1, 1\}$ and $\alpha \in [0.1, 0.9]$. All these parameters are tuned based on the mean accuracies over the 10 random experiments. The mean accuracies and their standard deviations are reported.

## 4.4    Experiments on Synthetic Data

In Fig. 2, we demonstrate the neighborhood graph structure with neighbor size equal to 10 and its affinity matrix used in LGC, anchors in AGR with two approaches (Gauss and LAE) for obtaining $Z$, and our proposed graph construction approach by optimizing high-density points and a tree structure. In comparing Fig. 2(e) with Fig. 2(f), we highlight the key differences between anchor points and high-density points: 1) high-density points locate in the high-density regions, so they are different from cluster centroids by the $k$-means method; 2) the additional tree structure shown in Fig. 2(f) is the unique feature compared to the existing methods. We notice that the matrices $Z$ obtained by AGR and HiDeGL are quite similar (see Fig. 2 (d), (g) and (h)). Hence, both methods are able to capture the relations between input data and latent points (anchor points in AGR and high-density points in HiDeGL).

Table 1 shows the average accuracies with standard deviations over 10 randomly drawn labeled data obtained by the compared methods in terms of the varying number of labeled data. From Table 1, we have the following observations: 1) our proposed HiDeGL outperforms other methods over all varying number of labels; 2) With small numbers of labels such as $l \in \{3, 10\}$, HiDeGL performs significantly better than others; 3) the four variants of HiDeGL with two graph construction approaches and two inference approaches for unlabeled data achieve almost similar accuracies. These observations imply that our proposed methods are effective for SSL, especially with very small amount of labeled data.

## 4.5    Experiments on Real Data of Varied Sizes

We evaluate four variants of our HiDeGL on varying sizes of datasets by comparing with baseline methods in terms of varying number of labeled data. The experimental setting same on synthetic data is applied. The average accuracies and their standard deviations are shown in Table 2, Table 3, and Table 4, for varied data sizes. Over all sizes of tested datasets, HiDeGL gives the best accuracy than other methods when the number of labeled data points is small. On benchmark datasets, SGT is the best for $l = 150$ on USPS-2. For medium-size data, the similar results can be observed for a small number of labeled data. With a large number of labeled data, HiDeGL also shows better performance than others. For two large-scale data sets, EMNIST-Digits and Extended MINIST, HiDeGL significantly outperforms AGR over all testing cases. These observations imply that our constructed graphs are effective for SSL.

We further show in Table 4 the CPU time of HiDeGL compared with AGR on EMNIST-Digits and Extended MINIST as the number of labeled data points varies and $k = 500$. It is clear that 1) AGR(Gauss) is the fastest method, while its performance in accuracy is the worst; 2) AGR(LAE) is the slowest since solving LAE for each point is time consuming; 3) HiDeGL with all four variants shows the similar CPU time but 10 times faster than AGR(LAE), and also demonstrates the best performance over all varying numbers of labeled data points.

**Table 4.** Average accuracies with standard deviations and CPU time of compared methods over 10 randomly drawn labeled data on EMNIST-digits and Extended MNIST in terms of varying number of labeled data points. Best results are in bold.

| Method | $l = 10$ | $l = 50$ | $l = 100$ | $l = 150$ |
|---|---|---|---|---|
| Accuracy on EMNIST-digits ($k = 500, n = 280000, c = 10, d = 784$) | | | | |
| AGR(Gauss) | $77.34 \pm 4.92$ | $86.46 \pm 1.44$ | $88.89 \pm 1.17$ | $90.07 \pm 0.79$ |
| AGR(LAE) | $77.93 \pm 5.22$ | $87.17 \pm 1.94$ | $89.43 \pm 1.22$ | $90.60 \pm 0.85$ |
| HiDeGL(L-approx) | $79.55 \pm 6.38$ | $89.34 \pm 1.34$ | $\mathbf{91.46 \pm 0.90}$ | $\mathbf{91.85 \pm 0.92}$ |
| HiDeGL(L-accurate) | $79.63 \pm 6.40$ | $89.36 \pm 1.36$ | $\mathbf{91.46 \pm 0.90}$ | $\mathbf{91.85 \pm 0.92}$ |
| HiDeGL(A-approx) | $\mathbf{79.84 \pm 6.45}$ | $\mathbf{89.55 \pm 1.52}$ | $91.46 \pm 0.89$ | $91.83 \pm 0.98$ |
| HiDeGL(A-accurate) | $\mathbf{79.86 \pm 6.52}$ | $\mathbf{89.55 \pm 1.52}$ | $91.46 \pm 0.89$ | $91.83 \pm 0.98$ |
| CPU Time on EMNIST-digits (in seconds) | | | | |
| AGR(Gauss) | $6.59 \pm 0.71$ | $6.45 \pm 0.28$ | $6.55 \pm 0.42$ | $6.56 \pm 0.24$ |
| AGR(LAE) | $8886 \pm 18.5$ | $8634 \pm 14.1$ | $8641 \pm 7.9$ | $8607 \pm 14.4$ |
| HiDeGL(L-approx) | $414.6 \pm 2.5$ | $414.4 \pm 3.9$ | $410.5 \pm 2.7$ | $415.1 \pm 3.2$ |
| HiDeGL(L-accurate) | $635.5 \pm 14.8$ | $647.7 \pm 11.5$ | $629.7 \pm 3.0$ | $648.7 \pm 6.9$ |
| HiDeGL(A-approx) | $403.8 \pm 2.1$ | $403.6 \pm 1.8$ | $403.7 \pm 1.4$ | $403.0 \pm 1.0$ |
| HiDeGL(A-accurate) | $402.9 \pm 1.9$ | $401.4 \pm 1.4$ | $401.6 \pm 1.5$ | $402.5 \pm 2.3$ |
| Accuracy on Extended MNIST ($k = 500, n = 630000, c = 10, d = 784$) | | | | |
| AGR (Gauss) | $64.59 \pm 7.36$ | $76.79 \pm 1.26$ | $79.88 \pm 0.80$ | $82.11 \pm 0.78$ |
| AGR (LAE) | $66.27 \pm 7.27$ | $78.72 \pm 1.44$ | $80.97 \pm 0.75$ | $83.13 \pm 0.65$ |
| HiDeGL (L-approx) | $68.10 \pm 7.81$ | $\mathbf{80.97 \pm 1.42}$ | $\mathbf{82.55 \pm 1.24}$ | $83.99 \pm 1.08$ |
| HiDeGL (L-accurate) | $68.13 \pm 7.80$ | $80.96 \pm 1.48$ | $\mathbf{82.55 \pm 1.24}$ | $\mathbf{84.00 \pm 1.08}$ |
| HiDeGL (A-approx) | $\mathbf{68.14 \pm 8.33}$ | $79.40 \pm 1.45$ | $81.41 \pm 1.02$ | $83.25 \pm 1.06$ |
| HiDeGL (A-accurate) | $\mathbf{68.14 \pm 8.33}$ | $79.40 \pm 1.45$ | $81.42 \pm 1.02$ | $83.25 \pm 1.06$ |
| CPU time on Extended MNIST (in seconds) | | | | |
| AGR (Gauss) | $13.81 \pm 1.50$ | $13.94 \pm 1.63$ | $14.49 \pm 1.84$ | $14.20 \pm 1.62$ |
| AGR (LAE) | $12120 \pm 3281$ | $12134 \pm 3331$ | $12242 \pm 3307$ | $12183 \pm 3413$ |
| HiDeGL (L-approx) | $1074 \pm 11.6$ | $1078 \pm 9.1$ | $1086 \pm 9.3$ | $1083 \pm 17.1$ |
| HiDeGL (L-accurate) | $1123 \pm 4.7$ | $1146 \pm 11.9$ | $1169 \pm 7.4$ | $1097 \pm 65.7$ |
| HiDeGL (A-approx) | $1049 \pm 15.4$ | $1050 \pm 10.8$ | $1055 \pm 11.8$ | $1059 \pm 7.5$ |
| HiDeGL (A-accurate) | $1047 \pm 10.0$ | $1055 \pm 6.1$ | $1052 \pm 7.8$ | $1053 \pm 8.3$ |

### 4.6   Parameter Sensitivity Analysis

We conduct the parameter sensitivity analysis of HiDeGL(L-accurate) as an illustrating example in terms of different amount of labeled data. Specifically, we report the best accuracy for the parameter over results obtained by tuning the others using $k = 500$. Figure 3 shows the accuracies of HiDeGL(L-accurate) by varying parameters. First, we notice that our method is quite robust with respect to $\lambda_1$ and $\eta$. Second, $\sigma$ and $\alpha$ can have large impact on the classification performance. It is clear to see that the accuracy changes as $\sigma$ varies more smoothly with a peak in $[0.05, 0.1]$. Third, the classification accuracies improve as $l$ increases. However, parameters are robust to $l$ due to similar trends.

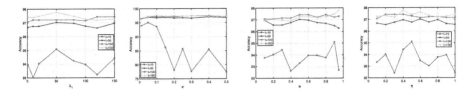

**Fig. 3.** Parameter sensitivity analysis of HiDeGL(L-accurate) on USPS-2 by varying the corresponding parameters $\lambda_1, \sigma, \alpha, \eta$ respectively with $k = 500$ and $\lambda_2 \in \{10^{-3}, 10^{-2}\}$ in terms of the number of labeled data points $l \in \{10, 50, 100, 150\}$.

## 5 Conclusion

We proposed a novel graph construction approach for graph-based SSL methods by learning a set of high-density points, the assignment of each input data point to these high-density points, and the relationships over these high-density points as represented by a spanning tree. Our theoretical results showed various useful properties about the constructed graphs, and that AGR is a special case of our approach. Our experimental results showed that our methods not only achieved competitive performance to baseline methods but also were more efficient for large-scale data. More importantly, we found that our methods outperformed all baseline methods on the datasets with extremely small amount of labeled data.

**Acknowledgements.** L. Wang was supported in part by NSF DMS-2009689. R. Chan was supported in part by HKRGC GRF Grants CUHK14301718, CityU11301120, and CRF Grant C1013-21GF. T. Zeng was supported in part by the National Key R&D Program of China under Grant 2021YFE0203700.

## References

1. Belkin, M., Niyogi, P., Sindhwani, V.: Manifold regularization: a geometric framework for learning from labeled and unlabeled examples. JMLR **7**, 2399–2434 (2006)
2. Bishop, C.M.: Pattern Recognition and Machine Learning, 1st edn. Springer, New York (2006)
3. Chang, X., Lin, S.B., Zhou, D.X.: Distributed semi-supervised learning with kernel ridge regression. JMLR **18**(1), 1493–1514 (2017)
4. Chapelle, O., Schölkopf, B., Zien, A. (eds.): Semi-Supervised Learning. MIT Press, Cambridge, MA (2006)
5. Chen, J., Liu, Y.: Locally linear embedding: a survey. Artif. Intell. Rev. **36**(1), 29–48 (2011)
6. Chen, Y.C.: A tutorial on kernel density estimation and recent advances. Biostat. Epidemiol. **1**(1), 161–187 (2017)
7. Cohen, G., Afshar, S., Tapson, J., van Schaik, A.: EMNIST: an extension of MNIST to handwritten letters. arXiv preprint arXiv:1702.05373 (2017)
8. Elhamifar, E., Vidal, R.: Sparse manifold clustering and embedding. In: NIPS, pp. 55–63 (2011)
9. Fergus, R., Weiss, Y., Torralba, A.: Semi-supervised learning in gigantic image collections. In: NIPS, pp. 522–530 (2009)

10. Franceschi, L., Niepert, M., Pontil, M., He, X.: Learning discrete structures for graph neural networks. In: ICML (2019)
11. Geršgorin, S.: Über die abgrenzung der eigenwerte einer matrix. Izv. Akad. Nauk SSSR Ser. Mat $\mathbf{1}$(7), 749–755 (1931)
12. Joachims, T.: Transductive learning via spectral graph partitioning. In: ICML, pp. 290–297 (2003)
13. Karlen, M., Weston, J., Erkan, A., Collobert, R.: Large scale manifold transduction. In: ICML, pp. 448–455 (2008)
14. Kipf, T.N., Welling, M.: Semi-supervised classification with graph convolutional networks. In: ICLR (2017)
15. Krishnapuram, R., Keller, J.M.: The possibilistic c-means algorithm: insights and recommendations. IEEE Trans. Fuzzy Syst. $\mathbf{4}$(3), 385–393 (1996)
16. Kruskal, J.B.: On the shortest spanning subtree of a graph and the traveling salesman problem. Proc. Am. Math. Soc. $\mathbf{7}$(1), 48–50 (1956)
17. Li, Q., Wu, X.M., Liu, H., Zhang, X., Guan, Z.: Label efficient semi-supervised learning via graph filtering. In: CVPR, pp. 9582–9591 (2019)
18. Liu, W., He, J., Chang, S.F.: Large graph construction for scalable semi-supervised learning. In: ICML, pp. 679–686 (2010)
19. Mao, Q., Wang, L., Tsang, I.W.: Principal graph and structure learning based on reversed graph embedding. IEEE TPAMI $\mathbf{39}$(11), 2227–2241 (2016)
20. Melacci, S., Belkin, M.: Laplacian support vector machines trained in the primal. JMLR $\mathbf{12}$(Mar), 1149–1184 (2011)
21. Nocedal, J., Wright, S.: Numerical optimization. Springer Series in Operations Research and Financial Engineering, 2 edn. Springer, New York (2006). https://doi.org/10.1007/978-0-387-40065-5
22. Roweis, S.T., Saul, L.K.: Nonlinear dimensionality reduction by locally linear embedding. Science $\mathbf{290}$(5500), 2323–2326 (2000)
23. Sivananthan, S., et al.: Manifold regularization based on nyström type subsampling. Appl. Comput. Harmonic Anal. $\mathbf{44}$, 1–200 (2018)
24. Subramanya, A., Bilmes, J.: Semi-supervised learning with measure propagation. JMLR $\mathbf{12}$(Nov), 3311–3370 (2011)
25. Taherkhani, F., Kazemi, H., Nasrabadi, N.M.: Matrix completion for graph-based deep semi-supervised learning. In: Proceedings of the AAAI Conference on Artificial Intelligence, vol. 33, pp. 5058–5065 (2019)
26. Wang, F., Zhang, C.: Label propagation through linear neighborhoods. IEEE TKDE $\mathbf{20}$(1), 55–67 (2007)
27. Wang, M., Li, H., Tao, D., Lu, K., Wu, X.: Multimodal graph-based reranking for web image search. IEEE TIP $\mathbf{21}$(11), 4649–4661 (2012)
28. Yin, K., Tai, X.C.: An effective region force for some variational models for learning and clustering. J. Sci. Comput. $\mathbf{74}$(1), 175–196 (2018)
29. Zhang, H., Zhang, Z., Zhao, M., Ye, Q., Zhang, M., Wang, M.: Robust triple-matrix-recovery-based auto-weighted label propagation for classification. arXiv preprint arXiv:1911.08678 (2019)
30. Zhang, K., Kwok, J.T., Parvin, B.: Prototype vector machine for large scale semi-supervised learning. In: ICML, pp. 1233–1240. ACM (2009)
31. Zhang, Z., Jia, L., Zhao, M., Liu, G., Wang, M., Yan, S.: Kernel-induced label propagation by mapping for semi-supervised classification. IEEE TBD $\mathbf{5}$(2), 148–165 (2019)
32. Zhang, Z., Zhang, Y., Liu, G., Tang, J., Yan, S., Wang, M.: Joint label prediction based semi-supervised adaptive concept factorization for robust data representation. In: IEEE TKDE (2019)

33. Zhang, Z., Zhao, M., Chow, T.W.: Marginal semi-supervised sub-manifold projections with informative constraints for dimensionality reduction and recognition. Neural Netw. **36**, 97–111 (2012)
34. Zhou, D., Bousquet, O., Lal, T.N., Weston, J., Schölkopf, B.: Learning with local and global consistency. In: NIPS, pp. 321–328 (2003)
35. Zhuang, L., Zhou, Z., Gao, S., Yin, J., Lin, Z., Ma, Y.: Label information guided graph construction for semi-supervised learning. IEEE TIP **26**(9), 4182–4192 (2017)

# Near Out-of-Distribution Detection for Low-Resolution Radar Micro-doppler Signatures

Martin Bauw[1,2]([✉]), Santiago Velasco-Forero[1], Jesus Angulo[1], Claude Adnet[2], and Olivier Airiau[2]

[1] Center for Mathematical Morphology, Mines Paris, PSL University, Paris, France
`martin.bauw@minesparis.psl.eu`
[2] Thales LAS France, Advanced Radar Concepts, Limours, France

**Abstract.** Near out-of-distribution detection (OODD) aims at discriminating semantically similar data points without the supervision required for classification. This paper puts forward an OODD use case for radar targets detection extensible to other kinds of sensors and detection scenarios. We emphasize the relevance of OODD and its specific supervision requirements for the detection of a multimodal, diverse targets class among other similar radar targets and clutter in real-life critical systems. We propose a comparison of deep and non-deep OODD methods on simulated low-resolution pulse radar micro-doppler signatures, considering both a spectral and a covariance matrix input representation. The covariance representation aims at estimating whether dedicated second-order processing is appropriate to discriminate signatures. The potential contributions of labeled anomalies in training, self-supervised learning, contrastive learning insights and innovative training losses are discussed, and the impact of training set contamination caused by mislabelling is investigated.

**Keywords:** Anomaly detection · Out-of-distribution detection · Micro-doppler · Radar target discrimination · Deep learning · Self-supervised learning

## 1 Introduction

Near out-of-distribution detection (OODD) aims at distinguishing one or several data classes from semantically similar data points. For instance, identifying samples from one class of CIFAR10 among samples of the other classes of the same dataset solves a near OODD task. On the other hand, separating CIFAR10 samples from MNIST samples is a far OODD task: there is no strong semantic proximity between the data points being separated. OODD defines a kind of anomaly detection (AD) since OODD can be seen as separating a normal class from infinitely diverse anomalies, with a training set only or mostly composed of normal samples, and anomalies being possibly semantically close to normal samples [24]. This training paradigm relies on lower supervision requirements compared to supervised classification, for which each class calls for a representative set of samples in the training data.

© The Author(s), under exclusive license to Springer Nature Switzerland AG 2023
M.-R. Amini et al. (Eds.): ECML PKDD 2022, LNAI 13716, pp. 384–399, 2023.
https://doi.org/10.1007/978-3-031-26412-2_24

This work considers both unsupervised and semi-supervised AD (SAD). Unsupervised AD trains the model with a representative set of normal data samples, while semi-supervised AD also benefits from labeled anomalies [15,26] that can not be representative, since anomalies are by definition infinitely diverse. A distinction can however be observed between benefiting from far and near anomalies, in analogy with far and near OODD, to refine the discrimination training. The contribution of self-supervision will be taken into account through the supply of far artificial anomalies for additional supervision during training.

Near OODD constitutes an ideal mean to achieve radar targets discrimination, where an operator wants an alarm to be raised everytime specific targets of interest are detected. This implies discriminating between different kinds of planes, or ships, sometimes being quite similar from a radar perspective. For example, two ships can have close hull and superstructure sizes, implying close radar cross-sections, even though their purpose and equipment on deck are completely different. Analogous observations could be made for helicopters, planes and drones. In an aerial radar context, whereas separating aerial vehicles would constitute a near OODD task, spotting weather-related clutter would define a far OODD. Such an OODD-based detection setup is directly applicable to other sensors.

The motivation behind the application of OODD methods to low-resolution pulse Doppler radar (PDR) signatures stems from the constraints of some air surveillance radars. Air surveillance PDRs with rotating antennas are required to produce very regular updates of the operational situation and to detect targets located at substantial ranges. The regular updates dictate the rotation rate and limit the number of pulses, and thus the number of Doppler spectrum bins, over which to integrate and refine a target characterization. The minimum effective range restricts the pulse repetition frequency (PRF), which in turns diminishes the range of velocities covered by the Doppler bins combined. The operating frequencies of air surveillance radars are such that they can not make up for this Doppler resolution loss [18]. This work aims at exploring the potential of machine learning to discriminate targets within these air surveillance radars limitations, using the targets Doppler spectrums. Refining radar targets discrimination with limited supervision is critical to enable the effective detection of targets usually hidden in cluttered domains, such as small and slow targets.

The AD methods examined will take a series of target Doppler spectrums as an input sample. This series is converted into a second-order representation through the computation of a covariance matrix to include an AD method adapted to process symmetric positive definite (SPD) inputs in our comparison. Radar Doppler signatures with sufficient resolution to reveal micro-doppler spectrum modulations is a common way to achieve targets classification in the radar literature, notably when it comes to detecting drones hidden in clutter [3,9,13]. The processing of second-order representations is inspired by their recent use in the machine learning literature [16,31], including in radar processing [6], and is part of the much larger and very active research on machine learning on Riemannian manifolds [5,7].

This paper first details the simulation setup which generates the micro-doppler dataset, then describes the OODD methods compared. Finally, an experimental section compares quantitatively various supervision scenarios involving SAD and self-supervision. The code for both the data generation and the OODD experiments is available[1]. The code made available does not restrict itself to the experiments put forward in the current document, pieces of less successful experiments being kept for openness and in case they help the community experiment on the data with similar approaches.

## 2    Micro-doppler Dataset

A PDR is a radar system that transmits bursts of modulated pulses, and after each pulse transmission waits for the pulse returns. The pulse returns are sampled and separated into range bins depending on the amount of time observed between transmission and reception. The spectral content of the sampled pulses is evaluated individually in each range bin, as depicted on Fig. 1. This content translates into the Doppler information which amounts to a velocity descriptor: the mean Doppler shift reveals the target bulk speed, and the spectrum modulation its rotating blades. These Doppler features are available for each burst, under the assumption that the velocities detected in a given range bin change negligibly during a burst. The number of pulses in a burst, equating the number of samples available to compute a spectrum, determines the resolution of the Fourier bins or Doppler bins. The PRF sampling frequency defines the range of speeds covered by the spectrum. PDR signatures are generated by a MATLAB [20] simulation. The Doppler signatures are a series of periodograms, i.e. the evolution of spectral density over several bursts, one periodogram being computed per burst. The samples on which the discrete Fourier transform is computed are sampled at the PRF frequency, i.e. one sample is available per pulse return for each range bin.

The main parameters of the simulation are close to realistic radar and target characteristics. A carrier frequency of 5 GHz was selected, with a PRF of 50 KHz. An input sample is a Doppler signature extracted from 64 bursts of 64 pulses, i.e. 64 spectrums of 64 samples, ensuring the full rank of the covariance matrix computed over non-normalized Doppler, i.e. Fourier, bins. The only simulation parameter changing across the classes of helicopter-like targets is the number of rotating blades: Doppler signatures are associated with either one, two, four or six rotating blades, as can be found on drones and radio-controlled helicopters. The quality of the dataset is visually verified: a non-expert human is easily able to distinguish the four target classes, confirming the discrimination task is feasible. The classes intrinsic diversity is ensured by receiver noise, blade size and revolutions-per-minute (RPM) respectively uniformly sampled in [4.5, 7] and [450, 650], and a bulk speed uniformly sampled so that the signature central frequency changes while staying approximately centered. The possible

---

[1] https://github.com/Blupblupblup/Doppler-Signatures-Generation
https://github.com/Blupblupblup/Near-OOD-Doppler-Signatures.

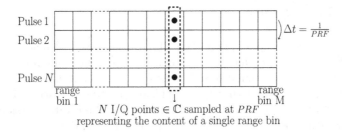

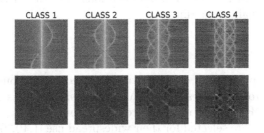

**Fig. 1.** Each pulse leads to one complex-valued I/Q sample per range bin, while each burst is composed of several pulses. Each range bin is thus associated with a complex-valued discrete signal with as many samples as there are pulses. Air surveillance radars with rotating antennas are required to provide regular situation updates in every direction, severely constraining the number of pulses per burst acceptable.

**Fig. 2.** One sample of each target class: the varying number of rotating blades defines the classes, the modulation pattern being easily singled out. The first line of images shows Doppler signatures, i.e. the time-varying periodogram of targets over 64 bursts of 64 pulses. On those images, each row is the periodogram computed over one burst, and each column a Fourier i.e. a Doppler bin. The second line contains the covariance SPD representation of the first line samples. The width of the Doppler modulations around the bulk speed on the periodograms varies within each class, as well as the bulk speed, the latter being portrayed by the central vertical illumination of the signature.

bulk speeds and rotor speeds are chosen in order for the main Doppler shift and the associated modulations to remain in the unambiguous speeds covered by the Doppler signatures [18]. Example signatures and their covariance representations are depicted for each class on Fig. 2. For each class, 3000 samples are simulated, thus creating a 12000-samples dataset. While small for the deep learning community, possessing thousands of relevant and labeled real radar detections would not be trivial in the radar industry, making larger simulated datasets less realistic for this use case.

# 3    OODD Methods

This work compares deep and non-deep OODD methods, called shallow, including second-order methods harnessing the SPD representations provided by the covariance matrix of the signatures. The extension of the deep learning architectures discussed to SAD and self-supervised learning (SSL) is part of the comparison. The use of SSL here consists in the exploitation of a rotated version of every training signature belonging to the normal class in addition to its non-rotated version, whereas SAD amounts to the use of a small minority of actual anomalies taken in one of the other classes of the dataset. In the first case one creates artificial anomalous samples from the already available samples of a single normal class, whereas in the second case labeled anomalies stemming from real target classes are made available. No SSL or SAD experiments were conducted on the SPD representations, since the SSL and SAD extensions of the deep methods are achieved through training loss modifications, and the SPD representations were confined to shallow baselines.

## 3.1    Non-deep Methods

Common non-deep anomaly detection methods constitute our baselines: one-class support vector machines (OC-SVM) [28], isolation forests (IF) [19], local outlier factor (LOF) [4] and random projections outlyingness (RPO) [12]. The three first methods are selected for their widespread use [1,10,27], and the diversity of the underlying algorithms. OC-SVM projects data points in a feature space where a hyperplane separates data points from the origin, thus creating a halfspace containing most samples. Samples whose representation lies outside of this halfspace are then considered to be anomalies. IF evaluates how easy it is to isolate data points in the feature space by recursively partitioning the representation space. The more partitions are required to isolate a data point, the more difficult it is to separate this point from other samples, and the less anomalous this point is. LOF uses the comparison of local densities in the feature space to determine whether a point is anomalous or not. Points that have local densities similar to the densities of their nearest neighbors are likely to be inliers, whereas an outlier will have a much different local density than its neighbors. RPO combines numerous normalized outlyingness measures over 1D projections with a $max$ estimator in order to produce a unique and robust multivariate outlyingness measure, which translates into the following quantity:

$$O(x; p, X) = \max_{u \in \mathbb{U}} \frac{|u^T x - MED(u^T X)|}{MAD(u^T X)} \tag{1}$$

where $x$ is the data point we want to compute the outlyingness for, $p$ the number of random projections (RP) $u$ of unit norm gathered in $\mathbb{U}$, and $X$ the training data matrix. $MED$ stands for median and $MAD$ for median absolute deviation. This outlyingness actually leads to the definition of a statistical depth approximation [12,17].

## 3.2   Deep Methods

The deep AD methods experimented on in this work are inspired by the deep support vector data description (SVDD) original paper [27]. Deep SVDD achieves one-class classification by concentrating latent space representations around a normality centroid with a neural network trained to minimize the distance of projected data samples to the centroid. The centroid is defined by the average of the initial forward pass of the training data, composed of normal samples. The intuition behind the use of Deep SVDD for AD is similar to the way one detects anomalies with generative models: whereas generative models detect outliers because they are not as well reconstructed as normal samples, deep SVDD projects outliers further away from the normality centroid in the latent space. One can note that Deep SVDD is a deep learning adaptation of SVDD [29], which can be equivalent to the OC-SVM method in our comparison if one uses a Gaussian kernel. The training loss of Deep SVDD for a sample of size $n$ with a neural network $\Phi$ with weights $W$ distributed over $L$ layers is as follows:

$$\min_{W} \left[ \frac{1}{n} \sum_{i=1}^{n} ||\Phi(x_i; W) - c||^2 + \frac{\lambda}{2} \sum_{l=1}^{L} ||W^l||^2 \right] \quad (2)$$

where $c$ is the fixed normality centroid. The second term is a weights regularization adjusted with $\lambda$. Deep SVDD naturally calls for a latent multi-sphere extension. An example of such an extension is Deep multi-sphere SVDD (MSVDD) [14], which is part of our comparison. Deep MSVDD initializes numerous latent normality hyperspheres using k-means and progressively discards the irrelevant centroids during training. The relevance of latent hyperspheres is determined thanks to the cardinality of the latent cluster they encompass. The deep MSVDD training loss is:

$$\min_{W, r_1 \ldots r_K} \left[ \frac{1}{K} \sum_{k=1}^{K} r_k^2 + \frac{1}{\nu n} \sum_{i=1}^{n} max(0, ||\Phi(x_i; W) - c_j||^2 - r_j^2) + \frac{\lambda}{2} \sum_{l=1}^{L} ||W^l||^2 \right] \quad (3)$$

The first term minimizes the volume of hyperspheres of radius $r_k$, while the second is controlled by $\nu \in [0, 1]$ and penalizes points lying outside of their assigned hypersphere, training samples being assigned to the nearest hypersphere of center $c_j$. A second Deep SVDD variant considered here is Deep RPO [2], which replaces the latent Euclidean distance to the normality centroid with a RPs-based outlyingness measure in the latent space. This outlyingness measure ensures normality is described by a latent ellipsoid instead of a latent hypersphere, and leads to the following loss:

$$\min_{W} \left[ \frac{1}{n} \sum_{i=1}^{n} \left( \underset{u \in \mathbb{U}}{mean} \frac{|u^T \Phi(x_i; W) - MED(u^T \Phi(X; W))|}{MAD(u^T \Phi(X; W))} \right) + \frac{\lambda}{2} \sum_{l=1}^{L} ||W^l||^2 \right] \quad (4)$$

This training loss uses the outlyingness defined in Eq. 1, with a *max* estimator transformed into a *mean* as suggested in [2] for better integration with the deep learning setup.

SAD is achieved through outlier exposure [15,26], which adds supervision to the training of the model thanks to the availability of few and non representative labeled anomalies. To take into account anomalies during training, Deep SAD [26] repels the outliers from the normality centroid by replacing the minimization of the distance to the centroid with the minimization of its inverse in the training loss. Outliers could not globally be gathered around a reference point since they are not concentrated. This adaptation can be repeated for both Deep RPO and Deep MSVDD, although in Deep MSVDD the multiplicity of normality centers calls for an additional consideration on how to choose from which centroid the labeled anomalies should be repelled. The experiments implementing Deep MSVDD adapted to SAD with an additional loss term for labeled anomalies were inconclusive, such an adaptation will therefore not be part of the presented results. The reunion of normal latent representations achieved through the deep one-class classification methods mentioned is analogous to the alignment principle put forward in [30], which also argued for a latent uniformity. The extension of the Deep SVDD loss to encourage such latent uniformity using the pairwise distance between normal samples during training was investigated without ever improving the baselines.

### 3.3   Riemannian Methods for Covariance Matrices

Two SPD-specific AD approaches were considered. The first approach consists in replacing the principal component analysis (PCA) dimensionality reduction preceding shallow AD with an SPD manifold-aware tangent PCA (tPCA). The tPCA projects SPD points on the tangent space of the Fréchet mean, a Riemannian mean which allows to compute an SPD mean, keeping the computed centroid on the Riemannian manifold naturally occupied by the data. Using tPCA offers the advantage of being sensible to the manifold on which the input samples lie, but implies that input data is centered around the Riemannian mean. This makes tPCA a questionable choice when the objective set is AD with multimodal normality [23], something that is part of the experiments put forward in this work. Nonetheless, the Euclidean PCA being a common tool in the shallow AD literature, tPCA remains a relevant candidate for this study since it enables us to take a step back with respect to non-deep dimensionality reduction.

The second SPD-specific approach defines a Riemannian equivalent to Deep SVDD: inspired by recent work on SPD neural networks, which learn representations while keeping them on the SPD matrices manifold, a Deep SVDD SPD would transform input covariance matrices and project the latter into a latent space comprised within the SPD manifold. Taking into account SAD and SSL labeled anomalies during training was expected to be done as for the semi and self-supervised adaptations of Deep SVDD described earlier, where labeled anomalies are pushed away from the latent normality centroid thanks to an inverse distance term in the loss. Despite diverse attempts to make such a Deep SVDD SPD model work, with and without geometry-aware non-linearities in the neural network architecture, no effective learning was achieved on our dataset. This second approach will therefore be missing from the reported experimental results. Since this approach defined the ReEig [16] non-linearity rectifying small eigenvalues of SPD representations, the related shallow AD approach using the norm of the last PCA components as an anomaly score was also considered. This *negated PCA* is motivated by the possibility that, in one-class classification where fitting occurs on normal data only, the first principal components responsible for most of the variance in normal data are not the most discriminating ones when it comes to distinguishing normal samples from anomalies [21,25]. This approach was applied to both spectral and covariance representations, with the PCA and tPCA last components respectively, but was discarded as well due to poor performances. The latter indicate that anomalous samples are close enough to the normal ones for their information to be carried in similar components, emphasizing the near OODD nature of the discrimination pursued.

## 4    Experiments

AD experiments are conducted for two setups: a first setup where normality is made of one target class, and a second setup where normality is made of two target classes. When a bimodal normality is experimented on, the normal classes are balanced. Moreover, the number of normal modes is not given in any way to the AD methods, making the experiments closer to the arbitrary and, to a certain extent, unspecified one-class classification useful to a radar operator. Within the simulated dataset, 90% of the samples are used to create the training set, while the rest is equally divided to create the validation and test sets. All non-deep AD methods include a preliminary PCA or tPCA dimensionality reduction.

*Preprocessing.* This work is inspired by [26], which experimented on Fashion-MNIST, a dataset in which samples are images of objects without background or irrelevant patterns. In order to guarantee a relevant neural architecture choice, this kind of input format is deliberately reproduced. The series of periodograms, i.e. non-SPD representations are therefore preprocessed such that only the columns with top 15% values in them are kept, this operation being done after a switch to logarithmic scale. This results in periodograms where only the active Doppler bins, portraying target bulk speed and micro-doppler modulations, have non-zero value. Only a grayscale region of interest (ROI) remains in the input matrix with various Doppler shifts and modulation widths, examples of which are shown on Fig. 3. This preprocessing leads to the "(SP)" input format as indicated in the results tables, and is complementary to the covariance representation. Covariance matrices are computed without such preprocessing, except for the switch to logarithmic scale which precedes the covariance computation. Comparing covariance-based OODD to OODD on spectral representations is fair since both representations stem from the same inputs, the covariance only implying an additional transformation of the input before training the AD. All input data is min-max normalized except for the covariance matrices used by tPCA.

*Deep Learning Experiments.* The test AUC score of the best validation epoch in terms of AUC is retained, in line with [11]. All experiments were conducted with large 1000 samples batches, which stabilizes the evolution of the train, validation and test AUCs during training. The training is conducted during 300 epochs, the last 100 epochs being fine-tuning epochs with a reduced learning rate, a setup close to the one in [27]. A relatively small learning rate of $10^{-4}$ is chosen to help avoid the latent normality hypersphere collapse, i.e. the convergence to a constant projection point in the latent space, in the non-SAD and non-SSL cases, with $\lambda = 10^{-6}$. Hyperparameters are kept constant across all experiments conducted, in order to ensure fair comparisons. In the results tables, the second and third columns indicate whether SAD and SSL samples were used for additional supervision during training, and describe how such samples affected the training loss if present. When the SAD or SSL loss term is defined by a centroid, it means that the distance to the mentioned centroid is minimized during training, whereas "away" implies the projection of the SAD or SSL samples are repelled from the normality centroid thanks to an inverse distance as described previously. For example, the first line of the second part of Table 2 describes an experiment where SAD samples are concentrated around the SAD samples latent centroid, and SSL samples concentrated around the SSL samples latent centroid. Centroids are computed, as for the normal training samples, with the averaging of an initial forward pass, therefore yielding the average latent representation.

*Non-deep Learning Experiments.* Shallow AD conducted on the covariance representation after a common PCA uses the upper triangular part of the min-max normalized input as a starting point, avoiding redundant values. This contrasts with the Riemannian approach replacing PCA with the tPCA, the latter requiring the raw SPD representation. Furthermore, shallow approaches were also tried on the periodograms individually, where each row of an input signature, i.e. one vector of Doppler bins described for one burst, was given a score, the complete signature being then given the mean score of all its periodograms. This ensemble method did not yield relevant results and is therefore missing from our comparison. Such an approach ignores the order of periodograms in signatures.

*Neural Network Architecture.* While the Fashion-MNIST input format is thus replicated, the 2D features remain specific to radar signal processing and may therefore benefit from a different neural network architecture. Several neural networks architectures were considered, including architectures beginning with wider square and rectangular convolutions extended along the (vertical) bursts input axis, with none of the investigated architectures scoring systematically higher than the Fashion-MNIST architecture from the original Deep SAD work [26], which was only modified in order to handle the larger input size. The latter was consequently selected to produce the presented results. This architecture projects data with two

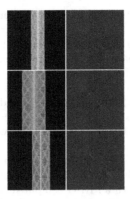

**Fig. 3.** Random samples of the fourth class after the preprocessing erasing the irrelevant background. One can notice the varying modulation width and central shift. The fourth class has the highest number of rotating blades on the helicopter-like target, hence the higher complexity of the pattern.

convolutional layers followed by two dense layers, each layer being separated from the next one by a batch normalization and a leaky ReLU activation. The outputs of the two convolutional layers are additionally passed through a 2D max-pooling layer.

*Riemannian AD.* The tPCA was computed thanks to the dedicated Geomstats [22] function, while experiments implementing a Riemannian equivalent of Deep SVDD were conducted using the SPD neural networks library torchspdnet [8]. The AD experiments based on a SPD neural network ending up inconclusive, they are not part of the results tables.

### 4.1   Unsupervised OODD with Shallow and Deep Learning

Unsupervised AD results, for which the training is only supervised by normal training samples, are presented in Table 1. These results indicate the superiority of deep learning for the OODD task considered, while demonstrating the substantial contribution of geometry-aware dimensionality reduction through the use of tPCA for non-deep AD. RPO is kept in Table 1 even though it does not achieve useful discrimination because it is the shallow equivalent of Deep RPO, one of the highlighted deep AD methods, deprived of the neural network encoder and with a *max* estimator instead of a *mean*, as was previously justified. Deep MSVDD does not lead to the best performances, and is as effective as Deep SVDD and Deep RPO, which could seem surprising at least when normality is made of two target classes.

**Table 1.** Unsupervised AD experiments results (average test AUCs in % ± StdDevs over ten seeds). These machine learning methods are trained on fully normal training sets, without labeled anomalies for SAD or self-supervision transformations. The four last methods are our deep AD baselines, trained on normalized spectral representations only. Deep MSVDD "mean best" indicates the neural network was trained using a simpler loss, analogous to the Deep SVDD loss, where only the distance to the best latent normality centroid is minimized, thus discarding the radius loss term. One should note that whereas Deep SVDD uses the Euclidean distance to the latent normality centroid as a test score, Deep MSVDD replaces this score with the distance to the nearest latent centroid remaining after training, from which the associated radius is subtracted. Very often in our experiments, even with multimodal normality during training, only one latent sphere remains at the end of Deep MSVDD training. Deep RPO replaces the Euclidean distance score with an RPO computed in the encoding neural network output space. PCA and tPCA indicate that the AD model is trained after an initial dimensionality reduction, which is either PCA or tangent PCA. RPO, with or without prior neural network encoding, is always implemented with 1000 random projections.

| AD method (input format) | SAD loss | SSL loss | Mean test AUC (1 mode) | Mean test AUC (2 modes) |
|---|---|---|---|---|
| OC-SVM (SP-PCA) | / | / | 49.16 ± 26.69 | 45.48 ± 27.53 |
| OC-SVM (SPD-PCA) | / | / | 64.68 ± 9.10 | 58.23 ± 15.12 |
| OC-SVM (SPD-tPCA) | / | / | 57.59 ± 3.91 | 55.33 ± 9.48 |
| IF (SP-PCA) | / | / | 50.96 ± 17.37 | 48.50 ± 18.76 |
| IF (SPD-PCA) | / | / | 52.36 ± 22.47 | 47.50 ± 20.32 |
| IF (SPD-tPCA) | / | / | 66.91 ± 9.65 | 61.23 ± 12.65 |
| LOF (SP-PCA) | / | / | 56.80 ± 2.38 | 61.55 ± 10.29 |
| LOF (SPD-PCA) | / | / | 66.44 ± 21.37 | 65.83 ± 19.52 |
| LOF (SPD-tPCA) | / | / | 78.38 ± 8.86 | 73.56 ± 10.09 |
| RPO (SP-PCA) | / | / | 49.61 ± 6.89 | 50.43 ± 7.13 |
| RPO (SPD-PCA) | / | / | 51.08 ± 19.66 | 54.95 ± 17.58 |
| RPO (SPD-tPCA) | / | / | 33.97 ± 7.36 | 38.08 ± 14.58 |
| Deep SVDD (SP) | no SAD | no SSL | 83.03 ± 6.83 | **78.29 ± 6.68** |
| Deep MSVDD (SP) | no SAD | no SSL | 82.27 ± 9.67 | **78.30 ± 8.28** |
| Deep MSVDD "mean best" (SP) | no SAD | no SSL | 82.29 ± 7.20 | 78.02 ± 6.80 |
| Deep RPO (SP) | no SAD | no SSL | **83.60 ± 5.35** | 78.13 ± 6.02 |

## 4.2 Potential Contribution of SAD and SSL

The contribution of additional supervision during training through the introduction of SAD samples and SSL samples is examined in Table 2. Regarding SAD experiments, labeled anomalies will be taken from a single anomalous class for simplicity, and because only four classes are being separated, this avoids unrealistic experiments where labeled anomalies from every anomalous class are seen during training. When SAD samples are used during training, labeled anomalies represent one percent of the original training set size. This respects the spirit of SAD, for which labeled anomalies can only be a minority of training samples, which is not representative of anomalies. This is especially realistic in the radar processing setup initially described where labeled detections would rarely be available. SSL samples are generated thanks to a rotation of the spectral input format, rendering the latter absurd but encouraging better features extraction since the network is asked to separate similar patterns with different orientations. SSL samples are as numerous as normal training samples, implying they don't define a minority of labeled anomalies for training as SAD samples do, when they are taken into account.

Individually, SAD samples lead to better performances than SSL ones, but the best results are obtained when combining the two sets of samples for maximal training supervision. Deep SVDD appears to be substantially better at taking advantage of the additional supervision provided by SAD and SSL samples. Quite surprisingly for a radar operator, the best test AUC is obtained when SSL samples are concentrated around a specialized centroid while SAD samples are repelled from the normality centroid. Indeed, SSL samples being the only absurd samples considered in our experiments radarwise, it could seem

**Table 2.** Experiments with additional supervision provided by SAD and/or SSL labeled samples during training (average test AUCs in % ± StdDevs over ten seeds). When available, SAD samples are the equivalent of one percent of the normal training samples in quantity. The first half of the Table reports performances where only one of the two kinds of additional supervision is leveraged, while the second half describes the performances for setups where both SAD and SSL labeled samples contribute to the model training. Each couple of lines compares Deep SVDD and Deep RPO in a shared AD supervision setup, thus allowing a direct comparison. *c.* stands for centroid.

| AD method (input format) | SAD loss | SSL loss | Mean test AUC (1 mode) | Mean test AUC (2 modes) |
|---|---|---|---|---|
| Deep SVDD (SP) | no SAD | SSL c | 86.79 ± 6.54 | 83.91 ± 7.92 |
| Deep RPO (SP) | no SAD | SSL c | 88.70 ± 5.10 | 84.59 ± 8.54 |
| Deep SVDD (SP) | no SAD | away | 81.43 ± 8.62 | 77.01 ± 8.20 |
| Deep RPO (SP) | no SAD | away | 80.21 ± 9.06 | 78.93 ± 9.39 |
| Deep SVDD (SP) | SAD c | no SSL | 86.79 ± 8.94 | 87.65 ± 6.44 |
| Deep RPO (SP) | SAD c | no SSL | 81.38 ± 6.09 | 76.45 ± 6.30 |
| Deep SVDD (SP) | away | no SSL | 93.93 ± 4.82 | 93.50 ± 7.61 |
| Deep RPO (SP) | away | no SSL | 84.19 ± 5.32 | 80.37 ± 7.22 |
| Deep SVDD (SP) | SAD c | SSL c | 91.00 ± 6.45 | 90.51 ± 7.38 |
| Deep RPO (SP) | SAD c | SSL c | 87.79 ± 5.81 | 82.69 ± 8.51 |
| Deep SVDD (SP) | SAD c | away | 89.98 ± 7.79 | 91.03 ± 6.71 |
| Deep RPO (SP) | SAD c | away | 78.86 ± 9.10 | 79.11 ± 9.64 |
| Deep SVDD (SP) | away | SSL c | **95.06 ± 4.20** | 93.91 ± 7.31 |
| Deep RPO (SP) | away | SSL c | 89.82 ± 5.21 | 87.17 ± 8.17 |
| Deep SVDD (SP) | away | away | 94.63 ± 4.31 | **94.02 ± 7.30** |
| Deep RPO (SP) | away | away | 90.91 ± 5.94 | 92.69 ± 7.98 |

**Table 3.** Contamination experiments results (average test AUCs in % ± StdDevs over ten seeds): the SAD labeled anomalies are integrated within the training samples and taken into account as normal samples during training, thus no SAD loss term is used for SAD samples. The contamination rate is one percent, i.e. the equivalent of one percent of the normal training samples in labeled anomalies is added to confuse the AD.

| AD method (input format) | SAD loss | SSL loss | Mean test AUC (1 mode) | Mean test AUC (2 modes) |
|---|---|---|---|---|
| Deep SVDD (SP) | no SAD | no SSL | 80.76 ± 7.11 | 76.02 ± 6.66 |
| Deep MSVDD (SP) | no SAD | no SSL | 78.31 ± 11.18 | 74.49 ± 9.13 |
| Deep MSVDD "mean best" (SP) | no SAD | no SSL | 79.84 ± 7.82 | 74.89 ± 7.01 |
| Deep RPO (SP) | no SAD | no SSL | 81.29 ± 5.92 | 74.82 ± 5.89 |
| Deep SVDD (SP) | no SAD | SSL c | 85.34 ± 6.85 | 81.36 ± 7.47 |
| Deep RPO (SP) | no SAD | SSL c | **86.66 ± 6.41** | **82.78 ± 8.25** |
| Deep SVDD (SP) | no SAD | away | 79.62 ± 9.02 | 75.38 ± 8.28 |
| Deep RPO (SP) | no SAD | away | 76.16 ± 9.87 | 76.56 ± 8.69 |

more intuitive to project SAD samples, which remain valid targets, next to a dedicated centroid while repelling SSL samples. Likewise, on an ideal outlyingness scale, SSL samples should be further away from normality than SAD samples. This counter-intuitive performance could stem from the test set which only evaluates the separation of targets in a near OODD context. No invalid target representation, like the SSL samples are, is present in the test set, only valid representation from the four targets classes make up the latter. This is consistent with the application put forward in this study: use OODD to discriminate between various kinds of radar detections.

### 4.3 Training with a Contaminated Training Set

Unsupervised AD refers to the experiments of Table 1 where only training samples assumed to be normal supervise the training of the neural network. Real-life datasets, labeled by algorithms or experts, are unlikely to respect that assumption and will suffer from contamination of normal samples with unlabeled anomalies. The results in Table 3 depict how sensible the deep AD methods previously introduced are to training set contamination. The contamination is carried out using the one percent SAD samples already used for SAD experiments. While in the SAD experiments SAD samples were repelled from the normality centroid or concentrated next to their dedicated latent reference point, here they will be processed as normal samples. SSL samples again appear to better contribute to improving AD when concentrated next to a specialized centroid, while the performance drop due to contamination does not seem to be particularly stronger for one of the approaches considered (Fig. 4).

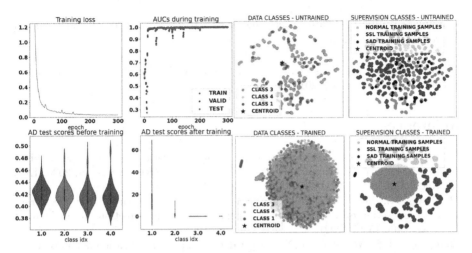

**Fig. 4.** Left - Training metrics of a successful run where normal samples are concentrated around their average initial projection, and SAD and SSL samples are pushed away thanks to a loss term using the inverse of the distance with respect to the normality latent centroid. This is one of the most successful setups in Table 2, and one of the easiest AD experiments since the two classes defining normality here are class 3 (four blades are responsible for the modulation pattern around the central Doppler shift) and class 4 (six blades are responsible for the modulation pattern around the central Doppler shift), meaning the separation with the other classes deemed anomalous is actually a binary modulation complexity threshold. One of the contributions of the SAD and SSL supervisions can be observed on the evolution of AUCs during training: no AUC collapse can be seen during training. Experiments showed that large training batches contributed to stable AUCs growth. Spikes in the training loss match the drops in AUCs. **Right** - Latent distribution of the training samples visualized in 2D using t-SNE after projection by the untrained (top) and the trained neural network (bottom). One can notice that normal training samples from both normal classes are completely mixed up with the minority of SAD labeled anomalies from class 1 in red (one blade), semantically similar, whereas SSL samples which are rotated normal training samples are already gathered in their own latent subclusters. SAD labeled anomalies end up well separated after training. (Color figure online)

## 5    Conclusion

The near OODD performances of various deep and non-deep AD methods were compared on a radar Doppler signatures simulated dataset. Deep AD approaches were evaluated in various supervision setups, which revealed the relevance of combining a minority of labeled anomalies with transformed normal training samples to improve near OODD performances, and avoid latent normality hypersphere collapse. Among the limitations of our study, one can note the lack of OODD experiments on a multimodal normal training set with unbalanced normal classes, which would make the OODD task more realistic. The benefits of deep learning clearly showed, and while not leading to the best overall performances, geometry-aware processing proved to be the source of a substantial improvement for non-deep AD.

**Acknowledgements.** This work was supported by the French Defense Innovation Agency (Cifre-Défense 001/2019/AID).

# References

1. Bauw, M., Velasco-Forero, S., Angulo, J., Adnet, C., Airiau, O.: From unsupervised to semi-supervised anomaly detection methods for HRRP targets. In: 2020 IEEE Radar Conference (RadarConf20), pp. 1–6. IEEE (2020)
2. Bauw, M., Velasco-Forero, S., Angulo, J., Adnet, C., Airiau, O.: Deep random projection outlyingness for unsupervised anomaly detection. arXiv preprint arXiv:2106.15307 (2021)
3. Björklund, S., Wadströmer, N.: Target detection and classification of small drones by deep learning on radar micro-doppler. In: 2019 International Radar Conference (RADAR), pp. 1–6. IEEE (2019)
4. Breunig, M.M., Kriegel, H.P., Ng, R.T., Sander, J.: LOF: identifying density-based local outliers. In: Proceedings of the 2000 ACM SIGMOD international conference on Management of data, pp. 93–104 (2000)
5. Bronstein, M.M., Bruna, J., LeCun, Y., Szlam, A., Vandergheynst, P.: Geometric deep learning: going beyond Euclidean data. IEEE Signal Process. Mag. **34**(4), 18–42 (2017)
6. Brooks, D., Schwander, O., Barbaresco, F., Schneider, J.Y., Cord, M.: A Hermitian positive definite neural network for micro-doppler complex covariance processing. In: 2019 International Radar Conference (RADAR), pp. 1–6. IEEE (2019)
7. Brooks, D., Schwander, O., Barbaresco, F., Schneider, J.Y., Cord, M.: Riemannian batch normalization for SPD neural networks. In: Advances in Neural Information Processing Systems 32 (2019)
8. Brooks, D., Schwander, O., Barbaresco, F., Schneider, J.-Y., Cord, M.: Second-order networks in PyTorch. In: Nielsen, F., Barbaresco, F. (eds.) GSI 2019. LNCS, vol. 11712, pp. 751–758. Springer, Cham (2019). https://doi.org/10.1007/978-3-030-26980-7_78
9. Brooks, D.A., Schwander, O., Barbaresco, F., Schneider, J.Y., Cord, M.: Temporal deep learning for drone micro-doppler classification. In: 2018 19th International Radar Symposium (IRS), pp. 1–10. IEEE (2018)
10. Chandola, V., Banerjee, A., Kumar, V.: Anomaly detection: a survey. ACM Comput. Surv. (CSUR) **41**(3), 1–58 (2009)
11. Chong, P., Ruff, L., Kloft, M., Binder, A.: Simple and effective prevention of mode collapse in deep one-class classification. In: 2020 International Joint Conference on Neural Networks (IJCNN), pp. 1–9. IEEE (2020)
12. Donoho, D.L., Gasko, M., et al.: Breakdown properties of location estimates based on halfspace depth and projected outlyingness. Ann. Stat. **20**(4), 1803–1827 (1992)
13. Gérard, J., Tomasik, J., Morisseau, C., Rimmel, A., Vieillard, G.: Micro-doppler signal representation for drone classification by deep learning. In: 2020 28th European Signal Processing Conference (EUSIPCO), pp. 1561–1565. IEEE (2021)
14. Ghafoori, Z., Leckie, C.: Deep multi-sphere support vector data description. In: Proceedings of the 2020 SIAM International Conference on Data Mining, pp. 109–117. SIAM (2020)

15. Hendrycks, D., Mazeika, M., Dietterich, T.: Deep anomaly detection with outlier exposure. In: Proceedings of the International Conference on Learning Representations (2019)
16. Huang, Z., Van Gool, L.: A Riemannian network for SPD matrix learning. In: Thirty-First AAAI Conference on Artificial Intelligence (2017)
17. Huber, P.J.: Projection pursuit. Ann. Statist. **13**(2), 435–475 (1985)
18. Levanon, N., Mozeson, E.: Radar signals. John Wiley & Sons (2004)
19. Liu, F.T., Ting, K.M., Zhou, Z.H.: Isolation forest. In: 2008 Eighth IEEE International Conference On Data Mining, pp. 413–422. IEEE (2008)
20. MATLAB: version 9.11.0 (R2021b). The MathWorks Inc., Natick, Massachusetts (2021)
21. Miolane, N., Caorsi, M., Lupo, U., et al.: ICLR 2021 challenge for computational geometry & topology: design and results. arXiv preprint arXiv:2108.09810 (2021)
22. Miolane, N., Guigui, N., Brigant, A.L., et al.: Geomstats: a python package for Riemannian geometry in machine learning. J. Mach. Learn. Res. **21**(223), 1–9 (2020). http://jmlr.org/papers/v21/19-027.html
23. Pennec, X.: Barycentric subspace analysis on manifolds. Ann. Stat. **46**(6A), 2711–2746 (2018)
24. Ren, J., Fort, S., Liu, J., et al.: A simple fix to Mahalanobis distance for improving near-ood detection. arXiv preprint arXiv:2106.09022 (2021)
25. Rippel, O., Mertens, P., Merhof, D.: Modeling the distribution of normal data in pre-trained deep features for anomaly detection. In: 2020 25th International Conference on Pattern Recognition (ICPR), pp. 6726–6733. IEEE (2021)
26. Ruff, L., Vandermeulen, R.A., Görnitz, N., et al.: Deep semi-supervised anomaly detection. In: International Conference on Learning Representations (2020). https://openreview.net/forum?id=HkgH0TEYwH
27. Ruff, L., Vandermeulen, R.A., Görnitz, N., et al.: Deep one-class classification. In: Proceedings of the 35th International Conference on Machine Learning, vol. 80, pp. 4393–4402 (2018)
28. Schölkopf, B., Platt, J.C., Shawe-Taylor, J., Smola, A.J., Williamson, R.C.: Estimating the support of a high-dimensional distribution. Neural Comput. **13**(7), 1443–1471 (2001)
29. Tax, D.M., Duin, R.P.: Support vector data description. Mach. Learn. **54**(1), 45–66 (2004)
30. Wang, T., Isola, P.: Understanding contrastive representation learning through alignment and uniformity on the hypersphere. In: International Conference on Machine Learning, pp. 9929–9939. PMLR (2020)
31. Yu, K., Salzmann, M.: Second-order convolutional neural networks. arXiv preprint arXiv:1703.06817 (2017)

# SemiITE: Semi-supervised Individual Treatment Effect Estimation via Disagreement-Based Co-training

Qiang Huang[1,3,4,5], Jing Ma[2], Jundong Li[2], Huiyan Sun[1,3,5(✉)],
and Yi Chang[1,3,4,5(✉)]

[1] School of Artificial Intelligence, Jilin University, Changchun, China
huangqiang18@mails.jlu.edu.cn, {huiyansun,yichang}@jlu.edu.cn
[2] University of Virginia, Charlottesville, USA
{jm3mr,jundong}@virginia.edu
[3] International Center of Future Science, Jilin University, Changchun, China
[4] Key Laboratory of Symbolic Computation and Knowledge Engineering,
Jilin University, Changchun, China
[5] Innovation Group of Marine Engineering Materials and Corrosion Control,
Southern Marine Science and Engineering Guangdong Laboratory, Zhuhai, China

**Abstract.** Recent years have witnessed a surge of interests in Individual Treatment Effect (ITE) estimation, which aims to estimate the causal effect of a treatment (e.g., job training) on an outcome (e.g., employment status) for each individual (e.g., an employee). Various machine learning based methods have been proposed recently and have achieved satisfactory performance of ITE estimation from observational data. However, most of these methods overwhelmingly rely on a large amount of data with labeled treatment assignments and corresponding outcomes. Unfortunately, a significant amount of labeled observational data can be difficult to collect in real-world applications due to time and expense constraints. In this paper, we propose a Semi-supervised Individual Treatment Effect estimation (*SemiITE*) framework with a disagreement-based co-training style, which aims to utilize massive unlabeled data to better infer the factual and counterfactual outcomes of each instance with limited labeled data. Extensive experiments on two widely used real-world datasets validate the superiority of our *SemiITE* over the state-of-the-art ITE estimation models.

**Keywords:** Treatment effect estimation · Semi-supervised learning

## 1 Introduction

Estimating individual treatment effect (ITE) is an important problem in causal inference, which aims to estimate the causal effect of a treatment on an outcome for each individual, e.g., "how would participating in a job training would influence the employment status of an employee?". ITE estimation plays an important role in a wide range of areas, such as decision making and policy evaluation

© The Author(s), under exclusive license to Springer Nature Switzerland AG 2023
M.-R. Amini et al. (Eds.): ECML PKDD 2022, LNAI 13716, pp. 400–417, 2023.
https://doi.org/10.1007/978-3-031-26412-2_25

regarding healthcare [9,20], education [15], and economics [27]. A traditional solution for this problem is to conduct randomized controlled trials (RCTs), which randomly divide individuals into treatment group and control group with different treatment assignments (e.g., participating in the job training or not), and then estimate the causal effect of treatment assignment with the outcome difference over these two groups. However, performing RCTs could be costly, time-consuming, and even unethical [5,13]. To overcome these issues, different from RCTs, many machine learning based methods [6,18,26,30] have been proposed to estimate individual treatment effect directly from observational data and have achieved great success in recent years.

Despite the great success the aforementioned machine learning based models have achieved in causal effect estimation, most of them often require a large amount of *labeled observational data* (i.e., instances that come with treatment assignments and corresponding factual outcomes) in the training process. To show how the amount of *labeled observational data* affects the ITE estimation performance, we conduct an initial exploration by training two ITE estimation models CFR [26] and TARNet [17] on the IHDP dataset [4] with different proportions of labeled observational data. The ITE estimation prediction on the test data is shown in Fig. 1. Here, we adopt the widely used metrics of $\sqrt{\epsilon_{PEHE}}$ and $\epsilon_{ATE}$ [13] (more details can be seen in Sect. 4) to evaluate the performance of the two models on ITE estimation. We can observe that the performance of these two models is poor when the proportion of training data is low, but improves significantly as the proportion increases to a large percentage. This example demonstrates the indispensability of the large amount of labeled observational data for existing state-of-the-art ITE estimation models. However, in many domains such as health care, the labeled observational data is often very scarce. The process of collecting such labeled observational data could take years and be extremely expensive, or it may face serious ethical issues [14]. Fortunately, *unlabeled observational data* (i.e., instances only with covariates) is easy to obtain, and many studies [8,34] have shown that unlabeled data is also beneficial to the performance of machine learning models. Exploiting unlabeled data for ITE estimation can greatly reduce the cost of collecting labeled observational data. Therefore, how to estimate ITE from limited labeled observational data by using unlabeled data is a pressing issue in causal inference.

To tackle the above problem, in this paper, we propose to use co-training framework [3] to harness the power of unlabeled observational data to aid ITE estimation. Co-training is a popular semi-supervised learning framework that has achieved great success in many problems. It first trains multiple diverse base learners on the limited labeled data, then the trained base learners are used to predict unlabeled data. At last, the most confident predictions of the base learners on the unlabeled data are iteratively added into the labeled data set. However, such co-training frameworks cannot be directly grafted into the ITE estimation problem, mainly because of the following difficulties. First, the existence of hidden confounders (i.e., the unobserved variables that influence both the treatment and the outcome) may result in confounding bias in ITE

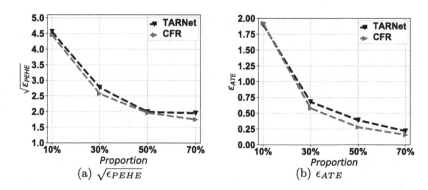

**Fig. 1.** The performance of CFR [26] and TARNet [17] for ITE estimation on the IHDP dataset with different proportions of labeled observational data (lower is better).

estimation, hence how to control the confounding bias is an issue to be addressed. Second, traditional co-training framework relies on multiple views of data (e.g., the acoustic attribute view and pictorial attribute view for a movie sample) to train multiple diverse base learners, otherwise the co-training degrades to self-training [21]. However, it is hard to collect such observational data with multiple views in causal inference, hence generating base causal models with diversity is critical for learning the individual treatment effect. Third, most of existing co-training frameworks [2,8,32] are mainly designed for classification problems. When it comes to ITE estimation, which is naturally a regression problem in most cases, base learners usually need to be re-trained to check whether the candidate instance prediction reduces their error rate in each instance selection round, which would increase the cost of computation and time. Thus designing an appropriate co-training strategy for ITE estimation to avoid re-training issue is also pressing.

To address the aforementioned difficulties, we propose a novel Semi-supervised Individual Treatment Effect estimation framework (*SemiITE*) via disagreement-based co-training. *SemiITE* builds a shared module to capture the hidden confounders so as to alleviate the confounding bias. To effectively enhance the ITE estimation performance under the semi-supervised setting, *SemiITE* generates three base potential outcome prediction models with diversity in a variety of ways. Moreover, to better utilize the unlabeled observational data, we design a novel co-training strategy in *SemiITE* based on the disagreement information of the three base models, which can select the most confident unlabeled instance predictions directly without re-training in each instance selection round. The following are the main contributions of our work:

- We formulate a novel research problem to utilize unlabeled observational data in a co-training manner for better individual treatment effect estimation.

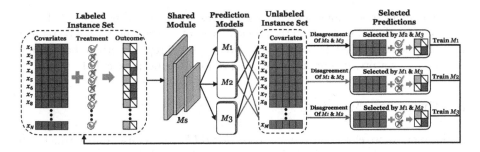

**Fig. 2.** An overview of the proposed semi-supervised ITE estimation framework *Semi-ITE* with disagreement-based co-training that utilizes unlabeled instances.

- We design a novel disagreement-based co-training framework *SemiITE* for semi-supervised individual treatment effect estimation, which can make use of unlabeled instances effectively, eliminate confounding bias, and avoid the re-training issue in each instance selection round.
- We perform extensive experiments and the results show that the proposed ITE estimation framework *SemiITE* is superior to existing state-of-the-art methods for ITE estimation when labeled observational data is limited.

## 2 Preliminaries

**Notations.** Let $L = \{(\boldsymbol{x}_1, t_1, y^{t_1}), ..., (\boldsymbol{x}_N, t_N, y^{t_N})\}$ denote the set of labeled observational instances with covariates, treatment assignments, and corresponding outcomes, where $\boldsymbol{x}_i \in \mathbb{R}^d$, $t_i \in \{0, 1\}$, $y^{t_i}$ represent the covariates, treatment assignment, observed factual outcome given treatment assignment $t_i$ of instance $i$, respectively; and $N$ is the number of labeled instances. Let $U = \{\boldsymbol{x}_1, ..., \boldsymbol{x}_M\}$ denote the set of unlabeled instances only with covariates, where $M$ is the number of unlabeled instances.

**Problem Statement.** We develop our framework based on the potential outcome framework [24, 25], which is widely used in causal inference. The individual treatment effect of instance $i$ is defined as $\tau_i = y_i^1 - y_i^0$. Noting that in real-world scenarios, only one of the potential outcomes can be observed for each instance, and the remaining unobserved potential outcome is also known as the counterfactual outcome. Inferring the counterfactual outcome from observational data is one of the most challenging tasks in causal inference [25]. Using the above, we provide the formal problem statement as follows: given the set of labeled observational instances $L = \{(\boldsymbol{x}_i, t_i, y^{t_i})\}_{i=1}^N$ and the set of unlabeled instances $U = \{\boldsymbol{x}_1, ..., \boldsymbol{x}_M\}$, our goal is to learn ITE $\tau_i$ for each instance $i$ from limited labeled observational data by making use of massive unlabeled data.

**Co-training.** Co-training [3] is a widely used solution to utilize unlabeled data to aid prediction in semi-supervised learning. In co-training, multiple base learners will be trained on the limited labeled data. Then for each base learner, the most confident predictions of unlabeled samples predicted by its peer base learners would be chosen to add into the labeled data set and the model will be refined using the newly labeled data set. The above steps will repeat until no base learners update or a preset number of learning rounds has been executed.

# 3   Proposed Method

In this section, we elaborate the proposed *SemiITE*, a novel framework which can utilize unlabeled instances for ITE estimation. Figure 2 depicts an overview of *SemiITE*. The framework mainly contains three components: a shared module, triple base potential outcome prediction models, and a co-training strategy.

More specifically, we first build a shared module to capture deep information for each instance and to balance the latent representations of treatment group and control group, then build three backbone neural network based models with different structures and initializations to infer potential outcomes, noting that the shared module and the three backbone models are integrated as an ensemble model. We first train the ensemble model using labeled observational instance set $L$, then in each round of the co-training, we select some instance(s) with its predicted potential outcomes by fitting the unlabeled instances to the trained ensemble model according to the disagreement information of the three backbone prediction models and add them to the labeled instance set, until all the instances in the unlabeled set $U$ are selected or the number of training rounds reaches the preset maximum.

## 3.1   Model Structure of *SemiITE*

First, we illustrate the model structure of the proposed framework. Generally, the model structure of SemiITE contains a shared module which aims to capture the hidden confounders and three potential outcome prediction models for inferring individual treatment effect.

**Shared Module.** To conduct unbiased ITE estimation [18], we capture the hidden confounders in the proposed framework by building a shared module with a multi-layer neural network. This shared module maps the original covariates to latent space and generates shared latent representation of each instance for the following base potential outcome prediction models. Furthermore, as proved in [26], the representations with closer distance between treatment and control groups can help mitigate the biases in causal effect estimation, thus we refine the representations generated by the shared module to obtain balanced representations between treatment group and control group towards more unbiased ITE estimation, which will be introduced later. For the shared module which is denoted as $M_s$, we aim to learn a representation learning function $f_s$ : $\mathcal{X} \in \mathbb{R}^d \to \mathbb{R}^m$, which maps the observed covariates to an $m$-dimensional latent space. Specifically, we parameterize the representation learning function $f_s$ by stacking $L_s$ neural network layers. The representations generated by the shared module for instance $i$ can be formulated as follows:

$$h_i = f_s(x_i) = \varphi(W_{L_s}...\varphi(W_1 x_i + b_1) + b_{L_s}), \tag{1}$$

where $x_i$ is the original covariates of instance $i$ and $h_i \in \mathbb{R}^m$ is the learned representation of instance $i$ by function $f_s$, $\varphi(\cdot)$ denotes the ReLU activation function, $W_S$ and $b_S$ $(S = 1, 2, ..., L_s)$ are the learning weight matrix and bias term of the $S$-th layer, respectively.

**Triple Base Prediction Models with Diversity.** To ensure the effectiveness of the co-training strategy, the base models should be diverse, which can help address the limited label issue [8,21], because if all of the base learners are identical, the training process with multiple learners will degrade to self-training with a single learner. Conventional co-training based methods typically require sufficient and redundant views of data to train the base learners in order to diversify them [19,22]. Given that the requirement for data with sufficient views is too stringent to meet in causal inference due to the expensive and time-consuming nature of observational data collection, in this work we build three outcome prediction models $M_1$, $M_2$, and $M_3$ by stacking multiple neural network layers to infer individual treatment effect for each instance and achieve the diversity of the three outcome prediction models from the following several aspects. First, we let the network structures of the three potential outcome prediction models differ. We assign different values of the depth and width, i.e., the number of hidden layers and neurons in each layer for each base model. Second, we let the learning weights of network architecture of each base model to be initialized by different methods. We take Gaussian random initialization, uniform random initialization, and Glorot initialization [10] on $M_1$, $M_2$, and $M_3$, respectively. Third, we use different optimization methods to update the three potential outcome prediction models. In particular, we use stochastic gradient descent to optimize $M_1$ and $M_2$, Adam optimization to optimize $M_3$. Fourth, we let $M_s$ and $M_1$ to be updated together after adding selected instances with outcomes predicted by $M_2$ and $M_3$ to the set of labeled instances while fixing the other modules. $M_2$ and $M_3$ will be updated separately while fixing the other modules, including the shared module $M_s$.

Specifically, we aim to learn a prediction function $f_v : \mathbb{R}^m \to \mathbb{R}$ $(v = 1, 2, 3)$ for each potential outcome prediction model $M_v$ and parameterize the prediction function $f_v$ by stacking multiple layers of neural network. With the representation $h_i$ of instance $i$ by the aforementioned representation learning function $f_s(\cdot)$ and the corresponding treatment assignment $t_i \in \{0, 1\}$, the predicted outcome $y_i$ with treatment $t_i$ of instance $i$ in model $M_v$ $(v = 1, 2, 3)$ can be computed by the function $f_v$ as:

$$f_v(h_i, t_i) = \begin{cases} \hat{y}_i^{t_i=0} = f_v^0(h_i) & \text{if } t_i = 0 \\ \hat{y}_i^{t_i=1} = f_v^1(h_i) & \text{if } t_i = 1, \end{cases} \tag{2}$$

where $f_v^0$ and $f_v^1$ are parameterized by $L_v$ fully connected layers followed by an output layer for $t_i = 0$ and $t_i = 1$, respectively. More specifically, the formulation of function $f_v^t$ $(t = 0, 1)$ can be written as:

$$f_v^t(\boldsymbol{h}_i) = \boldsymbol{w}^t \varphi(\boldsymbol{W}_{L_v}^t \cdots \varphi(\boldsymbol{W}_1^t \boldsymbol{h}_i + c_1) + c_{L_v}) + c^t, \tag{3}$$

where $\varphi(\cdot)$ denotes the ReLU activation function, $\boldsymbol{W}_K^t$ and $c_K$ $(K = 1, 2 \ldots, L_v)$ are the weight matrix and bias term for the $K$-th hidden layer, respectively. $\boldsymbol{w}^t$ is the weight vector and $c^t$ is the corresponding bias term of the final prediction layer.

Besides, we propose to utilize HSIC (Hilbert-Schmidt Independence Criterion) [12] to enhance and quantify the diversity of the three base models by measuring the dependence of the predicted factual outcomes from the three prediction models. Assuming that the predicted factual outcomes by $M_1$, $M_2$, and $M_3$ for $n$ samples are $\boldsymbol{Y}_1$, $\boldsymbol{Y}_2$, and $\boldsymbol{Y}_3 \in \mathbb{R}^n$, respectively, the diversity of the three base models can be calculated as follows:

$$\mathcal{L}_{div} = \sum_{i,j \in \{1,2,3\}, i<j} HSIC(\boldsymbol{Y}_i, \boldsymbol{Y}_j),$$

$$\text{where} \quad HSIC(\boldsymbol{Y}_i, \boldsymbol{Y}_j) = \frac{1}{(n-1)^2} \boldsymbol{K}_i \boldsymbol{J} \boldsymbol{K}_j \boldsymbol{J}. \tag{4}$$

Here, $\boldsymbol{K}_i$ and $\boldsymbol{K}_j$ are kernel matrices for prediction vectors $\boldsymbol{Y}_i$ and $\boldsymbol{Y}_j$, respectively. In this work, we use Gaussian kernel for computing the kernel matrix. $\boldsymbol{J} = \boldsymbol{I} - \frac{1}{n}\boldsymbol{1}$, where $\boldsymbol{I}$ is identity matrix and $\boldsymbol{1}$ is the vector with all elements of 1. Smaller HSIC value indicates stronger independence between the two variables. More details about the derivation can be found in [12]. Furthermore, based on the three potential outcome prediction models with diversity, we can use the disagreement information between them to design a novel co-training strategy to avoid the re-training issue in each instance selection round. More details can be found in Sect. 3.3.

## 3.2 Loss Function

With the above network structure including $M_s$, $M_1$, $M_2$, and $M_3$, we design a loss term to combine these components for inferring potential outcomes in an end-to-end manner.

**Loss for Predicted Potential outcomes of $M_1$, $M_2$, and $M_3$.** First we need to minimize the difference between the inferred factual outcome by function $f_v$ $(v = 1, 2, 3)$ and the observed factual outcome. We use mean squared error (MSE) function to evaluate the predicted outcomes to approximate the observed factual outcomes. Given the set of labeled observational instances $L = \{(\boldsymbol{x}_i, t_i, y^{t_i})\}_{i=1}^N$, the loss term for the three outcome prediction models can be written as:

$$\mathcal{L}_{mse} = \sum_{v=1}^{3} \sum_{i=1}^{N} MSE(f_v(f_s(\boldsymbol{x}_i), t_i), y^{t_i}), \tag{5}$$

**Loss for Representation Balancing.** Due to the fact that we only minimize the error of factual outcomes in Eq. (5), whereas the counterfactual distribution generally differs from the factual distribution, which lead to a biased ITE estimation [17]. Therefore, it is necessary to balance the distributions of representations generated by the shared module $M_s$ between the treatment group and the control group, which will help with the unbiased ITE estimation. Here, following previous work [14], we use integral probability metric (IPM) to measure the difference between the representation distributions of instances in the treatment group and those in the control group. We denote the balance term as $\mathcal{L}_{IPM}$.

**Diversity Term.** Besides, we add the HSIC term $\mathcal{L}_{div}$ shown in Eq. (4) into the overall loss function to control and ensure the diversity between the three prediction base models in the training process.

**Overall Loss.** To sum up, the final loss function of our proposed framework *SemiITE* can be written as follows:

$$\mathcal{L} = \mathcal{L}_{mse} + \alpha\mathcal{L}_{IPM} + \beta\mathcal{L}_{div} + \gamma\|\theta\|_2^2, \tag{6}$$

where $\alpha$, $\beta$, and $\gamma$ are hyperparameters to control the trade-off between corresponding loss term and other terms. $\|\theta\|_2^2$ is the regular term imposed on all learning parameters $\theta$ to avoid over-fitting.

## 3.3   Co-training Strategy via Disagreement

In this subsection, we introduce the co-training strategy of *SemiITE*, in which the framework chooses the predictions of unlabeled instances predicted by the three different potential outcome prediction models based on their disagreement information, and the three models will be refined by these chosen unlabeled instances iteratively. In addition, we illustrate that *SemiITE* avoids re-training issue in each instance selection round, while such issue exists in previous work [22,35] for regression with co-training.

Before introducing the proposed co-training strategy of *SemiITE*, we present the re-training issue in traditional co-training for regression problem, which increases the computational cost greatly. We take an example of traditional co-training regression method [35] to present the issue. Two different regression models are used in this method, which estimates the prediction confidence for each unlabeled sample based on the following principle: whether the error rate of regression model is reduced after adding new predicted sample from unlabeled data set to the training data set. Thus, the method needs to calculate the error reduction rate $\Delta_{x_u} = \sum_{x_i \in L}(y_i - h(x_i))^2 - (y_i - h'(x_i))^2$ on the training set for each candidate unlabeled instance $x_u$, where $h$ is the original learner and $h'$ is the newly trained learner by adding $x_u$ to the training set. Finally, the instance with the largest positive $\Delta_{x_u}$ would be chosen to add into the training set. One can see that we need to re-train the model $|U|$ times in each instance selection round when the two regression models are parameterized (e.g., SVM, Neural networks), which demands a lot of computational resources.

To address the above problem, we utilize the disagreement information of the three outcome prediction models to select unlabeled instance without re-training the models in each instance selection round. Next we introduce the co-training

---

**Algorithm 1.** *SemiITE*

---

**Input:** The labeled set $L = \{(\boldsymbol{x}_i, t_i, y^{t_i})\}_{i=1}^{N}$, unlabeled set $U = \{\boldsymbol{x}_i\}_{j=1}^{M}$, and the maximum number of instance selection rounds $T$

**Output:** The shared module $M_s$ and three outcome prediction models $M_1$, $M_2$, and $M_3$

  1: **Initialization:**
  2: Build network modules $M_s$, $M_1$, $M_2$, and $M_3$
  3: $L_1, L_2, L_3 = L$
  4: Train $M_s$, $M_1$, $M_2$, and $M_3$ based on $\mathcal{L}$ in Eq. (6)
  5: **Training:**
  6: **for** $t = 1 \rightarrow T$ **do**
  7:    **for** $v = 1 \rightarrow 3$ **do**
  8:      $CL_v = \emptyset$
  9:      $\boldsymbol{x}_u \leftarrow$ chosen unlabeled instance based on Eq. (7)
10:      $CL_v = CL_v \cup (\boldsymbol{x}_u, 1, f_k(f_s(\boldsymbol{x}_u), 1)) \ (k \neq v)$
11:      $\boldsymbol{x}_w \leftarrow$ chosen unlabeled instance based on Eq. (8)
12:      $CL_v = CL_v \cup (\boldsymbol{x}_w, 0, f_h(f_s(\boldsymbol{x}_w), 0)) \ (h \neq k, v)$
13:      $\widetilde{L}_v = L_v \cup CL_v$
14:      **if** $v = 1$ **then**
15:        Train $M_s$ and $M_1$ on $\widetilde{L}_v$
16:      **else**
17:        Train $M_v$ on $\widetilde{L}_v$
18:      **end if**
19:    **end for**
20: **end for**
21: **return** $M_s$, $M_1$, $M_2$ and $M_3$

---

strategy of *SemiITE* to avoid the re-training issue when estimating individual treatment effect. First, we utilize the set of labeled observational instances $L$ with treatment assignments and corresponding factual outcomes to train the initialized holistic model, then we can obtain a trained inference model denoted as $\mathcal{M}$. Then we begin to conduct the co-training procedure to choose instances from unlabeled set $U$ for $T$ rounds. In each round, we first fit the unlabeled instances only with covariates into the trained inference model $\mathcal{M}$, then we can obtain the potential outcomes $y_i^{t_i=1}|v$ and $y_i^{t_i=0}|v$ $(v = 1, 2, 3)$ by the function $f_v(\boldsymbol{h}_i, t_i)$ for each unlabeled instance $i$. After that, we choose unlabeled instances by the disagreement information between $M_1$, $M_2$, and $M_3$ and add the selected instances with their corresponding treatment assignment and predicted outcome into the training set. More specifically, we take a strategy that if the disagreement (i.e., the difference) between two outcome prediction models (e.g., $M_2$ and $M_3$) on the prediction of instance $i$ from unlabeled set $U$ is minimal, then the two outcome prediction models will teach the third outcome prediction model (e.g., $M_1$) on this instance. Then we add the instance $i$ with its covariates, treatment assignment, and the corresponding outcome predicted by $M_2$ and $M_3$ into the training set of instances to train the third prediction model $M_1$. One can see that there are two predicted outcomes (e.g., by $M_2$ and $M_3$) for an instance $i$ with $t_i$, here we randomly choose one of the two outcomes as the prediction to add into the trained set $L$. We take an example that chooses unlabeled instances for training prediction model $M_1$ by the disagreement of $M_2$ and $M_3$. The selection formulation can be written as:

$$\arg\min_{\boldsymbol{x}_u \in U} \| f_2(f_s(\boldsymbol{x}_u), t = 1) - f_3(f_s(\boldsymbol{x}_u), t = 1) \|^2 \tag{7}$$

$$\arg\min_{\boldsymbol{x}_v \in U} \| f_2(f_s(\boldsymbol{x}_v), t = 0) - f_3(f_s(\boldsymbol{x}_v), t = 0) \|^2, \tag{8}$$

then we will add the selected instances with pseudo-labels and treatment assignments $(\boldsymbol{x}_u, t_u = 1, f_2(f_s(\boldsymbol{x}_u), 1))$ and $(\boldsymbol{x}_v, t_v = 0, f_3(f_s(\boldsymbol{x}_v), 0))$ into the trained set $L$ to train $M_s$ and $M_1$ based on the loss function in Eq. (6) without MSE losses of $M_2$ and $M_3$. Similarly, $M_2$ and $M_3$ will be updated by such instance selection procedure while the shared module would be maintained when training $M_2$ and $M_3$.

The summary of the proposed framework $SemiITE$ is shown in Algorithm 1. Noting that the proposed framework $SemiITE$ avoids the re-training issue by utilizing the disagreement information of the three outcome prediction models in each instance selection round, which can greatly reduce the computational cost of the co-training framework. After finishing the co-training procedure, we can infer the individual treatment effect for an unseen instance by any of the three potential outcome prediction models. We use $M_1$ as an example to infer the ITE of an unseen instance $i$: $\hat{\tau}_i = f_1(f_s(\boldsymbol{x}_i), 1) - f_1(f_s(\boldsymbol{x}_i), 0)$.

## 4    Experiments

In this section, we present the experimental results of the proposed framework $SemiITE$, including ITE performance evaluation, ablation study, and hyperparameter study.

### 4.1    Experimental Setting

**Datasets.** We conduct the experiments on two benchmark real-world datasets IHDP [4] and Job training [7], which have been widely used in previous works of causal inference [17,26]. In IHDP, each instance's covariates include 25 variables that measure various aspects of children and their mothers. The treatment group's infants receive intensive high-quality childcare and specialist home visits, while the control group's infants do not, and the outcome is the infants' cognitive test scores. In Job training, each instance is an employee with 17 covariates such as age, education, and ethnicity. Instances in the treatment group participate in job training, while those in the control group do not. The outcome of each instance is the employment status.

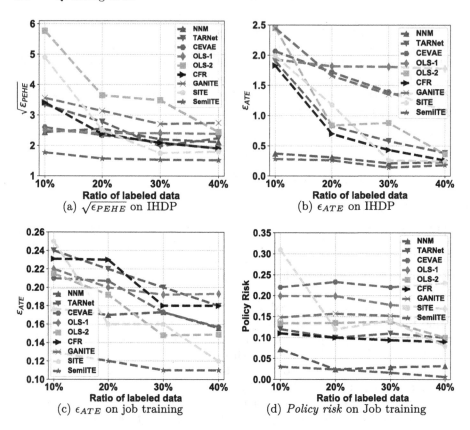

**Fig. 3.** ITE estimation performance comparison results for different methods on the IHDP data and the Job training data. The horizontal axis represents the proportion of labeled data, and the vertical one denotes the values of metrics (the lower the better).

**Baselines.** We compare the performance of the proposed framework *SemiITE* on ITE estimation with the following state-of-the-art causal inference models: (1) OLS-1 [26] is the ordinary linear regression model which treats the treatment assignment as a covariate of an instance. (2) OLS-2 [26] are two separated linear regression models for treatment ($t = 1$) and control ($t = 0$) instances. (3) Nearest neighbor matching (NNM) [6] is a matching-based method that infers the potential outcomes of an instance by using its nearby instances. Here, we use the Euclidean distance to measure the similarity between two instances. (4) Causal Effect Variational Autoencoder (CEVAE) [18] follows the causal structure of inference with proxies and builds deep latent variable model to estimate the unknown latent space that summarizes the confounders and the causal effect. (5) Counterfactual Regression (CFR) [26] is a multilayer perceptron based method to infer the counterfactual outcome and minimize the imbalance between treatment and control group. Here we use the Wasserstein-1 distance. (6) TARNet

[17] is a variant of the CFR model, which does not have a built-in representation balancing component. (7) GANITE [31] is a generative adversarial net based model to infer ITE. (8) SITE [30] infers the ITE by capturing hidden confounders and preserving local similarity of data. Due to the fact that the baselines are full-supervised methods, to ensure the fairness of the comparison, we adopt the manifold assumption [16] (i.e., the samples with similar inputs should get similar outputs) and design a corresponding term to add into the loss function for each baseline (except NNM). Assuming that the function $f(\boldsymbol{x}_i, t)$ denotes the predicted outcomes for a certain causal inference model (e.g., CFR) where $\boldsymbol{x}_i, t$ denote the original covariates and treatment assignment of unit $i$ respectively, then a corresponding term based on manifold assumption is added into the loss function of the model:

$$\mathcal{L}_m = \sum_t \sum_{i,j=1}^{N+M} w_{ij}(f(\boldsymbol{x}_i, t) - f(\boldsymbol{x}_j, t))^2, \tag{9}$$

where $N$ and $M$ represent the number of labeled and unlabeled samples, respectively. $w_{ij}$ denotes the similarity between two units $\boldsymbol{x}_i$ and $\boldsymbol{x}_j$. Here, we use the Gaussian kernel to compute the similarity for each unit pair.

**Evaluation Metrics.** For the IHDP dataset, we adopt two widely used metrics in causal inference for ITE estimation: (1) Rooted Precision in Estimation of Heterogeneous Effect $\sqrt{\epsilon_{PEHE}} = \sqrt{\frac{1}{n}\sum_{i=1}(\tau_i - \hat{\tau}_i)^2}$, where $\tau_i = y_i^{t_i=1} - y_i^{t_i=0}$ and $\hat{\tau}_i = \hat{y}_i^{t_i=1} - \hat{y}_i^{t_i=0}$ are the ground truth ITE and the inferred ITE, respectively; and (2) Mean Absolute Error on ATE $\epsilon_{ATE} = \frac{1}{n}|\sum_{i=1}\hat{\tau}_i - \sum_{i=1}\tau_i|$. For the Job training dataset, we use $\epsilon_{ATE}$ and *policy risk*, which is detailed in previous work [17]. Lower values of all metrics denote better performance.

We spilt the data into training set, unlabeled data set, validation set, and test set, where the size of training set is limited. For both IHDP and Job training, we use different ratios of training data to evaluate the performance. We use {10%, 20%, 30%, 40%} of the whole data for labeled set of instances, {70%, 60%, 50%, 40%} for unlabeled set of instances to be selected in co-training rounds, and the rest of the data is used for validation and test set (5% for validation and 15% for test). Regarding the hyperparameters of the proposed framework, we utilize the grid search strategy to find the optimal hyperparameters combination based on the results of validation set. Specifically, for the shared module $M_s$, we set the number of hidden layer as 3, and the dimension of each hidden layer as 100. For the potential outcome prediction models $M_v$ ($v = 1, 2, 3$), the number of hidden layer varies in {2, 3, 4} and the dimension of each layer ranges in {200, 300, 400, 500}. The trade-off hyperparameters $\alpha$, $\beta$, and $\gamma$ are set in range {$10^{-1}, 10^{-2}, 10^{-3}, 10^{-4}$}. The maximum number of unlabeled instance selection rounds $T = 500$. We use the predictions of $M_1$ as the final inferred potential outcomes and run the experiments 10 times and the average performance of each method is reported. Besides, all codes are implemented by Python and we use Intel(R) Xeon(R) CPU E5-2690 v4 @ 2.60GHz 264G Memory, and NVIDIA Corporation GP100GL.

## 4.2   ITE Estimation Performance

We compare the proposed framework *SemiITE* against the aforementioned baseline methods with respect to the ITE estimation performance. The results are shown in Fig. 3. By analyzing the experiment results we can conclude that:

- **(i)** The supervised state-of-the-art baselines have unsatisfactory performance of ITE estimation when the proportion of labeled observation is low (e.g., below 20%), but their performance gradually improves with the increasing of proportion of labeled data and achieve satisfactory level when the proportion of labeled data is over 40%, which demonstrates that the existing methods require a plenty of labeled observational instances to support their effectiveness for ITE estimation.
- **(ii)** *SemiITE* clearly outperforms several supervised causal inference models with different ratios of labeled instances. And it is worthy to note that the lower the proportion of labeled observational data is, the greater the superiority of the proposed *SemiITE* over other causal inference models will exhibit, which illustrates that *SemiITE* can utilize those unlabeled instances effectively and extract the useful information from them for ITE estimation.
- **(iii)** The performance of *SemiITE* does not change significantly when the proportion of labeled data changes, indicating that *SemiITE* is stable, because *SemiITE* can utilize unlabeled instances to infer potential outcomes more precisely with limited labeled data. Its performance is not greatly affected by the proportion of labeled data, which illustrates that *SemiITE* would still be effective even if the labeled observational data is limited.

## 4.3   Ablation Study

Here we develop the following three variants of *SemiITE* to explore three components of the framework for the individual treatment effect estimation.

- **SemiITE w/o Shared Module**: This variant does not contain the shared module $M_s$ in the framework, which means that we directly train the three potential outcome models without capturing some hidden confounders. We denote this variant as *SemiITE w/o SM*.
- **SemiITE w/o representation balance**: This variant does not balance the distribution of representations generated by shared module $M_s$ for the treatment and control groups, i.e., the variant does not add the loss $\mathcal{L}_{IPM}$ into the overall loss function. We denote this variant as *SemiITE w/o RB*.
- **SemiITE w/o model diversity**: This variant does not consider the diversity of the three prediction models and the diversity term $\mathcal{L}_{div}$ is not added into the loss function. The same network structure, initialization, and optimization method are adopted for the three prediction models. We denote this variant as *SemiITE w/o MD*.

We conduct the ablation study experiment to compare the performance of the proposed *SemiITE* with the aforementioned variants. The results are shown in

Fig. 4. Due to the page limit, we only report the results on the IHDP dataset with labeled instance proportion $p = \{10\%, 20\%\}$, but similar observations can also be found in the other datasets and other settings. We have the following observations:

- *SemiITE w/o SM* performs the worst, which demonstrates the importance of capturing hidden confounders in individual treatment effect estimation.
- The performance of *SemiITE w/o RB* is also degraded, because it fails to control the confounding bias, which is a common problem in causal inference.
- *SemiITE w/o MD* also performs worse than the original framework *SemiITE* because the diversity of the three prediction models cannot be achieved, which is important in co-training based semi-supervised learning.

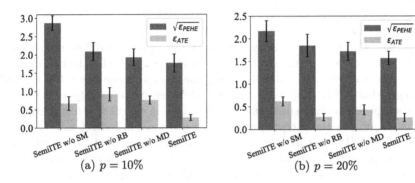

**Fig. 4.** Ablation study of *SemiITE* on the IHDP dataset.

### 4.4  Hyperparameter Study

We further explore the impact of two important hyperparameters $\alpha$ and $\beta$ in Eq. (6) on the performance of the proposed co-training based semi-supervised ITE estimation framework. We set the range of the two hyperparameters as $\{10^{-1}, 10^{-2}, 10^{-3}, 10^{-4}\}$ and the hyperparameter study results are shown in Fig. 5. Due to the page limit, we only report the results for $\sqrt{\epsilon_{PEHE}}$ and $\epsilon_{ATE}$ on IHDP with labeled data proportion $p = 10\%$. We have similar results for other datasets with different settings of $p$. We can observe that the performance is generally stable when the two hyperparameters vary, and the performance is relatively better when $\alpha$ and $\beta$ range in $\{0.001, 0.01\}$, which demonstrates the robustness of the proposed framework.

## 5  Related Work

**Causal Inference with Machine Learning.** Machine learning based causal inference methods have been shown to be effective in observational studies [13,29]. Among them, $k$-NN [23] is adopted as a matching strategy to find the

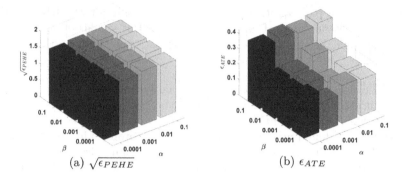

(a) $\sqrt{\epsilon_{PEHE}}$                    (b) $\epsilon_{ATE}$

**Fig. 5.** Hyperparameter study of *SemiITE* on IHDP with labeled data proportion $p = 10\%$.

instance pair with the closest distance in covariates space but different treatment assignments to obtain causal effect. OLS-1 and OLS-2 [28] infer the causal effect by predicting the potential outcomes using linear regression models. Counterfactual Regression (CFR) [26] casts counterfactual inference as a type of domain adaptation problem and estimates individual treatment effect using neural network by learning balanced representations for instances in control and treatment groups. CEVAE [18] captures the hidden confounders to estimate unbiased causal effect by mapping the original covariates to latent space with variational autoencoder [1]. Yao et al. [30] proposed a local similarity preserved individual treatment effect (SITE) estimation method based on deep representation learning, which can capture the hidden confounders and preserve local similarity of data. However, all these methods are supervised in nature and require massive labeled observational data.

**Semi-supervised Learning with Co-training.** The original co-training algorithm was proposed by [3], which assumes that there are two independent, sufficient, and redundant natural views in the sample space, then two separated models can be trained on these two views. Without multiple views of data sample space, many co-training algorithms are proposed to mine useful information from unlabeled data by a single view. For example, Goldman and Zhou [11] proposed to train two different decision tree models from a single view. Zhou and Li [33] adopted a re-sampling strategy to generate three sub-datasets from the original dataset and train three diverse classification models on each generated dataset. Zhou et al. [35] trained two K-NN regressors with different distance orders and chose the predicted unlabeled samples by making the error rate of the regressor reduced most after adding the unlabeled samples into training set. Chen et al. [8] proposed to train three different CNN models with model initialization, diversity augmentation, and pseudo-label editing in a co-training framework. However, most co-training methods are designed for classification and cannot be directly applied to the causal inference problem.

# 6    Conclusion

In this work, we propose a semi-supervised individual treatment effect estimation framework *SemiITE* via disagreement based co-training, which can effectively utilize unlabeled instances to aid ITE estimation. *SemiITE* chooses the most confident unlabeled instance predictions and then add them into the labeled instance set by the disagreement information of the base prediction models via a co-training strategy, which can avoid the re-training issue in each unlabeled instance selection round. Finally, extensive experiments on two public real-world datasets show the superiority of *SemiITE* on ITE estimation over the existing ITE estimation methods when the labeled data is limited.

**Acknowledgements.** Qiang Huang, Huiyan Sun, and Yi Chang are supported in part by the National Natural Science Foundation of China (No.U19A2065, No.61902144, No.61976102).

# References

1. Auto-encoding variational Bayes (2014)
2. Appice, A., Guccione, P., Malerba, D.: A novel spectral-spatial co-training algorithm for the transductive classification of hyperspectral imagery data. Pattern Recogn. **63**, 229–245 (2017)
3. Blum, A., Mitchell, T.: Combining labeled and unlabeled data with co-training. In: Proceedings of the Eleventh Annual Conference on Computational Learning Theory, pp. 92–100 (1998)
4. Brooks-Gunn, J., Liaw, F.R., Klebanov, P.K.: Effects of early intervention on cognitive function of low birth weight preterm infants. J. Pediatr. **120**(3), 350–359 (1992)
5. Cheng, L., Guo, R., Moraffah, R., Candan, K.S., Raglin, A., Liu, H.: A practical data repository for causal learning with big data. In: Proceedings of the 2019 International Symposium on Benchmarking, Measuring and Optimization, pp. 234–248 (2019)
6. Cui, P., et al.: Causal inference meets machine learning. In: Proceedings of the ACM SIGKDD International Conference on Knowledge Discovery and Data Mining, pp. 3527–3528 (2020)
7. Dehejia, R.H., Wahba, S.: Causal effects in nonexperimental studies: Reevaluating the evaluation of training programs. J. Am. Stat. Assoc. **94**(448), 1053–1062 (1999)
8. Dong-DongChen, W., WeiGao, Z.H.: Tri-net for semi-supervised deep learning. In: Proceedings of Twenty-seventh International Joint Conference on Artificial Intelligence, pp. 2014–2020 (2018)
9. Glass, T.A., Goodman, S.N., Hernán, M.A., Samet, J.M.: Causal inference in public health. Annu. Rev. Public Health **34**, 61–75 (2013)
10. Glorot, X., Bengio, Y.: Understanding the difficulty of training deep feedforward neural networks. In: Proceedings of the Thirteenth International Conference On Artificial Intelligence and Statistics, pp. 249–256. JMLR Workshop and Conference Proceedings (2010)
11. Goldman, S., Zhou, Y.: Enhancing supervised learning with unlabeled data. In: International Conference on Machine Learning, pp. 327–334. CiteSeer (2000)

12. Gretton, A., Bousquet, O., Smola, A., Schölkopf, B.: Measuring statistical dependence with Hilbert-Schmidt norms. In: Jain, S., Simon, H.U., Tomita, E. (eds.) ALT 2005. LNCS (LNAI), vol. 3734, pp. 63–77. Springer, Heidelberg (2005). https://doi.org/10.1007/11564089_7
13. Guo, R., Cheng, L., Li, J., Hahn, P.R., Liu, H.: A survey of learning causality with data: problems and methods. ACM Comput. Surv. (CSUR) **53**(4), 1–37 (2020)
14. Guo, R., Li, J., Liu, H.: Learning individual causal effects from networked observational data. In: Proceedings of the 13th International Conference on Web Search and Data Mining, pp. 232–240 (2020)
15. Hill, J.L.: Bayesian nonparametric modeling for causal inference. J. Comput. Graph. Stat. **20**(1), 217–240 (2011)
16. Iscen, A., Tolias, G., Avrithis, Y., Chum, O.: Label propagation for deep semi-supervised learning. In: Proceedings of the IEEE/CVF Conference on Computer Vision and Pattern Recognition, pp. 5070–5079 (2019)
17. Johansson, F., Shalit, U., Sontag, D.: Learning representations for counterfactual inference. In: International Conference on Machine Learning, pp. 3020–3029. PMLR (2016)
18. Louizos, C., Shalit, U., Mooij, J., Sontag, D., Zemel, R., Welling, M.: Causal effect inference with deep latent-variable models. arXiv preprint arXiv:1705.08821 (2017)
19. Ma, F., Meng, D., Dong, X., Yang, Y.: Self-paced multi-view co-training. J. Mach. Learn. Res. **21**(57), 1–38 (2020)
20. Ma, J., Dong, Y., Huang, Z., Mietchen, D., Li, J.: Assessing the causal impact of COVID-19 related policies on outbreak dynamics: a case study in the US. In: Proceedings of the ACM Web Conference 2022, pp. 2678–2686 (2022)
21. Nigam, K., Ghani, R.: Analyzing the effectiveness and applicability of co-training. In: Proceedings of the Ninth International Conference on Information and Knowledge Management, pp. 86–93 (2000)
22. Ning, X., et al.: A review of research on co-training. Concurrency and Computation: Practice and Experience, p. e6276 (2021)
23. Peterson, L.E.: K-nearest neighbor. Scholarpedia **4**(2), 1883 (2009)
24. Rubin, D.B.: Estimating causal effects of treatments in randomized and nonrandomized studies. J. Educ. Psychol. **66**(5), 688 (1974)
25. Rubin, D.B.: Causal inference using potential outcomes: design, modeling, decisions. J. Am. Stat. Assoc. **100**(469), 322–331 (2005)
26. Shalit, U., Johansson, F., et al.: Estimating individual treatment effect: generalization bounds and algorithms. In: International Conference on Machine Learning, pp. 3076–3085 (2017)
27. Varian, H.R.: Causal inference in economics and marketing. Proc. Natl. Acad. Sci. **113**(27), 7310–7315 (2016)
28. Weisberg, S.: Applied linear regression, vol. 528. John Wiley & Sons (2005)
29. Yao, L., Chu, Z., Li, S., Li, Y., Gao, J., Zhang, A.: A survey on causal inference. arXiv preprint arXiv:2002.02770 (2020)
30. Yao, L., Li, S., Li, Y., Huai, M., Gao, J., Zhang, A.: Representation learning for treatment effect estimation from observational data. In Advances in Neural Information Processing Systems 31 (2018)
31. Yoon, J., Jordon, J., Van Der Schaar, M.: GANITE: estimation of individualized treatment effects using generative adversarial nets. In: International Conference on Learning Representations (2018)
32. Zhang, X., Song, Q., Liu, R., Wang, W., Jiao, L.: Modified co-training with spectral and spatial views for semisupervised hyperspectral image classification. IEEE J. Select. Top. Appl. Earth Observ. Remote Sens. **7**(6), 2044–2055 (2014)

33. Zhou, Z.H., Li, M.: Tri-training: exploiting unlabeled data using three classifiers. IEEE Trans. Knowl. Data Eng. **17**(11), 1529–1541 (2005)
34. Zhou, Z.H., Li, M.: Semi-supervised learning by disagreement. Knowl. Inf. Syst. **24**(3), 415–439 (2010)
35. Zhou, Z.H., Li, M., et al.: Semi-supervised regression with co-training. In: Proceedings of International Joint Conference on Artificial Intelligence, vol. 5, pp. 908–913 (2005)

# Multi-task Adversarial Learning for Semi-supervised Trajectory-User Linking

Sen Zhang[1,2], Senzhang Wang[2,3], Xiang Wang[4(✉)], Shigeng Zhang[3], Hao Miao[5], and Junxing Zhu[4]

[1] Nanjing University of Aeronautics and Astronautics, Nanjing, China
zhangsen@nuaa.edu.cn
[2] Shenzhen Research Institute, Nanjing University of Aeronautics and Astronautics, Shenzhen, China
[3] Central South University, Changsha, China
{szwang,sgzhang}@csu.edu.cn
[4] National University of Defense Technology, Changsha, China
{xiangwangcn,zhujunxing}@nudt.edu.cn
[5] Aalborg University, Aalborg, Denmark
haom@cs.aau.dk

**Abstract.** Trajectory-User Linking (TUL), which aims to link the trajectories to the users who have generated them, is critically important to many real applications. Existing approaches generally consider TUL as a supervised learning problem which requires a large number of labeled trajectory-user pairs. However, in real scenarios users may not be willing to make their identities publicly available due to data privacy concerns, leading to the scarcity of labeled trajectory-user pairs. In addition, the trajectory data are usually sparse as users will not always check-in when they go to POIs. To address these issues, in this paper we propose a multi-task adversarial learning model named TULMAL for semi-supervised TUL with spare trajectory data. Specifically, TULMAL first conducts sparse trajectory completion through a proposed seq2seq model. Kalman filter is also coupled into the decoder of the seq2seq model to calibrate the generated new locations. The completed trajectories are next input into a generative adversarial learning model for semi-supervised TUL. The insight is that we consider all the users and their trajectories as a whole and perform TUL in the data distribution level. We first project users and trajectories into the common latent feature space through learning a projection function (generator) to minimize the distance between the user distribution and the trajectory distribution. Then each unlabeled trajectory will be linked to the user who is closest to it in the latent feature space without much guidance of labels. The two tasks are jointly conducted and optimized under a multi-task learning framework. Extensive experimental results on two real-world trajectory datasets demonstrate the superiority of our proposal by comparison with existing approaches.

**Keywords:** Trajectory completion · Trajectory-user linking · Adversarial learning · Multi-task learning

---

S. Zhang and S. Wang—Equal contribution.

M.-R. Amini et al. (Eds.): ECML PKDD 2022, LNAI 13716, pp. 418–434, 2023.
https://doi.org/10.1007/978-3-031-26412-2_26

# 1 Introduction

With the rapid development of satellite positioning technology and location-based services, many location-related mobile data such as human trajectory data and taxi OD data are ubiquitous nowadays. The large volume of human mobility data can facilitate us to have a better understanding on human behavior patterns and provide great opportunities for various trajectory mining tasks [11,13,33]. For example, the user mobility data collected from location-based social networks (LBSNs) such as Foursquare, can be used for POI recommendation [5,17], trajectory classification [2,4], traffic prediction [20–24,29] and human mobility prediction [6,9,14].

In many online applications, users usually are not willing to make their personal identity information associated with their trajectories publicly available due to privacy concerns. In such a case the platforms can only collect the trajectory data, but the users who have generated them are unknown. Linking the trajectories to the users who have generated them, which is also called Trajectory User Linking (TUL), is fundamentally important to many tasks such as personalized POI recommendation and terrorists/criminal identification [7].

Existing works generally model TUL as a supervised learning task, which use RNN-based methods to learn a projection function between trajectories and users based on a large number of labeled trajectory-user pairs. TULER [7] is the first model proposed to address the TUL problem, which uses RNN to model the trajectory sequences and learn the dependencies between the location points to the users for TUL. Zhou et al. [33] proposed to use variational autoencoder to learn the hierarchical semantic features of user trajectories. The unlabeled data was also incorporated to deal with the data sparsity issue to improve TUL performance. DeepTUL [15] focused on the TUL task by learning the multi-periodic nature of user mobility from their historical trajectories and exploiting both spatial and temporal features of the trajectory data.

However, the performance of existing works may not be promising in real application scenarios due to the following two major challenges. First, in many LBSN platforms like Foursquare, users will not always check in and share their locations when going to a POI. It is common that users are not willing to check in due to data privacy concerns, which results in huge amount of sparse and incomplete trajectories. Existing works mostly consider the user trajectories are complete, and thus they may not be applicable in real applications. Second, existing supervised learning based methods need a large number of annotated trajectory-user pairs, which is extreamly time consuming and costly. How to conduct TUL with a few labeled trajectory-user pairs under a semi-supervised learning framework is challenging and less explored.

To address the above challenges, this paper proposes a multi-task adversarial learning model named TULMAL to perform sparse trajectory completion and semi-supervised TUL simultaneously. Specifically, TULMAL first completes the sparse raw trajectory data through a proposed seq2seq model. The sparse trajectory data are first input the encoder of the seq2seq model to learn the latent feature representations. Then motivated by the effectiveness of Kalman filter

[10] in calibrating noise estimation for temporal data, we adopt Kalman filtering to calibrate the estimated new locations of the completed trajectory, which is coupled into the decoder of the seq2seq model. The completed trajectories are then input into an adversarial learning model for TUL. Instead of matching the trajectory-user pairs one by one, we consider all the users and trajectories as a whole and perform TUL from the data distribution level. We aim to learn a projection function $\Phi$ to embed users and trajectories into a common latent space. With the assumption that two users are similar if their trajectories are similar and vice versa, the projection $\Phi$ should make the trajectory close to the user who has generated it in the feature space. To this end, TULMAL uses an adversarial learning framework to learn the projection function $\Phi$. Specifically, TULMAL contains an encoder $E$, a decoder $O$ and a discriminator $D$. Encoder $E$ maps the feature vectors of trajectories into a shared latent space, and decoder $O$ projects the latent space features into the user space as the generated samples. The encoder and decoder together work as a projection function $\Phi$. The discriminator $D$ aims to distinguish the real instances of users from the samples generated by the decoders. Through adversarial learning, the discriminator essentially estimates the approximate Wasserstein distance between the user distribution and the projected trajectory distribution. Through the competition with discriminators, the projection function $\Phi$ will be updated to minimize the estimated Wasserstein distance. Given a new unlabeled trajectory, it will be projected into the user space by $\Phi$ first and then be linked to the user who is closest to it. In summary, our main contributions are as follows:

- To the best of our knowledge, we are the first to study the semi-supervised TUL problem with sparse and incomplete trajectory data.
- A novel model TULMAL is proposed to effectively address the studied problem. TULMAL first conducts sparse trajectory completion with a Kalman filter enhanced seq2seq model, and then performs semi-supervised TUL with an adversarial learning framework. The two tasks are jointly conducted under a multi-task learning framework.
- We conduct extensive experiments on two real-world datasets to evaluate the effectiveness of TULMAL. The experimental results show that our model provides significant performance improvement over existing state-of-the-arts.

## 2   Related Work

### 2.1   Trajectory Completion

Existing works for trajectory completion can be roughly categorized into the following two types. The first type of works is to directly complete the trajectories with missing locations, and the second type aims to recovery the trajectories through the next step or short-term POI prediction. For the direct location completion approach, Zheng et al. [31] proposed to infer the missing part of sparse trajectories by comparing the similarity of historical trajectories with the sparse trajectories. An attentional neural network model AttnMove [26] was proposed to complete individual trajectories by recovering unobserved locations based on historical trajectories. Xi et al. [25] proposed a bidirectional spatial and temporal dependency and a dynamic preference model of users to identify missing POI check-ins. MTrajRec [16] used a seq2seq multi-task learning model to patch missing trajectory points while mapping them to the road network.

For next location prediction works, STRNN proposed by Liu et al. [12] tried to model the temporal and spatial context of each layer. Specifically, STRNN used a specific excess matrix for different time intervals and geographical distances to predict the next POI. Feng et al. [3] proposed a recurrent neural network with multimodal embedding named DeepMove to capture the complex sequential transitions by jointly embedding multiple factors that governed human movement. DeepMove also used a historical attention model with two mechanisms to capture multi-level periodicity, effectively exploiting the nature of periodicity to enhance recurrent neural networks' mobility prediction. However, a significant drawback of existing works is that they are not effective to reduce data noise in trajectory data completion.

## 2.2  Trajectory-User Linking

Existing TUL models are mostly supervised, which use machine learning models especially RNN-based approaches to learn a projection function between users and trajectories through a large number of labeled trajectory-user pairs. TULER [7] is the first model proposed for TUL. It used RNN to model the trajectory sequence for capturing the dependencies between location points. DeepTUL [15] learned the multi-periodic nature of user mobility from the user's historical trajectory and used both spatial and temporal features of the trajectory data for user-trajectory matching. Considering the large number of users, TULSN [28] was proposed to model the trajectory data by linking networks, and only a small amount of trajectory data were needed for training the model. TULVAE [33] was a novel semi-supervised variational autoencoder framework, which used variational autoencoder to learn the hierarchical semantic features of user trajectories and incorporated unlabeled data to solve the data sparsity problem for TUL. However, the data sparsity issue in many real scenarios is largely ignored by existing works. Existing works usually need a large number of annotated trajectory-user pairs, which is labor intensive and costly thus infeasible in many applications.

## 3  Problem Definition

In this section, we will first give definitions of some terminologies, and then formally define the studied problem.

**Definition 1.** *(Cell region). We divide a city under study into a set of equal-sized grid cells, denoted as $R$. Each cell region $r \in R$ is a square region. The coordinates of a cell region $r$ are denoted by its latitude and longitude $\langle x_r, y_r \rangle$.*

**Definition 2.** *(Trajectory).* *A trajectory* $\widetilde{T} = \langle s_1, s_2, \cdots, s_n \rangle$ *is defined as a sequence of geographically located points with time order, where* $s_i = \langle x, y, t, r \rangle$ *represents a location point consisting of latitude* $x$, *longitude* $y$, *timestamp* $t$ *and the cell region* $r$ *where* $s_i$ *is located.*

Note that $\widetilde{T}$ is sparse with some locations missing. We denote the complete trajectory as $T = (p_1, p_2, ..., p_n)$, where $T$ includes all visited locations and $\widetilde{T} \in T$. The sparse trajectory dataset $\widetilde{\mathbf{T}} = \left\{ \widetilde{T}_1, \widetilde{T}_2, \cdots, \widetilde{T}_m \right\}$ contains a small number of labeled or linked trajectories $\widetilde{\mathbf{T}}^l$ and a large number of unlabeled or unlinked ones $\widetilde{\mathbf{T}}^u$. Let $\mathbf{U} = \{u_1, u_2, \cdots, u_n\}$ denote the user set. We assume that each user has some linked trajectories, and the linked trajectory set associated with each user is denoted as $\widetilde{\mathbf{T}}^l = \left\{ (\widetilde{T}^{u_1}, u_1), (\widetilde{T}^{u_2}, u_2), \cdots, (\widetilde{T}^{u_n}, u_n) \right\}$, where $(\widetilde{T}^{u_i}, u_i)$ means trajectories $\widetilde{T}^{u_i}$ belongs to $u_i$, and thus $\widetilde{T}^{u_i}$ can be used to represent $u_i$. The studied problem is formally defined as follows.

**Problem Statement.** Given the sparse trajectory set $\widetilde{\mathbf{T}}$, the user set $\mathbf{U}$ and some trajectory-user pairs $\widetilde{\mathbf{T}}^l = \left\{ (\widetilde{T}^{u_1}, u_1), (\widetilde{T}^{u_2}, u_2), \cdots, (\widetilde{T}^{u_n}, u_n) \right\}$, our goal is to complete the trajectories in $\widetilde{\mathbf{T}}$ to obtain the complete trajectories $\mathbf{T} = \{\mathbf{T}^l, \mathbf{T}^u\}$, and then learn a projection function $\Phi$ to link all the trajectories in $\mathbf{T}^u$ to the users in $\mathbf{U}$.

## 4   Methodology

Figure 1 shows the model framework, which contains the trajectory completion step and the adversarial learning based TUL step. In the first step, we propose SeqKF model which integrates the Seq2Seq model with Kalman filter to accomplish the trajectory completion task. We first obtain the cell region where the coordinates are located by Seq2Seq. Then Kalman filtering is used for fine-grained calibration to obtain the exact coordinate values. In the adversarial learning step, we aim to learn a projection function that minimizes the distance between the generated trajectory distribution and the user distribution.

### 4.1   SeqKF for Trajectory Completion

The proposed SeqKF model for trajectory completion consists of the encoder and the decoder with Kalman filter. The encoder learns the spatio-temporal dependency of the trajectories, while the decoder generates the completed trajectory. Inspired by [32], instead of directly predicting the coordinate values of the missing trajectory points, we predict the grid cells where it is located. This approach allows for easier modeling than using coordinate values directly. Then we predict the cell region where the location point is located by the Seq2Seq model and use the center of the cell as the predicted coordinate value of the location point. Finally, we correct the predicted coordinate values by a Kalman filter to obtain an accurate prediction.

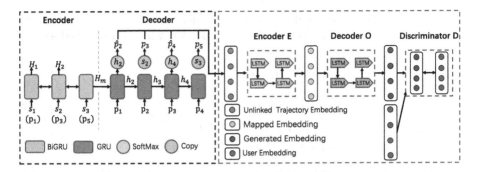

**Fig. 1.** The TULMAL model. The SeqKF step is in the black dashed box, and the adversarial learning based TUL step is in the red dashed box. (Color figure online)

**Encoder.** We encode the input sparse trajectory as a fixed vector and feed it into the Bidirectional Gated Recurrent Unit (BiGRU). GRU is a variant of Long Short Term Memory (LSTM) network, which is able to learn long-term dependencies of continuous data without performance degradation, while BiGRU can capture both forward and backward temporal dependencies.

BiGRU is actually two GRUs processing the data into forward and backward paths, and then combining the outputs in these two directions to obtain the final hidden state. The forward and backward hidden states in time step $i$ are denoted as $\overrightarrow{H_i}$ and $\overleftarrow{H_i}$. Then the output $H_i$ in time step $i$ is the combination of $\overrightarrow{H_i}$ and $\overleftarrow{H_i}$, i.e. $H_i = \overrightarrow{H_i} + \overleftarrow{H_i}$.

**Decoder with Kalman Filter.** The GRU decoder is used to recover the sparse trajectories $\widetilde{\mathbf{T}} = \langle s_1, s_2, \cdots, s_m \rangle$. Unlike the standard Seq2Seq, in our model, for the points presenting in the raw trajectory, we use the idea of replication operation that is widely used in NLP tasks [8,27]. We replicate these points directly from the output slot of the decoder as follows

$$p_i = \begin{cases} \hat{p}_i, & \text{if } j_k < i < j_{k+1}, \\ s_k, & \text{if } i = j_k, \end{cases} \tag{1}$$

where $\hat{p}_i$ represents the cell where the missing points are predicted by decoder.

To take global relevance into account, we add an attention mechanism so that the hidden unit $h_i$ in the decoder is updated by

$$h_i = GRU(h_{i-1}, p_{i-1}, a_i, H_m), \tag{2}$$

where $H_m$ is the output state from the encoder and $a_i$ is the weighted sum computed from all output vectors $H$ in the encoder, which is expressed as

$$
\begin{aligned}
a_i &= \sum_{i=1}^{m} \alpha_{i,k} H_i, \\
\alpha_{i,k} &= \frac{exp(u_{i,k})}{\sum_{k'=1}^{m} exp(u_{i,k'})}, \\
u_{i,k} &= v^T \cdot tanh(W_h h_i + W_H H_k),
\end{aligned}
\tag{3}
$$

where $v$, $W_h$ and $W_H$ are learnable parameters and $h_i$ denotes the current state of the decoder. When we obtain the hidden state $h_i$ from the decoder, for points that are not in the trajectory, we apply the softmax function to generate the corresponding cells of the missing trajectory points conditional on the probability of $p(c|h_i)$ as

$$
pro(c|h_i) = \frac{exp(h_i^T \cdot w_c)}{\sum_{c' \in C} exp(h_i^T \cdot w_{c'})},
\tag{4}
$$

where $w_c$ is the $c$-th column vector of the trainable parameter matrix $\mathbf{W}_c$.

Now we have the cell regions corresponding to the missing locations, we next combine the Kalman filter (KF) with the decoder to estimate the exact locations. KF is essentially an optimal state estimator under the assumption of linear and Gaussian noise. It is used to calibrate the coarse-grained predictions generated by the Seq2Seq output. In the KF model, we denote the state of the object at timestep $k$ as $g_k$, which is denoted as

$$
g_k = \mathbf{A} g_{k-1} + d_g, \quad d_g \sim \mathbf{N}(0, \mathbf{P}),
\tag{5}
$$

where $\mathbf{A}$ is the state update matrix, $d_g$ denotes Gaussian noise, and $\mathbf{P}$ denotes the covariance matrix of $d_g$. The current state can be obtained from the measurement value $z_k$, which is denoted as

$$
z_k = \mathbf{B} g_k + e_z, \quad e_z \sim \mathbf{N}(0, \mathbf{Q}),
\tag{6}
$$

where $\mathbf{B}$ is the measurement matrix, $e_z$ denotes the measurement Gaussian noise, and $\mathbf{Q}$ denotes the covariance matrix of $e_z$. The KF model mainly estimates the true state value $g$ based on the predicted value $\hat{g}^-$ and the measured value $z$, and it is divided into two steps: prediction and calibration.

In the prediction phase, the KF model predicts the prior value $\hat{g}^-$ and the prior error covariance matrix $\mathbf{R}^-$ at time step $k$ by the following equations

$$
\hat{g}_k^- = \mathbf{A} \hat{g}_{k-1},
\tag{7}
$$

$$
\mathbf{R}_k^- = \mathbf{A} \mathbf{R}_{k-1} \mathbf{A}^T + \mathbf{P}.
\tag{8}
$$

In the calibration phase, the KF model obtains the posteriori estimated state value $\hat{g}$ and updates the covariance matrix $\mathbf{R}$ by using the measured value $z$ and

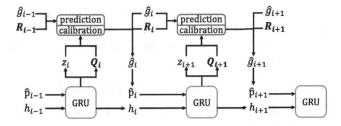

**Fig. 2.** Details of the combination of the decoder and Kalman filter, where the red part shows the two steps of updating and calibration of the Kalman filter. (Color figure online)

the a priori value $\hat{g}^-$ by the following equations

$$\hat{g}_k = \hat{g}_k^- + K_k \left( z_k - \mathbf{B}\hat{g}_k^- \right), \tag{9}$$

$$\mathbf{R_k} = (\mathbf{I} - K_k\mathbf{B}) \mathbf{R_k^-}, \tag{10}$$

where $K_k$ is the optimal Kalman gain, which combines the priori value $\hat{g}^-$ and the measured value $z$ for final estimation. $K_k$ is given by the following equation

$$K_k = \mathbf{R_k^-}\mathbf{B^T} \left( \mathbf{B}\mathbf{R_k^-}\mathbf{B^T} + \mathbf{Q} \right)^{-1}. \tag{11}$$

$K_k$ is used to denote the importance of the estimation error covariance matrix $\mathbf{R_k}$ and the measurement error covariance matrix $\mathbf{Q}$.

KF has two inputs $z$ and $\mathbf{Q}$. For the measurement $z$, we take out the center coordinates $(x_{c_i}, y_{c_i})$ of the prediction unit $c_i$ output from decoder as the value of $z_i$. For $\mathbf{Q}$, the traditional KF model sets $\mathbf{Q}$ as a fixed prior parameter, but intuitively, the uncertainty of the measurements should be constantly changing at different time periods. Therefore, we introduce a dynamic covariance matrix. For a given set of grid cells $R$, we aggregate the central coordinates of all grids $r$ into a matrix $\mathbf{V}$ of size $2 * |R|$. At each timestamp $i$, we calculate the currently estimated expected coordinate vector by

$$\bar{v}_i = \sum_{c' \in G} pro \left( c'|h_i \right) \cdot v_{c'}, \tag{12}$$

where $pro\left(c'|h_i\right)$ is the predicted probability of $c'$ calculated by the softmax function in Eq. (4), and $u_{c'}$ represents the $c$-th column vector in the matrix $\mathbf{V}$. We then calculate the covariance matrix $\mathbf{Q}$ by

$$\mathbf{Q_i} = \sum_{r' \in R} pro \left( r'|h_i \right) \cdot \left( v_{r'} - \bar{v}_i \right) \cdot \left( v_{r'} - \bar{v}_i \right)^T, \tag{13}$$

where $\{pro\left(r'|h_i\right)\}_{r' \in R}$ is used to combine the covariance matrix of each cell as the expected covariance of the measurement $z$.

As shown in Fig. 2, at each timestep $i$, the decoder cell feeds the center coordinates of the predicted cell into the KF component, and KF calibrates the observation $z_i$ through two procedures, prediction and correction, to obtain the final prediction $g_i$. Then $g_i$ is further discredited into a grid cell $\hat{p}_i$, which is used as the input of the next decoder cell. By this combination, the prediction noise can be effectively reduced.

**Loss Function.** Given a training set $\mathbf{D} = \left\{\mathbf{T}, \widetilde{\mathbf{T}}\right\}$ containing a set of sparse trajectories $\widetilde{\mathbf{T}}$ and the corresponding completed trajectories $\mathbf{T}$, we use the cross entropy as the loss function.

$$L_1 = \sum_{(\mathbf{T},\widetilde{\mathbf{T}})\in\mathbf{D}} -log(pro(\mathbf{T}|\widetilde{\mathbf{T}})). \tag{14}$$

To optimize the KF component, we use the mean squared error as the loss function, which is defined as

$$L_2 = \frac{1}{2} \sum_{(\mathbf{T},\widetilde{\mathbf{T}})\in\mathbf{D}} \sum_{p_i\in\mathbf{T} \ and \ p_i\notin\widetilde{T}} \left(\begin{bmatrix} p_i.x \\ p_i.y \end{bmatrix} - \hat{g}_i\right)^2. \tag{15}$$

Then the loss function of the trajectory completion task can be expressed as

$$L_{coml} = L_1 + \lambda L_2, \tag{16}$$

where $\lambda$ is a parameter to balance the importance of the two terms.

### 4.2 Adversarial Learning for Semi-supervised TUL

**Generator.** The generator aims to generate the representations of the trajectories from the raw feature space to the user space, and it consists of an encoder and a decoder. The encoder is responsible for mapping the input trajectories into a latent space, and the decoder is responsible for projecting the latent embedding in the latent space into the target user space. We use LSTM as the encoder and decoder. After mapping the trajectories to the target user space by the generator, we can identify the real instances from the users by the discriminator $D$. Next we will derive the discriminator needed for the TUL task starting from the objective function.

**Objective Function.** Given the distribution $\mathbb{D}^{\mathbf{T}}$ represented by the set of trajectories and the distribution $\mathbb{D}^{\mathbf{U}}$ represented by the set of users, the objective of TULMAL can be defined as follows:

$$\min_{\Phi} WD(\mathbb{D}^{\mathbf{U}}, \mathbb{D}^{\Phi(\mathbf{T})}) = \inf_{\gamma\in\Upsilon(\mathbb{D}^{\mathbf{U}},\mathbb{D}^{\Phi(\mathbf{T})})} \mathbb{E}_{(\mathbf{U},\Phi(\mathbf{T}))\sim\gamma}[d(\mathbf{U}, \Phi(\mathbf{T}))]. \tag{17}$$

The right-hand side of Eq. (17) is a representation of the Wasserstein distance, which measures the distance between two probability distributions $\mathbb{D}^{\mathbf{U}}$ and $\mathbb{D}^{\Phi(\mathbf{T})}$. $\Upsilon(\mathbb{D}^{\mathbf{U}}, \mathbb{D}^{\Phi(\mathbf{T})})$ is the set of all possible joint probability distributions for the combination of distributions $\mathbb{D}^{\mathbf{U}}$ and $\mathbb{D}^{\Phi(\mathbf{T})}$. $d$ represents the distance between two points (set as Euclidean distance in this paper). WD aims to find the ideal joint distribution $\varUpsilon$ to reach the expectation infimum. However, it is difficult to compute $inf_{\gamma \in \varUpsilon(\mathbb{D}^{\mathbf{U}}, \mathbb{D}^{\varPhi(\mathbf{T})})}$ [30] by traversing all joint distributions. The work [18] presents a simple version of WD, which can be formulated as follows when Kantorovich-Rubinstein duality is satisfied:

$$WD = \frac{1}{K} \sup_{\|f\|_L \leq K} \mathbb{E}_{\mathbf{U} \sim \mathbb{D}^{\mathbf{U}}} f(\mathbf{U}) - \mathbb{E}_{\varPhi(\mathbf{T}) \sim \mathbb{D}^{\varPhi(T)}} f(\varPhi(\mathbf{T})). \qquad (18)$$

where the function $f$ is required to be K-Lipschitz continuous. For Eq. (18), we have to learn an ideal K-Lipschitz function $f$ to implement it. Since the neural network itself has a powerful approximation capability, we choose a multilayer feedforward network to find $f$. It can be regarded as a discriminator $D$ that distinguishes between the target and generated samples, and the loss of the discriminator can then be expressed as follows:

$$\min_{\alpha} L_D = \mathbb{E}_{\varPhi(\mathbf{T}) \sim \mathbb{D}^{\varPhi(T)}} D(\varPhi(\mathbf{T})) - \mathbb{E}_{\mathbf{U} \sim \mathbb{D}^{\mathbf{U}}} D(\mathbf{U}), \qquad (19)$$

where $\alpha$ is the set of parameters of the feedforward network $f$ (i.e., the discriminator $D$). To satisfy the K-Lipschitz restriction, we use the clipping trick by sandwiching the weights $\alpha$ in a small window [-c,c] after each gradient update.

The generator $\varPhi$ is designed to minimize WD. For Eq. (19), $\varPhi$ exists only in the second term on the left-hand side of the equation, so we can learn the ideal $\varPhi$ by minimizing the following loss:

$$\min_{\varPhi} L_\varPhi = \mathbb{E}_{\varPhi(\mathbf{T}) \sim \mathbb{D}^{\varPhi(T)}} D(\varPhi(\mathbf{T})). \qquad (20)$$

As the loss of the generator $\varPhi$ gradually decreases, the loss of the discriminator $D$, i.e., WD, also decreases, so that trajectories belonging to the same user are grouped together in the latent space. Meanwhile, we also incorporate a small number of annotations $\mathbf{T}^l$. For a matched pair of trajectories $(T^{u_i}, u_i)$ in $\mathbf{T}^l$, our goal is to minimize the distance between the trajectory and the user as follows:

$$\min_{\varPhi} L_W = \frac{\theta_w}{|T^l|} \sum_{(\mathbf{T}^{u_i}, u_i) \in \mathbf{T}^l} dis(\varPhi(T^{u_i}), u_i), \qquad (21)$$

where $\theta_w$ is the hyper-parameter controlling the weight of the loss $L_W$.

**Adversarial Loss.** The loss function for adversarial learning is as follows

$$L_{al} = L_\varPhi + L_D + L_W, \qquad (22)$$

where $L_\varPhi$ represents the loss of generator $\varPhi$, $L_D$ represents the loss of discriminator $D$, and $L_W$ represents the loss of labeled trajectory-user pairs.

### 4.3  Final Objective Function

The sparse trajectory completion and TUL tasks are optimized simultaneously, and the overall loss $L$ of the two tasks is as follows

$$L = L_{al} + \mu L_{coml}, \qquad (23)$$

**Table 1.** Dataset Description.$|U|$: number of users; $|T_r|/|T_{te}|$: number of trajectories for training and testing; $|P|$: number of trajectory point; $|R|$: average length of trajectories (before segmentation).

| Dataset | $|U|$ | $|T_{tr}|/|T_{te}|$ | $|P|$ | $|R|$ |
|---------|-------|---------------------|-------|-------|
| NYC | 113 | 9122/2280 | 3561 | 214 |
| TKY | 258 | 15843/3871 | 4126 | 188 |

where $L_{coml}$ is the loss for trajectory completion, $\mu$ is the hyperparameter, and $L_{al}$ is the loss for TUL. The work is implemented with the Huawei MindSpore AI computing framework.

## 5 Experiment

### 5.1 Dataset and Experiment Setup

**Dataset.** Two datasets collected from Foursquare are used in our experiment. **NYC dataset** records about 10 months of check-ins in New York City. Each check-in includes its timestamp, GPS coordinates and semantic information (represented by fine-grained venue-categories). **TKY dataset** contains about ten months of check-in records in Tokyo.

Following [34], we randomly select $|U|$ users and their generated trajectories from both datasets for evaluation. For each trajectory, we randomly sample $r\%$ points and remove the others to simulate the sparse trajectory. In our experiments we keep 50% and 70% points for each trajectory, respectively. 10% of the entire trajectories and their users are used as annotations for supervision. Table 1 shows the details of the two datasets.

**Evaluation Metrics.** $ACC@K$ and $macro\text{-}F1$ are used to evaluate the performance of the model. In addition, to verify the effectiveness of trajectory completion, we use three metrics $RMSE$, $NDTW$, and $EDR$ for evaluation. $ACC@K$ is to evaluate the accuracy of the TUL problem, which is defined as

$$ACC@K = \frac{correctly\ linked\ trajectories@K}{the\ number\ of\ trajectories}.$$

$Macro\text{-}F1$ is the harmonic mean of accuracy ($macro\text{-}P$) and recall ($macro\text{-}R$), averaged over all categories (users in TUL).

$$macro\text{-}F1 = \frac{2 \times macro\text{-}P \times marco\text{-}R}{marco\text{-}P + macro\text{-}R}$$

$RMSE$ is the root mean square error between the actual and predicted values of the coordinates of the missing trajectory points that is formulated as

$$RMSE = \sqrt{\frac{1}{m}\sum_{j=1}^{m}(dis\,(p_j, \hat{p}_j))^2},$$

where $dis\left(p_j, \hat{p}_j\right)$ represents the Euclidean distance between the true value $p_j$ and the predicted value $\hat{p}_j$.

$NDTW$ is the normalized dynamic time warping distance between two trajectories, which is modified from dynamic time warping distance (DTW),

$$NDTW\left(T, \widetilde{T}\right) = \frac{DTW\left(T, \widetilde{T}\right)}{length\left(T\right)}.$$

$EDR$ is the edit distance on real sequence. Specifically, given two trajectories $T$ and $\widetilde{T}$, the edit distance between them is the number of operations required to transform $T$ into $\widetilde{T}$ through insert, delete and replace operations.

**Baselines.** To evaluate the effectiveness of TULMAL, we first compare it with the following baseline methods: DTW [19] and LCSS [1], which compute the distance between two trajectories by using different criteria for comparing similarity; TULER [7], TULER-LSTM, TULER-GRU, BiTULER, TULVAE [34] and DeepTUL [15], which use deep learning classification models to learn the projection function between trajectories and users.

To further test the power of adversarial learning in the TUL problem, we compare TULMAL with its variant model TULMAL-NoSeqKF, which removes the completion module. We also compare TULMAL with the baseline methods plus our proposed trajectory completion model SeqKF to study the effectiveness of SeqKF. We also select two trajectory completion baseline methods, Deep-Move [3] and STRNN [12], to compare with SeqKF to verify its effectiveness in trajectory completion.

## 5.2    Experimental Results

Tables 2 and 3 show the performance of various methods on the two datasets in terms of $ACC@K$ and $macro$-$F1$. Table 2 shows the comparison result at a sampling rate of 70% and Table 3 shows the comparison result at a sampling rate of 50%, where the best values are highlighted in bold. In Table 2, one can observe that on the dataset NYC, TULMAL-NoSeqKF improves by 2.86%, 3.15% and 1.79% in terms of $ACC@1$, $ACC@5$ and $macro$-$F1$, respectively, compared to the best baseline method DeepTUL. This superior result is due to its ability to exploit the multi-period nature of user mobility and to address the data sparsity issue. After adding SeqKF to the baseline methods, the performance of all the methods improve, which indicates that the trajectory completion component is effective to improve the TUL performance. Our method TULML achieves the best performance, improving the three metrics by 4.94%, 3.51% and 2.72%, respectively, compared with the best baseline method SeqKF+DeepTUL. This indicates that our adversarial learning based TUL component is more effective than the baselines.

In Table 3 one can see that TULML also achieves the best performance when the sampling rate is 50%. Compared with DeepTUL, TULMAL-NoSeqKF improves by 2.39%, 4.9% and 2.35% in terms of the three metrics, respectively. Compared with SeqKF+DeepTUL, TULMAL improves the three metrics by

**Table 2.** Performance comparison over two datasets under a sampling rate (SR) of 70%

| | NYC | | | TKY | | |
|---|---|---|---|---|---|---|
| | acc@1 | acc@5 | macro-F1 | acc@1 | acc@5 | macro-F1 |
| | u=113, SR=70% | | | u=258,SR=70% | | |
| DTW | 13.74% | 24.58% | 6.14% | 10.08% | 20.65% | 5.41% |
| LCSS | 15.21% | 28.63% | 7.54% | 12.38% | 23.86% | 6.44% |
| TULER-LSTM | 19.31% | 46.65% | 11.99% | 16.14% | 38.42% | 9.18% |
| TULER-GRU | 20.25% | 46.82% | 13.87% | 17.31% | 39.41% | 9.68% |
| BiTULER | 22.63% | 50.42% | 16.55% | 18.52% | 42.35% | 12.94% |
| TULVAE | 23.32% | 50.25% | 16.83% | 18.65% | 43.28% | 13.76% |
| DeepTUL | 25.62% | 53.16% | 19.00% | 21.44% | 45.83% | 15.07% |
| TULMAL-NoSeqKF | 28.48% | 56.31% | 20.79% | 23.83% | 48.11% | 15.97% |
| SeqKF+DTW | 15.43% | 30.74% | 7.62% | 12.62% | 25.83% | 6.28% |
| SeqKF+LCSS | 20.12% | 37.57% | 9.45% | 18.07% | 31.46% | 8.10% |
| SeqKF+TULER-LSTM | 23.37% | 48.63% | 16.20% | 20.75% | 42.96% | 12.42% |
| SeqKF+TULER-GRU | 25.14% | 48.86% | 16.40% | 22.43% | 42.75% | 12.95% |
| SeqKF+BiTULER | 26.84% | 52.49% | 17.69% | 23.87% | 45.62% | 15.05% |
| SeqKF+TULVAE | 27.89% | 53.78% | 19.28% | 25.16% | 48.37% | 16.77% |
| SeqKF+DeepTUL | 30.48% | 57.86% | 21.18% | 27.45% | 50.38% | 18.78% |
| TULMAL | **35.42%** | **61.37%** | **23.90%** | **32.84%** | **51.48%** | **22.58%** |

**Table 3.** Performance comparison over two datasets under a sampling rate (SR) of 50%

| | NYC | | | TKY | | |
|---|---|---|---|---|---|---|
| | acc@1 | acc@5 | macro-F1 | acc@1 | acc@5 | macro-F1 |
| | u = 113, SR = 50% | | | u = 258, SR = 50% | | |
| DTW | 9.58% | 16.83% | 4.43% | 7.62% | 13.27% | 3.19% |
| LCSS | 11.32% | 22.67% | 6.17% | 8.51% | 18.47% | 4.62% |
| TULER-LSTM | 15.38% | 26.77% | 7.65% | 12.85% | 22.66% | 6.41% |
| TULER-GRU | 16.19% | 30.68% | 8.74% | 14.67% | 25.14% | 6.97% |
| BiTULER | 17.84% | 32.73% | 9.42% | 15.98% | 27.31% | 7.39% |
| TULVAE | 19.18% | 35.96% | 11.37% | 17.22% | 30.37% | 8.14% |
| DeepTUL | 22.47% | 41.85% | 13.21% | 20.05% | 34.99% | 11.65% |
| TULMAL-NoSeqKF | 24.86% | 46.75% | 15.56% | 21.96% | 38.41% | 12.87% |
| SeqKF+DTW | 12.54% | 21.48% | 6.21% | 9.63% | 15.85% | 4.31% |
| SeqKF+LCSS | 13.79% | 25.31% | 7.34% | 10.95% | 21.07% | 5.56% |
| SeqKF+TULER-LSTM | 18.46% | 31.24% | 9.17% | 14.16% | 27.60% | 7.47% |
| SeqKF+TULER-GRU | 20.15% | 34.58% | 10.83% | 16.52% | 29.89% | 8.51% |
| SeqKF+BiTULER | 20.97% | 36.27% | 11.68% | 17.83% | 31.04% | 9.33% |
| SeqKF+TULVAE | 22.49% | 40.62% | 13.15% | 19.57% | 34.41% | 11.60% |
| SeqKF+DeepTUL | 25.91% | 47.36% | 15.91% | 23.12% | 38.77% | 12.84% |
| TULMAL | **30.03%** | **53.64%** | **19.58%** | **26.70%** | **43.41%** | **15.49%** |

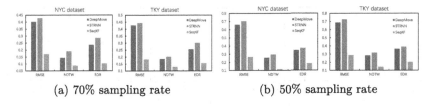

(a) 70% sampling rate                    (b) 50% sampling rate

**Fig. 3.** Performance comparison on SeqKF at sampling rates of 70% and 50%

4.12%, 6.28% and 3.67%, respectively. From the two tables one can also see that the results are generally better on the NYC dataset than that on the TKY dataset. This is mainly because there are fewer users in NYC than in TKY, resulting in denser trajectories in NYC than in TKY. Denser trajectories provide more information and thus the models can achieve better performance.

### 5.3 Effectiveness of Trajectory Completion

Figure 3 shows the performance comparison of the proposed SeqKF with other baseline methods under the sampling rates of 70% and 50%, respectively. In Fig. 3(a) one can observe that DeepMove is better than STRNN because it considers more history information. SeqKF model is better than DeepMove. This is because SeqKF can effectively reduce the effect of noise by using Kalman filter. One can also observe that the performance of the methods on NYC is better than that on TKY. This is mainly because the trajectories in TKY are sparser than those in NYC. In Fig. 3(b), SeqKF also significantly outperforms the two baselines DeepMove and STRNN. One can also see that the performance drop of SeqKF is much smaller than the other two methods when the sparsity rate changes from 70% to 50%. It further verifies SeqKF can effectively reduce the effect of noise and improve the robustness of the model by combining Seq2Seq with Kalman filter. The performance of the two baselines degrades quickly because the smaller sparsity rate leads to sparser trajectories.

### 5.4 Parameter Sensitivity Study

We finally investigate the performance sensitivity of TULMAL on three parameters: the weight $\mu$ of the final loss function, the annotation guidance weight $\theta$ in TUL, and the weight $\lambda$ of the loss function in SeqKF. We let $\mu$ increase from 0 to 0.5, $\theta$ increase from 0.1 to 0.5, and $\lambda$ increase from 0 to 0.25. In our experiment, the sampling rate is set to 70% for both datasets. As can be seen in Fig. 4, the performance of the model first increases and then decreases as $\mu$ keeps increasing, which indicates that an appropriate SeqKF loss can improve the performance of TUL, but a too large SeqKF loss can overwhelm the adversarial loss and thus hurt the performance. Similar result is produced for the annotation guidance weight $\theta$. The performance of the model increases first and then decreases as $\theta$ increases. $\theta = 0.3$ is a suitable setting for both datasets. As $\lambda$

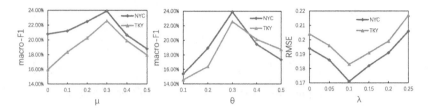

**Fig. 4.** The effect of three parameters $\mu$, $\theta$ and $\lambda$ on the model performance.

increases, the RMSE first decreases and then increases. The best performance of SeqKF is achieved when $\lambda$ is 0.1, which proves that KF is effective in improving the performance of trajectory completion.

## 6 Conclusion

In this paper, we proposed a multi-task adversarial learning model for semi-supervised TUL with sparse trajectory data. TULMAL first used the trajectory completion component SeqKF to effectively complete the sparse trajectories. SeqKF combined Seq2Seq model and Kalman filter effectively to alleviate the noise for data completion. Then the TUL problem was solved by capturing the multi-periodicity of user movement through a proposed adversarial learning model. As TUL was conducted in the data distribution level rather than the trajectory-user pair data instance level, the proposed model required only a small number of annotations. We conducted extensive experiments on two real datasets. The experimental results showed that our proposal outperformed previous approaches in the two tasks.

**Acknowledgments.** This research was funded by National Science Foundation of China (No. 62172443 and 61802424), Guangdong Basic and Applied Basic Research Foundation (No.20 21A1515012239), CAAI-Huawei MindSpore Open Fund and the Fundamental Research Funds for the Central Universities (No.: NZ2020014).

## References

1. Berndt, D.J., Clifford, J.: Using dynamic time warping to find patterns in time series. In: Proceedings of KDD Workshop (1994)
2. Damiani, M.L., Guting, R.H.: Semantic trajectories and beyond. In: Proceedings of MDM (2014)
3. Feng, J., et al.: Deepmove: predicting human mobility with attentional recurrent networks. In: Proceedings of WWW (2018)
4. Freitas, N., Silva, T., Macêdo, J., Junior, L.M., Cordeiro, M.: Using deep learning for trajectory classification. In: Proceedings of ICAART (2021)
5. Gao, Q., Trajcevski, G., Zhou, F., Zhang, K., Zhong, T., Zhang, F.: Deeptrip: adversarially understanding human mobility for trip recommendation. In: Proceedings of ACM SIGSPATIAL (2019)

6. Gao, Q., Zhou, F., Trajcevski, G., Zhang, K., Zhong, T., Zhang, F.: Predicting human mobility via variational attention. In: Proceedings of WWW (2019)
7. Gao, Q., Zhou, F., Zhang, K., Trajcevski, G., Luo, X., Zhang, F.: Identifying human mobility via trajectory embeddings. In: Proceedings of IJCAI (2017)
8. Gu, J., Lu, Z., Li, H., Li, V.O.: Incorporating copying mechanism in sequence-to-sequence learning. arXiv preprint arXiv:1603.06393 (2016)
9. Huang, L., Yang, Y., Chen, H., Zhang, Y., Wang, Z., He, L.: Context-aware road travel time estimation by coupled tensor decomposition based on trajectory data. Knowl.-Based Syst. **245**, 108596 (2022)
10. Kalman, R.E.: A new approach to linear filtering and prediction problems. Fluids Eng. **82D**, 35–45 (1959)
11. Li, J., Wang, S., Zhang, J., Miao, H., Zhang, J., Yu, P.: Fine-grained urban flow inference with incomplete data. IEEE Trans. Knowl. Data Eng. 1–14 (2022)
12. Liu, Q., Wu, S., Wang, L., Tan, T.: Predicting the next location: a recurrent model with spatial and temporal contexts. In: Proceedings of AAAI (2016)
13. Mahajan, R., Mansotra, V.: Predicting geolocation of tweets: using combination of CNN and BILSTM. Data Sci. Eng. **6**(4), 402–410 (2021)
14. Miao, C., Luo, Z., Zeng, F., Wang, J.: Predicting human mobility via attentive convolutional network. In: Proceedings of WSDM (2020)
15. Miao, C., Wang, J., Yu, H., Zhang, W., Qi, Y.: Trajectory-user linking with attentive recurrent network. In: Proceedings of AAMAS (2020)
16. Ren, H., et al.: MTRAJREC: map-constrained trajectory recovery via seq2seq multi-task learning. In: Proceedings of ACM SIGKDD (2021)
17. Sun, K., Qian, T., Chen, T., Liang, Y., Nguyen, Q.V.H., Yin, H.: Where to go next: modeling long-and short-term user preferences for point-of-interest recommendation. In: Proceedings of AAAI (2020)
18. Villani, C.: Optimal Transport: Old and New. Springer, Heidelberg (2009). https://doi.org/10.1007/978-3-540-71050-9
19. Vlachos, M., Kollios, G., Gunopulos, D.: Discovering similar multidimensional trajectories. In: Proceedings of ICDE (2002)
20. Wang, S., Miao, H., Chen, H., Huang, Z.: Multi-task adversarial spatial-temporal networks for crowd flow prediction. In: Proceedings of CIKM (2020)
21. Wang, S., Miao, H., Li, J., Cao, J.: Spatio-temporal knowledge transfer for urban crowd flow prediction via deep attentive adaptation networks. IEEE Trans. Intell. Transp. Syst. **23**(5), 4695–4705 (2021)
22. Wang, S., Zhang, J., Li, J., Miao, H., Cao, J.: Traffic accident risk prediction via multi-view multi-task spatio-temporal networks. IEEE Trans. Knowl. Data Eng. 1–14 (2021)
23. Wang, S., Zhang, M., Miao, H., Peng, Z., Yu, P.S.: Multivariate correlation-aware SPATIO-temporal graph convolutional networks for multi-scale traffic prediction. ACM Trans. Intell. Syst. Technol. **13**(3), 1–22 (2022)
24. Wang, S., Zhang, M., Miao, H., Yu, P.S.: Mt-STNets: Multi-task spatial-temporal networks for multi-scale traffic prediction. In: Proceedings of SDM (2021)
25. Xi, D., Zhuang, F., Liu, Y., Gu, J., Xiong, H., He, Q.: Modelling of bi-directional SPATIO-temporal dependence and users' dynamic preferences for missing poi check-in identification. In: Proceedings of AAAI (2019)
26. Xia, T., et al.: ATTNmove: history enhanced trajectory recovery via attentional network. arXiv preprint arXiv:2101.00646 (2021)
27. Yang, Z., Ma, J., Chen, H., Zhang, J., Chang, Y.: Context-aware attentive multilevel feature fusion for named entity recognition. IEEE Trans. Neural Networks Learn. Syst. 1–12 (2022)

28. Yu, Y., et al.: TULSN: SIAMESE network for trajectory-user linking. In: Proceedings of IJCNN (2020)
29. Yuan, H., Li, G.: A survey of traffic prediction: from SPATIO-temporal data to intelligent transportation. Data Sci. Eng. **6**(1), 63–85 (2021)
30. Zhang, M., Liu, Y., Luan, H., Sun, M.: Earth mover's distance minimization for unsupervised bilingual lexicon induction. In: Proceedings of EMNLP (2017)
31. Zheng, K., Zheng, Y., Xie, X., Zhou, X.: Reducing uncertainty of low-sampling-rate trajectories. In: Proceedings of ICDE (2012)
32. Zheng, S., Yue, Y., Hobbs, J.: Generating long-term trajectories using deep hierarchical networks. In: Proceedings of NeurIPS (2016)
33. Zheng, Y.: Trajectory data mining: an overview. ACM Trans. Intell. Syst. Technol. **6**(3), 1–41 (2015)
34. Zhou, F., Gao, Q., Trajcevski, G., Zhang, K., Zhong, T., Zhang, F.: Trajectory-user linking via variational autoencoder. In: Proceedings of IJCAI (2018)

# Consistent and Tractable Algorithm for Markov Network Learning

Vojtech Franc[(✉)] [iD], Daniel Prusa [iD], and Andrii Yermakov [iD]

Czech Technical University in Prague, Prague, Czech Republic
{xfrancv,prusa,yermaand}@fel.cvut.cz

**Abstract.** Markov network (MN) structured output classifiers provide a transparent and powerful way to model dependencies between output labels. The MN classifiers can be learned using the M3N algorithm, which, however, is not statistically consistent and requires expensive fully annotated examples. We propose an algorithm to learn MN classifiers that is based on Fisher-consistent adversarial loss minimization. Learning is transformed into a tractable convex optimization that is amenable to standard gradient methods. We also extend the algorithm to learn from examples with missing labels. We show that the extended algorithm remains convex, tractable, and statistically consistent.

## 1 Introduction

Structured output classification aims at the prediction of a set of statistically interdependent labels. A transparent way to model dependencies between the labels provides the Markov Network (MN) classifier, formally defined as follows. Let $\mathcal{X}$ be a set of observations. Let $\mathcal{V}$ be a finite set of objects, and let $\mathcal{E} \subseteq \binom{\mathcal{V}}{2}$ be a set of interacting objects. An object $v \in \mathcal{V}$ is characterized by a label $y \in \mathcal{Y}_v$ of a finite set $\mathcal{Y}_v$. Let $\mathcal{Y} = \times_{v \in \mathcal{V}} \mathcal{Y}_v$ be the structured output space, and let $\boldsymbol{y} = (y_y \in \mathcal{Y}_v \mid v \in \mathcal{V}) \in \mathcal{Y}$ denote the labeling of all objects in $\mathcal{V}$. The match between observation $x$ and a label $y_v \in \mathcal{Y}$ assigned to object $v \in \mathcal{V}$ is scored by a function $f_v \colon \mathcal{X} \times \mathcal{Y} \to \mathbb{R}$. The match between the labels $(y_v, y_{v'})$ assigned to the interacting objects $\{v, v'\} \in \mathcal{E}$ is scored by a function $f_{vv'} \colon \mathcal{Y} \times \mathcal{Y} \to \mathbb{R}$. Given an observation $x \in \mathcal{X}$, the MN classifier $\boldsymbol{h} \colon \mathcal{X} \to \mathcal{Y}$ returns the labeling $\boldsymbol{y} \in \mathcal{Y}$ with the maximum total score:

$$h(x) \in \operatorname*{Argmax}_{y \in \mathcal{Y}} \left[ \sum_{v \in \mathcal{V}} f_v(x, y_v) + \sum_{v, v' \in \mathcal{E}} f_{vv'}(y_v, y_{v'}) \right] \tag{1}$$

Inference (1) requires solving a valued constrained satisfaction problem which is an NP-hard in general. There are subclasses solvable efficiently; e.g., when $(\mathcal{V}, \mathcal{E})$ is acyclic, it can be solved by dynamic programming. In a general setup, the problem can be addressed using linear programming (LP) relaxation [19].

---

**Supplementary Information** The online version contains supplementary material available at https://doi.org/10.1007/978-3-031-26412-2_27.

Linearly parameterized score functions can be learned efficiently by Maximum Margin Markov Network (M3N) algorithms [14,15,17] even if the graph $(\mathcal{V}, \mathcal{E})$ is generic and the inference is not tractable [4,5]. The original M3N algorithm requires fully annotated examples; however, an extension for learning from partially annotated examples, when some labels can be missing, was proposed in [6]. M3N algorithms translate learning into convex and tractable minimization of the margin rescaling loss and its variants, which serve as a surrogate of the target loss we would like to actually optimize. An unsettling issue is the statistical properties. Namely, algorithms based on margin rescaling loss minimization are not statistically consistent [10]; that is, they are not guaranteed to learn the optimal Bayes classifier even if fed in with an unlimited amount of data.

Recently, [2,3] proposed an adversarial loss whose minimization yields a statistically consistent algorithm. Unfortunately, the evaluation of the adversarial loss requires solving a Min-Max problem whose size scales with the number of labels, and thus it is not tractable for structured prediction. In [12] an algorithm was proposed that minimizes adversarial loss instantiated for structured predictors. However, the algorithm relies on an oracle that solves a Min-Max problem of the same complexity as the one in the definition of adversarial loss. Therefore, the algorithm is applicable only for MN classifiers when the neighborhood graph $(\mathcal{V}, \mathcal{E})$ is restricted to be acyclic, and even then the oracle is not guaranteed to solve the problem optimally.

In this paper, we contribute to the problem of learning MN classifiers by:

1. We propose a novel surrogate loss, named MArkov Network Adversarial (MANA) loss, for learning MN classifiers. The MANA loss is defined by a convex optimization which is tractable for general neighborhood graph $(\mathcal{V}, \mathcal{E})$. In Theorem 3 we prove that the MANA loss is equivalent to the adversarial loss. The MANA loss is, to our knowledge, the first surrogate for learning generic MN classifiers which is simultaneously statistically consistent, convex and tractable. Minimization of the MANA loss is amenable to standard gradient methods.
2. We extend the MANA loss for learning MN classifiers on partially annotated examples when the labels are missing at random. The extended loss, named partial MANA loss, has the same computational complexity as its supervised counterpart. In Theorem 5 we prove that the partial MANA loss is Fisher consistent.
3. We evaluate the algorithms minimizing margin-rescaling loss and the proposed MANA loss using both fully annotated and partially annotated data sets. We show that the empirical performance of both losses is similar. This find is not that surprising because we also show that the margin rescaling loss is a close approximation of the consistent MANA loss, although both surrogates were developed from completely different principles.

The necessary background and state-of-the-art is given in Sect. 2. The contributions of this paper are presented in Sect. 3. Section 4 provides an empirical evaluation of the proposed and existing methods, and Sect. 5 concludes the paper. Proofs of the novel Theorems 3, 4 and 5 are deferred to the supplementary material.

## 2    State-of-the-Art

In this section, we describe the state-of-the-art in risk minimization approaches applicable to learning MN classifiers. We survey existing surrogate losses and describe which are Fisher-consistent and which are tractable when applied for learning the MN classifier. Section 2.1 focuses on supervised learning, and Sect. 2.2 on learning from partially annotated examples.

### 2.1    Supervised Learning

Assume that instances $(x, y)$ are generated from a distribution $p_{XY}(x, y)$ on $\mathcal{X} \times \mathcal{Y}$. Let $\ell \colon \mathcal{Y} \times \mathcal{Y} \to \mathbb{R}$ be a target loss penalizing the predictions of the labeling. In this paper, we focus on additive losses, that is,

$$\ell(\boldsymbol{y}, \hat{\boldsymbol{y}}) = \sum_{v \in \mathcal{V}} \ell_v(y_v, \hat{y}_{v'}) \tag{2}$$

where $\ell_v \colon \mathcal{Y}_v \times \mathcal{Y}_v \to \mathbb{R}$ are some single-label losses. The goal is to find a classifier $\boldsymbol{h} \colon \mathcal{X} \to \mathcal{Y}$ that minimizes the expected risk[1]:

$$R_\ell(\boldsymbol{h}, p_{XY}) = \mathbb{E}_{x,y \sim p_{XY}} \ell(\boldsymbol{y}, \boldsymbol{h}(x)) \,.$$

At best we achieve the Bayes risk $R_\ell^*(p_{XY}) = \inf_{h \colon \mathcal{X} \to \mathcal{Y}} R_\ell(\boldsymbol{h}, p_{XY})$. The classifier is usually modeled as a composed function $\boldsymbol{h}(x) = \boldsymbol{T} \circ \boldsymbol{f}(x)$, where $\boldsymbol{f} \colon \mathcal{X} \to \mathbb{R}^d$ is a score map, and $\boldsymbol{T} \colon \mathbb{R}^d \to \mathcal{Y}$ is a fixed label decoding. For example, the most common prediction model, also considered in this paper, assigns labels based on maximization of a score function $f \colon \mathcal{X} \times \mathcal{Y} \to \mathbb{R}$, i.e., $\boldsymbol{h}(x) \in \operatorname*{Argmax}_{y \in \mathcal{Y}} f(x, y)$. This corresponds to $d = |\mathcal{Y}|$ and $\boldsymbol{T}(\boldsymbol{f}) \in \operatorname*{Argmax}_{y \in \mathcal{Y}} f_y$ where $\boldsymbol{f}(x) = (f(x, y), y \in \mathcal{Y}) \in \mathbb{R}^{|\mathcal{Y}|}$. The MN classifier (1) is obtained when $f(x, y)$ decomposes over objects $\mathcal{V}$ and edges $\mathcal{E}$, i.e.,

$$f(x, y) = \sum_{v \in \mathcal{V}} f_v(x, y_v) + \sum_{v, v' \in \mathcal{E}} f_{vv'}(y_v, y_{v'}) \,. \tag{3}$$

The finding of the classifier can be posed as a minimization of $R_\ell(\boldsymbol{T} \circ \boldsymbol{f})$ w.r.t. $\boldsymbol{f}$. However, direct minimization of the $\ell$-risk is difficult due to the discrete nature of the commonly used losses $\ell$. Therefore, $\ell$ is replaced by a surrogate loss $\psi \colon \mathbb{R}^d \times \mathcal{Y} \to \mathbb{R}$, which evaluates the score map $\boldsymbol{f}$ on $p_{XY}(x, y)$ using a $\psi$-risk $R_\psi(\boldsymbol{f}, p_{XY}) = \mathbb{E}_{x,y \sim p_{XY}} \psi(\boldsymbol{f}(x), y)$. The optimal score w.r.t. the $\psi$-risk is then $\boldsymbol{f}_\psi \in \operatorname*{Argmin}_{f \colon \mathcal{X} \to \mathbb{R}^d} R_\psi(\boldsymbol{f}, p_{XY})$. There are two requirements on the surrogate loss $\psi$. First, optimization of the surrogate should be tractable, hence, $\psi(\boldsymbol{f}, \boldsymbol{y})$ is designed to be convex in $\boldsymbol{f}$ and cheap to evaluate. Second, the resulting classifier $\boldsymbol{h}(x) = \boldsymbol{T} \circ \boldsymbol{f}_\psi(x)$ should achieve low $\ell$-risk, being our true objective. Ideally, we require the surrogate $\psi$ to be Fisher consistent [9,20]:

$$R_\ell^*(p_{XY}) = R_\ell(\boldsymbol{T} \circ \boldsymbol{f}_\psi, p_{XY}) \,, \tag{4}$$

i.e., the classifier found by minimizing the $\psi$-risk achieves the Bayes $\ell$-risk.

---

[1] We refer to the expectation of a loss LOSS as the LOSS-risk.

In the common ML setup, the distribution $p_{XY}(x, y)$ is unknown; however, we have a training sequence $\mathcal{T}_{XY} = ((x^i, y^i) \in \mathcal{X} \times \mathcal{Y} \mid i = 1, \ldots, m)$ drawn from i.i.d. random variables with distribution $p_{XY}(x, y)$. Training data $\mathcal{T}_{XY}$ are used to approximate $p_{XY}(x, y)$ by the empirical distribution $\hat{p}_{XY}^m(x, y) = \frac{1}{m} \sum_{i=1}^{m} [\![x = x^i \wedge y = y^i]\!]$, and the classifier is found using the empirical risk minimization (ERM)

$$ f_\psi^m \in \underset{f \in \mathcal{F}}{\text{Argmin}} \, R_\psi(f, \hat{p}_{XY}^m) \,, \tag{5} $$

where $\mathcal{F} \subseteq \{f \colon \mathcal{X} \to \mathbb{R}^d\}$ is an a priori chosen class of functions. Under suitable conditions, with an increasing number of examples $m$, the population $\psi$-risk converges in probability to the minimal attainable $\psi$-risk, i.e., $R_\psi(f_\psi^m, p_{XY}) \overset{p}{\to} R_\psi(f_\psi^*, p_{XY})$. In this case, a Fisher-consistent surrogate $\psi$ (i.e., the surrogate satisfying (4)), which is also continuous and bounded from below, guarantees the convergence of the $\ell$-risk to the Bayes $\ell$-risk, i.e. $R_\ell(T \circ f_\psi^m, p_{XY}) \overset{p}{\to} R_\ell^*(p_{XY})$ [20].

**Structured Output Support Vector Machines.** [15–17] (SO-SVM) is an instance of the ERM, that is designed to learn the linear classifier and the surrogate is a certain convex piecewise linear function.

Let $\phi \colon \mathcal{X} \times \mathcal{Y} \to \mathbb{R}^n$ be an input-output feature map that embeds $\mathcal{X} \times \mathcal{Y}$ in a parameter space $\mathbb{R}^n$. Let $f(x, y) = \phi(x)^T \theta$ be the score function parameterized by $\theta \in \mathbb{R}^n$. We will use $\Phi(x) = (\phi(x, y), y \in \mathcal{Y}) \in \mathbb{R}^{n \times |\mathcal{Y}|}$ to denote a matrix that for a given $x \in \mathcal{X}$ contains the feature maps of all labelings $y \in \mathcal{Y}$. Let $T(f) \in \text{Argmax}_{y \in \mathcal{Y}} f_y$ be the label decoding and $f(x) = (f(x, y), y \in \mathcal{Y}) \in \mathbb{R}^{|\mathcal{Y}|}$ be the score map. The linear classifier can be written in a compact way as

$$ h(x) \in \underset{y \in \mathcal{Y}}{\text{Argmax}} \, \phi(x, y)^T \theta = T \circ f(x) = T \circ \Phi(x)^T \theta \,. \tag{6} $$

In this paper, we concentrate on a linear MN classifier, obtained when the input-output feature map decomposes over objects $\mathcal{V}$ and edges $\mathcal{E}$ as

$$ \phi(x, y) = \sum_{v \in \mathcal{V}} \phi_v(x, y_v) + \sum_{v, v' \in \mathcal{E}} \phi_{vv'}(y_v, y_{v'}) \,, \tag{7} $$

where $\phi_v \colon \mathcal{X} \times \mathcal{Y}_v \to \mathbb{R}^n$, $v \in \mathcal{V}$, and $\phi_{vv'} \colon \mathcal{Y}_v \times \mathcal{Y}_{v'} \to \mathbb{R}^n$, $\{v, v'\} \in \mathcal{E}$.

Marginal rescaling loss is the most widely used surrogate in structured output classification [17], and it is defined as

$$ \psi_{\text{mr}}(f, y) = \max_{y' \in \mathcal{Y}} \left[ \ell(y, y') + f(x, y') \right] - f(x, y) \,. \tag{8} $$

Given training examples $\mathcal{T}_{XY}$, the SO-SVM algorithm finds parameters $\theta$ of the linear classifier (6) by solving a convex unconstrained problem

$$ \theta_{mr}^m = \underset{\theta \in \mathbb{R}^n}{\text{argmin}} \left[ \frac{\lambda}{2} \|\theta\|^2 + \frac{1}{m} \sum_{i=1}^{m} \psi_{\text{mr}}(\Phi(x^i)^T \theta, y^i) \right] \,, \tag{9} $$

where $\lambda > 0$ is the regularization constant. The SO-SVM problem (9) corresponds to the ERM (5) with proxy $\psi_{\mathrm{mr}}$ and a class of linear scores $\mathcal{F} = \{f(x) = \boldsymbol{\Phi}(x)^T\boldsymbol{\theta} \mid \|\boldsymbol{\theta}\| \leqslant r(\lambda)\}$, where $r\colon \mathbb{R} \to \mathbb{R}$ is a monotonic function of $\lambda$.

There are two issues with the SO-SVM algorithm. First, the margin-rescaling loss $\psi_{\mathrm{mr}}$ is not Fisher-consistent, in general. It is Fisher-consistent in the binary case $|\mathcal{Y}| = 2$, when $\psi_{\mathrm{mr}}$ becomes the hinge loss of the binary SVM [9]. In the multiclass case, $|\mathcal{Y}| > 2$, it is Fisher-consistent if only if $\max_{y \in \mathcal{Y}} p_{Y|X}(\boldsymbol{y} \mid x) > 0.5$, $\forall x \in \mathcal{X}$ [10]. Second, evaluating the margin-rescaling loss requires an oracle solving the loss-augmented prediction

$$\hat{\boldsymbol{y}} \in \operatorname*{Argmax}_{\boldsymbol{y} \in \mathcal{Y}} \left[ \ell(\hat{\boldsymbol{y}}, \boldsymbol{y}) + f(x, \boldsymbol{y}) \right]. \tag{10}$$

In the case of the MN classifier, (10) is intractable, in general. It is tractable when the target loss is additive and the neighborhood graph $(\mathcal{V}, \mathcal{E})$ is restricted to be acyclic, in which case (10) can be solved by dynamic programming. The intractability of loss-augmented prediction can be resolved by replacing the intractable maximization problem (10) with a linear programming (LP) upper bound [13,19], which was done for a generic MN classifier in [4]. The linear programming margin-rescaling (LP-MR) loss for the MN classifier (1) reads

$$\psi_{\mathrm{lp}}(\boldsymbol{f}, \hat{\boldsymbol{y}}) = \min_{\boldsymbol{\alpha} \in \mathbb{R}^{n_\alpha}} \left[ \sum_{v \in \mathcal{V}} \max_{y \in \mathcal{Y}_v} \left[ f_v(x, y) - \sum_{v' \in \mathcal{N}(v)} \alpha_{vv'}(y) + \ell_v(\hat{y}_v, y) \right] \right.$$
$$\left. + \sum_{\{v,v'\} \in \mathcal{E}} \max_{(y,y') \in \mathcal{Y}_v \times \mathcal{Y}_{v'}} \left[ f_{vv'}(y, y') + \alpha_{vv'}(y) + \alpha_{v'v}(y') \right] \right] - f(x, \hat{\boldsymbol{y}}), \tag{11}$$

where $\boldsymbol{\alpha} = (\alpha_{vv'} \colon \mathcal{Y}_v \times \mathcal{Y}_{v'} \to \mathbb{R}, \alpha_{v'v} \colon \mathcal{Y}_{v'} \times \mathcal{Y}_v \to \mathbb{R}, \{v, v'\} \in \mathcal{E})$ is a vector of $n_\alpha = 2\sum_{\{v,v'\} \in \mathcal{E}}(|\mathcal{Y}_v| + |\mathcal{Y}_{v'}|)$ auxiliary variables. Evaluating $\psi_{\mathrm{lp}}(\boldsymbol{f}, \hat{\boldsymbol{y}})$ requires solving a convex unconstrained problem, which can be done using gradient methods simultaneously with learning the score function. In contrast to the original margin-rescaling loss, the optimization of $\psi_{\mathrm{lp}}$ is tractable for an arbitrary neighborhood graph $(\mathcal{V}, \mathcal{E})$. In the case of the acyclic graph $(\mathcal{V}, \mathcal{E})$, the bound is tight and $\psi_{\mathrm{lp}}(\boldsymbol{f}, \boldsymbol{y}) = \psi_{\mathrm{mr}}(\boldsymbol{f}, \boldsymbol{y})$, $\forall \boldsymbol{f}, \boldsymbol{y}$. Therefore, $\psi_{\mathrm{lp}}$ is not Fisher consistent in general.

**Adversarial Loss.** [2,3] posed the prediction as an adversarial problem between the predictor minimizing the risk and an adversarial maximizing the risk with respect to the posterior distribution that matches the statistics computed on the examples. They show that adversarial prediction is an example of the risk minimization approach. In this case, the adversarial surrogate loss is expressed as a Min-Max problem:

$$\psi_{\mathrm{adv}}(\boldsymbol{f}, \hat{\boldsymbol{y}}) = \max_{q \in \Delta} \min_{p \in \Delta} \mathbb{E}_{\boldsymbol{y} \sim p, \boldsymbol{y'} \sim q} \left[ \ell(\boldsymbol{y}, \boldsymbol{y'}) + f(x, \boldsymbol{y}) - f(x, \hat{\boldsymbol{y}}) \right] \tag{12}$$

where $\Delta = \{\boldsymbol{q} \in \mathbb{R}_+^{|\mathcal{Y}|} \mid \sum_{\boldsymbol{y} \in \mathcal{Y}} q(\boldsymbol{y}) = 1\}$ is a probability simplex on $\mathcal{Y}$. The adversarial loss is Fisher-consistent (Theorem 15 in [3]):

**Theorem 1.** *Let $R_{adv}(f, p_{XY}) = \mathbb{E}_{x,y \sim p_{XY}} \psi_{adv}(f(x), y)$ be $\psi_{adv}$-risk given by the adversarial loss (12), induced from a target loss $\ell: \mathcal{Y} \times \mathcal{Y} \rightarrow \mathbb{R}$ satisfying $\ell(y, y) < \ell(y, y')$, $\forall y \neq y'$. Let the set of optimal predictions $\hat{\mathcal{Y}}(x) = \text{Argmin}_{y' \in \mathcal{Y}} \mathbb{E}_{y \sim p_{Y|X}} \ell(y, y')$ be a singleton, $|\hat{\mathcal{Y}}(x)| = 1$, for all inputs $x \in \mathcal{X}$. Then, we have*

$$R_\ell^*(p_{XY}) = R_\ell(T \circ f_{adv}, p_{XY}),$$

*where $T(x) \in \text{Argmax}_{y \in \mathcal{Y}} f_y$ and $f_{adv} \in \text{Argmin}_{f: \mathcal{X} \rightarrow \mathbb{R}^{|\mathcal{Y}|}} R_{adv}(f, p_{XY})$ is a minimizer of the $\psi_{adv}$-risk with respect to all measurable functions.*

Nice statistical properties of the adversarial loss are paid off by the computational issues, i.e., evaluating the loss (12) requires solving the Min-Max problem with $2|\mathcal{Y}|$ variables, which in case of the structured prediction is intractable. A generalized Block Coordinate Frank-Wolfe (GBCFW) algorithm to learn structured output linear classifiers by regularized ERM with the adversarial loss was recently proposed in [12]. The GBCFW relies on an oracle solving a Min-Max problem of as similar complexity as (12). [12] propose an alternating procedure to solve the Min-Max approximately which, however, has no guarantee to reach a global optimum and, in case of the MN classifier it is tractable only when the neighbourhood graph $(\mathcal{V}, \mathcal{E})$ is restricted to be acyclic.

## 2.2   Learning from Partially Annotated Examples

Assume that we do not have access to full labeling $y \in \mathcal{Y} = \times_{v \in \mathcal{V}} \mathcal{Y}_v$ but instead we obtain an annotation $a \in \mathcal{A} = \times_{v \in \mathcal{V}} \mathcal{A}_v$ where $\mathcal{A}_v = \{\mathcal{Y}_v \cup \{?\}\}$. That is, for an object $v \in \mathcal{V}$ we either know the true label, $a_v = y_v$, or the label is not given, $a_v = ?$. Given the instance $(x, y) \in \mathcal{X} \times \mathcal{Y}$ generated from $p_{XY}(x, y)$, the annotation $a \in \mathcal{A}$ is generated from $p_{A|XY}(a \mid x, y)$. A partially annotated training set $\mathcal{T}_{XA} = \{(x^i, a^i) \in \mathcal{X} \times \mathcal{A} \mid i = 1, \ldots, m\}$ contains examples drawn from i.i.d. random variables with distribution

$$p_{XA}(x, a) = \sum_{y \in \mathcal{Y}} p_{A|XY}(a \mid x, y) p_{XY}(x, y). \tag{13}$$

The goal is to use $\mathcal{T}_{XA}$ to learn a classifier with $\ell$-risk $R_\ell(h, p_{XY})$ close to the Bayes $\ell$-risk $R_\ell^*(p_{XY})$. That is, the goals of supervised learning and learning from partially annotated examples are the same, but the training sets are different.

To apply the risk minimization approach, we need a surrogate loss $\psi^p: \mathbb{R}^d \times \mathcal{A} \rightarrow \mathbb{R}$, whose value $\psi^p(f, a)$ evaluates the score map $f: \mathcal{X} \rightarrow \mathbb{R}^d$ based on the partial annotation $a \in \mathcal{A}$. Let us define the $\psi^p$-risk $R_{\psi^p}(f, p_{XA}) = \mathbb{E}_{x,a \sim p_{XA}} \psi^p(f(x), a)$. An optimal score under $\psi^p$-risk is obtained by solving

$$f_{\psi^p} \in \underset{f: \mathcal{X} \rightarrow \mathbb{R}^d}{\text{Argmin}} R_{\psi^p}(f, p_{XA}).$$

As in the supervised case, we require the surrogate $\psi^p$ to be tractable and Fisher-consistent:

$$R_\ell^*(p_{XY}) = R_\ell(T \circ f_{\psi^p}, p_{XY}),$$

i.e., the classifier found by minimizing the $\psi^p$-risk on $p_{XA}(x, a)$, achieves the Bayes $\ell$-risk on $p_{XY}(x, y)$.

**Labels Missing at Random.** The distribution $p_{A|XY}(\boldsymbol{a} \mid \boldsymbol{x}, \boldsymbol{y})$ governing the annotation process cannot be arbitrary to make the learning possible. For example, when $p_{A|XY}(\boldsymbol{a} \mid \boldsymbol{x}, \boldsymbol{y}) = p_A(\boldsymbol{a})$, the annotation $\boldsymbol{a}$ is useless as it carries no information about the labeling $\boldsymbol{y}$. In this paper, we consider labels Missing At Random (MAR) annotation process [1,6] defined by

$$p_{A|XY}(\boldsymbol{a} \mid \boldsymbol{x}, \boldsymbol{y}) = \sum_{\boldsymbol{z} \in \{0,1\}^{\mathcal{V}}} p_{Z|X}(\boldsymbol{z} \mid \boldsymbol{x}) \prod_{v \in \mathcal{V}} [\![a_v = c(y_v, z_v)]\!], \tag{14}$$

where $c(y_v, z_v) = y_v$ if $z_v = 1$, $c(y_v, z_v) =?$ if $z_v = 0$, and $p_{Z|X}(\boldsymbol{z} \mid \boldsymbol{x})$, $\boldsymbol{x} \in \mathcal{X}$, are conditional distributions on $\boldsymbol{z} \in \{0,1\}^{\mathcal{V}}$ such that $p_{Z_v|X}(z_v \mid \boldsymbol{x}) > 0$, $\forall v \in \mathcal{V}$. The MAR process implies that the annotation in $\mathcal{T}_{XA}$ is generated as follows. Nature generates $(\boldsymbol{x}, \boldsymbol{y})$ from $p_{XY}(\boldsymbol{x}, \boldsymbol{y})$. The annotator decides the objects to label based on the observation of the input $\boldsymbol{x}$. His decision is stochastic, represented by a binary vector $\boldsymbol{z} \in \{0,1\}^{\mathcal{V}}$ generated from $p_{Z|X}(\boldsymbol{z} \mid \boldsymbol{x})$. The annotator reveals the labels of the objects $\mathcal{V}_{lab} = \{v \in \mathcal{V} \mid z_v = 1\}$, i.e., he sets $a_v = y_v$, $v \in \mathcal{V}_{lab}$, while the labels of the remaining objects are not provided, i.e., $a_v =?$, $v \in \mathcal{V} \setminus \mathcal{V}_{lab}$.

**Ramp-Loss.** SO-SVM was extended to learn from partially annotated examples in [11]. The method uses the Ramp loss defined as

$$\psi_{\text{ramp}}^p(\boldsymbol{f}, \boldsymbol{a}) = \max_{\boldsymbol{y} \in \mathcal{Y}} \left[ \ell^p(\boldsymbol{a}, \boldsymbol{y}) + f(\boldsymbol{x}, \boldsymbol{y}) \right] - \max_{\boldsymbol{y} \in \mathcal{Y}} f(\boldsymbol{x}, \boldsymbol{y})$$

where $\ell^p(\boldsymbol{a}, \boldsymbol{y}) = \sum_{v \in \mathcal{V}} [\![a_v \neq ?]\!] \ell_v(a_v, y_v)$ is the partial additive loss. In case of the MAR annotation, the ramp-loss is Fisher-consistent [1]. However, the ramp-loss is non-convex, and in case of the score of the MN-classifier even its evaluation is not tractable in general. Unlike the margin-rescaling loss, the LP upper bound is not applicable here.

**Partial LP Margin-Rescaling Loss.** Partial LP margin-rescaling loss for learning linear MN classifiers from partially annotated examples was proposed in [6]. The loss reads[2]

$$\psi_{\text{lp}}^p(\boldsymbol{x}, \boldsymbol{\theta}, \boldsymbol{a}) = \min_{\boldsymbol{\alpha} \in \mathbb{R}^{n_\alpha}} \left[ \sum_{v \in \mathcal{V}} \max_{y \in \mathcal{Y}_v} \left[ \boldsymbol{\phi}_v(\boldsymbol{x}, y)^T \boldsymbol{\theta} - \sum_{v' \in \mathcal{N}(v)} \alpha_{vv'}(y) + \ell_v(\hat{y}_v, y) \right] \right.$$
$$\left. + \sum_{\{v,v'\} \in \mathcal{E}} \max_{(y,y') \in \mathcal{Y}_v \times \mathcal{Y}_{v'}} \left[ \boldsymbol{\phi}_{vv'}(y, y')^T \boldsymbol{\theta} + \alpha_{vv'}(y) + \alpha_{v'v}(y') \right] \right] - \boldsymbol{\phi}^p(\boldsymbol{x}, \boldsymbol{a})^T \boldsymbol{\theta}, \tag{15}$$

where $\boldsymbol{\phi}^p \colon \mathcal{X} \times \mathcal{A} \to \mathbb{R}^n$ is input-annotation feature map defined as

$$\boldsymbol{\phi}^p(\boldsymbol{x}, \boldsymbol{a}) = \sum_{v \in \mathcal{V}} \frac{[\![a_v \neq ?]\!]}{p_{Z_v|X}(1 \mid \boldsymbol{x})} \boldsymbol{\phi}_v(\boldsymbol{x}, y_v) + \sum_{v,v' \in \mathcal{E}} \frac{[\![a_v \neq ? \wedge a_{v'} \neq ?]\!]}{p_{Z_v, Z_{v'}|X}(1,1 \mid \boldsymbol{x})} \boldsymbol{\phi}_{vv'}(y_v, y_{v'}). \tag{16}$$

---

[2] To emphasize that $\psi_{\text{lp}}^p$ is applicable only for the linear MN classifier, we use the notation $\psi_{\text{lp}}^p(\boldsymbol{x}, \boldsymbol{\theta}, \boldsymbol{a})$ instead of $\psi_{\text{lp}}^p(\boldsymbol{f}, \boldsymbol{a})$ with $\boldsymbol{f}(\boldsymbol{x}) = \boldsymbol{\Phi}(\boldsymbol{x})^T \boldsymbol{\theta}$.

The loss $\psi_{\mathrm{lp}}^p$ is obtained from the LP-MR loss (11) by replacing the correct labeling score $f(x, y) = \phi(x, y)^T \theta$, which cannot be computed since the complete labeling $y$ is unknown, by the score $\phi^p(x, a)^T \theta$ which can be computed on the partial annotation $a$. The replacement is justified by the fact that the expectation of the input-output features equals the expectation of the input-annotation features as stated by the following theorem (Theorem 1 in [6]):

**Theorem 2.** *Let $\phi \colon \mathcal{X} \times \mathcal{Y} \to \mathbb{R}^d$ be the input-output feature map defined by (7) and $\phi^p \colon \mathcal{X} \times \mathcal{A} \to \mathbb{R}^d$ the input-annotation feature map defined by (16). Let both $\phi$ and $\phi^p$ be constructed from the same set of $\phi_v \colon \mathcal{X} \times \mathcal{Y} \to \mathbb{R}^d$, $v \in \mathcal{V}$, and $\phi_{vv'} \colon \mathcal{Y} \times \mathcal{Y} \to \mathbb{R}^d$, $\{v, v'\} \in \mathcal{E}$. Let $p_{A|XY}(a \mid x, y)$ be the MAR annotation process (14). Then, we have*

$$\mathbb{E}_{a \sim p_{A|X}} \phi^p(x, a) = \mathbb{E}_{y \sim p_{Y|X}} \phi(x, y), \qquad \forall x \in \mathcal{X},$$

*where $p_{A|X}(a \mid x)$ and $p_{Y|X}(y \mid x)$ are conditional distributions derived from $p_{XYA}(x, y, a) = p_{XY}(x, y) p_{A|XY}(a \mid x, y)$.*

Note that computation of the input-annotation feature map (16) requires the unary marginals $p_{Z_v|X}(z_v \mid x)$, $v \in \mathcal{V}$, and pair-wise marginals $p_{Z_v, Z_{v'}|X}(z_v, z_{v'} \mid x)$, $\{v, v'\} \in \mathcal{E}$, of the distribution $p_{Z|X}(z \mid x)$ describing the label missingness. The marginals can be easily estimated from the partially annotated examples $\mathcal{T}_{XA}$ using the maximum likelihood method [6].

The partial LP-MR loss $\psi_{\mathrm{lp}}^p$ is convex, and it can be efficiently optimized by gradient methods. In the limit case, when no labels are missing, it coincides with the supervised margin-rescaling loss, and hence, it is not Fisher-consistent.

## 3    Contributions

### 3.1    Tractable Adversarial Loss for the MN Classifier

The additive loss (2) and the score $f(x, y)$ of the MN classifier (3), both decompose as a sum of functions with arity at most two. We noticed that in this case the Min-Max problem that defines adversarial loss (12) can be converted to a linear program whose dual form is tractable. This leads to a novel surrogate loss, termed the MArkov Network Adversarial (MANA) loss, which is defined as a tractable convex optimization

$$\psi_{\mathrm{mana}}(f, \hat{y}) = \min_{\substack{\alpha \in \mathbb{R}^{n_\alpha} \\ \mu \in \mathcal{M}}} \left[ \sum_{v \in \mathcal{V}} \max_{y \in \mathcal{Y}_v} \left[ f_v(x, y) - \sum_{v' \in \mathcal{N}(v)} \alpha_{vv'}(y) + \sum_{y' \in \mathcal{A}} \mu_v(y') \ell_v(y_v, y') \right] \right.$$
$$\left. + \sum_{\{v, v'\} \in \mathcal{E}} \max_{(y, y') \in \mathcal{Y}_v \times \mathcal{Y}_{v'}} \left[ f_{vv'}(y, y') + \alpha_{vv'}(y) + \alpha_{v'v}(y') \right] \right] - f(x, \hat{y}),$$
$$\tag{17}$$

where the vector $\alpha = (\alpha_{vv'} \colon \mathcal{Y}_v \times \mathcal{Y}_{v'} \to \mathbb{R}, \alpha_{v'v} \colon \mathcal{Y}_{v'} \times \mathcal{Y}_v \to \mathbb{R}, \{v, v'\} \in \mathcal{E})$ has $n_\alpha = 2 \sum_{\{v, v'\} \in \mathcal{E}} (|\mathcal{Y}_v| + |\mathcal{Y}_{v'}|)$ variables, the vector $\mu = (\mu_v \in \Delta_v, v \in \mathcal{V}) \in \mathcal{M} \subset \mathbb{R}^{n_\mu}$ is composed of vectors $\mu_v \in \Delta_v$, $v \in \mathcal{V}$, from the probability simplex on $\mathcal{Y}_v$ and it has $n_\mu = \sum_{v \in \mathcal{V}} |\mathcal{Y}_v|$ variables in total. Note that evaluating the objective of the minimization problem (17) for fixed $(\alpha, \mu)$ does not require any oracle to solve an intractable problem, unlike the algorithm of [12]. The following theorem, one of the main results of this paper, ensures that the MANA loss (17) coincides with the Fisher consistent adversarial loss (12).

**Theorem 3.** *Let $\ell \colon \mathcal{Y} \times \mathcal{Y} \to \mathbb{R}$ be an additive loss (2). Let $\mathcal{F} = \{f \colon \mathcal{X} \to \mathbb{R}^{|\mathcal{Y}|}\}$ be a set composed of the MN classifier score maps given by (3). Then, we have*

$$\psi_{\mathrm{adv}}(f, y) = \psi_{\mathrm{mana}}(f, y), \quad \forall f \in \mathcal{F}, \forall y \in \mathcal{Y}.$$

**Comparison with the LP Margin-Rescaling Loss.** We would like to point out a striking similarity between the MANA loss (17) and the LP-MR loss (11) although they were derived from completely different principles. The MANA loss can be obtained from the LP-MR loss by replacing the ground truth labels in the maximization terms of (11) by their one-hot encodings, and minimizing the value of the loss w.r.t those encodings. Or, equivalently, fixing the values of $\mu_v(y)$, $v \in \mathcal{V}$, in (17) to one-hot encoding of ground truth labels $\hat{y}_v$, $v \in \mathcal{V}$, instead of minimizing them, makes MANA loss equal to the LP-MR loss. This subtle change makes the inconsistent LP-MR loss to Fisher-consistent MANA loss without significantly increasing the computational complexity. However, the LP-MR loss can be seen as a close approximation of the consistent MANA loss which also provides additional explanation for its good empirically observed performance.

**MANA as Unconstrained Convex Optimization.** Most frequently, the single-label losses $\ell_v \colon \mathcal{Y}_v \times \mathcal{Y}_v \to \mathbb{R}$, $v \in \mathcal{V}$, defining the additive loss $\ell(y, y')$ are normalized 0/1 losses, e.g., when $\ell(y, \hat{y}) = \frac{1}{|\mathcal{V}|} \sum_{v \in \mathcal{V}} [\![y_v \neq \hat{y}_v]\!]$ is the Hamming loss. In this case, the MANA loss (17) can be simplified by eliminating the variables $\mu$ as stated in the following theorem.

**Theorem 4.** *Let $\ell \colon \mathcal{Y} \times \mathcal{Y} \to \mathbb{R}$ be an additive loss (2) composed of $\ell_v(y, y') = K_v [\![y \neq y']\!]$, $v \in \mathcal{V}$, where $K_v > 0$, $v \in \mathcal{V}$, are positive scalars. Then*

$$\psi_{\mathrm{mana}}(f, \hat{y}) = \min_{\alpha \in \mathbb{R}^{n_\alpha}} \left[ \sum_{v \in \mathcal{V}} \max_{S \subseteq \mathcal{Y}_v, |S| > 0} \left[ \frac{1}{|S|} \sum_{y \in S} \left[ f_v(x, y) - \sum_{v' \in \mathcal{N}(v)} \alpha_{vv'}(y) \right] \right.\right.$$
$$\left. + K_v - \frac{K_v}{|S|} \right] + \sum_{\{v, v'\} \in \mathcal{E}} \max_{(y, y') \in \mathcal{Y}_v \times \mathcal{Y}_{v'}} \left[ f_{vv'}(y, y') + \alpha_{vv'}(y) + \alpha_{v'v}(y') \right] \bigg] - f(x, \hat{y}).$$

$$(18)$$

*Remark 1.* The inner maximization in (18) is of type

$$\max_{S \subseteq \mathcal{Y}_v, |S| > 0} \left[ \frac{1}{|S|} \sum_{y \in S} g_v(x, y) + K_v - \frac{K_v}{|S|} \right]$$

and it can be solved in $\mathcal{O}(|\mathcal{Y}_v| \log |\mathcal{Y}_v|)$ time as follows. First, sort the values $g_v(x, y)$, $y \in \mathcal{Y}_v$, in non-increasing order into a sequence $a_1, \ldots, a_{|\mathcal{Y}_v|}$. Second, compute $k^* \in \text{Argmax}_{k \in \{1, \ldots, |\mathcal{Y}_v|\}} [\frac{1}{k} \sum_{i=1}^k a_i + K_v - \frac{K_v}{k}]$. Third, construct the optimal set $S$ from the first $k^*$ labels in the sorted order.

Using the MANA loss (18) as a surrogate in the regularized ERM problem (9), leads to an unconstrained convex optimization with $\mathcal{O}(m \cdot n_\alpha + n)$ variables. The optimization problem can be solved efficiently using standard sub-gradient methods.

## 3.2   Fisher-Consistent Surrogate for Partially Annotated Examples

In this section, we extend the MANA for learning the linear MN classifier on partially annotated examples. In particular, we assume the linear predictor (6) with the score map $\boldsymbol{f}(x) = \boldsymbol{\Phi}(x)^T \boldsymbol{\theta}$ given by the parameters $\boldsymbol{\theta} \in \mathbb{R}^n$ and the input-output feature map (7). We further assume that the partial annotations are generated by the MAR process (14). For this setting, we propose the partial MANA loss defined as

$$
\begin{aligned}
\psi_{\text{mana}}^p(x, \boldsymbol{\theta}, \boldsymbol{a}) = \min_{\substack{\alpha \in \mathbb{R}^{n_\alpha} \\ \mu \in \mathcal{M}}} & \left[ \sum_{v \in \mathcal{V}} \max_{y \in \mathcal{Y}_v} \left[ \boldsymbol{\theta}^T \boldsymbol{\phi}_v(x, y) - \sum_{v' \in \mathcal{N}(v)} \alpha_{vv'}(y) + \sum_{y' \in \mathcal{A}} \mu_v(y') \ell_v(y_v, y') \right] \right. \\
& \left. + \sum_{\{v, v'\} \in \mathcal{E}} \max_{(y, y') \in \mathcal{Y}_v \times \mathcal{Y}_{v'}} \left[ \boldsymbol{\theta}^T \boldsymbol{\phi}_{vv'}(y, y') + \alpha_{vv'}(y) + \alpha_{v'v}(y') \right] \right] - \boldsymbol{\theta}^T \boldsymbol{\phi}^p(x, \boldsymbol{a})
\end{aligned}
\tag{19}
$$

where $\boldsymbol{\phi}^p(x, \boldsymbol{a})$ is the input-annotation feature map (16). The partial MANA loss (19) is obtained from the (supervised) MANA loss (17) after substituting the linear score $f(x, \boldsymbol{y}) = \boldsymbol{\theta}^T \boldsymbol{\phi}(x, \boldsymbol{y})$ and replacing the correct labeling score (the last term in (17)), which cannot be evaluated as the complete labeling $\boldsymbol{y}$ is unknown, by $\boldsymbol{\theta}^T \boldsymbol{\phi}^p(x, \boldsymbol{a})$, which can be evaluated on a partial annotation $\boldsymbol{a}$. Note that the partial MANA loss has the exact same computational complexity as the (supervised) MANA loss. In the case of fully annotated examples, it follows from (16) that $\boldsymbol{\phi}^p(x, \boldsymbol{a}) = \boldsymbol{\phi}(x, \boldsymbol{y})$, and hence both losses coincide.

The following theorem, another main contribution of this paper, ensures that the partial MANA loss is Fisher-consistent.

**Theorem 5.** *Assume the same setup as in Theorem 1. In addition, assume that:*

1. *The partially annotated examples $(x, \boldsymbol{a}) \in \mathcal{X} \times \mathcal{A}$ are generated from $p_{XA}(x, \boldsymbol{a}) = \sum_{\boldsymbol{y} \in \mathcal{Y}} p_{A|XY}(\boldsymbol{a} \mid x, \boldsymbol{y}) p_{XY}(x, \boldsymbol{y})$ where $p_{A|XY}(\boldsymbol{a} \mid x, \boldsymbol{y})$ is a MAR annotation process (14).*
2. *The set $\mathcal{F} = \{\boldsymbol{f}(x) = \boldsymbol{\Phi}(x)^T \boldsymbol{\theta} \mid \boldsymbol{\theta} \in \Theta \subseteq \mathbb{R}^n\}$ contains score maps of linear MN classifier (7), and $\mathcal{F} \supset \text{Argmin}_{\boldsymbol{f} : \mathcal{X} \to \mathbb{R}^{|\mathcal{Y}|}} R_{\text{adv}}(\boldsymbol{f}, p_{XY})$.*

*Then, we have*
$$
R_*^\ell(p_{XY}) = R^\ell(\boldsymbol{T} \circ \boldsymbol{\Phi}(x)^T \boldsymbol{\theta}_{\text{mana}}^p, p_{XY}),
$$

*where $\boldsymbol{T}(x) \in \text{Argmax}_{y \in \mathcal{Y}} f_y$ and $\boldsymbol{\theta}_{\text{mana}}^p \in \text{Argmin}_{\boldsymbol{\theta} \in \Theta} R_{\text{mana}}^p(\boldsymbol{\theta}, p_{XA})$ is a minimizer of $R_{\text{mana}}^p(\boldsymbol{\theta}, p_{XA}) = \mathbb{E}_{x, a \sim p_{XA}} \psi_{\text{mana}}^p(x, \boldsymbol{\theta}, \boldsymbol{a})$.*

The theorem guarantees that the linear MN classifier $h(x) = T \circ \Phi(x)^T \theta^p_{\text{mana}}$ with parameters found minimizing $\psi^p_{\text{mana}}$-risk on $p_{XA}(x, a)$ achieves the Bayes $\ell$-risk on $p_{XY}(x, y)$. The theorem requires the annotations to be generated from MAR process (14), and that the class of linear scores $\mathcal{F}$ is sufficiently rich to contain a minimizer of the $\psi_{\text{adv}}$-risk.

# 4 Experiments

We evaluate the precision of a linear MN classifier trained by solving the regularized ERM problem (9) with different surrogate losses. In all experiments, we use the normalized Hamming loss, $\ell(y, y') = \frac{1}{|\mathcal{V}|} \sum_{v \in \mathcal{V}} [\![ y_v \neq y'_v ]\!]$, as the target loss plugged into the surrogates. We evaluate the following algorithms:

1. The baseline, referred to as the M3N algorithm, solves (9) with the MR-LP surrogate (11) when learning from fully annotated examples, and the partial MR-LP surrogate (15) when learning from the partial annotations. When $(\mathcal{V}, \mathcal{E})$ is a chain and the examples are fully annotated, the algorithm becomes the standard Maximum Margin Markov network algorithm [16,17]. The generalization for an arbitrary graph $(\mathcal{V}, \mathcal{E})$ was proposed in [4]. The generalization for partially annotated examples comes from [6]. In all cases, the surrogates are derived from the margin rescaling loss (8), hence we use the M3N algorithm for all the variants.
2. The proposed algorithm, referred to as MANA algorithm, solves (9) with the MANA surrogate (18) when learning from fully annotated examples and partial loss of MANA (19) when learning from partial annotations.

As benchmark problems, in Sect. 4.1 we consider the prediction of sequences generated from the hidden Markov chain, and in Sect. 4.2 the prediction of the solution of the Sudoku puzzle [6].

**Inference.** The inference of the MN classifier (1) is solved by the dynamic programming when $(\mathcal{V}, \mathcal{E})$ is a chain. For general $(\mathcal{V}, \mathcal{E})$, we use the Augmented Directed Acyclic Graph solver [13,19].

**Optimization.** Regardless of the surrogate used, ERM (9) leads to convex unconstrained optimization with the same number of variables. We solve ERM (9) using ADAM [7] with $\beta_1 = 0.9$, $\beta_2 = 0.999$, 5000 passes through all $m$ training examples, and the learning rate $\frac{1}{100\,t}$, $t \in \{1, \ldots, 5000\,m\}$ .

**Computation of Partial Losses.** Partial loss of LP-MR (11) and Partial MANA loss (19) require knowing the marginals of the distribution $p_{Z|X}(z \mid x)$ that govern the missingness of the labels. Following [6], we assume that the distribution is homogeneous and the labels are missing completely at random, that is,

$$p_{Z|X}(z \mid x) = \prod_{v \in \mathcal{V}} \tau^t (1 - \tau)^t \tag{20}$$

where $t = \sum_{v \in \mathcal{V}} z_v$ and $\tau \in [0, 1]$ is the probability that a randomly chosen object $v \in \mathcal{V}$ is annotated. Under this assumption, marginals can be estimated from the partially annotated examples $\mathcal{T}_{XA}$ using the maximum likelihood approach:

$$\begin{aligned} \hat{p}_{Z_v|X}(1 \mid x) &= \tau, \quad v \in \mathcal{V}, \\ \hat{p}_{Z_v,Z_{v'}|X}(1,1 \mid x) &= \tau^2, \quad \{v, v'\} \in \mathcal{E}, \end{aligned} \quad \text{where} \quad \tau = \frac{1}{m|\mathcal{V}|} \sum_{i=1}^{m} \sum_{v \in \mathcal{V}} \llbracket a_v^i \neq ? \rrbracket. \quad (21)$$

The estimated marginals are used to calculate $\phi^P(x, a)$ defined by (16).

**Evaluation Protocol.** For each data set, we generate $K$ random divisions of the examples into training, validation, and testing parts. The training part is used to learn the parameters $\theta$. The optimal regularization constant $\lambda \in \{0, 1, 10, 100\}$ selected based on the minimal Hamming loss evaluated on the validation= part. We report the mean and standard deviation of the Hamming loss and the $0/1$ loss of the model with the optimal $\lambda$ calculated on the K example divisions.

## 4.1  Synthetic Data: Hidden Markov Chain

The input and output are sequences of symbols $x = (x_1, \ldots, x_{100}) \in \{1, \ldots, 30\}^{100}$ and $y = (y_1, \ldots, y_{100}) \in \{1, \ldots, 30\}^{100}$ generated from the hidden Markov chain:

$$p_{XY}(x, y) = p(y_1) \prod_{i=2}^{100} p(y_i \mid y_{i-1}) p(x_i \mid y_i). \quad (22)$$

The initial state distribution $p(y_1)$ is randomly generated, the emission probability is $p(x_i \mid y_i) = 7/10$ if $x_i = y_i$ and $p(x_i \mid y_i) = 3/290$ otherwise, and the transition probability is $p(y_i \mid y_{i-1}) = 7/10$ if $y_i = y_{i-1}$ and $p(y_i \mid y_{i-1}) = 3/290$ otherwise. The known model allows us to construct the Bayes classifier, optimal for the Hamming loss. The Bayes risk estimated from 100,000 examples is 0.2013.

We generate the partial annotation $a \in (\{1, \ldots, 30\} \cup \{?\})^{100}$ using $p(a \mid x, y)$ given by (14), and the missingness distribution $p(z \mid x)$ given by (20). We vary the probability $\tau \in \{0, 0.1, 0.2\}$ to generate the complete annotation and partial annotations with 10% and 20% labels missing at random. The graph $(\mathcal{V}, \mathcal{E})$ is a chain. The feature maps $\psi_v(x, y) = \mathbf{1}_{x_v, y}$, and $\psi_{vv'}(y, y') = \mathbf{1}_{y, y'}$, are one-hot encodings of the symbols $(x_v, y)$ and $(y, y')$, respectively. We used $K = 5$ random divisions of the data. The test set has 10,000 examples, the validation 5000 examples, and the size of the training set was $m \in \{10, 100, 1000\}$.

The test error of the MN classifier for different sizes of the training set and different amounts of missing labels is summarized in Table 1. The errors obtained for the M3N and MANA algorithms are very similar. The M3N performs slightly better when the number of training examples is small, while the MANA performs slightly better when the number of training examples is high. Differences become more pronounced with a greater number of missing labels. The best test risk $0.2050 \pm 0.0005$, obtained with the MANA algorithm on 1,000 fully annotated examples, is close to the Bayes risk 0.2013 estimated from 100,000 examples and using the ground truth model (22) to construct the Bayes predictor.

**Table 1.** Test error of linear MN classifiers predicting sequences of symbols generated by hidden Markov chain. The classifiers are trained by M3N and MANA algorithm on varying number of training examples with varying amount of missing labels.

|  |  |  | M3N | MANA |
|---|---|---|---|---|
|  |  |  | Test error | Test error |
|  |  | #trn | Hamming loss | Hamming loss |
| missing labels | 0% | 10 | $0.3500 \pm 0.0124$ | $0.3764 \pm 0.0152$ |
|  |  | 100 | $0.2351 \pm 0.0007$ | $0.2319 \pm 0.0016$ |
|  |  | 1000 | $0.2094 \pm 0.0008$ | $0.2050 \pm 0.0005$ |
|  | 10% | 10 | $0.3495 \pm 0.0189$ | $0.3528 \pm 0.0154$ |
|  |  | 100 | $0.2441 \pm 0.0023$ | $0.2420 \pm 0.0018$ |
|  |  | 1000 | $0.2117 \pm 0.0008$ | $0.2065 \pm 0.0007$ |
|  | 20% | 10 | $0.3409 \pm 0.0200$ | $0.3423 \pm 0.0177$ |
|  |  | 100 | $0.2547 \pm 0.0017$ | $0.2526 \pm 0.0017$ |
|  |  | 1000 | $0.2135 \pm 0.0008$ | $0.2078 \pm 0.0007$ |

## 4.2  Sudoku Solver

**Symbolic Inputs.** The Sudoku is made up of $9 \times 9$ cells $\mathcal{V} = \{(i,j) \in \mathcal{N} \mid 1 \leqslant i \leqslant 9, 1 \leqslant j \leqslant 9\}$ filled with numbers 1 to 9 or kept empty $\square$. The puzzle assignment is $\boldsymbol{x} = (x_v \in \{\square, 1, \ldots, 9\} \mid v \in \mathcal{V})$. The task is to fill the empty cells so that the rows, columns, and non-overlapping subgrids $3 \times 3$ contain all numbers from 1 to 9. The puzzle solution is $\boldsymbol{y} = (y_v \in \{1, \ldots, 9\} \mid v \in \mathcal{V})$. Prior knowledge is encoded by revealing the algorithm that cells in rows, columns, and $3 \times 3$ sub-grids are related, that is, by setting $\mathcal{E} = \{\{(v, v'), (u, u')\} \mid v = v' \vee v' = u' \vee (\lceil v/3 \rceil = \lceil u/3 \rceil \wedge \lceil v'/3 \rceil = \lceil u'/3 \rceil)\}$. The feature maps $\boldsymbol{\psi}_v(\boldsymbol{x}, y) = \mathbf{1}_{x_v, y}$, and $\boldsymbol{\psi}_{vv'}(y, y') = \mathbf{1}_{y, y'}$, are one-hot encodings of the pair of symbols $(x_v, y)$ and $(y, y')$, respectively.

We use a database of Sudoku assignments and their correct solutions to create a training set. The partial annotation/solution was generated using the MAR process (14) with $p_{Z_v|X}(1 \mid \boldsymbol{x}) = 1$ if $x_v \in \{1, \ldots, 9\}$ and $p_{Z_v|X}(1 \mid \boldsymbol{x}) = 1 - \tau$ if $x_v = \square$, where $\tau \in \{0, 0.1, 0.2\}$ is the probability that the empty cell is not annotated. We generate three training sets with a complete solution, with 10% and 20% of the empty cells left empty, respectively. We varied the number of training examples $m \in \{10, 100, 1000\}$. We tested on 100 puzzles; note that it involves the prediction of $9 \cdot 9 \cdot 100 = 8{,}100$ labels. In addition to Hamming loss, we also evaluated the prediction using the 0/1 loss, in which case the test error corresponds to the portion of puzzles that were not solved perfectly.

The results are summarized in Table 2. The precisions obtained for the M3N and MANA algorithm are similar. It was enough to use $m = 100$ training examples to reach zero test error regardless of the amount of missing labels used. In the case of $m = 10$ training examples, the differences are at the level of the standard deviation for both 0/1 loss and Hamming loss.

**Table 2.** Test error of linear MN classifiers predicting solution of Sudoku puzzle from either symbolic assignment or visual assignment composed of the MNIST digits. The classifiers are trained by M3N and the proposed MANA algorithm on varying number of training examples with varying amount of missing labels.

Symbolic Sudoku

| | | | M3N | | MANA | |
|---|---|---|---|---|---|---|
| | | | Test error | | Test error | |
| | | #trn | 0/1-loss [%] | Hamming loss | 0/1-loss [%] | Hamming loss |
| missing labels | 0% | 10 | 0.6±0.9 | 0.0021±0.0029 | 0.6±0.5 | 0.0021±0.0020 |
| | | 100 | 0.0±0.0 | 0.0000±0.0000 | 0.0±0.0 | 0.0000±0.0000 |
| | 10% | 10 | 0.6±0.9 | 0.0018±0.0028 | 0.4±0.5 | 0.0013±0.0018 |
| | | 100 | 0.0±0.0 | 0.0000±0.0000 | 0.0±0.0 | 0.0000±0.0000 |
| | 20% | 10 | 0.6±0.9 | 0.0018±0.0028 | 1.0±1.2 | 0.0031±0.0038 |
| | | 100 | 0.0±0.0 | 0.0000±0.0000 | 0.0±0.0 | 0.0000±0.0000 |

Visual Sudoku

| | | | M3N | | MANA | |
|---|---|---|---|---|---|---|
| | | | Test error | | Test error | |
| | | #trn | 0/1-loss [%] | Hamming loss | 0/1-loss [%] | Hamming loss |
| missing labels | 0% | 10 | 96.2±1.8 | 0.4407± 0.0070 | 96.8±1.3 | 0.4475± 0.0201 |
| | | 100 | 19.2±4.3 | 0.0625± 0.0153 | 20.4±3.9 | 0.0710± 0.0160 |
| | | 1000 | 5.8±1.3 | 0.0149± 0.0035 | 5.8±0.8 | 0.0155± 0.0037 |
| | 10% | 10 | 95.6±2.6 | 0.4402±0.0155 | 96.2±2.4 | 0.4512±0.0106 |
| | | 100 | 36.2±4.9 | 0.1254±0.0205 | 42.6±4.7 | 0.1467±0.0238 |
| | | 1000 | 37.2±4.8 | 0.0928±0.0195 | 40.6±3.8 | 0.0952±0.0160 |
| | 20% | 10 | 97.6±2.5 | 0.4557±0.0213 | 98.0±1.9 | 0.4643± 0.0129 |
| | | 100 | 46.2±2.3 | 0.1593±0.0120 | 50.4±3.4 | 0.1706± 0.0150 |
| | | 1000 | 52.4±3.2 | 0.1260±0.0184 | 52.8±3.3 | 0.1261± 0.0180 |

**MNIST Digits Used as Input.** We replace the input symbols $\{1, \ldots, 9\}$ with $28 \times 28$ images of handwritten digits from the MNIST data set [8]. The empty cells are replaced by all-black images. As a feature map of the unary scores, we use $\boldsymbol{\psi}_v(\boldsymbol{x}, y) = (\bar{\boldsymbol{\psi}}_1, \ldots, \bar{\boldsymbol{\psi}}_9)$, where $\bar{\boldsymbol{\psi}}_{y'} \in \mathbb{R}^{2000}$, $y' \neq y$, are all-zero vectors, $\bar{\boldsymbol{\psi}}_y = (k(\boldsymbol{x}_v, \boldsymbol{\mu}_1), \ldots, k(\boldsymbol{x}_v, \boldsymbol{\mu}_{2000})) \in \mathbb{R}^{2000}$ is a vector of RBF kernels $k(\boldsymbol{x}_v, \boldsymbol{\mu}_i) = \exp(-2\|\boldsymbol{x}_v - \boldsymbol{\mu}_i\|^2)$ evaluated for the image $\boldsymbol{x}_v$ of the $v$-th cell and 2,000 randomly sampled training images. All other settings are the same as for the symbolic Sudoku experiment. The results are summarized in Table 2. The M3N algorithm achieves slightly better results when the number of training examples is small. For $m = 1000$, the differences are at the standard deviation level.

**Comparison with Neural Architectures.** Learning deep NN to solve Sudoku was considered in [18]. They used both the symbolic and the MNIST digits as inputs. They trained the SATNet architecture, which is a CNN with a maximum satisfiability (MAXSAT) solver as the last layer. SATNet can better learn hard interactions between output variables than canonical neural architectures (ConvNet) used as a baseline. They use $9,000$ fully annotated training examples and $1,000$ test examples. Table 3 presents the portion of incorrectly predicted solutions of the test Sudoku puzzles. For comparison, we include the performance of linear MN classifiers trained with the M3N and MANA algorithm on 1,000 completely annotated examples. Although the MN classifier is trained on a smaller number of examples, it significantly outperforms both neural architectures in both symbolic and visual Sudoku.

**Table 3.** Comparison of the MN classifier trained from 1,000 examples, and neural architectures trained from 9,000 examples on the problem of predicting Sudoku solution from symbolic and visual assignments composed of MNIST digits.

| Method | Test error, 0/1-loss [%] | |
| --- | --- | --- |
| | Symbolic | Visual |
| MN classifier - M3N | $0.0\pm0.0$ | $5.8\pm1.3$ |
| MN classifier - MANA | $0.0\pm0.0$ | $5.8\pm0.8$ |
| ConvNet | 84.9 | 99.9 |
| SATNet | 1.7 | 63.8 |

# 5   Conclusions

We proposed a novel surrogate loss, the MANA loss, to train MN classifiers. Minimizing MANA loss leads to tractable convex optimization that is amenable to standard gradient methods. We prove that the MANA loss is equivalent to the adversarial loss defined by the Min-Max problem, which is Fisher consistent but intractable in the context of the structure prediction. To our knowledge, the proposed MANA loss is the first surrogate for learning MN classifiers with a generic neighborhood graph that is simultaneously statistically consistent, convex, and tractable. This is not an obvious result because even an evaluation of a generic MN classifier leads to discrete optimization, which is intractable, in general.

We also proposed a partial MANA loss applicable to learning linear MN classifiers on partially annotated examples when the labels are missing at random. The partial MANA loss has the same computational complexity as its supervised counterpart, and we prove that the partial MANA loss is also Fisher-consistent.

The experiments show that the empirical performance of the ERM algorithms minimizing the MANA loss, which is consistent, and the LP margin scaling loss, which is not consistent, are comparable. The deviations are usually at the level of the estimation error. The comparable performance is not that surprising, because we have also shown that the LP margin rescaling loss is a close approximation of the MANA loss, although both surrogates were originally developed from completely different principles.

The code and data are available at: https://github.com/xfrancv/manet

**Acknowledgments.** The research was supported by the Czech Science Foundation project GACR GA19-21198S and OP VVV project CZ.02.1.01\0.0\0.0\16 019\0000765 Research Center for Informatics.

# References

1. Antoniuk, K., Franc, V., Hlaváč, V.: Consistency of structured output learning with missing labels. In: Asian Conference on Machine Learning (ACML) (2015)
2. Fathony, R., Liu, A., Asif, K., Ziebart, B.: Adversarial multiclass classification: A risk minimization perspective. In: NIPS (2016)
3. Fathony, R., et al.: Consistent robust adversarial prediction for general multiclass classification (2018). https://arxiv.org/abs/1812.07526
4. Franc, V., Laskov, P.: Learning maximal margin Markov networks via tractable convex optimization. Control Syst. Comput. **2**, 25–34 (2011)
5. Franc, V., Savchynskyy, B.: Discriminative learning of max-sum classifiers. J. Mach. Learn. Res. **9**(1), 67–104 (2008)
6. Franc, V., Yermakov, A.: Learning maximum margin Markov networks from examples with missing labels. In: Asian Conference on Machine Learning (2021)
7. Kingma, D.P., Ba, J.: Adam: a method for stochastic optimization. In: Proceedings of International Conference on Learning Representations (ICLR) (2015)
8. LeCun, Y., Cortes, C.: MNIST handwritten digit database (2010). http://yann.lecun.com/exdb/mnist/
9. Lin, Y.: A note on margin-based loss functions in classification. Stat. Probab. Lett. **68**(1), 73–82 (2004)
10. Liu, Y.: Fisher consistency of multicategory support vector machines. In: International Conference on Artificial Intelligence and Statistics, pp. 291–298 (2007)
11. Lou, X., Hamprecht, F.A.: Structured learning from partial annotations. In: International Conference on Machine Learning (ICML), pp. 1519–1526 (2012)
12. Nowak, A., Bach, F., Rudi, A.: Consistent structured prediction with max-min margin Markov networks. In: International Conference on Machine Learning (2020)
13. Schlesinger, M.: Syntactic analysis of two-dimensional visual signals in noisy conditions. Kibernetika **4**, 113–130 (1976). in Russian
14. Taskar, B., Chatalbashev, V., Koller, D.: Learning associative Markov networks. In: International Conference on Machine Learning (ICML) (2004)
15. Taskar, B., Guestrin, C., Koller, D.: Maximum-margin markov networks. In: Proceedings of Neural Information Processing Systems (NIPS) (2004)
16. Taskar, B., Guestrin, C., Koller, D.: Max-margin Markov networks. In: NIPS (2003)
17. Tsochantaridis, I., Joachims, T., Hofmann, T., Altun, Y.: Large margin methods for structured and interdependent output variables. J. Mach. Learn. Res. **6**(50), 1453–1484 (2005)
18. Wang, P., Donti, P., Wilder, B., Kolter, J.: SATnet: bridging deep learning and logical reasoning using a differential satisfiability solver. In: ICML (2019)

19. Werner, T.: A linear programming approach to max-sum problem: a review. IEEE Trans. Pattern Anal. Mach. Intell. **29**(7), 1165–1179 (2007)
20. Yhang, T.: Statistical analysis of some multi-category large margin classification methods. J. Mach. Learn. Res. **5** (2004)

# Deep Active Learning for Detection of Mercury's Bow Shock and Magnetopause Crossings

Sahib Julka[1]([✉]), Nikolas Kirschstein[1], Michael Granitzer[1],
Alexander Lavrukhin[2], and Ute Amerstorfer[3]

[1] University of Passau, Passau, Germany
{sahib.julka,michael.granitzer}@uni-passau.de
[2] Skobeltsyn Institute of Nuclear Physics, LMSU, Moscow, Russian Federation
lavrukhin@physics.msu.ru
[3] Space Research Institute, Austrian Academy of Sciences, Graz, Austria
ute.amerstorfer@oeaw.ac.at

**Abstract.** Accurate and timely detection of bow shock and magnetopause crossings is essential for understanding the dynamics of a planet's magnetosphere. However, for Mercury, due to the variable nature of its magnetosphere, this remains a challenging task. Existing approaches based on geometric equations only provide average boundary shapes, and can be hard to generalise to environments with variable conditions. On the other hand, data-driven methods require large amounts of annotated data to account for variations, which can scale up the costs quickly. We propose to solve this problem with machine learning. To this end, we introduce a suitable dataset, prepared by processing raw measurements from NASA's MESSENGER (**ME**rcury **S**urface, **S**pace **E**nvironment, **GE**ochemistry, and **R**anging) mission and design a five-class supervised learning problem. We perform an architectural search to find a suitable model, and report our best model, a Convolutional Recurrent Neural Network (CRNN), achieves a macro F1 score of 0.82 with accuracies of approximately 80% and 88% on the bow shock and magnetopause crossings, respectively. Further, we introduce an approach based on active learning that includes only the most informative orbits from the MESSENGER dataset measured by Shannon entropy. We observe that by employing this technique, the model is able to obtain near maximal information gain by training on just two Mercury years worth of data, which is about 10% of the entire dataset. This has the potential to significantly reduce the need for manual labeling. This work sets the ground for future machine learning endeavors in this direction and may be highly relevant to future missions such as BepiColombo, which is expected to enter orbit around Mercury in December 2025.

**Keywords:** Active learning · Neural networks · Magnetosphere

© The Author(s), under exclusive license to Springer Nature Switzerland AG 2023
M.-R. Amini et al. (Eds.): ECML PKDD 2022, LNAI 13716, pp. 452–467, 2023.
https://doi.org/10.1007/978-3-031-26412-2_28

# 1   Introduction

The *magnetosphere* of a planet is the region surrounding it where its magnetic field dominates over the magnetic field of the interplanetary space. The *magnetopause* marks the outer boundary of the magnetosphere. Above the magnetopause, lies the *magnetosheath*, which is the region between the magnetopause and the *bow shock*—a shock wave that slows down the approaching supersonic solar wind, and deflects it around the planet's magnetospheric cavity. Principally, the locations and characteristics of these regions around a planet are affected by the varying solar wind conditions [9]. This is particularly the case for Mercury (C.f. Fig. 1 (a)), the innermost planet in our solar system. Adding to it, its weak magnetic field—only about 1% of the Earth's [6], makes the magnetic conditions around the planet even more dynamic, and thus interesting to study. Studying such magnetospheres can yield valuable insights into understanding more complex magnetospheres, such as that of our planet Earth.

It has long been of scientific interest in the planetary science community to study Mercury's bow shock and magnetopause signatures. To this end, NASA launched a space-probe called MESSENGER orbiting Mercury for a long-term empirical study. The relatively small size of Mercury's magnetosphere, an order of magnitude less than the Earth's, allowed the collection of large amounts of data in a significantly shorter time. During the four years of its voyage from 2011 to 2015, the spacecraft completed over 4000 orbits around the planet. As sketched in Fig. 1(b), it passed through all the magnetic regions, yielding more than 8000 incidences of bow shock and magnetopause crossings.

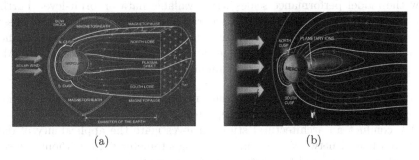

(a)                                            (b)

**Fig. 1.** (a) Schematic view of Mercury's magnetic conditions [22].The bow shock slows down the approaching solar wind to subsonic speeds. The magnetopause further acts as an obstacle. (b) A typical MESSENGER orbit path: the spacecraft passed from the *interplanetary magnetic field* (IMF) through bow shock, magnetosheath, magnetopause and magnetosphere regions of Mercury and then through the same sequence in reverse [26].

Based on the data from the MESSENGER magnetometer, several studies proposed geometric models of Mercury's magnetosphere [12,17,24,25]. However, due to their global and static nature, they could only provide an average shape of the bow shock and magnetopause boundaries. The respective authors found that the models struggle to capture the many fluctuations and nuances necessary to generalise to all events. This issue may successfully be tackled by employing data-driven statistical machine learning techniques. Given sufficient data, deep neural networks have shown increasing promise in approximately modelling any distribution, and have successfully been applied to complex tasks relating to event detection, including but not limited to rare event detection in audio signals [4,5] and images [13]. The problem of detecting boundary crossings in a continuous stream of magnetic flux data could be viewed similarly.

The planetary science community recognises the importance of this paradigm shift [16]. We follow suit and propose to solve this problem, as a first step, in a supervised deep learning setting. However, supervised learning requires a suitable dataset and expert annotations. As this effort can get very costly given the usually large amounts of unlabelled data in planetary sciences, it is prudent to only annotate the most useful samples. Active learning can facilitate efficient manual labeling by taking classifier specifics into account. However, it may not necessarily be useful in all domain contexts. In planetary science, however, the problem domain and data gathering context could provide an important frame for devising a domain-specific active learning strategy.

In particular, it is reasonable to assume that different orbits may exhibit similarities in their magnetic field structure, yet at the same time at least one entire Mercury year would be necessary to capture all seasonal nuances. It remains, however, unknown how the inter-orbital year distributions vary, and thus questions such as what is the lower bound on number of orbits required to obtain a near maximum informational gain remain open. In this regard, we examine how the model performance scales with available data on orbit-level. Further, we consider it necessary not only to accurately classify and localise the crossing timestamps, but also to classify ahead in time, which would be highly beneficial for tasks such as instrument parameter adjustment, during real time use. More precisely, our contributions can be summarised as follows:

1. We introduce a dataset suited to machine learning tasks and make it available open source: https://github.com/epn-ml/messenger-prep
2. We conduct an architectural study to investigate the applicability of data-driven neural networks by using just magnetometer data, without the solar wind conditions, and to identify some best practices.
3. We devise a domain-specific active learning strategy and investigate how many Mercury years' worth of data are required for a sufficiently representative model.
4. We provide a high-quality codebase that may be used as a framework for further studies on neural detection of bow shock and magnetopause crossings. It is publicly available at: https://github.com/epn-ml/Freddie

## 2    Related Work

The task of modelling the boundary crossings is not new. Naturally, Earth has had the lion's share of related work as evidenced by [14,20,21,23]. This enabled subsequent studies investigating various structural and statistical properties of the magnetopause [10]. The empirical and statistical studies require that a consistent catalogue of boundary crossings is available from the in situ data. This process has been recognised to be time-consuming, ambiguous and poorly reproducible, and one that would significantly benefit from automation.

To this end, [11] proposed a threshold-based method. However that turned out to be hard to generalise given the different scales and distributions from different missions [15]. In another line of work, models using paraboloids of revolution with variable flaring angles are explored for Mercury [2], Earth [1], Jupiter [7], and Saturn [3]. These models were obtained by parabolic parameterisation of the magnetopause and bow shock crossing shapes. The averaged boundary shapes can be used as initial parameter values for magnetospheric magnetic field modeling. In this vein, [12] attempted to model Mercury's boundary crossings using such a model. This was followed by [24], where the authors explored the applicability of hyperboloids and a figure similar to the Earth's magnetopause shape, and also [25] whose authors modelled it as a three dimensional non-axially symmetric shape. Philpott et al. [17] extended the aforementioned studies using a combination of an axisymmetric shape and a three-dimensional shape with indentations in the cusp regions and a magnetotail that is wider in the north-south versus east-west direction.

All these approaches share the drawback of applying static models that cannot capture variable conditions in the environment, since they propose a fixed geometric shape cemented for all times. We utilise the boundary crossing catalogue provided by Philpott et. al as approximate guides for supervised deep learning. This is particularly useful to test our active learning strategy so in the future works this is suited to a semi-supervised setup, where only the most necessary samples are required to be annotated by the domain expert.

## 3    Dataset

As the Fast Imaging Plasma Spectrometer (FIPS) on MESSENGER was not equipped to capture the solar wind data, there are no in situ estimates of solar wind parameters controlling the solar wind dynamic pressure which in part determines the position and the flaring angle of the bow shock and magnetopause boundaries. Thus, we are limited to features based on magnetic field measurements only. We chiefly use Reduced Data Record (RDR) data products of the MESSENGER MAG magnetometer instrument, obtained from the NASA PDS PPI repository, and process it in the follwing manner: First, we remove the calibration signals, in order to not be biased by them. Next, we enrich the dataset with Mercury position information. Then, to prepare the data indexed on orbit boundaries, we split them based on UTC-based day boundaries, and MESSEN-GER orbit apoapsis[1] points as markers to separate individual orbits. To simplify subsequent analysis, we also include the estimated planetary dipole magnetic field contribution for each point, planetocentric distance of the spacecraft, and recalculate position and magnetic field data in the aberrated MSO coordinate system which accounts for the non-negligible orbital velocity of Mercury relative to the speed of the solar wind. For more details and links to original sources, please refer to the dataset repository.

Consequently, we obtain a prepared dataset comprising of 4049 orbits. Additionally, we perform a few more removal steps, specific to our pipeline in this work: (a) Missing values: Some of the orbits lack individual measurements or even entire time steps. We conveniently remove those orbits, instead of correcting or filling with interpolation. (b) Overhanging crossings: Some orbits have crossings that extend into neighboring orbits or vice versa. After the cleaning step, there remain 2776 orbits. We randomly split these orbits into *training*, *validation* and *test* sets with a 70–20–10% split, and normalise using Z-score standardisation.

Finally, we leverage the crossing annotations by Philpott et al. [17] visualised in Fig. 2, to assign each time step a magnetic region. This yields the class distribution shown in Table 1, which exhibits a significant imbalance that we address later.

**Table 1.** Class labels with their abbreviations and frequency of occurence. The boundary classes are highly underrepresented.

| Label | Magnetic region | Share |
|-------|-----------------|-------|
| 0 | Interplanetary magnetic field (IMF) | 64.8% |
| 1 | Bow shock crossing (SK) | 3.7% |
| 2 | Magnetosheath (MSh) | 14.8% |
| 3 | Magnetopause crossing (MP) | 2.3% |
| 4 | Magnetosphere (MSp) | 14.4% |

## 4    Methodology

### 4.1    Problem Formulation

To obtain aggregates, and have an augmented set of fixed shaped input vectors, we use a sliding window. It has a stride of one, which ensures each time step of the original series is contained in multiple windows, such that the crossings can be presented to the model in all possible arrangements, to account for translation-equivariance[2]. Hence, the model's input is a window of $w \in \mathbb{N}$ successive time

---

[1] The apoapsis of an elliptic orbit is the point farthest away from the planet.
[2] The position of the event in the window should not matter.

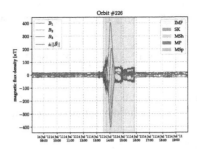

**Fig. 2.** Example annotation for orbit #226 (best viewed in colour). Annotations mark the start and end of a magnetic region. We label the entire region inside as belonging to the respective crossing. Each crossing appears twice in an orbit. (Color figure online)

steps. Each of these time steps consists of $d \in \mathbb{N}$ scalar features. We abstractly represent the input window as follows:

$$X := \left[ x^{(1)} \; x^{(2)} \; \cdots \; x^{(w)} \right] \in \mathbb{R}^{d \times w}$$

Consider the last time step $x^{(w)}$ in a window as representing the 'present'. Instead of merely classifying the 'past' time steps within a window, we also seek to compute predictions on the magnetic region for $f \in \mathbb{N}$ future time steps. Therefore, we expect the output per time step as *one-hot* vectors. As we expect the model to predict a class per time step, we pack multiple of these one-hot vectors next to each other into a matrix. Thus, the target output matrix of one-hot vectors is $Y \in \mathbb{R}^{5 \times (w+f)}$.

Formally, the task can be framed as a multi-dimensional multi-class classification with a future component: Given the window $X$, we predict a sequence of magnetic region probabilities, where each column sums up to one:

$$\hat{Y} := \begin{bmatrix} p_{1,1} & \cdots & p_{1,w} & p_{1,w+1} & \cdots & p_{1,w+f} \\ \vdots & & \vdots & \vdots & & \vdots \\ p_{5,1} & \cdots & p_{5,w} & p_{5,w+1} & \cdots & p_{5,w+f} \end{bmatrix} \in [0,1]^{5 \times (w+f)}$$

With the setup ready, the normalised vectors are then passed through the neural networks, and the final activations can be represented as: $\gamma_\theta : p_{ij} = \gamma_\theta(X)$ where $\gamma$ is a chosen model, with $\theta$ as its parameters. We experimentally find a window size of two minutes, i.e., $w = 120$, to be both practical and computationally kind, and fix the future size to $f = 20$ s. The selected features include the 3 three-dimensional features, namely MSO position, flux density and measurement errors, chosen via manual tuning on the evaluation split, resulting in dimensionality $d = 9$.

To measure the error between the prediction $\hat{Y}$ and the ground truth $Y$, we employ the standard categorical cross-entropy loss. Counteracting the considerable class imbalance inherent in the dataset, we weight each class inversely proportional to its frequency $f_c \in \mathbb{N}$ in the dataset by virtue of $w_c := (\sum_{i=1}^{5} f_i)/f_c$.

The resulting weighted loss for a single time step $j$ in a window is then

$$\mathcal{L}_j(\hat{\boldsymbol{Y}}, \boldsymbol{Y}) := -\sum_{i=1}^{5} w_i Y_{ij} \log(\hat{Y}_{ij})$$

Note that the sum is only a formal construct, since exactly one of the $Y_{ij}$ for fixed $j$ is non-zero. By averaging across all time steps in a window, we straightforwardly obtain the window loss:

$$\mathcal{L}(\hat{\boldsymbol{Y}}, \boldsymbol{Y}) := \frac{1}{w+f} \sum_{j=1}^{w+f} \mathcal{L}_j(\hat{\boldsymbol{Y}}, \boldsymbol{Y}) = -\frac{1}{w+f} \sum_{i=1}^{5} w_i \sum_{j=1}^{w+f} Y_{ij} \log(\hat{Y}_{ij})$$

Finally, the average loss over all windows extracted from the training set forms the overall optimisation target.

## 4.2    Model Architectures

As a first step in a feasibility study for model selection, we consider a total of six architecture categories, namely: Multi Layer Perceptron (MLP), Convolutional Neural Network (CNN), Fully Convolutional Neural Network (FCNN), Recurrent Neural Network (RNN), Convolutional Recurrent Neural Network (CRNN), and Convolutional Attentional Neural Network (CANN). The reader is encouraged to refer to the code repository for specific implementation details. While our architecture search space is biased towards shallower models, we are only concerned with their relative performance, and by interpreting Table 2 find the CRNN to be a suitable candidate for further experimentation.

## 4.3    Active Learning

For our active learning experiment, we exploit the domain specific data gathering properties, particularly that bow shock characteristics differ between orbits. Consequently, we ask the question whether an orbit-level informativeness measure can be constructed to reduce the amount of manual labelling. To evaluate the impact of this orbit-level informativeness measure, we compare the model performance when adding orbits to the training process.

We use an instance of *pool-based* active learning [19]: Initially untrained, the model repeatedly selects samples from a pool of yet unlabeled samples, obtains the labels, and trains incrementally on them. To address our performance scaling question, we increment the training set not by individual windows but on the level of entire orbits. In order to choose the next orbit(s) to add, it is needed that we rank all yet unused orbits according to an *informativeness* measure. Although solely relying on the top uncertain samples could sometimes lead to overfitting [18], since we always add an entire orbit covering all classes, we find this to be non-issue in our study, and for convenience, resort to it.

As our model has a series of Multinoulli distributions for output, we may measure uncertainty as a function of the output probabilities. *Shannon entropy* is a mathematically well-funded measure of uncertainty in a probability distribution that we utilise as the basis of our active learning strategy: Consider the training set $\mathcal{D} \subseteq \mathbb{R}^{d \times w} \times \{\text{IMF}, \text{SK}, \text{MSh}, \text{MP}, \text{MSp}\}^{w+f}$ with number of features $d \in \mathbb{N}$, window size $w \in \mathbb{N}$ and future size $f \in \mathbb{N}$. Given a model prediction $\hat{\boldsymbol{Y}} = [\hat{\boldsymbol{y}}^{(1)}, \ldots, \hat{\boldsymbol{y}}^{(w+f)}] \in [0,1]^{5 \times (w+f)}$, we define its uncertainty as

$$\mathfrak{u}(\hat{\boldsymbol{Y}}) := \max_j H(\hat{\boldsymbol{y}}^{(j)}) = -\min_j \sum_{i=1}^{5} y_i^{(j)} \log(y_i^{(j)}),$$

$$\mathfrak{u}(\hat{\boldsymbol{Y}}) := \max_j H(\hat{\boldsymbol{y}}^{(j)}) = -\min_j \sum_{i=1}^{5} y_i^{(j)} \log(y_i^{(j)}),$$

where $H : \triangle^4 \to \mathbb{R}$ is the Shannon entropy on the standard 4-simplex[3].

To achieve this on the orbit level, we must reduce the individual window uncertainties to a single orbit score. As we are only interested in the crossings, we can argue that the most uncertain windows of an orbit will usually overlap with a crossing region, and thus for simplicity, only consider the uncertainty of such windows for the overall orbit uncertainty. Let hence $\mathcal{D}_o \subseteq \mathcal{D}$ be the windows belonging to the orbit $o \in \mathbb{N}$ and

$$\widetilde{\mathcal{D}}_o := \{(\boldsymbol{X}, \boldsymbol{y}) \in \mathcal{D}_o \mid \boldsymbol{y} \cap \{\text{SK}, \text{MP}\} \neq \emptyset\}$$

be only those samples that overlap with a bow shock or magnetopause boundary region. The average uncertainty over these windows then defines the integrated orbit uncertainty of a model $\hat{f}_\theta : \mathbb{R}^{d \times w} \to \mathbb{R}^{d \times (w+f)}$ for our task:

$$\mathfrak{U}_{\hat{f}_\theta}(\widetilde{\mathcal{D}}_o) := \frac{1}{|\widetilde{\mathcal{D}}_o|} \sum_{(\boldsymbol{X}, \boldsymbol{y}) \in \widetilde{\mathcal{D}}_o} \mathfrak{u}(\hat{f}_\theta(\boldsymbol{X}))$$

Using this uncertainty measure, we formulate our active learning procedure in Algorithm 1. Instead of strictly adding orbits one-by-one, we more generally allow for an *increment function* $\partial : \mathbb{N}_0 \to \mathbb{N}$ that dictates the number of most uncertain orbits to add, depending on the number of already seen orbits.

## 5 Experiments

### 5.1 Model Evaluation

We compare the six models listed in Sect. 4.2 as to their classification performance on the test set. To this end, we employ the following metrics: Macro F1, overall accuracy, and the class-wise accuracy for the critical bow shock and magnetopause classes, respectively. Table 2 illustrates results for all the models, with their respective number of trainable parameters as an indicator of their size. We see a clear improvement between the variants with and without the recurrent component. The combination of convolutional and recurrent does noticeably

---

[3] $\triangle^{n-1} := \{(p_1, p_2, \ldots, p_n) \in \mathbb{R}^n \mid \forall i : p_i \geq 0, \sum_{i=1}^n p_i = 1\} \subseteq [0,1]^n.$

```
/* actively trains the given model on an incrementally
   growing subset of the training data                        */
active_learning(f̂_θ : ℝ^{d×w} → ℝ^{5×(w+f)} : model,
                Ω ⊆ 𝒫(𝒟) : set of all training orbits,
                ∂ : ℕ_0 → ℕ : increment function):
```

1   $\mathcal{T} := \emptyset$  // current training orbits
2   **while** $|\mathcal{T}| < |\Omega|$ :
3      $U := \mathsf{hash\_table}()$  // empty hash map
4      **for** $\mathcal{D}_o \in \Omega \setminus \mathcal{T}$ **do**
5         $U[\mathcal{D}_o] := \mathfrak{U}_{\hat{f}_\theta}(\widetilde{\mathcal{D}}_o)$  // determine orbit uncertainty
6      $\mathcal{T} := \mathcal{T} \cup \mathsf{top\_k}(U, \partial(|\mathcal{T}|))$  // add $\partial(|\mathcal{T}|)$ most uncertain orbits
7      $\hat{f}_\theta := \mathsf{train}(\hat{f}_\theta, \mathcal{T})$  // retrain model on updated set
8   **return** $\hat{f}_\theta$

Algorithm 1: Active learning scheme for incrementally adding orbits to the training procedure in a flexible manner.

better than either alone, however the contribution is marginal compared to that of RNN alone. The CRNN however achieves the highest overall scores and the highest magnetopause accuracy. Our experimental CANN, with an attention mechanism, accomplishes almost the same magnetopause performance but lags slightly behind on the overall metrics. Although the CANN achieves a higher bow shock accuracy than the CRNN, we continue our experiments with the latter for its best overall performance.

**Table 2.** Comparison of the model architectures.

| Model | Macro F1 | Accuracy | SK accur. | MP accur. | # params |
|-------|----------|----------|-----------|-----------|----------|
| MLP   | 74.73%   | 86.60%   | 73.87%    | 84.05%    | 245180   |
| CNN   | 77.80%   | 89.29%   | 74.75%    | 84.62%    | 1413372  |
| FCNN  | 78.97%   | 90.88%   | 78.83%    | 89.08%    | 1444796  |
| RNN   | 79.93%   | 92.03%   | **81.50%** | 91.75%   | 237701   |
| CRNN  | **81.21%** | **93.04%** | 79.22%  | **92.22%** | 267333 |
| CANN  | 80.20%   | 92.46%   | 81.30%    | **92.23%** | 246469   |

Further, upon evaluating the best model on the test set, we see no real evidence of overfitting (C.f. Table 3), which is a good sign. The model performs better on the relatively easier classes of IMF, magnetosheath and magnetosphere. The confusion matrices in Fig. 3 show that the model consistently does better on recall over precision.

**Table 3.** CRNN performance on the test versus evaluation set.

| Set | Macro F1 | Accuracy | SK accur. | MP accur. |
|---|---|---|---|---|
| Eval. | 81.21% | 93.04% | 79.22% | 92.22% |
| Test | 81.95% | 93.13% | 79.93% | 87.51% |

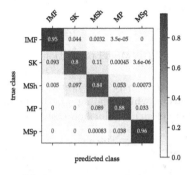

(a) Row-wise normalisation.
Diagonal corresponds to recall.

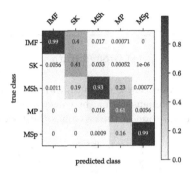

(b) Column-wise normalisation.
Diagonal corresponds to precision.

**Fig. 3.** Normalised confusion matrices for the CRNN. The results indicate applicability for real-time predictions.

**Qualitative Evaluation.** To confirm our findings, we evaluate the CRNN qualitatively, and utilise its past-only classifications in a window to infer predictions for an entire orbit. Each time step receives distinct predictions from all sliding windows it is contained in, which we integrate by averaging to obtain an overall class probability distribution, and arg max yields a class prediction for the time step. We do this for all orbits in the test set and plot their magnetic flux density along with the predictions.[4] Upon visually inspecting all orbits in the test set, we can confirm that the model overall predicts contiguous magnetic regions. Further, we notice that in some cases, the crossings, albeit exaggerated w.r.t existing ground truth, correctly predicts the boundaries (Fig. 4), indicating that it might be learning associations not available explicitly in labels. Although more work needs to be done in this regard, this is very promising as it might lead to explanations that could benefit the physical understanding of certain phenomena. Nevertheless, we also identify some major qualitative issues that still remain, such as scattered predictions, and boundary exaggerations (C.f. Fig. 5), some of which may possibly be tackled by solutions that we identify in Sect. 6.

---

[4] All plots for the entire test set are made available in the code repository linked in Sect. 1.

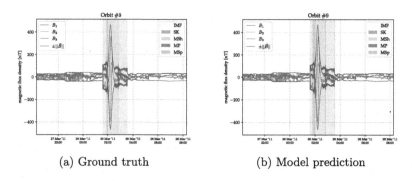

(a) Ground truth                    (b) Model prediction

**Fig. 4.** An example prediction where boundaries are slightly exaggerated. The network tries to compensate for the conservative annotation, while yielding a better prediction on the duration of the crossings (best viewed in colour). (Color figure online)

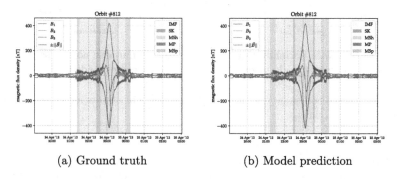

(a) Ground truth                    (b) Model prediction

**Fig. 5.** An example prediction with significantly exaggerated and scattered bow shock crossing.

## 5.2  Active Learning

For our active learning experiments, we run Algorithm 1 with two different choices for the increment function: one leading to a constantly growing training set and one leading to an linearly growing training set. In this manner, we explore how the classification performance scales with available data and determine the order of magnitude of orbits required for a sufficiently informed model.

**Constant Increment** A straightforward choice of increment function would be to add orbits one by one. Due to computational concerns and observed overfitting on single orbits, we found $\partial(n) := 10$ to be more suitable.

**Linear Increment** Due to some problems we identified with a constant increment, we conduct another active learning experiment with the linear increment function $\partial(n) := \max\{\lfloor n/2, 10 \rfloor\}$. This choice ensures a constant proportion of 'new' vs 'old' training orbits while preventing overfitting to a single orbit.

Figure 6 plots the evaluation metrics discussed previously over the number of already included orbits for both increment function choices. The learning curve for the constant increment shows only the first 1000 orbits, as the experiment could not run until completion, but the evolution is clearly evident. In both cases, we observe a rapid increase of all metrics in the beginning, followed by a period of flattening. After no more than 500 orbits, the performance metrics are comparable to those of the passively trained model.

In the constant case, the class accuracies for bow shock and magnetopause later decrease, while the overall metrics continue to rise. This divergence implies that the model focusses more on the majority classes and increasingly ignores the two boundary classes we are concerned with. We suspect the constant increment causes this mediocre development. Since the number of orbits added in each iteration does not depend on the number of already seen orbits, their relative proportion becomes increasingly skewed towards the known orbits. As a result, the marginal returns diminish while learning from new orbits but continues to optimise over the familiar ones repeatedly.

These observations explain our choice of the linear increment function. Indeed, it leads to a much better development while at the same time requiring substantially less iterations and hence computational cost. Due to the latter reason, the improvement is slower in the beginning but reaches far higher scores in the long run. However, they do not surpass the performance of the passively trained model. This indicates that the lower bound on number of orbits required is not too high, further emphasising the need for clever data sampling approaches.

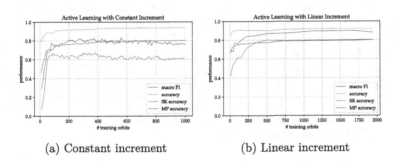

(a) Constant increment        (b) Linear increment

**Fig. 6.** Performance metric development during active learning.

Besides the performance metrics, we also evaluate the development of our uncertainty measure during the active learning process. After all, it is the very measure by which orbits are selected for training in the active learning scheme and indicates the model's confidence about its decisions. We are interested in the point from where the uncertainty does not significantly decrease anymore, implying that the model has nearly saturated its learning capabilities. In a sense, the model has 'seen enough' until that point. Figure 7 plots the worst occurring orbit uncertainty at each iteration as a function of the number of orbits included in the process. Both start in the beginning with a value of just under $\log(5) \approx 1.609$, which is the entropy of a uniform distribution on the five outcomes. This is not a coincidence but rather results from the definition of our orbit uncertainty measure in Sect. 4.3 and the model's random parameter initialisation. Then, the orbit uncertainty decreases rapidly during the subsequent iterations. Analogously to the performance metrics in Fig. 6, the uncertainty eventually flattens out and seems to almost asymptotically approach values of 0.5 and 0.6 respectively. Again, the maximum marginal improvement appears during the first half, until about 500 orbits are included.

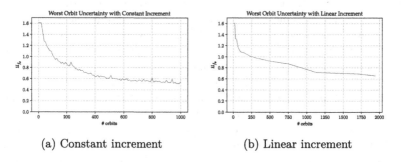

(a) Constant increment                    (b) Linear increment

**Fig. 7.** Orbit uncertainty development during active learning.

Taking all insights together, we conclude that the model's learning capacity saturates after 450 to 500 orbits. This constitutes an upper bound for the number of orbits required for a representative model. When summing the duration spanned by the concrete orbits chosen by the model, this equates to roughly two full Mercury years' worth of MESSENGER orbits. We may therefore claim that two Mercury years make for a sufficient set of observations for the model to learn from. On the other hand, revisiting Fig. 6 and Fig. 7 confirms our intuition that one complete Mercury year (around 230 orbits in this case) is at least required. It remains for future work, hence, to explore the range in between. With the improvements we propose in Sect. 6, it might even be possible to lower this bound to just one Mercury year.

## 6   Conclusion

In this work, we built a discriminative end-to-end deep learning model for detecting Mercury's bow shock and magnetopause crossing signatures based on raw measurements from NASA's MESSENGER mission. Additionally, we devised an active learning scheme to address the question of how many orbits worth of measurement data is required for a representative model. To this end, we prepared a dataset suited to machine learning tasks, which we make available publicly to facilitate future research in this direction. To inspect the applicability of machine learning to this problem we formulated a five class supervised machine learning task, that given a window of measurements, predicts the classes for each time step in the current window, and at the same time predicts the classes the next specified number of time steps. We applied various architectures and configurations of neural networks to determine a suitable fit for the architecture, and observed that the CRNN performs relatively better, which is consistent with findings in other types of signals too, where both spatial and temporal features are of relevance. We also observe that the neural networks are capable of predicting ahead in time, which is a good indication that there might be presence of autoregressive characteristics in the signal. Our best CRNN model achieves a macro F1 of about 82% and consistently predicts magnetopause crossings better than the bow shock crossings. This is no surprise since the magnetopause crossings are also better discernible to the human eye. Further, the recall scores of 78% and 86% on the bow shock and magnetopause crossings respectively, are significantly and consistently better than the precision scores of 39% and 61% respectively. There can be several explanations for this: first, the model clearly prefers not missing a boundary at the cost of false positives. Given the use case, it is more important that a boundary is not missed, over exaggerated crossings. Second, the annotations we used are clearly too conservative in many instances, so the network tries to compensate for those based on the learned statistical associations.

Based on the best model, we approached the central question underlying this work with an active learning scheme. It employs the uncertainty sampling strategy with a custom orbit-level measure based on Shannon entropy, by which we iteratively determine the next orbits to include in the training set. After a preliminary experiment with a constantly growing training set, we conducted our main experiment with a linear increment, and observed it to be significantly better. It likely ensures a constant portion of unseen orbits throughout all iterations. Although these strategies might suffer slightly from overfitting on the very first set of orbits, we were able to derive that at least one and at most two Mercury years' worth of measurement data may be sufficient for a representative model that performs reasonably. Finally we recognise while our work yields comprehensive insights into the structure of the MESSENGER magnetometer data and hence the magnetic dynamics around Mercury, it can only provide a starting point in machine learning endeavors.

As part of future work, it would be worthwhile to improve quantitative evaluation by employing metrics that are more sensitive to temporal onsets and offsets. It would also be interesting to investigate if it suffices to let the model predict only one class per window. For inference, this would result in one prediction for each time step instead of multiple votes. This may tackle some of the issues where some crossings are scattered. Likewise, the future classification output may be compressed to a single value. For instance, this could use a binary flag that indicates whether the class predicted for the present time step changes in the near future. It would be useful to explore how concept drift detection techniques help in this regard.

By and large, this work reveals two insights on a broader level: First, deep learning can be used to build sophisticated models of the bow shock and magnetopause, a favourable alternative to the existing geometric models that suffer the downside of being static, and often do not accurately predict the duration of the crossings. Second, active learning serves not only for enhancing labeling efficiency but also for addressing data representativeness questions. We strongly encourage future work to continue and improve our study, taking note of the suggestions made above. The outcomes might become relevant for the upcoming Mercury mission BepiColombo [8], which with its twin-aircraft probe will collect significantly more data.

**Acknowledgements.** The authors acknowledge support from *Europlanet 2024 RI* that has received funding from the European Union's *Horizon 2020* research and innovation programme under grant agreement No. 871149.

# References

1. Alexeev, I.I., Belenkaya, E.S., Bobrovnikov, S.Y., Kalegaev, V.V.: Modelling of the electromagnetic field in the interplanetary space and in the earth's magnetosphere. Space Sci. Rev. **107**(1), 7–26 (2003)
2. Alexeev, I.I., et al.: Mercury's magnetospheric magnetic field after the first two messenger flybys. Icarus **209**(1), 23–39 (2010)
3. Alexeev, I., et al.: A global magnetic model of saturn's magnetosphere and a comparison with cassini soi data. Geophys. Res. Lett. **33**(8), 1–4 (2006)
4. Amiriparian, S., et al.: Recognition of echolalic autistic child vocalisations utilising convolutional recurrent neural networks (2018)
5. Amiriparian, S., Cummins, N., Julka, S., Schuller, B.: Deep convolutional recurrent neural network for rare acoustic event detection. In: Proceedings of DAGA, pp. 1522–1525 (2018)
6. Anderson, B.J., et al.: The magnetic field of mercury. Space Sci. Rev. **152**(1), 307–339 (2010)
7. Belenkaya, E., Bobrovnikov, S.Y., Alexeev, I., Kalegaev, V., Cowley, S.: A model of jupiter's magnetospheric magnetic field with variable magnetopause flaring. Planet. Space Sci. **53**(9), 863–872 (2005)
8. Benkhoff, J., et al.: Bepicolombo-comprehensive exploration of mercury: mission overview and science goals. Planet. Space Sci. **58**(1–2), 2–20 (2010)
9. Fairfield, D.H.: Average and unusual locations of the earth's magnetopause and bow shock. J. Geophys. Res. **76**(28), 6700–6716 (1971)
10. Haaland, S., et al.: Characteristics of the flank magnetopause: mms results. J. Geophys. Res.: Space Phys. **125**(3), e2019JA027623 (2020)
11. Jelínek, K., Němeček, Z., Šafránková, J.: A new approach to magnetopause and bow shock modeling based on automated region identification. J. Geophys. Res.: Space Phys. **117**(A5) (2012)
12. Johnson, C.L., et al.: Messenger observations of mercury's magnetic field structure. J. Geophys. Res.: Planets **117**(E12) (2012)
13. Kraeft, S.K.: Detection and analysis of cancer cells in blood and bone marrow using a rare event imaging system. Clin. Cancer Res. **6**(2), 434–442 (2000)
14. Lin, R., Zhang, X., Liu, S., Wang, Y., Gong, J.: A three-dimensional asymmetric magnetopause model. J. Geophys. Res.: Space Phys. **115**(A4) (2010)
15. Nguyen, G., Aunai, N., Michotte de Welle, B., Jeandet, A., Fontaine, D.: Automatic detection of the earth bow shock and magnetopause from in-situ data with machine learning. In: Annales Geophysicae Discussions, pp. 1–22 (2019)

16. Nikolaou, N., et al.: Lessons learned from the 1st ariel machine learning challenge: correcting transiting exoplanet light curves for stellar spots. arXiv preprint arXiv:2010.15996 (2020)
17. Philpott, L.C., Johnson, C.L., Anderson, B.J., Winslow, R.M.: The shape of mercury's magnetopause: the picture from messenger magnetometer observations and future prospects for bepicolombo. J. Geophys. Res.: Space Phys. **125**(5), e2019JA027544 (2020)
18. Ren, P., et al.: A survey of deep active learning. ACM Comput. Surv. (CSUR) **54**(9), 1–40 (2021)
19. Settles, B.: Active learning. Synth. Lect. Artif. Intell. Mach. Learn. **6**(1), 1–114 (2012)
20. Shue, J.H., et al.: A new functional form to study the solar wind control of the magnetopause size and shape. J. Geophys. Res.: Space Phys. **102**(A5), 9497–9511 (1997)
21. Sibeck, D.G., Lopez, R., Roelof, E.C.: Solar wind control of the magnetopause shape, location, and motion. J. Geophys. Res.: Space Phys. **96**(A4), 5489–5495 (1991)
22. Slavin, J.A.: Mercury's magnetosphere. Adv. Space Res. **33**(11), 1859–1874 (2004)
23. Wang, Y., et al.: A new three-dimensional magnetopause model with a support vector regression machine and a large database of multiple spacecraft observations. J. Geophys. Res.: Space Phys. **118**(5), 2173–2184 (2013)
24. Winslow, R.M., et al.: Mercury's magnetopause and bow shock from messenger magnetometer observations. J. Geophys. Res.: Space Phys. **118**(5), 2213–2227 (2013)
25. Zhong, J.: Mercury's three-dimensional asymmetric magnetopause. J. Geophys. Res.: Space Phys. **120**(9), 7658–7671 (2015)
26. Zurbuchen, T.H., et al.: Messenger observations of the spatial distribution of planetary ions near mercury. Science **333**(6051), 1862–1865 (2011)

# A Stopping Criterion for Transductive Active Learning

Daniel Kottke[1]([⊠]) [ID], Christoph Sandrock[1], Georg Krempl[2] [ID],
and Bernhard Sick[1] [ID]

[1] University of Kassel, Wilhelmshöher Allee 73, 34121 Kassel, Germany
{daniel.kottke,christoph.sandrock,bsick}@uni-kassel.de
[2] Utrecht University, Princetonplein 5, 3584 CC Utrecht, The Netherlands
g.m.krempl@uu.nl

**Abstract.** In transductive active learning, the goal is to determine the correct labels for an unlabeled, known dataset. Therefore, we can either ask an oracle to provide the right label at some cost or use the prediction of a classifier which we train on the labels acquired so far. In contrast, the commonly used (inductive) active learning aims to select instances for labeling out of the unlabeled set to create a generalized classifier, which will be deployed on unknown data. This article formally defines the transductive setting and shows that it requires new solutions. Additionally, we formalize the theoretically cost-optimal stopping point for the transductive scenario. Building upon the probabilistic active learning framework, we propose a new transductive selection strategy that includes a stopping criterion and show its superiority.

**Keywords:** Stopping criteria · Active learning · Transduction

## 1 Introduction

In classification, the goal is to create a classifier that predicts the true labels for unlabeled instances. Therefore, the classifier needs a set of instance-label pairs (i.e., the training set) which is often not directly available. Fortunately, unlabeled data is usually available at a low cost. However, labeling data is often expensive. Thus, active learning may reduce the annotation cost by selecting instances for labeling that help the classifier in its training progress the most [24].

In this article, we propose to distinguish inductive and transductive active learning. To visualize the difference between both scenarios, we give the following examples: (1) We aim to train a general model to identify protected animals on high-resolution satellite images to surveil their population. In this inductive learning example, we aim to build a general classifier as we want to use it periodically and not only on the images of the initial set (i.e., the test data is unknown). (2) After a natural disaster destroyed some buildings, we search

---

**Supplementary Information** The online version contains supplementary material available at https://doi.org/10.1007/978-3-031-26412-2_29.

M.-R. Amini et al. (Eds.): ECML PKDD 2022, LNAI 13716, pp. 468–484, 2023.
https://doi.org/10.1007/978-3-031-26412-2_29

for survivors. Hence, we take satellite images to find collapsed buildings across the affected regions. In that transductive context, it is important to classify the collected images correctly as their evaluation decides between life and death. In such a transductive scenario, the performance on the collected data is important. Hence, it might be beneficial to use the classifier mainly for simple cases and annotate difficult cases manually even if they do not improve the classifier's performance much. Mixed inductive-transductive scenarios are also possible, where the generalization of the performance beyond the collected data might be relevant. However, to highlight the characteristics and consequences of each scenario, and due to space limitations, this paper will focus on disjoint scenarios.

Up until now, almost all literature refers to inductive active learning and only a few works exist that mention the transductive scenario. Tong [26, p. 15] even argued that the transductive scenario is a special case of inductive active learning and, therefore, solving the inductive case is sufficient. Recently, some articles [16,23] consider transductive active learning but they did not mention its distinct difference to the standard inductive setting in detail.

When deploying classifiers that have been trained with active learning, it is crucial to decide when to stop acquiring more labels [11]. Therefore, cost-sensitive stopping criteria balance misclassification and annotation costs [6,19]. In the inductive scenario, it is difficult to reliably estimate the misclassifications cost because the number of instances to be classified after deployment is often unknown. As we already know the instances to be classified in the transductive scenario, it is straightforward to define and evaluate stopping criteria.

Within this article, our contributions are:

1. We formally define and describe transductive active learning and show that it is beneficial to develop transductive selection strategies (Hypothesis A).
2. We propose a new transductive selection strategy and show its superiority (Hypothesis B). Therefore, we additionally introduce the minimum aggregated cost score, which is a new transductive, cost-based evaluation measure that considers annotation and misclassification costs.
3. We propose a new cost-based stopping criterion for transductive active learning which outperforms its competitors (Hypothesis C).

Next, we discuss the related work, followed by the problem definition, the probabilistic active learning framework, the extension to the transductive case, and our new stopping criterion. Our evaluation is based on three hypotheses.

## 2    Background and Related Work

In the early 1970 s, Vapnik introduced the concept of transductive inference, which he discussed in more detail in his later publications, e. g. [29, pp. 339ff.]. Both concepts mainly differ in the availability of an evaluation set. In inductive inference, the evaluation set is unknown, whereas it is known for transductive inference. The concept of transduction became especially relevant in the area of *semi-supervised learning* [4, pp. 453ff.]. Here, labels are only partially available, and the assumption is that incorporating the unlabeled instances can improve the classifier's performance. One approach is to successively label the most certain unlabeled instances based on the current classification results. Thereby, the approaches incorporate the structure of the data to build more realistic classification hypotheses [25,27]. In this paper, we extend this idea to active learning.

The main idea of active learning is to actively ask for information that helps best to improve the classifier's predictions [24]. In general, the active learning cycle starts with an initially unlabeled set of instances. A selection strategy successively selects some of these instances and then, an oracle provides the corresponding class labels for these instances. After updating the classifier, the cycle restarts. The main focus of active learning research is on finding an appropriate selection strategy. The most commonly used is uncertainty sampling [14], which selects instances where the classifier is most uncertain. These uncertainty scores are mainly based on probabilistic predictions. Query-by-committee [15] builds a classifier ensemble and selects instances where its members disagree the most. Expected error reduction [22] optimizes the generalization error by simulating potential label acquisitions and thereby provides a decision-theoretic score. Chapelle [3] observed that the used probabilities can be unreliable for only a few labels. Hence, he introduced a prior on the classes for regularization. Value of information [9] differs from expected error reduction in the way that it evaluates the generalization error only on the unlabeled instances and assumes that an unlabeled instance is correct after labeling. In probabilistic active learning [12], the generalization error for both, the current and the simulated (with the additional label) classifier, is evaluated on the same probability distribution.

The term transduction also appears in different contexts in active learning literature. Varying from our definition of transductive active learning, the authors of [7,20] use the term transduction as a technique of propagating labels to the remaining unlabeled data by using the predictions of the classifier. This self-labeling approach is used to create a more robust classifier as it is known from semi-supervised learning. Yu et al. [31] propose a transductive experimental design. Instead of using discrete classes as in classification tasks, they train a model for noisy, continuous targets. Balasubramanian et al. [1] present a selection strategy in the online-based setting. New instances are labeled if the current estimated performance of the classifier is insufficient. As they know this new instance when evaluating it, they use the term transductive learning.

Ishibashi and Hino [8] recently summarized existing stopping criteria for active learning. They divide them into three categories: (1) Accuracy-based approaches (e. g., [13]) evaluate the predictive error of the classifier on unlabeled data or already queried data. (2) Confidence-based approaches (e. g., [30]) use the uncertainty of the model on the remaining unlabeled data to determine the stopping point. (3) Stability-based approaches (e. g., [2]) consider the changes in the model parameters and stop if the model does not change much anymore.

In their survey, Pullar-Strecker et al. [19] compare different stopping criteria and define a cost measure based on the combined cost from annotation and misclassification. Their results indicate that previously proposed stopping criteria based on the accuracy per label tend to stop learning early, while stopping criteria based on classification changes tend to stop late. They conclude that criteria should consider the trade-off between annotation and misclassification costs.

Dimitrakakis et al. [6] introduce a cost-sensitive scenario with a parameter balancing the annotation and misclassification cost. They propose two stopping criteria that compare the expected performance gain and the annotation cost caused by querying an instance. The first one uses convergence properties to estimate the performance gain, while the second one builds on a probabilistic classifier serving this purpose. This idea uses the generalization error of expected error reduction [22] which has been extended in [9,10]. The stopping criterion proposed in [8] compares the performance gain of a parameterized model with the acquisition cost of new labels. As shown in [19], the balancing parameter used by [6,8] is not directly applicable in real-world applications. This is because both articles consider an inductive setting where the size of the evaluation set is implicitly included in their parameters. However, even parameterizing the evaluation set size directly, as proposed by [19], may not solve the problem as it is hard to be estimated. In transduction, the evaluation set is given, which allows us to define a more intuitive and general cost function. To our knowledge, the transductive setting has not been investigated in a cost-sensitive scenario.

## 3   Problem Definition

For this section, we use a slightly adapted version of Vapnik's [29, p. 15] definition of "learning from examples". A learning task consists of: (1) a generator of random vectors (the instances) $x \in \mathbb{R}^D$, drawn independently from a fixed but unknown probability distribution function $p(x)$, (2) an oracle that returns an output value (the label) $y \in \mathcal{Y}$, according to a conditional distribution function $p(y|x)$, also fixed but unknown, and (3) a classifier $f$ that aims to predict the oracle's outputs.

In pool-based active learning, we have a dataset $\mathcal{D} = \{(x_1, y_1), ..., (x_N, y_N)\}$, where all instances $x_i$ but only a few/no labels $y_i$ are known to the learner, and $\mathcal{D} \overset{\text{i.i.d.}}{\sim} p(x, y) = p(y|x) \cdot p(x)$. Specifically, the learner has access to[1]:

1. A small or empty set of initially labeled instances $\mathcal{L}_0 \subseteq \mathcal{D}$.
2. A set of initially unlabeled instances $\mathcal{U}_0 = \{x : (x, y) \in \mathcal{D} \setminus \mathcal{L}_0\}$.
3. An oracle $o$ that returns the label $y = o(x)$ for every $(x, y) \in \mathcal{D}$.

In each iteration $i \geq 1$, a *selection strategy* selects one instance from the candidate pool $\tilde{x} \in \mathcal{U}_{i-1}$ with the goal to improve the performance of the classifier. The selected instance $\tilde{x}$ is labeled by the oracle with $\tilde{y} = o(\tilde{x})$, added to the set of labeled instances and removed from the candidate pool.

$$\mathcal{L}_i = \mathcal{L}_{i-1} \cup \{(\tilde{x}, \tilde{y})\} \tag{1}$$
$$\mathcal{U}_i = \mathcal{U}_{i-1} \setminus \{\tilde{x}\} \tag{2}$$

---

[1] We assume that the instances are unique to simplify the notation. This is not a limitation as one can easily drop this assumption by addressing instance-label pairs through their index.

After each iteration, the classifier is updated on the current labeled set which we denote by $f^{\mathcal{L}_i}$. Note that $\mathcal{U}_i$ only contains instances, whereas $\mathcal{D}$ and $\mathcal{L}_i$ consist of instance-label pairs. For readability purposes, we write $\mathcal{U}$ and $\mathcal{L}$ without the indices if possible.

In *transductive active learning*, the goal is to determine the correct labels for all instances in $\mathcal{D}$. As we assume that the oracle provided the true labels for instances in $\mathcal{L}$, we only need the classifier to predict the labels for instances in $\mathcal{U}$. To simplify the notation, we define a meta-classifier $g_f^{\mathcal{L}}$ that returns the known labels for instances in the labeled set and uses the classifier $f^{\mathcal{L}}$ to predict the unknown labels. This is necessary as we cannot be sure that $f^{\mathcal{L}}(\boldsymbol{x}) = y$ for all $(\boldsymbol{x}, y) \in \mathcal{L}$.

$$g_f^{\mathcal{L}}(\boldsymbol{x}) = \begin{cases} y & \text{if } (\boldsymbol{x}, y) \in \mathcal{L} \\ f^{\mathcal{L}}(\boldsymbol{x}) & \text{else} \end{cases} \tag{3}$$

We define the *transductive risk* as the sum of classification losses $L$ over $\mathcal{D}$. As stated above, it is sufficient to evaluate over $\mathcal{U}$.

$$R_{\mathcal{D}}^{\text{tr}}(f^{\mathcal{L}}) = \sum_{(\boldsymbol{x},y)\in\mathcal{D}} L(y, g_f^{\mathcal{L}}(\boldsymbol{x})) = \sum_{\boldsymbol{x}\in\mathcal{U}} L(o(\boldsymbol{x}), f^{\mathcal{L}}(\boldsymbol{x})) = R_{\mathcal{U}}^{\text{tr}}(f^{\mathcal{L}}) \tag{4}$$

Throughout this article, we use the zero-one loss that compares the true label $y$ with the prediction $f^{\mathcal{L}}(\boldsymbol{x})$ label and returns 0 if the prediction is correct and 1 otherwise.

$$L(y, f^{\mathcal{L}}(\boldsymbol{x})) = \begin{cases} 0 & y = f^{\mathcal{L}}(\boldsymbol{x}) \\ 1 & \text{otherwise} \end{cases} \tag{5}$$

In *inductive active learning*, we aim to train a classifier for every (possibly unknown) instance $\boldsymbol{x} \overset{\text{i.i.d.}}{\sim} p(\boldsymbol{x})$ with the goal of generalization. Consequently, we do not know the evaluation instances during training in the inductive setting. The distribution $p(\boldsymbol{x}, y)$ is usually approximated with a labeled validation set. As in [29], the (inductive) risk is defined as follows.

$$R(f^{\mathcal{L}}) = \underset{p(\boldsymbol{x},y)}{\mathbb{E}} \left[ L(y, f^{\mathcal{L}}(\boldsymbol{x})) \right] = \underset{p(\boldsymbol{x})}{\mathbb{E}} \left[ \underset{p(y|\boldsymbol{x})}{\mathbb{E}} \left[ L(y, f^{\mathcal{L}}(\boldsymbol{x})) \right] \right] \tag{6}$$

The transductive active learning setting differs from the inductive one in two ways: (1) One knows the data used to evaluate the model beforehand, and one does not need to build a generalized model. (2) One can exclude data from being predicted by the classifier by asking for the label from the oracle.

## 4   From Inductive to Transductive Active Learning

We build our selection strategy for transductive active learning upon the probabilistic active learning framework [12] that estimates the expected risk reduction when a candidate instance is selected for label acquisition. In the first subsection, we summarize the existing method for the inductive scenario and derive the equations for the transductive case in the second subsection.

## 4.1   The Probabilistic Active Learning Framework

To estimate the inductive risk, we need to estimate the unknown distributions $p(\boldsymbol{x})$ and $p(y|\boldsymbol{x})$ in Eq. 6. As suggested by [12,21], we approximate $p(\boldsymbol{x})$ using a Monte Carlo approach with an unlabeled set $\mathcal{E} \overset{\text{i.i.d.}}{\sim} p(\boldsymbol{x})$. Here, we use $\mathcal{E} = \{\boldsymbol{x} : (\boldsymbol{x}, y) \in \mathcal{L}\} \cup \mathcal{U}$. We estimate $p(y|\boldsymbol{x})$ with $p([)\mathcal{L}]y\boldsymbol{x}$ using the data in $\mathcal{L}$ [3, 12,17]. The probability is based on a kernel frequency estimate $\boldsymbol{k}_{\boldsymbol{x}}^{\mathcal{L}}$ that contains the number of samples for every class near $\boldsymbol{x}$ using the similarity/kernel $K(\cdot, \cdot)$. By using a Bayesian approach that introduces a prior $\boldsymbol{\epsilon} \in \mathbb{R}_+^{|\mathcal{Y}|}$, the probability $\mathrm{p}^{\mathcal{L}}(y|\boldsymbol{x})$ is given by the $y$-th element of the normalized vector $\boldsymbol{k}_{\boldsymbol{x}}^{\mathcal{L}} + \boldsymbol{\epsilon}$.

$$\mathrm{p}^{\mathcal{L}}(y|\boldsymbol{x}) = \frac{(\boldsymbol{k}_{\boldsymbol{x}}^{\mathcal{L}} + \boldsymbol{\epsilon})_y}{||\boldsymbol{k}_{\boldsymbol{x}}^{\mathcal{L}} + \boldsymbol{\epsilon}||_1} \qquad k_{\boldsymbol{x},y}^{\mathcal{L}} = \sum_{\substack{(\boldsymbol{x}',y') \in \mathcal{L} \\ y'=y}} K(\boldsymbol{x}, \boldsymbol{x}') \tag{7}$$

The inductive risk of a classifier is estimated as follows.

$$\hat{R}_{\mathcal{E},p^{\mathcal{L}}}(f^{\mathcal{L}}) = \frac{1}{|\mathcal{E}|} \sum_{\boldsymbol{x} \in \mathcal{E}} \sum_{y \in \mathcal{Y}} \mathrm{p}^{\mathcal{L}}(y|\boldsymbol{x}) L(y, f^{\mathcal{L}}(\boldsymbol{x})) \approx R(f^{\mathcal{L}}) \tag{8}$$

For a given candidate $\tilde{\boldsymbol{x}} \in \mathcal{U}$, we calculate the probabilistic gain (xgain) as the expectation value over all possible labeling outcomes $\tilde{y} \in \mathcal{Y}$ of the estimated inductive risk reduction. Therefore, we compare the inductive risks (estimated on $\mathcal{E}$ and $p^{\mathcal{L}^+}$) of the current classifier $f^{\mathcal{L}}$ and the simulated classifier $f^{\mathcal{L}^+}$ that includes the candidate with $\mathcal{L}^+ = \mathcal{L} \cup (\tilde{\boldsymbol{x}}, \tilde{y})$. Since we want to maximize the gain, we consider the negative risk reduction.

$$\mathrm{xgain}(\tilde{\boldsymbol{x}}, \mathcal{L}, \mathcal{E}) = - \underset{p^{\mathcal{L}}(\tilde{y}|\tilde{\boldsymbol{x}})}{\mathbb{E}} \left[ \hat{R}_{\mathcal{E},p^{\mathcal{L}^+}}(f^{\mathcal{L}^+}) - \hat{R}_{\mathcal{E},p^{\mathcal{L}^+}}(f^{\mathcal{L}}) \right] \tag{9}$$

$$= - \sum_{\tilde{y} \in \mathcal{Y}} \mathrm{p}^{\mathcal{L}}(\tilde{y}|\tilde{\boldsymbol{x}}) \left[ \frac{1}{|\mathcal{E}|} \sum_{\boldsymbol{x} \in \mathcal{E}} \sum_{y \in \mathcal{Y}} \mathrm{p}^{\mathcal{L}^+}(y|\boldsymbol{x}) \left( L\left(y, f^{\mathcal{L}^+}(\boldsymbol{x})\right) - L\left(y, f^{\mathcal{L}}(\boldsymbol{x})\right) \right) \right] \tag{10}$$

$$= - \sum_{\tilde{y} \in \mathcal{Y}} \frac{(\boldsymbol{k}_{\tilde{\boldsymbol{x}}}^{\mathcal{L}} + \boldsymbol{\beta})_{\tilde{y}}}{||\boldsymbol{k}_{\tilde{\boldsymbol{x}}}^{\mathcal{L}} + \boldsymbol{\beta}||_1} \cdot \frac{1}{|\mathcal{E}|} \sum_{\boldsymbol{x} \in \mathcal{E}} \sum_{y \in \mathcal{Y}} \frac{(\boldsymbol{k}_{\boldsymbol{k}_{\boldsymbol{x}}}^{\mathcal{L}^+} + \boldsymbol{\alpha})_y}{||\boldsymbol{k}_{\boldsymbol{k}_{\boldsymbol{x}}}^{\mathcal{L}^+} + \boldsymbol{\alpha}||_1} \left( L(y, f^{\mathcal{L}^+}(\boldsymbol{x})) - L(y, f^{\mathcal{L}}(\boldsymbol{x})) \right) \tag{11}$$

The vectors $\boldsymbol{\alpha}$ and $\boldsymbol{\beta}$ are the priors of the label distribution of the evaluation sample $\boldsymbol{x}$ and the candidate $\tilde{\boldsymbol{x}}$, respectively. They can be interpreted as the number of pseudo-labels added to each region of the dataset. High numbers lead to high regularization of the probabilities and vice versa. As proposed in [12], we set $\boldsymbol{\alpha} = \boldsymbol{\beta} = (10^{-3}, \ldots, 10^{-3})$.

The selection strategy chooses the candidate instance $\tilde{\boldsymbol{x}}^*$ that maximizes the probabilistic gain.

$$\tilde{\boldsymbol{x}}^* = \arg\max_{\tilde{\boldsymbol{x}} \in \mathcal{U}} \{\mathrm{xgain}(\tilde{\boldsymbol{x}}, \mathcal{L}, \mathcal{E})\} \tag{12}$$

## 4.2  Transductive Probabilistic Active Learning

The goal of transductive active learning is to determine the correct label for all instances in the dataset $\mathcal{D}$. As we assume that the oracle is omniscient, we know that the labels in $\mathcal{L}$ are already correct. To get the label of the remaining instances in $\mathcal{U}$, we can either ask the oracle (and be certain that it is correct) or use the classifier's predictions $f^{\mathcal{L}}(x)$. In the latter case, we run into the risk of making mistakes.

Due to these specific characteristics of the transductive scenario, we need to adapt the estimate in Eq. 7 such that the probability for the correct label $y$ for labeled instances $x$ with $(x, y) \in \mathcal{L}$ is 1.

$$
\mathrm{p}_{\mathrm{tr}}^{\mathcal{L}}(y|x) = \begin{cases} 1 & (x, y) \in \mathcal{L} \\ 0 & (x, y') \in \mathcal{L} \wedge y \neq y' \\ \mathrm{p}^{\mathcal{L}}(y|x) & \text{otherwise} \end{cases} \tag{13}
$$

To calculate the probabilistic gain in the transductive setting, we use the same estimation idea as before, but with the transductive risk. The first step follows the simplification in Eq. 4.

$$
\hat{R}_{\mathcal{D}, \mathrm{p}_{\mathrm{tr}}^{\mathcal{L}}}^{\mathrm{tr}}(f^{\mathcal{L}}) = \hat{R}_{\mathcal{U}, \mathrm{p}_{\mathrm{tr}}^{\mathcal{L}}}^{\mathrm{tr}}(f^{\mathcal{L}}) = \sum_{x \in \mathcal{U}} \sum_{y \in \mathcal{Y}} \mathrm{p}_{\mathrm{tr}}^{\mathcal{L}}(y|x) \cdot L(y, g_f^{\mathcal{L}}(x)) \approx R_{\mathcal{U}}^{\mathrm{tr}}(f^{\mathcal{L}}) \tag{14}
$$

This estimate allows us to define the estimated risk reduction in the transductive setting as follows:

$$
\Delta \hat{R}_{\mathcal{D}, \mathrm{p}_{\mathrm{tr}}^{\mathcal{L}^+}}^{\mathrm{tr}}(f^{\mathcal{L}^+}, f^{\mathcal{L}}) = \hat{R}_{\mathcal{U}, \mathrm{p}_{\mathrm{tr}}^{\mathcal{L}^+}}^{\mathrm{tr}}(f^{\mathcal{L}^+}) - \hat{R}_{\mathcal{U}, \mathrm{p}_{\mathrm{tr}}^{\mathcal{L}^+}}^{\mathrm{tr}}(f^{\mathcal{L}}) \tag{15}
$$

$$
= \sum_{x \in \mathcal{U}} \sum_{y \in \mathcal{Y}} \mathrm{P}_{\mathrm{tr}}^{\mathcal{L}^+}(y|x) \left( L(y, g_f^{\mathcal{L}^+}(x)) - L(y, g_f^{\mathcal{L}}(x)) \right) \tag{16}
$$

$$
= \sum_{x \in \mathcal{U} \setminus \{\tilde{x}\}} \sum_{y \in \mathcal{Y}} \mathrm{P}_{\mathrm{tr}}^{\mathcal{L}^+}(y|x) \left( L(y, f^{\mathcal{L}^+}(x)) - L(y, f^{\mathcal{L}}(x)) \right)
$$
$$
- \sum_{y \in \mathcal{Y}} \mathrm{P}_{\mathrm{tr}}^{\mathcal{L}^+}(y|\tilde{x}) \left( L(y, \tilde{y}) - L(y, f^{\mathcal{L}}(\tilde{x})) \right) \tag{17}
$$

$$
= \sum_{x \in \mathcal{U} \setminus \{\tilde{x}\}} \sum_{y \in \mathcal{Y}} \mathrm{P}_{\mathrm{tr}}^{\mathcal{L}^+}(y|x) \left( L(y, f^{\mathcal{L}^+}(x)) - L(y, f^{\mathcal{L}}(x)) \right) - L(\tilde{y}, f^{\mathcal{L}}(\tilde{x})) . \tag{18}
$$

In Eq. 17, we separate $\tilde{x}$ from $\mathcal{U}$ as the candidate serves two purposes. In the first part of the equation, we estimate the *inductive* risk reduction for the remaining unlabeled instances resulting from the improvement of the model with the additional label. In the second part, we assume that the label $\tilde{y}$ is correct. Therefore, we only need to consider the case $y = \tilde{y}$ as $\mathrm{P}_{\mathrm{tr}}^{\mathcal{L}^+}(\tilde{y}|\tilde{x}) = 1$ and $\mathrm{P}_{\mathrm{tr}}^{\mathcal{L}^+}(y|\tilde{x}) = 0$ for $y \neq \tilde{y}$. Hence, we simplify that term to $L(\tilde{y}, f^{\mathcal{L}}(\tilde{x}))$.

Analogous to Eq. 9, the transductive probabilistic gain is calculated as follows:

$$\text{xgain}^{\text{tr}}(\tilde{x}, \mathcal{L}, \mathcal{D}) = - \underset{p_{\text{tr}}^{\mathcal{L}}(\tilde{y}|\tilde{x})}{\mathbb{E}} \left[ \Delta \hat{R}_{\mathcal{D}, p([])\mathcal{L}^+]}^{\text{tr}} (f^{\mathcal{L}^+}, f^{\mathcal{L}}) \right] \qquad (19)$$

$$= - \sum_{\tilde{y} \in \mathcal{Y}} \frac{(k_{\tilde{x}}^{\mathcal{L}} + \beta)_{\tilde{y}}}{\|k_{\tilde{x}}^{\mathcal{L}} + \beta\|_1} \cdot \sum_{x \in \mathcal{U} \setminus \{\tilde{x}\}} \sum_{y \in \mathcal{Y}} \frac{(k_{k_x}^{\mathcal{L}^+} + \alpha)_y}{\|k_{k_x}^{\mathcal{L}^+} + \alpha\|_1} \left( L(y, f^{\mathcal{L}^+}(x)) - L(y, f^{\mathcal{L}}(x)) \right)$$

$$+ \sum_{\tilde{y} \in \mathcal{Y}} \frac{(k_{\tilde{x}}^{\mathcal{L}} + \beta)_{\tilde{y}}}{\|k_{\tilde{x}}^{\mathcal{L}} + \beta\|_1} \cdot L(\tilde{y}, f^{\mathcal{L}}(\tilde{x})) \qquad (20)$$

The first part is equal to the inductive probabilistic gain evaluated on $\mathcal{U} \setminus \{\tilde{x}\}$ multiplied by the number of instances in that set. This factor is necessary as the transductive risk is defined as the sum over all losses whereas the inductive risk uses the average loss. We call the second part of the equation the *candidate gain* (cgain) as it results from acquiring the correct label from the candidate instance. In summary, we can write the transductive probabilistic gain as the sum of the inductive and the candidate gain:

$$\text{xgain}^{\text{tr}}(\tilde{x}, \mathcal{L}, \mathcal{U}) = |\mathcal{U} \setminus \{\tilde{x}\}| \cdot \text{xgain}(\tilde{x}, \mathcal{L}, \mathcal{U} \setminus \{\tilde{x}\}) + \text{cgain}(\tilde{x}, \mathcal{L}, \{\tilde{x}\}) . \qquad (21)$$

### 4.3 Illustrative Example

Figure 1 shows the inductive and the candidate gain for a synthetic 2-dimensional dataset with two classes. The 7 already labeled instances are marked with a gray circle. The classifier's decision boundary is given as a black line and the

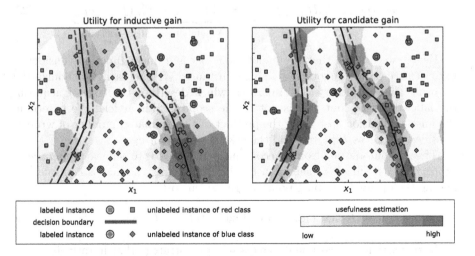

**Fig. 1.** Utility plots for the inductive and the candidate gain on a synthetic 2-dimensional dataset with 7 labels. (Color figure online)

dashed lines mark its confidence. The utilities are calculated for every unlabeled instance and are given as green surfaces (the color refers to the utility of the nearest instance). We see that the candidate gain (right plot) focuses on difficult instances in regions of high Bayesian error (near the decision boundary). Hence, it does not explore the data space but aims to ask the oracle to prevent the classifier from making wrong predictions. In contrast, the inductive gain (left plot) aims at improving the performance of the classifier. Therefore, it explores regions that are not yet covered with labels (upper left and lower right) and exploits the labels that already are available by refining the decision boundary. Moreover, we observe that regions of higher density (lower right) are preferred over regions with lower density (upper left) as labels have more impact on the classifier's performance there.

## 5   A Transductive Stopping Criterion

To define a stopping criterion for transductive active learning, we introduce a performance metric using an economic rationale. Therefore, we consider the most relevant kinds of costs involved in an active learning scenario: (1) The *annotation cost* $c_{AN} \in \mathbb{R}^{\geq 0}$ describes the cost of acquiring one label from an oracle, and (2) the *misclassification cost* $c_{ER} \in \mathbb{R}^{\geq 0}$ describes the cost induced by one wrong prediction of the classifier. Intuitively, the annotation cost is dependent on the number of acquired labels, whereas the misclassification cost usually decreases as more labels become available.

We define the *aggregated cost* as the sum of annotation and misclassification costs. Consequently, the aggregated cost can be written as follows for the $i$-th iteration of the active learning cycle.

$$\text{aggcost}(f, \mathcal{L}_i, \mathcal{U}_i, c_{AN}, c_{ER}) = \underbrace{|\mathcal{L}_i| \cdot c_{AN}}_{\substack{\text{Annotation} \\ \text{Cost}}} + \underbrace{R^{\text{tr}}_{\mathcal{U}_i}(f^{\mathcal{L}_i}) \cdot c_{ER}}_{\substack{\text{Misclassification} \\ \text{Cost}}} \tag{22}$$

Hence, we assume that the annotation cost is a linear function considering fixed costs $c_{AN}$ for annotating a single instance. We can easily generalize this by using some arbitrary cost function, which describes the cost of acquiring the labeled set $\mathcal{L}_i$, but this is not in the scope of this article. We determine the misclassification cost using the product of the estimated number of wrongly classified instances $R^{\text{tr}}_{\mathcal{U}_i}(f^{\mathcal{L}_i})$ and the cost for one error $c_{ER}$.

The optimal solution from an economic perspective is to achieve the *minimum aggregated cost* (mac), as shown in Eq. 23. Calculating the mac is equivalent to finding the optimal stopping point for the given costs.

$$\text{mac}(f, c_{AN}, c_{ER}) = \min_i \left( \text{aggcost}(f, \mathcal{L}_i, \mathcal{U}_i, c_{AN}, c_{ER}) \right) \tag{23}$$

In this article, we assume to have a selection strategy that iteratively selects one sample. In each iteration of the active learning cycle, we have to decide whether to acquire the label of another instance or to stop querying new labels.

Consequently, we stop the acquisition as soon as the annotation cost $c_{AN}$ exceeds the estimated cost reduction, based on the transductive probabilistic gain:

$$\text{Stop when } \Delta c_{ER} < c_{AN} \quad \text{with} \quad \Delta c_{ER} = \text{xgain}^{\text{tr}}(\tilde{x}^*, \mathcal{L}_i, \mathcal{U}_i) \cdot c_{ER}. \quad (24)$$

## 6  Experimental Evaluation

This section presents our experimental evaluation and starts by describing the experimental setup including the used datasets, competitors, and visualizations. Our evaluation approach is based on three hypotheses as motivated in the introduction. For each contribution, we formulate one hypothesis, present the key findings, and provide a detailed discussion with plots and/or tables.

### 6.1  Setup, Datasets, and Competitors

All experiments have been implemented in Python using scikit-learn and scikit-activeml[2]. We conduct experiments with the following selection strategies: random sampling (rand), least confidence uncertainty sampling (lc) [14], epistemic uncertainty sampling (epis) [17], query by committee (qbc) [15] with the Kullback-Leibler divergence as a disagreement measure and bootstrapping to generate a committee of 10 classifiers, Monte Carlo expected error reduction (mc) [21] including the extension of Chapelle with $\epsilon = 10^{-3}$ (chap) [3], and value of information (voi) [9]. To show the benefits of the new transductive probabilistic active learning (xpal_tr), we also compare it to the inductive (standard) variant (xpal) [12]. The expected error based strategies mc, chap (with [6]), voi, and xpal_tr implement a cost-based stopping criterion. Whereas voi already evaluates only on the unlabeled instances, we use the unlabeled set as the evaluation set for mc and chap to ensure comparability in the transductive setting.

We use a Parzen window classifier [18] with an RBF kernel as the classifier (similar to [3,12,17]). The main advantages of this classifier are the low number of parameters, the deterministic character, its probabilistic nature, and the fact that it is generic in a way that all methods can be used with that classifier. Using the same classifier for comparison is important as doing otherwise could induce additional biases. The bandwidth parameter of the kernel is set by the mean criterion [5].

We use 10 datasets from OpenML [28]. For simplicity, we remove all samples that contain missing values and standardize all features independently to zero mean and a standard deviation of one. We repeatedly (25 times) split all datasets randomly into two subsets. The first one, which contains 67% of the samples, is used for the active learning circle and builds the initially unlabeled set $\mathcal{U}_0$ according to Sect. 3. This set is used for evaluating the transductive setting. The remaining samples (33%) build the test set for the inductive setting.

---

[2] https://github.com/dakot/stopTransAL, https://github.com/scikit-activeml.

## 6.2  Visualization Techniques

To visualize the results, we provide learning curves (e. g., Fig. 2) showing the transductive (resp. inductive) risk. For each dataset and selection strategy, we averaged the risks after every iteration over the 25 repetitions. The goal is to achieve a low error fast.

We summarize these results in ranking tables (e. g., Fig. 3). There, we show the rank of each strategy for every dataset with respect to the area under the performance curve. We calculate the rank for each of the 25 repetitions independently and average these ranks into the final score. Depending on the evaluation goal, we define a baseline strategy that will be compared to all other competitors using a paired Wilcoxon signed-rank test. We identify if the evaluation score of the competitor is significantly higher (arrow up), significantly lower (arrow down), or not significantly different (no sign) than the baseline strategy ($p$-value .05). These are summarizes as win/tie/loss statistics.

Moreover, we evaluate the transductive scenario by plotting the aggregated cost (e. g., Fig. 4). There, we evaluate the aggregated cost (i. e., the sum of annotation and misclassification costs) for different cost ratios. Depending on the application this ratio might differ and the practitioner can find a suitable algorithm. In Fig. 4, we show the minimum aggregated cost as we identify the optimal stopping point for every selection strategy. Hence, we can assess the quality of selection strategies without the bias of a stopping criterion. In Fig. 5 and Fig. 6 (dashed lines), the aggregated cost is determined based on the proposed stopping point of a stopping criterion. The black lines in the aggregated cost plots show the naive baselines which are determined by the minimum cost between classifying all instances as one class without acquiring any label and acquiring all labels.

Due to the large variety of plots, we only show the most interesting results. You can find all plots in the supplemental material on github.

## 6.3  Results

**Hypothesis A: It is beneficial to develop specific selection strategies for transductive active learning.**

*Key Findings:* When comparing inductive and transductive probabilistic active learning, we show that xpal (inductive) wins when evaluated on the inductive risk, and xpal_tr wins for the transductive risk. Hence adapting the selection strategy is beneficial and solving the inductive case (considering generalization capabilities) is not sufficient to solve transductive active learning.

*Detailed Discussion:* In Fig. 2, we exemplary selected three datasets to show the inductive and the transductive risk for all selection strategies. We see that the transductive risk finishes at zero risk as there are no errors when all labels are acquired. In contrast, the inductive risk converges at the Bayesian error rate. In Fig. 3, we show the ranking statistics based on the area under the inductive/transductive risk curve as described in the previous subsection. Please note that epis is only valid for 2-class problems. The results show the superiority of xpal in the inductive case (rank 2.56 vs. rank 2.95) and of xpal_tr in the transductive case (rank 1.87 vs. 2.14). The reason for that is that xpal_tr specifically incorporated the acquisition of difficult instances into the target function through the candidate gain as discussed in Subsect. 4.3.

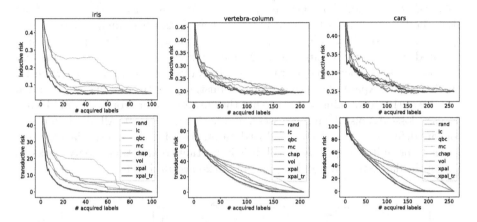

**Fig. 2.** Learning curves of selection strategies with respect to the inductive (upper) and the transductive (lower) risk.

| transductive | *iris* | *prnn_crabs* | *cpu* | *vertebra-column* | *ecoli* | *autoMpg* | *user-knowledge* | *cars* | *chscase_vine2* | *irish* | *mean* | *win/tie/loss* | inductive | *mean* | *win/tie/loss* |
|---|---|---|---|---|---|---|---|---|---|---|---|---|---|---|---|
| rand | 5.1↗ | 6.4↗ | 7.5↗ | 6.1↗ | 7.5↗ | 7.6↗ | 5.4↗ | 7.0↗ | 4.7↗ | 6.9↗ | 6.43 | 10/0/0 | rand | 5.69 | 10/0/0 |
| lc | 5.6↗ | 4.2↗ | 3.2↗ | 1.4 | 3.0↗ | 2.2 | 4.0↗ | 2.9↗ | 6.5↗ | 5.9↗ | 3.89 | 8/2/0 | lc | 5.19 | 9/1/0 |
| qbc | 4.3↗ | 6.4↗ | 4.4↗ | 4.4↗ | 4.6↗ | 4.3↗ | 5.0↗ | 4.8↗ | 4.9↗ | 5.0↗ | 4.81 | 10/0/0 | qbc | 4.81 | 8/2/0 |
| epis | — | 2.2 | 5.6↗ | — | — | 4.1↗ | — | — | 1.3↘ | 3.6↗ | 3.36 | 3/1/1 | epis | 3.52 | 3/0/2 |
| mc | 6.7↗ | 7.6↗ | 8.3↗ | 7.4↗ | 7.2↗ | 8.2↗ | 7.2↗ | 6.9↗ | 8.0↗ | 7.8↗ | 7.54 | 10/0/0 | mc | 6.36 | 10/0/0 |
| chap | 5.4↗ | 6.6↗ | 6.0↗ | 7.4↗ | 5.6↗ | 7.9↗ | 4.4↗ | 7.0↗ | 7.2↗ | 6.9↗ | 6.44 | 10/0/0 | chap | 5.40 | 9/1/0 |
| voi | 5.5↗ | 7.2↗ | 6.3↗ | 4.5↗ | 5.0↗ | 5.7↗ | 7.0↗ | 3.1↗ | 7.0↗ | 5.8↗ | 5.71 | 10/0/0 | voi | 5.78 | 9/1/0 |
| xpal | 1.7 | 1.7↘ | 1.7 | 3.0↗ | 2.0↗ | 3.2↗ | 1.5 | 3.0↗ | 2.4 | 1.0↘ | 2.14 | 4/4/2 | xpal | 2.56 | baseline |
| xpal_tr | 1.7 | 2.7 | 1.9 | 1.7 | 1.1 | 1.8 | 1.6 | 1.3 | 3.0 | 2.0 | 1.87 | baseline | xpal_tr | 2.95 | 4/6/0 |

**Fig. 3.** Ranking statistics with respect to the area under the transductive (left) and inductive (right) risk.

**Hypothesis B: Our selection strategy xpal_tr performs best for the transductive risk and the minimum aggregated cost.**

*Key Findings:* We show that transductive probabilistic active learning outperforms the other competitors in the transductive scenario on average when evaluated on the transductive risk and the minimum aggregated cost, i. e., the sum of the annotation and misclassification cost for the optimal stopping point.

*Detailed Discussion:* To evaluate this hypothesis, we consider the figures from Hypothesis A to evaluate the transductive risk and Fig. 4 to evaluate the minimum aggregated cost. The results show: (1) For the transductive risk, xpal_tr is only defeated significantly in three cases (2 times by xpal and once by epis). Whereas epis performs mediocre on cpu (rank 5.6), the ranks of xpal_tr are all between 1.1 and 3.0. Hence, xpal_tr seems to be fairly robust. (2) For the minimum aggregated cost, we see in the ranking statistics that the hardest competitors are xpal (4 wins, 4 ties, 2 losses), epis (3 wins, 2 losses), and lc (7 wins, 3 ties). All other competitors are defeated significantly on all 10 datasets. Hereby, epis is a special case as it seems to be quite competitive. Still, it is important to note that it only works on half of the datasets as it is only applicable to 2-class problems.

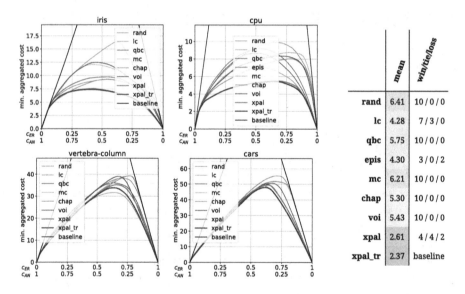

**Fig. 4.** Minimum aggregated cost curves (left) and ranking statistics with respect to the area under the mac curve (right).

**Hypothesis C: Our new stopping criterion performs best compared to existing methods.**

*Key Findings:* The selection strategy xpal_tr with the new stopping criterion outperforms the existing selection strategies that implement a stopping criterion (mc, chap, voi). To evaluate these stopping criteria independently from the selection strategy, we tested their performance together with random sampling to ensure comparability and show the superiority of our method.

*Detailed Discussion:* To evaluate the stopping criteria, we show the aggregated cost for the chosen stopping point with respect to the given cost ratios (left) and the ranking statistics (right): In Fig. 5, we evaluated the proposed combinations of a selection strategy and a stopping criterion. Figure 6 shows the results based on a random selection. We use random for the comparison as it induces the smallest bias on the selection. In this scenario, we cannot assume that the best candidate is always selected. Hence, we average the estimated misclassification cost reduction instead of choosing the one from the selected candidate to decide about stopping. Our method xpal_tr significantly outperforms all competitors on all datasets for both cases with only one exception (1 tie).

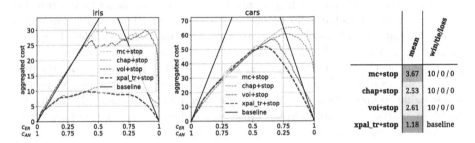

**Fig. 5.** Aggregated cost curves for selection strategies that implement a stopping criterion (left) and their ranks based on the area under these curves (right).

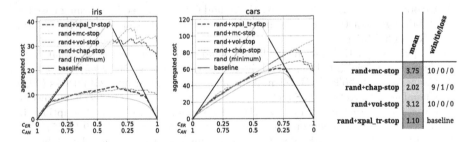

**Fig. 6.** Aggregated cost curves for different stopping criteria using rand as a selection strategy (left) and their ranks based on the area under these curves (right).

# 7  Conclusion and Outlook

In this article, we introduced and formalized the transductive active learning scenario. We showed that this scenario is not just a special case of the inductive one and that it requires new methods for instance selection. To address this problem, we proposed a novel transductive selection strategy based on the probabilistic active learning framework and experimentally showed that it performs better than the inductive version in the transductive setting. We introduced and motivated a target function for stopping criteria for transductive active learning that considers the misclassification and the annotation costs. Based on this target function, we introduced the minimum aggregated cost that evaluates stopping criteria based on how well they perform for different cost ratios. We used our strategy to derive a novel cost-based stopping criterion. The empirical evaluation showed that it outperforms existing criteria.

In the future, we aim to investigate how the prior influences the proposed methods (here set to 0.001 following [3,12]). In this article, we only considered fixed annotation and misclassification costs and omniscient oracles. However, it is often more realistic that instances have different annotation costs (e. g., dependent on the annotation time, or quality) or that instances have different misclassification costs (e. g., dependent on the instance's importance). Moreover, considering computational cost for the selection might be beneficial. Finally, we want to analyze how our stopping criterion can be used also with other active learning strategies such as uncertainty sampling.

# References

1. Balasubramanian, V., Chakraborty, S., Panchanathan, S.: Generalized query by transduction for online active learning. In: International Conference on Computer Vision (Workshops), pp. 1378–1385 (2009)
2. Bloodgood, M., Vijay-Shanker, K.: A method for stopping active learning based on stabilizing predictions and the need for user-adjustable stopping. arXiv preprint arXiv:1409.5165 (2014)
3. Chapelle, O.: Active learning for Parzen window classifier. In: International Workshop on Artificial Intelligence and Statistics, vol. 5, pp. 49–56 (2005)
4. Chapelle, O., Schölkopf, B., Zien, A.: Semi-supervised learning. MIT Press (2010)
5. Chaudhuri, A., Kakde, D., Sadek, C., Gonzalez, L., Kong, S.: The mean and median criteria for kernel bandwidth selection for support vector data description. In: International Conference on Data Mining (Workshops), pp. 842–849 (2017)
6. Dimitrakakis, C., Savu-Krohn, C.: Cost-minimising strategies for data labelling: optimal stopping and active learning. In: International Symposium on Foundations of Information and Knowledge Systems, pp. 96–111 (2008)
7. Güttler, F.N., Ienco, D., Poncelet, P., Teisseire, M.: Combining transductive and active learning to improve object-based classification of remote sensing images. Remote Sens. Lett. **7**(4), 358–367 (2016)
8. Ishibashi, H., Hino, H.: Stopping criterion for active learning based on error stability. arXiv preprint arXiv:2104.01836 (2021)

9. Joshi, A.J., Porikli, F., Papanikolopoulos, N.: Multi-class active learning for image classification. In: Conference on Computer Vision and Pattern Recognition, pp. 2372–2379 (2009)
10. Kapoor, A., Horvitz, E., Basu, S.: Selective supervision: guiding supervised learning with decision-theoretic active learning. In: Int. Joint Conference on Artificial Intelligence, pp. 877–882 (2007)
11. Kottke, D., Calma, A., Huseljic, D., Krempl, G., Sick, B.: Challenges of reliable, realistic and comparable active learning evaluation. In: Workshop on Interactive Adaptive Learning, pp. 2–14 (2017)
12. Kottke, D., Herde, M., Sandrock, C., Huseljic, D., Krempl, G., Sick, B.: Toward optimal probabilistic active learning using a Bayesian approach. Mach. Learn. **110**(6), 1199–1231 (2021)
13. Laws, F., Schätze, H.: Stopping criteria for active learning of named entity recognition. In: International Conference on Computational Linguistics, pp. 465–472 (2008)
14. Lewis, D.D.: A sequential algorithm for training text classifiers. In: International ACM SIGIR Conference on Research and Development in Information Retrieval (1995)
15. McCallumzy, A.K., Nigamy, K.: Employing EM and pool-based active learning for text classification. In: International Conference on Machine Learning, pp. 359–367 (1998)
16. Min, F., Liu, F.L., Wen, L.Y., Zhang, Z.H.: Tri-partition cost-sensitive active learning through kNN. Soft. Comput. **23**(5), 1557–1572 (2019)
17. Nguyen, V.-L., Shaker, M.H., Hüllermeier, E.: How to measure uncertainty in uncertainty sampling for active learning. Mach. Learn. **111**, 89–122 (2021). https://doi.org/10.1007/s10994-021-06003-9
18. Parzen, E.: On estimation of a probability density function and mode. Ann. Math. Stat. **33**(3), 1065–1076 (1962)
19. Pullar-Strecker, Z., Dost, K., Frank, E., Wicker, J.: Hitting the target: stopping active learning at the cost-based optimum. arXiv preprint arXiv:2110.03802 (2021)
20. Reitmaier, T., Calma, A., Sick, B.: Transductive active learning-a new semi-supervised learning approach based on iteratively refined generative models to capture structure in data. Inf. Sci. **293**, 275–298 (2015)
21. Roy, N., McCallum, A.: Toward optimal active learning through Monte Carlo estimation of error reduction. In: International Conference on Machine Learning, pp. 441–448 (2001)
22. Roy, N., Mccallum, A., Com, M.W.: Toward optimal active learning through Monte Carlo estimation of error reduction. In: Proceedings of the International Conference on Machine Learning (ICML), p. 8. San Francisco, CA, USA (2001)
23. Scharei, K., Herde, M., Bieshaar, M., Calma, A., Kottke, D., Sick, B.: Automated active learning with a robot. Arch. Data Science, Ser. A **5**(1), 16 (2018)
24. Settles, B.: Active learning literature survey. Technical report, University of Wisconsin, Department of Computer Science (2010)
25. Sun, S., Hardoon, D.R.: Active learning with extremely sparse labeled examples. Neurocomputing **73**(16–18), 2980–2988 (2010)
26. Tong, S.: Active learning: theory and applications, Ph. D. thesis, Stanford (2001)
27. Triguero, I., García, S., Herrera, F.: Self-labeled techniques for semi-supervised learning: taxonomy, software and empirical study. Knowl. Inf. Syst. **42**(2), 245–284 (2015)
28. Vanschoren, J., van Rijn, J.N., Bischl, B., Torgo, L.: OpenML: networked science in machine learning. SIGKDD Explor. **15**(2), 49–60 (2013)

29. Vapnik, V.N.: Statistical learning theory. John Wiley & Sons, Inc. (1998)
30. Vlachos, A.: A stopping criterion for active learning. Comput. Speech Lang. **22**(3), 295–312 (2008)
31. Yu, K., Bi, J., Tresp, V.: Active learning via transductive experimental design. In: International Conference on Machine learning, pp. 1081–1088 (2006)

# Multi-domain Active Learning
# for Semi-supervised Anomaly Detection

Vincent Vercruyssen$^{(\boxtimes)}$, Lorenzo Perini, Wannes Meert, and Jesse Davis

KU Leuven, Leuven, Belgium
{vincent.vercruyssen,lorenzo.perini,wannes.meert,jesse.davis}@kuleuven.be

**Abstract.** Active learning aims to ease the burden of collecting large amounts of annotated data by intelligently acquiring labels during the learning process that will be most helpful to learner. Current active learning approaches focus on learning from a *single* dataset. However, a common setting in practice requires simultaneously learning models from *multiple* datasets, where each dataset requires a separate learned model. This paper tackles the less-explored multi-domain active learning setting. We approach this from the perspective of multi-armed bandits and propose the *active learning bandits* (ALBA) method, which uses bandit methods to both explore and exploit the usefulness of querying a label from different datasets in subsequent query rounds. We evaluate our approach on a benchmark of 7 datasets collected from a retail environment, in the context of a real-world use case of detecting anomalous resource usage. ALBA outperforms existing active learning strategies, providing evidence that the standard active learning approaches are less suitable for the multi-domain setting.

**Keywords:** Anomaly detection · Active learning · Semi-supervised learning · Multi-armed bandits

## 1 Introduction

Active learning (AL) attempts to alleviate the time and monetary cost of acquiring labeled data by intelligently deciding exactly which unlabeled instances require a label [20]. One task where active learning can be particularly helpful is in *anomaly detection* (AD), where the goal is to learn a model that can identify anomalous instances in a dataset. While anomaly detection was typically treated an unsupervised learning problem, there is growing evidence that in practice AD algorithms benefit from small amounts of labeled data [15,25]. In particular, labels can help overcome the assumptions encoded in unsupervised AD approaches (e.g., all rare behavior is anomalous) by providing examples of infrequent normal behavior such as maintenance. However, anomalies are rare by nature, making it costly to find

**Supplementary Information** The online version contains supplementary material available at https://doi.org/10.1007/978-3-031-26412-2_30.

and label them. Thus, AL can help select those instances whose label would be most *informative* to the underlying anomaly detector [24].

A drawback to classic AL approaches is that they focus on learning from a single dataset. This contrasts with a scenario that often arises in practice, particularly in anomaly detection, where it is necessary to simultaneously model data from a fleet of similar yet slightly different entities. As an illustrative example, consider trying to detect anomalous resource usage in a chain of retail stores. Each store is different in terms of its location, size, opening hours, services offered, etc. Thus, each store's resource usage, and the resulting dataset, is characterized by a different marginal distribution [16]. Consequently, what constitutes anomalous behavior is store-dependent, which necessitates a separate detection model per store. Or consider building classifiers to detect the occurrence of blade icing in different wind turbines [27]. Each turbine generates its own data and is different from the other turbines in terms of position, size, etc. Hence, training a single model for use in all stores/turbines would not work. Unfortunately, classic AL strategies are not optimized to deal with multiple datasets simultaneously.

In this paper, we focus on adapting active learning to the multi-domain setting in the context of AD. Given multiple datasets and a global fixed budget for the number of labels that can be acquired across all datasets, our objective is to employ active learning to learn one model for each dataset. The key challenge is to decide how to best divide this budget across the different datasets. Naively spending an equal budget on each dataset is likely to be suboptimal as some datasets will require fewer labeled instances to learn an accurate model than others. Hence, one needs to estimate the *marginal gain* of acquiring another label in each dataset. This is challenging as the marginal gain is diminishing: as more labels are actively acquired in a dataset, each one will have a smaller effect on the learned model's performance. This can be viewed through the prism of the exploration-exploitation trade-off. One needs to spend some labeling effort in each dataset to estimate this marginal gain while simultaneously trying to mostly label the datasets with high gains. We address this challenge from the perspective of *multi-armed bandits* (MAB) and propose the *active learning bandits* (ALBA) method. ALBA maintains an estimate of the marginal gain of querying a label from each dataset over the course of multiple query rounds and queries those labels that optimize the exploration-exploitation trade-off. This yields three differences with the classic AL techniques. First, ALBA can handle multiple datasets. Second, ALBA tracks the marginal gain of acquiring a label from *groups of instances* whereas classic AL tends to estimate the marginal gain of a *single* instance. Third, ALBA computes the marginal gain of an instance's label *after* the oracle has been queried and has provided the instance's true label. To summarize, this paper makes the following contributions:[1]

1. We identify and show how multi-domain active learning and multi-armed bandits are related;
2. We propose an approach to multi-domain active learning that uses *rotting bandits* to cope with the diminishing returns of acquiring labels;

---

[1] Appendix & Code: https://github.com/Vincent-Vercruyssen/ALBA-paper.

3. We explore theoretically and experimentally how the integration of either a heuristic or a random active learning strategy in ALBA impacts its performance;

4. We empirically demonstrate that ALBA outperforms multiple baselines on 7 real-world datasets about water usage where the task is anomaly detection.

## 2    Preliminaries

***Multi-domain Dataset.*** A *domain* consists of an input space $\mathcal{X}$, a label space $\mathcal{Y}$, and a joint probability distribution over the input-label space pair. By sampling observations from a domain's distribution, we obtain a dataset $D$.

A *multi-domain dataset* $\mathcal{M} = \{D^k\}_{k=1}^K$ consists of $K$ datasets, each sampled from a different underlying domain's distribution. A *multi-domain instance* is denoted as $x_i^k$ and its label as $y_i^k$, i.e., instance $i$ of the dataset $D^k$. We assume the input and label space are the same for each of the $K$ domains.

***Pool-based Active Learning.*** Each dataset $D^k \in \mathcal{M}$ contains both labeled and unlabeled instances. In pool-based active learning, one tries to construct a classifier $f^k$ for a dataset $D^k$. Initially, no labels are available. Over the course of subsequent iterations, one unlabeled instance in $D^k$ is chosen to be labeled by the oracle, its label is added to $D^k$, and the classifier is retrained [20].

***Multi-armed Bandits.*** The origin of the MAB problem stems from clinical trials [23]. An MAB algorithm is typically given a fixed set of actions $\mathcal{A} = \{1, ..., K\}$ (the arms) and a fixed number of rounds to play $T$ (the budget). In each round $t$, the algorithm has to choose one action $j \in \mathcal{A}$ and receives a single random payoff $r_j$ from the corresponding unknown payoff distribution [2]. When action $j$ is taken for the $n^{\text{th}}$ time, the mean of the payoff distribution is $\mu_j(n)$.

In our active learning setting, the payoff distribution is non-stationary and the expected payoff of an action decreases over time. Thus, for all actions, $\mu_j(n)$ is assumed to be positive and non-increasing in $n$. This corresponds to the *rotting bandit* setting [12]. Let $N_j(t)$ be the number of times action $j$ is taken at round $t$, let $\pi$ be a policy (i.e., an infinite sequence of actions), and let $\pi(t)$ denote the action chosen by policy $\pi$ in round $t$. Then, the goal of the MAB algorithm is to maximize the expected sum of payoffs after round $T$ which equals $\mathbb{E}\left[\sum_{t=1}^T \mu_{\pi(t)}\left(N_{\pi(t)}(t)\right)\right]$. MAB algorithms embody the *exploration-exploitation* trade-off, exploiting the best action while spending some time exploring the payoff of each action [2]. In this paper, we use the recent *sliding-window average* (SWA) algorithm as a solver for the non-parametric rotting bandit setting, which comes with strong performance guarantees [12].

## 3    Multi-domain Active Learning

The multi-domain active learning (MDAL) problem for anomaly detection is:

**Given:** A multi-domain dataset $\mathcal{M}$ consisting of $K$ unlabeled datasets, a fixed label budget $T$, and an oracle $\mathcal{O}$ that can provide one label at a time;

**Do:** maximize the performance of each domain's anomaly detector $f^k$ by querying additional labels to the oracle.

The two key challenges are (1) figuring out the marginal benefit of acquiring
labels in each dataset of $\mathcal{M}$ (= *labeling payoff*), and (2) dealing with the diminishing returns of labeling additional instances. Some datasets in $\mathcal{M}$ will require
less labels to learn an accurate detector than other datasets, i.e., their labeling
payoff is higher, but we do not know beforehand which ones.

We propose the *active learning bandits* (ALBA) approach which leverages an
MAB algorithm to solve the trade-off between exploration (figuring out which
datasets have a high labeling payoff) and exploitation (focusing our labeling efforts
on the high-payoff datasets). First, ALBA defines several groups of instances for
which to track the labeling payoff (Sect. 3.1). Second, ALBA updates its estimate
of the average labeling payoff of each group using a reward function that measures
the impact of labeling an instance from that group on the corresponding anomaly
detector (Sect. 3.2). Third, in each query round ALBA picks a group from which to
query an instance using the SWA *rotting bandit* algorithm which can handle non-
stationary rewards (Sect. 3.3). Finally, after choosing a group, ALBA still needs
to decide which individual instance of that group to query (Sect. 3.4). ALBA maintains one detector $f^k$ per dataset in $\mathcal{M}$ that is retrained upon receiving a new label
from oracle $\mathcal{O}$. We assume that the cost of querying and labeling is the same and
constant for all instances in $\mathcal{M}$, and that the oracle is queried one instance at a
time. See Sect. 3.5 for the algorithm's pseudocode.

## 3.1    Choosing an Action Set

We define a group of instances $G^j$ such that $\forall j$ there exists a $k$ such that $G^j \subset D^k$
and $\forall i \neq j : G^j \cap G^i = \varnothing$. Then, we define the action set $\mathcal{A}$ such that each action
$j \in \mathcal{A}$ corresponds to choosing a particular group of instances $G^j$ from which
one instance will be queried. The most straightforward idea is to let each of the
$K$ datasets in $\mathcal{M}$ be its own group such that $G^j = D^k$. Thus, $|\mathcal{A}| = K$.

However, the distribution of informative instances likely varies substantially
within each dataset. Therefore, we propose to first divide each dataset into
smaller groups using a clustering algorithm, obtaining a set of $C$ clusters for
each dataset. Each action now corresponds to choosing a particular cluster and
$|\mathcal{A}| = K \times C$. Note that $\forall j : G^j \subset D^k$. This approach gives the MAB algorithm
(Sect. 3.3) more fine-grained control in selecting different groups of instances and
learning their labeling payoff. However, it comes at the cost of increased exploration because the algorithm now has to figure out the reward structure for a
larger set of actions.

## 3.2    MAB Reward Function

ALBA keeps track of the labeling payoff for each group of instances $G^j$. After
choosing action $j$, and receiving the label for the single selected instance $x_i^j \in G^j$
(Sect. 3.4), the challenge is to design a reward function that reflects the updated

payoff of querying the label for one of the remaining unlabeled instances from this group. We consider two possibilities.

*Entropy of the Predictions.* One measure of the labeling payoff of instance $x_i^j$ is its ability to decrease the overall prediction uncertainty of anomaly detector $f^k$ trained on the dataset to which $x_i^j$ belongs. Thus, the payoff $r_j$ of action $j$ that results in querying $x_i^j$ is computed as the decrease in prediction entropy of the detector after retraining:

$$r_j = \sum_{x \in D^k} \left[ H_{f_+^k}(x) - H_{f^k}(x) \right] \tag{1}$$

where $f_+^k$ represents the detector after retraining with the label of $x_i^j$ provided by the oracle. $H_f(x)$ is the Shannon entropy of the predicted label probability (by detector $f$) that $x$ belongs to one of the classes in $\mathcal{Y}$.

*Cosine Similarity of the Predictions.* A more direct measurement of the labeling payoff of $x_i^j$ looks at how many instances the anomaly detector changes its predicted label for after retraining. If this number is large, the model changed a lot and we can say that labeling $x_i^j$ had a large impact. The payoff $r_j$ of action $j$ that results in querying $x_i^j$ is computed as the cosine similarity between the predicted-label vectors:

$$r_j = 1 - \frac{Y_{f_+^k} \cdot Y_{f^k}}{\|Y_{f_+^k}\| \|Y_{f^k}\|} \tag{2}$$

where $Y_f$ is the vector of predicted labels for a dataset by detector $f$ and contains all 0's or 1's (in our experiments, 0 signifies "normal" and 1 "anomalous"). $f_+^k$ and $f^k$ represent the anomaly detectors trained respectively with and without the queried label.

### 3.3   MAB Algorithm

The MAB algorithm chooses an action $j$ from $\mathcal{A}$ in each query round $t$. In our setting, the number of query rounds is fixed and equal to the label budget $T$, the set of possible actions is fixed, only the payoff of the chosen action is observed at each round ($r_j = 0$ if $j$ is not chosen), and the observed payoffs are bounded to the interval $[0, 1]$. The labeling payoff of a group decreases as more instances from that group are queried and labeled. Intuitively, if most of the instances in a group are labeled, acquiring yet another label will have little effect on the anomaly detector, so the labeling payoff will be close to zero. In contrast, if few or no instances are labeled, observing even one label might greatly improve the detector. Hence, given the non-stationary rewards, ALBA uses the SWA rotting bandit algorithm [12] to choose between different actions in each query round $t$. During the AL loop, SWA tracks the decreasing labeling payoff of each action by estimating a sliding-window average of the obtained rewards with window size $W$, i.e., $\bar{\mu}_j(N_j(t)) = \frac{1}{W} \sum_{n=N_j(t)-W}^{N_j(t)} r_j(n)$ where $r_j(n)$ is the reward obtained when action $j$ is chosen for the $n^{\text{th}}$ time. Initially, this estimate is 0 as ALBA only obtains information about an action's reward distribution *after* querying instances.

### 3.4   Query Selection Strategy

We can query one instance per query round to the oracle. Although the MAB algorithm tells us from which group $G^j$ we should query, it does not inform us which particular instance $x_i^j \in G^j$ to query. The solution is to select query instance $x_i^j$ either *randomly* from $G^j$ (RAND) or *heuristically*, using uncertainty sampling (UC) or another suitable AL method.

Using a random selection strategy results in a better regret bound than using a heuristic strategy, because random payoffs produce a less biased estimate of the average payoff $\mu_j(N_j(t))$ for any action $j \in \mathcal{A}$ at any round $t \leq T$. Let us first assume that the rate of change of the true labeling payoff of an action is not affected by which instance from the corresponding group is queried and labeled in any given round $t$.[2] Then, in a theoretical scenario with infinite instances and infinite budget, randomly collecting labels from each group, computing the rewards, and estimating the average labeling payoff with a sliding window average will result in gradually more accurate estimates of each group's *true* average labeling payoff: $|\bar{\mu}_j(N_j(t)) - \mu_j(N_j(t))| \to 0$ for $t \to \infty$, $j = 1, \ldots, K \times C$. This means that, for a given tolerance error $\varepsilon_j > 0$, there exists a certain necessary cost $c_j \in \mathbb{N}$ such that it is guaranteed that the estimate error is smaller than the tolerance: $|\bar{\mu}_j(N_j(t)) - \mu_j(N_j(t))| < \varepsilon_j$, for all $t \geq c_j$. By taking the total cost $c = \sum_{j=1}^{K \times C} c_j$, and the minimum tolerance $\varepsilon = \min\{\varepsilon_j : j \leq K \times C\}$, we can claim that all estimates of the average payoffs are accurate enough, independently of which group is considered: $|\bar{\mu}_j(N_j(t)) - \mu_j(N_j(t))| < \varepsilon$ for all $t \geq c$ and $j = 1, \ldots, K \times C$. After paying cost $c$ (i.e., some number of query rounds), the MAB algorithm begins to pick the optimal actions (exploitation). In contrast, as long as $t < c$, it will sometimes pick sub-optimal actions (exploration). Hence, the lower $c$ (i.e., the faster the estimate of all actions' payoff converges), the lower the expected average regret.

The previous statement is true when any unbiased estimator of the average payoff is used. We now show that estimating the average payoff by *heuristically* selecting the instances, e.g., using the UC selection strategy, results in a biased estimate of the average payoff. Intuitively, this is because in each round $t$ the heuristic strategy picks the instance that yields the largest (potential) reward, resulting in an overly optimistic estimate of the true labeling payoff of each group, forcing the MAB algorithm to spend more time exploring ($c$ is larger).

In the following paragraphs, we fix the index $j$ that refers to a specific action and denote with $\mathcal{W}$ the maximum budget that can be spent on the group $G^j$, corresponding to the number of available payoffs. Additionally, $n$ denotes the number of rounds spent on the given group $G^j$, i.e., for any $n$ there exists a round $t$ such that $n = N_j(t)$. For example, $\mu_j(N_j(t))$ would become $\mu(n)$.

**Proposition 1.** *Let $R_1^n, \ldots, R_{\mathcal{W}-n}^n$ be i.i.d. random variables that take the payoffs as values, such that $\mathbb{E}[R_q^n] = \mu(n)$, for $q = 1, \ldots, \mathcal{W} - n$, $n \in \mathcal{W} + 1, \ldots, \mathcal{W}$.*

---

[2] Because AL typically operates with a small budget and large datasets, this assumption is reasonable as the marginal benefit of labeling each additional instance is small.

*Assume that $\mu(n)$ is a positive non-decreasing function and that $\mu(n) - \mu(n+1)$ does not depend on which instance is queried at round $n$. Then,*

$$\mu(n) \leq \mathbb{E}[\bar{\mu}_{rand}] < \mathbb{E}[\bar{\mu}_{heur}].$$

*Proof.* Without loss of generality, let us assume that the random selection strategy picks the random variable with index $p$. For any $n > W$,

$$\mathbb{E}[\bar{\mu}_{rand}] = \mathbb{E}\left[ \frac{1}{W} \sum_{i=n-W}^{n} R_p^i \right] = \frac{1}{W} \sum_{i=n-W}^{n} \mu(i) \geq \mu(n),$$

where the last inequality is due to $\mu(n)$ being non-decreasing. This proves the first inequality.

The heuristic selection strategy always queries the instance that yields the maximum of the available payoffs. Now, $R_*^n = \max(R_1^n, \ldots, R_{W-n}^n)$ is the random variable getting the maximum payoff at each round $n \in W+1, \ldots, W$. Such a random variable has a different distribution with respect to $R_1^n, \ldots, R_{W-n}^n$. With $F$ being the cumulative density function of the payoff random variables, for any value $z \in [0, 1]$,

$$\mathbb{P}(R_*^n \leq z) = \prod_{q=1}^{W-n} \mathbb{P}(R_q^n \leq z) \Rightarrow F_{R_*^n}(z) = (F(z))^{W-n},$$

which means that the cdf of the maximum payoff is not the same as the cdf of any payoff. Given that $F(z) \leq 1$, with $F$ non-constantly equal to 1, and that it is a non-decreasing function, there exists a minimum value $\hat{z} \in [0, 1]$ for which the value of the cdf $F$ equals 1. At the same time, for all $0 \leq z < \hat{z}$, $F(z) < 1$. Because the power of positive values lower than 1 returns smaller values, $F_{R_*^n}(z) = F(z)^{W-n} < F(z)$ for all $z < \hat{z}$, i.e. $1 - F_{R_*^n}(z) > 1 - F(z)$. Finally we can apply Cavalieri's principle to derive the expected value from the cdf,

$$\mathbb{E}[\bar{\mu}_{heur}] = \int_0^1 \left(1 - F_{R_*^n}\right) dz > \int_0^1 (1 - F) \, dx = \mathbb{E}[\bar{\mu}_{rand}]$$

which proves the second inequality.

The previous proposition states that taking the maximum rewards in a decreasing fashion results in a biased estimate of the average payoff that is strictly greater than the random selection estimate. Therefore, given $\varepsilon$ and $c$ such that $|\bar{\mu}_{rand}(t) - \mu(t)| < \varepsilon$ for all $t \geq c$, the estimate obtained by $\bar{\mu}_{heur}$ has not accurately estimated $\mu(t)$ yet.[3]

---

[3] Section 5 provides empirical evidence that the random selection strategy indeed leads to better results than the heuristic strategy. Note that the proof relies on the heuristic strategy being able to rank the instances correctly according to their informativeness. In reality, this ranking is approximate.

**Algorithm 1.** ALBA: Active Learning Bandits

---

1: **Input:** Multi-domain dataset $\mathcal{M}$, label budget $T$, oracle $\mathcal{O}$, number of clusters $C$
2: **Output:** Set of trained anomaly detectors $\mathcal{F}$
3: $\mathcal{A} = \varnothing, \mathcal{F} = \varnothing, t = 0$
4: **for** $D^k \in \mathcal{M}$ **do**
5:     $\mathcal{A} = \mathcal{A} \cup \text{CLUSTERDATA}(D^k, C)$
6:     $\mathcal{F} = \mathcal{F} \cup \text{TRAINDETECTOR}(D^k)$
7: **end for**
8: $\overrightarrow{r} = [0]^{j \in \mathcal{A}}$                       ▷ initialize the payoff vector
9: **while** $t < T$ **do**
10:     $j = \text{SWA}(\mathcal{A}, \overrightarrow{r}, t)$                       ▷ choose an action
11:     $x_i^j = \text{RAND}(G^j)$                 ▷ choose instance to be queried
12:     $y_i^j = \text{QUERY}(\mathcal{O}, x_i^j)$                 ▷ query the label to the oracle
13:     $\mathcal{F}^k = \text{TRAINDETECTOR}(D^k \cup y_i^j)$
14:     $\overrightarrow{r}^j = \text{ESTIMATEPAYOFF}(\mathcal{F}_{t-1}^k, \mathcal{F}_t^k)$
15:     $t = t + 1$
16: **end while**

---

### 3.5  ALBA Algorithm

Algorithm 1 details the full ALBA algorithm. On lines 4–7 the action set is instantiated by first clustering each dataset in $\mathcal{M}$ into $C$ clusters, and an initial anomaly detector is trained for each dataset. On line 9 the payoff vector that stores the obtained rewards, is initialized to zero. Lines 10–17 contain ALBA's active learning loop. ALBA proceeds in five steps: (i) it selects the group of instances from which to query a label according to the current MAB reward estimate (line 11), (ii) it selects an instance from that group to query (line 12), (iii) it queries the instance's label to the oracle (line 13), (iv) it retrains the model (line 14), and (v) it computes the actual labeling payoff using Eq. 1 or 2 to update the MAB reward estimate for the selected group in step (i) (line 15).

ALBA is computationally more time-efficient in the multi-domain setting than some AL strategies, such as uncertainty sampling. The cost of retraining the detector after an instance has been labeled is identical for both ALBA and every AL technique. However, while most AL techniques, such as uncertainty sampling, use heuristics to estimate the potential labeling payoff of each instance upfront (resulting in $\sum_{k=1}^{K} |D^k|$ computations each query round), ALBA only has to keep track of the labeling payoff of the different groups. This results in only $K \times C$ computations in total per query round.

**Table 1.** Classification of the *problem dimensions* tackled in AL related work. A check mark (✓) and dash (-) signify what was part of the original problem description.

| Reference | No. tasks | | No. datasets | | No. classes | | No. views | | Paradigm |
|---|---|---|---|---|---|---|---|---|---|
| | 1 | ≥ 2 | 1 | ≥ 2 | 2 | ≥ 3 | 1 | ≥ 2 | |
| [20,24] | ✓ | - | ✓ | - | ✓ | - | ✓ | - | Classic AL |
| [1,17,28] | - | ✓ | ✓ | - | ✓ | - | ✓ | - | Multi-task AL |
| [9,19,22,29] | ✓ | - | ✓ | - | - | ✓ | ✓ | - | Multi-class AL |
| [26] | ✓ | - | ✓ | - | ✓ | - | - | ✓ | Multi-view AL |
| [14,31] | ✓ | - | - | ✓ | ✓ | - | ✓ | - | MDAL |
| ALBA | ✓ | - | - | ✓ | ✓ | - | ✓ | - | MDAL |

# 4 Related Work

## 4.1 Active Learning

This work only considers sequential, pool-based active learning, where labels are queried one-by-one until the budget is spent. To see how our work fits within the vast body of research on AL, we roughly divide the spectrum of AL techniques along four axis: the *number of tasks* solved, the *number of datasets* considered, the *number of classes to predict* in each dataset, and whether multiple classifiers are learned on *different views* (i.e., feature subsets) of the data. Table 1 summarizes how the related works discussed below, fit these four axis. The classic AL techniques, such as uncertainty sampling [13], query-by-committee [21], expected-error reduction [18], density-based approaches [20], were originally designed for single-task, single-dataset scenario's with the features treated as a single set (view). In-depth surveys on these AL techniques are [20,24].

ALBA differs from these classic AL techniques in two important ways. First, ALBA is explicitly designed for multiple datasets, each of which requires the training of a separate classifier (or anomaly detector). Second, ALBA's use of an MAB strategy fundamentally changes how an acquired label informs subsequent query rounds, because it tracks the marginal gain of acquiring a label of groups of instances and not of single instances. An instance's "true labeling payoff" can *only* be measured by comparing classifier performances before and after retraining with said labeled instance. ALBA can measure this payoff exactly while the classic AL techniques have to resort to heuristics to estimate it upfront. To see why, consider the order of operations. In classic AL we (1) estimate the payoff of getting each instance's label, (2) get the label, and (3) retrain the model. In contrast, ALBA (1) selects a group of instances from a dataset according to the current reward estimate, (2) selects an unlabeled instance from the chosen group, (3) gets the instance's label, (4) retrains the model, and then (5) computes the actual payoff to update the reward estimate for the selected group in step (1).

Most related to our work, are [14,31]. [14] proposes a method for MDAL for classification. Their work properly conforms with the MDAL setting. However, their AL method is fully integrated with the underlying SVM classifier and geared towards text classification. This makes it difficult to use for our experiments without significant alterations to the proposed method. [31] designed a strategy for active learning for multi-domain recommendation. Recommendation is quite different from classification as the optimization target is distinct, rendering the proposed approach unsuitable for our task. The online Appendix 7.1 to this paper provides further details on how our work relates and differs from multi-task, multi-view, and multi-class active learning.

## 4.2 MAB Strategies and Active Learning

Since there is no single best AL algorithm [5], some researchers look at *learning active learning* [10,11]. The idea is to learn how to select the best AL strategy from a pool of strategies, potentially using an MAB approach [5,10], or learning which instances in a dataset are likely to improve the classifier [11]. ALBA differs from all these approaches as it is not concerned with finding the best AL strategy among $K$ strategies for 1 dataset, but rather with identifying the labeling payoff of $K$ different datasets using 1 strategy.

The work of [8] proposes the use of MAB strategies to sequentially select instances presented to the oracle. The main difference with ALBA is that they conceptually view each learned hypothesis as an action, while ALBA equates each action with a group of instances. Moreover, ALBA works for multiple datasets that require potentially different classifiers (hypotheses). Finally, Fang et al. also use an MAB approach to decide which instance to query [6]. There are two key differences with our work. First, ALBA equates *actions* with groups of instances and not with different learned tasks. Second, ALBA explicitly accounts for the diminishing payoffs of labeling additional instances.

## 5   Experiments

We evaluate multi-domain active learning in the context of *anomaly detection* where we have access to real-world multi-domain data consisting of nearly four years of water consumption data from 7 different retail stores. Anomaly detection naturally fits this paper's problem setting for two reasons. First, many real-world anomaly detection problems consist of multiple distinct datasets where an anomaly detector has to be learned for each dataset. Second, while anomaly detection problems typically were posed as unsupervised problems due to difficulties of obtaining labeled data, there is growing evidence that in practice achieving good performance requires labeling some data data [25].

We try to answer following questions empirically:

**Q1.** Does ALBA outperform the classic AL baselines when dealing with multi-domain datasets where $K > 1$?

**Q2.** How does the choice of the action set $\mathcal{A}$ impact ALBA's performance?

**Q3.** How do the choice of the MAB reward function and the query selection strategy impact ALBA's performance?

### 5.1    Experimental Setup[4]

*Compared Methods.* For all approaches, we use the same semi-supervised anomaly detector: SSDO with the Isolation Forest algorithm as prior to generate the unsupervised anomaly scores [25]. One could use other detectors in theory. The unsupervised prior of SSDO allows us to exploit information in the unlabeled data. For active learning, we compare to uncertainty and random sampling as they (1) can be used with any underlying detector, and (2) have been shown to consistently perform well against more complicated AL strategies [9,24].

We compare ALBA to seven baselines, divided into two categories. Category 1 baselines **combine all the datasets into one big dataset and learn a single anomaly detector** using the random (C-RAND) or uncertainty sampling (C-UC) active learning strategy. Though necessary baselines [9], learning a single detector for the combined datasets is likely suboptimal as each dataset has distinct marginal and conditional distributions. This would result in difficult-to-learn regions in the combined instance space for the detector.

Category 2 baselines **treat each domain independently and learn a separate model for each one.** I-U learns a completely unsupervised anomaly detector for each dataset. I-RAND or I-UC use the random and uncertainty sampling strategies in the following way in each query round. First, they apply their AL strategy to each dataset to select the most informative instance within each dataset. Given this set of identified instances, they again apply their respective AL strategy to select the single most informative instance to be labeled. I-RAND and I-UC do not attempt to ensure any balance in terms of how many instances are queried from a given a dataset. Finally, I-R-RAND or I-R-UC impose an additional *restriction* on I-RAND or I-UC. Given a fixed total query budget $T$ and $K$ datasets, at most $\lfloor T/K \rfloor$ instances can be sampled from a given dataset.

*Evaluation Metrics.* The performance of the *anomaly detector* on a single dataset is evaluated using the *area under the receiver operating characteristic* (AUROC) as is standard in anomaly detection [3]. The performance on a multi-domain dataset is obtained by averaging the AUROC scores on the $K$ individual datasets ($\text{AUROC}_K$). The performance of an *active learning* strategy is revealed by the progress curve which captures how the $\text{AUROC}_K$ evolves as a function of the number of labeled instances (i.e., the spent label budget) [24]. As the number of experiments increases, these curves are summarized using the *area under the active learning curve* (AULC) [24]. AULC scores are $\in [0, 1]$ and higher scores are better.

---

[4] Online Appendix 7.2 has detailed information on the (choice of) evaluation metrics, benchmark data, and hyperparameters. It also has additional results on the impact of the dataset characteristics on ALBA's performance.

***Benchmark Data.*** We use 7 real-world datasets that each track the water consumption of a different retail store, measured every 5 min, over the course of 4 years. Each of the 7 datasets is further divided into 24 datasets by grouping the data per hour-of-the-day, yielding 7 (stores) × 24 (hours) = 168 fully labeled water datasets. This division is necessary as the hour-of-the-day strongly influences the observed water consumption. The binary labels are "normal" or "anomalous" usage (e.g., water leaks). Then, we transform each hour-long segment into a feature-vector[5], and train a separate anomaly detector per dataset.

We now construct an appropriate multi-domain AL benchmark as follows. First, we compute for each of the 168 datasets a *labeling payoff score* which is simply the difference in AUROC obtained by (1) SSDO trained with 20% of the data labeled and (2) SSDO trained without labels. Then, we construct a *multi-domain dataset* by selecting $K$ datasets from the set of 168 water datasets. A fraction $\psi$ of these $K$ datasets are selected to have a high labeling payoff score, while the remaining datasets $(1 - \psi K)$ have a low labeling payoff scores. By varying $K$ in $[2, 10]$ and $\psi$ in $[0.1, 1]$, we obtain the full benchmark of 54 unique multi-domain datasets.

***Setup.*** Given a multi-domain dataset $\mathcal{M}$ from the benchmark, the anomaly detector, and an AL method, each experiment proceeds in four steps. First, each of the $K$ datasets in $\mathcal{M}$ is randomly divided into 2/3 train and 1/3 test set. Second, we simulate an oracle iteratively labeling one training instance at a time selected by the AL method across the $K$ datasets, until the label budget $T = 500$ is spent. Third, each iteration, the appropriate anomaly detector is retrained and we recompute the $\text{AUROC}_K$ on the test data. Each experiment is repeated 5 times to average out any random effects, resulting in a total of 7 (methods) × 5 × 54 = 1890 experiments where each experiment has $T + K$ training and $T \times K$ evaluation runs. The baselines have no hyperparameters, ALBA has three. These are $C = 5$ using KMEANS, the SWA bandit algorithm with the *cosine* reward function, and a RAND query selection strategy.

## 5.2 Experimental Results

***Q1:* ALBA *versus the baselines*** Fig. 1 plots the progress curves for 5 multi-domain datasets randomly selected from the benchmark of 54 datasets. The plots[6] reveals four insights. One, all approaches (except C-UC) outperform the unsupervised baseline after about 100 query rounds. Two, learning a separate anomaly detector per dataset clearly outperforms combining all datasets and learning a single detector, as evidenced by the lagging performances of C-RAND and C-UC versus the other baselines. Three, a completely random AL strategy surprisingly outperforms the heuristic strategy in the multi-domain setting, as

---

[5] We use 8 statistical (average, standard deviation, max, min, median, sum, entropy, skewness, and Kurtosis) and 2 binary features (whether its a Friday or a Sunday), 10 in total.

[6] See online Appendix 7.3 for the plots for all 54 benchmark datasets.

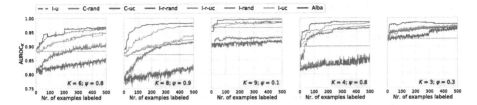

**Fig. 1.** The figure shows the progress curves of ALBA and the baselines for 5 multi-domain datasets randomly selected from the full benchmark. Each progress curve shows how the $AUROC_K$ evolves as a function of the total number of instances labeled by the oracle. The characteristics of each selected multi-domain dataset ($K$ and $\psi$) are shown on the corresponding plot.

evidenced by C-RAND, I-RAND, and I-R-RAND outperforming respectively C-UC, C-RAND, and C-R-RAND on all but one benchmark dataset. Four, I-RAND and I-R-RAND perform similarly.

After enough query rounds, all approaches (except the unsupervised baseline) will converge to the performance of a fully supervised classifier. Better AL strategies, however, converge faster. To investigate each method's convergence, we look at their performance after both 100 query rounds and 500 query rounds. Table 2 shows for the 54 benchmark multi-domain datasets how many times ALBA wins/draws/loses in terms of AULC versus each baseline[7] as well as the average AULC rank for each method [4]. Table 2a shows the results after 100 queries, while Table 2b shows the results after 500 queries. After 100 query rounds, the Friedman test rejects the null-hypothesis that all methods perform similarly (p-value < 1e-8). The post-hoc Bonferroni-Dunn test [4] with $\alpha = 0.05$ finds that ALBA is significantly better than every baseline. After 500 query rounds, ALBA is still significantly better than most baselines, except I-RAND and I-R-RAND. This illustrates that the methods start converging. However, ALBA converges faster (this is also illustrated by the progress curves), which is especially useful in scenario's where labeling is costly, such as anomaly detection.

*Q2: Impact of the Choice of Action Set.* We explore the effect of the granularity of the action set considered by ALBA on its performance. We do so by varying the number of clusters $C$ per dataset. When $C = 1$, the action set is coarse-grained as each action corresponds to a full dataset. As $C$ increases, the action set becomes more fine-grained as each dataset is further partitioned into groups using KMEANS. Figure 2 points to a correlation between $C$ and ALBA's performance. As $C$ increases, the MAB method can make a more fine-grained estimate of the usefulness of different groups of instances.[8]

---

[7] All the experimental evaluations maintain a precision of 1e–4 and a threshold of 0.001 (e.g., to determine the similarity of two AULC scores).

[8] See online Appendix 7.3 for a more detailed discussion.

**Table 2.** The table shows the number of AULC wins/draws/losses of ALBA versus each baseline, and the average AULC rank (± Standard Deviation) of each method on the full benchmark, both after 100 and 500 query rounds.

| Method | Nr. of times ALBA: wins | draws | loses | Ranking Avg. ± SD |
|---|---|---|---|---|
| ALBA | - | - | - | **1.315 ± 0.894** |
| I-RAND | 48 | 2 | 4 | 2.639 ± 0.573 |
| I-R-RAND | 48 | 2 | 4 | 2.639 ± 0.573 |
| I-U | 48 | 2 | 4 | 4.333 ± 1.656 |
| I-UC | 53 | 0 | 1 | 5.157 ± 0.551 |
| I-R-UC | 53 | 0 | 1 | 5.231 ± 0.497 |
| C-RAND | 54 | 0 | 0 | 6.741 ± 0.865 |
| C-UC | 54 | 0 | 0 | 7.944 ± 0.404 |

(a) Results @ 100 query rounds

| Method | Nr. of times ALBA: wins | draws | loses | Ranking Avg. ± SD |
|---|---|---|---|---|
| ALBA | - | - | - | **1.306 ± 0.710** |
| I-RAND | 45 | 2 | 7 | 2.417 ± 0.507 |
| I-R-RAND | 46 | 1 | 7 | 2.417 ± 0.507 |
| I-R-UC | 54 | 0 | 0 | 4.537 ± 0.686 |
| I-UC | 53 | 1 | 0 | 4.546 ± 0.512 |
| I-U | 54 | 0 | 0 | 6.370 ± 0.818 |
| C-RAND | 54 | 0 | 0 | 6.491 ± 0.717 |
| C-UC | 54 | 0 | 0 | 7.917 ± 0.382 |

(b) Results @ 500 query rounds

***Q3: Impact of the MAB Algorithm, Reward Function, and Query Selection Strategy.*** We explore the effect of the choice of MAB algorithm, reward function, and query selection strategy on ALBA's performance. Table 3 shows the resulting average AULC ranks ($C = 5$). The best-performing version of ALBA uses a *cosine* reward function and *random* instance selection strategy. Generally, the RAND versions of ALBA outperform their UC counterparts $\sim 63\%$ of time on the full benchmark, and they draw $\sim 18\%$ of the time. Repeating the analysis with $C = 1$, the RAND versions outperform their UC counterparts $\sim 88\%$ of time and they draw $\sim 5.5\%$ of the time. This aligns with the theoretical results. When fixing the query strategy, our proposed cosine reward function outperforms the entropy reward function. See online Appendix 7.3 for linear regression analyses on these results, as well as further analyses on the impact of the dataset characteristics, $K$ and $\psi$, on the performance of ALBA.

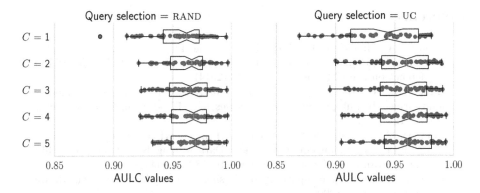

**Fig. 2.** Box plot overlaying a scatter plot of the AULCs obtained by ALBA with different values for $C$ on the 54 benchmark datasets. Results are shown for two versions of ALBA (with different instance selection strategies).

**Table 3.** Average AULC rank ($\pm$ Standard Deviation) of versions of ALBA with different settings for the query selection strategy and reward function ($C = 5$).

| Reward function | Query sel. strategy | Ranking Avg. $\pm$ SD |
|---|---|---|
| Cosine | Rand | **1.806 $\pm$ 0.813** |
| Cosine | Uc | 2.944 $\pm$ 0.926 |
| Entropy | Rand | 2.241 $\pm$ 0.843 |
| Entropy | Uc | 3.009 $\pm$ 0.825 |

## 6 Conclusion

This paper tackled the multi-domain active learning problem for anomaly detection, which often arises in practice. The key challenge was to determine from which dataset an instance should be queried as labels are not equally beneficial in all domains. To cope with this problem, we proposed a method (ALBA) that exploits multi-armed bandit strategies to track the label-informativeness of groups of instances over time and decides which instances are optimal to query to an oracle. Empirically, ALBA outperformed existing active learning strategies on a benchmark of 7 real-world water consumption datasets.

**Acknowledgements.** This work is supported by the Flemish government "Onderzoeksprogramma Artificiële Intelligentie Vlaanderen", "Agentschap Innoveren & Ondernemen (VLAIO)" as part of the innovation mandate HBC.2020.2297, the FWO-Vlaanderen aspirant grant 1166222N, and Leuven.AI, B-3000 Leuven, Belgium.

# References

1. Acharya, A., Mooney, R.J., Ghosh, J.: Active multi-task learning using both latent and supervised shared topics. In: Proceedings of the 2014 SIAM International Conference on Data Mining, pp. 190–198 (2014)
2. Bubeck, S., Cesa-Bianchi, N., et al.: Regret analysis of stochastic and nonstochastic multi-armed bandit problems. Foundations and Trends® in Machine Learning 5(1), 1–122 (2012)
3. Campos, G.O., et al.: On the evaluation of unsupervised outlier detection: measures, datasets, and an empirical study. Data Mining Knowl. Discovery **30**(4), 891–927 (2016). https://doi.org/10.1007/s10618-015-0444-8
4. Demšar, J.: Statistical comparisons of classifiers over multiple data sets. J. Mach. Learn. Res. **7**, 1–30 (2006)
5. Desreumaux, L., Lemaire, V.: Learning active learning at the crossroads? evaluation and discussion. arXiv preprint arXiv:2012.09631 (2020)
6. Fang, M., Tao, D.: Active multi-task learning via bandits. In: Proceedings of the 2015 SIAM International Conference on Data Mining, pp. 505–513 (2015)
7. Ganin, Y., et al.: Domain-adversarial training of neural networks. J. Mach. Learn. Res. **17**(1), 1–35 (2016)
8. Ganti, R., Gray, A.: Building bridges: viewing active learning from the multi-armed bandit lens. arXiv preprint arXiv:1309.6830 (2013)
9. He, R., He, S., Tang, K.: Multi-domain active learning: a comparative study. arXiv preprint arXiv:2106.13516 (2021)
10. Hsu, W.N., Lin, H.T.: Active learning by learning. In: Proceedings of the 29th AAAI Conference on Artificial Intelligence (2015)
11. Konyushkova, K., Sznitman, R., Fua, P.: Learning active learning from data. arXiv preprint arXiv:1703.03365 (2017)
12. Levine, N., Crammer, K., Mannor, S.: Rotting bandits. In: Advances in Neural Information Processing Systems, vol. 30 (2017)
13. Lewis, D.D., Catlett, J.: Heterogeneous uncertainty sampling for supervised learning. In: Machine Learning Proceedings, pp. 148–156. Morgan Kaufmann (1994)
14. Li, L., Jin, X., Pan, S.J., Sun, J.T.: Multi-domain active learning for text classification. In: Proceedings of the 18th ACM SIGKDD International Conference on Knowledge Discovery and Data Mining, pp. 1086–1094 (2012)
15. Perini, L., Vercruyssen, V., Davis, J.: Class prior estimation in active positive and unlabeled learning. In: Proceedings of the 29th International Joint Conference on Artificial Intelligence and the 17th Pacific Rim International Conference on Artificial Intelligence (IJCAI-PRICAI 2020), pp. 2915–2921. IJCAI-PRICAI (2020)
16. Perini, L., Vercruyssen, V., Davis, J.: Transferring the contamination factor between anomaly detection domains by shape similarity. In: Proceedings of the Thirty-Sixth AAAI Conference on Artificial Intelligence (2021)
17. Reichart, R., Tomanek, K., Hahn, U., Rappoport, A.: Multi-task active learning for linguistic annotations. In: Proceedings of ACL-08: HLT, pp. 861–869 (2008)
18. Roy, N., McCallum, A.: Toward optimal active learning through monte-carlo estimation of error reduction. In: Proceedings of the 18th International Conference on Machine Learning, pp. 441–448 (2001)
19. Sener, O., Savarese, S.: Active learning for convolutional neural networks: a core-set approach. arXiv preprint arXiv:1708.00489 (2017)
20. Settles, B.: Active learning. Synth. Lect. Artif. Intell. Mach. Learn. **6**(1), 1–114 (2012)

21. Seung, H., Opper, M., Sompolinsky, H.: Query by committee. In: Proceedings of the 5th Annual Workshop on Computational Learning Theory, pp. 287–294 (1992)
22. Sinha, S., Ebrahimi, S., Darrell, T.: Variational adversarial active learning. In: Proceedings of the IEEE International Conference on Computer Vision, pp. 5972–5981 (2019)
23. Thompson, W.R.: On the likelihood that one unknown probability exceeds another in view of the evidence of two samples. Biometrika $25(3/4)$, 285–294 (1933)
24. Trittenbach, H., Englhardt, A., Böhm, K.: An overview and a benchmark of active learning for outlier detection with one-class classifiers. Expert Syst. Appl. **168**, 114372 (2021)
25. Vercruyssen, V., Meert, W., Verbruggen, G., Maes, K., Bäumer, R., Davis, J.: Semi-supervised anomaly detection with an application to water analytics. In: Proceedings of the IEEE International Conference on Data Mining, pp. 527–536 (2018)
26. Wang, W., Zhou, Z.H.: On multi-view active learning and the combination with semi-supervised learning. In: Proceedings of the 25th International Conference on Machine learning, pp. 1152–1159 (2008)
27. Wei, K., Yang, Y., Zuo, H., Zhong, D.: A review on ice detection technology and ice elimination technology for wind turbine. Wind Energy **23**(3), 433–457 (2020)
28. Xiao, Y., Chang, Z., Liu, B.: An efficient active learning method for multi-task learning. Knowledge-Based Syst. **190**, 105137 (2020)
29. Yoo, D., Kweon, I.S.: Learning loss for active learning. In: Proceedings of the IEEE Conference on Computer Vision and Pattern Recognition, pp. 93–102 (2019)
30. Zhang, Y., Yang, Q.: A survey on multi-task learning. arXiv preprint arXiv:1707.08114 (2017)
31. Zhang, Z., Jin, X., Li, L., Ding, G., Yang, Q.: Multi-domain active learning for recommendation. In: 30th AAAI Conference on Artificial Intelligence (2016)

# CMG: A Class-Mixed Generation Approach to Out-of-Distribution Detection

Mengyu Wang[1], Yijia Shao[1], Haowei Lin[1], Wenpeng Hu[1], and Bing Liu[2(✉)]

[1] Wangxuan Institute of Computer Technology, Peking University, Beijing, China
{wangmengyu,shaoyj,linhaowei,wenpeng.hu}@pku.edu.cn
[2] University of Illinois at Chicago, Chicago, USA
liub@uic.edu

**Abstract.** Recently, contrastive learning with data and class augmentations has been shown to produce markedly better results for out-of-distribution (OOD) detection than previous approaches. However, a major shortcoming of this approach is that it is extremely slow due to the significant increase in data size and in the number of classes and the quadratic pairwise similarity computation. This paper shows that this heavy machinery is unnecessary. A novel approach, called CMG (*Class-Mixed Generation*), is proposed, which generates pseudo-OOD data by mixing class embeddings as abnormal conditions to CVAE (conditional variational Auto-Encoder) and then uses the data to fine-tune a classifier built using the given in-distribution (IND) data. To our surprise, the obvious approach of using the IND data and the pseudo-OOD data to directly train an OOD model is a very poor choice. The fine-tuning based approach turns out to be markedly better. Empirical evaluation shows that CMG not only produces new state-of-the-art results but also is much more efficient than contrastive learning, at least 10 times faster (Code is available at: https://github.com/shaoyijia/CMG).

**Keywords:** Out-of-distribution detection · Data generation

## 1 Introduction

*Out-of-distribution* (OOD) detection aims to detect novel data that are very different from the training distribution or *in-distribution* (IND). It has a wide range of applications, e.g., autonomous driving [40] and medical diagnosis [5]. So far, many approaches have been proposed to solve this problem, from distance-based methods [2,3,12,20], to generative models [36,38,41,60] and self-supervised learning methods [4,11,17,19]. Recently, *contrastive learning* with *data augmentation* has produced state-of-the-art (SOTA) OOD detection results [45,53].

---

M. Wang and Y. Shao—Equal contribution.

---

**Supplementary Information** The online version contains supplementary material available at https://doi.org/10.1007/978-3-031-26412-2_31.

However, data augmentation-based contrastive learning has a major drawback. It is extremely inefficient and resource-hungry due to a large amount of augmented data and quadratic pairwise similarity computation during training. For example, CSI [53] creates 8 augmented instances for each original image. Furthermore, every 2 samples in the augmented batch is treated as a pair to calculate contrastive loss. The performance is also poor if the batch size is small, but a large batch size needs a huge amount of memory and a very long time to train. It is thus unsuitable for edge devices that do not have the required resources. In Sect. 4.4, we will see that even for a moderately large dataset, CSI has difficulty to run.

In this paper, we propose a novel and yet simple approach, called CMG (*Class-Mixed Generation*), that is both highly effective and efficient, to solve the problem. CMG consists of two stages. The first stage trains a pseudo-OOD data generator. The second stage uses the generated pseudo-OOD data and the IND training data to fine-tune (using an *energy function*) a classifier already trained with the IND data. We discuss the first stage first.

OOD detection is basically a classification problem but there is no OOD data to use in training. This paper proposes to generate pseudo-OOD data by working in the latent space of a Conditional Variational Auto-Encoder (CVAE). The key novelty is that the pseudo-OOD data generation is done by manipulating the CVAE's conditional information using *class-mixed embeddings*. CVAE generates instances from the training distribution on the basis of latent representations consisting of the conditional information and variables sampled from a prior distribution of CVAE, normally the Gaussian distribution. If the latent space features or representations are created with some *abnormal conditions*, the CVAE will generate "*bad*" instances but such instances can serve as pseudo-OOD samples. Our abnormal conditions are produced by mixing embeddings of class labels in the IND data, which ensures the generated pseudo-OOD data to be similar but also different from any existing IND data.

With the pseudo-OOD data generated, the conventional approach is to use the IND data and the pseudo-OOD data to build a classifier for OOD detection. However, to our surprise, *this approach is a poor choice*. We will discuss the reason in Sect. 3.3 and confirm it with experimental results in Sect. 4.5. We discovered that if we build a classifier first using the IND data and then fine-tune only the final classification layer using both the IND and pseudo-OOD data, the results improve dramatically. This is another novelty of this work. The paper further proposes to use an *energy function* to fine-tune the final classification layer, which produces even better results.

Our contributions can be summarized as follows:

(1) We propose a novel method using CVAE to generate pseudo-OOD samples by providing *abnormal conditions*, which are mixed embeddings of different class labels. To our knowledge, this has not been done before.[1]

---

[1] By no means do we claim that this CVAE method is the best. Clearly, other generators may be combined with the proposed class-mixed embedding approach too. It is also known that CVAE does not generate high resolution images, but our experiments show that low resolution images already work well.

(2) We discovered that the obvious and conventional approach of using the IND and pseudo-OOD data to train a classifier in one stage performed very poorly. Our two-stage framework with fine-tuning performs dramatically better. Again, to our knowledge, this has not been reported before.

(3) Equally importantly, since the classifier in CMG is not specified, CMG can be applied to existing OOD detection models to improve them too.

Extensive experiments show that the proposed CMG approach produces new SOTA results and is also much more efficient than contrastive learning for OOD detection, requiring only one-tenth of the execution time.

## 2  Related Work

Early ideas for solving the OOD detection problem focused on modifying softmax scores to obtain calibrated confidences for OOD detection [3,13]. Many other score functions have also been proposed, e.g., likelihood ratio [46], input complexity [50] and typicality [37]. A recent work utilizes Gram matrices to characterize activity patterns and identify OOD samples [48].

Methods that use anomalous data to improve detection [16,35] are more closely related to our work. Generative models have been used to anticipate novel data distributions. In some of these methods, generated data are treated as OOD samples to optimize the decision boundary and calibrate the confidence [54,60]. In some other methods, generative models such as auto-encoders [43,59] and generative adversarial networks (GAN) are used to reconstruct the training data [8,42]. During GAN training, low quality samples acquired by the generator are used as OOD data [44]. Their reconstruction loss can also help detect OOD samples. There are also works using given OOD data to train a model [33]. It has been shown recently that using pre-trained representations and few-shot outliers can improve the results [9]. Self-supervised techniques have been applied to OOD detection too. They focus on acquiring rich representations through training with some pre-defined tasks [10,25]. Self-supervised models show outstanding performance [4,25]. CSI [53] is a representative method (see more below), which uses contrastive learning and data augmentation to produce SOTA results. However, it is extremely slow and memory demanding. Our CMG method for generating pseudo-OOD data is much more efficient. Some researchers also tried to improve contrastive learning based methods [49] and proposed distance-based methods [34]. However, our experiments show that CSI outperforms them. Our CMG method is a generative approach. But unlike existing methods that use perturbations to anticipate OOD data, CMG uses synthetic conditions and CVAE to obtain effective and diverse pseudo-OOD data.

Auto-Encoder (AE) is a family of unsupervised neural networks [1,47]. A basic AE consists of an encoder and a decoder. The encoder encodes the input data into a low-dimensional hidden representation and the decoder transforms the representation back to the reconstructed input data [7,18,55]. Variational auto-encoder is a special kind of AE [23]. It encodes the input as a given probability distribution (usually Gaussian) and the decoder reconstructs data instances according to variables sampled from that distribution. CVAE is an extension of

VAE [24]. It encodes the label or conditional information into the latent representation so that a CVAE can generate new samples from specified class labels. CVAE makes it easy to control the generating process, i.e., to generate samples with features of specified classes. We make use of this property of CVAE to generate high quality pseudo-OOD data.

## 3    Proposed CMG Method

OOD detection is commonly formulated as a classification problem without OOD data/class available in training. To effectively train an OOD detection model, an intuitive idea is to generate pseudo-OOD data and use them together with the IND data to jointly build a OOD detection model. We take this approach. We propose a method using Conditional Variational Auto-Encoder (CVAE) to generate pseudo-OOD data and a new fine-tuning framework based on an energy function to leverage the generated pseudo-OOD data to produce a highly effective and efficient OOD detection model.

### 3.1    Conditional Variational Auto-encoder

Conditional Variational Auto-Encoder (CVAE) is derived from Variational auto-encoder (VAE). We first introduce VAE which is a conditional directed graphical model consisting of three main parts, an encoder $q_\phi(\cdot)$ with parameters $\phi$, a decoder $p_\theta(\cdot)$ with parameters $\theta$ and a loss function $\mathcal{L}(\mathbf{x}; \theta, \phi)$, where $\mathbf{x}$ represents an input sample. The loss function is as follows:

$$\mathcal{L}(\mathbf{x}; \theta, \phi) = -\mathbb{E}_{\mathbf{z} \sim q_\phi(\mathbf{z}|\mathbf{x})}[\log p_\theta(\mathbf{x}|\mathbf{z})] + KL(q_\phi(\mathbf{z}|\mathbf{x})||p_\theta(\mathbf{z})) \qquad (1)$$

where $q_\phi(\mathbf{z}|\mathbf{x})$ is a proposal distribution to approximate the prior distribution $p_\theta(\mathbf{z})$, $p_\theta(\mathbf{x}|\mathbf{z})$ is the likelihood of the input $\mathbf{x}$ with a given latent representation $\mathbf{z}$, and $KL(\cdot)$ is the function to calculate Kullback-Leibler divergence. $q_\phi(\mathbf{z}|\mathbf{x})$ is the encoder and $p_\theta(\mathbf{x}|\mathbf{z})$ is the decoder. In Eq. (1), the expected negative log-likelihood term encourages the decoder to learn to reconstruct the data with samples from the latent distribution. The KL-divergence term forces the latent distribution to conform to a specific prior distribution such as the Gaussian distribution, which we use. After training, a VAE can generate data using the decoder $p_\theta(\mathbf{x}|\mathbf{z})$ with a set of latent variables $\mathbf{z}$ sampled from the prior distribution $p_\theta(\mathbf{z})$. Commonly, the prior distribution is the centered isotropic multivariate Gaussian $p_\theta(\mathbf{z}) = \mathcal{N}(\mathbf{z}; \mathbf{0}, \mathbf{I})$.

However, VAE does not consider the class label information which is available in classification datasets and thus has difficulty generating data of a particular class. Conditional variational Auto-Encoder (CVAE) was introduced to extend VAE to address this problem. It improves the generative process by adding a conditional input information into latent variables so that a CVAE can generate samples with some specific characteristics or from certain classes. We use $c$ to denote the prior class information. The loss function for CVAE is as follows:

$$\mathcal{L}(\mathbf{x}; \theta, \phi) = -\mathbb{E}_{\mathbf{z} \sim q_\phi(\mathbf{z}|\mathbf{x})}[\log p_\theta(\mathbf{x}|\mathbf{z}, c)] + KL(q_\phi(\mathbf{z}|\mathbf{x}, c)||p_\theta(\mathbf{z}|c)) \qquad (2)$$

One CVAE implementation uses a one-hot vector to represent a class label $y_c$, and a weight matrix is multiplied to it to turn the one-hot vector to a class embedding $\mathbf{y}_c$. Then a variable $\mathbf{z}$, generated from the prior distribution $p_\theta(\mathbf{z})$, is concatenated with $\mathbf{y}_c$ to construct the whole latent variable. Finally, the generated instance $p_\theta(\mathbf{x}|\mathbf{z}, c)$ of class $c$ is produced. We can formulate the process as:

$$p_\theta(\mathbf{x}|\mathbf{z}, c) = p_\theta(\mathbf{x}|[\mathbf{y}_c, \mathbf{z}]) \tag{3}$$

## 3.2   Generating Pseudo-OOD Data

CVAE's ability to control the generating process using the conditional information (e.g. class label in our case) inspired us to design a method to generate pseudo-OOD samples. This is done by the conditional decoder using atypical prior information $c$ in $p_\theta(\mathbf{x}|\mathbf{z}, c)$. As introduced before, OOD data need to be different from in-distribution (IND) data but also resemble them. The continuity property of CVAE, which means that two close points in the latent space should not give two completely different contents once decoded [6], ensures that we can manipulate CVAE's latent space features to generate high quality pseudo-OOD data. Since we have no information of the future OOD data, we have to make use of the existing training data (i.e., IND data) to construct pseudo-OOD data. We can provide it with pseudo label information to generate pseudo-OOD data.

Specifically, we propose to construct pseudo class embedding by combining the embeddings of two existing classes in the IND training data. The formulation is as follows:

$$p_\theta(\mathbf{x}|\mathbf{z}, \mathbf{k}, c_i, c_j) = p_\theta(\mathbf{x}|[\mathbf{k} * \mathbf{y}_{c_i} + (1 - \mathbf{k}) * \mathbf{y}_{c_j}, \mathbf{z}]) \tag{4}$$

where $\mathbf{k}$ is a vector generated from Bernoulli distribution $\mathcal{B}(0.5)$ with the same length as the class label embedding. $\mathbf{k}$ is basically for the system to randomly select the vector components of the two class embeddings with equal probability. Such a generated sample $p_\theta(\mathbf{x}|\mathbf{z}, \mathbf{k}, c_i, c_j)$ will not likely to be an instance of either class $c_i$ or $c_j$ but still keep some of their characteristics, which meets the need of the pseudo-OOD data. Furthermore, the pseudo class embedding has a great variety, owing to the diverse choices of classes and the vector $\mathbf{k}$. To generate pseudo-OOD samples, we also need to sample $\mathbf{z}$ from the encoder. In CVAE training, we ensure that $\mathbf{z}$ fits the Gaussian $\mathcal{N}(\mathbf{0}, \mathbf{I})$. To sample $\mathbf{z}$ for generating, we use another flatter Gaussian distribution $\mathcal{N}(\mathbf{0}, \sigma^2 * \mathbf{I})$, where $\sigma > 1 \in \mathbb{Z}$, to make the generated samples highly diverse.

## 3.3   Fine-Tuning for OOD Detection

As discussed earlier, using the generated pseudo-OOD data and the original in-distribution (IND) training data to directly train a classifier for OOD detection is not a good approach. Here, we propose a fine-tuning method that uses the generated pseudo-OOD samples and the IND training data to learn an OOD detection model in two stages.

**Stage 1 (IND classifier building and CVAE training):** Only the original IND data is used to train a classification model $\mathcal{C}$. The classification model can be decomposed into two functions $f(\cdot)$ and $h(\cdot)$, where $f(\cdot)$ is the final linear *classifier* and $h(\cdot)$ is the *feature extractor*. $f(h(\mathbf{x}))$ is the classification output. A separate CVAE model is also trained for generating pseudo-OOD data.

**Stage 2 (fine-tuning the classifier):** We keep the trained feature extractor $h(\cdot)$ fixed (or frozen) and fine-tune only the classification/linear layer $f(\cdot)$ using both the IND and the pseudo-OOD data for OOD detection.

The CMG approach is in fact a framework, which is illustrated in Fig. 1 with the 2-stage training process. The framework is flexible as the classifier in the first stage can use any model. Our pseudo-OOD data can help various classifiers improve the ability of OOD detection. Stage 2 is also flexible and can use different approaches. Here we introduce two specific approaches, which are both highly efficient. The second approach CMG-Energy produces new SOTA OOD detection results. As we will see in Sect. 4.4, fine-tuning an existing OOD detection model also enables it to improve.

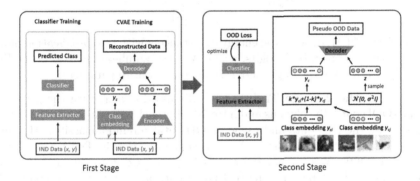

First Stage                    Second Stage

**Fig. 1.** CMG framework and its training process. The OOD loss can be cross entropy in CMG-softmax, cross-entropy+energy in CMG-energy, or other possible losses. Although we put Classifier Training and CVAE Training in First Stage, they are independent.

**CMG-Softmax Fine-Tuning.** In this approach to fine-tuning, we simply add an additional class (which we call the OOD class) in the classification layer to accept the pseudo-OOD data. If the IND data has $N$ classes, we add parameters to the classifier to make it output $N + 1$ logits. These added parameters related to the $(N + 1)$th OOD class are randomly initialized. We then train the model by only fine-tuning the classification layer using the cross entropy loss with feature extractor trained in Stage 1 fixed. Finally, we use the softmax score of the $(N + 1)$th class as the OOD score.

**CMG-Energy Fine-Tuning.** This approach adds an energy loss to the cross entropy loss ($\mathcal{L}_{ent} + \lambda\mathcal{L}_{energy}$) to fine-tune the classification layer using the IND and the pseudo-OOD data. No OOD class is added. The energy loss is,

$$\mathcal{L}_{energy} = \mathbb{E}_{\mathbf{x}_{ind} \sim \mathcal{D}_{ind}}(\max(0, E(\mathbf{x}_{ind}) - m_{ind}))^2$$
$$+ \mathbb{E}_{\mathbf{x}_{ood} \sim \mathcal{D}_{ood}}(\max(0, m_{ood} - E(\mathbf{x}_{ood})))^2 \tag{5}$$

where $\mathcal{D}_{ind}$ denotes the IND training data, $\mathcal{D}_{ood}$ denotes generated pseudo-OOD data, and $m_{ind}$ and $m_{ood}$ are margin hyper-parameters. The idea of this loss is to make the OOD data get similar values for all $N$ logits so that they will not be favored by any $N$ IND data classes. Here $N$ is the number of classes of the IND data. As the loss function shows, the OOD data are necessary. This loss was used in [33], which has to employ some *real* OOD data but such OOD data are often not available in practice. This loss cannot be used by other OOD methods since they have no OOD data available [3,21,53]. However, this is not an issue for us as we have pseudo data to replace real OOD training data.

Stage 2 produces an energy score calculated from a classification model for OOD detection for a test instance $\mathbf{x}$:

$$E(\mathbf{x}; f(h)) = -T \cdot \log \sum_{i=1}^{N} e^{f_i(h(\mathbf{x}))/T} \tag{6}$$

where $E(\mathbf{x}; f(h(\cdot)))$ denotes the energy of instance $\mathbf{x}$ with the classification model $f(h(\cdot))$, which maps $\mathbf{x}$ to $N$ logits, where $N$ is the number of classes in the IND data, $f_i(h(\mathbf{x}))$ is the $i$-th logit and $T$ is the temperature parameter.

**Reason for the 2-Stage Training.** As discussed earlier, with the generated pseudo-OOD data, the obvious and intuitive approach to training an OOD detector is to use the IND data and pseudo-OOD data to build a classifier. However, as the results in Sect. 4.5 show, this is a very poor choice. The reason is that the pseudo-OOD data are not the real OOD data used in testing and their difference can be large because the real OOD data are completely unpredictable and can be anywhere in the space. This combined one-stage training fits only the pseudo-OOD data and may still perform poorly for the real OOD data. The proposed fine-tuning is different. Its Stage 1 training uses only the IND data to learn features for the IND data. In Stage 2, with the feature extractor $h(\cdot)$ fixed, we fine-tune only the final classification layer $f(\cdot)$ using the IND data and the pseudo-OOD data. Since the feature extractor $h(\cdot)$ is not updated by the pseudo-OOD data, the final model $f(\cdot)$ will not overfit the pseudo-OOD data and can give the model more generalization power for OOD detection. Experimental results will demonstrate the extreme importance of the proposed two-stage training.

## 4   Experiments

We construct OOD detection tasks using benchmark datasets and compare the proposed CMG with the state-of-the-art existing methods.

## 4.1  Experiment Settings and Data Preparation

We use two experimental settings for evaluation.

**Setting 1 - Near-OOD Detection on the Same Dataset:** In this setting, IND (in-distribution) and OOD instances are from different classes of the same dataset. This setting is often called *open-set detection.* We use the following 4 popular datasets for our experiments in this setting.

(1) **MNIST** [29]: A handwritten digit classification dataset of 10 classes. The dataset has 70,000 examples/instances, with the splitting of 60,000 for training and 10,000 for testing.
(2) **CIFAR-10** [26]: A 10-class classification dataset consisting of 60,000 $32 \times 32$ color images with the splitting of 50,000 for training and 10,000 for testing.
(3) **SVHN** [39]: A colorful street view house number classification dataset of 10 classes. It contains 99289 instances with the splitting of 73257 for training and 26032 for testing.
(4) **TinyImageNet** [28]: A classification dataset of 200 classes. Each class contains 500 training samples and 50 testing samples of resolution $64 \times 64$.

We follow the data processing method in [51,58] to split known and unknown classes. For each dataset, we conduct 5 experiments using different splits of known (IND) and unknown (OOD) classes. These same 5 splits are used by all baselines and our system. Following [51], for MNIST, CIFAR-10 and SVHN, 6 classes are chosen as IND classes, and the other 4 classes are regarded as OOD classes. The following 5 fixed sets of IND classes, 0–5, 1–6, 2–7, 3–8, and 4–9, are used and they are called **partitions 1, 2, 3, 4**, and **5**, respectively. The rest 4 classes in each case serve as the OOD classes. For TinyImageNet, each set of IND data contains 20 classes and the sets of IND classes in the 5 experiments are 0–19, 40–59, 80–99, 120–139, and 160–189 respectively. The rest 180 classes are regarded as the OOD classes. The reason for using different partitions is discussed in the supplementary material.

**Setting 2 - Far-OOD Detection on Different Datasets:** The IND data and OOD data come from different datasets. We use CIFAR-10 and CIFAR100 as the IND dataset respectively and each of the following datasets as the OOD dataset. When CIFAR-10 is used as the IND dataset, the following are used as the OOD datsets (when CIFAR100 is used as the IND dataset, CIFAR-10 is also one of the OOD datasets).

(1) **SVHN** [39]: See above. All 26032 testing samples are used as OOD data.
(2) **LSUN** [56]: This is a large-scale scene understanding dataset with a testing set of 10,000 images from 10 different scenes. Images are resized to $32 \times 32$ in our experiment.
(3) **LSUN-FIX** [53]: To avoid artificial noises brought by general resizing operation, this dataset is generated by using a fixed resizing operation on LSUN to change the images to $32 \times 32$.
(4) **TinyImageNet** [28]: See above. All 10,000 testing samples are used as OOD data.

(5) **ImageNet-FIX** [28]: 10,000 images are randomly selected from the training set of ImageNet-30, excluding "airliner", "ambulance", "parkingmeter", and "schooner" to avoid overlapping with CIFAR-10. A resizing operation is applied to transform the images to 32 × 32.

(6) **CIFAR100** [27]: An image classification dataset with 60,000 32 × 32 color images of 100 classes. Its 10,000 test samples are used as the OOD data.

## 4.2  Baselines

We compare with 10 state-of-the-art baselines, including 2 generative methods and 2 contrastive learning methods.

(1) **Softmax**: This is the popular classification score model. The highest softmax probability is used as the confidence score for OOD detection.

(2) **OpenMax** [3]: This method combines the softmax score with the distance between the test sample and IND class centers to detect OOD data.

(3) **ODIN** [32]: This method improves the OOD detection performance of a pre-trained neural network by using temperature scaling and adding small perturbations to the input.

(4) **Maha** [31]: This method uses Mahalanobis distance to evaluate the probability that an instance belongs to OOD.

(5) **CCC** [30]: This is a GAN-based method, jointly training the classification model and the pseudo-OOD generator for OOD detection.

(6) **OSRCI** [38]: This method also uses GAN to generate pseudo instances and further improves the model to predict novelty (OOD) examples.

(7) **CAC** [34]: This is a distance-based method, using the Class Anchor Clustering loss to cluster IND samples tightly around the anchored centers.

(8) **SupCLR** [21]: This is a contrastive learning based method. It extends contrastive learning to fully-supervised setting to improve the quality of features

(9) **CSI** [53]: This is also a supervised contrastive learning method. It uses extensive data augmentations to generate shifted data instances. It also has a score function that benefits from the augmented instances for OOD detection.

(10) **React** [52]: This method exploits the internal activations of neural networks to find distinctive signature patterns for OOD distributions.

For Softmax, OpenMax and OSRCI, we use OSRCI's implementation[2]. For SupCLR and CSI, we use CSI's code[3]. For ODIN, Maha, CCC, CAC and React, we use their original code[4,5,6,7,8]. We also use their default hyper-parameters.

---

[2] https://github.com/lwneal/counterfactual-open-set.
[3] https://github.com/alinlab/CSI.
[4] https://github.com/facebookresearch/odin.
[5] https://github.com/pokaxpoka/deep_Mahalanobis_detector.
[6] https://github.com/alinlab/Confident_classifier.
[7] https://github.com/dimitymiller/cac-openset.
[8] https://github.com/deeplearning-wisc/react.

### 4.3   Implementation Details

For MNIST, we use a 9-layer CNN as the encoder (feature extractor) and a 2-layer MLP as the projection head. CVAE includes a 2-layer CNN as the encoder and a 2-layer deconvolution network [57] as the decoder, as well as two 1-layer MLPs to turn features into means and variations. For CIFAR100, the encoder is a PreActResnet [15] and the projection head is a 1-layer MLP. Its CVAE is the same as for the other datasets below. For the other datasets, the encoder is a ResNet18 [14] and the projection head is a 2-layer MLP. CVAE also uses ResNet18 as the encoder, and 2 residual blocks and a 3-layer deconvolution network as the decoder. The mean and variation projection are completed by two 1-layer MLPs. During the first stage of training, we use Adam optimizer [22] with $\beta_1 = 0.9$, $\beta_2 = 0.999$ and learning rate of 0.001. We train both the classification model and CVAE model for 200 epochs with batch size 512. In the second stage, the learning rate is set to 0.0001 and the fine-tuning process with the generated pseudo data are run for 10 epochs. The number of generated pseudo-OOD data is the same as the IND data (we will study this further shortly). Each batch has 128 IND samples and 128 generated OOD samples. There is no special hyper-parameter for CMG-softmax in stage 2. For CMG-energy, two special hyper-parameters of the energy loss $m_{ind}$ and $m_{ood}$ are decided at the beginning of stage 2 by IND and pseudo data. We calculate the energy of all training IND data and generated pseudo data. Then $m_{ind}$ and $m_{ood}$ are chosen to make 80% of IND data's energy larger than $m_{ind}$ and 80% of pseudo data's energy smaller than $m_{ood}$. This ensures that 80% of data get non-zero loss. We use a NVIDIA-GeForce-RTX-2080Ti GPU for the experiments of evaluating the running speed of different methods.

### 4.4   Results and Discussions

Table 1 shows the results of the two OOD detection settings on different datasets. Due to the large image size, numerous IND classes and a large batch size requirement, we were unable to run SupCLR and CSI using TinyImageNet on our hardware and thus do not have their results in Setting 1. In Setting 2, CAC crashes owing to too many IND data classes. On average, our CMG achieves the best results in both Setting 1 and Setting 2. Specifically, CMG greatly outperforms the two GAN-based generation methods, CCC and OSRCI, which shows the superiority of CMG in generating and utilizing pseudo-OOD data. We also notice that our CMG-s (CMG-softmax) is slightly weaker than our CMG-e (CMG-energy), which shows the energy function is effective.

Table 2 demonstrates that CMG's fine-tuning (stage 2) can improve the 4 best performing baselines in Table 1, i.e., GAN-based OSRCI and contrastive learning based SupCLR and CSI, and a newest work React. Here after each baseline finishes its training, we apply fine-tuning of CMG's stage 2 to fine-tune the trained model using CMG-energy. We can see that the baselines OSRCI, SupCLR, CSI and React are all improved.

Table 3 shows that CMG is much more efficient than the contrastive learning methods. With the best overall performances on OOD detection, CMG spends about only 10% of contrastive learning training time.

**Table 1.** AUC (Area Under the ROC curve) (%) on detecting IND and OOD samples in 2 experimental settings. For Setting 1, the results are averaged over the 5 partitions. CMG-s uses CMG-softmax fine-tuning and CMG-e uses CMG-energy fine-tuning. Every experiment was run 5 times. Each result in brackets () in the *average* row for Setting 1 is the mean of the first three datasets as SupCLR and CSI cannot run on TinyImageNet. CAC crashed when using CIFAR100 as the IND data.

| Datasets | Softmax | OpenMax | ODIN | Maha | CCC | OSRCI | CAC | SupCLR | CSI | React | CMG-s | CMG-e |
|---|---|---|---|---|---|---|---|---|---|---|---|---|
| Setting 1 - Near-OOD detection on the same dataset | | | | | | | | | | | | |
| MNIST | 97.6 | 98.1 | 98.1 | 98.4 | 94.2 | 98.3 | **99.2** | 97.1 | 97.2 | 98.6 | 98.3 | 99.0 |
| (std) | ±0.7 | ±0.5 | ±1.1 | ±0.4 | ±0.8 | ±0.9 | ±0.1 | ±0.2 | ±0.3 | ±0.1 | ±0.2 | ±0.2 |
| CIFAR-10 | 65.5 | 66.9 | 79.4 | 73.4 | 74.0 | 67.5 | 75.9 | 80.0 | 84.7 | 85.5 | **86.3** | 85.6 |
| (std) | ±0.5 | ±0.4 | ±1.6 | ±2.2 | ±1.4 | ±0.8 | ±0.7 | ±0.5 | ±0.3 | ±0.2 | ±1.2 | ±0.6 |
| SVHN | 90.3 | 90.7 | 89.4 | 91.5 | 64.6 | 91.7 | 93.8 | 93.8 | **93.9** | 92.8 | 91.8 | 92.1 |
| (std) | ±0.5 | ±0.4 | ±2.0 | ±0.6 | ±2.3 | ±0.2 | ±0.2 | ±0.2 | ±0.1 | ±0.1 | ±0.5 | ±0.4 |
| TinyImageNet | 57.5 | 57.9 | 70.9 | 56.3 | 51.0 | 58.1 | 71.9 | \ | \ | 51.9 | 72.2 | **73.7** |
| (std) | ±0.7 | ±0.2 | ±1.5 | ±1.9 | ±1.2 | ±0.4 | ±0.7 | \ | \ | ±0.0 | ±0.5 | ±0.6 |
| Average | 77.8 | 78.4 | 84.5 | 79.9 | 71.0 | 78.9 | 85.2 | (90.3) | (91.9) | 82.2 | 87.2 | **87.6** |
| Setting 2 - Far-OOD detection on different datasets | | | | | | | | | | | | |
| CIFAR-10 as IND | | | | | | | | | | | | |
| SVHN | 80.2 | 82.7 | 83.2 | 97.5 | 83.3 | 80.2 | 87.3 | 97.3 | **97.9** | 92.2 | 95.8 | 96.2 |
| (std) | ±1.8 | ±1.9 | ±1.5 | ±1.6 | ±0.8 | ±1.8 | ±4.6 | ±0.1 | ±0.1 | ±1.1 | ±0.6 | ±2.5 |
| LSUN | 70.1 | 72.2 | 82.1 | 61.5 | 85.6 | 79.9 | 89.1 | 92.8 | 97.7 | 96.5 | 96.8 | **97.7** |
| (std) | ±2.5 | ±1.8 | ±1.9 | ±5.0 | ±2.3 | ±1.8 | ±3.4 | ±0.5 | ±0.4 | ±0.7 | ±1.4 | ±0.9 |
| LSUN-FIX | 76.7 | 75.6 | 84.1 | 77.8 | 86.6 | 78.2 | 85.5 | 91.6 | 93.5 | 90.6 | **94.1** | 93.7 |
| (std) | ±0.8 | ±1.2 | ±1.7 | ±2.1 | ±1.6 | ±0.5 | ±0.7 | ±1.5 | ±0.4 | ±1.9 | ±0.9 | ±0.4 |
| TinyImageNet | 62.5 | 65.2 | 68.7 | 56.8 | 83.2 | 70.0 | 86.4 | 91.4 | **97.6** | 94.3 | 94.8 | 95.2 |
| (std) | ±3.6 | ±3.1 | ±2.2 | ±2.1 | ±1.8 | ±1.7 | ±4.6 | ±1.2 | ±0.3 | ±0.5 | ±1.6 | ±2.7 |
| ImageNet-FIX | 75.9 | 75.6 | 74.8 | 79.0 | 83.7 | 78.1 | 85.6 | 90.5 | **94.0** | 92.0 | 89.7 | 92.9 |
| (std) | ±4.6 | ±0.7 | ±0.6 | ±3.1 | ±1.1 | ±0.3 | ±0.3 | ±0.5 | ±0.1 | ±2.2 | ±0.3 | ±1.2 |
| CIFAR100 | 74.6 | 75.5 | 74.5 | 61.4 | 81.9 | 77.4 | 83.9 | 88.6 | **92.2** | 88.4 | 87.9 | 89.3 |
| (std) | ±0.5 | ±0.4 | ±0.8 | ±0.9 | ±0.5 | ±0.4 | ±0.2 | ±0.2 | ±0.1 | ±0.7 | ±0.4 | ±0.4 |
| CIFAR100 as IND | | | | | | | | | | | | |
| SVHN | 66.7 | 65.9 | 71.7 | **93.1** | 66.0 | 65.5 | \ | 83.4 | 88.2 | 88.6 | 90.0 | 90.2 |
| (std) | ±4.0 | ±4.5 | ±1.7 | ±0.6 | ±1.0 | ±1.1 | \ | ±0.5 | ±0.7 | ±1.3 | ±2.4 | ±2.5 |
| LSUN | 48.1 | 53.7 | 66.0 | **95.6** | 68.7 | 74.4 | \ | 81.6 | 80.9 | 88.1 | 85.9 | 88.3 |
| (std) | ±4.6 | ±6.7 | ±1.0 | ±0.1 | ±1.0 | ±0.8 | \ | ±0.5 | ±0.5 | ±2.8 | ±2.6 | ±3.6 |
| LSUN-FIX | 47.5 | 50.4 | 72.6 | 63.4 | 59.3 | 69.7 | \ | 70.9 | 74.0 | 69.7 | 76.5 | **77.4** |
| (std) | ±4.1 | ±5.8 | ±7.5 | ±3.4 | ±1.7 | ±0.6 | \ | ±0.1 | ±0.2 | ±0.5 | ±8.5 | ±2.4 |
| TinyImageNet | 62.5 | 62.5 | 73.5 | **93.3** | 69.7 | 63.9 | \ | 78.5 | 79.4 | 87.0 | 84.3 | 88.2 |
| (std) | ±2.9 | ±3.0 | ±3.3 | ±0.7 | ±1.6 | ±1.2 | \ | ±0.8 | ±0.2 | ±3.2 | ±4.9 | ±2.0 |
| ImageNet-FIX | 64.8 | 64.5 | 76.9 | 61.2 | 60.6 | 63.8 | \ | 75.0 | **79.2** | 78.9 | 72.5 | 76.5 |
| (std) | ±0.5 | ±0.6 | ±9.1 | ±0.9 | ±0.0 | ±0.9 | \ | ±0.5 | ±0.2 | ±0.3 | ±2.2 | ±1.3 |
| CIFAR-10 | 63.0 | 62.7 | 67.9 | 56.3 | 63.7 | 58.8 | \ | 72.2 | **78.2** | 74.4 | 68.8 | 71.5 |
| (std) | ±1.0 | ±1.0 | ±2.1 | ±0.5 | ±0.5 | ±0.8 | \ | ±0.6 | ±0.2 | ±1.3 | ±3.1 | ±2.9 |
| Average | 66.1 | 67.2 | 74.7 | 74.7 | 74.4 | 71.7 | \ | 84.5 | 87.7 | 86.7 | 86.4 | **88.1** |

**Table 2.** AUC (Area Under the ROC curve) (%) results of the original model (denoted by **original**) and the model plus fine-tuning using CMG-energy (denoted by **+CMG-e**). Almost every +CMG-e version of the baselines outperforms the original model. Every experiment was run 5 times.

| Datasets | OSRCI | | SupCLR | | CSI | | React | |
|---|---|---|---|---|---|---|---|---|
| | Original | +CMG-e | Original | +CMG-e | Original | +CMG-e | Original | +CMG-e |
| Setting 1 - Near-OOD detection on the same dataset | | | | | | | | |
| MNIST | 98.3±0.9 | 99.1±0.4 | 97.1±0.2 | 98.6±0.2 | 97.2±0.3 | 99.3±0.1 | 98.6±0.1 | 98.9±0.1 |
| CIFAR-10 | 67.5±0.8 | 72.3±0.6 | 80.0±0.5 | 88.9±0.5 | 84.7±0.3 | 89.8±0.6 | 85.5±0.2 | 85.8±0.1 |
| SVHN | 91.7±0.2 | 92.1±0.1 | 93.8±0.2 | 96.5±0.3 | 93.9±0.1 | 96.7±0.2 | 92.8±0.1 | 92.8±0.1 |
| TinyImageNet | 58.1±0.4 | 59.9±0.3 | \ | \ | \ | \ | 51.9±0.0 | 51.8±0.1 |
| Average | 78.9 | **80.9** | 90.3 | **94.7** | 91.9 | **95.3** | 82.2 | **82.3** |
| Setting 2 - Far-OOD detection on different datasets | | | | | | | | |
| CIFAR-10 as IND | | | | | | | | |
| SVHN | 80.2±1.8 | 79.3±2.5 | 97.3±0.1 | 93.0±1.3 | 97.9±0.1 | 97.8±0.6 | 92.1±1.1 | 98.2±0.8 |
| LSUN | 79.9±1.8 | 92.1±0.6 | 92.8±0.5 | 97.7±0.6 | 97.7±0.4 | 99.2±0.1 | 96.5±0.7 | 96.4±0.4 |
| LSUN-FIX | 78.2±0.5 | 81.2±1.0 | 91.6±1.5 | 94.1±0.3 | 93.5±0.4 | 96.2±0.3 | 90.6±1.9 | 91.9±0.2 |
| TinyImageNet | 70.0±1.7 | 83.2±1.7 | 91.4±1.2 | 96.3±0.8 | 97.6±0.3 | 98.7±0.3 | 94.3±0.5 | 94.0±0.6 |
| ImageNet-FIX | 78.1±0.3 | 78.5±0.2 | 90.5±0.5 | 92.9±0.3 | 94.0±0.1 | 95.7±0.1 | 92.0±2.2 | 91.3±0.1 |
| CIFAR100 | 77.4±0.4 | 77.4±0.6 | 88.6±0.2 | 90.3±0.2 | 92.2±0.1 | 92.0±0.2 | 88.4±0.7 | 89.2±0.2 |
| CIFAR-100 as IND | | | | | | | | |
| SVHN | 65.5±1.1 | 87.2±1.3 | 83.4±0.5 | 85.3±1.1 | 88.2±0.7 | 85.9±1.2 | 88.6±1.3 | 97.1±1.3 |
| LSUN | 74.4±0.8 | 76.5±0.7 | 81.6±0.5 | 84.3±0.9 | 80.9±0.5 | 89.9±0.8 | 88.1±2.8 | 89.3±0.8 |
| LSUN-FIX | 69.7±0.6 | 71.7±0.9 | 70.9±0.1 | 69.8±0.8 | 74.0±0.2 | 74.0±1.3 | 69.7±0.5 | 70.6±2.5 |
| TinyImageNet | 63.9±1.2 | 67.3±0.4 | 78.5±0.8 | 84.2±1.1 | 79.4±0.2 | 89.4±0.8 | 87.0±3.2 | 87.9±0.5 |
| ImageNet-FIX | 63.8±0.9 | 66.1±0.9 | 75.0±0.5 | 72.4±0.8 | 79.2±0.2 | 79.6±1.1 | 78.9±0.3 | 79.8±0.2 |
| CIFAR-10 | 58.8±0.8 | 61.9±0.4 | 72.2±0.6 | 75.9±0.7 | 78.2±0.2 | 72.2±0.4 | 74.4±1.3 | 72.9±0.5 |
| Average | 71.7 | **76.9** | 84.5 | **86.4** | 87.7 | **89.2** | 86.7 | **88.2** |

**Table 3.** Execution time (min) of each method spent in running the whole experiment on benchmark datasets for Setting 1.

| Datasets | Softmax | OpenMax | ODIN | Maha | CCC | OSRCI | CAC | SupCLR | CSI | React | CMG-e |
|---|---|---|---|---|---|---|---|---|---|---|---|
| MNIST | 6 | 6 | 71 | 54 | 133 | 49 | 13 | 1260 | 1728 | 65 | 24 |
| CIFAR-10 | 20 | 20 | 61 | 56 | 111 | 70 | 49 | 1110 | 1428 | 61 | 144 |
| SVHN | 20 | 20 | 142 | 140 | 196 | 71 | 37 | 1770 | 2471 | 79 | 249 |
| TinyImageNet | 22 | 22 | 64 | 54 | 46 | 79 | 64 | \ | \ | 65 | 131 |

## 4.5   Ablation Study

We now perform the ablation study with various options of CMG-e and report AUC scores on the 5 partitions of CIFAR-10 in Setting 1.[9]

---

[9] We also conducted some experiments using a pre-trained feature extractor. Using a pre-trained feature extractor can be controversial, which is discussed in the supplementary material.

**CMG's Two-Stage Training vs. One-Stage Direct Training.** As we stated in the introduction, one-stage *direct training* using the IND training and the generated pseudo-OOD data produces very poor results compared with CMG's 2-stage training with only fine-tuning of the classification layer in the second stage. Here we show the comparison results. We compare three training strategies: (1) *Direct Training*, (2) *Unfrozen Fine-tuning*, i.e., keeping stage 1 but fine-tuning the whole model in stage 2 without freezing the feature extractor, and (3) *CMG training* (CMG-e). Figure 2(a) shows that Direct Training produces very poor results and Unfrozen Fine-Tuning is also weak. CMG Training (CMG-e) performs considerably better. We explained the reason in Sect. 3.3. In these experiments, the energy function in Eq. 6 is used to compute the OOD score. In the supplementary material, we also show that in experiment Setting 2, the same trend applies.

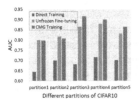
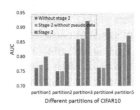
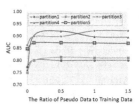

(a) Training Strategies.    (b) CMG-e Stage 2.    (c) Amount of Pseudo Data.

**Fig. 2.** Ablation studies: (a) different training strategies, (b) CMG-e stage 2, and (c) Amount of Pseudo-OOD Data.

**CMG Stage 2**. To verify the effect of different options of stage 2, we compare the results of CMG-e model with **(1)** *without stage 2*, i.e., we directly compute the energy score using Eq. (6) on the classification model from stage 1, **(2)** *stage 2 without using pseudo-OOD data*, i.e., we use only IND data to fine-tune the classifier with loss using Eq. (5), and **(3)** *full stage 2*. Figure 2(b) shows that without stage 2, stage 1 produces poor results. Stage 2 without the generated pseudo-OOD data only improves the performance slightly. The full stage 2 with the generated pseudo-OOD data greatly improves the performance of OOD detection. These experiments prove the necessity of stage 2 and the effectiveness of the generated pseudo-OOD data.

**Amount of Pseudo-OOD Data.** We run experiments of stage 2 with different numbers of generated pseudo-OOD samples to analyze their effectiveness. Figure 2(c) shows that the model benefits significantly from only a few pseudo-OOD samples. With only 10% of that of the IND data, the pseudo-OOD data can already improve the results markedly, which indicates the importance of the pseudo-OOD data. The results are similar when pseudo-OOD samples are more than a half of the IND samples. We use the same number of pseudo-OOD samples as the IND samples in all our experiments.

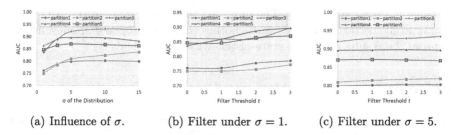

(a) Influence of $\sigma$.          (b) Filter under $\sigma = 1$.          (c) Filter under $\sigma = 5$.

**Fig. 3.** Ablation study on different options in generating pseudo-OOD data. Figure 3(a) shows AUC results of different $\sigma$ values of the sampling distribution. Figure 3(b) filters values near the center of the Gaussian distribution with $\sigma = 1$. Figure 3(c) filters values near the center of the Gaussian Distribution with $\sigma = 5$.

**Pseudo-OOD Data Distribution.** The CVAE generator is trained to make the latent variables or features conform to the Gaussian distribution $\mathcal{N}(\mathbf{0}, \mathbf{I})$ (see Sect. 3.2). To make pseudo data diverse and different from the training data, we sample the latent variables $\mathbf{z}$ from a pseudo data sampling distribution $\mathcal{N}(\mathbf{0}, \sigma^2 * \mathbf{I})$. We conduct experiments to study the effect of the distribution. First, we study the influence of $\sigma$. Note that $\sigma$ is 1 in training. With larger $\sigma$ values, the sampled values will be more likely to be far from $\mathbf{0}$ (which is the mean) to make the latent features different from those seen in training.[10] Fig. 3(a) shows the results, which indicate the necessity of using $\sigma > 1$ and results are similar for a large range of $\sigma$ values. We use $\sigma = 5$ in all our experiments.

Intuitively, we may only keep latent features $\mathbf{z}$ that are far from the Gaussian distribution mean by filtering out values that are close to $\mathbf{0}$ (or the mean). We use a filtering threshold $t$ to filter out the sampled $\mathbf{z}$ whose component values are within the range $[-t, t]$. Experimental results in Fig. 3(b) allow us to make the following observations. When $\sigma = 1$, as $t$ grows, the performance improves slightly. But comparing with Fig. 3(a), we see that a larger $\sigma$ improves the performance more. Figure 3(c) tells us that when $\sigma = 5$, the effect of filtering diminishes. For simplicity and efficiency, all our experiments employed $\sigma = 5$ without filtration. In the supplementary material, a visual analysis is done for our pseudo-OOD data to further illustrate their high quality.

## 5    Conclusion

This paper proposed a novel and yet simple method for OOD detection based on OOD data generation and classifier fine-tuning, which not only produces SOTA results but is also much more efficient than existing highly effective contrastive learning based OOD detection methods. Also importantly, we discovered that using the IND data and the generated pseudo-OOD data to directly train a

---

[10] We include images generated with different choices of $\sigma$ in the supplementary material. Images generated with larger $\sigma$'s are more different from the IND data and show a more comprehensive coverage of the OOD area.

classifier performs very poorly. The proposed fine-tuning framework CMG works dramatically better. It is also worth noting that the proposed framework can improve the results of diverse state-of-the-art OOD methods too.

# References

1. Baldi, P., Hornik, K.: Neural networks and principal component analysis: learning from examples without local minima. Neural Netw. **2**(1), 53–58 (1989)
2. Bendale, A., Boult, T.: Towards open world recognition. In: CVPR, pp. 1893–1902 (2015)
3. Bendale, A., Boult, T.E.: Towards open set deep networks. In: CVPR, pp. 1563–1572 (2016)
4. Bergman, L., Hoshen, Y.: Classification-based anomaly detection for general data. In: International Conference on Learning Representations (2019)
5. Caruana, R., Lou, Y., Gehrke, J., Koch, P., Sturm, M., Elhadad, N.: Intelligible models for healthcare: predicting pneumonia risk and hospital 30-day readmission. In: 21th ACM SIGKDD, pp. 1721–1730 (2015)
6. Cemgil, T., Ghaisas, S., Dvijotham, K.D., Kohli, P.: Adversarially robust representations with smooth encoders. In: ICLR (2019)
7. Chen, M., Xu, Z.E., Weinberger, K.Q., Sha, F.: Marginalized denoising autoencoders for domain adaptation. In: ICML (2012). https://icml.cc/2012/papers/416.pdf
8. Deecke, L., Vandermeulen, R., Ruff, L., Mandt, S., Kloft, M.: Image anomaly detection with generative adversarial networks. In: Berlingerio, M., Bonchi, F., Gärtner, T., Hurley, N., Ifrim, G. (eds.) ECML PKDD 2018. LNCS (LNAI), vol. 11051, pp. 3–17. Springer, Cham (2019). https://doi.org/10.1007/978-3-030-10925-7_1
9. Fort, S., Ren, J., Lakshminarayanan, B.: Exploring the limits of out-of-distribution detection. In: Advances in NeurlPS, vol. 34 (2021)
10. Gidaris, S., Singh, P., Komodakis, N.: Unsupervised representation learning by predicting image rotations. In: ICLR (2018)
11. Golan, I., El-Yaniv, R.: Deep anomaly detection using geometric transformations. In: Advances in NeurlPS, vol. 31 (2018)
12. Gunther, M., Cruz, S., Rudd, E.M., Boult, T.E.: Toward open-set face recognition. In: CVPR Workshops, pp. 71–80 (2017)
13. Guo, C., Pleiss, G., Sun, Y., Weinberger, K.Q.: On calibration of modern neural networks. In: ICML, pp. 1321–1330. PMLR (2017)
14. He, K., Zhang, X., Ren, S., Sun, J.: Deep residual learning for image recognition. In: CVPR, pp. 770–778 (2016)
15. He, K., Zhang, X., Ren, S., Sun, J.: Identity mappings in deep residual networks. In: Leibe, B., Matas, J., Sebe, N., Welling, M. (eds.) ECCV 2016. LNCS, vol. 9908, pp. 630–645. Springer, Cham (2016). https://doi.org/10.1007/978-3-319-46493-0_38
16. Hendrycks, D., Mazeika, M., Dietterich, T.: Deep anomaly detection with outlier exposure. In: ICLR (2019). https://openreview.net/forum?id=HyxCxhRcY7
17. Hendrycks, D., Mazeika, M., Kadavath, S., Song, D.: Using self-supervised learning can improve model robustness and uncertainty. In: Advances in NeurlPS, vol. 32 (2019)
18. Hinton, G.E., Osindero, S., Teh, Y.W.: A fast learning algorithm for deep belief nets. Neural Comput. **18**(7), 1527–1554 (2006)

19. Hu, W., Wang, M., Qin, Q., Ma, J., Liu, B.: HRN: a holistic approach to one class learning. Adv. Neural. Inf. Process. Syst. **33**, 19111–19124 (2020)

20. Júnior, P.R.M., et al.: Nearest neighbors distance ratio open-set classifier. Mach. Learn. **106**(3), 359–386 (2017)

21. Khosla, P., et al.: Supervised contrastive learning. In: Advances in NeurIPS, vol. 33, pp. 18661–18673 (2020)

22. Kingma, D.P., Ba, J.: Adam: a method for stochastic optimization. In: ICLR (2015)

23. Kingma, D.P., Welling, M.: Auto-encoding variational Bayes. arXiv preprint arXiv:1312.6114 (2013)

24. Kingma, D.P., Mohamed, S., Jimenez Rezende, D., Welling, M.: Semi-supervised learning with deep generative models. In: Advances in NeurIPS, vol. 27 (2014)

25. Kolesnikov, A., Zhai, X., Beyer, L.: Revisiting self-supervised visual representation learning. In: CVPR, pp. 1920–1929 (2019)

26. Krizhevsky, A., Hinton, G.: Convolutional deep belief networks on cifar-10. Unpublished Manuscript **40**(7), 1–9 (2010)

27. Krizhevsky, A., Hinton, G., et al.: Learning multiple layers of features from tiny images (2009)

28. Le, Y., Yang, X.: Tiny imagenet visual recognition challenge. CS 231N **7**, 7 (2015)

29. LeCun, Y., Cortes, C., Burges, C.: Mnist handwritten digit database (2010)

30. Lee, K., Lee, H., Lee, K., Shin, J.: Training confidence-calibrated classifiers for detecting out-of-distribution samples. In: ICLR (2018)

31. Lee, K., Lee, K., Lee, H., Shin, J.: A simple unified framework for detecting out-of-distribution samples and adversarial attacks. In: Advances in NeurIPS, vol. 31 (2018)

32. Liang, S., Li, Y., Srikant, R.: Enhancing the reliability of out-of-distribution image detection in neural networks. arXiv preprint arXiv:1706.02690 (2017)

33. Liu, W., Wang, X., Owens, J., Li, Y.: Energy-based out-of-distribution detection. In: Advances in NeurIPS, vol. 33, pp. 21464–21475 (2020)

34. Miller, D., Sunderhauf, N., Milford, M., Dayoub, F.: Class anchor clustering: a loss for distance-based open set recognition. In: Proceedings of the IEEE/CVF Winter Conference on Applications of Computer Vision, pp. 3570–3578 (2021)

35. Mohseni, S., Pitale, M., Yadawa, J., Wang, Z.: Self-supervised learning for generalizable out-of-distribution detection. In: Proceedings of the AAAI Conference on Artificial Intelligence, vol. 34, pp. 5216–5223 (2020)

36. Nalisnick, E., Matsukawa, A., Teh, Y.W., Gorur, D., Lakshminarayanan, B.: Do deep generative models know what they don't know? In: ICLR (2019)

37. Nalisnick, E., Matsukawa, A., Teh, Y.W., Lakshminarayanan, B.: Detecting out-of-distribution inputs to deep generative models using typicality (2020)

38. Neal, L., Olson, M., Fern, X., Wong, W.-K., Li, F.: Open set learning with counterfactual images. In: Ferrari, V., Hebert, M., Sminchisescu, C., Weiss, Y. (eds.) ECCV 2018. LNCS, vol. 11210, pp. 620–635. Springer, Cham (2018). https://doi.org/10.1007/978-3-030-01231-1_38

39. Netzer, Y., Wang, T., Coates, A., Bissacco, A., Wu, B., Ng, A.Y.: Reading digits in natural images with unsupervised feature learning (2011)

40. Nitsch, J., et al.: Out-of-distribution detection for automotive perception. In: 2021 IEEE ITSC, pp. 2938–2943. IEEE (2021)

41. Oza, P., Patel, V.M.: C2AE: class conditioned auto-encoder for open-set recognition. In: CVPR, pp. 2307–2316 (2019)

42. Perera, P., Nallapati, R., Xiang, B.: OcGAN: one-class novelty detection using GANs with constrained latent representations. In: CVPR, pp. 2898–2906 (2019)

43. Pidhorskyi, S., Almohsen, R., Doretto, G.: Generative probabilistic novelty detection with adversarial autoencoders. In: Advances in NeurIPS, vol. 31 (2018)
44. Pourreza, M., Mohammadi, B., Khaki, M., Bouindour, S., Snoussi, H., Sabokrou, M.: G2D: generate to detect anomaly. In: Proceedings of the IEEE/CVF Winter Conference on Applications of Computer Vision, pp. 2003–2012 (2021)
45. Qiu, C., Pfrommer, T., Kloft, M., Mandt, S., Rudolph, M.: Neural transformation learning for deep anomaly detection beyond images. In: International Conference on Machine Learning, pp. 8703–8714. PMLR (2021)
46. Ren, J., et al.: Likelihood ratios for out-of-distribution detection. In: Advances in NeurIPS, vol. 32 (2019)
47. Rumelhart, D.E., Hinton, G.E., Williams, R.J.: Learning representations by back-propagating errors. Nature $\mathbf{323}$(6088), 533–536 (1986)
48. Sastry, C.S., Oore, S.: Detecting out-of-distribution examples with gram matrices. In: ICML, pp. 8491–8501. PMLR (2020)
49. Sehwag, V., Chiang, M., Mittal, P.: SSD: a unified framework for self-supervised outlier detection. In: ICLR (2021). https://openreview.net/forum?id=v5gjXpmR8J
50. Serrà, J., Álvarez, D., Gómez, V., Slizovskaia, O., Núñez, J.F., Luque, J.: Input complexity and out-of-distribution detection with likelihood-based generative models. In: ICLR (2020)
51. Sun, X., Yang, Z., Zhang, C., Ling, K.V., Peng, G.: Conditional gaussian distribution learning for open set recognition. In: CVPR, pp. 13480–13489 (2020)
52. Sun, Y., Guo, C., Li, Y.: React: out-of-distribution detection with rectified activations. In: Advances in NeurIPS (2021)
53. Tack, J., Mo, S., Jeong, J., Shin, J.: CSI: novelty detection via contrastive learning on distributionally shifted instances. In: Advances in NeurIPS, vol. 33, pp. 11839–11852 (2020)
54. Vernekar, S., Gaurav, A., Abdelzad, V., Denouden, T., Salay, R., Czarnecki, K.: Out-of-distribution detection in classifiers via generation. arXiv preprint arXiv:1910.04241 (2019)
55. Vincent, P., Larochelle, H., Bengio, Y., Manzagol, P.A.: Extracting and composing robust features with denoising autoencoders. In: Proceedings of the 25th ICML, pp. 1096–1103 (2008)
56. Yu, F., Seff, A., Zhang, Y., Song, S., Funkhouser, T., Xiao, J.: LSUN: construction of a large-scale image dataset using deep learning with humans in the loop. arXiv preprint arXiv:1506.03365 (2015)
57. Zeiler, M.D., Taylor, G.W., Fergus, R.: Adaptive deconvolutional networks for mid and high level feature learning. In: 2011 ICCV, pp. 2018–2025. IEEE (2011)
58. Zhou, D.W., Ye, H.J., Zhan, D.C.: Learning placeholders for open-set recognition. In: CVPR, pp. 4401–4410, June 2021
59. Zong, B., et al.: Deep autoencoding Gaussian mixture model for unsupervised anomaly detection. In: ICLR (2018)
60. Zongyuan Ge, S.D., Garnavi, R.: Generative openmax for multi-class open set classification. In: Tae-Kyun Kim, Stefanos Zafeiriou, G.B., Mikolajczyk, K. (eds.) BMVC, pp. 42.1-42.12. BMVA Press, September 2017. https://doi.org/10.5244/C.31.42

# GraphMixup: Improving Class-Imbalanced Node Classification by Reinforcement Mixup and Self-supervised Context Prediction

Lirong Wu[(✉)], Jun Xia[(✉)], Zhangyang Gao, Haitao Lin, Cheng Tan, and Stan Z. Li

School of Engineering, Westlake University, Hangzhou 310030, China
{wulirong,xiajun,gaozhangyang,linhaitao,tancheng,
stan.zq.li}@westlake.edu.cn

**Abstract.** Data imbalance, i.e., some classes may have much fewer samples than others, is a serious problem that can lead to unfavorable node classification. However, most existing GNNs are based on the assumption that node samples for different classes are balanced. In this case, directly training a GNN classifier with raw data would under-represent samples from those minority classes and result in sub-optimal performance. This paper proposes GraphMixup, a novel mixup-based framework for improving class-imbalanced node classification on graphs. However, directly performing mixup in the input space or embedding space may produce out-of-domain samples due to the extreme sparsity of minority classes; hence we construct semantic relation spaces that allow *Feature Mixup* to be performed at the semantic level. Moreover, we apply two context-based self-supervised techniques to capture both local and global information in the graph structure and specifically propose *Edge Mixup* to handle graph data. Finally, we develop a *Reinforcement Mixup* mechanism to adaptively determine how many samples are to be generated by mixup for those minority classes. Extensive experiments on three real-world datasets have shown that GraphMixup yields truly encouraging results for the task of class-imbalanced node classification. Codes are available at: https://github.com/LirongWu/GraphMixup.

**Keywords:** Class-imbalance · Graph model · Self-supervised learning

## 1 Introduction

In many real-world applications, including social networks, chemical molecules, and citation networks, data can be naturally modeled as graphs. Recently, the emerging Graph Neural Networks (GNNs) have demonstrated their powerful capability to handle the task of semi-supervised node classification: inferring unknown node labels by using the graph structure and node features with partially known node labels. Despite all these successes, existing works are mainly based on the assumption that node samples for different classes are roughly balanced. However,

M.-R. Amini et al. (Eds.): ECML PKDD 2022, LNAI 13716, pp. 519–535, 2023.
https://doi.org/10.1007/978-3-031-26412-2_32

there exists serious class-imbalance in many real-world applications, i.e., some classes may have significantly fewer samples for training than other classes. For example, the majority of users in a transaction fraud network are benign users, while only a small portion of them are bots. Similarly, topic classification for citation networks also suffers from the class-imbalanced problem, as papers for some topics may be scarce, compared to those on-trend topics.

The class-imbalanced problems have been well studied in the image domain [6,7,11], and data-level algorithms can be summarized into two groups: down-sampling and over-sampling [13]. The down-sampling methods sample a representative sample set from the majority class to make its size close to the minority class, but this inevitably entails a loss of information. In contrast, the over-sampling methods aim to generate new samples for minority classes, which have been found to be more effective and stable. However, directly applying existing over-sampling strategies to graph data may lead to sub-optimal results due to the non-Euclidean property of graphs. Three key problems for the class-imbalanced problem are: *(1) How to generate new nodes and their node features for minority classes? (2) How to capture the connections between generated and existing nodes? (3) How to determine the upsampling ratio for minority classes?*

Mixup [24] is an effective method to solve *Problem (1)*, which performs feature interpolation for minority classes to generate new samples. However, most existing mixup methods are performed either in the input or embedding space, which may generate out-of-domain samples, especially for those minority classes due to their extreme sparsity. To alleviate this problem, we construct semantic relation spaces that allow *Feature Mixup* to be performed at the semantic level (see Sect. 3.3). To solve *Problem (2)*, a natural solution is to train an edge generator through the task of adjacency matrix reconstruction and then apply it to predict the existence of edges between generated nodes and existing nodes. However, MSE-based matrix reconstruction completely ignores graph topological information, making the edge generator over-emphasize the connections between nodes with similar features while neglecting the long-range dependencies between nodes. Therefore, we design two context-based self-supervised auxiliary tasks to consider both local and global graph structural information (see Sect. 3.4). Finally, unlike hand-crafted or heuristic estimation for *Problem (3)*, we develop a reinforcement mixup mechanism to adaptively determine the upsampling ratio for minority classes (see Sect. 3.5). Extensive experiments show that GraphMixup outperforms other leading methods at the low-to-high class-imbalance ratios.

## 2   Related Work

### 2.1   Class-Imbalanced Problem

The class-imbalanced problem is common in real-world scenarios and has become a popular research topic [7,14,15,25]. The mainstream algorithms for class-imbalanced problems can be mainly divided into two categories: algorithm-level and data-level. The algorithm-level methods seek to directly increase the importance of minority classes with suitable penalty functions or model regularization. For example, RA-GCN [3] proposes to learn a parametric penalty function through adversarial training, which re-weights samples to help the classifier fit better between majority and minority classes, and thus avoid bias towards either of the classes. Besides, DR-GCN [17] tackles the class-imbalanced problem by imposing two types of regularization, which adopts a conditional adversarial training together with latent distribution alignment.

Different from those algorithm-level methods, the data-level methods usually adjust class sizes through down-sampling or over-sampling. **In this paper, we mainly focus on solving the class-imbalanced problem for graph data with over-sampling algorithms.** The vanilla over-sampling is to replicate existing samples, which reduces the class imbalance but can lead to overfitting as no extra information is introduced. Instead, SMOTE [2] solves this problem by generating new samples by feature interpolation between samples of minority classes and their nearest neighbors, and many of its variants [1] have been proposed with promising results. Despite their great success, few attempts have been made on class-imbalanced problems for non-Euclidean graph data. GraphSMOTE [25] is the first work to consider the problem of node-class imbalance on graphs, but their contribution is only to extend SMOTE to graph settings without making full use of graph topological information. In addition, ImGAGN [14] proposes a generative adversarial model, which utilizes a generator to generate a set of synthetic minority nodes. Then a GCN discriminator is trained to discriminate between real nodes and fake (i.e., generated) nodes, and also between minority nodes and majority nodes.

### 2.2    Graph Self-Supervised Learning (SSL)

The primary goal of Graph SSL is to learn transferable prior knowledge from abundant unlabeled data with well-designed pretext tasks and then generalize the learned knowledge to downstream tasks. The existing graph SSL methods can be divided into three categories: contrastive, generative, and predictive [21]. The contrastive methods contrast the views generated from different augmentation by mutual information maximization. Instead, the generative methods focus on the (intra-data) information embedded in the graph, generally based on pretext tasks such as reconstruction. Moreover, the predictive methods generally self-generate labels by some simple statistical analysis or expert knowledge and then perform prediction-based tasks based on self-generated labels. In this paper, we mainly focus on predictive methods since it takes full account of the contextual information in the graph structure, both local and global, allowing us to better capture connections between generated and existing nodes.

## 3    Methodology

### 3.1    Notions

Given an graph $\mathcal{G} = (\mathcal{V}, \mathcal{E})$, where $\mathcal{V}$ is the set of $N$ nodes and $\mathcal{E} \subseteq \mathcal{V} \times \mathcal{V}$ is the set of edges. Each node $v \in \mathcal{V}$ is associated with a features vector $\mathbf{x}_v \in \mathcal{X}$, and each edge $e_{u,v} \in \mathcal{E}$ denotes a connection between node $u$ and node $v$. The graph structure can also be represented by an adjacency matrix $\mathbf{A} \in [0,1]^{N \times N}$ with $A_{u,v} = 1$ if $e_{u,v} \in \mathcal{E}$ and $A_{u,v} = 0$ if $e_{u,v} \notin \mathcal{E}$.

## 3.2  Problem Statement

We first define the concepts about node class-imbalance ratio as follows

**Definition 1.** *Suppose there are $M$ classes $\mathcal{C} = \{C_1, \ldots, C_M\}$, where $|C_i|$ is the sample number of $i$-th class. Class-Imbalance Ratio $h = \frac{\min_i(|C_i|)}{\max_j(|C_j|)}$ is defined as the ratio of the size of the smallest minority class to the largest majority class.*

Node classification is a typical node-level task where only a subset of node $\mathcal{V}_L$ with corresponding node features $\mathcal{X}_L$ and labels $\mathcal{Y}_L$ are known, and we denote the labeled set as $\mathcal{D}_L = (\mathcal{V}_L, \mathcal{X}_L, \mathcal{Y}_L)$ and unlabeled set as $\mathcal{D}_U = (\mathcal{V}_U, \mathcal{X}_U, \mathcal{Y}_U)$. The purpose of GraphMixup is to perform feature, label and edge mixups for minority classes $\mathcal{C}_S \subseteq \mathcal{C}$ to generate a synthetic set $\mathcal{D}_S = (\mathcal{V}_S, \mathcal{X}_S, \mathcal{Y}_S)$ and its corresponding edge set $\mathcal{E}_S = \{e_{v',u}|v' \in \mathcal{V}_S, u \in \mathcal{V}\}$. Then the synthesized set $\mathcal{D}_S$ can be moved into the labeled set $\mathcal{D}_L$ to obtain an updated labeled set $\mathcal{D}_O = \mathcal{D}_L \bigcup \mathcal{D}_S$. Similarly, we can obtain an updated edge set $\mathcal{E}_O = \mathcal{E} \bigcup \mathcal{E}_S$ as well as its corresponding adjacency matrix $\mathbf{A}_O$, where we set $\mathbf{A}_O[: N, : N] = \mathbf{A}$. Let $\Phi : \mathcal{V} \to \mathcal{Y}$ be a graph neural network trained on labeled data $\mathcal{D}_O$ so that it can be used to infer the labels $\mathcal{Y}_U$ of unlabeled data.

An overview of the proposed GraphMixup framework is shown in Fig. 1. The main idea of GraphMixup is to perform feature mixup to generate synthetic minority nodes $\mathcal{V}_S$ in the semantic relation spaces by a *Semantic Feature Mixup* module (see Sect. 3.3). Besides, two context-based self-supervised pretext tasks are applied to train a *Contextual Edge Mixup* module (see Sect. 3.4) that captures both local and global connections $\mathcal{E}_S$ between generated nodes $\mathcal{V}_S$ and existing nodes $\mathcal{V}$. Finally, we detail the *Reinforcement Mixup* mechanism (see Sect. 3.5), which can adaptively determine the number of samples to be generated (i.e., upsampling scale) by mixup for minority classes.

## 3.3  Semantic Feature Mixup

We propose a **Semantic Feature Extractor (SFE)** to learn semantic features and then perform feature mixup to generate minority nodes [22]. Specifically, we first construct several semantic relation spaces, then perform aggregation and transformation in each semantic space separately, and finally merge the semantic features from each space into a concatenated semantic feature.

**Semantic Relation Learning.** First, we transform the input node features to a low-dimensional hidden space, done by multiplying the features of nodes with a parameter matrix $\mathbf{W}_h \in \mathbb{R}^{F \times d}$, that is $\mathbf{h}'_i = \mathbf{W}_h \mathbf{x}_i$, where $d$ and $F$ are the dimensions of input and hidden spaces. The transformed features are then used to generate a semantic relation graph $\mathbf{G}_k$ with respect to semantic relation $k$

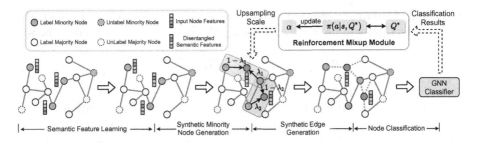

**Fig. 1.** Illustration of the proposed GraphMixp framework, which consists of the following four important steps: (1) learning semantic features by constructing semantic relation spaces; (2) generating synthetic minority nodes by semantic-level feature mixup; (3) generating synthetic edges by performing edge mixup with an edge predictor trained on two context-based self-supervised auxiliary tasks; (4) Classifying nodes with a GNN node classifier and feed the classification results back to the RL agent (reinforcement mixup module) to further update the upsampling scale.

$(1 \leq k \leq K)$. The weight coefficient $\mathbf{G}_{k,i,j}$ between node $i$ and node $j$ in the graph $\mathbf{G}_k$ can be difined as follows

$$G_{k,i,j} = \sigma\big(\Omega_k(\mathbf{h}'_i, \mathbf{h}'_j)\big), \forall e_{i,j} \in \mathcal{E} \tag{1}$$

where $\sigma = \mathrm{sigmoid}(\cdot)$ is an activation function, and $\Omega_k(\cdot)$ is a function that takes the concated features of node $i$ and $j$ as input and takes the form of a two-layer MLP. However, without any other constraints, some of the generated relation graphs may contain similar structures. More importantly, it is not easy to directly maximize the gap between different semantic relation graphs due to the non-Euclidean property of graph. Therefore, we first derive a graph descriptor $\mathbf{d}_k$ for each relation graph $\mathbf{G}_k$, as follows

$$\mathbf{d}_k = f\Big(\mathrm{Readout}\big(\mathcal{A}(\mathbf{G}_k, \mathbf{H}')\big)\Big), \tag{2}$$

where $\mathcal{A}(\cdot)$ is a two-layer graph autoencoder [10] which takes $\mathbf{H}' = \{\mathbf{h}'_1, \mathbf{h}'_2, \cdots, \mathbf{h}'_N\}$ as inputs and generates new features for each node, $\mathrm{Readout}(\cdot)$ performs global average pooling for all nodes, and $f(\cdot)$ is a fully connected layer. Note that all semantic relation graphs $\{\mathbf{G}_k\}_{k=1}^K$ share the same node features $\mathbf{H}'$, making sure that the information discovered by the feature extractor comes only from the differences between graph structures rather than node features. The loss used to train the extractor is defined as

$$\mathcal{L}_{dis} = \sum_{i=1}^{K-1} \sum_{j=i+1}^{K} \frac{\mathbf{d}_i \cdot \mathbf{d}_j^T}{\|\mathbf{d}_i\| \|\mathbf{d}_j\|}. \tag{3}$$

**Semantic Feature Learning.** Once the semantic relations are learned, the semantic features can be learned by taking the weighted sum of its neighbors,

$$\mathbf{h}_{i,k}^{(l)} = \sigma\Big( \sum_{j \in \mathcal{N}_{i,k}} \mathbf{G}_{k,i,j} \mathbf{W}^{(l,k)} \mathbf{h}_j^{(l-1)} \Big), \quad 1 \le l \le L \tag{4}$$

where $\mathbf{h}_j^{(0)} = \mathbf{x}_j$ and $\mathbf{h}_{i,k}^{(l)}$ is the semantic feature of node $i$ w.r.t relation $k$ in the $l$-th layer. $\mathcal{N}_{i,k}$ is the neighbours of node $i$ in the graph $\mathbf{G}_k$ and $\mathbf{W}^{(l,k)} \in \mathbb{R}^{F \times F}$ is a parameter matrix. Finally, the learned features from different semantic relation spaces can be **concated** to produce a final semantic feature, as follows

$$\mathbf{h}_i^{(l)} = \|_{k=1}^K \mathbf{h}_{i,k}^{(l)}. \tag{5}$$

The proposed semantic feature extractor is somewhat similar to GAT [18], i.e., learning features by considering multi-head attentions between nodes, but the difference is that we learn disentangled semantic features through imposing the constraint $\mathcal{L}_{dis}$ defined in Eq. (3). Moreover, the performance of learning with semantic feature extractor and GAT has been compared by qualitative and quantitative experiments provided in Sect. 4.4.

**Minority Node Generation.** Once features $\mathbf{H}_{\mathcal{V}}^{(L)} = \{\mathbf{h}_1^{(L)}, \mathbf{h}_2^{(L)}, \cdots, \mathbf{h}_N^{(L)}\}$ for nodes $\mathcal{V}$ have been learned by semantic feature extractor, we can perform semantic-level feature mixup to generate new samples $\mathcal{V}_S$ for minority classes. Specifically, we perform linear interpolation on sample $v$ from one target minority class with its nearest neighbor $nn(v)$ to generate a new minority node $v' \in \mathcal{V}_S$,

$$\mathbf{h}_{v'}^{(L)} = (1-\delta) \cdot \mathbf{h}_v^{(L)} + \delta \cdot \mathbf{h}_{nn(v)}^{(L)}, \quad nn(v) = \underset{u \in \{\mathcal{V}/v\}, y_u = y_v}{\arg\min} \left\| \mathbf{h}_u^{(L)} - \mathbf{h}_v^{(L)} \right\|. \tag{6}$$

where $\delta \in [0, 1]$ is a random variable, following uniform distribution. The final node embedding matrix $\mathbf{H}_O^{(L)} = [\mathbf{H}_{\mathcal{V}}^{(L)} \| \mathbf{H}_S^{(L)}]$ is obtained by concatenating the semantic features of original and generated nodes. Since node $v$ and $nn(v)$ belong to the same class and are close to each other, the generated node $v'$ should also belong to the same class. Therefore, the *label mixup* can be simplified to directly assign the same label as the source node $v$ to the newly generated node $v'$.

### 3.4 Contextual Edge Mixup

While we have generated synthetic nodes $\mathcal{V}_S$ and labels $\mathcal{Y}_S$, these new synthetic nodes are still isolated from the raw graph $\mathcal{G}$ and do not have any links with original nodes $\mathcal{V}$. Therefore, we introduce *edge mixup* to capture the connections between generated and existing nodes. To this end, we design an edge predictor that is trained on the raw node set $\mathcal{V}$ and edge set $\mathcal{E}$ and then used to predict the connectivity between generated nodes in $\mathcal{V}_S$ and existing nodes in $\mathcal{V}$. Specifically, the edge predictor is implemented as

$$\widehat{\mathbf{A}}_{v,u} = \sigma\left(\mathbf{z}_v \cdot \mathbf{z}_u^T\right); \mathbf{z}_u = \overline{\mathbf{W}}\mathbf{h}_u^{(L)}, \mathbf{z}_v = \overline{\mathbf{W}}\mathbf{h}_v^{(L)} \tag{7}$$

where $\widehat{\mathbf{A}}_{v,u}$ is the predicted connectivity between node $v$ and $u$, and $\overline{\mathbf{W}} \in \mathbb{R}^{F \times F}$ is a parameter matrix. The loss for training the edge predictor is defined as

$$\mathcal{L}_{rec} = \left\| \widehat{\mathbf{A}} - \mathbf{A} \right\|_F^2. \tag{8}$$

Since the above MSE-based matrix reconstruction in Eq. (8) only considers the connectivity between nodes based on feature similarity, it may ignore important graph structural information, so we employ two context-based self-supervised tasks to capture both local and global structural information for learning a better edge predictor.

**Context-based Self-supervised Prediction.** The first pretext task *Local-Path Prediction* is to predict the shortest path length between different node pairs. To prevent very noisy ultra-long pairwise distances from dominating the optimization, we truncate the shortest path longer than 4, which also *forces the model to focus on the local structure*. Specifically, it first randomly samples a certain amount of node pairs $\mathcal{S}$ from all node pairs $\{(v,u)|v,u \in \mathcal{V}\}$ and calculates the pairwise node shortest path length $d_{v,u} = d(v,u)$ for each node pair $(v,u) \in \mathcal{S}$. Furthermore, it groups the shortest path lengths into four categories: $C_{v,u} = 0, C_{v,u} = 1, C_{v,u} = 2$, and $C_{v,u} = 3$ corresponding to $d_{v,u} = 1, d_{v,u} = 2, d_{v,u} = 3$, and $d_{v,u} \geq 3$, respectively. The learning objective is then formulated as a multi-class classification problem, defined as follows

$$\mathcal{L}_{local} = \frac{1}{|\mathcal{S}|} \sum_{(v,u) \in \mathcal{S}} \mathcal{L}_{CE}\left( f_\omega^{(1)}\left(|\mathbf{z}_v - \mathbf{z}_u|\right), C_{v,u}\right) \tag{9}$$

where $\mathcal{L}_{CE}(\cdot,\cdot)$ denotes the cross-entropy loss and $f_\omega^{(1)}(\cdot)$ linearly maps the input to a 4-dimension vector. The second task *Global-Path Prediction* pre-obtains a set of clusters from raw node set $\mathcal{V}$ and then guides the model to *preserve global topology information* by predicting the shortest path from each node to the anchor nodes associated with cluster centers. Specifically, it first partitions the graph into $T$ clusters $\{O_1, O_2, \cdots, O_T\}$ by applying unsupervised graph partition or clustering algorithm [8,23]. Inside each cluster $O_t$ $(1 \leq t \leq T)$, the node with the highest degree is taken as corresponding cluster center, denoted as $o_t$. Then it calculates the distance $\mathbf{l}_i \in \mathbb{R}^T$ from node $v_i$ to cluster centers $\{o_k\}_{k=1}^T$. The learning objective is then formulated as a regression problem, defined as

$$\mathcal{L}_{global} = \frac{1}{|\mathcal{V}|} \sum_{v_i \in \mathcal{V}} \left\| f_\omega^{(2)}\left(\mathbf{z}_i\right) - \mathbf{l}_i \right\|^2 \tag{10}$$

where $f_\omega^{(2)}(\cdot)$ linearly maps the input to a $T$-dimension vector. Finally, the total loss to train the edge predictor is defined as

$$\mathcal{L}_{edge} = \mathcal{L}_{rec} + \mathcal{L}_{local} + \mathcal{L}_{global} \tag{11}$$

**Synthetic Edge Generation.** With the learned edge predictor, we can perform *Edge Mixup* in two different ways. The first is to directly use *continuous edges*,

$$\mathbf{A}_O[v', u] = \widehat{\mathbf{A}}_{v', u} \qquad (12)$$

where $v' \in \mathcal{V}_S$ and $u \in \mathcal{V}$. The second scheme is to obtain the *binary edges* by setting a fixed threshold value $\eta$, as follows

$$\mathbf{A}_O[v', u] = \begin{cases} 1, & \text{if } \widehat{\mathbf{A}}_{v', u} > \eta \\ 0, & \text{otherwise} \end{cases} \qquad (13)$$

The above two are both implemented in this paper, denoted as GraphMixup$_C$ and GraphMixup$_B$, and their performances are compared in the experiment part.

### 3.5   Reinforcement Mixup Mechanism

The upsampling scale, i.e., the number of synthetic samples to be generated, is important for model performance. A too-large scale may introduce redundant and noisy information, while a too-small scale is not efficient enough to alleviate the class-imbalanced problem. Therefore, instead of setting the upsampling scale $\alpha$ as a fixed hyperparameter for all minority classes and then estimating it heuristically, we use a novel reinforcement learning algorithm that adaptively updates the upsampling scale for minority classes. Formally, we model the updating process as a Markov Decision Process (MDP) [20], where the state, action, transition, reward, and termination are defined as:

- **State.** For minority class $\mathcal{C}_S$, the state $s_e$ at epoch $e$ is defined as the number of new samples, that is $s_e = \{|C_i| \cdot \alpha_i\}_{C_i \in \mathcal{C}_S}$ with $\alpha_i = \alpha_i^{init} + \kappa_i$, where $\alpha_i^{init}$ and $\kappa_i$ are the initial and cumulative values of class $C_i$.
- **Action.** RL agent updates $\{\kappa_i\}_{C_i \in \mathcal{C}_S}$ by taking action $a_e$ based on reward. We define action $a_e$ as add or minus a fixed value $\Delta\kappa$ from $\{\kappa_i\}_{C_i \in \mathcal{C}_S}$.
- **Transition.** We generate $|C_i| \cdot \alpha_i$ new synthetic nodes as defined in Eq. (6) for each minority class in the next epoch.
- **Reward.** Due to the black-box nature of GNN, it is hard to sense its state and cumulative reward. So we define a discrete reward function reward $(s_e, a_e)$ for each action $a_e$ at state $s_e$ directly based on the classification results,

$$\text{reward}(s_e, a_e) = \begin{cases} +1, & \text{if } cla_e > cla_{e-1} \\ 0, & \text{if } cla_e = cla_{e-1} \\ -1, & \text{if } cla_e < cla_{e-1} \end{cases} \qquad (14)$$

where $cla_e$ is the macro-F1 score on the validation set at epoch $e$. Therefore, Eq. (14) indicates that if the macro-F1 with action $a_e$ is higher than the previous epoch, the reward for $a_e$ is positive, otherwise it is negative.
- **Termination.** If the change of $\{\kappa_i\}_{C_i \in \mathcal{C}_S}$ among twenty consecutive epochs is no more than $\Delta\kappa$, the RL algorithm will stop, and $\{\kappa_i\}_{C_i \in \mathcal{C}_S}$ will remain fixed during the next training process. The terminal condition is defined as

$$\text{Range}\left(\{\kappa_i^{e-20}, \cdots, \kappa_i^e\}\right) \leq T_\kappa, \quad C_i \in \mathcal{C}_S \qquad (15)$$

The $Q$-learning algorithm [19] is applied to learn the above MDP. The $Q$-learning algorithm is an off-policy algorithm that seeks to find the best actions given the current state. It fits the Bellman optimality equation, as follows

$$Q^* (s_e, a_e) = \text{reward} (s_e, a_e) + \gamma \arg \max_{a'} Q^* (s_{e+1}, a') \qquad (16)$$

where $\gamma \in [0, 1]$ is a discount factor of future reward. Finally, we adopt a $\varepsilon$-greedy policy to solve Eq. (16) with an explore probability $\varepsilon$, as follows

$$\pi (a_e \mid s_e; Q^*) = \begin{cases} \text{random action} & \text{w.p. } \varepsilon \\ \arg \max_{a_e} Q^* (s_e, a) & \text{otherwise} \end{cases} \qquad (17)$$

This means that the RL agent explores new states by selecting an action at random with probability $\varepsilon$ instead of only based on the max future reward. Furthermore, the RL agent can be trained jointly with other proposed modules *in an end-to-end manner*. The results in Sect. 4.6 have validated the effectiveness of the reinforcement mixup mechanism.

### 3.6    Optimization Objective and Training Strategy

Finally, we apply a GNN classifier to obtain label prediction for node $v$, as

$$\mathbf{h}_v^{(L+1)} = \sigma \left( \widetilde{\mathbf{W}}^{(1)} \cdot [\mathbf{h}_v^{(L)} \| \mathbf{H}_O^{(L)} \cdot \mathbf{A}_O[:, v]] \right)$$
$$\widehat{\mathbf{y}}_v = \text{softmax}(\widetilde{\mathbf{W}}^{(2)} \cdot \mathbf{h}_v^{(L+1)}) \qquad (18)$$

where $\|$ denotes the concatation operation, $\widetilde{\mathbf{W}}^{(1)} \in \mathbb{R}^{F \times 2F}$ and $\widetilde{\mathbf{W}}^{(2)} \in \mathbb{R}^{M \times F}$ are parameter matrices. Besides, $\mathbf{H}_O^{(L)}$ and $\mathbf{A}_O$ are the node embedding matrix and adjacency matrix composed of both original and generated nodes. The above GNN classifier is optimized using cross-entropy loss on $\mathcal{V}_O = \mathcal{V}_L \bigcup \mathcal{V}_S$, as follows

$$\mathcal{L}_{\text{node}} = \frac{1}{|\mathcal{V}_O|} \sum_{v \in \mathcal{V}_O} \mathcal{L}_{CE}(\widehat{\mathbf{y}}_v, y_v) \qquad (19)$$

As the model performance heavily depends on the quality of embedding space and generated edges, we adopt a two-stage training strategy to make training phrase more stable. Let $\theta$, $\gamma$, $\phi$ be the parameters for semantic feature extractor, edge predictor, and node classifier, respectively. Firstly, the semantic feature extractor and edge predictor are pre-trained with loss $\mathcal{L}_{dis}$ and $\mathcal{L}_{edge}$, then the pre-trained parameters $\theta_{init}$ and $\gamma_{init}$ are used as the initialization. At the fine-tuning stage, the pre-trained encoder $\theta_{init}(\cdot)$ with a node classifier is trained under $\mathcal{L}_{node}$. The learning objective is defined as

$$\theta^*, \phi^* = \arg \min_{(\theta, \phi)} \mathcal{L}_{node}(\theta, \gamma, \phi) \qquad (20)$$

with initialization $\theta_{init}, \gamma_{init} = \arg \min_{(\theta, \gamma)} \mathcal{L}_{dis}(\theta) + \beta \mathcal{L}_{edge}(\gamma)$, where $\beta$ is the weight to balance these two losses. Since $\mathcal{L}_{dis}$ and $\mathcal{L}_{edge}$ are roughly on the same order of magnitude, without loss of generality, we set $\beta$ to 1.0 by default.

The pseudo-code of GraphMixup is shown in Algorithm 1. The computational burden of GraphMixup mainly comes from three parts: (1) semantic feature mixup $\mathcal{O}(|\mathcal{V}|dF + K|\mathcal{E}|F)$; (2) contextual edge mixup $\mathcal{O}(|\mathcal{V}|^2 F^2)$; (3) node classification $\mathcal{O}(|\mathcal{V}|F^2)$, where $d$ and $F$ are the input and hidden dimensions, and $K$ is the relation number. Since $K$ usually takes a value less than 10 in practice, the total complexity is nearly quadratic w.r.t the number of nodes $|\mathcal{V}|$ and linear w.r.t the edge number $|\mathcal{E}|$, which is in the same order as GraphSMOTE [25].

---

**Algorithm 1.** Algorithm for the proposed GraphMixup framework

---

**Input:** Feature Matrix: $\mathbf{X}$; Adjacency Matrix: $\mathbf{A}$.
**Output:** Predicted Labels $\mathcal{Y}_U$.
1: Initialize the semantic feature extractor, edge predictor and node classifier.
2: Initialize upsampling scale $\alpha_i^{init} = \frac{N}{M|C_i|}$ and $\kappa_i = 0$ for minority class $C_i \in \mathcal{C}_S$;
3: Train the feature extractor and edge predictor based on $L_{dis}$ and $L_{edge}$.
4: **while** Not Converged **do**
5:     Obtaining semantic node features $\mathbf{H}^{(L)}$ by Eq. (4) and Eq. (5);
6:     **for** class $i$ in minority classes set $\mathcal{C}_S$ **do**
7:         Calculating upsampling scale $\alpha_i = \alpha_i^{init} + \kappa_i$
8:         **for** $j \in \{0, 1, \cdots, |C_i| * \alpha_i\}$ **do**
9:             Generating new samples $j$ for minority class $i$ by Eq. (6);
10:        **end for**
11:    **end for**
12:    Training feature extractor and node classifier with $L_{node}$ by Eq. (19);

13:    **if** Eq. (15) is False **then**
14:        reward $(s_e, a_e) \leftarrow$ Eq. (14);
15:        $a_e \leftarrow$ Eq. (17);
16:        $\kappa_i \leftarrow a_e \cdot \Delta\kappa$ for $C_i \in \mathcal{C}_S$;
17:    **end if**
18: **end while**
19: **return** Predicted labels $\mathcal{Y}_U$ for unlabeled nodes $\mathcal{V}_U$.

---

## 4   Experiments

The experiments aim to answer five questions: **Q1.** How does GraphMixup perform for class-imbalanced node classification? **Q2.** Is GraphMixup robust enough to different class-imbalance ratios? **Q3.** How does semantic feature extractor (in Sect. 3.3) influence the model performance? Is GraphMixup robust to other bottleneck encoders? **Q4.** How do two self-supervised tasks (in Sect. 3.4) influence the model performance, and do they help improve performance across datasets? **Q5.** How well does the reinforcement mixup mechanism (in Sect. 3.5) work?

## 4.1    Experimental Setups

**Datasets and Hyperparameters.** The experiments are conducted on three commonly used real-world datasets. The first one is the BlogCatalog dataset [5], where 14 classes with fewer than 100 samples are taken as minority classes. The second one is the Wiki-CS dataset [12], where we consider classes with fewer than the average samples per class as minority classes. Finally, on the Cora dataset [16], we randomly selected three classes as minority classes and the rest as majority classes, where all majority classes have a training set of 20 samples and all minority classes have $20 \times h$ samples with class-imbalance ratio $h$ defaulted to 0.5. Furthermore, to evaluate the robustness of GraphMixup to different class imbalance ratios, we have varied $h \in \{0.1, 0.2, 0.3, 0.4, 0.5, 0.6\}$ in Sect. 4.3. The following hyperparameters are set for all datasets: Adam optimizer with learning rate $lr = 0.001$ and weight decay $decay = $ 5e-4; Maximum Epoch $E = 4000$; Layer number $L = 2$ with hidden dimension $F = 32$; Relation $K = 4$; Loss weights $\beta = 1.0$; Threshold $\eta = 0.5$. In the reinforcement mixup module, we set $\gamma = 1$, $\varepsilon = 0.9$, $\Delta \kappa = 0.05$. Besides, the initialization value $\alpha_i^{init} = \frac{N}{M|C_i|}$ is set class-wise for minority class $C_i \in \mathcal{C}_S$ on each dataset. Each set of experiments is run five times with different random seeds, and the average is reported as metric.

**Baselines and Metrics.** To demonstrate the power of GraphMixup, we compare it with five general baselines: (1) *Origin*: original implementation; (2) *Over-Sampling*: repeat samples directly from minority classes; (3) *Re-weight*: assign higher loss weights to samples from minority classes; (4) *SMOTE*: generate synthetic samples by interpolating in the input space, and the edges of newly generated nodes are set to be the same as source nodes; (5) *Embed-SMOTE*: an extension of SMOTE by interpolating in the embedding space. Besides, four graph-specific class-imbalanced methods: *DR-GCN*, *RA-GCN*, *GraphSMOTE*, and *ImGAGN* are also included in the comparison. Finally, based on strategies for setting edges, two varients of GraphMixup are considered, that is *GraphMixup$_B$*: the generated edges are set to binary values by thresholding as Eq. (13), and *GraphMixup$_C$*: the generated edges are set as continuous values as Eq. (12). Three evaluation metrics are adopted in this paper, including Accuracy (*Acc*), AUC-ROC, and Macro-F1. The accuracy is calculated on all test samples at once and thus may underestimate those minority classes. In contrast, both AUC-ROC and Macro-F1 are calculated for each class separately and then non-weighted average over them; thus, it can better reflect the performance on minority classes.

## 4.2    Class-Imbalanced Classification *(Q1)*

To evaluate the effectiveness of GraphMixup in class-imbalanced node classification tasks, we compare it with the other nine baselines on three real-world datasets. Table 1 shows that the improvements brought by GraphMixup are much larger than directly applying other over-sampling algorithms. For example, compared with GraphSMOTE, *GraphMixup$_C$* shows an improvement of 3.3% in *Acc* score and 3.4% in *Macro-F1* score on the Cora dataset. Moreover, both two variants of GraphMixup show significant improvements compared to almost all baselines on all datasets. Notably, we find that $GraphMixup_C$ exhibits slightly better performance than $GraphMixup_B$, which demonstrates the advantage of soft continuous edges over thresholded binary edges.

### 4.3   Robustness to Different Imbalance Ratio (Q2)

The performance under different imbalance ratios is reported in Table 2 to evaluate their robustness. Experiments are conducted in the Cora dataset by varying class imbalance ratio $h \in \{0.1, 0.2, 0.3, 0.4, 0.5, 0.6\}$. The ROC-AUC scores in Table 2 show that: (1) GraphMixup generalizes well to different imbalance ratios and achieves the best performance across all settings. (2) The improvement gain of GraphMixup is more significant when the imbalance ratio is extreme. For example, when $h = 0.1$, $GraphMixup_C$ outperforms SMOTE by 6.4%, and the gap reduces 1.5% when $h$ reaches 0.6.

### 4.4   Influence of Bottleneck Encoder (Q3)

To analyze the effectiveness of the *Semantic Feature Extractor (SFE)* proposed in Sect. 3.3 and the applicability of GraphMixup to different bottleneck encoders, we consider three common encoders: GCN [9], SAGE [4], and GAT [18]. Due to space limitations, only the performance of the AUC-ROC scores on the Cora dataset is reported. Table 3 shows that GraphMixup works well with all four bottleneck encoders, achieving the best performance. Moreover, results with SFE as the bottleneck encoder are slightly better than the other three across all methods, indicating the benefits of performing semantic-level feature mixup. Furthermore, Fig. 2 shows the correlation analysis of 128-dimensional latent features with $K = 4$ semantic relations, from which we find that only the correlation map of SFE exhibits four clear diagonal blocks, which demonstrates its excellent capability to extract highly independent semantic features.

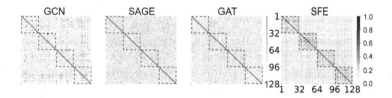

**Fig. 2.** Feature correlation analysis on the Cora dataset.

### 4.5   Self-Supervised Prediction Analysis (Q4)

To evaluate the influence of the two self-supervised prediction tasks on the performance of $GraphMixup_C$, we design four sets of experiments: the model without

**Table 1.** Performance comparison on three dataset, with best metrics <u>underline</u>.

| Methods | Cora | | | BlogCatlog | |
|---|---|---|---|---|---|
| | Acc | AUC-ROC | Macro-F1 | Acc | AUC-ROC |
| Origin | $0.718 \pm 0.002$ | $0.919 \pm 0.002$ | $0.715 \pm 0.003$ | $0.208 \pm 0.005$ | $0.583 \pm 0.004$ |
| Over-Sampling | $0.731 \pm 0.007$ | $0.927 \pm 0.006$ | $0.728 \pm 0.008$ | $0.202 \pm 0.004$ | $0.592 \pm 0.003$ |
| Re-weight | $0.728 \pm 0.009$ | $0.925 \pm 0.005$ | $0.724 \pm 0.006$ | $0.204 \pm 0.005$ | $0.785 \pm 0.004$ |
| SMOTE | $0.732 \pm 0.010$ | $0.925 \pm 0.007$ | $0.729 \pm 0.005$ | $0.206 \pm 0.004$ | $0.795 \pm 0.003$ |
| Embed-SMOTE | $0.722 \pm 0.006$ | $0.918 \pm 0.003$ | $0.721 \pm 0.004$ | $0.202 \pm 0.006$ | $0.781 \pm 0.004$ |
| DR-GNN | $0.748 \pm 0.002$ | $0.932 \pm 0.002$ | $0.744 \pm 0.004$ | $0.244 \pm 0.004$ | $0.650 \pm 0.005$ |
| RA-GNN | $0.754 \pm 0.004$ | $0.937 \pm 0.003$ | $0.755 \pm 0.003$ | $0.253 \pm 0.003$ | $0.655 \pm 0.004$ |
| GraphSMOTE | $0.742 \pm 0.003$ | $0.930 \pm 0.002$ | $0.739 \pm 0.002$ | $0.247 \pm 0.004$ | $0.644 \pm 0.005$ |
| ImGAGN | $0.757 \pm 0.002$ | $0.935 \pm 0.004$ | $0.760 \pm 0.003$ | $0.250 \pm 0.005$ | $0.657 \pm 0.004$ |
| GraphMixup$_B$ | $0.761 \pm 0.001$ | $0.934 \pm 0.002$ | $0.758 \pm 0.002$ | $0.255 \pm 0.003$ | $0.663 \pm 0.003$ |
| GraphMixup$_C$ | $\underline{0.775 \pm 0.003}$ | $\underline{0.942 \pm 0.002}$ | $\underline{0.773 \pm 0.001}$ | $\underline{0.268 \pm 0.003}$ | $\underline{0.673 \pm 0.001}$ |

| Methods | BlogCatlog | Wiki-CS | | |
|---|---|---|---|---|
| | Macro-F1 | Acc | AUC-ROC | Macro-F1 |
| Origin | $0.067 \pm 0.002$ | $0.767 \pm 0.001$ | $0.940 \pm 0.002$ | $0.735 \pm 0.001$ |
| Over-Sampling | $0.072 \pm 0.003$ | $0.779 \pm 0.002$ | $0.948 \pm 0.002$ | $0.744 \pm 0.002$ |
| Re-weight | $0.069 \pm 0.002$ | $0.761 \pm 0.002$ | $0.939 \pm 0.002$ | $0.738 \pm 0.002$ |
| SMOTE | $0.073 \pm 0.001$ | $0.780 \pm 0.004$ | $0.945 \pm 0.003$ | $0.745 \pm 0.003$ |
| Embed-SMOTE | $0.070 \pm 0.003$ | $0.750 \pm 0.005$ | $0.943 \pm 0.003$ | $0.721 \pm 0.004$ |
| DR-GNN | $0.119 \pm 0.004$ | $0.786 \pm 0.004$ | $0.950 \pm 0.003$ | $0.757 \pm 0.004$ |
| RA-GNN | $0.124 \pm 0.002$ | $0.790 \pm 0.004$ | $0.952 \pm 0.004$ | $0.764 \pm 0.004$ |
| GraphSMOTE | $0.123 \pm 0.002$ | $0.785 \pm 0.003$ | $0.955 \pm 0.004$ | $0.752 \pm 0.003$ |
| ImGAGN | $0.125 \pm 0.004$ | $0.789 \pm 0.003$ | $0.953 \pm 0.003$ | $0.761 \pm 0.004$ |
| GraphMixup$_B$ | $0.126 \pm 0.002$ | $0.792 \pm 0.002$ | $0.958 \pm 0.002$ | $0.764 \pm 0.002$ |
| GraphMixup$_C$ | $\underline{0.132 \pm 0.002}$ | $\underline{0.804 \pm 0.002}$ | $\underline{0.964 \pm 0.003}$ | $\underline{0.775 \pm 0.001}$ |

(A) Local-Path Prediction ($w/o\ LP$); (B) Glocal-Path Prediction ($w/o\ GP$); (C) both Local-Path and Global-Path Prediction ($w/o\ LP\ and\ GP$), and (D) the full model. Experiments are conducted on the Cora dataset, and ROC-AUC scores are reported as performance metrics. After analyzing the reported results in Fig. 3(a), we can observe that both Local-Path Prediction and Glocal-Path Prediction contribute to improving model performance. More importantly, applying these two tasks together can further improve performance on top of each of them, resulting in the best performance, which demonstrates the benefit of self-supervised tasks on capturing local and global topological information.

### 4.6   RL Process Analysis *(Q5)*

To demonstrate the importance of the reinforcement mixup mechanism, we remove it from GraphMixup to obtain a variant, GraphMixup-Fix, which sets a

**Table 2.** ROC-AUC under different imbalance ratios, with best metrics <u>underline</u>.

| Methods | Class-Imbalanced Ratio $h$ | | | | | |
|---|---|---|---|---|---|---|
| | 0.1 | 0.2 | 0.3 | 0.4 | 0.5 | 0.6 |
| Origin | 0.843 | 0.890 | 0.907 | 0.913 | 0.919 | 0.920 |
| Over-Sampling | 0.830 | 0.898 | 0.917 | 0.922 | 0.927 | 0.929 |
| Re-weight | 0.869 | 0.906 | 0.921 | 0.923 | 0.925 | 0.928 |
| SMOTE | 0.839 | 0.897 | 0.917 | 0.924 | 0.925 | 0.929 |
| Embed-SMOTE | 0.870 | 0.897 | 0.906 | 0.912 | 0.918 | 0.925 |
| DR-GNN | 0.890 | 0.908 | 0.921 | 0.925 | 0.929 | 0.934 |
| RA-GNN | 0.895 | 0.913 | 0.925 | 0.931 | 0.933 | 0.938 |
| GraphSMOTE | 0.887 | 0.912 | 0.923 | 0.927 | 0.930 | 0.932 |
| ImGAGN | 0.894 | 0.913 | 0.925 | 0.930 | 0.935 | 0.937 |
| GraphMixup$_B$ | 0.898 | 0.915 | 0.923 | 0.932 | 0.934 | 0.935 |
| GraphMixup$_C$ | <u>0.903</u> | <u>0.919</u> | <u>0.931</u> | <u>0.935</u> | <u>0.942</u> | <u>0.944</u> |

**Table 3.** AUC-ROC for different bottleneck encoders, with best metrics <u>underline</u>.

| Methods | Bottleneck Encoder | | | |
|---|---|---|---|---|
| | GCN | SAGE | GAT | SFE |
| Origin | 0.909 | 0.897 | 0.912 | 0.919 |
| Over-Sampling | 0.916 | 0.907 | 0.923 | 0.927 |
| Re-weight | 0.917 | 0.904 | 0.919 | 0.925 |
| SMOTE | 0.917 | 0.907 | 0.919 | 0.925 |
| Embed-SMOTE | 0.914 | 0.906 | 0.916 | 0.918 |
| DR-GNN | 0.919 | 0.915 | 0.925 | 0.929 |
| RA-GNN | 0.924 | 0.917 | 0.927 | 0.933 |
| GraphSMOTE | 0.920 | 0.914 | 0.923 | 0.930 |
| ImGAGN | 0.921 | 0.913 | 0.925 | 0.935 |
| GraphMixup$_B$ | 0.924 | 0.916 | 0.926 | 0.934 |
| GraphMixup$_C$ | <u>0.926</u> | <u>0.919</u> | <u>0.932</u> | <u>0.942</u> |

*fixed* upsampling scale $\alpha$ for all minority classes. Then, we plot the performance curves of GraphMixup-Fix and four baselines under different upsampling scales on the Cora dataset. It can be seen in Fig. 3(b) that generating more samples for minority classes helps achieve better performance when the upsampling scale $\alpha$ is smaller than 0.8 (or 1.0). However, when the upsampling scale becomes larger, keeping increasing it may result in the opposite effect, as excessive new synthesis nodes will only introduce redundant information.

Since the proposed RL agent is trained jointly with GNNs, its updating and convergence process is very important. As shown in Fig. 3(c), we visual-

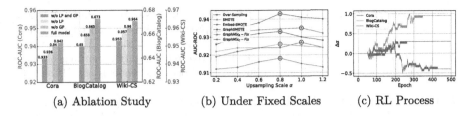

(a) Ablation Study        (b) Under Fixed Scales        (c) RL Process

**Fig. 3.** Ablation study on self-supervised tasks and analysis on reinforcement mixup.

ize the updating process of the cumulative change in upsampling ratio, i.e., $\Delta\alpha = \alpha_i - \alpha_i^{init}$. Since other modules are trained together with the RL agent in the proposed framework, the RL environment is not very stable at the beginning, so we make the RL algorithm start to run only after the first 50 epochs. When the framework gradually converges, $\Delta\alpha$ bumps for several rounds and meets the terminal condition. Finally, we find from Fig. 3(c) that $\Delta\alpha$ eventually converges to 0.3 on the Cora dataset, resulting in an upsampling scale $\alpha_i = \Delta\alpha + \alpha_i^{init} = 0.8$ with initialization value $\alpha_i^{init} = 0.5$. This corresponds to the circled result in Fig. 3(b), where GraphMixup$_C$ obtains the best performance when the upsampling scale is set as 0.8, which demonstrates the effectiveness of the reinforcement mixup mechanism, as it adaptively determines suitable upsampling scale without the need for heuristic estimation like the grid search in Fig. 3(b).

## 5    Conclusion

In this paper, we propose GraphMixup, a novel framework for improving class-imbalanced node classification on graphs. GraphMixup implements the feature, label, and edge mixup simultaneously in a unified framework in an end-to-end manner. Extensive experiments have shown that the GraphMixup framework outperforms other leading methods for the class-imbalanced node classification.

**Acknowledgement.** This work is supported by the Science and Technology Innovation 2030 - Major Project (No. 2021ZD0150100) and National Natural Science Foundation of China (No. U21A20427).

## References

1. Bunkhumpornpat, C., Sinapiromsaran, K., Lursinsap, C.: Safe-Level-SMOTE: safe-level-synthetic minority over-sampling technique for handling the class imbalanced problem. In: Theeramunkong, T., Kijsirikul, B., Cercone, N., Ho, T.-B. (eds.) PAKDD 2009. LNCS (LNAI), vol. 5476, pp. 475–482. Springer, Heidelberg (2009). https://doi.org/10.1007/978-3-642-01307-2_43
2. Chawla, N.V., Bowyer, K.W., Hall, L.O., Kegelmeyer, W.P.: Smote: synthetic minority over-sampling technique. J. Artif. Intell. Res. **16**, 321–357 (2002)

3. Ghorbani, M., Kazi, A., Baghshah, M.S., Rabiee, H.R., Navab, N.: Ra-GCN: graph convolutional network for disease prediction problems with imbalanced data. Med. Image Anal. **75**, 102272 (2022)
4. Hamilton, W., Ying, Z., Leskovec, J.: Inductive representation learning on large graphs. In: Advances in Neural Information Processing Systems, pp. 1024–1034 (2017)
5. Huang, X., Li, J., Hu, X.: Label informed attributed network embedding. In: Proceedings of the Tenth ACM International Conference on Web Search and Data Mining, pp. 731–739 (2017)
6. Japkowicz, N., Stephen, S.: The class imbalance problem: a systematic study. Intell. Data Anal. **6**(5), 429–449 (2002)
7. Johnson, J.M., Khoshgoftaar, T.M.: Survey on deep learning with class imbalance. J. Big Data **6**(1), 1–54 (2019). https://doi.org/10.1186/s40537-019-0192-5
8. Karypis, G., Kumar, V.: A fast and high quality multilevel scheme for partitioning irregular graphs. SIAM J. Sci. Comput. **20**(1), 359–392 (1998)
9. Kipf, T.N., Welling, M.: Semi-supervised classification with graph convolutional networks. arXiv preprint arXiv:1609.02907 (2016)
10. Kipf, T.N., Welling, M.: Variational graph auto-encoders. arXiv preprint arXiv:1611.07308 (2016)
11. Leevy, J.L., Khoshgoftaar, T.M., Bauder, R.A., Seliya, N.: A survey on addressing high-class imbalance in big data. J. Big Data **5**(1), 1–30 (2018). https://doi.org/10.1186/s40537-018-0151-6
12. Mernyei, P., Cangea, C.: Wiki-CS: a wikipedia-based benchmark for graph neural networks. arXiv preprint arXiv:2007.02901 (2020)
13. More, A.: Survey of resampling techniques for improving classification performance in unbalanced datasets. arXiv preprint arXiv:1608.06048 (2016)
14. Qu, L., Zhu, H., Zheng, R., Shi, Y., Yin, H.: ImGAGN: imbalanced network embedding via generative adversarial graph networks. arXiv preprint arXiv:2106.02817 (2021)
15. Rout, N., Mishra, D., Mallick, M.K.: Handling imbalanced data: a survey. In: Reddy, M.S., Viswanath, K., K.M., S.P. (eds.) International Proceedings on Advances in Soft Computing, Intelligent Systems and Applications. AISC, vol. 628, pp. 431–443. Springer, Singapore (2018). https://doi.org/10.1007/978-981-10-5272-9_39
16. Sen, P., Namata, G., Bilgic, M., Getoor, L., Galligher, B., Eliassi-Rad, T.: Collective classification in network data. AI Mag. **29**(3), 93 (2008)
17. Shi, M., Tang, Y., Zhu, X., Wilson, D., Liu, J.: Multi-class imbalanced graph convolutional network learning. In: Proceedings of the Twenty-Ninth International Joint Conference on Artificial Intelligence (IJCAI-20) (2020)
18. Veličković, P., Cucurull, G., Casanova, A., Romero, A., Lio, P., Bengio, Y.: Graph attention networks. arXiv preprint arXiv:1710.10903 (2017)
19. Watkins, C.J., Dayan, P.: Q-learning. Mach. Learn. **8**(3–4), 279–292 (1992)
20. White, C.C., III., White, D.J.: Markov decision processes. Eur. J. Oper. Res. **39**(1), 1–16 (1989)
21. Wu, L., Lin, H., Gao, Z., Tan, C., Li, S., et al.: Self-supervised on graphs: contrastive, generative, or predictive. arXiv preprint arXiv:2105.07342 (2021)
22. Wu, L., Lin, H., Xia, J., Tan, C., Li, S.Z.: Multi-level disentanglement graph neural network. Neural Comput. Appl. **34**(11), 9087–9101 (2022). https://doi.org/10.1007/s00521-022-06930-1

23. Wu, L., Yuan, L., Zhao, G., Lin, H., Li, S.Z.: Deep clustering and visualization for end-to-end high-dimensional data analysis. In: IEEE Transactions on Neural Networks and Learning Systems (2022)
24. Zhang, H., Cisse, M., Dauphin, Y.N., Lopez-Paz, D.: mixup: beyond empirical risk minimization. arXiv preprint arXiv:1710.09412 (2017)
25. Zhao, T., Zhang, X., Wang, S.: GraphSMOTE: imbalanced node classification on graphs with graph neural networks. In: Proceedings of the 14th ACM International Conference on Web Search and Data Mining, pp. 833–841 (2021)

# Private and Federated Learning

# Non-IID Distributed Learning
# with Optimal Mixture Weights

Jian Li[1], Bojian Wei[1,2], Yong Liu[3(✉)], and Weiping Wang[1]

[1] Institute of Information Engineering, Chinese Academy of Sciences, Beijing, China
{lijian9026,wangweiping}@iie.ac.cn
[2] School of Cyber Security, University of Chinese Academy of Sciences,
Beijing, China
[3] Gaoling School of Artificial Intelligence, Renmin University of China,
Beijing, China
liuyonggsai@ruc.edu.cn

**Abstract.** Distributed learning can well solve the problem of training model with large-scale data, which has attracted much attention in recent years. However, most existing distributed learning algorithms set uniform mixture weights across clients when aggregating the global model, which impairs the accuracy under Non-IID (Not Independently or Identically Distributed) setting. In this paper, we present a general framework to optimize the mixture weights and show that our framework has lower expected loss than the uniform mixture weights framework theoretically. Moreover, we provide strong generalization guarantee for our framework, where the excess risk bound can converge at $\mathcal{O}(1/n)$, which is as fast as centralized training. Motivated by the theoretical findings, we propose a novel algorithm to improve the performance of distributed learning under Non-IID setting. Through extensive experiments, we show that our algorithm outperforms other mainstream methods, which coincides with our theory.

**Keywords:** Distributed learning · Excess risk bound · Optimal mixture weights

## 1 Introduction

With the development of Internet of Things (IoT) technology and the popularity of intelligent terminal devices, it is difficult to continue the traditional centralized training of machine learning algorithms. Fortunately, distributed learning [2, 23, 29] provides an effective way for model training with large-scale data.

In standard distributed learning, many clients collaboratively train a global model under the coordination of a central server, where the training samples are splitted on clients to alleviate the storage and computing limitations of the

---

J. Li and B. Wei—Contribute equally to this work.

---

**Supplementary Information** The online version contains supplementary material available at https://doi.org/10.1007/978-3-031-26412-2_33.

server. Recently, there are many studies analyze the properties of distributed learning from different perspectives [3,18]. EasyASR [26] provides a distributed platform for training and serving large-scale automatic speech recognition models, which supports both pre-defined networks and user-customized networks. In high-dimensional settings, Acharya et al. [1] analyzed the communication problem in distributed learning, and obtained algorithms that enjoy optimal error with logarithmic communication by relaxing the boundedness assumptions. Random topologies [30] is applied to tackle the unreliable networks problem in distributed learning, which can achieve comparable convergence rate to centralized learning. MRE [24] aims to reduce the error in IID distributed learning, where the error bound meets the existing lower bounds up to poly-logarithmic factors. Deep Q-learning based synchronization policies [31] is used for parameter server-based distributed training, which can generalize to different cluster environments and datasets.

With the increasing attention paid to privacy-preserving, data sharing in distributed learning has been strictly limited. Thus, federated learning [19,28] was proposed to maintain or improve the performance of distributed learning while protecting users' privacy. However, local distributions on different clients may be different due to the personality of users, which brings us the Non-IID problem in distributed learning, where the global model is difficult to converge to the optimal solution. To solve the problem, FedAvg [19] runs multi-step SGD (stochastic gradient descent) on clients and aggregates local models by periodically communications. FedProx [16] introduces a proximal term to constrain the divergence between local models and the global model. There are many other studies try to tackle the Non-IID problem by different algorithms [11,22,27].

Although many related work has presented various methods to improve the performance of Non-IID distributed learning, the mixture weights for model aggregation are usually fixed as $\frac{n_k}{n}$, where $n_k$ denotes the sample size on the $k$-th client and $n$ denotes the total sample size among all clients. In fact, the uniform mixture weights is a good choice under IID settings, but it can not reflect the heterogeneous characteristics. When we use the uniform mixture weights to Non-IID distributed learning, the global model will shift to the local model with larger sample size, which impairs the performance of global model. Furthermore, most existing algorithms for distributed learning lack generalization guarantees, which restricts their portability to some extent.

In this paper, we present a general framework for Non-IID distributed learning, where the mixtures weights can be optimized to promote the global model to converge to the optimal solution. We also provide a strong generalization guarantee for our distributed learning framework based on local Rademacher complexity. The main contributions of this paper are summarized as follows.

- **A general framework.** We present a general framework for Non-IID distributed learning, where the mixtures weights are optimized together with model parameters by minimizing the objective. Theoretically, we demonstrate that our framework has lower expected loss than distributed learning with uniform mixture weights.
- **A strong generalization guarantee.** To our best knowledge, we derive a sharper excess risk bound for Non-IID distributed learning with convergence

rate of $\mathcal{O}(1/n)$ based on local Rademacher complexity for the first time, which meets the current bounds in centralized learning and is much faster than the existing bounds of distributed learning.

- **A novel algorithm.** Based on our general framework and theoretical findings, we propose a novel distributed learning algorithm DL-opt, which optimizes the mixture weights on server-side with validation samples and constrains local Rademacher complexity with an additional regularization term on the local objective. Through extensive experiments, we show that DL-opt significantly outperforms distributed learning with uniform mixture weights.
- **An effective extension to federated learning.** We extend DL-opt to federated learning, named FedOMW, which executes periodical communications to alternately optimize the mixture weights and model parameters. We illustrate that FedOMW performs better than FedAvg and FedProx with a clear margin through a series of experiments.

## 2    Preliminaries and Notations

In this section, we first introduce the Non-IID distributed learning scenario and then demonstrate the general notations used in this paper.

In a Non-IID distributed learning scenario, there are $K$ clients and a central server, where the local training samples $\mathcal{D}_k = \{(\boldsymbol{x}_{ik}, y_{ik})\}_{i=1}^{n_k}$ on the $k$-th client are drawn from a local distribution $\rho_k$ with size of $n_k$. The underlying local distribution is different on different clients: $\rho_i \neq \rho_j$. We denote $n = \sum_{k=1}^{K} n_k$ the total number of training samples across all clients.

Let $\mathcal{H}$ be the hypothesis space consisting of labeling functions $h : \mathcal{X} \to \mathcal{Y}$, where $\mathcal{X}$ denotes the input space and $\mathcal{Y}$ denotes the output space. The labeling function is formed as $h(\boldsymbol{x}) = \boldsymbol{w}^T \phi(\boldsymbol{x})$, where $\boldsymbol{w}$ denotes the vector of learnable parameters and $\phi(\cdot)$ denotes a fixed feature mapping. Let $\ell : \mathcal{Y} \times \mathcal{Y} \to \mathbb{R}_+$ be the loss function, we denote the loss space associated to $\mathcal{H}$ by $\mathcal{G} = \{\ell(h(\boldsymbol{x}), y) | h \in \mathcal{H}\}$. For the $k$-th client, we define the expected loss as

$$\mathcal{L}_k(h; \boldsymbol{w}) = \mathbb{E}_{(\boldsymbol{x}, y) \sim \rho_k} \left[ \ell(h(\boldsymbol{x}), y) \right],$$

and the corresponding empirical loss as

$$\widehat{\mathcal{L}}_k(h; \boldsymbol{w}) = \frac{1}{n_k} \sum_{i=1}^{n_k} \ell(h(\boldsymbol{x}_{ik}), y_{ik}).$$

The target of distributed learning is to obtain a global model. For traditional distributed learning, the global model $\boldsymbol{w}$ is obtained by aggregating local models ($\boldsymbol{w}_k$ denotes the local model on the $k$-th client) which are trained locally to converge on clients. For federated learning, the global model is obtained by alternately performing client-side local training and server-side model aggregating. The objective of distributed learning can be formed as

$$\min_{\boldsymbol{w} \in \mathcal{H}} \sum_{k=1}^{K} p_k \widehat{\mathcal{L}}_k(h; \boldsymbol{w}),$$

where $p_k$ is the mixture weight of the $k$-th client.

Many distributed learning algorithms use uniform mixture weights ($p_k = n_k/n$) to aggregate local models, which make use of local sample sizes, and they work well when local training samples are independently drawn from an identical distribution (IID situation). However, under Non-IID setting, the uniform mixture weights fails to capture the discrepancy among local distributions. To this end, we consider optimizing the mixture weights together with $\boldsymbol{w}$ to get an optimal solution, which can truely minimize the objective under Non-IID setting.

Therefore, we present a general framework for Non-IID distributed learning, and the general objective is defined as

$$\min_{\boldsymbol{w}\in\mathcal{H}}\min_{\boldsymbol{p}\in\mathcal{P}} \widehat{\mathcal{L}}(h;\boldsymbol{w},\boldsymbol{p}) = \sum_{k=1}^{K} p_k \widehat{\mathcal{L}}_k(h;\boldsymbol{w}), \tag{1}$$

where $\boldsymbol{p} = [p_1,\cdots,p_K]$ is the vector of mixture weights and $\mathcal{P}$ is the parameter space of $\boldsymbol{p}$. The above objective is the empirical general loss of Non-IID distributed learning, and the corresponding expected general loss is $\mathcal{L}(h;\boldsymbol{w},\boldsymbol{p}) = \sum_{k=1}^{K} p_k \mathcal{L}_k(h;\boldsymbol{w})$.

In our framework, we relax the constraint on $\boldsymbol{p}$: the sum of $K$ elements is 1 ($\sum_{k=1}^{K} p_k = 1$) and the value of each element $p_k$ is in $(0,1)$ and expands the range of feasibility, where each element $p_k$ can take any value under the assumption that $|p_k|$ is upper bounded by $\tau$ ($\tau < \infty$). Thus, many gradient-based algorithms can be applied to optimize the mixture weights to get the optimal solution.

## 3  Generalization Guarantee

We introduce two specific estimators in hypothesis space $\mathcal{H}$: The empirical estimator is defined as

$$\widehat{h} = \operatorname*{arg\,min}_{\boldsymbol{w}\in\mathcal{H},\boldsymbol{p}\in\mathcal{P}} \widehat{\mathcal{L}}(h;\boldsymbol{w},\boldsymbol{p}),$$

and the optimal estimator is defined as

$$h^* = \operatorname*{arg\,min}_{\boldsymbol{w}\in\mathcal{H},\boldsymbol{p}\in\mathcal{P}} \mathcal{L}(h;\boldsymbol{w},\boldsymbol{p}),$$

where $\widehat{h}$ minimizes the empirical general loss of Non-IID distributed learning and $h^*$ minimizes the corresponding expected general loss.

Excess risk is often used to represent the generalization performance of an estimator [7], which measures the gap between the empirical estimator and the optimal estimator. We define the excess risk of Non-IID distributed learning as follows:

$$\mathcal{L}(\widehat{h};\boldsymbol{w},\boldsymbol{p}) - \mathcal{L}(h^*;\boldsymbol{w},\boldsymbol{p}). \tag{2}$$

In the previous work, generalization error of centralized learning and distributed learning with uniform mixture weights has been widely studied, which is actually the upper bound of excess risk [7]. Through Rademacher complexity [6,12] and stability theory [8,9], the current generalization error bounds for centralized learning and distributed learning with uniform mixture weights converge at $\mathcal{O}(1/\sqrt{n})$. The convergence rate of generalization error bounds for centralized learning can be improved to $\mathcal{O}(1/n)$ by local Rademacher complexity [5] and some advanced techniques in stability. However, there is no existing work on the generalization error bounds for distributed learning with convergence rate of $\mathcal{O}(1/n)$.

In the following part, we will derive a sharper excess risk bound for Non-IID distributed learning to give a stronger generalization guarantee on the general framework defined in this paper.

## 3.1 Excess Risk Bound with Local Rademacher Complexity

We first introduce two important assumptions.

**Assumption 1.** *Assume that the loss function is $\lambda$-Lipschitz continuous and upper bounded by $M$ ($M > 0$), that is*

$$|\ell(h(\boldsymbol{x}), y) - \ell(h(\boldsymbol{x}'), y')| \leq \lambda \, |h(\boldsymbol{x}) - h(\boldsymbol{x}')|$$

*and*

$$|\ell(h(\boldsymbol{x}), y)| < M, \quad \forall (\boldsymbol{x}, y) \in \mathcal{X} \times \mathcal{Y}.$$

**Assumption 2.** *Assume that the loss function satisfies the Bernstein condition: For some $B > 0$, it holds that*

$$\mathbb{E}\left[\ell(h(\boldsymbol{x}), y) - \ell(h^*(\boldsymbol{x}), y)\right]^2 \leq B\left(\mathcal{L}(h; \boldsymbol{w}, \boldsymbol{p}) - \mathcal{L}(h^*; \boldsymbol{w}, \boldsymbol{p})\right).$$

Assumption 1 is a commonly used assumption in generalization analysis [4, 25], where many loss functions meet this condition, such as hinge loss, margin loss and their variants. Meanwhile, Assumption 2 is widely used in statistical learning theory, such as local Rademacher complexity [5,10,20] and stability [8,13,14].

The empirical Rademacher complexity of $\mathcal{G}$ is formed as

$$\widehat{\mathcal{R}}(\mathcal{G}) = \mathbb{E}_\epsilon\left[\sup_{h \in \mathcal{H}} \sum_{k=1}^{K} \frac{p_k}{n_k} \sum_{i=1}^{n_k} \epsilon_{ik} \ell(h(\boldsymbol{x}_{ik}), y_{ik})\right],$$

and the empirical Rademacher complexity of $\mathcal{H}$ is formed as

$$\widehat{\mathcal{R}}(\mathcal{H}) = \mathbb{E}_\epsilon\left[\sup_{h \in \mathcal{H}} \sum_{k=1}^{K} \frac{p_k}{n_k} \sum_{i=1}^{n_k} \epsilon_{ik} h(\boldsymbol{x}_{ik})\right],$$

where $\{\epsilon_{ik}\}_{i \in [n_k]}^{k \in [K]}$ are independent Rademacher variables sampling uniformly from $\{-1, +1\}$.

We define the empirical local Rademacher complexity of $\mathcal{G}$ and $\mathcal{H}$ on training samples as follows:

$$\widehat{\mathcal{R}}(\mathcal{G}, r) = \widehat{\mathcal{R}}\left(\{\ell_h | \ell_h \in \mathcal{G}, \mathbb{E}[\ell_h - \ell_{h^*}]^2 \leq r\}\right),$$
$$\widehat{\mathcal{R}}(\mathcal{H}, r) = \widehat{\mathcal{R}}\left(\{h | h \in \mathcal{H}, \mathbb{E}[\ell_h - \ell_{h^*}]^2 \leq r\}\right),$$

where $\ell_h = \ell(h(\boldsymbol{x}), y)$, for simplicity.

Without loss of generality, the feature mapping $\phi(\cdot)$ mentioned Sect. 2 is assumed to be upper bounded by $\kappa$: $\kappa = \sup_{\boldsymbol{x} \in \mathcal{X}} \|\phi(\boldsymbol{x})\| \leq \infty$, and it is often used in kernel methods. Moreover, the depth and structure of neural networks are becoming deeper and more diverse in order to model more complex tasks, so the value of hidden vector after feature mapping should be constrained by normalization or other techniques to avoid training problems such as no convergence. Thus, this condition also applies to current deep learning methods.

We present the excess risk bound for Non-IID distributed learning with local Rademacher complexity in the following theorem.

**Theorem 1 (Excess Risk Bound).** *Let $d$ be the VC dimension of hypothesis space $\mathcal{H}$, $\psi(r)$ be a sub-root function and $r^*$ be the fixed point of $\psi(r)$. Assume that $\|\boldsymbol{w}\|^2 \leq \frac{r}{\lambda^2 \kappa^2}$, under Assumption 1 and 2, $\forall \delta \in (0, 1]$ and $\forall r \geq r^*$, it holds that*

$$\psi(r) \geq \frac{\lambda \tau \sqrt{2dK \log\left[\frac{en}{d}\right]}}{n} \geq \lambda \mathbb{E}\left[\widehat{\mathcal{R}}(\mathcal{H})\right]. \tag{3}$$

*With probability at least $1 - \delta$, the following bound holds:*

$$\mathcal{L}(\widehat{h}; \boldsymbol{w}, \boldsymbol{p}) - \mathcal{L}(h^*; \boldsymbol{w}, \boldsymbol{p}) \leq \frac{705}{B} r^* + \frac{(11M + 27B) \log(1/\delta)}{n}. \tag{4}$$

The proof is in Appendix A.1 of the supplementary file.

In Theorem 1, we derive a sharper excess risk bound for Non-IID distributed learning related to our general framework, which provides strong generalization guarantee for algorithms under our framework.

According to Theorem 1, the fixed point $r^*$ dominates the excess risk of Non-IID distributed learning, which is affected by local Rademacher complexity with the sub-root function $\psi(r)$. In (3), we have proved that local Rademacher complexity can converge at $\mathcal{O}(1/n)$. Meanwhile, the rest part in (4) also has the convergence rate of $\mathcal{O}(1/n)$ due to the self-bounding property [5]:

$$\mathbb{E}\left[\widehat{\mathcal{R}}(\mathcal{H})\right] \leq \widehat{\mathcal{R}}(\mathcal{H}) + \sqrt{\frac{2\mathbb{E}\left[\widehat{\mathcal{R}}(\mathcal{H})\right] \log(1/\delta)}{n}}.$$

Therefore, if we ignore the constants and other unrelevant factors, the excess risk bound for Non-IID distributed learning can be rewritten as

$$\mathcal{L}(\widehat{h}; \boldsymbol{w}, \boldsymbol{p}) - \mathcal{L}(h^*; \boldsymbol{w}, \boldsymbol{p}) \leq \mathcal{O}\left(\frac{\sqrt{K}}{n}\right).$$

Note that $K$ is the total number clients, so the above result shows that the generalization performance can be worse if there are too many clients participated in a Non-IID distributed learning, which is consistent with the actual application. Moreover, when there is only one client, the above result degrades into $\mathcal{O}(1/n)$, which meets the best excess risk bound for centralized learning. Thus, our general framework for Non-IID distributed learning has strong generalization guarantee and we provide a sharper excess risk bound for Non-IID distributed learning with convergence rate of $\mathcal{O}(1/n)$ for the first time.

### 3.2  Comparison with Current Framework

In this part, we will demonstrate that our general framework has lower expected loss than current distributed learning framework.

Traditional distributed learning framework uses uniform mixture weights ($p_k = \frac{n_k}{n}$) to aggregate the global model, we denote the empirical estimator of this uniform framework by $\widehat{h}_{\mathrm{uf}}$. The expected loss of $\widehat{h}_{\mathrm{uf}}$ is formed as $\mathcal{L}(\widehat{h}_{\mathrm{uf}}; \boldsymbol{w}, \boldsymbol{p})$, we show that the expected loss of our framework is upper bounded by the expected loss of uniform framework in the following theorem.

**Theorem 2.** *Suppose that the distributed learning framework is applied to solve a binary-classification task, where $\mathcal{Y} = \{0,1\}$. Let the loss function $\ell$ be the cross-entropy loss. For some $\mathcal{P}$ and $\rho_k$ ($k \in [K]$), we have*

$$\mathcal{L}(\widehat{h}; \boldsymbol{w}, \boldsymbol{p}) \leq \mathcal{L}(\widehat{h}_{\mathrm{uf}}; \boldsymbol{w}, \boldsymbol{p}). \tag{5}$$

*Proof (Proof of Theorem 2).* For simplicity, we consider that there are only two clients and the sample sizes $n_k$ are equal. Given a single point $\boldsymbol{x}$, we assume that the local distribution on the first client satisfies $\rho_1(\boldsymbol{x}, 0) = 0$, $\rho_1(\boldsymbol{x}, 1) = 1$, and the local distribution on the second client satisfies $\rho_2(\boldsymbol{x}, 0) = \frac{1}{2}$, $\rho_2(\boldsymbol{x}, 1) = \frac{1}{2}$. We denote $\mathrm{Pr}_0$ the probability that $h$ assigns to class 0 and $\mathrm{Pr}_1 = 1 - \mathrm{Pr}_0$ that $h$ assigns to class 1.

Note that the objective is the weighted sum of local loss functions and the mixture weights are $[p_1, p_2] = [\frac{1}{2}, \frac{1}{2}]$ in the uniform framework. Then, the expected loss of $\widehat{h}_{\mathrm{uf}}$ is

$$
\begin{aligned}
\mathcal{L}(\widehat{h}_{\mathrm{uf}}; \boldsymbol{w}, \boldsymbol{p}) =& \mathbb{E}_{(\boldsymbol{x}, y)} \left[ -\log \mathrm{Pr}_y \right] = \frac{1}{4} \log \frac{1}{\mathrm{Pr}_0} + \frac{3}{4} \log \frac{1}{\mathrm{Pr}_1} \\
=& \frac{1}{4} \log 4 + \frac{3}{4} \log \frac{4}{3} + \mathrm{KL} \left( [\frac{1}{4}, \frac{3}{4}] \,\|\, [\mathrm{Pr}_0, \mathrm{Pr}_1] \right) \\
\geq& \frac{1}{4} \log 4 + \frac{3}{4} \log \frac{4}{3},
\end{aligned}
$$

where $\mathrm{KL}(\cdot)$ denotes the Kullback-Leibler divergence. Furthermore, we set $\mathrm{Pr}_0 = \frac{1}{4}$ and $\mathrm{Pr}_1 = \frac{3}{4}$. Thus, the expected loss of uniform framework becomes

$$\mathcal{L}(\widehat{h}_{\mathrm{uf}}; \boldsymbol{w}, \boldsymbol{p}) = \frac{1}{4} \log 4 + \frac{3}{4} \log \frac{4}{3} = \log \frac{4}{\sqrt[4]{27}}.$$

Under the same settings, the expected loss of our framework is

$$\mathcal{L}(\widehat{h}; \boldsymbol{w}, \boldsymbol{p}) = \min_{\boldsymbol{w} \in \mathcal{H}} \min_{\boldsymbol{p} \in \mathcal{P}} \left\{ \log \frac{1}{\text{Pr}_0}, \frac{1}{2} \log \frac{1}{\text{Pr}_0} + \frac{1}{2} \log \frac{1}{\text{Pr}_1} \right\}$$
$$= \min \left\{ \log \frac{4}{3}, \frac{1}{2} \log 4 + \frac{1}{2} \log \frac{4}{3} \right\}$$
$$= \log \frac{4}{3} \le \log \frac{4}{\sqrt[4]{27}}.$$

This completes the proof.

According to Theorem 2, our general framework for Non-IID distributed learning has lower expected loss than current uniform framework, which demonstrates that our framework has surpassed uniform framework theoretically.

## 4 Algorithm

Our general framework aims to optimize mixture weights $\boldsymbol{p}$ together with the global model $\boldsymbol{w}$, and we relax the constraints on $\boldsymbol{p}$, so we consider applying SGD to get the optimal mixture weights.

### 4.1 DL-opt: Distributed Learning with Optimal Mixture Weights

In Non-IID distributed learning, training samples are stored on $K$ clients, where local distribution vary across clients because of personal properties. Here, we use classic distributed learning method to train $\boldsymbol{w}_k$ on the $k$-th client locally until it converges. Meanwhile, we define an additional constraint $\|\boldsymbol{w}\|^2 \le \frac{r}{\lambda^2 \kappa^2}$ on $\|\boldsymbol{w}\|$ in Sect. 3 to provide strong generalization guarantee for our general framework, which indicates that the norm of $\boldsymbol{w}$ can not be very large. To this end, we add $\|\boldsymbol{w}\|$ to the local objective as a regularization term. Thus, the local objective on the $k$-th client is formed as

$$\min_{\boldsymbol{w}_k \in \mathcal{H}} L_k(\mathcal{D}_k) = \frac{1}{n_k} \sum_{i=1}^{n_k} \ell(h_k(\boldsymbol{x}_{ik}), y_{ik}) + \gamma \|\boldsymbol{w}_k\|, \tag{6}$$

where $\gamma$ is a tunable parameter and $h_k = \boldsymbol{w}_k^T \phi(\boldsymbol{x}_{ik})$ related to Sect. 2.

On the other hand, it is unwise to optimize $\boldsymbol{p}$ on client-side, because the properties of other clients can not be integrated on the $k$-th client, which may cause the global model to deviate from the global optima. In order to capture global information and improve the performance of the aggregated global model $\boldsymbol{w} = \sum_{k=1}^{K} p_k \boldsymbol{w}_k$, we optimize the mixture weights on the central server with a group of validation samples $\mathcal{D}_{\text{val}}$, where the validation samples $\mathcal{D}_{\text{val}}$ are randomly sampled from each client in a small proportion.

After local training, local models $\boldsymbol{w}_k$ are uploaded to the central server to aggregate the global model $\boldsymbol{w}$. Then, we use SGD to optimize $\boldsymbol{p}$ on the validation samples $\mathcal{D}_{\text{val}}$. We first get $K$ predictions $[h_1(\boldsymbol{x}^{\text{val}}), \cdots, h_K(\boldsymbol{x}^{\text{val}})]$ by $K$ local models. Next, we combine the mixture weights with the prediction vector as $\sum_{k=1}^{K} p_k h_k(\boldsymbol{x}^{\text{val}})$. Note that $\mathcal{D}_{\text{val}}$ is relevant to the task, so we can use the same loss function as local objectives to construct the central objective, that is

$$\min_{\boldsymbol{p} \in \mathcal{P}} L_{\boldsymbol{p}}(\mathcal{D}_{\text{val}}) = \frac{1}{n_{\text{val}}} \sum_{j=1}^{n_{\text{val}}} \ell \left( \sum_{k=1}^{K} p_k h_k(\boldsymbol{x}_j^{\text{val}}), y_j^{\text{val}} \right), \tag{7}$$

where $n_{\text{val}}$ is the sample size of $\mathcal{D}_{\text{val}}$.

The pseudo code of DL-opt is listed in Algorithm 1.

---

**Algorithm 1.** DL-opt (Distributed Learning with Optimal Mixture Weights)

---

**Input:** $\bigcup_{k=1}^{K} \mathcal{D}_k$ (local samples), $\mathcal{D}_{\text{val}}$ (validation samples), $\boldsymbol{w}^0 \in \mathcal{H}$ (model parameters), $\boldsymbol{p}^0 \in \mathcal{P}$ (mixture weights), $T_l, T_c$ (total iterations of local training and central training), $\eta_w, \eta_p$ (learning rates).

**Output:** $\boldsymbol{w}^{\text{global}}$.

**Client-side local training**

1: $K$ clients download the initial model: $\boldsymbol{w}_k^0 \leftarrow \boldsymbol{w}^0 \ (k = 1, 2, \cdots, K)$
2: **for** $k = 1, 2, \cdots, K$ **do**
3:    **for** $t = 1, 2, \cdots, T_l$ **do**
4:       $\boldsymbol{w}_k^t = \boldsymbol{w}_k^{t-1} - \eta_w \nabla_{\boldsymbol{w}_k} L_k(\mathcal{D}_k)$
5:    **end for**
6: **end for**

**Server-side central training and aggregating**

1: $K$ clients upload local models $\boldsymbol{w}_k^{T_l} \ (k = 1, 2, \cdots, K)$
2: **for** $t = 1, 2, \cdots, T_c$ **do**
3:    $\boldsymbol{p}^t = \boldsymbol{p}^{t-1} - \eta_p \nabla_{\boldsymbol{p}} L_{\boldsymbol{p}}(\mathcal{D}_{\text{val}})$
4: **end for**
5: $\boldsymbol{w}^{\text{global}} = \sum_{k=1}^{K} p_k^{T_c} \boldsymbol{w}_k^{T_l}$

---

### 4.2  FedOMW: Federated Learning Version of DL-opt

Federated learning is a new distributed learning paradigm preserving users' privacy, which is a rising star in recent years. In addition to the encryption and compression techniques, federated learning uses an alternating communication mechanism to train the global model (shown in FedAvg [19]). In this part, we extend DL-opt to federated learning and propose a novel Non-IID federated learning algorithm FedOMW (Federated Learning with Optimal Mixture Weights).

The local objective and central objective in FedOMW are the same as DL-opt, because both of them are induced from our general framework for Non-IID distributed learning. The main difference between FedOMW and DL-opt is that

FedOMW executes client-side local training and server-side central training periodically instead of training local models to converge. Moreover, the validation samples can not be sampled from clients for privacy issues. To this end, the feature vectors after feature mapping and encrypting can be uploaded to the central server in a small proportion. Alternatively, we can also train a generator locally on each client and use it to generator several samples on the central server to get $\mathcal{D}_{\text{val}}$. For some very common learning tasks such as sentiment analysis of comments and next-word prediction, the validation samples are easy to get without sharing or uploading from clients. For example, if we apply distributed learning to solve a flower recognition task, we can easily get many flower pictures from Internet, which are used to construct $\mathcal{D}_{\text{val}}$ after preprocessing, and the whole process of constructing $\mathcal{D}_{\text{val}}$ will not bring privacy issues. Thus, the strategy of introducing $\mathcal{D}_{\text{val}}$ is not difficult to implement in federated learning.

We list the pseudo code of FedOMW in Algorithm 2.

---

**Algorithm 2.** FedOMW (Federated Learning with Optimal Mixture Weights)

---

**Input:** $\bigcup_{k=1}^{K} \mathcal{D}_k$ (local samples), $\mathcal{D}_{\text{val}}$ (validation samples), $\boldsymbol{w}^0 \in \mathcal{H}$ (model parameters), $\boldsymbol{p}^0 \in \mathcal{P}$ (mixture weights), $\mathcal{T}$ (total communication rounds), $\mathcal{T}_l, \mathcal{T}_c$ (total iterations of local training and central training), $\eta_w, \eta_p$ (learning rates).
**Output:** $\boldsymbol{w}^{\text{global}}$.

**Server-side central training and aggregating**

1: $K$ clients download the initial model: $\boldsymbol{w}_k^0 \leftarrow \boldsymbol{w}^0$ $(k = 1, 2, \cdots, K)$
2: **for** $\nu = 1, 2, \cdots, \mathcal{T}$ **do**
3:    $\boldsymbol{w}_k^{\mathcal{T}_l} \leftarrow$ Client-side local training $(k = 1, 2, \cdots, K)$
4:    **for** $t = 1, 2, \cdots, \mathcal{T}_c$ **do**
5:       $\boldsymbol{p}^t = \boldsymbol{p}^{t-1} - \eta_p \nabla_p L_p(\mathcal{D}_{\text{val}})$
6:    **end for**
7:    $\boldsymbol{w}^\nu = \sum_{k=1}^{K} p_k^{\mathcal{T}_c} \boldsymbol{w}_k^{\mathcal{T}_l}$
8: **end for**
9: $\boldsymbol{w}^{\text{global}} = \boldsymbol{w}^{\mathcal{T}}$

**Client-side local training**

1: **for** $k = 1, 2, \cdots, K$ **do**
2:    **for** $t = 1, 2, \cdots, \mathcal{T}_l$ **do**
3:       $\boldsymbol{w}_k^t = \boldsymbol{w}_k^{t-1} - \eta_w \nabla_{\boldsymbol{w}_k} L_k(\mathcal{D}_k)$
4:    **end for**
5: **end for**
6: $K$ clients upload local models $\boldsymbol{w}_k^{\mathcal{T}}$ $(k = 1, 2, \cdots, K)$

---

## 5  Experiments

In this section, we evaluate all algorithms on various real-world datasets with Non-IID partitioning.

## 5.1 Experimental Setup

In the following experiments, we mainly focus on multi-classification task, so the input space and output space can be expressed as $\mathcal{X} \in \mathbb{R}^{d_x}$ and $\mathcal{Y} \in \mathbb{R}^C$, where $d_x$ denotes the input dimension and $C$ denotes the output dimension related to $C$ classes. We use cross-entry as the loss function. As shown in Sect. 2, the model $h$ is formed as $h(x) = w^T \phi(x)$. Here, we use random Fourier feature [21] as the feature mapping, that is $\phi(x) = \frac{1}{\sqrt{D}} \cos(\Omega^T x + b)$, where $\phi : \mathbb{R}^{d_x} \to \mathbb{R}^D$, $\Omega \in \mathbb{R}^{d_x \times D}$, $b \in \mathbb{R}^D$. According to [21], the entries in matrix $\Omega$ obey Gaussian distribution with $\Omega \sim \mathcal{N}(0, 1/\sigma^2)$ and the elements in vector $b$ are uniformly sampled from $[0, 2\pi]$. We set $D = 2000$ for the following datasets.

**Real-World Datasets.** The real-world datasets come from LIBSVM Data[1], which provides both training and testing data publicly. To construct a Non-IID partitioning setup [15,17], we first divide the original training datasets into training samples $\bigcup_{k=1}^K \mathcal{D}_k$ and validation samples $\mathcal{D}_{\text{val}}$ according to the ratio of $8 : 2$. Then, we split the training samples across 50 clients using a Dirichlet distribution [27] $\text{Dir}_K(0.01)$ to get the local training samples $\mathcal{D}_k$ for each client, the original testing datasets are used to evaluate the performance of the global model. The statistical information of all the datasets are listed in Table 1.

All the experiments are conducted on a Linux server equipped with two NVIDIA GeForce 2080ti, and all the algorithms are implemented by Pytorch[2]. We tune all the hyperparameters by grid search and list the best results in Table (Appendix B.1 of the supplementary file).

**Table 1.** Statistical information of datasets.

| Datasets | Training size | Testing size | Dimensions | Classes |
|---|---|---|---|---|
| *usps* | 7291 | 2007 | 256 | 10 |
| *pendigits* | 7494 | 3498 | 16 | 10 |
| *satimage* | 4435 | 2000 | 36 | 6 |
| *letter* | 15000 | 5000 | 16 | 26 |
| *dna* | 2000 | 1186 | 180 | 3 |
| *mnist* | 60000 | 10000 | $28 \times 28$ | 10 |

## 5.2 Experiments of Distributed Learning

In this part, we compare DL-opt to distributed learning with uniform mixture weights (abbreviated as DL-u). To ensure the fairness of comparison, we tune local learning rate $\eta_w$ for DL-u and apply the same value to DL-opt. We set the epoch of local training as 200 and the epoch of central training as 100. The initial mixture weights $p^0$ in DL-opt is the same as DL-u ($p_k = n_k/n$).

---

[1] Available at https://www.csie.ntu.edu.tw/~cjlin/libsvmtools/.

[2] Codes are available at https://github.com/Bojian-Wei/Non-IID-Distributed-Learning-with-Optimal-Mixture-Weights.

**Table 2.** Test accuracy (%) of distributed learning algorithms on Non-IID datasets.

| Algorithms | Datasets | | | | | |
|---|---|---|---|---|---|---|
| | usps | pendigits | satimage | letter | dna | mnist |
| DL-u | 91.53 ± 0.25 | 96.13 ± 0.18 | 80.05 ± 0.25 | 57.28 ± 0.50 | 91.03 ± 0.31 | 95.56 ± 0.24 |
| DL-opt | **92.49 ± 0.09** | **96.53 ± 0.35** | **83.25 ± 0.27** | **62.95 ± 1.31** | **92.80 ± 0.31** | **96.29 ± 0.13** |

We run each experiment with 3 random seeds and record the average and standard deviation in Table 2. In Table 2, we bold the result results and underline the results which are not significantly worse than the best one. As shown in Table 2, we observe that DL-opt is significantly better than DL-u with confidence level 95% and DL-opt generally outperforms DL-u with a clear margin (more than 5% on *letter*). This illustrates that our general framework is effective in dealing with Non-IID distributed learning, which is consistent with our theoretical findings.

### 5.3    Experiments of Federated Learning

We also propose a federated learning algorithm FedOMW based on our framework. It is well know that FedAvg [19] and FedProx [16] are two mainstream algorithms in federated learning with uniform mixture weights. Thus, we conduct a comparative experiment to compare FedOMW with the two methods.

**Table 3.** Test accuracy (%) of federated learning algorithms on Non-IID datasets.

| Algorithms | Datasets | | | | | |
|---|---|---|---|---|---|---|
| | usps | pendigits | satimage | letter | dna | mnist |
| FedAvg | 90.82 ± 0.26 | 95.45 ± 0.14 | 79.52 ± 0.19 | 51.17 ± 0.64 | 90.30 ± 0.01 | 95.15 ± 0.22 |
| FedProx | 90.73 ± 0.19 | 95.23 ± 0.23 | 79.35 ± 0.16 | 51.20 ± 0.63 | 89.32 ± 0.22 | 95.13 ± 0.21 |
| FedOMW | textbf92.81 ± 0.02 | **96.92 ± 0.04** | **81.97 ± 0.25** | **63.57 ± 0.46** | **90.98 ± 0.59** | **96.29 ± 0.05** |

To ensure the fairness of comparison, we tune local learning rate $\eta_w$ for FedAvg and apply the same value to FedProx and FedOMW. We set the epoch of local training as 2, the epoch of central training as 100 and the total communication round as 100. The initial mixture weights $p^0$ in FedOMW remains $p_k = n_k/n$.

We also run each experiment with 3 random seeds and record the average and standard deviation in Table 3, and we also bold the result results and underline the results which are not significantly worse than the best one. According to Table 3, it is obvious that FedOMW performs significantly (confidence level 95%) better than FedAvg and FedProx, and FedOMW yields a marginal improvement up to 12% (on *letter*) compared to the other algorithms. Moreover, in Fig 1, we find that FedOMW not only performs better than the other algorithms, but also converges much faster. More experimental results can be found in Appendix B.2

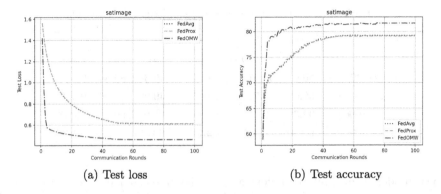

(a) Test loss                              (b) Test accuracy

**Fig. 1.** Results of federated learning algorithms on Non-IID *satimage*.

of the supplementary file. Therefore, `FedOMW` is an effective algorithm to tackle the Non-IID problem in federated learning, and our general framework is proved to be well applied in classic distributed learning and federated learning scenarios.

### 5.4   Ablation Study

There are two important components in our framework: the optimization of mixture weights $p$ and the regularization term of $\|w\|$, where the optimization of $p$ is the key strategy to improve the performance of distributed learning under Non-IID settings. In order to analyze the contribution of these two components to the proposed algorithms, we conduct an ablation experiment on both `DL-opt` and `FedOMW`. We report the results in Table 4 and Table 5, where -p denotes the algorithm only with the optimization of $p$, -w denotes the algorithm only with the regularization term of $\|w\|$ and -non denotes the algorithm without the two components.

**Table 4.** Ablation results of `DL-opt` on Non-IID datasets.

| Algorithms | Datasets | | | | | |
|---|---|---|---|---|---|---|
| | *usps* | *pendigits* | *satimage* | *letter* | *dna* | *mnist* |
| DL-opt-non | 91.61 | 94.51 | 80.25 | 57.02 | 90.13 | 95.68 |
| DL-opt-w | 91.69 | 94.80 | 80.45 | 57.04 | 90.14 | 95.70 |
| DL-opt-p | 92.42 | 95.94 | 82.80 | 63.28 | 93.41 | 96.35 |
| DL-opt | **92.58** | **96.28** | **83.20** | **63.42** | **93.59** | **96.40** |

As shown in Table 4 and Table 5, we find that the performance of `DL-u` (equal to `DL-opt-non`) and `FedAvg` (equal to `FedOMW-non`) can be improved markedly by only optimizing mixture weights $p$, which indicates the effectiveness of correcting

**Table 5.** Ablation results of `FedOMW` on Non-IID datasets.

| Algorithms | Datasets | | | | | |
|---|---|---|---|---|---|---|
| | usps | pendigits | satimage | letter | dna | mnist |
| FedOMW-non | 90.63 | 95.51 | 79.75 | 49.38 | 89.54 | 95.37 |
| FedOMW-w | 90.68 | 95.71 | 79.65 | 49.90 | 89.66 | 95.39 |
| FedOMW-p | 92.36 | 96.77 | 82.25 | 61.94 | 90.39 | 96.34 |
| FedOMW | **92.48** | **96.83** | **82.35** | **64.66** | **90.56** | **96.39** |

local models' contributions before getting the global model in our framework. Moreover, the performance can be further improved by constraining $\|w\|$, which coincides with our generalization theory.

### 5.5 Experiments of Mixture Weights

We visualize the mixture weights of 5 clients via central training in Fig 2. DL-u uses fixed uniform mixture weights, so $p$ won't change during training. DL-opt optimizes the mixture weights on $\mathcal{D}_{\mathrm{val}}$ through central training and the target is to minimize the classification loss. Thus, DL-opt adaptively assigns bigger mixture weights to the local model with smaller classification loss on $\mathcal{D}_{\mathrm{val}}$. Combined with the above experiments, we can conclude that our min-min framework improves the performance of Non-IID distributed learning by selecting the optimal mixture weights.

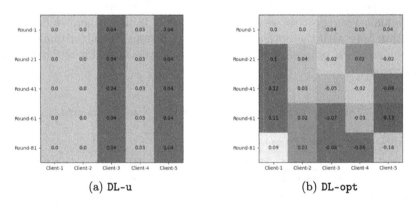

(a) DL-u                    (b) DL-opt

**Fig. 2.** Mixture weights via central training on Non-IID *letter*.

## 6   Conclusion

In this paper, we present a general framework for Non-IID distributed learning, which optimizes the mixture weights together with model parameters to obtain the optimal combination of local models. Compared to the classic distributed learning with uniform mixture weights, we demonstrate that our framework has lower expected loss theoretically. Furthermore, we provide a strong generalization guarantee for our framework based on local Rademacher complexity, where the excess risk bound can converge at $\mathcal{O}(1/n)$. Driven by our framework and theory, we propose an improved algorithm for Non-IID distributed learning and extend it to federated learning, where both of them perform significantly better than the current methods. The proof techniques in this paper may pave a way for studying generalization properties in other learning scenarios. Furthermore, we will study optimization errors and convergence guarantees in the future work.

**Acknowledgement.** This work was supported in part by the Excellent Talents Program of Institute of Information Engineering, CAS, the Special Research Assistant Project of CAS, the Beijing Outstanding Young Scientist Program (No. BJJWZYJH012019100020098), Beijing Natural Science Foundation (No. 4222029), and National Natural Science Foundation of China (No. 62076234, No. 62106257).

# References

1. Acharya, J., Sa, C.D., Foster, D.J., Sridharan, K.: Distributed learning with sublinear communication. In: ICML 2019, vol. 97, pp. 40–50 (2019)
2. Arjevani, Y., Shamir, O.: Communication complexity of distributed convex learning and optimization. In: NIPS 2015, pp. 1756–1764 (2015)
3. Aviv, R.Z., Hakimi, I., Schuster, A., Levy, K.Y.: Asynchronous distributed learning: adapting to gradient delays without prior knowledge. In: ICML 2021, vol. 139, pp. 436–445 (2021)
4. Bartlett, P.L., Boucheron, S., Lugosi, G.: Model selection and error estimation. Mach. Learn. **48**, 85–113 (2002)
5. Bartlett, P.L., Bousquet, O., Mendelson, S.: Local Rademacher complexities. Ann. Stat. **33**(4), 1497–1537 (2005)
6. Bartlett, P.L., Mendelson, S.: Rademacher and gaussian complexities: risk bounds and structural results. J. Mach. Learn. Res. **3**(Nov), 463–482 (2002)
7. Bottou, L., Bousquet, O.: The tradeoffs of large scale learning. In: Advances in Neural Information Processing Systems, vol. 21 (NIPS), pp. 161–168 (2008)
8. Bousquet, O., Elisseeff, A.: Stability and generalization. J. Mach. Learn. Res. **2**, 499–526 (2002)
9. Bousquet, O., Klochkov, Y., Zhivotovskiy, N.: Sharper bounds for uniformly stable algorithms. In: COLT, pp. 610–626 (2020)
10. Cortes, C., Kloft, M., Mohri, M.: Learning kernels using local Rademacher complexity. In: Advances in Neural Information Processing Systems, vol. 26 (NIPS), pp. 2760–2768 (2013)
11. Karimireddy, S.P., Kale, S., Mohri, M., Reddi, S.J., Stich, S.U., Suresh, A.T.: SCAFFOLD: stochastic controlled averaging for federated learning. In: ICML 2020, vol. 119, pp. 5132–5143 (2020)
12. Koltchinskii, V., Panchenko, D.: Empirical margin distributions and bounding the generalization error of combined classifiers. Ann. Stat. 1–50 (2002)
13. Kutin, S., Niyogi, P.: Almost-everywhere algorithmic stability and generalization error. In: Proceedings of the 18th Conference in Uncertainty in Artificial Intelligence (UAI), pp. 275–282 (2002)

14. Lange, T., Braun, M.L., Roth, V., Buhmann, J.M.: Stability-based model selection. In: Advances in Neural Information Processing Systems, vol. 15 (NIPS), pp. 617–624 (2002)
15. Li, T., Sahu, A.K., Talwalkar, A., Smith, V.: Federated learning: challenges, methods, and future directions. IEEE Sig. Process. Mag. **37**(3), 50–60 (2020)
16. Li, T., Sahu, A.K., Zaheer, M., Sanjabi, M., Talwalkar, A., Smith, V.: Federated optimization in heterogeneous networks. In: MLSys (2020)
17. Lin, S.B., Wang, D., Zhou, D.X.: Distributed kernel ridge regression with communications. J. Mach. Learn. Res. **21**(93), 1–38 (2020)
18. Liu, Y., Liu, J., Wang, S.: Effective distributed learning with random features: improved bounds and algorithms. In: ICLR 2021 (2021)
19. McMahan, B., Moore, E., Ramage, D., Hampson, S., Arcas, B.A.: Communication-efficient learning of deep networks from decentralized data. In: AISTATS 2017, vol. 54, pp. 1273–1282 (2017)
20. Oneto, L., Ghio, A., Ridella, S., Anguita, D.: Local Rademacher complexity: sharper risk bounds with and without unlabeled samples. Neural Netw. **65**, 115–125 (2015)
21. Rahimi, A., Recht, B.: Random features for large-scale kernel machines. In: NIPS 2007, pp. 1177–1184 (2007)
22. Reddi, S.J., et al.: Adaptive federated optimization. In: ICLR 2021 (2021)
23. Richards, D., Rebeschini, P., Rosasco, L.: Decentralised learning with random features and distributed gradient descent. In: ICML 2020, vol. 119, pp. 8105–8115 (2020)
24. Sharif-Nassab, A., Salehkaleybar, S., Golestani, S.J.: Order optimal one-shot distributed learning. In: NeurIPS 2019, pp. 2165–2174 (2019)
25. Vapnik, V.: The Nature of Statistical Learning Theory. Springer, New York (2000). https://doi.org/10.1007/978-1-4757-3264-1
26. Wang, C., Cheng, M., Hu, X., Huang, J.: EasyASR: a distributed machine learning platform for end-to-end automatic speech recognition. In: AAAI 2021, pp. 16111–16113 (2021)
27. Wang, J., Liu, Q., Liang, H., Joshi, G., Poor, H.V.: Tackling the objective inconsistency problem in heterogeneous federated optimization. In: NeurIPS 2020 (2020)
28. Wei, B., Li, J., Liu, Y., Wang, W.: Federated learning for non-IID data: from theory to algorithm. In: PRICAI 2021, vol. 13031, pp. 33–48 (2021)
29. Woodworth, B.E., Patel, K.K., Srebro, N.: Minibatch vs local SGD for heterogeneous distributed learning. In: NeurIPS 2020 (2020)
30. Yu, C., et al.: Distributed learning over unreliable networks. In: ICML 2019, vol. 97, pp. 7202–7212 (2019)
31. Zhu, R., Yang, S., Pfadler, A., Qian, Z., Zhou, J.: Learning efficient parameter server synchronization policies for distributed SGD. In: ICLR 2020 (2020)

# Marginal Release Under Multi-party Personalized Differential Privacy

Peng Tang[1,2], Rui Chen[3], Chongshi Jin[1,2], Gaoyuan Liu[1,2],
and Shanqing Guo[1,2(✉)]

[1] Key Laboratory of Cryptologic Technology and Information Security Ministry of
Education, Shandong University, Qingdao, China
{tangpeng,guoshanqing}@sdu.edu.cn,
{jinchongshi,gaoyuan_liu}@mail.sdu.edu.cn
[2] School of Cyber Science and Technology, Shandong University, Qingdao, China
[3] College of Computer Science and Technology, Harbin Engineering University,
Harbin, China
ruichen@hrbeu.edu.cn

**Abstract.** Given a set of local datasets held by multiple parties, we study
the problem of learning marginals over the integrated dataset while satisfying differential privacy for each local dataset. Different from existing works
in the multi-party setting, our work allows the parties to have different
privacy preferences for their data, which is referred to as the multi-party
personalized differential privacy (PDP) problem. The existing solutions
to PDP problems in the centralized setting mostly adopt sampling-based
approaches. However, extending similar ideas to the multi-party setting
cannot satisfactorily solve our problem. On the one hand, the data owned
by multiple parties are usually not identically distributed. Sampling-based
approaches will incur a serious distortion in the results. On the other hand,
when the parties hold different attributes of the same set of individuals, sampling at the tuple level cannot meet parties' personalized privacy
requirements for different attributes. To address the above problems, we
first present a mixture-of-multinomials-based marginal calculation approach, where the global marginals over the stretched datasets are formalized as a multinomial mixture model. As such, the global marginals
over the original datasets can be reconstructed based on the calculated
model parameters with high accuracy. We then propose a privacy budget segmentation method, which introduces a privacy division composition
strategy from the view of attributes to make full use of each party's privacy budget while meeting personalized privacy requirements for different
attributes. Extensive experiments on real datasets demonstrate that our
solution offers desirable data utility.

**Keywords:** Personalized differential privacy · Multiple party ·
Marginal release

---

S. Guo and R. Chen—Co-corresponding authors.

# 1 Introduction

In many real-life applications, a mass of data are stored among multiple distributed parties [28]. There are two typical multi-party settings: *horizontally partitioned* and *vertically partitioned*. In the former setting, it is assumed that all the local databases have the same schema and that the parties possess different individuals' information. In the latter one, all of the local datasets are over the same set of individuals, and each party observes a subset of attributes of the individuals. Calculating the marginals over such distributed data can lead to better decision-making. However, since such data may contain highly sensitive personal information, calculating the marginals in the multi-party setting needs to be conducted in a way that no private information is revealed to other participating entities or any other potential adversaries. In the multi-party setting, differential privacy has been widely used in the distributed data analysis [3,8,23,30,32]. All these above studies afford the same level of privacy protection for the individuals of all the local datasets. However, it is common that the parties have different expectations regarding their data's acceptable level of privacy. That is, users in different local datasets can have different privacy needs, where some users are extremely restrictive while others are relatively loose. As a real-world example, an analyst may want to do medical research on a hospital's data. This requires integrated data from different hospital departments. Some departments that treat sensitive diseases may require a higher level of privacy needs than others. As another practical example, medical researchers may want to study a potential correlation between travel patterns and certain types of illnesses. This requires integrated data from different sources, such as an airline reservation system and a hospital database. As medical data is usually more sensitive, the hospital may have a higher privacy need. In the above scenarios, the data analyst employing differential privacy has limited options. Setting high-level global privacy to satisfy all the local datasets will introduce a large amount of noise into the analysis outputs, resulting in poor utility. While, setting a lower privacy level may force the analyst to exclude the local datasets with strict privacy needs from analysis, which may also significantly harm utility.

This leads to the multi-party personalized differential privacy (PDP) problem. To achieve PDP in the centralized setting, Jorgensen et al. [18] propose an advanced method $\mathcal{PE}$. Analogous to the exponential mechanism [21], for each item in a marginal, $\mathcal{PE}$ calculates its noisy count by sampling from an output set. Specifically, the data owner first calculates the item's true count. Based on this count, the owner can compute the score of the true count and the score of the other noisy counts, and then sample one from these counts according to their scores. However, such a method cannot be extended to the multi-party setting. As $\mathcal{PE}$ requires knowing the true global count of each item. While in the multi-party setting, to guarantee differential privacy for each local dataset, it is not allowed for each party to learn the true global count. Other solutions [18,20] mostly adopt sampling-based approaches. Such approaches capture the additional randomness about the input data by employing non-uniform random sampling at the tuple level to yield the precise level of privacy required by each individual. Unfortunately, extending similar ideas to the multi-party setting cannot satisfactorily solve our problem. In the horizontally partitioned setting, the data owned by multiple parties are usually not identically distributed. Then there will be a serious distortion in the results calculated by sampling-based approaches, i.e., the marginals calculated over the non-uniformly sampled datasets are not equal to the marginals calculated over the original datasets. In the vertically partitioned setting, the parties hold different attributes of the same individuals, and the attributes in different local datasets have different privacy requirements. Then, sampling will be "invalid" to adjust privacy preference because sampling at the tuple level cannot meet the parties' personalized privacy requirements for different attributes.

### 1.1 Contributions

To address the above challenges, we first present a mixture-of-multinomials-based marginal calculation approach for the horizontally partitioned setting. In this approach, stretching [2] is used to adjust the parties' different privacy preferences while avoiding the error caused by sampling, and the global marginals can be formalized as a multinomial mixture model. Thus, it is possible first to calculate marginals over the stretched datasets and then accurately reconstruct the global marginals over the original datasets by calculating the model parameters.

For the vertically partitioned setting, we propose a privacy budget segmentation method, which can adjust privacy preferences from the view of the attribute. This method elaborately divides the privacy budget of each party into multiple parts, and let the parties assemble some different teams. Each team calculates an intermediate result by consuming part of the privacy budget. Based on these intermediate results, this method can reconstruct the marginal by employing consistency post-processing. Using such a privacy division composition strategy, this method can fully use each party's privacy budget while satisfying personalized privacy requirements for different attributes.

We conduct an extensive experimental study over several real datasets. The experimental results suggest that our methods are practical to offer desirable data utility.

## 2  Related Work

There exist three kinds of most relevant works, i.e., personalized privacy in the centralized setting, multi-party differential privacy, and local differential privacy.

In the centralized setting, personalized privacy allows the users have quite different expectations regarding the acceptable level of privacy for their data. A line of work, started by Xiao and Tao [36], introduce personalized privacy for $k$-anonymity, and present a new generalization framework called *personalized anonymity*. For differential privacy, Alaggan et al. [2] develop the privacy notion called *heterogeneous differential privacy*, which considers differential privacy with non-uniform privacy guarantees. Following, Jorgensen et al. [18] propose the privacy definition called *personalized differential privacy* (PDP), where users specify a personal privacy requirement for their data, and introduce an advanced method $\mathcal{PE}$ for achieving PDP. Recently, Kotsogiannis et al. [20] study the problem of privacy-preserving data sharing, wherein only a subset of the records in a database is sensitive. To pursue higher data utility while satisfying personalized differential privacy, Niu et al. [27] propose a utility-aware personalized Exponential mechanism. These approaches inspire us to initiate a new approach to solving our problem. However, unlike the centralized setting, in the distributed setting, each party is not allowed to reveal the sensitive personal information that contained in their local datasets to other parties. Besides, the data owned by multiple parties are usually not identically distributed, and the attributes of the same individuals may have different privacy requirements. These new challenges are exactly the focus of our work.

In the multi-party setting, differential privacy has been widely used in the distributed data analysis [4,7,13,15,25]. Besides, there are some works [3,8,23,30,32] for differentially privately data publishing in the multi-party setting. Using the published integrated data, the marginals can also be calculated. Different from our work, all these studies afford the same level of privacy protection for the individuals of all the local datasets. In contrast, our work aims to satisfy each party's different privacy preferences.

In the distributed scenario, another kind of differential privacy exists, i.e., Local Differential Privacy (LDP). There exist some studies of personalized differential privacy in the *local setting* [6,14,16,26,29,35,37]. Both multi-party differential privacy and LDP do not require a trusted data aggregate. However, as discussed in [34], in LDP, each user independently perturbs their own input before the aggregation on an untrusted server. This results in a large error of $O(\sqrt{N})$ in the output, where $N$ denotes the number of users. While in multi-party differential privacy, there is a complementary synergy between secure multiparty computation and differential privacy. Multi-party differential privacy can maintain the same level of accuracy as in centralized differential privacy. The final output has only an error of $O(1)$.

## 3    Preliminaries

Differential privacy [11] is a recent privacy definition that provides a strong privacy guarantee. Naturally, differential privacy is built upon the concept of *neighboring databases*. Two databases $D$ and $\hat{D}$ are neighbors if they differ on at most one record. Differential privacy can be defined as follows.

**Definition 1.** *A randomized algorithm $\varphi$ achieves $\varepsilon$-differential privacy, if for any pair of neighboring databases $D$ and $\hat{D}$, and all $\mathcal{O} \subseteq Range(\varphi)$,*

$$\Pr\left(\varphi(D) \in \mathcal{O}\right) \leq e^{\varepsilon} \times \Pr\left(\varphi(\hat{D}) \in \mathcal{O}\right), \tag{1}$$

*where the probability $\Pr\left(\cdot\right)$ is taken over coin tosses of $\varphi$.*

A fundamental concept for achieving differential privacy is *sensitivity* [11]. Let $F$ be a function that maps a database into a fixed-size vector of real numbers. For all neighboring databases $D$ and $\hat{D}$, the sensitivity of $F$ is: $S\left(F\right) = \max\limits_{D,\hat{D}} \left\| F\left(D\right) - F\left(\hat{D}\right) \right\|_1$, where $\|\cdot\|_1$ denotes the $L_1$ norm. For a function $F$ whose outputs are real, differential privacy can be achieved by the *Laplace mechanism* [11]. This mechanism works by adding random noise to the true outputs. The noise is drawn from a Laplace distribution with the probability density function $p(x) = \frac{1}{2\lambda} e^{-|x|/\lambda}$, where the scale $\lambda = S\left(F\right)/\varepsilon$ is determined by both the function's sensitivity $S\left(F\right)$ and the privacy budget $\varepsilon$.

# 4   Problem Formulation

## 4.1   System and Threat Models

Following the common convention [5,19,30] in the fields of privacy, we consider a *semi-trusted* curator in our setting. With the assistance of the curator, $K$ parties calculate the marginals over the integrated dataset collaboratively. Both the parties and the curator are semi-trusted (i.e., "honest-but-curious"). That is, the parties and the curator will correctly follow the designed protocols, but act in a "curious" fashion that they may infer private information other than what they are allowed to learn (e.g., sensitive information about the tuples in the local datasets). Our threat model also considers collusion attacks. In particular, there exist two kinds of collusion attacks. One kind is collusion attacks among the parties, and the other is collusion attacks between some parties and the curator.

In our problem, there is a complementary synergy between secure multiparty computation and differential privacy. Together they can prevent attackers from inferring sensitive information about the input local datasets using either intermediate results or outputs. Certainly, this requires an additional assumption of all parties and the curator being computationally bounded in the protocol. Therefore, in our privacy model, the overall scheme actually satisfies *computational differential privacy* [22,34].

## 4.2   Problem Definition

In the problem of *multi-party marginal calculation under personalized differential privacy*, there are $K$ parties (i.e., data owners), each of which $P_k$ ($1 \leq k \leq K$) holds a local dataset $D_k$ and specifies a privacy budget $\varepsilon_k$. The attributes contained in $D_k$ can be either numerical or categorical. Over the local datasets, the $K$ parties would like to jointly calculate the marginal of a given attribute set $\mathcal{X}$, while meeting multi-party personalized differential privacy. Multi-party personalized differential privacy is a kind of computational differential privacy, defined below.

**Definition 2.** *There are $K$ parties. All parties are assumed to be computationally bounded, and each of them specifies a privacy budget $\varepsilon_k$. A randomized* *algorithm $\varphi$ achieves multi-party personalized differential privacy, if the computing is secure according to secure multiparty computation, and for any two sets of datasets $\{D_1, \ldots, D_K\}$ and $\{\hat{D}_1, \ldots, \hat{D}_K\}$, where there exists a $k$ in $\{1, 2, \cdots, K\}$, $D_k$ and $\hat{D}_k$ are neighbors $(|D_k \oplus \hat{D}_k| = 1)$, and for any other $k' \neq k$ in $\{1, 2, \ldots, K\}$, $D_{k'} = \hat{D}_{k'}$, and for all $\mathcal{O} \subseteq Range(\varphi)$,*

$$\Pr\left(\varphi(\bigcup_{k=1}^{K} D_k) \in \mathcal{O}\right) \leq e^{\varepsilon_k} \times \Pr\left(\varphi(\bigcup_{k=1}^{K} \hat{D}_k) \in \mathcal{O}\right).$$

There are two typical multi-party settings: horizontally partitioned setting and vertically partitioned setting. In the former setting, it is assumed that all the local datasets have the same schema (i.e., attribute set) $\mathcal{A} = \{A_1, \ldots, A_n\}$ and that a single individual's information is exclusively possessed by a single party, and the given attribute set $\mathcal{X} \subseteq \mathcal{A}$. In the latter one, all of the local datasets are over the same set of individuals that are identified by a common identifier attribute. $\mathcal{A}_i$ denotes the set of attributes observed by $P_i$. It is assumed that, for any two local datasets $D_i$ and $D_j$, $\mathcal{A}_i \cap \mathcal{A}_j = \emptyset$. The attribute set $\mathcal{X} = \bigcup_{k=1}^{K} \mathcal{X}_k$ and $\mathcal{X}_k \subseteq \mathcal{A}_k$. In the vertically partitioned setting, it is common to assume that different parties share common identifiers of the users and hold mutually exclusive sets of attributes [17, 23, 24]. If the parties have overlapping attributes, they can send their data schemas to the curator to constructs exclusive sets of attributes as a preprocessing step of our solution. Since data schemas are considered public information, such a process does not lead to privacy breaches.

## 5    Baseline Solutions and Limitations

### 5.1    Horizontally Partitioned Setting

To solve the problem in the horizontally partitioned setting, there exist three kinds of baseline solutions. Firstly, a straightforward method lets each party add noise of different levels to the local marginals before sharing them with the curator. However, this will lead to the global marginals containing multiple noises, making the results useless (as discussed in Sect. 7.2). Secondly, in the centralized setting, Jorgensen et al. [18] propose an advanced method $\mathcal{PE}$ to achieve PDP. Analogous to the exponential mechanism [21], $\mathcal{PE}$ calculates its noisy count by sampling from an output set for each item in a marginal. However, such a method cannot be extended to the multi-party setting. As $\mathcal{PE}$ requires knowing the true global count of each item. While in the multi-party setting, to guarantee differential privacy for each local dataset, it is not allowed for each party to learn the true global count. Thirdly, a sampling-based solution can be proposed. Specifically, each party $P_k$ first takes sampling on the tuples in $D_k$ with the probability $p_k = \frac{e^{\varepsilon_k} - 1}{e^{\varepsilon_{\max}} - 1}$ to obtain a sampled dataset $\widetilde{D}_k$, where $\varepsilon_{\max} = \max\{\varepsilon_1, \varepsilon_2, \ldots, \varepsilon_K\}$. Based on these sampled datasets, the curator and the parties can calculate a noisy marginal of $\mathcal{X}$ using $\varepsilon_{max}$ as the privacy parameter. However, the data owned by multiple parties are usually not identically distributed. Sampling with different probabilities on multiple datasets will seriously distort the results calculated in the sampling-based approaches.

*Example 1.* There exist three parties $P_1, P_2, P_3$. Each $P_k$ ($1 \leq k \leq K$) holds a local medical dataset $D_k$ and specifies a privacy budget $\varepsilon_k$. Let the size of each local dataset be $|D_1| = |D_2| = |D_3| = 100$, and $\varepsilon_1 = 0.1, \varepsilon_2 = 0.3, \varepsilon_3 = 0.5$. Each local dataset contains some cancer patients, and the number of patients is 50, 30, 20, respectively. To calculate the probability of the patients over the global dataset $\bigcup_{k=1}^{3} D_k$ while satisfying personalized differential privacy for each local dataset, $P_1, P_2, P_3$ first take sampling on the tuples in their local dataset with the probability $p_1 = \frac{e^{0.1}-1}{e^{0.5}-1} \approx 0.2, p_2 \approx 0.6, p_3 = 1$ to obtaining sampled datasets $\widetilde{D}_1, \widetilde{D}_2, \widetilde{D}_3$. Based on these sampled datasets, the curator and the parties can calculate a probability of the patients, approximately equal to $\frac{0.2 \times 50 + 0.6 \times 30 + 1 \times 20}{0.2 \times 100 + 0.6 \times 100 + 1 \times 100} = \frac{48}{180} \approx 0.27$. However, the actual probability of the patients is $\frac{50+30+20}{100+100+100} = \frac{100}{300} \approx 0.33$. Thus, the result calculated on the sampled data set is far from the result calculated on the original data set.

## 5.2   Vertically Partitioned Setting

In the vertically partitioned setting, each party $P_k$ holds a local dataset $D_k$ with a set of attributes $\mathcal{A}_k$, and keeps a privacy budget $\varepsilon_k$, where $k \in \{1, 2, \ldots, K\}$. For a given attribute set $\mathcal{X}$, where $\mathcal{X} = \bigcup_{k=1}^{K} \mathcal{X}_k$ and $\mathcal{X}_k$ is from $D_k$, i.e., $\mathcal{X}_k \subseteq \mathcal{A}_k$, the curator and the parties want to calculate its marginal under personalized differential privacy. The intuitive idea is that, following the methods used in the centralized setting, each party $P_k$ first takes sampling on the tuple level in $D_k$ with the probability $p_k = \frac{e^{\varepsilon_k}-1}{e^{\varepsilon_{\max}}-1}$ to get a sampled dataset $\widetilde{D}_k$, where $\varepsilon_{\max} = \max\{\varepsilon_k | 1 \leq k \leq K\}$. Then the curator and the parties calculate the marginal distribution of $\mathcal{X}$ over the integrated sampled dataset $\bowtie_{k=1}^{K} \widetilde{D}_k$ with privacy budget $\varepsilon_{\max}$, where $\bowtie$ denotes the join of two datasets. However, such a sampling method cannot meet the personalized privacy preference of attributes in different local datasets. It would also lead to the sampled global dataset being too sparse, which will reduce the utility of the calculated marginal distribution. The reason lies in that all the local datasets are over the same set of individuals in the vertically partitioned setting. Employing sampling at the tuple level on multiple datasets is equivalent to sampling individuals with the same small probability. Specifically, for any individual with $ID = x$, we have $\Pr\left(x \in \bowtie_{k=1}^{K} \widetilde{D}_k\right) = \prod_{k=1}^{K} p_k$. With the increase of $K$, $\prod_{k=1}^{K} p_k$ becomes smaller and $\bowtie_{k=1}^{K} \widetilde{D}_k$ becomes sparser.

By careful analysis, we learn that the main cause of the issue is that, in the vertically partitioned setting, personalized privacy requirements are for different attributes, while sampling is working at the tuple level. Therefore, to guarantee personalized differential privacy for each local dataset while enjoying reduced noise in the vertically partitioned setting, we need to propose a privacy adjusting method from the view of the attribute.

## 6 Our Solution

This section proposes the mixture-of-multinomials-based approach for the horizontally partitioned setting and the privacy budget segmentation method for the vertically partitioned setting. Note that, in the multi-party setting, all the communications between the curator and parties must be secure to guarantee computational differential privacy for each local dataset. We will first focus on the noisy marginals computation methods and then describe their implementation details under encryption.

### 6.1 Horizontally Partitioned Setting

In this setting, we propose a mixture of multinomials based method. In this method, the global marginals of $\mathcal{X}$ can be seen as a mixture of multinomial distributions and calculated by maximizing a posterior. The details are as follows.

*1. Calculating the local counts.* Given the attribute set $\mathcal{X}$, for each $\mathcal{X} = x_i$, where $i \in \{1, 2, \ldots, l\}$ and $l$ denotes the size of the domain of $\mathcal{X}$, each party $P_k$ calculates its local count $c_{ik}$ over dataset $D_k$ and multiplies the count by a scaling factor $s_k = \frac{\varepsilon_k}{\varepsilon_{max}}$, i.e., $\widetilde{c_{ik}} = c_{ik} \cdot s_k$. This can be seen as performing statistics on a stretched dataset $\widetilde{D_k}$. In the stretched dataset, the count of each tuple is multiplied by $s_k$. Here stretching is used to adjust the parties' different privacy preferences to satisfy personalized differential privacy, while avoiding the error caused by sampling.

*2. Construction of the likelihood function.* Based on the local counts, the parties and the curator can obtain the number $\widetilde{c}_i$ of the tuples with $\mathcal{X} = x_i$ for each $i \in \{1, 2, \ldots, l\}$ that contained in the stretched datasets, i.e., $\widetilde{c}_i = \sum_{k=1}^{K} \widetilde{c_{ik}}$. As the distribution of $\mathcal{X}$ in each local stretched dataset $\widetilde{D_k}$ (referred to as $\Pr\left(\mathcal{X}|\widetilde{D_k}\right)$) follows a multinomial distribution with parameters $\{\mu_{1k}, \mu_{2k}, \ldots, \mu_{lk}\}$, where $\mu_{ik} = \Pr\left(x_i|\widetilde{D_k}\right)$. And the prior probability of each multinomial element is $\alpha_k = \Pr\left(\widetilde{D_k}\right) = (s_k \cdot |D_k|)/\sum_{j=1}^{K}(s_j \cdot |D_j|)$. Thus the curator can calculate the probability $\mathcal{L} = \prod_{i=1}^{l}\left(\Pr\left(x_i\right)\right)^{\widetilde{c}_i}$, where $\Pr\left(x_i\right) = \sum_{k=1}^{K} \Pr\left(x_i|\widetilde{D_k}\right) \cdot \Pr\left(\widetilde{D_k}\right) = \sum_{k=1}^{K} \mu_{ik} \cdot \alpha_k$. $\mathcal{L}$ can be referred to as the likelihood function. The corresponding logarithmic likelihood function is:

$$\log\left(\mathcal{L}\right) = \log\left(\prod_{i=1}^{l}\left(\Pr\left(x_i\right)\right)^{\widetilde{c}_i}\right) = \sum_{i=1}^{l} \widetilde{c}_i \cdot \log\left(\sum_{k=1}^{K} \mu_{ik} \cdot \alpha_k\right).$$

Note that, $\sum_{i=1}^{l} \mu_{ik} = 1$ and $\sum_{k=1}^{K} \alpha_k = 1$. Thus, it can be seen as a constrained maximization problem.

*3. Calculation of model parameters* $\mu_{ik}$. Given the local datasets $D_k$ and the scaled factors $s_k$ (where $1 \leq k \leq K$), $\alpha_k = (s_k \cdot |D_k|)/\sum_{j=1}^{K}(s_j \cdot |D_j|)$ can be seen as a constant. And $\mu_{ik}$ can be calculated by using the criterion of maximum likelihood estimation. We introduce Lagrange multipliers $\lambda_k$ $(1 \leq k \leq K)$ to enforce the normalization constraint and then reduce the constrained maximization problem to the unconstrained maximization problem:

$$L = \sum_{i=1}^{l} \tilde{c}_i \cdot \log \left( \sum_{k=1}^{K} \mu_{ik} \cdot \alpha_k \right) - \sum_{k=1}^{K} \left( \lambda_k \left( \sum_{i=1}^{l} \mu_{ik} - 1 \right) \right).$$

*4. Recalculation of the marginal of* $\mathcal{X}$. The local marginal of $\mathcal{X}$ in the stretched dataset $\widetilde{D_k}$ is equal to that in the original dataset $D_k$, i.e., $\Pr(x_i|D_k) = \Pr\left(x_i|\widetilde{D_k}\right) = \mu_{ik}$. Based on the calculated $\mu_{ik}$, the curator can recalculate the marginal distribution of $\mathcal{X}$ over the original datasets:

$$\widehat{\Pr(x_i)} = \frac{\sum_{k=1}^{K} \mu_{ik} \cdot |D_k|}{\sum_{k=1}^{K} |D_k|} \qquad (2)$$

### 6.2 Vertically Partitioned Setting

In the vertically partitioned setting, we propose a privacy budget segmentation method, which is a privacy adjusting method from the view of the attribute. The method mainly consists of the following 4 steps.

1. We first sort the parties $P_1, P_2, \ldots, P_K$ according to their privacy budget. The sorted result can be denoted as $P_{s_1}, P_{s_2}, \ldots, P_{s_K}$, where for any $1 \leq i < j \leq K$, $\varepsilon_{s_i} \leq \varepsilon_{s_j}$. Note that, $\varepsilon_{s_1} = \min\{\varepsilon_1, \varepsilon_2, \ldots, \varepsilon_K\}$.
2. For each $k \in \{1, 2, \ldots, K\}$, the party $P_{s_k}$ splits $\varepsilon_{s_k}$ into $\varepsilon_{s_k} - \varepsilon_{s_{k-1}}$ and $\varepsilon_{s_{k-1}}$, where $\varepsilon_{s_0} = 0$.
3. For each $i \in \{1, 2, \ldots, K\}$, the parties $P_{s_i}, P_{s_2}, \ldots, P_{s_K}$ calculate the noisy marginal distribution of $\bigcup_{k=i}^{K} X_{s_k}$ under $\left(\varepsilon_{s_i} - \varepsilon_{s_{i-1}}\right)$-differential privacy.
4. The curator takes consistency post-processing on the above calculated marginal distributions to obtain a more accurate marginal distribution of $\bigcup_{k=1}^{K} X_k$.

Using the above privacy division composition strategy, such a method can make full use of the privacy budget of each party while satisfying personalized privacy requirements for different attributes according to the composition property of differential privacy. In the above process, the core is Step 4.

*For ease of understanding, let us first consider two-party setting.* There exist two parties $P_1$ and $P_2$, who hold $D_1$ with attribute set $A_1$ and $D_2$ with attribute set $A_2$, respectively. $P_1$ and $P_2$ want to calculate the marginal distribution of $(X_1, X_2)$ while satisfying $\varepsilon_1$-differential privacy for $D_1$ and $\varepsilon_2$-differential privacy for $D_2$, where $X_1 \in A_1$ and $X_2 \in A_2$, the domain of $X_1$ and $X_2$ are assumed to be both $\{0, 1\}$, and $\varepsilon_1 < \varepsilon_2$. In addition, in order to show the advantages of the proposed method more intuitively, we choose Gaussian noise as the added noise. This is because Gaussian distribution satisfies additivity. In Gaussian mechanism, the function satisfies $(\varepsilon, \delta)$-differential privacy, if it injects a Gaussian noise with the mean $\mu = 0$ and standard deviation $\sigma \geq cs/\varepsilon$ into the output, where $c^2 > 2\ln(1.25/\delta)$, and $s$ denotes the sensitivity of the function and $\delta \in (0, 1)$ denotes the relaxation factor.

At the beginning, we split $\varepsilon_2$ into $\varepsilon_1$ and $\varepsilon_2 - \varepsilon_1$. The parties $P_1, P_2$ and the curator calculate the marginal distribution $\Pr(X_1, X_2)$ over $D_1 \cup D_2$ and inject Gaussian noises with $\sigma = cs/\varepsilon_1$ into each item of $\Pr(X_1, X_2)$, the noisy results can be denoted as $p'_{00}, p'_{10}, p'_{01}, p'_{11}$, respectively. Given the calculated $p'_{00}, p'_{10}, p'_{01}, p'_{11}$, the curator can calculate $\Pr(X_2 = 0) = p'_{00} + p'_{10} = p'_0$ and $\Pr(X_2 = 1) = p'_{01} + p'_{11} = p'_1$. Besides, the party $P_2$ calculates the marginal distribution $\Pr(X_2)$ over $D_2$ and injects Gaussian noise with $\sigma = cs/(\varepsilon_2 - \varepsilon_1)$ into each item of $\Pr(X_2)$. The noisy results are denoted as $p''_0, p''_1$, respectively. Thus, we have that, for the attribute $X_2$, we get two noisy marginals. In reality, there can only exist one marginal distribution of $X_2$. We recalculate the marginal distribution of $X_2$ by employing consistency post-processing and learn that:

$$\tilde{p}_0 = \frac{(cs/(\varepsilon_2 - \varepsilon_1))^2 \cdot p'_0 + 2(cs/\varepsilon_1)^2 \cdot p''_0}{2(cs/\varepsilon_1)^2 + (cs/(\varepsilon_2 - \varepsilon_1))^2}, \tilde{p}_1 = \frac{(cs/(\varepsilon_2 - \varepsilon_1))^2 \cdot p'_1 + 2(cs/\varepsilon_1)^2 \cdot p''_1}{2(cs/\varepsilon_1)^2 + (cs/(\varepsilon_2 - \varepsilon_1))^2}$$

As $\tilde{p}_0 + \tilde{p}_1 = 1$, $(\tilde{p}_0, \tilde{p}_1)$ can be an estimated marginal of $X_2$. Furthermore, based on the reconstructed of $\Pr(X_2)$, i.e., $\tilde{p}_0$ and $\tilde{p}_1$, we can reconstruct $\Pr(X_1, X_2)$:

$$\tilde{p}_{00} = \tilde{p}_0 \cdot \frac{p'_{00}}{p'_0}, \tilde{p}_{10} = \tilde{p}_0 \cdot \frac{p'_{10}}{p'_0}, \tilde{p}_{01} = \tilde{p}_1 \cdot \frac{p'_{01}}{p'_1}, \tilde{p}_{11} = \tilde{p}_1 \cdot \frac{p'_{11}}{p'_1}.$$

*The above conclusions can be extended to the multi-party setting*, where there exist $K$ parties, where $K \geq 3$. Specifically, after Step 3, the curator can obtain noisy marginals, $\Pr\left(\bigcup_{j=1}^{K} X_{s_j} \mid \bowtie_{j=1}^{K} D_{s_j}\right)$, $\Pr\left(\bigcup_{j=2}^{K} X_{s_j} \mid \bowtie_{j=2}^{K} D_{s_j}\right)$, ..., $\Pr(X_{s_K} \mid D_{s_K})$. Based on each of these marginals, the curator can calculate:

$$\omega_{ki} = \begin{cases} \sum_{X_{s_k}} \cdots \sum_{X_{s_{i-1}}} \Pr\left(\bigcup_{j=k}^{K} X_{s_j} \mid \bowtie_{j=k}^{K} D_{s_j}\right), & k < i \leq K \\ \Pr\left(\bigcup_{j=k}^{K} X_{s_j} \mid \bowtie_{j=k}^{K} D_{s_j}\right), & i = k \end{cases} \tag{3}$$

$$\beta_{ki} = \begin{cases} \prod_{j=k}^{i-1} \left|\Omega_{X_{s_j}}\right|, & k < i \leq K \\ 1, & i = k \end{cases} \tag{4}$$

At this time, for each $i \in \{1, \ldots, K\}$, the curator gets multiple noisy marginal distributions of attribute set $\bigcup_{j=i}^{K} X_{s_j}$, i.e., $w_{1i}, \ldots, w_{ii}$. Based on these results, the curator can calculate:

$$\Pr\left(\widetilde{\bigcup_{j=i}^{K} X_{s_j}}\right) = \left(\sum_{k=1}^{i} \frac{\omega_{ki}}{\beta_{ki}\sigma_k^2}\right) / \left(\sum_{k=1}^{i} \frac{1}{\beta_{ki}\sigma_k^2}\right). \tag{5}$$

where $\sigma_k = cs/(\varepsilon_k - \varepsilon_{k-1})$. Furthermore, for each $k \in \{1, 2, \ldots, K-1\}$, the curator can iteratively calculate:

$$\mathrm{Pr}\left(\widehat{\bigcup_{i=k}^{K} X_{s_i}}\right) = \mathrm{Pr}\left(\widehat{\bigcup_{i=k+1}^{K} X_{s_i}}\right) \cdot \mathrm{Pr}\left(\widetilde{\bigcup_{i=k}^{K} X_{s_i}}\right) \bigg/ \sum_{X_{s_k}} \mathrm{Pr}\left(\widetilde{\bigcup_{i=k}^{K} X_{s_i}}\right).$$

$\mathrm{Pr}\left(\widehat{\bigcup_{i=1}^{K} X_{s_i}}\right)$ is the final noisy marginal of the attribute set $\mathcal{X} = \bigcup_{i=1}^{K} X_{s_k}$.

## 6.3   Implementation Details

We first consider the problem in the horizontally partitioned setting. After stretching, the parties and the curator privately calculate the marginal distribution of a given attribute set $\mathcal{X} \subseteq \mathcal{A}$ over the stretched datasets by using the threshold Homomorphic encryption [9]. In particular, for each $\mathcal{X} = x_i$, where $x_i \in \Omega_X$ and $1 \leq i \leq |\Omega_X|$, the parties first jointly generate a Laplace noise $\eta_i$ with scale $\lambda = \frac{2}{\varepsilon_{max}}$ by employing the Distributed Laplace Noise Generation (DLNG) method proposed in [31]. DLNG can allow the parties jointly generate a Laplace noise $\eta_i$ while preventing any parties and the curator from learning the value of $\eta_i$ and facilitate subsequent calculation. Specifically, DLNG randomly divides $\eta_i$ into $K$ parts and shared among the parties, i.e., $\eta_i = \sum_{k=1}^{K} \eta_{ik}$, and $P_1, \ldots, P_K$ hold $\eta_{i1}, \ldots, \eta_{iK}$, respectively. In [31], it has be proven that the randomness of each $\eta_{ik}$ is greater than $\eta_i$ and the privacy $\eta_i$ cannot be violated even when there exist some (even $K-1$) colluding parties. Then, each party $P_k$ locally counts the number of tuples that have $\mathcal{X} = x_i$, which can be referred to as $\widetilde{c_{ik}}$. Next, $P_k$ calculates $\widetilde{c_{ik}} + \eta_{ik}$ and sends it to the curator. After that, the curator calculates $\widetilde{c(x_i)} = \sum_{k=1}^{K} (\widetilde{c_{ik}} + \eta_{ik}) = \sum_{k=1}^{K} \widetilde{c_{ik}} + \eta_i$. Based on the above results, the curator can construct the likelihood function and solve the model parameters, and then calculate the noisy marginal distribution of $\mathcal{X}$ over the original datasets.

Calculating the marginal distribution in the vertically partitioned setting is rather complicated because the attributes are in different local datasets. We need some other security protocols to solve the problem, e.g., the secure scalar product protocol [12]. In particular, after privacy budget sort and segmentation, for each $i \in \{1, 2, \ldots, K\}$, the parties $P_{s_i}, \ldots, P_{s_K}$ jointly calculate the marginal distribution of $\mathcal{X} = \bigcup_{k=i}^{K} \mathcal{X}_{s_k}$ over their local datasets, while satisfying $(\varepsilon_{s_i} - \varepsilon_{s_{i-1}})$-differential privacy, where $\mathcal{X}_{s_k} \subseteq \mathcal{A}_{s_k}$.

At the beginning, each party $P_{s_k}$ first locally generates a vector $v_{s_k} = \left\{v_{s_k 1}, \ldots, v_{s_k |D_{s_k}|}\right\}$ with length $|D_{s_k}|$ for each $\mathcal{X}_{s_k} = x_{s_k}$, where $|D_{s_k}|$ denotes the size of the local dataset $D_{s_k}$. Note that, all of the local datasets have the same size, which can be referred to as $|D|$. Each element $v_{s_k j}$ in $v_{s_k}$ is 1 if $X_{s_k} = x_{s_k}$ in the $t^{th}$ tuple of $D_{s_k}$, otherwise $v_{s_k j} = 0$. Then, the parties calculate the number of tuples that have $\bigcup_{k=i}^{K} X_{s_k} = (x_{s_i}, \ldots, x_{s_K})$ by computing $\sum_{j=1}^{|D|} \prod_{k=i}^{K} v_{s_k j}$ in a secure way, and divide the result into $K$ parts $r_1, \ldots, r_K$, and share among the parties. Next, by employing DLNG, the parties generate a Laplace noise

$\eta$ with scale $\lambda = \frac{2}{\varepsilon_{s_i} - \varepsilon_{s_{i-1}}}$, and divide $\eta$ into $K$ parts $\eta_{i1}, \ldots, \eta_{iK}$ and shared among the parties. After that, each party sends $r_1 + \eta_{i1}$ to the curator, and the curator calculates $\widetilde{c(x)} = \sum_{k=1}^{K} (r_k(x) + \eta_k) = \sum_{k=1}^{K} r_k(x) + \eta$. Finally, based on $\widetilde{c(x)}$'s, the curator can calculate:

$$\widetilde{\Pr(x)} = \widetilde{c(x)}/\sum_{x' \in \Omega_X} \widetilde{c(x')}.$$

### 6.4   Privacy Analysis

Combining secure multiparty computation with differential privacy, we can guarantee that both the mixture-of-multinomials-based method and the privacy budget segmentation method satisfy $\varepsilon_k-$multi-party personalized differential privacy for each local dataset $D_k$.

## 7   Experiments

### 7.1   Experimental Settings

*Datasets* . In our experiments, we use two real datasets, *NLTCS*[1] and *BR2000*[2]. NLTCS contains records of 21,574 individuals who participated in the National Long Term Care Survey. BR2000 consists of 38,000 census records collected from Brazil in the year 2000. Each of the two datasets contains both continuous and categorical attributes. For each continuous attribute, we discretize its domain into a fixed number $b$ of equi-width ranges (we use $d = 16$).

To simulate the horizontally partitioned setting, we employ two categories of sampling methods (i.e., uniform sampling and non-uniform sampling) on the input dataset to obtain multiple local datasets that follow identically distributed, and that not. In addition, the size of each local dataset can be flexibly set. To simulate the vertically partitioned setting, we vertically partition the attributes among different parties randomly. We observe similar trends under different random partitionings.

*Competitors* . We first demonstrate the utility of the mixture-of-multinomials-based approach (denoted *MM*) for the horizontally partitioned setting by comparing it with four main approaches:

- **Independent.** The parties first add different noise levels to the local marginals independently according to their privacy preferences before sharing it with the curator. Then the curator aggregates these noisy local marginals together.
- **Minimum (MH).** The parties and the curator jointly calculate marginals over the original datasets using $\varepsilon_{min}$ as the privacy parameter, where $\varepsilon_{min} = \min\{\varepsilon_k | 1 \le k \le K\}$ and $\varepsilon_k$ denotes the privacy preference specified by $P_k$.

---

[1] http://lib.stat.cmu.edu/.
[2] https://international.ipums.org.

- **Sampling-based method (SAH).** It works by first sampling $D_k$ with probability $\frac{e^{\varepsilon_k}-1}{e^{\varepsilon_{max}}-1}$ and then calculating marginals over the sampled datasets using $\varepsilon_{max}$ as the privacy parameter, where $\varepsilon_{max} = \max\{\varepsilon_k | 1 \le k \le K\}$.
- **Stretching-based method (STH).** It works by first multiplying the value of each tuple in $D_k$ by a scaling factor $\frac{\varepsilon_k}{\varepsilon_{max}}$ and then calculating over the stretched datasets using $\varepsilon_{max}$ as the privacy parameter.
- **Product.** In the Product method, the attributes are assumed to be independent and the $k$-way marginal is estimated with the product of $k$ 1-way marginals.

Then, we evaluate the utility of the privacy budget segmentation method (denoted *PBS*) for the vertically partitioned setting by comparing it with Minimum and sampling-based methods for the vertically partitioned setting (denoted as *MV* and *SAV*, respectively).

*Metrics* . To measure the accuracy of a noisy marginal obtained by each method, we calculate the *total variation distance* [33] between the noisy marginal and its noise-free version, i.e., half of the $L_1$ distance between the two distributions. For each task, we repeat the experiment 100 times and report the average.

*Parameters* . There are three key parameters involved in our solutions:

- **Privacy preferences.** Following the setting in [18], we provide three kinds of privacy preferences for the parties i.e., $\varepsilon_{min}$, $\varepsilon_{mid}$, and $\varepsilon_{max}$. In particular, $\varepsilon_{max}$ is always set to be 1, $\varepsilon_{min}$ varies from 0.1 to 0.5, and $\varepsilon_{mid}$ is set to be $\varepsilon_{mid} = \frac{\varepsilon_{min}+\varepsilon_{max}}{2}$.
- **Number of parties.** We let the number of parties vary from 2 to 10. According to the privacy preferences, the parties can be divided into three groups that with privacy preferences $\varepsilon_{min}$, $\varepsilon_{mid}$, and $\varepsilon_{max}$, respectively.
- **Fraction of users.** In the horizontally partitioned setting, the fraction of users that choose different privacy preferences will affect the utility of our solutions. We denote the fraction of each group to be $f_{min}$, $f_{mid}$, and $f_{max}$. The fraction can be set based on findings from several studies regarding user privacy attitudes (e.g., [1]). In particular, $f_{mid}$ is always set to be 0.4, $f_{min}$ varies from 0.1 to 0.5, and $f_{max}$ is set to be $f_{max} = 1 - f_{min} + f_{mid}$.

*Computing environment setup* . All methods were implemented in Python. All methods were evaluated in a distributed environment using a cluster of nodes, which are connected by a 100Mbit network. Each node with an Intel Core i5-8300H processor and 16 GB of memory acts as either a curator or a party. The number of curator is 1, and the number of parties is up to 10.

## 7.2    Utility of Methods in the Horizontally Partitioned Setting

We first evaluate the impact of differences between private preferences by varying the value of the minimum privacy preference $\varepsilon_{min}$, where $\varepsilon_{max} = 1$ and

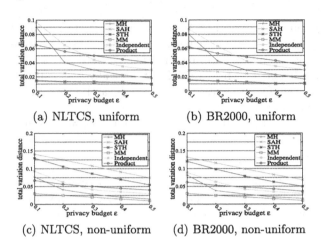

(a) NLTCS, uniform          (b) BR2000, uniform

(c) NLTCS, non-uniform     (d) BR2000, non-uniform

**Fig. 1.** Utility of methods in the horizontally partitioned setting.

$\varepsilon_{mid} = \frac{\varepsilon_{min}+\varepsilon_{max}}{2}$. Besides, the fraction of the group with privacy preference $\varepsilon_{max}/\varepsilon_{mid}/\varepsilon_{min}$ is set to be $f_{max} = 0.3$, $f_{mid} = 0.4$, $f_{min} = 0.3$, respectively.

Figure 1 shows the effects of differences between private preferences on each approach over NLTCS and BR2000, where the marginals are 3-way marginals. Figures 1(a)-1(b) show the total variation distance of each method when the local datasets follow identically distributed, and Figs. 1(c)-1(d) present the total variation distance of of each approach when the local datasets follow non-identically distributed. MM can always obtain the utility better than the others in all experiments. In particular, MM can obtain the utility better than Independent and Product. This is because Independent adds multiple shares of noise into the global marginals, reducing the utility of the results. In Product, it is assumed that the attributes are independent, this will incur a lot of precision loss. When the local datasets follow identically distributed, STH can obtain the utility as well as MM. But when the local datasets do not follow identically distributed, MM can obtain the utility better than STH. The reason lies in that, when the local datasets do not follow identically distributed, the calculated results in STH will be distorted. While by employing the Expectation-Maximization (EM) algorithm [10], MM can reconstruct the marginal distributions without distortion. Note that, $\varepsilon_{max}$ is fixed at 1, $\varepsilon_{mid}$ is set to be $\varepsilon_{mid} = \frac{\varepsilon_{min}+\varepsilon_{max}}{2}$. As $\varepsilon_{min}$ increases, the difference between $\varepsilon_{max}$, $\varepsilon_{mid}$, and $\varepsilon_{min}$ gradually narrows. The performance of MM, STH, and MH is all progressively close to that of the strategy that injects noise into the marginals according to a unified privacy budget. However, with the increase in the difference between private preferences, the superiority of the MM becomes more apparent, and MM can always obtain the utility better than the others. This is because, as the difference between private preferences increases, STH will stretch the data in the datasets that with the privacy preference $\varepsilon_{min}$ by a smaller scaling factor. This will incur more distortion. And $MH$ sets high-level global privacy to satisfy all the local datasets.

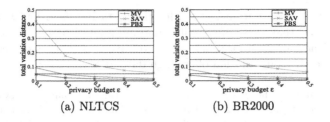

(a) NLTCS                          (b) BR2000

**Fig. 2.** Utility of methods in the vertically partitioned setting.

This will introduce a large amount of noise into the marginals, resulting in poor utility.

We also evaluate the impact of group fractions by varying the fraction $f_{min}$ of the local dataset, and find that, MM can obtain the utility better than the others, and with the increase in $f_{min}$, the superiority of the MM becomes more apparent. Due to the space limitation, we do not show the experiments.

### 7.3 Utility of Methods in the Vertically Partitioned Setting

In the vertically partitioned setting, we compare PBS with MV and SAV. We evaluate the impact of differences between private preferences by varying the value of the minimum privacy preference $\varepsilon_{min}$, where $\varepsilon_{max} = 1$ and $\varepsilon_{mid} = \frac{\varepsilon_{min} + \varepsilon_{max}}{2}$. Note that, all of the local datasets are over the same set of individuals, we need not consider the impact of group fractions in the vertically partitioned setting. In addition, to conveniently evaluate the performance of each method, in this experiment, we select one attribute from each local dataset and calculate 3-way marginals. Generally, for a given attribute set $\mathcal{X}$, if there exist multiple attributes contained in the local dataset $D_k$, denoted $X_1^k, \ldots, X_l^k$, then these attributes can be treated as a new attribute, whose domain is $\Omega_{X_1^k} \times \cdots \times \Omega_{X_l^k}$, where $\Omega_{X_i^k}$ denotes the domain of one attribute $X_i^k$.

Figure 2 shows the effects of differences between private preferences on each approach over NLTCS and BR2000. In all experiments, PBS can obtain the utility better than the others. The reason lies in that, by the privacy budget segmentation and composition, the PBS method can make full use of the privacy budget of each party. In addition, it is interesting that the sapling-based method SAV performs even worse than the Minimum method MV. This confirms our analysis in Sect. 5.2, i.e., the sampling-based method leads to a sparse sampled global dataset, which will reduce the utility of the calculated marginal distributions.

## 8   Conclusions

In this paper, we studied the problem of multi-party marginal distribution calculation under personalized differential privacy. We proposed the mixture-of-multinomials-based approach for the horizontally partitioned setting and the privacy budget segmentation method for the vertically partitioned setting. We formally proved that these approaches guarantee multi-party personalized differential privacy for each local dataset. Extensive experiments on real datasets demonstrated that our solution offers high data utility.

**Acknowledgment.** The work was supported by the National Key R&D Program of China under Grant No. 2020YFB1710200, National Natural Science Foundation of China under Grant No. 62002203, No. 61872105, No. 62072136, Shandong Provincial Natural Science Foundation No. ZR2020QF045, No. ZR2020MF055, No. ZR2021LZH007, No. ZR2020LZH002, the New Engineering Disciplines Research and Practice Project under Grant No. E-JSJRJ20201314, and Young Scholars Program of Shandong University.

# References

1. Acquisti, A., Grossklags, J.: Privacy and rationality in individual decision making. S&P 3(1), 26–33 (2005)
2. Alaggan, M., Gambs, S., Kermarrec, A.: Heterogeneous differential privacy. J. Priv. Confidentiality 7(2), 127–158 (2016)
3. Alhadidi, D., Mohammed, N., Fung, B.C.M., Debbabi, M.: Secure distributed framework for achieving $\varepsilon$-differential privacy. In: PETS (2012)
4. Bater, J., He, X., Ehrich, W., Machanavajjhala, A., Rogers, J.: Shrinkwrap: efficient SQL query processing in differentially private data federations. VLDB 12(3), 307–320 (2018)
5. Beimel, A., Nissim, K., Omri, E.: Distributed private data analysis: simultaneously solving how and what. In: CRYPTO (2008)
6. Chen, R., Li, H., Qin, A.K., Kasiviswanathan, S.P., Jin, H.: Private spatial data aggregation in the local setting. In: ICDE (2016)
7. Chen, R., Reznichenko, A., Francis, P., Gehrke, J.: Towards statistical queries over distributed private user data. In: NSDI (2012)
8. Cheng, X., Tang, P., Su, S., Chen, R., Wu, Z., Zhu, B.: Multi-party high-dimensional data publishing under differential privacy. TKDE 32(8), 1557–1571 (2020)
9. Cramer, R., Damgård, I., Nielsen, J.B.: Multiparty computation from threshold homomorphic encryption. In: EUROCRYPT (2001)
10. Do, C.B., Batzoglou, S.: What is the expectation maximization algorithm? Nat. Biotechnol. 26, 897–899 (2008)
11. Dwork, C., McSherry, F., Nissim, K., Smith, A.: Calibrating noise to sensitivity in private data analysis. In: TCC (2006)
12. Goethals, B., Laur, S., Lipmaa, H., Mielikäinen, T.: On private scalar product computation for privacy-preserving data mining. In: ICISC (2004)
13. Goryczka, S., Xiong, L.: A comprehensive comparison of multiparty secure additions with differential privacy. TDSC 14(5), 463–477 (2017)
14. Gu, X., Li, M., Xiong, L., Cao, Y.: Providing input-discriminative protection for local differential privacy. In: ICDE (2020)
15. Hardt, M., Nath, S.: Privacy-aware personalization for mobile advertising. In: CCS (2012)
16. Hong, D., Jung, W., Shim, K.: Collecting geospatial data with local differential privacy for personalized services. In: ICDE (2021)

17. Jiang, W., Clifton, C.: A secure distributed framework for achieving k-anonymity. VLDB J. **15**(4), 316–333 (2006)
18. Jorgensen, Z., Yu, T., Cormode, G.: Conservative or liberal? personalized differential privacy. In: ICDE (2015)
19. Kasiviswanathan, S.P., Lee, H.K., Nissim, K., Raskhodnikova, S., Smith, A.D.: What can we learn privately? In: FOCS (2008)
20. Kotsogiannis, I., Doudalis, S., Haney, S., Machanavajjhala, A., Mehrotra, S.: One-sided differential privacy. In: ICDE (2020)
21. McSherry, F., Talwar, K.: Mechanism design via differential privacy. In: FOCS (2007)
22. Mironov, I., Pandey, O., Reingold, O., Vadhan, S.P.: Computational differential privacy. In: CRYPTO (2009)
23. Mohammed, N., Alhadidi, D., Fung, B.C.M., Debbabi, M.: Secure two-party differentially private data release for vertically partitioned data. TDSC **11**(1), 59–71 (2014)
24. Mohammed, N., Fung, B.C.M., Debbabi, M.: Anonymity meets game theory: secure data integration with malicious participants. VLDB J. **20**(4), 567–588 (2011)
25. Narayan, A., Haeberlen, A.: DJoin: differentially private join queries over distributed databases. In: OSDI (2012)
26. Nie, Y., Yang, W., Huang, L., Xie, X., Zhao, Z., Wang, S.: A utility-optimized framework for personalized private histogram estimation. TKDE **31**(4), 655–669 (2019)
27. Niu, B., Chen, Y., Wang, B., Cao, J., Li, F.: Utility-aware exponential mechanism for personalized differential privacy. In: WCNC (2020)
28. Qardaji, W.H., Yang, W., Li, N.: PriView: practical differentially private release of marginal contingency tables. In: SIGMOD (2014)
29. Song, H., Luo, T., Wang, X., Li, J.: Multiple sensitive values-oriented personalized privacy preservation based on randomized response. TIFS **15**, 2209–2224 (2020)
30. Su, S., Tang, P., Cheng, X., Chen, R., Wu, Z.: Differentially private multi-party high-dimensional data publishing. In: ICDE (2016)
31. Tang, P., Chen, R., Su, S., Guo, S., Ju, L., Liu, G.: Differentially private publication of multi-party sequential data. In: ICDE (2021)
32. Tang, P., Cheng, X., Su, S., Chen, R., Shao, H.: Differentially private publication of vertically partitioned data. TDSC **18**(2), 780–795 (2021)
33. Tsybakov, A.B.: Introduction to Nonparametric Estimation. Springer, New York, NY (2009). https://doi.org/10.1007/b13794
34. Wagh, S., He, X., Machanavajjhala, A., Mittal, P.: DP-Cryptography: marrying differential privacy and cryptography in emerging applications. Commun. ACM **64**(2), 84–93 (2021)
35. Wu, D., et al.: A personalized preservation mechanism satisfying local differential privacy in location-based services. In: SPDE (2020)
36. Xiao, X., Tao, Y.: Personalized privacy preservation. In: SIGMOD (2006)
37. Xue, Q., Zhu, Y., Wang, J.: Mean estimation over numeric data with personalized local differential privacy. Front. Comput. Sci. **16**(3), 1–10 (2022). https://doi.org/10.1007/s11704-020-0103-0

# Beyond Random Selection: A Perspective from Model Inversion in Personalized Federated Learning

Zichen Ma[1,2(✉)] , Yu Lu[1,2] , Wenye Li[1,3] , and Shuguang Cui[1,3]

[1] The Chinese University of Hong Kong, Shenzhen, China
{zichenma1,yulu1}@link.cuhk.edu.cn, {wyli,shuguangcui}@cuhk.edu.cn
[2] JD AI Research, Beijing, China
[3] Shenzhen Research Institute of Big Data, Shenzhen, China

**Abstract.** With increasing concern for privacy issues in data, federated learning has emerged as one of the most prevalent approaches to collaboratively train statistical models without disclosing raw data. However, heterogeneity among clients in federated learning hinders optimization convergence and generalization performance. For example, clients usually differ in data distributions, network conditions, input/output dimensions, and model architectures, leading to the misalignment of clients' participation in training and degrading the model performance. In this work, we propose PFedRe, a personalized approach that introduces individual *relevance*, measured by Wasserstein distances among dummy datasets, into client selection in federated learning. The server generates dummy datasets from the inversion of local model updates, identifies clients with large distribution divergences, and aggregates updates from high relevant clients. Theoretically, we perform a convergence analysis of PFedRe and quantify how selection affects the convergence rate. We empirically demonstrate the efficacy of our framework on a variety of non-IID datasets. The results show that PFedRe outperforms other client selection baselines in the context of heterogeneous settings.

**Keywords:** Federated learning · Client selection · Personalization

## 1 Introduction

The ever-growing attention to data privacy has propelled the rise of federated learning (FL), a privacy-preserving distributed machine learning paradigm on decentralized data [24]. A typical FL system consists of a central server and multiple decentralized clients (e.g., devices or data silos). The training of an FL system is typically an iterative process, which has two steps: (i) each local client is synchronized by the global model and trained using its local data; (ii) the server updates the global model by aggregating the local models.

**Supplementary Information** The online version contains supplementary material available at https://doi.org/10.1007/978-3-031-26412-2_35.

However, as the number of clients and the complexity of the models grow, new challenges emerge concerning heterogeneity among clients [16]. For example, statistical heterogeneity in that data are not independent and identically distributed (IID) hinders the convergence of the model and is detrimental to its performance. Thus, methods to overcome the adverse effects of heterogeneity are proposed, including regularization [17,20], clustering [2,10], and personalization [6,30]. Despite these advances, client selection is a critical yet under-investigated topic.

In a cross-device FL training phase, it is plausible that not all of the client contributes to the learning objective [33]. Aggregating local updates from irrelevant clients to update the global model might degrade the system's performance. Moreover, McMahan et al. [24] show that only a fraction of clients should be selected by the server in each round, as adding more clients would diminish returns beyond a certain point. Hence, effective client selection schemes for heterogeneous FL are highly desired to achieve satisfactory model performances.

Thus far, some efforts have been devoted to selecting clients to alleviate heterogeneous issues and improve model performances, roughly grouped into two categories: (i) naive approaches to client selection identify and exclude irrelevant local model updates under the assumption that they are geometrically far from relevant ones [13,36]; (ii) another line of work assumes the server maintains a public validation dataset and evaluates local model updates using this dataset. Underperforming clients are identified as irrelevant and excluded from aggregation. [34,35].

Nevertheless, most of the existing client selection schemes have some limitations: (i) keeping a public validation dataset in the server and evaluating local updates on it disobeys the privacy principle of FL to some degree and might be impractical in real-world applications; (ii) current approaches are limited to the empirical demonstration without a rigorous analysis of how selection affects convergence speed.

Against this background, we propose a simple yet efficient personalized technique with client selection in heterogeneous settings. Clients with high relevance, measured by Wasserstein distances among dummy datasets, will be involved in the aggregation on the server, which boosts the system's efficiency.

**Contributions** of the paper are summarized as follows. First, we provide unique insights into client selection strategies to identify irrelevant clients. Specifically, the server derives dummy datasets from the inversion of local updates, excludes clients with large Wasserstein distances (large distribution divergences) among dummy datasets, and aggregates updates from high relevant clients. The proposed scheme has a crucial advantage: it uses dummy datasets from the inversion of local updates. Thus, there is no need for the server to keep a pubic validation dataset and the algorithm ensures the aggregation only involves highly relevant clients.

Second, we introduce a notion of individual relevance into FL, measured by Wasserstein distance among dummy datasets. As a motivating example, we examine two algorithms' (FedAvg [24] and FedProx [20]) performances with/without irrelevant clients on the MNIST dataset [19] in Fig. 1. The objective is to classify odd labeled digits, i.e., $\{1, 3, 5, 7, 9\}$. For the case with

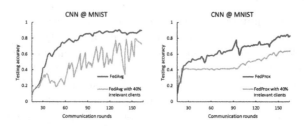

**Fig. 1.** Impacts of irrelevant clients in FL

irrelevant clients, odd labeled data are distributed to six clients, even labeled data are assigned to four clients, and even labels on four clients are randomly flipped to one of the odd labels. It is evident from the figure that irrelevant clients annihilate the stability of the training and incur lower accuracy, demonstrating the need to identify and exclude them from the system.

Finally, we explore the influences of client selection on the convergence of PFedRe. Theoretically, we show that, under some mild conditions, PFedRe will converge to an optimal solution for strongly convex function in non-IID settings. We illustrate that PFedRe can promote efficacy through extensive empirical evaluations while achieving superior prediction accuracy relative to recent state-of-the-art client selection algorithms.

## 2    Related Work

**Client Selection in FL.** Existing work in client selection focuses on (i) detecting and excluding irrelevant clients that are geometrically far from relevant ones. Blanchard et al. [1] explore the problem by choosing the local updates with the smallest distance from others and aggregating them to update the global model. Later, Trimmed Mean and Median [36] removes local updates with the largest and smallest $F$, and take the remaining mean and median as the aggregated model. In [4,32], authors alleviate the client selection issue while preserving efficient communication and boosting the convergence rate. However, some recently proposed work shows that irrelevant clients may be geometrically close to relevant ones [9,28]; (ii) another line of research needs to centralize a public validation dataset on the server and use it to evaluate local model updates in terms of test accuracy or loss. The error rate-based method [9] rejects local model updates that significantly negatively impact the global model's accuracy. Zeno [34,35] uses the loss decrease on the validation dataset to rank the model's relevance. Nevertheless, these schemes may violate the privacy-preserving principle of FL and may be challenging to implement in practice.

Recently, some work combines the two schemes and proposes hybrid client selection mechanisms. FLTrust [3] adopts a bootstrap on the server's validation dataset and uses the cosine similarity between the local and trained bootstrap models to rank the relevance. Later, DiverseFL [27] introduces a bootstrap method for each client using partial local data and compares this model with its updates to determine the selection. However, these approaches may also inherit limitations of two ways.

**Personalized FL.** Given the variability of data in FL, personalization is an approach used to improve accuracy, and numerous work has been proposed along this line. Particularly, Smith et al. [29] explore personalized FL via a primal-dual multi-task learning framework. As summarized in [6,22,31], the subsequent work has explored personalized FL through local customization [8,15,23], where models are built by customizing a well-trained global model. There are several ways to achieve personalization: (i) mixture of the global model and local models combines the global model with the clients' latent local models [6,14,23]; (ii) meta-learning approaches build an initial meta-model that can be updated effectively using Hessian or approximations of it, and the personalized models are learned on local data samples [7,8]; (iii) local fine-tuning methods customize the global model using local datasets to learn personalized models on each client [21,23].

# 3 Personalized Federated Learning with Relevance (PFedRe)

To explore client selection in personalized FL, we first formally define the personalized FL objective and introduce the system's workflow (Sect. 3.1). We then present PFedRe, a personalized algorithm that selects highly relevant clients to participate in training, and our proposed notion of individual relevance (Sect. 3.2). Finally, in Sect. 3.3, we analyze the influences of selection behaviors on training convergence.

## 3.1 Preliminaries and Problem Formulation

**Notations.** Suppose there are $M$ clients and a server in the system and denote by $X_k = \{x_{k,1}, x_{k,2}, ..., x_{k,m_k}\}$ the local data samples in the $k$-th client, where $x_{k,l}$ is the $l$-th sample and $l = 1, 2, ..., m_k$. Let $X = \cup_k X_k$ be the set of data among all clients, $\omega$ correspond to the global model, $\beta = (\beta_1, \beta_2, ..., \beta_M)$ with $\beta_k$ being the personalized local models on the $k$-th client, $F_k$ be the local objective function on the $k$-th client, and $E$ be the local epochs on clients, respectively. We denote $m = \sum_{k=1}^{M} m_k$ as the total number of samples.

In personalized FL, clients communicate with the server to solve the following problem:

$$\min_{\omega, \beta} F(\omega, \beta) = \frac{1}{m} \sum_{k=1}^{M} \sum_{l=1}^{m_k} f(\omega, \beta_k; x_{k,l}) = \sum_{k=1}^{M} \frac{m_k}{m} F_k(\omega, \beta_k) \qquad (1)$$

to find the global model $\omega$ and personalized model $\beta$. $f(\omega, \beta_k; x_{k,l})$ is the composite loss function for sample $x_{k,l}$ and model $\omega, \beta_k$. Generally, in clients, Eq. (1) is optimized w.r.t. $\omega$ and $\beta$ by stochastic gradient descent (SGD).

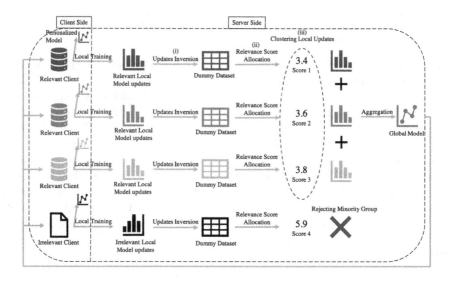

**Fig. 2.** Workflow of PFedRe

The communications in personalized FL only involve $\omega$, while the personalized models $\beta$ are stored locally and optimized without being sent to the server. Suppose the $k$-th client optimizes $F_k(\cdot)$ at most $T$ iterations. After a client receives the global model at the beginning of the $\tau$-th round $(0 \leq \tau < T)$, it updates the received model using $\omega_{\tau+1}^k = \omega_\tau^k - \eta_\tau \nabla_{\omega_k} F_k(\omega_\tau^k, \beta_\tau^k)$, and the personalized model is updated by $\beta_{\tau+1}^k = \beta_\tau^k - \delta_\tau \nabla_{\beta_k} F_k(\omega_\tau^k, \beta_\tau^k)$, where $\eta_\tau$ and $\delta_\tau$ are the learning rates.

After optimizing the personalized model $\beta_k$, each available client uploads $\omega_{\tau+1}^k$ every $E$ epochs. The server aggregates the received models by

$$\omega_{\tau+1}^G = \sum_{k=1}^{M} \frac{m_k}{m} \omega_{\tau+1}^k. \tag{2}$$

Personalized FL updates the global model with Eq. (2).

However, the server can only randomly select clients to participate in training due to the inaccessibility of clients' local training data and the uninspectable local training processes. As shown in Fig. 1, aggregating irrelevant clients' updates, in this case, hampers the stability and performance of the system. Hence, we introduce a client selection mechanism that facilitates the server to prune irrelevant clients.

Figure 2 elucidates the workflow of the proposed framework. It takes a different approach with three key differences compared with traditional methods. (i) local updates inversion: the received local model updates are inverted to generate corresponding dummy datasets on the server; (ii) relevance score allocation: Wasserstein distances among dummy datasets are calculated and recorded as the relevance score; (iii) relevant client clustering: local updates are clustered into two groups based on their relevance scores, and updates in the majority group are aggregated on the server.

## 3.2  PFedRe: Algorithm

In PFedRe, the $k$-th client performs $E$ epochs of the local model updates via mini-batch SGD with a size of $B$. Then, it submits local model update $\omega_{\tau+1}^k$ in the $\tau$-th round. The server works with the updates it receives from the clients. It first inverts local updates to generate dummy datasets using

$$x'^*_k = \arg\min_{x'_k} ||\frac{\partial F_k((x'_k, y'_k); \omega_\tau^k)}{\partial \omega_\tau^k} - \frac{a_\tau(\omega_{\tau+1}^k - \omega_\tau^k)}{Em_k/B}||_2^2, \tag{3}$$

where $(x'_k, y'_k)$ are the dummy data to be optimized, and $\omega_\tau^k = \omega_\tau^G$ is the current global model. PFedRe performs inversion by matching the dummy gradient with the equivalent gradient $\frac{(\omega_{\tau+1}^k - \omega_\tau^k)}{Em_k/B}$, where the term starts to cancel out, i.e., $\frac{(\omega_{\tau+1}^k - \omega_\tau^k)}{Em_k/B} \to 0$, as the global model converges. To signify the differences among gradients such that the server can identify the differences among clients' data distributions, PFedRe adds a scale factor $a_\tau$, i.e., $a$ to the power of $\tau$, where $a$ is a hyperparameter. The optimum, $x'^*_k$, substracts the initialization of the dummy data, respectively.

After generating dummy datasets, the server employs the Wasserstein distance metric to derive the relevance scores and the distribution divergences among dummy datasets. The divergence $\mathcal{D}_W$ between $(x'_k, y'_k)$ and $(x'_l, y'_l)$ is given by

$$\mathcal{D}_W(x'_k, x'_l) = \sum_{i=1}^{p}\sum_{j=1}^{q} \text{Wasserstein}[x'^{i,j}_k, x'^{i,j}_l], \tag{4}$$

where $x'^{i,j}_k$ is the vector composed of the $j$-th features of samples with label $i$, $p$ and $q$ are the numbers of labels and features of dummy datasets.

It is natural that the dummy datasets derived from irrelevant clients' updates have more considerable distribution divergences than relevant ones. However, this may not hold when statistical heterogeneity exists, i.e., data among clients are non-IID. Thus, instead of simply removing local updates with more significant distribution divergences, PFedRe collects those updates that have moderate distribution divergences.

Denote by $\mathcal{H} = \{\mathcal{D}_W(x'_k, x'_l)|l = 1, ..., M\}$ the set of $k$-th client's Wasserstein distances with all clients. More formally, individual relevance can be defined as

**Definition 1.** *Let $r_{\tau+1}^k$ be the relevance score of model $\omega_{\tau+1}^k$. For two models $\omega_{\tau+1}^k$, and $\omega_{\tau+1}^n$ in an FL system, we say $\omega_{\tau+1}^k$ is more relevant than $\omega_{\tau+1}^n$ if $r_{\tau+1}^k < r_{\tau+1}^n$, where*

$$r_{\tau+1}^k = \sum_{\mathcal{H}} |\mathcal{D}_W(x'_k, x'_l)|. \tag{5}$$

**Algorithm 1.** PFedRe. $M$ clients are indexed by $k$; $a_\tau$ represents the scaling factor; $\eta$ and $\delta$ denote the learning rates; $\gamma$ is the decay factor; $T$ is the maximal number of communication rounds, and $B$ denotes mini-batch size.

**Server executes:**
    initialize $\omega^0$, $\beta_k^0$, and dummy datasets $x'$. $\mathcal{H} \leftarrow \emptyset$, $\mathcal{R} \leftarrow \emptyset$.
    **for** $\tau = 0, 1, ..., T-1$ **do**
        **for** each client $k \in \{1, 2, ..., M\}$ **in parallel do**
            $\omega_{\tau+1}^k \leftarrow \text{ClientUpdate}(k, \omega_\tau^G)$
            $x'^*_k = \arg\min_{x'_k} ||\frac{\partial F_k((x'_k, y'_k); \omega_\tau^G)}{\partial \omega_\tau^G} - \frac{a_\tau(\omega_{\tau+1}^k - \omega_\tau^G)}{Em_k/B}||_2^2$
            $x'_k = x'^*_k - x'$
        **end for**
        **for** each client $k \in \{1, 2, ..., M\}$ **do**
            $\mathcal{H} \leftarrow \emptyset$
            **for** $l \in \{1, 2, ..., M\}\backslash\{k\}$ **do**
                $\mathcal{H} \leftarrow \mathcal{H} \cup \{\mathcal{D}_W(x'_k, x'_l)\}$
            **end for**
            $r_{\tau+1}^k = \sum_{\mathcal{H}} |\mathcal{D}(x'_k, x'_l)|$
            $\mathcal{R} \leftarrow \mathcal{R} \cup \{r_{\tau+1}^k\}$
        **end for**
        $\mathcal{R} \leftarrow \textbf{2-Median}(\mathcal{R})$
        $\Lambda \leftarrow \{k | r_{\tau+1}^k \in \mathcal{R}\}$
        $\omega_{\tau+1}^G = \sum_{k \in \Lambda} \frac{m_k}{\sum_{k \in \Lambda} m_k} \omega_{\tau+1}^k$
        **return** $\omega_{\tau+1}^G$ to participants.
    **end for**
**ClientUpdate**$(k, \omega_\tau^G)$:
    $B \leftarrow$ split local data into batches.
    **for** $i = 0, ..., E-1$ **do**
        **for** batch $\xi \in B$ **do**
            $\omega_{\tau+i+1}^k = \omega_{\tau+i}^k - \frac{\eta}{1+\gamma\tau} \frac{\partial F_k(\xi; \omega_{\tau+i}^k)}{\partial \omega_{\tau+i}^k}$
            $\beta_{\tau+i+1}^k = \beta_{\tau+i}^k - \frac{\delta}{1+\gamma\tau} \frac{\partial F_k(\xi; \beta_{\tau+i}^k)}{\partial \beta_{\tau+i}^k}$
        **end for**
    **end for**
    **return** $\omega_{\tau+E}^k$ to the server.

Let $\mathcal{R} = \{r_{\tau+1}^k | k = 1, ..., M\}$. PFedRe selects the majority group of $\mathcal{R}$ using the 2-Median clustering. Updates from the majority group are aggregated in the server. The details of the algorithm are summarized in Algorithm 1.

The inherent benefits of the proposed selection scheme are that (i) the individual relevance of a client is not the same across communication rounds, i.e., the value of a client's relevance score changes according to the state of the system that varies across rounds. Further, as the global model converges to the optimum, value of individual relevance also converges. Thus, the selection mechanism adapts to the dynamics of the heterogeneous settings that change over time; (ii) the framework is highly modular and flexible, i.e., we can readily use prior art developed for FL along with the client selection add-on, where the new methods still inherit the convergence benefits, if any.

### 3.3  PFedRe: Theoretical Analysis

In this section, we analyze the convergence behavior of PFedRe as described in Algorithm 1. We show that the proposed client selection scheme benefits to the convergence rate, albeit at the risk of incorporating a non-vanishing gap between the global optimum $\omega^{G*} = \arg\min_\omega F_k(\omega, \beta)$ and personalized optimum $\beta_k^* = \arg\min_{\omega,\beta} F_k(\omega, \beta)$.

**Assumption 1** *(L-smoothness). $F_k$ is L-smooth with constant $L > 0$ for $k = 1, 2, ..., M$, i.e. for all $v, w$,*

$$||\nabla F_k(v) - \nabla F_k(w)|| \le L||v - w||.$$

**Assumption 2** *(μ-strongly convexity). $F_k$ is μ-strongly convex with constant $\mu > 0$ for $k = 1, 2, ..., M$, i.e. for all $v, w$,*

$$F_k(w) - F_k(v) - \nabla F_k(v)||w - v|| \ge \frac{\mu}{2}||w - v||^2.$$

**Assumption 3** *(Unbiased gradient and bounded gradient discrepancy). For the mini-batch $\xi$ uniformly sampled at random from B, the resulting stochastic gradient is unbiased, i.e.,*

$$\mathbb{E}[g_k(\omega_\tau^k, \xi)] = \nabla_{\omega_\tau^k} F_k(\omega_\tau^k). \tag{6}$$

*Also, the discrepancy of model gradients is bounded by*

$$\mathbb{E}||g_k(\omega_\tau^k, \xi) - \nabla_{\omega_\tau^k} F_k(\omega_\tau^k)||^2 \le \chi^2, \tag{7}$$

*where $\chi$ is a scalar.*

**Assumption 4** *(Bounded model discrepancy). Denote by $\beta_k^* = \arg\min_{\omega,\beta} F(\omega, \beta)$ the optimal model in the k-th client, and $\omega^0$ the initialization of the global model. For a given ratio $q \gg 1$, the discrepancy between $\omega^0$ and $\omega^{G*}$ is sufficiently larger than the discrepancy between $\beta_k^*$ and $\omega^{G*}$, i.e. $||\omega^0 - \omega^{G*}|| > q||\beta_k^* - \omega^{G*}||$.*

Two metrics are introduced, i.e., the personalized-global objective gap, and the selection skew, to help the convergence analysis.

**Definition 2** *(Personalized-global objective gap). For the global optimum $\omega^{G*} = \arg\min_\omega F(\omega, \beta)$ and personalized optimum $\beta_k^* = \arg\min_{\omega,\beta} F(\omega, \beta)$, we define the personalized-global objective gap as*

$$\Gamma = F^* - \sum_{k=1}^M \frac{m_k}{m} F_k^* = \sum_{k=1}^M \frac{m_k}{m}(F_k(\omega^{G*}) - F_k(\beta_k^*)) \ge 0. \tag{8}$$

$\Gamma$ is an inherent gap between the personalized and global objective functions and is independent of the selection strategy. A more significant $\Gamma$ indicates higher data heterogeneity in the system. When $\Gamma = 0$, the personalized and global optimal values are the same, and no solution bias results from the selection.

The selection skew that captures the effect of the client selection strategy on the personalized-global objective gap can be defined as

**Definition 3** *(Selection skew) Let a client selection strategy $\pi$ be a function that maps the local updates to a selected set of clients $S(\pi, \omega_k)$, we define*

$$\rho(S(\pi, \omega_k), \beta_k) = \frac{\mathbb{E}_{S(\pi, \omega_k)}[\sum_{k \in S(\pi, \omega_k)} \frac{m_k}{m}(F_k(\omega_k) - F_k(\beta_k))]}{F_k(\omega_k) - \sum_{k=1}^{M} \frac{m_k}{m} F_k(\beta_k)} \geq 0, \quad (9)$$

*where $\mathbb{E}_{S(\pi, \omega_k)}$ is the expectation over the randomness from the selection strategy $\pi$.*

*We further define two related metrics independent of the global updates and personalized model to obtain a conservative error bound, where*

$$\overline{\rho} = \min_{\omega, \beta_k} \rho(S(\pi, \omega_k), \beta_k) \quad (10)$$

*and*

$$\widetilde{\rho} = \max_{\omega} \rho(S(\pi, \omega_k), \beta_k^*). \quad (11)$$

Equation (9) formulates the skew of a selection $\pi$. $\rho(S(\pi, \omega_k), \beta_k)$ is a function of versions of the global model's updates $\omega_k$ and personalized model $\beta_k$. According to Eq. (10) and (11), $\overline{\rho} \leq \widetilde{\rho}$ for a client selection strategy $\pi$.

For the client selection strategy $\pi_{random}$, we have $\rho(S(\pi_{random}, \omega_k), \beta_k) = 1$ for all $\omega_k$ and $\beta_k$ since the numerator and denominator of Eq. (9) become equal, and $\overline{\rho} = \widetilde{\rho} = 1$. For the proposed client selection strategy, $\pi$ chooses clients' updates within the majority group of individual relevance, where $\overline{\rho}$ and $\widetilde{\rho}$ will be more significant. The following analysis shows that a more substantial $\overline{\rho}$ leads to a faster convergence with a potential error gap proportional to $(\frac{\widetilde{\rho}}{\overline{\rho}-1})$.

The convergence results for a selection strategy $\pi$ with personalized-global objective gap $\Gamma$ and selection skew $\widetilde{\rho}, \overline{\rho}$ is presented in Theorem 1.

**Theorem 1.** *Given Assumptions 1 to 4, for learning rate $\eta_\tau = \frac{1}{\mu(\tau + \frac{4L}{\mu})}$, and any client selection strategy $\pi$, the error after $T$ rounds satisfies*

$$\mathbb{E}[F_k(\beta_T^k)] - F_k(\beta_k^*) \leq \frac{\mu}{\mu T + 4L} [\frac{4L(32E^2 q^2 + \frac{\chi^2}{|S|})}{3\mu^2 \overline{\rho}} + \frac{8L^2 \Gamma}{\mu^2} + \frac{2L^2 \|\beta_0^k - \beta_k^*\|^2}{\mu}] + Q(\overline{\rho}, \widetilde{\rho}),$$
$$(12)$$

*where $Q(\overline{\rho}, \widetilde{\rho}) = \frac{8L\Gamma}{3\mu}(\frac{\widetilde{\rho}}{\overline{\rho}} - 1)$.*

Theorem 1 provides the first convergence analysis of personalized FL with a biased client selection strategy $\pi$. It shows that a more significant selection skew $\overline{\rho}$ leads to faster convergence rate $O(\frac{1}{T\overline{\rho}})$. Since $\overline{\rho}$ is obtained by taking a minimum of the selection skew $\rho(S(\pi, \omega_k), \beta_k)$ over $\omega_k$ and $\beta_k$, the conservative bound on

**Table 1.** Statistics of datasets. The number of devices, samples, the mean and the standard deviation of data samples on each device are summarized.

| Dataset | # Devices | # Samples | Mean | SD |
|---|---|---|---|---|
| CIFAR100 | 100 | 59,137 | 591 | 32 |
| Shakespeare | 132 | 359,016 | 2,719 | 204 |
| Sentiment140 | 1,503 | 90,110 | 60 | 41 |
| EMNIST | 500 | 131,600 | 263 | 93 |

the actual convergence rate is obtained. If the selection skew $\rho(S(\pi, \omega_k), \beta_k)$ changes in training, the convergence rate can be improved by a more significant or at least a factor equal to $\overline{\rho}$.

The second term $Q(\overline{\rho}, \widetilde{\rho})$ in Eq. (12) represents the solution bias, depending on the selection strategy, and $Q(\overline{\rho}, \widetilde{\rho}) \geq 0$ according to the definitions of $\overline{\rho}$ and $\widetilde{\rho}$, if the selection strategy is unbiased, e.g., random selection, $\overline{\rho} = \widetilde{\rho} = 1$, and $Q(\overline{\rho}, \widetilde{\rho}) = 0$. If $\overline{\rho} > 1$, the method has faster convergence by $\overline{\rho}$ and $Q(\overline{\rho}, \widetilde{\rho}) \neq 0$. The proof of Theorem 1 is presented in Appendix.

## 4 Experiments

We evaluate the efficacy of our approach PFedRe on multiple datasets by considering various heterogeneous settings.

### 4.1 Setup

Both convex and non-convex models are evaluated on several benchmark datasets. Specifically, we adopt the EMNIST [5] dataset with Resnet50, CIFAR100 dataset [18] with VGG11, Shakespeare dataset with an LSTM [24] to predict the next character, and Sentiment140 dataset [11] with an LSTM to classify sentiment. Statistics of datasets are summarized in Table 1.

To demonstrate the effectiveness of PFedRe, we experiment with both vanilla and irrelevant clients, where two ways are adopted to simulate irrelevant clients. The first method flips data samples' labels to other classes on clients, and the second assigns out-of-distribution samples to clients and labels them randomly. Furthermore, three baselines are compared with PFedRe: (i) standard federated averaging (FedAvg) algorithm [24]; (ii) Selecting clients using the Shapely-based valuation (S-FedAvg) method [25]; (iii) Dynamic filtering of clients according to their cumulative losses (AFL) [12].

All experiments are implemented using PyTorch [26] and run on a cluster where each node is equipped with 4T P40 GPUs and 64 Intel(R) Xeon(R) CPU E5-2683 v4 cores @ 2.10 GHz. For reference, details of datasets partition and implementation settings are summarized in Appendix.

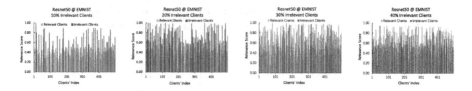

**Fig. 3.** Normalized relevance scores of clients on EMNIST dataset obtained by PFedRe after 1000 communication rounds. The irrelevant client percentage is 10%, 20%, 30%, and 40%. PFedRe identifies irrelevant clients (in blue) in training and excludes them from aggregation. (Color figure online)

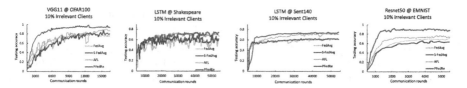

**Fig. 4.** The evolution of the testing accuracy is presented, where irrelevant clients have out-of-distribution samples. PFedRe outperforms other baselines in this case.

### 4.2 Detection of Irrelevant Clients

In this experiment, data samples of the EMNIST dataset are partitioned among 500 clients. To introduce the irrelevant clients, we flip data samples' labels to other classes on clients. The irrelevant client percentage in the system is 10%, 20%, 30%, and 40%. Figure 3 shows the normalized relevance score of clients learned using PFedRe after 1000 communication rounds. The yellow bars correspond to relevant clients, whereas the blue bars correspond to irrelevant clients. It is evident from the figure that the relevance scores of relevant clients are lower than that of irrelevant clients. Hence, using PFedRe, the server can differentiate between relevant and irrelevant clients. We further note that due to the dynamic nature of the generated dummy datasets in training, the magnitude of relevance keeps changing across communication rounds. However, the trend between relevant and relevant clients remains consistent.

### 4.3 Performance Comparison: Assigning Out-of-Distribution Data Samples

This experiment shows the impact of irrelevant clients with out-of-distribution data samples on the system's performance. We use four datasets where 10% of clients are irrelevant. Out of-distribution data samples with random labels are assigned to irrelevant clients. Figure 4 shows that even in the presence of out-of-distribution samples at irrelevant clients, the performance of PFedRe is significantly better than that of other baselines, indicating the efficacy of the proposed method. A close competitor to PFedRe is AFL, underlining the need for dynamic client filtering in heterogeneous settings.

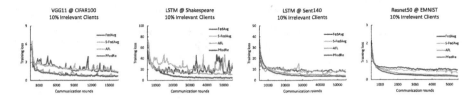

**Fig. 5.** The evolution of the training loss is presented, where the labels of samples are flipped to other classes on irrelevant clients. PFedRe exhibits better efficacy compared with baselines in this case.

### 4.4   Performance Comparison: Flipping Labels to Other Classes

In this experiment, we demonstrate the impact of the proposed client selection strategy on the performance (w.r.t. training loss) of algorithms. We implement PFedRe and the other three baselines independently on four datasets. For irrelevant clients, labels of samples are flipped to different classes on clients, and we set 10% of clients to be irrelevant in the system. Ideally, if PFedRe detects irrelevant clients correctly, the server would aggregate updates derived only from relevant clients, leading to better performance (lower training loss). Figure 5 shows that the system trained using PFedRe outperforms the models trained by other baselines. These results signify that identifying relevant clients and then aggregating updates from them is essential for building an efficient FL system.

### 4.5   Impact of Removing Clients with High/Low Relevance Score

This experiment shows that removing clients with high relevance scores deteriorates the system's performance, whereas removing clients who usually have low relevance scores helps improve it. We partition datasets to all clients and randomly flip 20% of samples' labels on 10% of clients. Then we run PFedRe for $\tau_0$ rounds ($\tau_0 \ll T$) on datasets. After $\tau_0$, the evolution of three testing accuracy is recorded, where PFedRe (i) keeps all clients in the system; (ii) removes clients determined as relevant more than 50% of rounds before $\tau_0$ and keeps others in training; (iii) removes clients who are judged as irrelevant more than 50% of rounds before $\tau_0$ and keeps others.

As shown in Fig. 6, it is evident that removing clients with high relevance scores indeed affects the system's performance adversely. On the contrary, eliminating clients with low relevance scores improves its performance. We can consistently observe that removing as many as 10% of clients with low relevance scores will enhance the system's efficacy. Whereas removing clients with high relevance scores has a noticeable negative impact.

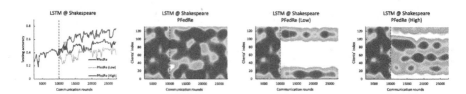

**Fig. 6.** The evolution of the testing accuracy is presented in the left figure. PFedRe trains models for $\tau_0 = 10000$ rounds. After $\tau_0$, PFedRe (i) keeps all clients and continues training (blue curve); (ii) removes clients determined as relevant for more than 5000 rounds and keeps others (yellow curve); (iii) removes clients judged as irrelevant for more than 5000 rounds and keeps the remaining clients (red curve). The heatmaps record the clients' participation in three methods. (Color figure online)

## 5    Conclusion and Future Work

This paper presents PFedRe, a novel personalized FL framework with client selection to mitigate heterogeneous issues in the system. By introducing the individual relevance into the algorithm, we extend the server to identify and exclude irrelevant clients via local updates' inversion, showing that dynamic client selection is instrumental in improving the system's performance. Both the analysis and empirical evaluations show the ability of PFedRe to achieve better performances in heterogeneous settings. In future work, we will explore potential competing constraints of client selection such as privacy and robustness to attacks and consider the applicability of PFedRe to other notions of the distributed system.

**Acknowledgements.** The work was supported in part by the Key Area R&D Program of Guangdong Province with grant No. 2018B030338001, by the National Key R&D Program of China with grant No. 2018YFB1800800, by Shenzhen Outstanding Talents Training Fund, and by Guangdong Research Project No. 2017ZT07X152 and 2021A1515011825.

## References

1. Blanchard, P., El Mhamdi, E.M., Guerraoui, R., Stainer, J.: Machine learning with adversaries: byzantine tolerant gradient descent. In: Advances in Neural Information Processing Systems, vol. 30 (2017)
2. Briggs, C., Fan, Z., Andras, P.: Federated learning with hierarchical clustering of local updates to improve training on non-iid data. In: 2020 International Joint Conference on Neural Networks (IJCNN), pp. 1–9. IEEE (2020)
3. Cao, X., Fang, M., Liu, J., Gong, N.Z.: Fltrust: byzantine-robust federated learning via trust bootstrapping. In: ISOC Network and Distributed System Security Symposium (NDSS) (2021)

4. Cho, Y.J., Gupta, S., Joshi, G., Yağan, O.: Bandit-based communication-efficient client selection strategies for federated learning. In: 2020 54th Asilomar Conference on Signals, Systems, and Computers, pp. 1066–1069. IEEE (2020)

5. Cohen, G., Afshar, S., Tapson, J., Van Schaik, A.: EMNIST: extending MNIST to handwritten letters. In: 2017 International Joint Conference on Neural Networks (IJCNN), pp. 2921–2926. IEEE (2017)

6. Deng, Y., Kamani, M.M., Mahdavi, M.: Adaptive personalized federated learning. arXiv preprint arXiv:2003.13461 (2020)

7. Fallah, A., Mokhtari, A., Ozdaglar, A.: On the convergence theory of gradient-based model-agnostic meta-learning algorithms. In: International Conference on Artificial Intelligence and Statistics, pp. 1082–1092. PMLR (2020)

8. Fallah, A., Mokhtari, A., Ozdaglar, A.: Personalized federated learning with theoretical guarantees: a model-agnostic meta-learning approach. Adv. Neural. Inf. Process. Syst. **33**, 3557–3568 (2020)

9. Fang, M., Cao, X., Jia, J., Gong, N.: Local model poisoning attacks to byzantine-robust federated learning. In: 29th USENIX Security Symposium (USENIX Security 20), pp. 1605–1622 (2020)

10. Ghosh, A., Chung, J., Yin, D., Ramchandran, K.: An efficient framework for clustered federated learning. Adv. Neural. Inf. Process. Syst. **33**, 19586–19597 (2020)

11. Go, A., Bhayani, R., Huang, L.: Twitter sentiment classification using distant supervision. CS224N project report, Stanford 1(12), 2009 (2009)

12. Goetz, J., Malik, K., Bui, D., Moon, S., Liu, H., Kumar, A.: Active federated learning. arXiv preprint arXiv:1909.12641 (2019)

13. Guerraoui, R., Rouault, S., et al.: The hidden vulnerability of distributed learning in byzantium. In: International Conference on Machine Learning, pp. 3521–3530. PMLR (2018)

14. Hanzely, F., Richtárik, P.: Federated learning of a mixture of global and local models. arXiv preprint arXiv:2002.05516 (2020)

15. Jiang, Y., Konečný, J., Rush, K., Kannan, S.: Improving federated learning personalization via model agnostic meta learning. arXiv preprint arXiv:1909.12488 (2019)

16. Kairouz, P., et al.: Advances and open problems in federated learning. Found. Trends® Mach. Learn. **14**(1–2), 1–210 (2021)

17. Karimireddy, P., Kale, S., Mohri, M., Reddi, S., Stich, S., Suresh, A.T.: Scaffold: stochastic controlled averaging for federated learning. In: International Conference on Machine Learning, pp. 5132–5143. PMLR (2020)

18. Krizhevsky, A., Hinton, G., et al.: Learning multiple layers of features from tiny images (2009)

19. LeCun, Y., Bottou, L., Bengio, Y., Haffner, P.: Gradient-based learning applied to document recognition. Proc. IEEE **86**(11), 2278–2324 (1998)

20. Li, T., Sahu, A., Zaheer, M., Sanjabi, M., Talwalkar, A., Smith, V.: Federated optimization in heterogeneous networks. Proc. Mach. Learn. Syst. **2**, 429–450 (2020)

21. Liang, P.P., et al.: Think locally, act globally: federated learning with local and global representations. arXiv preprint arXiv:2001.01523 (2020)

22. Ma, Z., Lu, Y., Li, W., Yi, J., Cui, S.: PFEDATT: attention-based personalized federated learning on heterogeneous clients. In: Asian Conference on Machine Learning, pp. 1253–1268. PMLR (2021)

23. Mansour, Y., Mohri, M., Ro, J., Suresh, A.T.: Three approaches for personalization with applications to federated learning. arXiv preprint arXiv:2002.10619 (2020)

24. McMahan, B., Moore, E., Ramage, D., Hampson, S., Arcas, B.A.: Communication-efficient learning of deep networks from decentralized data. In: Artificial Intelligence and Statistics, pp. 1273–1282. PMLR (2017)
25. Nagalapatti, L., Narayanam, R.: Game of gradients: mitigating irrelevant clients in federated learning. In: Proceedings of the AAAI Conference on Artificial Intelligence, vol. 35, pp. 9046–9054 (2021)
26. Paszke, A., et al.: Pytorch: an imperative style, high-performance deep learning library. In: Advances in Neural Information Processing Systems, vol. 32 (2019)
27. Prakash, S., Avestimehr, A.S.: Mitigating byzantine attacks in federated learning. arXiv preprint arXiv:2010.07541 (2020)
28. Shejwalkar, V., Houmansadr, A.: Manipulating the byzantine: optimizing model poisoning attacks and defenses for federated learning. In: NDSS (2021)
29. Smith, V., Chiang, C.K., Sanjabi, M., Talwalkar, A.: Federated multi-task learning. arXiv preprint arXiv:1705.10467 (2017)
30. T Dinh, C., Tran, N., Nguyen, T.: Personalized federated learning with MOREAU envelopes. In: Advances in Neural Information Processing Systems, vol. 33 (2020)
31. Tan, A.Z., Yu, H., Cui, L., Yang, Q.: Towards personalized federated learning. arXiv preprint arXiv:2103.00710 (2021)
32. Tang, M., Ning, X., Wang, Y., Wang, Y., Chen, Y.: FEDGP: correlation-based active client selection for heterogeneous federated learning. arXiv preprint arXiv:2103.13822 (2021)
33. Wang, H., Kaplan, Z., Niu, D., Li, B.: Optimizing federated learning on non-IID data with reinforcement learning. In: IEEE INFOCOM 2020-IEEE Conference on Computer Communications, pp. 1698–1707. IEEE (2020)
34. Xie, C., Koyejo, S., Gupta, I.: Zeno: distributed stochastic gradient descent with suspicion-based fault-tolerance. In: International Conference on Machine Learning, pp. 6893–6901. PMLR (2019)
35. Xie, C., Koyejo, S., Gupta, I.: Zeno++: Robust fully asynchronous SGD. In: International Conference on Machine Learning, pp. 10495–10503. PMLR (2020)
36. Yin, D., Chen, Y., Kannan, R., Bartlett, P.: Byzantine-robust distributed learning: Towards optimal statistical rates. In: International Conference on Machine Learning, pp. 5650–5659. PMLR (2018)

# Noise-Efficient Learning of Differentially Private Partitioning Machine Ensembles

Zhanliang Huang$^{(\boxtimes)}$, Yunwen Lei, and Ata Kabán

School of Computer Science, University of Birmingham, Birmingham, UK
ZXH898@cs.bham.ac.uk, {Y.Lei,A.Kaban}@bham.ac.uk

**Abstract.** Differentially private decision tree algorithms have been popular since the introduction of differential privacy. While many private tree-based algorithms have been proposed for supervised learning tasks, such as classification, very few extend naturally to the semi-supervised setting. In this paper, we present a framework that takes advantage of unlabelled data to reduce the noise requirement in differentially private decision forests and improves their predictive performance. The main ingredients in our approach consist of a median splitting criterion that creates balanced leaves, a geometric privacy budget allocation technique, and a random sampling technique to compute the private splitting-point accurately. While similar ideas existed in isolation, their combination is new, and has several advantages: (1) The semi-supervised mode of operation comes for free. (2) Our framework is applicable in two different privacy settings: when label-privacy is required, and when privacy of the features is also required. (3) Empirical evidence on 18 UCI data sets and 3 synthetic data sets demonstrate that our algorithm achieves high utility performance compared to the current state of the art in both supervised and semi-supervised classification problems.

**Keywords:** Differential privacy · Noise reduction · Ensembles

## 1 Introduction

Differential privacy (DP) [9] is a notion of privacy that provides a rigorous information-theoretic privacy guarantee. DP algorithms allow their outputs to be shared across multiple parties and used for analysis by introducing randomization in critical steps of the learning algorithm. However, DP algorithms often require a larger training set to achieve good performance due to the added noise. Moreover, large labelled training sets are expensive and sometimes impractical to obtain, especially for a sensitive data set (e.g. HIV positive data). On the other hand, unlabelled data can be less privacy-sensitive and in many cases much more cost-effective to obtain at large scale in comparison to accurately labelled data. Due to these reasons, it is extremely valuable if we can make use of unlabelled data to improve the performance of a learning algorithm in the private setting.

**Supplementary Information** The online version contains supplementary material available at https://doi.org/10.1007/978-3-031-26412-2_36.

In this paper, we present a framework that builds machine ensembles in both supervised and semi-supervised settings. The framework takes advantage of unlabelled data to reduce the noise requirements and hence it only requires a small number of labelled samples to achieve high performance. Our framework invokes constructing a tree structure that uses a density-informed splitting criterion to create balanced leaves and naturally extends to semi-supervised learning with different privacy settings. Current private tree-based algorithms in the literature either use a greedy-decision approach, or a random-tree approach. Methods with greedy approaches take the route of classical decision tree construction [5], and compute the optimal splits at each node privately. Popular splitting techniques such as the Gini index [11,27] or the information gain criterion [13] are applied in conjunction with a privatized algorithm. The drawback for this approach is that it cannot be naturally extended to semi-supervised learning as they greedily estimate the optimal split using the labels. On the other hand, random tree approaches construct the tree by randomized splits at each node [12,14,17]. Randomization is beneficial from the privacy perspective as it is data-independent and leaks zero information about individuals in the data set. However, a fully randomized split creates high variance and requires a large ensemble of trees to perform well. We cannot afford a large ensemble due to the privacy constraint. Furthermore, random-tree approaches do not take advantage of the unlabelled data as the splits are chosen fully randomly. Since these approaches do not naturally extend to the semi-supervised setting, we need to assign labels to the unlabelled set using a trained model if we want to make any use of the unlabelled data [16,20]. While this method can help to improve accuracy in some cases, it requires the predicted labels to be accurate for the output data to be useful, which cannot be guaranteed in general. Furthermore, since the output data contains the original features of the unlabelled set, it can only be applied where we do not need the privacy of the features at all.

Instead of the previous approaches, our approach proposes a semi-greedy median splitting criterion that uses the features to make formative splits. Median splitting has been used to build trees mostly in spatial decomposition where we partition data sets to allow quick access to different parts of the data [2, 7]. However it also can be used for classification and regression problems with good utility as shown in [4,18] – even though classical decision tree methods are better in general without privacy. A main intuition of median split is that it creates density-balanced nodes, the concept matches with the density-based dissimilarities in the work of Aryal *et al* [1]. That is, two points are more similar if they lie in a sparse region than two other points in a dense region with equal geometric distance. Each leaf comes with a similar amount of sample points. Hence we avoid empty leaves and the noise level of each leaf is balanced. To achieve high utility, we also employ several techniques to optimize each step of the privatized model as follows. 1) We use a geometric-scaling privacy budget allocation strategy to ensure accurate splits at each level. 2) We use a random sampling technique to compute the private median effectively. 3) We use disjoint subsets to create ensembles for both labelled and unlabelled sets to reduce noise effects. While these techniques have pre-existed in isolation as parts of other algorithms, the combination appears novel and it leads to a novel framework that achieves high performance in both the supervised and semi-supervised setting.

The remainder of the paper is organized as follows. We discuss the related works in Sect. 2, and introduce some background in Sect. 3. In Sect. 4 we present key steps of our strategy and the construction of a supervised algorithm. Section 5 demonstrates our framework of creating ensembles in semi-supervised learning. Finally, we present our experimental analysis of our method for both supervised and unsupervised learning in Sect. 6.

## 2    Related Work

There is a vast amount of research in machine learning on differential privacy since its introduction ( [10,15]). Tree-based methods are certainly one of the most popular research topics, with early works on private random trees [17] and greedy trees [13]. In [11] the authors proposed the use of local sensitivity [24] to reduce the randomness in greedy trees. The idea has been extended in [12] to improve private random trees using smooth sensitivity, which utilize an upper bound on the local sensitivity. More recently, [27] proposed a greedy approach that takes advantage of a notion of smooth sensitivity with the exponential mechanism for both Gini index and label output. Another DP forest algorithm by [26] considers $(\epsilon, \delta)$-differential privacy which is a weaker notion of the pure $\epsilon$-differential privacy that we are concerned with here. As discussed in [12], it is possible to obtain high utility while guaranteeing pure differential privacy. Other differentially private algorithms such as [23] consider private data release – a different problem setting from what we study here.

The construction of a tree structure using a median split has long lasted in spatial trees [2] and private spatial decomposition [7]. For classification problems it was initially used by [4], it then gradually gained attention and was analyzed theoretically by [3]. Similar idea is also used in spatial decomposition such as kd-trees [7]. More recently, [18] extended the idea to a median splitting random forest in the non-private setting. The authors demonstrated a random forest using a median-based splitting along with its theoretical analysis. Furthermore, a recent work by [6] proposed a private random forest using median splits however their method uses a greedy approach to compute each split-attribute and does not extend to semi-supervised learning. For private semi-supervised classifiers, the random forest by [17] can be extended to take advantage of unlabelled data [16]. Furthermore, work by [20] took advantage of unlabelled samples by providing predicted labels using a kNN classifier and then training a linear predictor using the predicted set. However both methods apply only when the privacy of the features is not a concern. Other semi-supervised methods [25] make extra assumptions on the data set and differ from our setting here.

## 3    Preliminaries: Differential Privacy

In this section, we use $\mathcal{X}$ to denote a universal set that contains all possible data points. We denote by $S$ a set of observations from $\mathcal{X}$. The privacy budget will be denoted by $\epsilon > 0$. Furthermore, we define the distance between two sample sets $S, S'$ to be the Hamming distance denoted as $\|S - S\|_H$, which equals the number of points to be added and removed from $S$ until $S = S'$.

**Definition 1 (Differential privacy [9]).** *A randomized algorithm $\mathcal{A}$ is said to satisfy $\epsilon$-differential privacy if for any $S, S' \in \mathcal{X}^n$ with $\|S - S'\|_H \leq 1$, we have*

$$\sup_{B \in \mathcal{B}} \frac{\mathbb{P}[\mathcal{A}(S) \in B]}{\mathbb{P}[\mathcal{A}(S') \in B]} \leq \exp(\epsilon), \tag{1}$$

*where $\mathcal{B}$ is the collection of all measurable sets in $Range(\mathcal{A})$.*

An immediate consequence of the definition is that DP algorithms are immune to *post-processing*. That is, if $\mathcal{A}$ is $\epsilon$-DP, then the composition $f \circ \mathcal{A}$ is also $\epsilon$-DP for an arbitrary mapping $f$ [10]. We note that some other literature on differential privacy considers the case where $S$ and $S'$ (of the same size) differ by at most one sample point. In contrast, we consider adding/removing a point, which is the setting considered in most of the related works. The two settings only differ by a constant factor on the sensitivity analysis.

**Definition 2 (Global $\ell_1$-sensitivity [9]).** *The global $\ell_1$-sensitivity $GS(f)$ of a function $f : \mathcal{X}^n \to \mathbb{R}^m$ is defined as*

$$GS(f) = \max_{S, S' \subset \mathcal{X}: \|S - S\|_H = 1} \|f(S) - f(S')\|_1. \tag{2}$$

The global sensitivity captures the maximum difference in the output when swapping a data set with a neighbouring one that differs by at most one point. There are other notions of sensitivity, such as the local sensitivity that considers the particular data set $S$. Local sensitivity can be significantly smaller than global sensitivity in many cases [11], however local sensitivity in itself does not guarantee differential privacy. Next, we present two well-known mechanisms for private algorithm design – the Laplace and the Exponential mechanisms.

**Definition 3 (Laplace mechanism of [10]).** *Given any function $f : \mathcal{X}^n \to \mathbb{R}^m$, the Laplace mechanism is defined as*

$$\mathcal{M}(S, f, \epsilon) = f(S) + (Y_1, \ldots, Y_m), \tag{3}$$

*where $Y_i$ are i.i.d. random variables drawn from $Lap(GS(f)/\epsilon)$.*

**Definition 4 (Exponential mechanism of [21]).** *Let $\mathcal{R}$ be an arbitrary set of output candidates. Given a utility function $u : \mathcal{X}^n \times \mathcal{R} \to \mathbb{R}$ that computes the quality of a candidate $r \in \mathcal{R}$, the exponential mechanism $\mathcal{M}(S, u, \mathcal{R})$ selects and outputs one of these with probability proportional to the following*

$$\mathbb{P}[\mathcal{M}(S, u, \mathcal{R}) = r] \propto \exp\left(\frac{\epsilon u(S, r)}{2 GS(u)}\right). \tag{4}$$

It is well-known that the Laplace and the Exponential mechanisms both satisfy $\epsilon$-DP. The reader can refer to [10] for detailed proofs of privacy guarantees. The following composition theorems allow us to combine multiple private mechanisms.

**Theorem 1 (Sequential composition** [21]**).** *Let* $\{f_i\}_{i=1}^N$ *be a sequence of queries on a data set* $S$ *each satisfying* $\{\epsilon_i\}_{i=1}^N$ *differential privacy. Then the output sequence* $\{f_i(S)\}_{i=1}^N$ *of all queries satisfies* $\sum_{i=1}^N \epsilon_i$ *differential privacy.*

**Theorem 2 (Parallel composition** [22]**).** *Let* $\{S_i\}_{i=1}^N$ *be disjoint subsets of* $S$, *and* $f$ *a query applied on each of the subsets* $S_i$ *while satisfying* $\epsilon$-DP. *Then the output sequence* $\{f(S_i)\}_{i=1}^N$ *satisfies* $\epsilon$-DP.

Sequential composition states that the more queries we send to the original data set, the less privacy guarantee we have. Parallel composition says that we do not lose the independent privacy guarantees if we query disjoint subsets independently.

# 4    A Density-Based Decision Tree

In this section, we describe the procedures and the key steps of our tree construction. Overall, the algorithm is broken down into the following steps, where the details are outlined in Algorithm 1.

1. Decide the parameters (number of trees and maximum depth).
2. At each tree node, uniformly randomly select an attribute to split on.
3. Calculate a private median for the selected attribute using the exponential mechanism.
4. When reaching a leaf node, use the Laplace mechanism to store the privatized counts for each class.
5. For a test point, collect together all the label-counts from all trees and output the label that has the majority count.

Note that, at each split we randomly select an attribute from the whole set of attributes, which avoids the label-dependent greedy computation of an optimal attribute as done in [6,27], and will allow us to construct the tree using only an unlabelled sample. Furthermore, random selection of the splitting feature can improve the diversity of trees while protecting privacy. After selection of a splitting attribute, we compute privately the median of the values for the selected attribute. This splitting method allows the feature space to be partitioned into even density regions. A key property of median splits is that it only depends on the features of the data, not the labels. Hence it does not overfit the data set easily, which is a concern with classic decision trees. Due to this property, we do not require any pruning process that label-dependent tree construction requires to avoid over-fitting. Other similar techniques involve the centred random forest described in [18] and mean-based rather than median-based splitting. However in the private setting, private mean estimation is usually more expensive than median estimation, as a single out-liar can largely affect the mean value – hence more noise is required to guarantee privacy. We terminate the splitting process of a branch either when we have reached our maximum depth or we only have a few points left (by default we set 10 as the minimum). The setting of a default minimum value prevents further splitting a node that has very few points.

---

**Algorithm 1.** BuildTree
---
1: Inputs: Sample set $S$, maximum depth $k$, privacy budget $\epsilon$.
2: **procedure** BUILDTREE$(S, k, \epsilon)$
3:     **if** k $\leq 0$ or $|S| \leq 10$ **then**
4:         Return a Leaf node
5:     **end if**
6:     LB, RB $= PrivateSplit(S, \epsilon)$
7:     BuildTree(LB,k-1, $\epsilon$), BuildTree(RB,k-1, $\epsilon$) # build subtrees
8:     Return a decision node that holds the split criteria and left/right branch.
9: **end procedure**
10: **procedure** PRIVATEMEDIAN$(V, \epsilon)$
11:     $a = \min V, b = \max V$
12:     $\mathcal{R} = \{$ set of random i.i.d. draws from $Uniform(a, b)\}$;
13:     **for** each $r$ in $\mathcal{R}$ **do**
14:         computes the quality score $u(V, r)$
15:     **end for**
16:     Return $\tilde{r} \in \mathcal{R}$ with exponential mechanism with budget $\epsilon$.
17: **end procedure**
18: **procedure** PRIVATESPLIT$(S, \epsilon)$
19:     Choose a splitting dimension $i$ uniformly from data dimension $[d]$
20:     $V = $ sorted$\{$set of values in dimension $i\}$
21:     Choose private median $p$ by $PrivateMedian(V, \epsilon)$
22:     **for** each sample $X$ in $S$ **do**
23:         **if** $X_i \leq p$ **then**
24:             add to left branch
25:         **else**
26:             add to right branch
27:         **end if**
28:     **end for**
29:     Return LeftBranch, RightBranch
30: **end procedure**

---

One of the key steps in our private tree construction is being able to compute the median accurately while preserving privacy. This is crucial for the final performance of the tree. A demonstration of the effect of the median estimation on accuracy is shown in Fig. 4. There are different methods of private median computation as discussed in [7], the most common approach is using the exponential mechanism. However, a direct application of the exponential mechanism considers all the feature values occurring in the data as candidates, some of which will be far away from the median. Each of these have low utility, but they accumulate a large fraction of the selection probability. We instead sample the candidates $\mathcal{R}$ uniformly from the interval defined by the minimum and maximum of the feature values occurring in the data – this significantly reduces the computations, and in our experience it also results in a more accurate estimate.

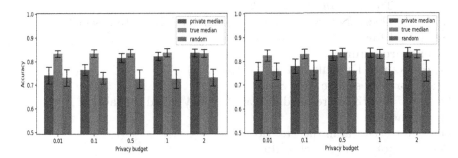

**Fig. 1.** Effect of median estimation on accuracy when we vary the privacy budget. Synthetic data sets of dimension 5 (left) and 10 (right) are used to compare the accuracy of Algorithm 1 with its variation that uses the true median, and random splitting. Observe, the accuracy increases as we make better estimates of the true median due to larger privacy budget.

We employ the exponential mechanism with a rank-based utility function as follows. Let $\epsilon_s$ denote the desired privacy parameter for tree construction. Denote a subset of the data as $S_i \subset S$ and let $V$ denote the set of all values of the $j$-th attribute for points in $S_i$. Then the utility of a candidate $r \in \mathcal{R}$ for attribute $j$ is defined as $u(V,r) = -|rank_j(r) - |V|/2|$, where $rank_j(r)$ denotes the number of points in $V$ that are no larger than $r$.

This utility function assigns a negative quality score to all values except for the median of the sample, which will have quality score zero. Values $r \in \mathcal{R}$ will have decreasing utilities the further away they are from the median. For categorical variables we use the same utility function, except that we let $\mathcal{R}$ to range over all categories for attribute $j$, and we define $rank_j(r)$ to be the number of points in $V$ that are equal to $r$. Note that the sensitivity of $u$ is $1/2$. Indeed, adding a data point into $V$ increases $|V|/2$ by $1/2$ and $rank_j(r)$ either increases by 1 or remains the same; removing a data point has a similar effect. Now by the exponential mechanism, for any given $\epsilon$ we can guarantee $\epsilon$-DP by outputting $r \in \mathcal{R}$ with probability

$$\mathbb{P}[\mathcal{M}(V, u, \mathcal{R}) = r] \propto \exp\left(\epsilon u(V,r)\right), \tag{5}$$

where the actual probability will be obtained through dividing the sum of proportional probabilities over all $r \in \mathcal{R}$.

---

**Algorithm 2.** Supervised Private Ensemble

---
1: Inputs: labelled set $S$, size of ensemble $N$, maximum depth $k$, privacy budget $\epsilon$
2: **procedure** SUPERVISEDENSEMBLE($S, N, k, \epsilon$)
3:     Split the total budget $\epsilon$ into $\epsilon_s, \epsilon_l$. # even split by default
4:     Randomly partition $S$ into $N$ disjoint subsets $\{S_i\}_{i=1}^N$.
5:     **for** $i$ in $1, \ldots, N$ **do**
6:         Tree $i = BuildTree(S_i, k, \epsilon_s)$
7:         $DistributeLabels$(Tree $i$, $S_i$, $\epsilon_l$) and add Tree $i$ to ensemble
8:     **end for**
9: **end procedure**
10: **procedure** DISTRIBUTELABELS(Tree, $S, \epsilon$)
11:     **for** each sample $X$ in $S$ **do**
12:         find the corresponding leaf of $X$ and record the label of $X$
13:     **end for**
14:     **for** each leaf in the Tree **do**
15:         **for** each label class **do**
16:             add $noise \sim Lap(1/\epsilon)$ to the label count.
17:         **end for**
18:     **end for**
19: **end procedure**

---

Note that we have not queried the sample labels in our construction of the tree. By partitioning the training set into $N$ disjoint subsets, and distributing the labels to the leaves of the trees privately (line 10 in Algorithm 2), we obtain a private supervised ensemble model for classification and regression tasks. The full algorithm is given in Algorithm 2.

### 4.1   Privacy Analysis

There are two steps in Algorithms 1 and 2 where we have used a privacy mechanism: (i) *PrivateSplit*, and (ii) *DistributeLabels*. In this section, we analyze the privacy guarantee of each mechanism. For the *PrivateSplit* procedure we note that the only computation required to query the data set is private median estimation. The splitting process afterwards only partitions the data set by the split condition, which guarantees the same privacy by the post-processing property of DP [10]. To guarantee $\epsilon_s$-DP over the whole sequence of splits along the tree construction, we need to split $\epsilon_s$ into a sequence of privacy budgets

$$\epsilon_0 + \epsilon_1 + \cdots + \epsilon_{k-1} = \sum_{i=0}^{k-1} \epsilon_i = \epsilon_s, \tag{6}$$

where $\epsilon_i$ is the privacy budget for splits at depth $i$, and $k$ is the maximum depth. A node with depth $k$ corresponds to a leaf, and hence no split is required. For splits at the same depth (say depth $i$), we can assign $\epsilon_i$ budget to each split because the data set held at the nodes of the same depth are disjoint. Therefore we have $\epsilon_i$-DP guaranteed simultaneously by the parallel composition theorem. Furthermore, by the sequential composition theorem we sum all splits at different depths and obtain $(\epsilon_0 + \cdots + \epsilon_{k-1}) = \epsilon_s$-DP.

Now, for *DistributeLabels* we will use the Laplace mechanism to output a private count of the classes while guaranteeing $\epsilon_l$-DP, where $\epsilon_l$ is the desired privacy parameter for leaf construction. We note that the sensitivity of the class count is 1 as adding or removing a point changes the count by at most 1. Hence, by the Laplace mechanism, it suffices to add random noise drawn from $Lap(1/\epsilon_l)$ to achieve $\epsilon_l$-DP. Moreover, since the data set in each leaf is disjoint from each other, by guaranteeing $\epsilon_l$-DP for each leave we can obtain $\epsilon_l$-DP for all leaves simultaneously by parallel composition. Thus, we have shown that, for any given $\epsilon_s$ and $\epsilon_l$, the ensemble construction in Algorithm 2 achieves $(\epsilon_s + \epsilon_l)$-DP.

## 4.2  Privacy Budget Allocation

In this section we discuss our strategy of privacy budget allocation for the construction of ensembles in Algorithm 2. For a total privacy budget $\epsilon$, since we have partitioned the sample set $S$ into $N$ disjoint subsets $\{S_i\}_{i=1}^N$, we can allocate the whole privacy budget $\epsilon$ to every tree by the parallel composition theorem. We further split $\epsilon$ to $\epsilon = \epsilon_s + \epsilon_l$ for privacy budget used in node splits and label predictions, respectively. We will use an equal share between the two as default, since both procedures are important to the final performance, i.e. $\epsilon_s = \epsilon_l = \epsilon/2$ (except in semi-supervised setting which we will discuss in Sect. 5).

We now discuss the budget allocation of $\epsilon_s$ along the nodes as follows. As a general intuition, the optimal budget allocation should depend on the difficulty of performing the private median computation. Based on this idea, we propose that the privacy budget allocation follows a geometrically-scaling sequence along the depths of the nodes.

Let $r, r' \in \mathcal{R}$ be any two potential outputs drawn uniformly from $[a, b], a, b \in \mathbb{R}$. The private estimation of the median is easier if we can distinguish the utility of $r, r'$ and output the better option with higher probability. This means that we want the utility difference $|u(r) - u(r')|$ to be large, as calculated as the number of values between $r$ and $r'$. We observe that the expected difference between two randomly chosen points from a uniform distribution is equal to $|a - b|/3$ (we cannot make any assumption on the values of the input samples). This observation implies that for every point added or removed, the probability that $|u(r) - u(r')|$ will change due to the added/removed point equals to $1/3$. Since any parent node is expected to receive 2 times the number of samples compared to its child nodes, the median estimation problem will be $2 \times (1/3)$ times easier comparing to its child node. Hence, for any node at depth $i$ that received $\epsilon_i$ budget, we assign $\epsilon_{i+1} = (3/2) * \epsilon_i$ privacy budget to its child nodes at depth $i + 1$. Furthermore, we must full-fill the condition that the sum of privacy budgets over all depths equals $\epsilon_s$. Hence we scale each $\epsilon_i$ by a constant $C$ so that we have $\sum_{i=0}^{k-1} \epsilon_i = \epsilon_s$, WLOG we assume $\epsilon_0 = C\epsilon_s$.

$$\epsilon_s = \sum_{i=0}^{k-1} \epsilon_i = \epsilon_0 + (3/2)\epsilon_0 + \cdots + (3/2)^{k-1}\epsilon_0$$

$$= C\epsilon_s(1 + (3/2) + \cdots + (3/2)^{k-1}) = C\epsilon_s \left( \frac{(3/2)^k - 1}{(3/2) - 1} \right), \quad (7)$$

which implies $\epsilon_i = C\epsilon_s \left(\frac{3}{2}\right)^i$ and $C = \frac{1}{2(3/2)^k - 2}$.

To illustrate the effectiveness of our privacy budget allocation strategy, we run simulations of Algorithm 1 to analyse the quality of the estimated median by comparing our allocation strategy with the uniform allocation baseline. We run our experiments on a synthetic data set generated from a normal distribution and we kept all other parameters fixed except the allocation strategy (Fig. 2).

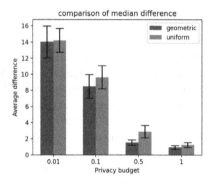 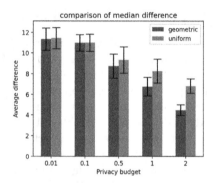

**Fig. 2.** Comparison between geometric and uniform allocation at different privacy levels with the maximum depth 5 (left) and 10 (right). Error bars indicate ($\pm 1$) standard deviation from the mean, and each experiment is repeated 50 times. We observe that our strategy achieves smaller average distance to the true median over all sets of experiments. This shows that the proposed allocation strategy has a significant effect on the median estimation while guaranteeing the same level of privacy.

# 5    Differentially Private Semi-supervised Ensembles

In this section we present our framework of private semi-supervised learning that creates private ensembles in the following private settings: (I) privacy of both the features and labels are required; (II) only privacy of labels is required. For the first case, existing work can only train on a labelled set, and cannot take advantage of a separate unlabelled set. In contrast, the method of our framework can build the tree using the unlabelled set only; as a result we can assign all privacy budget to the label predictions and reduce the noise. Moreover, we significantly reduce the number of labelled samples needed while achieving a good accuracy level. We present the procedures of the private ensembles in Algorithm 3 and 4, Algorithm 3 applied to both private settings (I) and (II) where Algorithm 4 is applicable only in setting (II). For setting (II) in Algorithm 3 we use $\infty$-privacy to build trees, meaning that we can compute the true median for each split, and no partitioning of the unlabelled set is required.

---

**Algorithm 3.** Semi-Supervised Private Ensemble

---

1: Inputs: labelled set $S$, unlabelled set $D$, maximum depth $k$, size of ensemble $N$, privacy budget $\epsilon$
2: **procedure** SS-ENSEMBLE$(S, D, k, N, \epsilon)$
3:     **if** need privacy for features and labels **then**
4:         Partition $S, D$ into $N$ disjoint portions: $\{S_i\}_{i=1}^N, \{D_i\}_{i=1}^N$
5:         **for** $i$ in range $1, \ldots, N$ **do**
6:             Tree $i$ = BuildTree$(D_i, k, \epsilon)$
7:             $DistributeLabels(Tree\ i, S_i, \epsilon)$ and add Tree $i$ to ensemble
8:         **end for**
9:     **else** #privacy for labels only
10:         $S' = S$ with labels removed
11:         Partition $S$ into $N$ disjoint portions: $\{S_i\}_{i=1}^N$
12:         **for** $i$ in range $1, \ldots, N$ **do**
13:             Tree $i$ = BuildTree$(S' \cup D, k, \infty)$
14:             $DistributeLabels(Tree\ i, S_i, \epsilon)$ and add Tree $i$ to ensemble
15:         **end for**
16:     **end if**
17:     Return ensemble
18: **end procedure**

---

In setting (II) we can perform computations with the features as many times as needed and release the output without privacy concern. A transductive approach can be applied in this case to take further advantage of the unlabelled data [20]. The transductive approach trains a small ensemble using a labelled set and then predicts labels for each sample in the unlabelled set using the trained ensemble. The newly-labelled set can then be used to train a larger ensemble. Our framework also takes advantage of this approach in the label-only privacy setting. A draw-back of this technique is that the newly-labelled set can have noisy labels due to inaccurate predictions which can decrease the accuracy of the final model.

# 6   Experimental Analysis

To illustrate the performance of the proposed algorithm, we perform a series of experiments using synthetic data sets as well as real data sets from the UCI [8]. The synthetic data sets are generated by forming normally distributed clusters

---

**Algorithm 4.** Private Transductive Ensemble

---

1: Inputs: labelled set $S$, unlabelled set $D$, maximum depth $k$, size of first ensemble $N1$, size of second ensemble $N2$, privacy budget $\epsilon$
2: **procedure** TRANSDUCTIVEENSEMBLE($S, D, k, N1, N2, \epsilon$)
3:     Partition $S$ into $N1$ disjoint portions: $\{S_i\}_{i=1}^{N}$
4:     **for** $i$ in range $1, \ldots, N1$ **do**
5:         Tree $i$ = BuildTree($S' \cup D$, $k$, $\epsilon$)
6:         $DistributeLabels(Tree\ i, S_i, \epsilon)$ and add Tree $i$ to ensemble1
7:     **end for**
8:     Assign labels to samples in $D$ using ensemble1 and denote by $D_l$
9:     **for** $i$ in range $1, \ldots, N2$ **do**
10:         Tree $i$ = BuildTree($S' \cup D$, $k$, $\epsilon$)
11:         $DistributeLabels(Tree\ i, D_l, \epsilon)$ and add Tree $i$ to ensemble2
12:     **end for**
13:     Return $ensemble1 \cup ensemble2$
14: **end procedure**

---

with random centers using the python package *sklearn.make_classification*. We generate three synthetic data sets each with 3000 samples with 5, 10 and 15 attributes, each data set contains 2 classes. The UCI data sets cover a wide range of real data with size ranging from 150 to 32561 and dimensions ranging from 4 to 33. We use 90% of the data for training and the remaining 10% for testing in all of our experiments. Each data set is randomly shuffled before training. Each experiment is repeated on the same data set 50 times, and the average and standard deviation of the prediction accuracy is reported.

### 6.1 Varying the Parameters

We demonstrate the effect of parameters ($N$ trees and max-depth) on the accuracy in Fig. 3 with three UCI data sets using the supervised ensemble (Algorithm 2). We see the accuracy is high with small $N$ and decreases as we add more trees into the forest, as we expected, since under privacy constraints each tree works on a disjoint subset of the data, leading to weak learning of the individual trees. From the plots we see that most of the accuracy curves decline after 10 trees, hence we have set $N = 10$ as our default. Furthermore, we see that the choice of max-depth=$d$ is a reasonable default as it reaches high accuracy across data sets. We also experimented with deeper ($2d$) trees (features to split on are sampled with replacement) and see in Fig. 1 that this may win in some cases. However, we must be cautious in general, as this increases the complexity of the function class and we run the risk of over-fitting as the leaf nodes become too small.

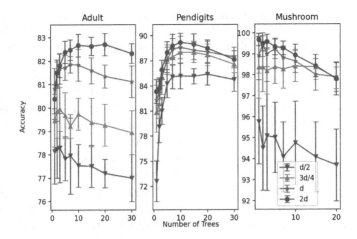

**Fig. 3.** Effect of number of trees (x-axis) and maximum depth (legend) on accuracy, where $d$ denotes the dimension of the data. Error bars indicate $\pm 1$ standard deviation from the mean, and $\epsilon$ is fixed to 2.

## 6.2    Comparison with Other Supervised Algorithms

We analyse the prediction accuracy in the supervised setting (no unlabelled set) by comparing our Supervised Private Ensemble (SPE) with three state-of-the-art differentially private tree-based algorithms: Smooth random trees (SRT) by [12], Random Decision Trees (RDT) by [17] and a version of greedy decision trees (MaxTree) by [19]. In the experiments, we build 10 trees with maximum depth $d$ as a default value where $d$ is the dimension. For competitors we used parameters recommended by the authors. For a non-private reference we use a random forest classifier with 100 trees as a benchmark for the best performance achievable on the particular data set without privacy constraints. The data sets used are described in Table 1.

To further evaluate if the reported result is statistically significant, we perform a Mann-Whitney U test (Wilcoxon rank sum test) at a 95% confidence level. From Table 1 we see that our method achieves higher accuracy for the majority of data sets with a statistical significance. The largest improvement is more than 25%, obtained on the *Robot* data set. Out of 22 total data sets tested, the only data set where our method is significantly worse is the *Nursery* data set.

## 6.3    Comparisons in Private Semi-supervised Learning

To assess our framework in the semi-supervised setting, we perform experiments with a reduced number of labelled samples and a separate unlabelled set. The data sets will be the same as our analysis in Sect. 6.2 except *Iris* and *Wine*, which have less than 200 points, hence too small for semi-supervised learning. For each of the remaining data sets we only use 20% of the training set as

**Table 1.** Comparisons with other private tree ensemble methods in supervised learning. The privacy budget is fixed to 2. The best result is highlighted in **bold**. We use the symbol $\star$ to indicate if the best result is statistically significant compared to others.

| | Size | Attributes | RF | SPE | SRT | RDT | MaxTree |
|---|---|---|---|---|---|---|---|
| Adults | 32561 | 14 | 85.47% | **82.05%** $\star$ | 76.89% | 76.16% | 81.80% |
| Bank | 4520 | 16 | 89.39% | 88.24% | **88.73%** | 88.21% | 88.31% |
| Banknotes | 1372 | 4 | 99.38% | **93.54%** $\star$ | 77.39% | 70.59% | 86.93% |
| Blood Transfusion | 747 | 4 | 74.99% | 78.00% | 75.71% | **78.05%** | 75.68% |
| Car | 1728 | 7 | 97.88% | **72.71%** $\star$ | 71.16% | 69.38% | 71.42% |
| Claves | 10787 | 16 | 80.82% | **73.84%** $\star$ | 70.80% | 73.02% | 73.22% |
| Credit Card | 30000 | 24 | 82.48% | **80.22%** $\star$ | 77.80% | 78.28% | 78.61% |
| Dry Bean | 13611 | 17 | 92.61% | **90.74%** $\star$ | 84.42% | 75.28% | 89.99% |
| GammaTele | 19020 | 10 | 87.99% | **82.34%** $\star$ | 71.62% | 66.84% | 78.58% |
| Iris | 150 | 4 | 95.33% | **81.87%** | 80.53% | 81.20% | 31.07% |
| Letter | 20000 | 17 | 96.13% | **67.86%** $\star$ | 47.61% | 40.25% | 62.99% |
| Mushroom | 8124 | 7 | 100.00% | **99.15%** $\star$ | 98.36% | 93.89% | 97.51% |
| Nursery | 12960 | 8 | 98.56% | 79.19% | 80.64% | 65.57% | **88.29%** $\star$ |
| Occupancy | 8142 | 7 | 99.71% | **98.19%** $\star$ | 90.26% | 82.87% | 97.07% |
| Pendigits | 7494 | 16 | 99.22% | **91.72%** $\star$ | 89.03% | 81.61% | 47.12% |
| Robot | 5456 | 4 | 99.46% | **87.43%** $\star$ | 61.50% | 56.86% | 47.70% |
| Student | 648 | 33 | 78.58% | 65.29% | 60.77% | 64.28% | **65.60%** |
| Wine | 178 | 4 | 100.00% | 73.00% | 72.56% | **74.11%** | 36.00% |
| Syn5d | 3000 | 5 | 92.49% | **90.41%** $\star$ | 86.52% | 79.78% | 88.34% |
| Syn10d | 3000 | 10 | 94.52% | **87.64%** | 80.55% | 78.57% | 86.71% |
| Syn15d | 3000 | 15 | 94.08% | **87.11%** | 80.67% | 80.24% | 86.83% |

the labelled training set, the other 80% will be used as an unlabelled set with their labels removed. We fix the privacy budget to 2 as before. We compare the methods of our framework in semi-supervised learning with two state-of-the-art competitors - Semi-supervised RDT (SSRDT) by [16] and Transductive Output Perturbation (TOP) by [20], where SSRDT is a tree-structured non-parametric method and TOP is a combination of kNN and linear predictors. Both competitors apply in the case where feature privacy is not required, hence we compare them with our second setting of Algorithm 3 (DPE-2) and our Private Transductive Ensemble (PTE) as they are in the same setting. We also include the first setting of Algorithm 3 in our comparison however the result can be worse since it guarantees a stronger privacy (both features and labels). From Table 2 we observe that our methods SSPE-2 and PTE achieve a better accuracy over SSRDT and TOP for the majority of data sets tested, and most improvements are statistically significant. Note that the results are in general worse than the figures in the supervised case, since we only have access to 20% of the labels. For SSPE-1, despite it guarantees the same level of privacy for both the features and the labels, the result has shown its performance remains on a same level

**Table 2.** Comparison with existing private semi-supervised methods. **Bold** indicates if the result is better than its competitors and ⋆ indicates if the difference is statistically significant. For SSPE-1, we use *underline* to indicate when it performs not significantly worse than SSRDT and TOP.

|                  | SSPE - 1         | SSPE - 2      | PTE           | SSRDT         | TOP         |
|------------------|------------------|---------------|---------------|---------------|-------------|
| Adults           | **80.53** % ⋆    | **80.88%** ⋆  | **80.29%**⋆   | 75.97%        | 76.18%      |
| Bank             | 87.97%           | 88.02%        | **88.84%**    | 88.58%        | 88.44%      |
| Banknotes        | 53.86%           | 89.67%        | **90.41%**⋆   | 88.52%        | 55.30%      |
| Blood_transfusion| 76.16%           | 75.84%        | 76.11%        | 75.68%        | **77.36%**  |
| Car              | 70.23%           | **72.55%** ⋆  | 71.11%        | 70.40%        | 33.88%      |
| Claves           | 67.88%           | 67.70%        | 57.68%        | **74.16%**⋆   | 9.49%       |
| Credit Card      | **78.79** % ⋆    | **78.38%** ⋆  | 77.89%        | 77.82%        | 77.80%      |
| Dry Bean         | 86.84% ⋆         | 87.76% ⋆      | **88.61%**⋆   | 81.41%        | 72.30%      |
| GammaTele        | **74.77** % ⋆    | **77.02%**⋆   | 74.11% ⋆      | 65.50%        | 64.83%      |
| Letter           | 42.27%           | 53.80%        | **56.91%** ⋆  | 16.72%        | 53.42%      |
| Mushroom         | 90.09%           | **95.96%** ⋆  | 95.46% ⋆      | 92.69%        | 51.65%      |
| Nursery          | 74.31 % ⋆        | **77.17%** ⋆  | 76.37% ⋆      | 66.30%        | 50.66%      |
| Occupancy        | 96.14 % ⋆        | 94.20%        | 92.77%        | **95.16%** ⋆  | 79.08%      |
| Pendigits        | 72.04%           | 85.22% ⋆      | **87.00%** ⋆  | 70.10%        | 84.26%      |
| Robot            | **77.30%** ⋆     | **87.01%** ⋆  | 82.56% ⋆      | 64.33%        | 61.43%      |
| Student          | 64.77%           | **65.14%**    | 65.05%        | 63.05%        | 64.98%      |
| Syn5d            | 86.82%           | **91.59%** ⋆  | 91.54% ⋆      | 90.09%        | 90.29%      |
| Syn10d           | 86.30%           | 90.51%        | 91.11%        | 83.93%        | **91.31%**  |
| Syn15d           | 76.40%           | 85.94%        | **86.45%**    | 79.59%        | 86.29%      |

(no significant difference detected) in comparison with the competitors which only guarantee privacy for the labels. Moreover, it has significant improvements over SSRDT and TOP on some data sets despite the additional added noise as shown in Table 2. Hence we conclude that the methods in our framework achieve a significant improvement over the state of the art in utility performance and privacy guarantee.

## 7 Conclusions

We have proposed a framework of differentially private classification for supervised and semi-supervised learning with high utility. This is based on a novel combination of techniques to build a new private machine ensemble for supervised learning, which naturally extends to a semi-supervised setting. We proposed a novel privacy budget allocation scheme that increases the usage efficiency of the available privacy budget and improves accuracy. Our experimental analysis over a wide range of data sets demonstrates that our method provides significantly better performance than the state of the art. In the semi-supervised setting, we proposed private ensembles that can be trained efficiently using a small number of labelled samples while achieving high utility, which allows us to reduce labelling efforts in data sets with sensitive information. In particular, we proposed the first semi-supervised private ensemble that is applicable in two privacy settings (feature and label privacy, and just label privacy). Empirically we found that our method provides high performance in both settings.

**Acknowledgements.** The last author is funded by EPSRC grant EP/P004245/1.

# References

1. Aryal, S., Ting, K.M., Washio, T., Haffari, G.: Data-dependent dissimilarity measure: an effective alternative to geometric distance measures. Knowl. Inf. Syst. **63**, 479–506 (2017)
2. Beniley, J.: Multidimensional binary search trees used for associative searching. ACM Commun. **18**(9), 509–517 (1975)
3. Biau, G.: Analysis of a random forests model. J. Mach. Learn. Res. **13**(1), 1063–1095 (2012)
4. Breiman, L.: Consistency for a simple model of random forests (2004)
5. Breiman, L., Friedman, J., Stone, C.J., Olshen, R.A.: Classification and regression trees. CRC Press (1984)
6. Consul, S., William, S.A.: Differentially private random forests for regression and classification. Association for the Advancement of Artificial Intelligence (2021)
7. Cormode, G., Procopiuc, C., Srivastava, D., Shen, E., Yu, T.: Differentially private spatial decompositions. In: 2012 IEEE 28th International Conference on Data Engineering. IEEE (2012)
8. Dua, D., Graff, C.: UCI machine learning repository (2017)
9. Dwork, C.: Differential privacy. Automata, Languages and Programming, pp. 1–12 (2006)
10. Dwork, C., Roth, A.: The algorithmic foundations of differential privacy. Found. Trends Theor. Comput. Sci. **9**(3–4), 211–407 (2014)
11. Fletcher, S., Islam, M.Z.: A differentially private decision forest. In: Proceedings of the 13-th Australasian Data Mining Conference (2015)
12. Fletcher, S., Islam, M.Z.: Differentially private random decision forests using smooth sensitivity. Expert Syst. Appl. **78**, 16–31 (2017)
13. Friedman, A., Schuster, A.: Data mining with differential privacy. In: Proceedings of the 16th SIGKDD Conference on Knowledge Discovery and Data Mining, pp. 493–502 (2010)
14. Geurts, P., Ernst, D., Wehenkel, L.: Extremely randomized trees. Mach. Learn. **63**(1), 3–42 (2006)
15. Gong, M., Xie, Y., Pan, K., Feng, K., Qin, A.: A survey on differentially private machine learning. IEEE Comput. Intell. Mag. **15**(2), 49–64 (2020)
16. Jagannathan, G., Monteleoni, C., Pillaipakkamnatt, K.: A semi-supervised learning approach to differential privacy. In: 2013 IEEE 13th International Conference on Data Mining Workshops, pp. 841–848. IEEE (2013)
17. Jagannathan, G., Pillaipakkamnatt, K., Wright, R.N.: A practical differentially private random decision tree classifier. In: 2009 IEEE International Conference on Data Mining Workshops, pp. 114–121. IEEE (2009)

18. Klusowski, J.: Sharp analysis of a simple model for random forests. In: International Conference on Artificial Intelligence and Statistics (2021)
19. Liu, X., Li, Q., Li, T., Chen, D.: Differentially private classification with decision tree ensemble. Appl. Soft Comput. **62**, 807–816 (2018)
20. Long, X., Sakuma, J.: Differentially private semi-supervised classification. In: 2017 IEEE International Conference on Smart Computing (SMARTCOMP), pp. 1–6. IEEE (2017)
21. McSherry, F., Talwar, K.: Mechanism design via differential privacy. In: 48th Annual IEEE Symposium on Foundations of Computer Science (FOCS2007), pp. 94–103. IEEE (2007)
22. McSherry, F.D.: Privacy integrated queries: an extensible platform for privacy-preserving data analysis. In: Proceedings of the 2009 ACM SIGMOD International Conference on Management of Data, pp. 19–30 (2009)
23. Mohammed, N., Chen, R., Fung, B.C., Yu, P.S.: Differentially private data release for data mining. In: Proceedings of the 17th ACM SIGKDD International Conference on Knowledge Discovery and Data Mining, pp. 493–501 (2011)
24. Nissim, K., Raskhodnikova, S., Smith, A.: Smooth sensitivity and sampling in private data analysis. In: Proceedings of the Thirty-ninth Annual ACM Symposium on Theory of Computing, pp. 75–84 (2007)
25. Pham, A.T., Xi, J.: Differentially private semi-supervised learning with known class priors. In: 2018 IEEE International Conference on Big Data (Big Data), pp. 801–810. IEEE (2018)
26. Rana, S., Gupta, S.K., Venkatesh, S.: Differentially private random forest with high utility. In: 2015 IEEE International Conference on Data Mining, pp. 955–960. IEEE (2015)
27. Xin, B., Yang, W., Wang, S., Huang, L.: Differentially private greedy decision forest. In: ICASSP 2019–2019 IEEE International Conference on Acoustics, Speech and Signal Processing (ICASSP), pp. 2672–2676 (2019)

# Differentially Private Bayesian Neural Networks on Accuracy, Privacy and Reliability

Qiyiwen Zhang, Zhiqi Bu, Kan Chen, and Qi Long[(✉)]

University of Pennsylvania, Philadelphia, USA
qiyiwen.zhang@pennmedicine.upenn.edu, {zbu,kanchen,qlong}@upenn.edu

**Abstract.** Bayesian neural network (BNN) allows for uncertainty quantification in prediction, offering an advantage over regular neural networks that has not been explored in the differential privacy (DP) framework. We fill this important gap by leveraging recent development in Bayesian deep learning and privacy accounting to offer a more precise analysis of the trade-off between privacy and accuracy in BNN. We propose three DP-BNNs that characterize the weight uncertainty for the same network architecture in distinct ways, namely DP-SGLD (via the noisy gradient method), DP-BBP (via changing the parameters of interest) and DP-MC Dropout (via the model architecture). Interestingly, we show a new equivalence between DP-SGD and DP-SGLD, implying that some non-Bayesian DP training naturally allows for uncertainty quantification. However, the hyperparameters such as learning rate and batch size, can have different or even opposite effects in DP-SGD and DP-SGLD.

Extensive experiments are conducted to compare DP-BNNs, in terms of privacy guarantee, prediction accuracy, uncertainty quantification, calibration, computation speed, and generalizability to network architecture. As a result, we observe a new tradeoff between the privacy and the reliability. When compared to non-DP and non-Bayesian approaches, DP-SGLD is remarkably accurate under strong privacy guarantee, demonstrating the great potential of DP-BNN in real-world tasks.

**Keywords:** Deep learning · Bayesian neural network · Differential privacy · Uncertainty quantification · Optimization · Calibration

## 1 Introduction

Deep learning has exhibited impressively strong performance in a wide range of classification and regression tasks. However, standard deep neural networks do not capture the model uncertainty and fail to provide the information available in statistical inference, which is crucial to many applications where poor

---

Q. Zhang and Z. Bu—Equal contribution.

---

**Supplementary Information** The online version contains supplementary material available at https://doi.org/10.1007/978-3-031-26412-2_37.

M.-R. Amini et al. (Eds.): ECML PKDD 2022, LNAI 13716, pp. 604–619, 2023.
https://doi.org/10.1007/978-3-031-26412-2_37

decisions are accompanied with high risks. As a consequence, neural networks are prone to overfitting and being overconfident about their prediction, reducing their generalization capability and more importantly, their reliability. From this perspective, Bayesian neural network (BNN) [7,22,23,27] is highly desirable and useful as it characterizes the model's uncertainty, which on one hand offers a reliable and calibrated prediction interval that indicates the model's confidence [4,15,19,35], and on the other hand reduces the prediction error through the model averaging over multiple weights sampled from the learned posterior distribution. For example, networks with the dropout [31] can be viewed as a Bayesian neural network by [13]; the dropout improves the accuracy from 57% [37] to 63% [31] on CIFAR100 image dataset and 69.0% to 70.4% on Reuters RCV1 text dataset [31]. In another example, on a genetics dataset where the task is to predict the occurrence probability of three alternative-splicing-related events based on RNA features. The performance of 'Code Quality' (a measure of the KL divergence between the target and the predicted probability distributions) can be improved from 440 on standard network to 623 on BNN [36].

In a long line of research, much effort has been devoted to making BNNs accurate and scalable. These approaches can be categorized into three main classes: (i) by introducing random noise into gradient methods (e.g. SG-MCMC [34]) to quantify the weight uncertainty; (ii) by considering each weight as a distribution, instead of a point estimate, so that the uncertainty is described inside the distribution; (iii) by introducing randomness on the network architecture (e.g. the dropout) that leads to a stochastic training process whose variability characterizes the model's uncertainty. To be more specific, we will discuss these methods including the Stochastic Gradient Langevin Descent (SGLD) [21], the Bayes By Backprop (BBP) [4] and the Monte Carlo Dropout (MC Dropout) [13].

Another natural yet urgent concern on the standard neural networks is the privacy risk. The use of sensitive datasets that contain information from individuals, including medical records, email contents, financial statements, and photos, has incurred serious risk of privacy violation. For example, the sale of Facebook user data to Cambridge Analytica [8] leads to the $5 billion fine to the Federal Trade Commission for its privacy leakage. As a gold standard to protect the privacy, the differential privacy (DP) has been introduced by [11] and widely applied to deep learning [1,5,6,30], due to its mathematical rigor.

Although both uncertainty quantification and privacy guarantee have drawn increasing attention, most existing work studied these two perspectives separately. Previous arts either studied DP Bayesian linear models [34,38] or studied DP-BNN using SGLD [20] but only for the accuracy measure without uncertainty quantification. In short, to the best of our knowledge, no existing deep learning models have equipped with the differential privacy and the Bayesian uncertainty quantification simultaneously.

Our proposal *contributes* on several fronts. First, We propose three distinct DP-BNNs that all use the DP-SGD (stochastic gradient descent) but characterize the weight uncertainty in fundamentally distinct ways, namely DP-SGLD (via the noisy gradient method), DP-BBP (via changing the parameters of interest), and DP-MC Dropout (via the model architecture). Our DP-BNNs are essentially DP Bayesian training procedures, as summarized in Fig. 10, while the inference procedures of DP-BNNs are the same as regular non-private BNNs.

Second, we establish the precise connection between the Bayesian gradient method, DP-SGLD and the non-Bayesian method, DP-SGD. Through a rigorous analysis, we show that DP-SGLD is a sub-class of DP-SGD yet the training hyperparameters (e.g. learning rate and batch size) have very different impacts on the performance of these two methods.

Finally, We empirically evaluate DP-BNNs through the classification and regression tasks. Notice that although all three DP-BNNs are equally private and capable of uncertainty quantification, their performance can be significantly different under various measures, as discussed in Sect. 4.

## 2  Differentially Private Neural Networks

In this work, we consider $(\epsilon, \delta)$-DP and also use $\mu$-GDP as a tool to compose the privacy loss $\epsilon$ iteratively. We first introduce the definition of $(\epsilon, \delta)$-DP in [12].

**Definition 1.** *A randomized algorithm M is $(\varepsilon, \delta)$-differentially private (DP) if for any pair of datasets $S, S'$ that differ in a single sample, and any event $E$,*

$$\mathbb{P}[M(S) \in E] \leqslant e^{\varepsilon} \mathbb{P}[M(S') \in E] + \delta. \tag{1}$$

A common approach to learn a DP neural network (NN) is to use DP gradient methods, such as DP-SGD (see Algorithm 1; possibly with the momentum and weight decay) and DP-Adam [5], to update the neural network parameters, i.e. weights and biases. In order to guarantee the privacy, DP gradient methods differ from its non-private counterparts in two steps. For one, the gradients are *clipped* on a per-sample basis, by a pre-defined clipping norm $C$. This is to ensure the sum of gradients has a bounded *sensitivity* to data points (this concept is to be defined in Appendix A). We note that in non-neural-network training, DP gradient methods may apply without the clipping, for instance, DP-SGLD in [34] requires no clipping and is thus different from our DP-SGLD in Algorithm 2 (also our DP-SGLD need not to modify the noise scale). For the other, some level of random Gaussian noises are added to the clipped gradient at each iteration. This is known as the *Gaussian mechanism* which has been rigorously shown to be DP by [12, Theorem 3.22].

---

**Algorithm 1:** Differentially private SGD (DP-SGD) with regularization

---

**Input:** Examples $\{(\boldsymbol{x}_i, y_i)\}$, loss $\ell(\cdot; \boldsymbol{w})$, regularization $r(\boldsymbol{w})$.
**for** $t = 1$ *to* $T$ **do**
  Randomly sample $B_t \subset \{1, 2, \ldots, N\}$;
  **for** $i \in B_t$ **do**
    Compute $g_i = \nabla_{\boldsymbol{w}} \ell(x_i, y_i; \boldsymbol{w}_{t-1})$
    Clip $\tilde{g}_i = \min\{1, \frac{C_t}{\|g_i\|_2}\} \cdot g_i.$ ;
  Add noise $\hat{g} = \frac{1}{|B_t|} \sum_{i \in B_t} \tilde{g}_i + \frac{\sigma \cdot C_t}{|B_t|} \cdot \mathcal{N}(0, I_d).$
  Update $\boldsymbol{w}_t \leftarrow \boldsymbol{w}_{t-1} - \eta_t (\hat{g} + \nabla_{\boldsymbol{w}} r(\boldsymbol{w}_{t-1}))$ ;
**Output:** $\boldsymbol{w}_1, \boldsymbol{w}_2, \cdots, \boldsymbol{w}_T$

---

In the training of neural networks, the Gaussian mechanism is applied multiple times and the privacy loss $\epsilon$ accumulates, indicating the model becomes increasingly vulnerable to privacy risk though more accurate. To compute the total privacy loss, we leverage the recent privacy accounting methods: Gaussian differential privacy (GDP) [5,10] and Moments accountant [1,9,18]. Both methods give valid though different upper bounds of $\epsilon$ as a consequence of using different composition theories. Notably, the rate at which the privacy compromises depends on the certain hyperparameters, such as the number of iterations $T$, the learning rate $\eta$, the noise scale $\sigma$, the batch size $|B|$, the clipping norm $C$. In the following sections, we exploit how these training hyperparameters influences DP and the convergence, and subsequently the uncertainty quantification.

## 3  Bayesian Neural Networks

BNNs have achieved significant success recently, by incorporating expert knowledge and making statistical inference through uncertainty quantification. On the high level, BNNs share the same architecture as regular NNs $f(x; \boldsymbol{w})$ but are different in that BNNs treat weights as a probability distribution instead of a single deterministic value. Learned properly, these weight distributions can characterize the uncertainty in prediction and improve the generalization behavior. For example, suppose we have obtained the weight distribution $W$, then the prediction distribution of BNNs is $f(x; W)$, which is unavailable by regular NNs. We now describe three popular yet distinct approaches to learn BNNs, leaving the algorithms in Sect. 4, which has the DP-BNNs but reduces to non-DP BNNs when $\sigma = 0$ (no noise) and $C_t = \infty$ (no clipping). We highlight that all three approaches are heavily based on SGD (though other optimizers can also be used): the difference lies in how SGD is applied. The Pytorch implementation is available at github.com/JavierAntoran/Bayesian-Neural-Networks.

### 3.1  Bayesian Neural Networks via Sampling

**Stochastic Gradient Langevin Dynamics (SGLD).** SGLD [21,35] is a gradient method that applies on the weights $\boldsymbol{w}$ of NN, and the weight uncertainty arises from the random noises injected into the training dynamics. Therefore, SGLD works on regular NN without any modification. However, unlike SGD, SGLD makes $\boldsymbol{w}$ to converge to a posterior distribution rather than to a point estimate, from which SGLD can sample and characterize the uncertainty of $\boldsymbol{w}$. In details, SGLD takes the following form

$$\boldsymbol{w}_t = \boldsymbol{w}_{t-1} + \eta_t \left( \nabla \log p(\boldsymbol{w}_{t-1}) + \frac{n}{|B_t|} \sum_{i \in B_t} \nabla \log p(\boldsymbol{x}_i, y_i | \boldsymbol{w}_{t-1}) \right) + \mathcal{N}(0, \eta_t)$$

where $p(\boldsymbol{w})$ is the pre-defined prior distribution of weights and $p(\boldsymbol{x}, y | \boldsymbol{w})$ is the likelihood of data. In the literature of empirical risk minimization, SGLD can be viewed as SGD with random Gaussian noise in the updates:

$$\boldsymbol{w}_t = \boldsymbol{w}_{t-1} - \eta_t \left( \nabla r(\boldsymbol{w}_{t-1}) + \frac{n}{|B_t|} \sum_{i \in B_t} \nabla \ell(\boldsymbol{x}_i, y_i; \boldsymbol{w}_{t-1}) \right) + \mathcal{N}(0, \eta_t),$$

where $r(w)$ is the regularization and $\ell(x, y; w)$ is loss. We summarize in Footnote 1 an one-to-one correspondence between the regularization $r(w)$ and the prior $p(w)$, as well as between the loss $\ell(x, y; w)$ and the likelihood $p(x, y|w)$. Writing the penalized loss as $\mathcal{L}_{\text{SGLD}}(\boldsymbol{x}_i, y_i; \boldsymbol{w}) := n \cdot \ell(\boldsymbol{x}_i, y_i; \boldsymbol{w}) + r(\boldsymbol{w})$, we obtain

$$\boldsymbol{w}_t = \boldsymbol{w}_{t-1} - \frac{\eta_t}{|B_t|} \sum_{i \in B_t} \frac{\partial \mathcal{L}_{\text{SGLD}}(\boldsymbol{x}_i, y_i; \boldsymbol{w}_{t-1})}{\partial \boldsymbol{w}_{t-1}} + \mathcal{N}(0, \eta_t).$$

Interestingly, although SGLD adds an isotropic Gaussian noise to the gradient (similar to DP-SGD in Algorithm 1), it is not guaranteed as DP without the per-sample gradient clipping. Nevertheless, while SGLD is different from SGD, we show in Theorem 1 that DP-SGLD is indeed a sub-class of DP-SGD.

## 3.2 Bayesian Neural Networks via Optimization

In contrast to SGLD, which is considered as a sampling approach that modifies the updating algorithm, we now introduce two optimization approaches of BNNs that use the regular optimizers like SGD, but modify the objective of minimization or the network architecture instead.

**Bayes by Backprop (BBP).** BBP [4] uses the standard SGD except it is applied on the hyperparameters of pre-defined weight distributions, rather than on the weights $\boldsymbol{w}$ directly. For example, suppose we assume that $w \sim \mathcal{N}(\mu, \sigma^2)$. Then BBP updates hyperparameters $(\mu, \sigma)$ while the regular SGD updates $w$.

This approach is known as the 'variational inference' or the 'variational Bayes', where a *variational distribution* $q(\boldsymbol{w}|\theta)$ is learned through its governing hyperparameter $\theta$. Consequently, the weight uncertainty is included in such variational distribution from which we can sample during the inference time.

In order to update the hyperparameter $\theta$, the objective of minimization requires highly non-trivial transformation from $\ell(x, y; w)$ and is derived as follows. Given data $D = \{(\boldsymbol{x}_i, y_i)\}$, the likelihood is $p(D|w) = \Pi_i p(y_i|\boldsymbol{x}_i, \boldsymbol{w})$ under some probabilistic model $p(y|\boldsymbol{x}, \boldsymbol{w})$. By the Bayes theorem, the posterior distribution $p(w|D)$ is proportional to the likelihood and the prior distribution $p(w)$,

$$p(w|D) \propto p(D|w)p(w) = \Pi_i p(y_i|x_i, w)p(w).$$

Within a pre-specified variational distribution $q(w|\theta)$, we seek the distributional parameter $\theta$ such that $q(w|\theta) \approx p(w|D)$. Conventionally, the variational distribution is restricted to be Gaussian and we learn its mean and standard deviation $\theta = (\mu, \sigma)$ through minimizing the KL divergence:

$$\min_\theta KL\left(q(w|\theta)\|p(w|D)\right) \equiv \mathbb{E}\log q(w|\theta) - \mathbb{E}\log p(w) - \mathbb{E}\log p(D|w). \quad (2)$$

This objective function is analytically intractable but can be approximated by drawing $w^{(j)}$ from $q(w|\theta)$ for $N$ independent times:

$$\mathcal{L}_{\text{BBP}}(D; \theta) := \frac{1}{N} \sum_{j \in [N]} \log q(w^{(j)}|\theta) + r(w^{(j)}) + \ell(D; \boldsymbol{w}^{(j)})$$

This approximated KL divergence is the actual objective to optimize instead of $\ell(D; w)$ used by the non-Bayesian NN (see the derivation of $\mathcal{L}_{BBP}$ in Appendix B.1). It follows that in BBP, the SGD updating rule for the reparameterization $\theta = (\mu, \rho)$ with $\sigma = \log(1 + \exp(\rho))$ is

$$\mu_t = \mu_{t-1} - \frac{\eta_t}{|B_t|} \sum_{i \in B_t} \frac{d\mathcal{L}_{BBP}(\boldsymbol{x}_i, y_i)}{d\mu}, \rho_t = \rho_{t-1} - \frac{\eta_t}{|B_t|} \sum_{i \in B_t} \frac{d\mathcal{L}_{BBP}(\boldsymbol{x}_i, y_i)}{d\rho}.$$

**Monte Carlo Dropout (MC Dropout).** MC Dropout is proposed by [13] that establishes an interesting connection: optimizing the loss with $L_2$ penalty in regular NNs with dropout layers is equivalent to learning Bayesian inference approximately. From this perspective, the weight uncertainty is described by the randomness of the dropout operation. We refer to Appendix B.2 for an in-depth review of MC dropout.

In more details, denoting $\mathcal{L}_{Dropout}(\boldsymbol{x}_i, y_i; \boldsymbol{w}) := \ell(\boldsymbol{x}_i, y_i; \boldsymbol{w}) + r(\boldsymbol{w})$, such connection claims equivalence between the problem $\min_w \frac{1}{n} \sum_{i=1}^{n} \mathcal{L}_{Dropout}(\boldsymbol{x}_i, y_i; \boldsymbol{w})$ and the variational inference problem (2), when the prior distribution is a zero mean Gaussian one. This equivalence makes MC Dropout similar to BBP in the sense of minimizing the same KL divergence. Nevertheless, while BBP directly minimizes the KL divergence, MC Dropout in practice leverages the equivalence to minimize the regular loss $\ell(D; w)$ via the empirical risk minimization. Hence MC Dropout also shares similarity with SGD or SGLD. From the algorithmic perspective, suppose $\boldsymbol{w}_t$ is the remaining weights after the dropout in the $t$-th iteration, then the updating rule for MC Dropout with SGD is

$$\boldsymbol{w}_t = \boldsymbol{w}_{t-1} - \frac{\eta_t}{|B_t|} \sum_{i \in B_t} \frac{\partial \mathcal{L}_{Dropout}(\boldsymbol{x}_i, y_i; \boldsymbol{w}_{t-1})}{\partial \boldsymbol{w}_{t-1}}.$$

## 4 Differentially Private Bayesian Neural Networks

To prepare the development of DP-BNNs, we summarize how to transform a regular NN to be Bayesian and to be DP, respectively. To learn a BNN, we need to establish the relationship between the Bayesian quantities (likelihood and prior) and the optimization loss and regularization. Under the Bayesian regime, $\ell$ is the negative log-likelihood $-\log p(x, y|\theta)$ and $\log p(\theta)$ is the log-prior. Under the empirical risk minimization regime, $\ell$ is the loss function and we view $-\log p(\theta)$ as the regularization or penalty[1]. To learn a DP network, we simply apply DP gradient methods that guarantee DP via the Gaussian mechanism (see Appendix A). Therefore, we can privatize each BNN to gain DP guarantee by applying DP gradient methods to update the parameters, as shown in Fig. 10 and Fig. 11 in Appendix E.

---

[1] For example, if the prior is $\mathcal{N}(0, \sigma^2)$, then $-\log p(\theta) \propto \frac{\|\theta\|^2}{2\sigma^2}$ is the $L_2$ penalty; if the prior is Laplacian, then $-\log p(\theta)$ is the $L_1$ penalty; additionally, the likelihood of a Gaussian model corresponds to the mean squared error loss..

Although the high-level ideas of DP-BNNs are easy to understand, we emphasize that different DP-BNNs vary significantly in terms of generalizability, computation efficiency, and uncertainty quantification (see Table 5).

## 4.1 Differentially Private Stochastic Gradient Langevin Dynamics

---

**Algorithm 2:** Differentially private SGLD (DP-SGLD)

---

**Input:** Examples $\{(\boldsymbol{x}_i, y_i)\}$, loss $\ell(\cdot; \boldsymbol{w})$, regularization $r(\boldsymbol{w})$.

**for** $t = 1$ *to* $T$ **do**

     Randomly sample a batch $B_t \subset \{1, 2, \ldots, n\}$;

     **for** $i \in B_t$ **do**

         Compute $g_i = \nabla_{\boldsymbol{w}} \ell(x_i, y_i; \boldsymbol{w}_{t-1})$

         Clip $\widetilde{g}_i = \min\{1, \frac{C_t}{\|g_i\|_2}\} \cdot g_i.$ ;

     Update $\boldsymbol{w}_t \leftarrow \boldsymbol{w}_{t-1} - \eta_t \left( \frac{n}{|B_t|} \sum_{i \in B_t} \widetilde{g}_i + \nabla_{\boldsymbol{w}} r(\boldsymbol{w}_{t-1}) \right) + \mathcal{N}(0, \eta_t)$ ;

**Output:** $\boldsymbol{w}_1, \boldsymbol{w}_2, \ldots, \boldsymbol{w}_T$

---

We first prove in Theorem 1 (with proof in Appendix D) that DP-SGLD is a sub-class of DP-SGD: every DP-SGLD is equivalent to some DP-SGD; however, only DP-SGD with $\sigma = \frac{|B|}{\sqrt{n}\eta C}$ is a DP-SGLD. In fact, DP-SGLD with non-informative prior is a special case of vanilla DP-SGD; DP-SGLD with Gaussian prior is equivalent to some DP-SGD with weight decay (i.e. with $L_2$ penalty).

**Theorem 1.** *For DP-SGLD with some prior assumption and DP-SGD with the corresponding regularization,*

$$\text{DP-SGLD}_{(\eta_{\text{SGLD}}=\eta, C_{\text{SGLD}}=C)} = \text{DP-SGD}_{(\eta_{\text{SGD}}=\eta n, \sigma_{\text{SGD}}=|B|/(n\sqrt{\eta}C), C_{\text{SGD}}=C)},$$

$$\text{DP-SGD}_{(\eta_{\text{SGD}}=\eta, \sigma_{\text{SGD}}=\sigma, C_{\text{SGD}}=C)} = \text{DP-SGLD}_{(\eta_{\text{SGLD}}=\eta/n, C_{\text{SGLD}}=C=|B|/(\sqrt{n\eta}\sigma))}.$$

In Fig. 1, we empirically observe that DP-SGLD is indeed a sub-class in the family of DP-SGD and is superior to other members of this family as it occupies the top left corner of the graph. In fact, it has been suggested by [34] in the non-deep learning that, training a Bayesian model using SGLD automatically guarantees DP.

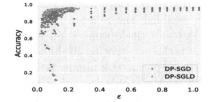

**Fig. 1.** Performance of DP-SGLD within DP-SGD family, on MNIST with CNN. Here $\delta = 10^{-5}, |B| = 256, \eta_{\text{SGD}} = 0.25, \eta_{\text{SGLD}} = 10^{-5}$, epoch $\leq 15$, $C \in [0.5, 5]$, $\sigma_{\text{SGD}} \in [0.5, 3]$.

In contrast, Theorem 1 is established in the deep learning regime and brings in a new perspective: training a regular NN using DP-SGD may automatically allow Bayesian uncertainty quantification.

Furthermore, DP-SGLD is generalizable to any network architecture (whenever DP-SGD works) and to any weight prior distribution (via different regularization terms); DP-SGLD does not require the computation of the complicated KL divergence. Computationally speaking, DP-SGLD enjoys fast computation speed (i.e. low computation complexity) since the per-sample gradient clipping can be very efficiently calculated using the outer product method [14, 29], the fastest acceleration technique of DP deep learning implemented in Opacus library. For example, on MNIST in Sect. 5, DP-SGLD requires only 10 sec/epoch, while DP-BBP takes 480 sec/epoch since it is incompatible with outer product.

However, DP-SGLD only offers empirical weight distribution $\{w_t\}$, which is not analytic and requires large memory for storage in order to give sufficiently accurate uncertainty quantification (e.g. we record 100 iterations of $w_t$ in Fig. 7 and 1000 iterations in Fig. 4). The memory burden can be too large to scale to large models that have billions of parameters, such as GPT-2.

### 4.2   Differentially Private Bayes by BackPropagation

Our DP-BBP can be viewed as DP-SGD working on the distributional hyperparameters such as the mean and the variance. In fact, it is the only method that does not works on weights directly, and thus requires to work with KL divergence via the variational inference problem (2).

---

**Algorithm 3:** Differentially private Bayes by BackPropagation (DP-BBP)

---

**Input:** Examples $\{(x_i, y_i)\}$, loss $\mathcal{L}_{\text{BBP}}(\cdot; \theta)$.
**for** $t = 1$ *to* $T$ **do**
  Randomly sample a batch $B_t \subset \{1, 2, \dots, n\}$;
  **for** $i \in B$ **do**
    **for** $j = 1$ *to* $N$ **do**
      Sample $w^{(j)}$ from $q(w|\theta_{t-1})$ and compute
$g_i^{(j)} = \nabla_\theta \mathcal{L}_{\text{BBP}}(x_i, y_i; w^{(j)}, \theta)$ ;
    Define $\bar{g}_i = \frac{1}{N} \sum_j g_i^{(j)}$ and clip $\widetilde{g}_i = \min\{1, \frac{C_t}{\|\bar{g}_i\|_2}\} \cdot \bar{g}_i$ ;
    Add noise $\hat{g} = \frac{1}{|B_t|} \sum_{i \in B} \widetilde{g}_i + \frac{\sigma \cdot C_t}{|B_t|} \cdot \mathcal{N}(0, I_d)$.
    Update $\theta_t \leftarrow \theta_{t-1} - \eta_t \hat{g}$ ;
**Output:** $\theta_T$

---

There are three major drawbacks of DP-BBP due to the KL divergence approach. Firstly, the updating rule needs significant modification for each type of network layers, e.g. convolutional layers and embedding layers. Therefore DP-BBP cannot work flexibly on general NNs. Secondly, DP-BBP suffers from high computation complexity. Under Gaussian variational distributions, DP-BBP needs to compute two hyperparameters (mean and standard deviation) for a single parameter (weight), which doubles the complexity of DP-SGLD, DP-MC Dropout and DP-SGD. The computational issue is further exacerbated due to the $N$ samplings of $\boldsymbol{w}^{(j)}$ from $q(\boldsymbol{w}|\theta_t)$, which means the number of back-propagation is $N$ times that of DP-SGLD and DP-MC Dropout. This introduces an inevitable tradeoff: when $N$ is larger, DP-BBP tends to be more accurate but its computational complexity is also higher, leading to the overall inefficiency of DP-BBP. Thirdly, DP-BBP cannot be accelerated by the outer product method as it violates the supported network layers[2]. Since the per-sample gradient clipping is the computational bottleneck for acceleration, DP-BBP can be too slow to be practically useful if the computation consideration overweighs its utility.

As for its advantages, DP-BBP is compatible to general DP optimizers such as DP-Adam. Similar to DP-SGLD, the DP-BBP can flexibly work under various priors by using different regularization terms $r(\boldsymbol{w})$ inside $\mathcal{L}_{\text{BBP}}$. Moreover, in sharp contrast to DP-SGLD and DP-MC Dropout, which only describe the weight distribution empirically, the distributional hyperparameters updated by DP-BBP directly characterize an analytic weight distribution for inference.

### 4.3    Differentially Private Monte Carlo Dropout

We can view our DP-MC Dropout as applying DP-SGD (or any other DP optimizers) on any NN with dropout layers, and thus DP-MC Dropout enjoys the low computation costs provided by the outer product acceleration in Opacus. Regarding the uncertainty quantification, DP-MC Dropout offers the empirical weight distribution at low storage costs since only $\boldsymbol{w}_T$ is stored, which means that its posterior is not analytic and will not be accurate if the number of training iterations is not sufficiently large.

---

**Algorithm 4:** Differentially private MC Dropout (DP-MC Dropout)

---

**Input:** Examples $\{(\boldsymbol{x}_i, y_i)\}$, loss $\mathcal{L}_{\text{Dropout}}(\cdot; \boldsymbol{w})$, regularization $r(\boldsymbol{w})$.
**for** $t = 1$ *to* $T$ **do**
 Randomly sample a batch $B_t \subset \{1, 2, \ldots, n\}$.
 Randomly drop out some weights and denote the remained as $\boldsymbol{w}_{t-1}$;
 **for** $i \in B_t$ **do**
  Compute $g_i = \nabla_{\boldsymbol{w}} \mathcal{L}_{\text{Dropout}}(\boldsymbol{x}_i, y_i; \boldsymbol{w}_{t-1})$
  Clip $\widetilde{g}_i = \min\{1, \frac{C_t}{\|g_i\|_2}\} \cdot g_i.$ ;
  Add noise $\hat{g} = \frac{1}{|B_t|} \sum_{i \in B_t} \widetilde{g}_i + \frac{\sigma \cdot C_t}{|B_t|} \cdot \mathcal{N}(0, I_d)$
  Update $\boldsymbol{w}_t \leftarrow \boldsymbol{w}_{t-1} - \eta_t(\hat{g} + \nabla_{\boldsymbol{w}} r(\boldsymbol{w}_{t-1}))$ ;
 **Output:** $\boldsymbol{w}_T$

---

[2] Since DP-BBP does not optimize the weights, the back-propagation is much different from using $\frac{\partial \ell}{\partial w}$ (see Appendix B) and thus requires new design that is currently not available. See https://github.com/pytorch/opacus/blob/master/opacus/supported_layers_grad_samplers.py.

A limitation to the theory of MC Dropout [13] is that the equivalence between the empirical risk minimization of $\mathcal{L}_{\text{Dropout}}$ and the KL divergence minimization (2) no longer holds beyond the Gaussian weight prior. Nevertheless, algorithmically speaking, DP-MC Dropout also works with other priors by using different regularization terms.

### 4.4 Analysis of Privacy

The following theorem gives the privacy loss $\epsilon$ by the GDP accountant [5,10].

**Theorem 2 (Theorem 5 in [5]).** *For both DP-MC Dropout and DP-BBP, under any DP-optimizers (e.g. DP-SGD, DP-Adam, DP-HeavyBall) with the number of iterations $T$, noise scale $\sigma$ and batch size $|B|$, the resulting neural network is $\sqrt{T(e^{1/\sigma^2} - 1)}|B|/n$-GDP.*

We remark that, from [10, Corollary 2.13], $\mu$-GDP can be mapped to $(\epsilon, \delta)$-DP via $\delta(\varepsilon; \mu) = \Phi\left(-\varepsilon/\mu + \mu/2\right) - e^{\varepsilon}\Phi\left(-\varepsilon/\mu - \mu/2\right)$. As alternatives to GDP, other privacy accountants such as the Moments Accountant (MA) [1,2,9,26] can be applied to characterize $\epsilon$, though implicitly (see Appendix A). Since DP-MC Dropout and DP-BBP do not quantify the uncertainty via optimizers, all privacy accountants give the same $\epsilon$ as training DP-SGD on regular NNs. We next give the privacy of DP-SGLD by writing it as DP-SGD.

**Theorem 3.** *For DP-SGLD with the number of iterations $T$, learning rate $\eta$, batch size $|B|$ and clipping norm $C$, the resulting neural network is $\sqrt{T(e^{n^2\eta C^2/|B|^2} - 1)}|B|/n$-GDP.*

The proof follows from Theorem 1 and [5, Theorem 5], given in Appendix D. We observe sharp contrast between Theorem 2 and Theorem 3: (1) while the clipping norm $C$ and learning rate $\eta$ have no effect on the privacy guarantee of DP-MC Dropout and DP-BBP, these hyperparameters play important roles in DP-SGLD. For instance, the learning rate triggers a tradeoff: larger $\eta$ converges faster but smaller $\eta$ is more private; see Fig. 2. (2) To get stronger privacy guarantee, DP-MC Dropout and DP-BBP need smaller $T$ and larger $\sigma$; however, DP-SGLD needs smaller $T, C$ and $\eta$. (3) Surprisingly, the batch size $|B|$ has *opposite* effects in DP-SGLD and in other methods: DP-SGLD with larger $|B|$ is more private, while smaller $|B|$ amplifies the privacy for DP-SGD [3,10,17,33].

## 5 Experiments

We further evaluate the proposed DP-BNNs on the classification (MNIST) and regression tasks, based on performance measures including uncertainty quantification, computational speed and privacy-accuracy tradeoff. In particular, we observe that DP-SGLD tends to outperform DP-MC Dropout, DP-BBP and DP-SGD, with little reduction in performance compared to non-DP models. All experiments (except BBP) are run with Opacus library under Apache License 2.0 and on Google Colab with a P100 GPU. A detailed description of the experiments can be found in Appendix C. Code of our implementation is available at https://github.com/littlekii/DPBBP.

## 5.1 Classification on MNIST

We first evaluate three DP-BNNs on the MNIST dataset, which contains $n = 60000$ training samples and 10000 test samples of $28 \times 28$ grayscale images of hand-written digits.

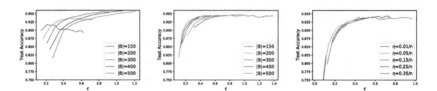

**Fig. 2.** Effects of batch size and learning rate on DP-SGD (left) and DP-SGLD (middle & right) with CNNs on MNIST. See Appendix C.3 for settings.

**Table 1.** Test accuracy and running time of DP-BBP, DP-SGLD, DP-MC Dropout, DP-SGD, and their non-DP counterparts. We use a default two-layer MLP and additionally a four-layer CNN by Opacus in parentheses.

| Methods | Weight prior | DP time/epoch | DP accuracy | Non-DP accuracy |
|---------|-------------|---------------|-------------|-----------------|
| SGLD | Gaussian | 10 s | 0.90 (0.95) | 0.95 (0.96) |
|  | Laplacian | 10 s | 0.89 (0.89) | 0.90 (0.89) |
| BBP | Gaussian | 480 s | 0.80 (——) | 0.97 (——) |
|  | Laplacian | 480 s | 0.81 (——) | 0.98 (——) |
| MC dropout | Gaussian | 9 s | 0.78 (0.77) | 0.98 (0.97) |
| SGD (non-Bayesian) | —— | 10 s | 0.77 (0.95) | 0.97 (0.99) |

**Accuracy and Privacy.** While all of non-DP methods have similar high test accuracy, in the DP regime in Table 1, DP-SGLD outperforms other Bayesian and non-Bayesian methods under almost identical privacy budgets (for details, see Appendix C). For the multilayer perceptron (MLP), all BNNs (DP or non-DP) do not lose much accuracy when gaining the ability to quantify uncertainty, compared to the non-Bayesian SGD. However, DP comes at high cost of accuracy, except for DP-SGLD which does not deteriorate comparing to its non-DP version, while other methods experience an accuracy drop $\approx 20\%$. Furthermore, DP-SGLD enjoys clear advantage in accuracy on more complicated convolutional neural network (CNN).

**Uncertainty and Calibration.** Regarding uncertainty quantification, we visualize the empirical prediction posterior of Bayesian MLPs in Fig. 7 over 100 predictions on a single image. Note that at each probability (x-axis), we plot a cluster of bins each of which represents a class[3]. For example, the left-most

---

[3] Within each cluster, the bins can interchange the ordering. Thus the bin's x-coordinate is not meaningful and only the cluster's x-coordinate represents the prediction probability.

cluster represents *not predicting* a class. As a measure of the reliability, the calibration [16,28] measures the distance between a classification model's accuracy and its prediction probability, i.e. confidence. Formally, denoting the vector of prediction probability for the $i$-th sample as $\pi_i$, the *confidence* for this sample is $\mathrm{conf}_i = \max_k[\pi_i]_k$ and the prediction is $\mathrm{pred}_i = \mathrm{argmax}_k[\pi_i]_k$. Two commonly applied calibration errors are the expected calibration error (ECE) and the maximum calibration error (MCE) [16].

Concretely, in the left-bottom plot, DP-SGLD has low red (class 3) and brown (class 5) bins on the left-most cluster, meaning it will predict 3 or 5. We see that non-DP BNNs usually predict correctly (with a low red bin in the left-most cluster), though the posterior probabilities of the correct class are different across three BNNs. Obviously, DP changes the empirical posterior probabilities significantly in distinct ways. First, all DP-BNNs are prone to make mistakes in prediction, e.g. both DP-SGLD and DP-BBP tend to predict class 5. In fact, DP-SGLD are equally likely to predict class 3 and 5 yet DP-BBP seldom predicts class 3 anymore, when DP is enforced. Additionally, DP-SGLD is less confident about its mistake compared to DP-BBP. This is indicated by the small x-coordinate of the right-most bins, and implies that DP-SGLD can be more calibrated, as discussed in the next paragraph. For MC Dropout, DP also reduces the confidence in predicting class 3 but the mistaken prediction spreads over several classes. Hence the quality of uncertainty quantification provided by DP-MC Dropout lies between that by DP-SGLD and DP-BBP.

Ideally, a reliable classifier should be calibrated in the sense that the accuracy matches the confidence. When a model is highly confident in its prediction yet it is not accurate, such classifier is over-confident; otherwise it is under-confident. It is well-known that the regular NNs are over-confident [16,25] and (non-DP) BNNs are more calibrated [24]. In Table 2 and Table 4, we again test the two-layer MLP and four-layer CNN on MNIST, with or without Gaussian prior under DP-BNNs regime. Notice that in the BNN regime, training with weight decay is equivalent to adopting a Gaussian prior, while training without weight decay is equivalent to using a non-informative prior.

On MLP, the Gaussian prior (or weight decay) significantly improves the MCE, in the non-DP regime and furthermore in the DP regime (see Fig. 8). However, on CNN, while the Gaussian prior helps in the non-DP regime, this may not hold true in the DP regime. For both neural network structures, DP exacerbates the mis-calibration: leading to worse MCE when the non-informative prior is used. See lower panel of Fig. 8 and Fig. 9. However, this is usually not the case when DP is guaranteed under the Gaussian prior. Additionally, BNNs often enjoy smaller MCE than the regular MLP but may have larger MCE than the regular CNN. In the case of SGLD, the effect of DP-BNN and prior distribution is visualized in Fig. 3.

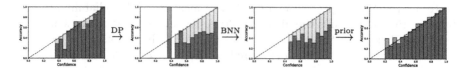

**Fig. 3.** Reliability diagram on MNIST with two-layer MLP, using SGD, DP-SGD and DP-SGLD (right two).

**Table 2.** Calibration errors of SGLD, BBP, MC Dropout, SGD, and their DP counterparts on MNIST with two-layer MLP, with or without Gaussian prior.

| Methods | DP-ECE | DP-MCE | Non-DP ECE | Non-DP MCE |
|---|---|---|---|---|
| BBP (w/ prior) | 0.204 | 0.641 | 0.024 | 0.052 |
| BBP (w/o prior) | 0.167 | 0.141 | 0.166 | 0.166 |
| SGLD (w/ prior) | 0.007 | 0.175 | 0.035 | 0.175 |
| SGLD (w/o prior) | 0.126 | 0.465 | 0.008 | 0.289 |
| MC Dropout (w/ prior) | 0.008 | 0.080 | 0.030 | 0.041 |
| MC Dropout (w/o prior) | 0.078 | 0.225 | 0.002 | 0.725 |
| SGD (w/ prior) | 0.013 | 0.089 | 0.016 | 0.139 |
| SGD (w/o prior) | 0.106 | 0.625 | 0.005 | 0.299 |

## 5.2   Heteroscedastic Synthetic Data Regression

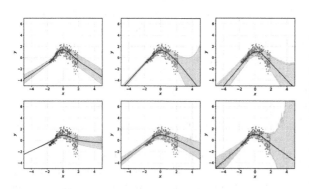

**Table 3.** Mean square error of heteroscedasticity regression with Gaussian prior. The reported error is the median over 20 independent simulations.

| Methods | DP | Non-DP |
|---|---|---|
| SGLD | 0.510 | 0.523 |
| BBP | 1.276 | 0.562 |
| MC Dropout | 0.682 | 0.591 |

**Fig. 4.** Prediction uncertainty on heteroscedasticity regression with Gaussian priors. Left to right: SGLD, BBP, MC Dropout. Upper: non-DP BNNs. Lower: DP-BNNs. Orange region refers to the posterior uncertainty. Blue region refers to the data uncertainty. Black line is the mean prediction. (Color figure online)

We compare the prediction uncertainty of BNNs on the heteroscedastic data generated from Gaussian process (see details in Appendix C). Here, the prediction uncertainty for each data point is estimated by the empirical posterior over 1000 predictions. Specifically, the prediction uncertainty can be decomposed into the posterior uncertainty (also called epistemic uncertainty, the blue region) and the data uncertainty (also called aleatoric uncertainty, the orange region), whose mathematical formulation is delayed in Appendix C. In Fig. 4, all three non-DP BNNs (upper panel) characterize similar prediction uncertainty, regarded as the ground truth.

In our experiments, we train all BNNs with DP-GD for 200 epochs and noise multiplier such that the DP is $\epsilon = 4.21, \delta = 1/250$. As shown in Table 3, SGLD is surprisingly accurate in both DP and non-DP scenarios while BBP and MC Dropout suffer notably from DP, even though their non-DP versions are accurate.

Clearly, the prediction uncertainty of SGLD and BBP are barely affected by DP; additionally, given that DP-SGLD has much better mean squared error, this experiment confirms that DP-SGLD is more desirable for uncertainty quantification with DP guarantee. Unfortunately, for MC Dropout, DP leads to substantially greater posterior uncertainty and unstable mean prediction. The resulting wide out-of-sample predictive intervals provide little information.

## 6  Discussion

This work proposes three DP-BNNs, namely DP-SGLD, DP-BBP and DP-MC Dropout, to both quantify the model uncertainty and guarantee the privacy in deep learning. Our work also provides valuable insights about the connection between DP-SGLD, a method often applied in the Bayesian settings, and DP-SGD, which is widely used without the consideration of Bayesian inference. This connection reveals novel findings about the impact of training hyperparameters on DP-SGLD, e.g. larger batch size enhances the privacy. All three DP-BNNs are evaluated through multiple metrics and demonstrate their advantages and limitations, supported by both theoretical and empirical analyses. For instance, as a sampling method, DP-SGLD outperforms the optimization methods, DP-BBP and DP-MC Dropout, on classification and regression tasks, at little expense of performance in comparison to the non-Bayesian or non-DP counterparts. However, DP-SGLD requires a possibly long period of burn-in to converge and its uncertainty quantification requires storing hundreds of weights, making the method less scalable.

For future directions, it is of interest to extend the connection between DP-SGD and DP-SGLD to a more general class, i.e. DP-SG-MCMC. Particularly, the convergence and generalization behaviors of DP-BNNs needs more investigation, similar to the analysis of different DP linear regression [32].

**Acknowledgment.** This research was supported by the NIH grants RF1AG063481 and R01GM124111.

# References

1. Abadi, M., et al.: Deep learning with differential privacy. In: Proceedings of the 2016 ACM SIGSAC Conference on Computer and Communications Security, pp. 308–318 (2016)
2. Asoodeh, S., Liao, J., Calmon, F.P., Kosut, O., Sankar, L.: A better bound gives a hundred rounds: enhanced privacy guarantees via f-divergences. In: 2020 IEEE International Symposium on Information Theory (ISIT), pp. 920–925. IEEE (2020)
3. Balle, B., Barthe, G., Gaboardi, M.: Privacy amplification by subsampling: tight analyses via couplings and divergences. arXiv preprint arXiv:1807.01647 (2018)
4. Blundell, C., Cornebise, J., Kavukcuoglu, K., Wierstra, D.: Weight uncertainty in neural network. In: International Conference on Machine Learning, pp. 1613–1622. PMLR (2015)
5. Bu, Z., Dong, J., Long, Q., Su, W.J.: Deep learning with gaussian differential privacy. Harvard Data Sci. Rev. 2020(23) (2020)
6. Bu, Z., Gopi, S., Kulkarni, J., Lee, Y.T., Shen, J.H., Tantipongpipat, U.: Fast and memory efficient differentially private-SGD via JL projections. arXiv preprint arXiv:2102.03013 (2021)
7. Buntine, W.L.: Bayesian backpropagation. Complex Syst. 5, 603–643 (1991)
8. Cadwalladr, C., Graham-Harrison, E.: Revealed: 50 million Facebook profiles harvested for Cambridge Analytica in major data breach. Guardian 17, 22 (2018)
9. Canonne, C., Kamath, G., Steinke, T.: The discrete Gaussian for differential privacy. arXiv preprint arXiv:2004.00010 (2020)
10. Dong, J., Roth, A., Su, W.J.: Gaussian differential privacy. arXiv preprint arXiv:1905.02383 (2019)
11. Dwork, C., McSherry, F., Nissim, K., Smith, A.: Calibrating noise to sensitivity in private data analysis. In: Halevi, S., Rabin, T. (eds.) TCC 2006. LNCS, vol. 3876, pp. 265–284. Springer, Heidelberg (2006). https://doi.org/10.1007/11681878_14
12. Dwork, C., Roth, A., et al.: The algorithmic foundations of differential privacy. Found. Trends Theor. Comput. Sci. 9(3–4), 211–407 (2014)
13. Gal, Y., Ghahramani, Z.: Dropout as a Bayesian approximation: Representing model uncertainty in deep learning. In: International Conference on Machine Learning, pp. 1050–1059. PMLR (2016)
14. Goodfellow, I.: Efficient per-example gradient computations. arXiv preprint arXiv:1510.01799 (2015)
15. Graves, A.: Practical variational inference for neural networks. In: Advances in Neural Information Processing Systems, vol. 24 (2011)
16. Guo, C., Pleiss, G., Sun, Y., Weinberger, K.Q.: On calibration of modern neural networks. In: International Conference on Machine Learning, pp. 1321–1330. PMLR (2017)
17. Kasiviswanathan, S.P., Lee, H.K., Nissim, K., Raskhodnikova, S., Smith, A.: What can we learn privately? SIAM J. Comput. 40(3), 793–826 (2011)
18. Koskela, A., Jälkö, J., Honkela, A.: Computing tight differential privacy guarantees using FFT. In: International Conference on Artificial Intelligence and Statistics, pp. 2560–2569. PMLR (2020)
19. Kuleshov, V., Fenner, N., Ermon, S.: Accurate uncertainties for deep learning using calibrated regression. In: International Conference on Machine Learning, pp. 2796–2804. PMLR (2018)
20. Li, B., Chen, C., Liu, H., Carin, L.: On connecting stochastic gradient MCMC and differential privacy. In: The 22nd International Conference on Artificial Intelligence and Statistics, pp. 557–566. PMLR (2019)

21. Li, C., Chen, C., Carlson, D., Carin, L.: Preconditioned stochastic gradient Langevin dynamics for deep neural networks. In: Proceedings of the AAAI Conference on Artificial Intelligence, vol. 30 (2016)
22. MacKay, D.J.: A practical bayesian framework for backpropagation networks. Neural Comput. **4**(3), 448–472 (1992)
23. MacKay, D.J.: Probable networks and plausible predictions - a review of practical Bayesian methods for supervised neural networks. Netw. Comput. Neural Syst. **6**(3), 469–505 (1995)
24. Maroñas, J., Paredes, R., Ramos, D.: Calibration of deep probabilistic models with decoupled Bayesian neural networks. Neurocomputing **407**, 194–205 (2020)
25. Minderer, M., et al.: Revisiting the calibration of modern neural networks. arXiv preprint arXiv:2106.07998 (2021)
26. Mironov, I., Talwar, K., Zhang, L.: R\'enyi differential privacy of the sampled Gaussian mechanism. arXiv preprint arXiv:1908.10530 (2019)
27. Neal, R.M.: Bayesian Learning for Neural Networks, vol. 118. Springer, New York (2012). https://doi.org/10.1007/978-1-4612-0745-0
28. Niculescu-Mizil, A., Caruana, R.: Predicting good probabilities with supervised learning. In: Proceedings of the 22nd International Conference on Machine Learning, pp. 625–632 (2005)
29. Rochette, G., Manoel, A., Tramel, E.W.: Efficient per-example gradient computations in convolutional neural networks. arXiv preprint arXiv:1912.06015 (2019)
30. Ryffel, T., et al.: A generic framework for privacy preserving deep learning. arXiv preprint arXiv:1811.04017 (2018)
31. Srivastava, N., Hinton, G., Krizhevsky, A., Sutskever, I., Salakhutdinov, R.: Dropout: a simple way to prevent neural networks from overfitting. J. Mach. Learn. Res. **15**(1), 1929–1958 (2014)
32. Wang, Y.X.: Revisiting differentially private linear regression: optimal and adaptive prediction & estimation in unbounded domain. arXiv preprint arXiv:1803.02596 (2018)
33. Wang, Y.X., Balle, B., Kasiviswanathan, S.P.: Subsampled rényi differential privacy and analytical moments accountant. In: The 22nd International Conference on Artificial Intelligence and Statistics, pp. 1226–1235. PMLR (2019)
34. Wang, Y.X., Fienberg, S., Smola, A.: Privacy for free: posterior sampling and stochastic gradient Monte Carlo. In: International Conference on Machine Learning, pp. 2493–2502. PMLR (2015)
35. Welling, M., Teh, Y.W.: Bayesian learning via stochastic gradient Langevin dynamics. In: Proceedings of the 28th International Conference on Machine Learning (ICML 2011), pp. 681–688. Citeseer (2011)
36. Xiong, H.Y., Barash, Y., Frey, B.J.: Bayesian prediction of tissue-regulated splicing using RNA sequence and cellular context. Bioinformatics **27**(18), 2554–2562 (2011)
37. Zeiler, M.D., Fergus, R.: Stochastic pooling for regularization of deep convolutional neural networks. arXiv preprint arXiv:1301.3557 (2013)
38. Zhang, Z., Rubinstein, B., Dimitrakakis, C.: On the differential privacy of Bayesian inference. In: Proceedings of the AAAI Conference on Artificial Intelligence, vol. 30 (2016)

# Differentially Private Federated Combinatorial Bandits with Constraints

Sambhav Solanki$^{(\boxtimes)}$, Samhita Kanaparthy, Sankarshan Damle,
and Sujit Gujar

Machine Learning Lab, International Institute of Information Technology (IIIT),
Hyderabad, India
{sambhav.solanki,s.v.samhita,sankarshan.damle}@research.iiit.ac.in,
sujit.gujar@iiit.ac.in

**Abstract.** There is a rapid increase in the cooperative learning paradigm in online learning settings, i.e., *federated learning* (FL). Unlike most FL settings, there are many situations where the agents are competitive. Each agent would like to learn from others, but the part of the information it shares for others to learn from could be sensitive; thus, it desires its *privacy*. This work investigates a group of agents working concurrently to solve similar combinatorial bandit problems while maintaining quality constraints. Can these agents collectively learn while keeping their sensitive information confidential by employing differential privacy? We observe that communicating can reduce the *regret*. However, differential privacy techniques for protecting sensitive information makes the data noisy and may deteriorate than help to improve regret. Hence, we note that it is essential to decide *when to communicate* and *what shared data to learn* to strike a functional balance between regret and privacy. For such a federated combinatorial MAB setting, we propose a Privacy-preserving Federated Combinatorial Bandit algorithm, P-FCB. We illustrate the efficacy of P-FCB through simulations. We further show that our algorithm provides an improvement in terms of regret while upholding quality threshold and meaningful privacy guarantees.

**Keywords:** Combinatorial multi-armed bandits · Differential privacy · Federated learning

## 1 Introduction

A large portion of the manufacturing industry follows the Original Equipment Manufacturer (OEM) model. In this model, companies (or aggregators) that design the product usually procure components required from an available set of OEMs. Foundries like TSMC, UMC, and GlobalFoundries handle the production of components used in a wide range of smart electronic offerings [1]. We also observe a similar trend in the automotive industry [2].

However, aggregators are required to maintain minimum *quality* assurance for their products while maximizing their revenue. Hence, they must judicially

© The Author(s), under exclusive license to Springer Nature Switzerland AG 2023
M.-R. Amini et al. (Eds.): ECML PKDD 2022, LNAI 13716, pp. 620–637, 2023.
https://doi.org/10.1007/978-3-031-26412-2_38

procure the components with desirable quality and cost from the OEMs. For this, aggregators should learn the quality of components provided by an OEM. OEM businesses often have numerous agents engaged in procuring the same or similar components. In such a setting, one can employ *online learning* where multiple aggregators, referred henceforth as *agents*, cooperate to learn the qualities [8,24]. Further, decentralized (or federated) learning is gaining traction for large-scale applications [20,33].

In general, an agent needs to procure and utilize the components from different OEMs (referred to as *producers*) to learn their quality. This learning is similar to the exploration and exploitation problem, popularly known as *Multi-armed Bandit* (MAB) [13,15]. It needs sequential interactions between sets of producers and the learning agent. Further, we associate qualities, costs, and capacities with the producers for each agent. We model this as a combinatorial multi-armed bandit (CMAB) [5] problem with assured qualities [15]. Our model allows the agents to maximize their revenues by communicating their history of procurements to have better estimations of the qualities. Since the agents can benefit from sharing their past quality realizations, we consider them engaged in a *federated* learning process. Federated MAB often improves performance in terms of *regret* incurred per agent [16,25][1].

Such a federated exploration/exploitation paradigm is not just limited to selecting OEMs. It is useful in many other domains such as stocking warehouse/distribution centres, flow optimization, and product recommendations on e-commerce websites [21,27]. However, agents are competitive; thus, engaging in federated learning is not straightforward. Agents may not be willing to share their private experiences since that could negatively benefit them. For example, sharing the exact procurement quantities of components specific to certain products can reveal the market/sales projections. Thus, we desire (or many times even it is necessary) to maintain privacy when engaged in federated learning. This paper aims to design a privacy-preserving algorithm for federated CMAB with quality assurances.

**Our Approach and Contributions.** Privacy concerns for sensitive information pose a significant barrier to adopting federated learning. To preserve the privacy of such information, we employ the strong notion of *differential privacy* (DP) [9]. Note that naive approaches (e.g., Laplace or Gaussian Noise Mechanisms [10]) to achieve DP for CMAB may come at a high privacy cost or outright perform worse than non-federated solutions. Consequently, the primary challenge is carefully designing methods to achieve DP that provide meaningful privacy guarantees while performing significantly better than its non-federated counterpart.

To this end, we introduce P-FCB, a Privacy-preserving Federated Combinatorial Bandit algorithm. P-FCB comprises a novel communication algorithm among agents, while each agent is learning the qualities of the producers to cooperate in the learning process. Crucially in P-FCB, the agent only communicates within a specific time frame – since it is not beneficial to communicate

---

[1] Regret is the deviation of utility gained while engaging in learning from the utility gained if the mean qualities were known.

in (i) earlier rounds (estimates have high error probability) or (ii) later rounds (value added by communicating is minimal). While communicating in each round reduces per agent regret, it results in a high privacy loss. P-FCB strikes an effective balance between learning and privacy loss by limiting the number of rounds in which agents communicate. Moreover, to ensure the privacy of the shared information, the agents add calibrated noise to sanitize the information a priori. P-FCB also uses error bounds generated for UCB exploration [3] to determine if shared information is worth learning. We show that P-FCB allows the agents to minimize their regrets while ensuring strong privacy guarantees through extensive simulations.

In recent times, research has focused on the intersection of MAB and DP [19,32]. Unlike P-FCB, these works have limitations to single-arm selections. To the best of our knowledge, this paper is the first to simultaneously study federated CMAB with assured quality and privacy constraints. In addition, as opposed to other DP and MAB approaches [8,12], we consider the sensitivity of attributes specific to a producer-agent set rather than the sensitivity of general observations. In summary, our contributions in this work are as follows:

1. We provide a theoretical analysis of improvement in terms of regret in a non-private homogeneous federated CMAB setting (Theorem 1, Sect. 4).
2. We show that employing privacy techniques naively is not helpful and has information leak concerns (Claim 1, Sect. 5.2).
3. We introduce P-FCB to employ privacy techniques practically (Algorithm 1). P-FCB includes selecting the information that needs to be perturbed and defining communication rounds to provide strong privacy guarantees. The communicated information is learned selectively by using error bounds around current estimates. Selective communication helps minimize regret.
4. P-FCB's improvement in per agent regret even in a private setting compared to individual learning is empirically validated through extensive simulations (Sect. 6).

## 2  Related Work

*Multi-armed bandits* (MAB) and their variants are a well studied class of problems [3,6,15,17,22,23] that tackle the exploration vs. exploitation trade-off in online learning settings. While the classical MAB problem [3,28] assumes single arm pull with stochastic reward generation, our work deals with combinatorial bandits (CMAB) [5,11,26,31], whereby the learning agent pulls a subset of arms. We remark that our single-agent (non-federated) MAB formulation is closely related to the MAB setting considered in [7], but the authors there do not consider federated learning.

*Federated MAB.* Many existing studies address the MAB problem in a federated setting but restrict themselves to single-arm pulls. The authors in [24,25] consider a federated extension of the stochastic single player MAB problem, while

Huang et al. [14] considers the linear contextual bandit in a federated setting. Kim et al. [16] specifically considers the federated CMAB setting. However, none of these works address privacy.

*Privacy-Preserving MAB.* The authors in [19,32] consider a differentially private MAB setting for a single learning agent, while the works in [4,18] consider differentially private federated MAB setting. However, these works focus only on the classical MAB setting, emphasising the communication bottlenecks. There also exists works that deal with private and federated setting for the contextual bandit problem [8,12]. However, they do not consider pulling subsets of arms. Further, Hannun et al. [12] consider privacy over the context, while Dubey and Pentland [8] consider privacy over context and rewards. Contrarily, this paper considers privacy over the procurement strategy used.

To the best of our knowledge, we are the first to propose a solution for combinatorial bandits (CMAB) in a federated setting with the associated privacy concerns.

# 3 Preliminaries

In this section, we formally describe the combinatorial multi-armed bandit setting and its federated extension. We also define differential privacy in our context.

## 3.1 Federated Combinatorial Multi Armed Bandits

We consider a combinatorial MAB (CMAB) setting where there are $[m]$ producers and $[n]$ agents. Each producer $i \in [m]$ has a cost $k_{ij}$ and capacity $c_{ij}$ for every agent $j \in [n]$. At any round $t \in \{1, 2, \ldots, T\}$, agents procure some quantity of goods from a subset of producers under given constraint(s). We denote the procurement of an agent $j$ by $\mathbf{s}_j = (l_{1j}, l_{2j}, \ldots, l_{mj})$ where $l_{ij} \in [0, k_{ij}]$ is the quantity procured from producer $i$.

*Qualities.* Each agent observes a quality realisation for each unit it procured from producers. Since the quality of a single unit of good may not be easily identifiable, we characterize it as a Bernoulli random variable. The expected realisation of a unit procured from a producer $i$ is referred to as its quality, $q_i$. In other words, $q_i$ denotes the probability with which a procured unit of good from producer $i$ will have a quality realisation of one. While the producer's cost and capacity vary across agents, the quality values are indifferent based on agents.

*Regret.* We use $r_{ij}$ to denote expected utility gain for the agent $j$ by procuring a single unit from producer $i$, where $r_{ij} = \rho q_i - c_{ij}$ (where $\rho > 0$, is a proportionality constant). Further, the expected revenue for a procurement vector $\mathbf{s}_j$, is given by $r_{\mathbf{s}_j} = \sum_{i \in [m]} l_{ij} r_{ij}$.

The goal for the agent is to maximise its revenue, under given constraints. We consider a constraint of maintaining a minimum expected quality threshold $\alpha$ (quality constraint), for our setting. To measure the performance of an a given algorithm $A$, we use the notion of regret which signifies the deviation of the algorithm from the procurement set chosen by an Oracle when mean qualities are known. For any round $t \in \{1, 2, \ldots, T\}$, we use the following to denote the regret for agent $j$ given an algorithm $A$,

$$\mathcal{R}^t_{Aj} = \begin{cases} r_{\mathbf{s_j^*}} - r_{\mathbf{s}^t_{Aj}}, & \text{if } s^t_{Aj} \text{ satisfies the quality constraint} \\ L, & \text{otherwise} \end{cases}$$

where $\mathbf{s_j^*}$ denotes the procurement set chosen by an Oracle, with the mean qualities known. $\mathbf{s}^t_A$ is the set chosen by the algorithm $A$ in round $t$. $L = \max_{r_{\mathbf{s}}}(r_{\mathbf{s_j^*}} - r_{\mathbf{s}})$ is a constant that represents the maximum regret one can acquire. The overall regret for algorithm $A$ is given by $\mathcal{R}_A = \sum_{j \in [n]} \sum_{t \in [T]} \mathcal{R}^t_{Aj}$.

*Federated Regret Ratio (FRR).* We introduce FRR to help quantify the reduction in regret brought on by engaging in federated learning. FRR is the ratio of the regret incurred by an agent via a federated learning algorithm $A$ over agent's learning individually via a non-federated algorithm $NF$, i.e., $FRR = \frac{\mathcal{R}_A}{\mathcal{R}_{NF}}$. We believe, *FRR* is a comprehensive indicator of the utility gained by engaging in federated learning, compared to direct regret, since it presents a normalised value and performance comparison over different data sets/algorithms is possible.

Observe that, $FRR \approx 1$ indicates that there is not much change in terms of regret by engaging in federated learning. If $FRR > 1$, it is detrimental to engage in federated learning, whereas if $FRR < 1$, it indicates a reduction in regret. When $FRR \approx 0$, there is almost complete reduction of regret in federated learning.

In our setting, we consider that agents communicate with each other to improve their regret. But in general, agents often engage in a competitive setting, and revealing true procurement values can negatively impact them. For instance, knowing that a company has been procuring less than their history can reveal their strategic plans, devalue their market capital, hinder negotiations etc. We give a formalisation of the notion of privacy used in our setting in the next subsection.

## 3.2 Differential Privacy (DP)

As opposed to typical federated models, we assume that the agents in our setting may be competing. Thus, agents will prefer the preservation of their sensitive information. Specifically, consider the history of procurement quantities $\mathbf{H}_{ij} = (l^t_{ij})_{t \in [T]}$ for any producer $i \in [m]$ is private to agent $j$. To preserve the privacy of $\mathbf{H}_{ij}$ while having meaningful utilitarian gains, we use the concept of Differential Privacy (DP). We tweak the standard DP definition in [9,10] for our setting. For this, let $\mathbf{S}_j = (\mathbf{s}^t_j)_{t \in [T]}$ be complete history of procurement vectors for agent $j$.

**Definition 1 (Differential Privacy).** *In a federated setting with $n \geq 2$ agents, a combinatorial MAB algorithm $A = (A_j)_{j=1}^n$ is said to be $(\epsilon, \delta, n)-$differentially private if for any $u, v \in [n], s.t., u \neq v$, any $t_o$, any set of adjacent histories*

Producers                                          Agent Communication

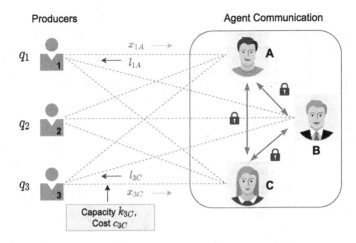

**Fig. 1.** Overview of the communication model for P-FCB: Agents interact with producers as part of the exploration and exploitation process. Agents also communicate among themselves to learn the qualities of producers. However, they share noisy data to maintain the privacy of their sensitive information.

$\mathbf{H}_{iu} = (l_{iu}^t)_{t \in [T]}, \mathbf{H}'_{iu} = (l_{iu}^t)_{t \in [T] \setminus \{t_o\}} \cup \bar{l}_{iu}^{t_o}$ *for producer i and any complete history of procurement vector* $\mathbf{S}_v$,

$$\Pr(A_v(\mathbf{H}_{iu}) \in \mathbf{S}_v) \le e^\epsilon \Pr(A_v(\mathbf{H}'_{iu}) \in \mathbf{S}_v) + \delta$$

Our concept of DP in a federated CMAB formalizes the idea that the selection of procurement vectors by an agent is insusceptible to any single element $l_{ij}^t$ from another agent's procurement history. Note that the agents are not insusceptible to their own histories here.

Typically, the "$\epsilon$" parameter is referred to as the *privacy budget*. The *privacy loss* variable $\mathcal{L}$ is often useful for the analysis of DP. More formally, given a randomised mechanism $\mathcal{M}(\cdot)$ and for any output $o$, the privacy loss variable is defined as,

$$\mathcal{L}^o_{\mathcal{M}(\mathbf{H}) \| \mathcal{M}(\mathbf{H}')} = \ln \left( \frac{\Pr[\mathcal{M}(\mathbf{H}) = o]}{\Pr[\mathcal{M}(\mathbf{H}') = o]} \right). \tag{1}$$

*Gaussian Noise Mechanism* [10]. To ensure DP, often standard techniques of adding noise to values to be communicated are used. The Gaussian Noise mechanism is a popular mechanism for the same. Formally, a randomised mechanism $\mathcal{M}(x)$ satisfies $(\epsilon, \delta)$-DP if the agent communicates $\mathcal{M}(x) \triangleq x + \mathcal{N}\left(0, \frac{2\Delta(x)^2 \ln(1.25/\delta)}{\epsilon^2}\right)$. Here, $x$ is the private value to be communicated with *sensitivity* $\Delta(x)$, and $\mathcal{N}(0, \sigma^2)$ the Gaussian distribution with mean zero and variance $\sigma^2$.

In summary, Fig. 1 provides an overview of the model considered. Recall that we aim to design a differentially private algorithm for federated CMAB with assured qualities. Before this, we first highlight the improvement in regret using

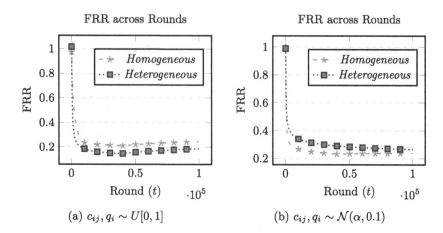

**Fig. 2.** Comparing $FRR$ values for Homogeneous and Heterogeneous Federated CMAB ($n = 10$, $m = 30$)

the federated learning paradigm. Next, we discuss our private algorithm, P-FCB, in Sect. 5.

## 4    Non-private Federated Combinatorial Multi-armed Bandits

We now demonstrate the advantage of federated learning in CMAB by highlighting the reduction in regret incurred compared to agents learning individually. We first categorize Federated CMAB into the following two settings: (i) *homogeneous*: where the capacities and costs for producers are the same across agents, and (ii) *heterogeneous*: where the producer's capacity and cost varies across agents.

**Homogeneous Setting.** The core idea for single-agent learning in CMAB involves using standard $UCB$ exploration [3]. We consider an Oracle that uses the $UCB$ estimates to return an optimal selection subset. In this paper, we propose that to accelerate the learning process and for getting *tighter* error bound for quality estimations, the agents communicate their observations with each other in every round. In a homogeneous setting, this allows all agents to train a shared model locally without a central planner since the Oracle algorithm is considered deterministic. We present the formal algorithm in the extended version [29]. It's important to note that in such a setting, each agent has the same procurement history and the same expected regret.

Further, the quality constraint guarantees for the federated case follow trivially from the single agent case ([7, Theorem 2]). Additionally, in Theorem 1, we prove that the upper bound for regret incurred by each agent is $\mathcal{O}(\frac{\ln(nT)}{n})$; a significant improvement over $\mathcal{O}(\ln T)$ regret the agent will incur when playing individually. The formal proof is provided in the extended version [29].

**Theorem 1.** *For Federated CMAB in a homogeneous setting with $n$ agents, if the qualities of producers satisfy $\gamma$-seperatedness, then the individual regret incurred by each of the agents is bounded by $\mathcal{O}(\frac{\ln(nT)}{n})$.*

**Heterogeneous Setting.** In real-world, the agents may not always have the same capacities. For such a heterogeneous setting, the regret analysis is analytically challenging. For instance, we can no longer directly use Hoeffding's inequality, needed for proving Theorem 1, since the procurement histories will differ across agents. Still, the intuition for regret reduction from cooperative learning carries over.

Even in a heterogeneous setting, communicating the observations allows the agent to converge their quality estimations to the mean faster and provide tighter error bounds. Even with shared quality estimates, Oracle may return different procurement vectors for different agents based on different capacities. Thus, a weighted update in estimation is essential, and the procurement vector would also need to be communicated.

We empirically demonstrate that using federated learning in heterogeneous setting shows similar $FRR$ (ratio of regret incurred in federated setting compared to non federated setting) trend compared to homogeneous setting, over 100000 rounds for two scenarios: (i) Costs and qualities are sampled from uniform distributions, i.e. $c_{ij} \sim U[0,1]$, $q_i \sim U[0,1]$, (ii) Costs and qualities are sampled from normal distributions around the quality threshold, i.e., $c_{ij} \sim \mathcal{N}(\alpha, 0.1)$, $q_i \sim \mathcal{N}(\alpha, 0.1)$.

Figure 2 depicts the results. From Fig. 2 we observe that the trend for both homogeneous and heterogeneous settings are quite similar. This shows that, similar to the homogeneous setting, employing federated learning reduces regret even in the heterogeneous setting.

## 5   P-FCB: **Privacy-Preserving Federated Combinatorial Bandit**

From Sect. 3.2, recall that we identify the procurement history of an agent-producer pair as the agent's sensitive information. We believe that the notion of DP w.r.t. the agent-producer procurement history is reasonable. A differentially private solution ensures that the probability with which other agents can distinguish between an agent's adjacent procurement histories is upper bounded by the privacy budget $\epsilon$.

Section Outline: In this section, we first argue that naive approaches for DP are not suitable due to their lack of meaningful privacy guarantees. Second, we show that all attributes dependent on the sensitive attribute must be sanitised before sharing to preserve privacy. Third, we define a privacy budget algorithm scheme. Fourth, we formally introduce P-FCB including a selective learning procedure. Last, we provide the $(\epsilon, \delta)$-DP guarantees for P-FCB.

### 5.1    Privacy Budget and Regret Trade-Off

Additive noise mechanism (e.g., Gaussian Noise mechanism [10]) is a popular technique for ensuring $(\epsilon, \delta)$-DP. To protect the privacy of an agent's procurement history within the DP framework, we can build a naive algorithm for heterogeneous federated CMAB setting by adding noise to the elements of the procurement vectors being communicated in each round.

However, such a naive approach does not suitably satisfy our privacy needs. Using the Basic Composition theorem [10], which adds the $\epsilon$s and $\delta$s across queries, it is intuitive to see that communicating in every round results in a high overall $\epsilon$ value which may not render much privacy protection in practice [30]. Consider the agents interacting with the producers for $10^6$ rounds. Let $\epsilon = 10^{-2}$ for each round they communicate the perturbed values. Using Basic Composition, we can see that the overall privacy budget will be bounded by $\epsilon = 10^4$, which is practically not acceptable. The privacy loss in terms of overall $\epsilon$ grows at worst linearly with the number of rounds.

It is also infeasible to solve this problem merely by adding more noise (reducing $\epsilon$ per round) since if the communicated values are too noisy, they can negatively affect the estimates. This will result in the overall regret increasing to a degree that it may be better to not cooperatively learn. To overcome this challenge, we propose to decrease the number of rounds in which agents communicate information.

Secondly, if the sample size for the local estimates is too small, noise addition can negatively effect the regret incurred. On the other hand, if the sample size of local estimate is too large, the local estimate will have tight error bounds and deviating from the local estimate too much may result in the same.

**When to Learn.** Based on the above observations, we propose the following techniques to strike an effective trade-off between the privacy budget and regret.

1. To limit the growth of $\epsilon$ over rounds, we propose that communication happens only when the current round number is equal to a certain threshold (denoted by $\tau$) which doubles in each communication round. Thus, there are only $\log(T)$ communications rounds, where density of communication rounds decrease over rounds.
2. We propose to communicate only for a specific interval of rounds, i.e., for each round $t \in [\underline{t}, \overline{t}]$. No communication occurs outside these rounds. This ensures that agent communication only happens in rounds when it is useful and not detrimental.

### 5.2    Additional Information Leak with Actual Quality Estimates and Noisy Weights

It is also important to carefully evaluate the way data is communicated every round since it may lead to privacy leaks. For example, consider that all agents communicate their local estimates of the producer qualities and perturbation of the total number of units procured from each producer to arrive at the estimation. We now formally analyse the additional information leak in this case.

**Procedure 1.** CheckandUpdate($W, \tilde{w}, Y, \tilde{y}, \omega_1, \omega_2, n, t$)

1: $\hat{q} \longleftarrow \frac{Y}{W}$

2: **if** $\frac{\tilde{y}}{\tilde{w}} \in \left[ \hat{q} - \omega_1 \sqrt{\frac{3ln(nt)}{2W}}, \hat{q} + \omega_1 \sqrt{\frac{3ln(nt)}{2W}} \right]$ **then**

3:     $W \longleftarrow W + \omega_2 \tilde{w}$

4:     $Y \longleftarrow Y + \omega_2 \tilde{y}$

5: **end if**

6: **return** $W, Y$

---

W.l.o.g. our analysis is for any arbitrarily picked producer $i \in [m]$ and agent $j \in [n]$. As such, we omit the subscripts "$i$" for producer and "$j$" for the agent. We first set up the required notations as follows.

Notations: Consider $\hat{q}^t, W^t$ as *true* values for the empirical estimate of quality and total quantity procured till the round $t$ (not including $t$). Next, let $\tilde{W}^t$ denote *noisy* value of $W^t$ (with the noise added using any additive noise mechanism for DP [10]). We have $w^t$ as the quantity procured in round $t$. Last, let $\hat{q}^{obsv_t}$ denote the quality estimate based on just round $t$. Through these notations, we can compute $\hat{q}^{t+1}$ for the successive round $t + 1$ as follows: $\hat{q}^{t+1} = \frac{W^t \times \hat{q}^t + w^t \times \hat{q}^{obsv_t}}{W^t + w^t}$.

**Claim 1.** *Given* $\hat{q}^t, W^t, \tilde{W}^t, w^t$ *and* $\hat{q}^{obsv_t}$, *the privacy loss variable* $\mathcal{L}$ *is not defined if* $\hat{q}^t$ *is also not perturbed.*

We present the formal proof in the extended version [29]. With Claim 1, we show that $\epsilon$ may not be bounded even after sanitising the sensitive data due to its dependence on other non-private communicated data. This is due to the fact that the local mean estimates are a function of the procurement vectors and the observation vectors. Thus, it becomes insufficient to just perturb the quality estimates.

We propose that whenever communication happens, only procurement and observation values based on rounds since last communication are shared. Additionally, to communicate weighted quality estimates, we use the Gaussian Noise mechanism to add noise to *both* the procurement values and realisation values. The sensitivity ($\Delta$) for noise sampling is equal to the capacity of the producer-agent pair.

### 5.3 Privacy Budget Allocation

Since the estimates are more sensitive to noise addition when the sample size is smaller, we propose using monotonically decreasing privacy budget for noise generation. Formally, let total privacy budget be denoted by $\epsilon$ with $(\epsilon^1, \epsilon^2, \ldots)$ corresponding to privacy budgets for communication rounds $(1, 2, \ldots)$. Then, we have $\epsilon^1 > \epsilon^2 > \ldots$. Specifically, we denote $\epsilon^z$ as the privacy budget in the $z^{th}$ communication round, where $\epsilon^z \longleftarrow \frac{\epsilon}{2 \times \log(T)} + \frac{\epsilon}{2^{z+1}}$.

---

**Algorithm 1.** P-FCB

---

1: **Inputs** : Total rounds $T$, Quality threshold $\alpha$, $\epsilon$, $\delta$, Cost set $\{\mathbf{c}_j\} = \{(c_{i,j})_{i\in[m]}\}$, Capacity set $\{\mathbf{k}_j\} = \{(k_{i,j})_{i\in[m]}\}$, Start round $\underline{t}$, Stop round $\bar{t}$

2: /* Initialisation Step */

3: $t \longleftarrow 0$, $\tau \longleftarrow 1$

4: $[\forall i \in [m], \forall j \in [n]]$ Initialise total and uncommunicated procurement $(W_{i,j}, w_{i,j})$ and realisations $(Y_{i,j}, y_{i,j})$

5: **while** $t \leq \frac{3ln(yT)}{2n\zeta^2}$ (**Pure Explore Phase**) **do**

6:     **for** all the agents $j \in [n]$ **do**

7:         Pick procurement vector $\mathbf{s}_j^t = (1)^m$ and observe quality realisations $\mathbf{X}_{\mathbf{s}_j^t,j}^t$.

8:         $[\forall i \in [m]]$ Update $W_{i,j}^{t+1}, w_{i,j}^{t+1}, Y_{i,j}^{t+1}, y_{i,j}^{t+1}$ using Eq. 2

9:         **if** $t \in [\underline{t}, \bar{t}]$ and $t \geq \tau$ **then**      ▷ Communication round

10:            $[\forall i \in [m]]$ Calculate $\tilde{w}_{i,j}, \tilde{y}_{i,j}$ according to Eq. 3,4

11:            **for** each agent $z \in [n]/j$ **do**

12:                Send $\{\tilde{w}_{i,j}, \tilde{y}_{i,j}\}$ to agent $z$

13:                $[\forall i \in [m]]$   $W_{i,z}^{t+1}, Y_{i,z}^{t+1}$  $\longleftarrow$  CheckandUpdate$(W_{i,z}^{t+1}, \tilde{w}_{i,j}, Y_{i,z}^{t+1}, \tilde{y}_{i,j}, .)$

14:            **end for**

15:            $[\forall i \in [m]]$ $w_{i,j}^{t+1} \longleftarrow 0$, $y_{i,j}^{t+1} \longleftarrow 0$

16:            $\tau \longleftarrow 2 \times \tau$

17:         **end if**

18:         Update quality estimate

19:         $t \longleftarrow t + 1$

20:     **end for**

21: **end while**

22: **while** $t \leq T$, $\forall j \in [n]$ (**Explore-Exploit Phase**) **do**

23:     $[\forall i \in [m]]$ Calculate the upper confidence bound of quality estimate, $(\hat{q}_{i,j}^t)^+$

24:     Pick procurement vector using $\mathbf{s}_j^t = \mathbf{Oracle}((\hat{q}_{i,j}^t)^+, \mathbf{c}_j, \mathbf{k}_j, .)$ and observe its realisations $\mathbf{X}_{\mathbf{s}_j^t,j}^t$.

25:     $[\forall i \in [m]]$ Update $W_{i,j}^{t+1}, w_{i,j}^{t+1}, Y_{i,j}^{t+1}, y_{i,j}^{t+1}$ using Eq. 2

26:     **if** $t \in [\underline{t}, \bar{t}]$ and $t \geq \tau$ **then**      ▷ Communication round

27:         $[\forall i \in [m]]$ Calculate $\tilde{w}_{i,j}, \tilde{y}_{i,j}$ according to Eq. 3,4

28:         **for** each agent $z \in [n]/j$ **do**

29:            Send $\{\tilde{w}_{i,j}, \tilde{y}_{i,j}\}$ to agent $z$

30:            $[\forall i \in [m]]$   $W_{i,z}^{t+1}, Y_{i,z}^{t+1}$  $\longleftarrow$  CheckandUpdate$(W_{i,z}^{t+1}, \tilde{w}_{i,j}, Y_{i,z}^{t+1}, \tilde{y}_{i,j}, .)$

31:         **end for**

32:         $[\forall i \in [m]]$ $w_{i,j}^{t+1} \longleftarrow 0$, $y_{i,j}^{t+1} \longleftarrow 0$

33:         $\tau \longleftarrow 2 \times \tau$

34:     **end if**

35:     Update quality estimate

36:     $t \longleftarrow t + 1$

37: **end while**

## 5.4   P-FCB: Algorithm

Based on the feedback from the analysis made in previous subsections, we now present a private federated CMAB algorithm for the heterogeneous setting, namely P-FCB. Algorithm 1 formally presents P-FCB. Details follow.

**Algorithm 1 Outline.** The rounds are split into two phases. During the initial pure exploration phase (Lines 6–22), the agents explore all the producers by procuring evenly from all of them. The length of the pure exploration phase is carried over from the non-private algorithm. In this second phase (Lines 23–38), explore-exploit, the agents calculate the $UCB$ for their quality estimates. Then the Oracle is used to provide a procurement vector based on the cost, capacity, $UCB$ values as well as the quality constraint ($\alpha$). Additionally, the agents communicate their estimates as outlined in Sects. 5.1 and 5.2. The agents update their quality estimates at the end of each round using procurement and observation values (both local and communicated), Lines 19 and 36.

$$w_{i,j}^{t+1} \longleftarrow w_{i,j}^t + l_{i,j}^t \; ; \; W_{i,j}^{t+1} \longleftarrow W_{i,j}^t + l_{i,j}^t$$
$$y_{i,j}^{t+1} \longleftarrow y_{i,j}^t + x_{i,j}^t \; ; \; Y_{i,j}^{t+1} \longleftarrow Y_{i,j}^t + x_{i,j}^t \tag{2}$$
$$q_{i,j}^{t+1} \longleftarrow \frac{Y_{i,j}^{t+1}}{W_{i,j}^{t+1}}$$

**Noise Addition.** From Sect. 5.2, we perturb both uncommunicated procurement and realization values for each agent-producer pair using the Gaussian Noise mechanism. Formally, let $w_{i,j}^t, y_{i,j}^t$ be the uncommunicated procurement and realization values. Then $\tilde{w}_{i,j}, \tilde{y}_{i,j}$ are communicated, which are calculated using the following privatizer,

$$\tilde{w}_{i,j} = w_{i,j}^t + \mathcal{N}(0, \frac{2k_{i,j}^2 \log(1.25/\delta)}{(\epsilon^z)^2}) \tag{3}$$

$$\tilde{y}_{i,j} = y_{i,j}^t + \mathcal{N}(0, \frac{2k_{i,j}^2 \log(1.25/\delta)}{(\epsilon^z)^2}) \tag{4}$$

where $\epsilon^z$ is the privacy budget corresponding to the $z^{th}$ communication round.

**What to Learn.** To minimise the regret incurred, we propose that the agents selectively choose what communications to learn from. Weighted confidence bounds around local estimates are used to determine if a communication round should be learned from. Let $\xi_{i,j}^t = \sqrt{\frac{3ln(t)}{2\sum_{z \in \{1,2,\dots,t\}} l_{i,j}^z}}$ denote the confidence interval agent $j$ has w.r.t. local quality estimate of producer $i$. Then, the agents only selects to learn from a communication if $\hat{q}_{i,j}^t - \omega_1 \xi_{i,j}^t < q_{(communicated)i,j} < \hat{q}_{i,j}^t + \omega_1 \xi_{i,j}^t$ where $\omega_1$ is a weight factor and $q_{(communicated)i,j} = \frac{\tilde{y}_{i,j}}{\tilde{w}_{i,j}}$.

The local observations are weighed more compared to communicated observations for calculating overall estimates. Specifically, $\omega_2 \in [0,1]$ is taken as the weighing factor for communicated observations.

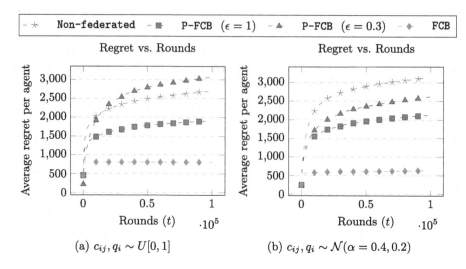

(a) $c_{ij}, q_i \sim U[0,1]$    (b) $c_{ij}, q_i \sim \mathcal{N}(\alpha = 0.4, 0.2)$

**Fig. 3.** EXP1: Regret Comparison across rounds ($n = 10$, $m = 30$)

### 5.5  P-FCB: $(\epsilon, \delta)$-DP Guarantees

In each round, we perturb the values being communicated by adding Gaussian noises satisfying $(\epsilon', \delta')$-DP to them. It is a standard practice for providing DP guarantees for group sum queries. Let $\mathcal{M}$ be a randomised mechanism which outputs the sum of values for a database input $d$ using Gaussian noise addition. Since Oracle is deterministic, each communication round can be considered a post-processing of $\mathcal{M}$ whereby subset of procurement history is the database input. Thus making individual communication rounds satisfy $(\epsilon', \delta')$-DP.

The distinct subset of procurement histories used in each communication round can be considered as independent DP mechanisms. Using the Basic Composition theorem, we can compute the overall $(\epsilon, \delta)$-DP guarantee. In P-FCB, we use a target privacy budget, $\epsilon$, to determine the noise parameter $\sigma$ in each round based on Basic composition. Thus, this can be leveraged as a tuning parameter for privacy/regret optimisation.

## 6  Experimental Results

In this section, we compare P-FCB with non-federated and non-private approaches for the combinatorial bandit (CMAB) setting with constraints. We first explain the experimental setup, then note our observations and analyze the results obtained.

### 6.1  Setup

For our setting, we generate costs and qualities for the producers from: (a) uniform distributions, i.e., $q_i, c_{ij} \sim U[0,1]$ (b) normal distributions, i.e., $q_i$,

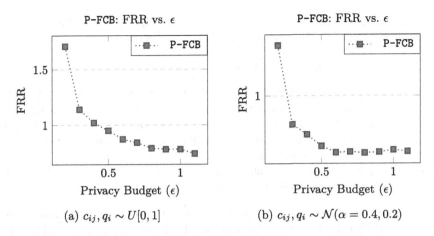

Fig. 4. EXP2: FRR for P-FCB while varying privacy budget $\epsilon$ (with $n = 10$, $m = 30$, $t = 100000$)

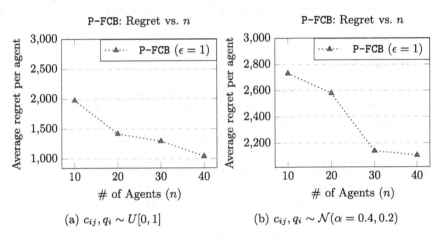

Fig. 5. EXP3: Average regret per agent with P-FCB by varying the number of learners $n$ (with $\epsilon = 1$, $t = 100000$)

$c_{ij} \sim \mathcal{N}(\alpha, 0)$. For both cases, the capacities are sampled from a uniform distribution, $k_{ij} \sim U[1, 50]$. We use the following tuning parameters in our experiments: $\alpha = 0.4$, $\delta = 0.01$ (i.e., $\delta < 1/n$), $\underline{t} = 200$, $\bar{t} = 40000$, $\omega_1 = 0.1$, $\omega_2 = 10$. For our Oracle, we deploy the *Greedy SSA* algorithm presented in Deva et al. [7]. Further, to compare P-FCB's performance, we construct the following two *non-private* baselines:

1. Non-Federated. We use the single agent algorithm for subset selection under constraints proposed in Deva et al. [7]. It follows $UCB$ exploration similar to P-FCB but omits any communication done with other agents.

2. FCB. This is the non-private variant of P-FCB. That is, instead of communicating $\tilde{w}_{ij}$ and $\tilde{y}_{ij}$, the true values $w_{ij}^t$ and $y_{ij}^t$ are communicated.

We perform the following experiments to measure P-FCB's performance:

- EXP1: For fixed $n = 10$, $m = 30$, we observe the regret growth over rounds ($t$) and compare it to non-federated and non-private federated settings.
- EXP2: For fixed $n = 10$, $m = 30$, we observe $FRR$ (ratio of regret incurred in federated setting compared to non federated setting) at $t = 100000$ while varying $\epsilon$ to see the regret variance w.r.t. privacy budget.
- EXP3: For fixed $\epsilon = 1$, $m = 30$, we observe average regret at $t = 100000$ for varying $n$ to study the effect of number of communicating agents.

For EXP1 and EXP2, we generate 5 instances by sampling costs and quality from both Uniform and Normal distributions. Each instance is simulated 20 times and we report the corresponding average values across all instances. Likewise for EXP3, instances with same producer quality values are considered with costs and capacities defined for different numbers of learners. For each instance, we average across 20 simulations.

## 6.2 Results

- EXP1. P-FCB shows significant improvement in terms of regret (Fig. 3) at the cost of relatively low privacy budget. Compared to FCB, P-FCB ($\epsilon = 1$) and Non-federated incurs 136%, 233% more regret respectively for uniform sampling and 235%, 394% more regret respectively for normal sampling. This validates efficacy of P-FCB.
- EXP2. We study the performance of the algorithm with respect to privacy budget (Fig. 4). We observe that according to our expectations, the regret decreases as privacy budget is increased. This decrease in regret is sub-linear in terms of increasing $\epsilon$ values. This is because as privacy budget increases, the amount of noise in communicated data decreases.
- EXP3. We see (Fig. 5) an approximately linear decrease in per agent regret as the number of learning agents increases. This reinforces the notion of reduction of regret, suggested in Sect. 4, by engaging in federated learning is valid in a heterogeneous private setting.

Discussion: Our experiments demonstrate that P-FCB, through selective learning in a federated setting, is able to achieve a fair regret and privacy trade-off. P-FCB achieves reduction in regret (compared to non-federated setting) for low privacy budgets.

With regards to hyperparameters, note that lower $\omega_2$ suggests tighter bounds while selecting what to learn, implying a higher confidence in usefulness of the communicated data. Thus, larger values for $\omega_1$ can be used if $\omega_2$ is decreased. In general, our results indicate that it is optimal to maintain the value $\omega_1 \cdot \omega_2$ used in our experiments. Also, the communication start time, should be such that the sampled noise is at-least a magnitude smaller than the accumulated uncommunicated data (e.g., $\underline{t} \approx 200$). This is done to ensure that the noisy data is not detrimental to the learning process.

The DP-ML literature suggests a privacy budget $\epsilon < 1$ [30]. From Fig. 4, we note that P-FCB performs well within this privacy budget. While our results achieve a fair regret and privacy trade-off, in future, one can further fine tune these hyperparameters through additional experimentation and/or theoretical analysis.

## 7 Conclusion and Future Work

This paper focuses on learning agents which interact with the same set of producers ("arms") and engage in federated learning while maintaining privacy regarding their procurement strategies. We first looked at a non-private setting where different producers' costs and capacities were the same across all agents and provided theoretical guarantees over optimisation due to federated learning. We then show that extending this to a heterogeneous private setting is non-trivial, and there could be potential information leaks. We propose P-FCB which uses *UCB* based exploration while communicating estimates perturbed using Gaussian method to ensure differential privacy. We defined a communication protocol and a selection learning process using error bounds. This provided a meaningful balance between regret and privacy budget. We empirically showed notable improvement in regret compared to individual learning, even for considerably small privacy budgets.

Looking at problems where agents do not share exact sets of producers but rather have overlapping subsets of available producers would be an interesting direction to explore. It is also possible to extend our work by providing theoretical upper bounds for regret in a differentially private setting. In general, we believe that the idea of when to learn and when not to learn from others in federated settings should lead to many interesting works.

## References

1. Foundry model. https://en.wikipedia.org/w/index.php?title=Foundry_model&oldid=1080269386
2. Original equipment manufacturer. https://en.wikipedia.org/w/index.php?title=Original_equipment_manufacturer&oldid=1080228401
3. Auer, P., Cesa-Bianchi, N., Fischer, P.: Finite-time analysis of the multiarmed bandit problem. Mach. Learn. **47**, 235–256 (2004)
4. Chen, S., Tao, Y., Yu, D., Li, F., Gong, B., Cheng, X.: Privacy-preserving collaborative learning for multiarmed bandits in IoT. IEEE Internet Things J. **8**(5), 3276–3286 (2021)
5. Chen, W., Wang, Y., Yuan, Y.: Combinatorial multi-armed bandit: general framework and applications. In: ICML. PMLR, 17–19 June 2013
6. Chiusano, F., Trovò, F., Carrera, G.D., Boracchi, Restelli, M.: Exploiting history data for nonstationary multi-armed bandit. In: ECML/PKDD (2021)

7. Deva, A., Abhishek, K., Gujar, S.: A multi-arm bandit approach to subset selection under constraints. In: AAMAS 2021, pp. 1492–1494. AAMAS (2021)

8. Dubey, A., Pentland, A.: Differentially-private federated linear bandits. Adv. Neural. Inf. Process. Syst. **33**, 6003–6014 (2020)

9. Dwork, C.: Differential privacy. In: Proceedings of the 33rd International Conference on Automata, Languages and Programming - Volume Part II. ICALP 2006, pp. 1–12 (2006)

10. Dwork, C., Roth, A.: The algorithmic foundations of differential privacy. Found. Trends Theor. Comput. Sci. **9**(3–4), 211–407 (2014)

11. Gai, Y., Krishnamachari, B., Jain, R.: Learning multiuser channel allocations in cognitive radio networks: a combinatorial multi-armed bandit formulation. In: DySPAN 2010 (2010)

12. Hannun, A.Y., Knott, B., Sengupta, S., van der Maaten, L.: Privacy-preserving contextual bandits. CoRR abs/1910.05299 (2019), http://arxiv.org/abs/1910.05299

13. Ho, C.J., Jabbari, S., Vaughan, J.W.: Adaptive task assignment for crowdsourced classification. In: ICML, pp. 534–542 (2013)

14. Huang, R., Wu, W., Yang, J., Shen, C.: Federated linear contextual bandits. In: Advances in Neural Information Processing Systems, vol. 34. Curran Associates, Inc. (2021)

15. Jain, S., Gujar, S., Bhat, S., Zoeter, O., Narahari, Y.: A quality assuring, cost optimal multi-armed bandit mechanism for expertsourcing. Artif. Intell. **254**, 44–63 (2018)

16. Kim, T., Bae, S., Lee, J., Yun, S.: Accurate and fast federated learning via combinatorial multi-armed bandits. CoRR (2020). https://arxiv.org/abs/2012.03270

17. Li, L., Chu, W., Langford, J., Schapire, R.E.: A contextual-bandit approach to personalized news article recommendation. In: International Conference on World Wide Web (2010)

18. Li, T., Song, L.: Privacy-preserving communication-efficient federated multi-armed bandits. IEEE J. Sel. Areas Commun. **40**(3), 773–787 (2022)

19. Malekzadeh, M., Athanasakis, D., Haddadi, H., Livshits, B.: Privacy-preserving bandits. In: Proceedings of Machine Learning and Systems, vol. 2, pp. 350–362 (2020)

20. McMahan, B., Moore, E., Ramage, D., Hampson, S., Arcas, B.A.: Communication-efficient learning of deep networks from decentralized data. In: Artificial Intelligence and Statistics. PMLR (2017)

21. Mehta, D., Yamparala, D.: Policy gradient reinforcement learning for solving supply-chain management problems. In: Proceedings of the 6th IBM Collaborative Academia Research Exchange Conference (I-CARE) on I-CARE 2014, pp. 1–4 (2014)

22. Roy, K., Zhang, Q., Gaur, M., Sheth, A.: Knowledge infused policy gradients with upper confidence bound for relational bandits. In: Oliver, N., Pérez-Cruz, F., Kramer, S., Read, J., Lozano, J.A. (eds.) ECML PKDD 2021. LNCS (LNAI), vol. 12975, pp. 35–50. Springer, Cham (2021). https://doi.org/10.1007/978-3-030-86486-6_3

23. Saber, H., Saci, L., Maillard, O.A., Durand, A.: Routine bandits: minimizing regret on recurring problems. In: ECML-PKDD 2021. Bilbao, Spain, September 2021

24. Shi, C., Shen, C.: Federated multi-armed bandits. In: Proceedings of the AAAI Conference on Artificial Intelligence, vol. 35, no. 11, pp. 9603–9611 (2021)

25. Shi, C., Shen, C., Yang, J.: Federated multi-armed bandits with personalization. In: Proceedings of The 24th International Conference on Artificial Intelligence and Statistics, pp. 2917–2925 (2021)
26. Shweta, J., Sujit, G.: A multiarmed bandit based incentive mechanism for a subset selection of customers for demand response in smart grids. In: Proceedings of the AAAI Conference on Artificial Intelligence, vol. 34, no. 02, pp. 2046–2053 (2020)
27. Silva, N., Werneck, H., Silva, T., Pereira, A.C., Rocha, L.: Multi-armed bandits in recommendation systems: a survey of the state-of-the-art and future directions. Expert Syst. Appl. **197**, 116669 (2022)
28. Slivkins, A.: Introduction to multi-armed bandits. CoRR abs/1904.07272 (2019). http://arxiv.org/abs/1904.07272
29. Solanki, S., Kanaparthy, S., Damle, S., Gujar, S.: Differentially private federated combinatorial bandits with constraints (2022). https://arxiv.org/abs/2206.13192
30. Triastcyn, A., Faltings, B.: Federated learning with Bayesian differential privacy. In: 2019 IEEE International Conference on Big Data (Big Data), pp. 2587–2596. IEEE (2019)
31. Wang, S., Chen, W.: Thompson sampling for combinatorial semi-bandits. In: Proceedings of the 35th International Conference on Machine Learning, pp. 5114–5122 (2018)
32. Zhao, H., Xiao, M., Wu, J., Xu, Y., Huang, H., Zhang, S.: Differentially private unknown worker recruitment for mobile crowdsensing using multi-armed bandits. IEEE Trans. Mob. Comput. (2021)
33. Zheng, Z., Zhou, Y., Sun, Y., Wang, Z., Liu, B., Li, K.: Applications of federated learning in smart cities: recent advances, taxonomy, and open challenges. Connect. Sci. (2021)

# Author Index

M.-R. Amini et al. (Eds.): ECML PKDD 2022, LNAI 13716, pp. 639–641, 2023.
https://doi.org/10.1007/978-3-031-26412-2

Printed in the United States
by Baker & Taylor Publisher Services